Linear Mathematical Models in Chemical Engineering

Second Edition

Linear Mathematical Models in Chemical Engineering

Second Edition

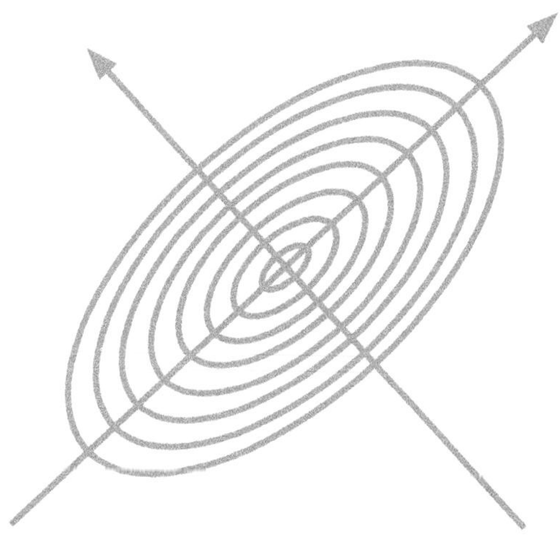

Martin A Hjortsø

Peter Wolenski

Louisiana State University, USA

World Scientific

NEW JERSEY • LONDON • SINGAPORE • BEIJING • SHANGHAI • HONG KONG • TAIPEI • CHENNAI • TOKYO

Preface

Like the first edition of this book, this second edition is focused on just chemical engineering models and their solution by analytical means. Several changes have been made to the first edition. In particular, the modelling chapter has been rewritten and partitioned into three chapters to separate model formulation from model simplifications and model examination. Similarly, the chapter on partial differential equations have been partitioned into three chapters, one on self-adjoint problems, one on first order equations and one on solution by similarity transformations. Substantial new material has been added to these chapters. The remaining chapters are quite similar to the comparable chapters in the first edition, with only minor changes and corrections of known errors.

In this text, only a small amount of effort will be spent on nonlinear systems. When nonlinear systems are mentioned, it is primarily to illustrate how complicated and different these systems are from linear systems and to make the reader appreciate the simplicity and beauty of linear systems.

PC solutions

Many of the mathematical methods described in this book entails the use of straight forward but tedious mathematical manipulations. We urge readers to look past all these tedious steps and focus on the structure of the solution methods and algorithms. The best way to do this is to have a PC carry out all the tedious steps and readers who are not already familiar with symbol manipulating software are urged to become acquainted with one of the many available software packages. In the first edition of the book, we gave the address of a web page with PC solutions of many of the examples in the text. We intend to continue to provide these solutions but maintaining a fixed web address in an academic environment has turned out to be difficult. Readers are therefore advised to check the web pages of either author to find a link to the page with PC solutions of problems and examples from the book (and to the errata for the book which will undoubtedly need to be penned). Readers of the this book are also invited to submit their own PC solutions to examples from the book, as was done by readers of the first edition, and we plan post them as we receive them.

Acknowledgment

During the years that this book first took shape, and through the years when the first edition was used, many students have contributed with suggestions and corrections. They have all helped to improve the text and we are grateful for them all. We have also received many suggestions from colleagues and are particularly grateful to our colleagues and former teachers who have given us permission to use their old homework problems in this book.

Nomenclature

The material covered in these notes is so diverse that a consistent nomenclature that does not clash with established nomenclature in some area is virtually impossible. The first three chapters, in particular, make use of a lot of different physical quantities and these are indicated using common chemical engineering nomenclature, such as C for concentrations, ρ for density etc. This nomenclature is given in Table 0.1 below. The remaining chapters are primarily abstract mathematics and the nomenclature used in these is given in Tables 0.2, 0.3 and 0.4.

Table 0.1 Nomenclature for physical (dimensioned) quantities.

Symbol	Variable	Units
A	Area	length^2
C	Concentration	moles/volume
c_p	Heat capacity	energy/(mass degree)
D	Diffusion coefficient	length^2/time
E	Activation energy	energy/mole
F	Molar flowrate	moles per time
h	Heat transfer coefficient	energy/(length^2 time degree)
ΔH_r	Heat of reaction	energy/mole
k	Reaction rate constant	Depends on the rate expression
k, k_T	Thermal conductivity[a]	energy/(length time degree)
k_m	Overall mass transfer coefficient	length/time
$k_{m,I}$	Mass transfer coefficients for phase I	length/time
k_0	Frequency factor of reaction rate constant	Depends on the rate expression
M_n	Molecular weight of compound n	
P	Pressure	force/area
$\dot{q}$	Specific rate of heat generation	energy/(time volume)
Q	Volumetric flow rate	volume/time
μ	Viscosity	mass/(length time)
$r(C)$	Specific reaction rate	moles/(time volume)
R	Gas constant	energy/(mole temperature)
t	time	time
T	Temperature	temperature
U	Overall heat transfer coefficient	energy/(time length^2 temperature)
V	Volume	length^3
ω	Mole fraction	number between 0 and 1
ρ	Density	mass/volume

[a]The simpler notation k is used whenever possible. However, in problems that include chemical reactions the k_T-notation is used to distinguish thermal conductivity from reaction rate constants.

Table 0.2 Latin letters.

Symbol	Common use
$a_{n,m}$	Element in row n, column m of the matrix A. Matrices themselves are indicated by capital letters and their elements by the same letter in lower case. The size of a matrix, i.e. the number of rows and columns, may be indicated as a subscript, e.g. $A_{N\times M}$.
B_x	Basis, identified by x, of a finite dimensional vector space.
C_n	Arbitrary constants.
$\{e_n\}$	Usual set of basis vectors in $\mathbb{R}^N$. Thus e_n is an N-dimensional vector for which all elements equals zero, except for element n which equals 1.
J	Jordan canonical form matrix.
$\{f_n\}$	Set of basis vectors in $\mathbb{R}^N$. Not necessarily same as $\{e_n\}$.
K	Matrix of eigenvectors and/or generalized eigenvector chains.
n, m	Counters, e.g. on vector or matrix elements or on finite sums. In general, $n = 1..N$ and $m = 1..M$.
N, M	Upper limits on counters n and m.
s	Dummy variable, such as dummy variable of integration or parameter in a parametric representation.
t	Free variable in ordinary differential equations. This usage works fine in differential equations that are models of time dependent phenomena, but may give rise to confusion in differential equations that are typically associated with variation along a spacial coordinate.
v, w	Vectors in finite dimensional vector spaces. Specific elements are indicated v_n and w_n.
$\tilde{v}_n$	Eigenvectors or rows of the eigenvalue λ_n. Or, generalized eigenvector of rank n if the eigenvalue is understood.
$\tilde{v}_{n,m}$	Generalized eigenvector of rank m of λ_n.
V, W	Finite dimensional vector spaces.

Table 0.3 Greek letters.

Symbol	Common use
Λ	Diagonal matrix of eigenvalues.
λ	Eigenvalue.
ϕ	Longitudinal coordinate in spherical system.
τ	Integer counter in difference equations.
θ	Angular coordinate in cylindrical coordinate system or latitudinal coordinate in spherical system or any angle.
$\theta_{n,m}$	Direction cosines.
Θ	Rotation matrix.

Table 0.4 Other math notation.

Symbol	Common use
$\tilde{}$	Eigenvector or generalized eigenvector.
$\hat{}$	Indicates either a different coordinate representation or parametric representation of an object. Thus v and $\hat{v}$ indicate the coordinate representation of the same vector and s and $\hat{s}$ indicate the parameter in different parametric representation of a vector space.
$\overline{C}$	Complex conjugate of C.
$*$	Adjoint operator.
$\mathbb{C}$	The field of complex numbers.
$\mathbb{F}$	Algebraic field.
i	$\sqrt{-1}$
$\mathbb{Q}$	The field of rational numbers.
$\mathbb{R}$	The field of real numbers.
$\mathbb{Z}$	The set of integers.
$\forall$	Short for "For all".
$\exists$	Short for "There exists".
$\ni$	Short for "Such that".

Contents

Preface v

1. Model Formulation 1

1.1 Point balances 6
1.2 Macroscopic balances 9
1.2.1 Balances in which fluxes are process streams 9
1.2.2 Balances with surface fluxes 18
1.2.3 Balances with chemical kinetics 24
1.3 Differential balances 30
1.3.1 Coordinate systems 30
1.3.2 Constitutive equations 33
1.3.3 Balances in one coordinate direction 35
1.3.4 Balances in several coordinate directions 47
1.3.5 Operator notation 56
1.3.6 Common boundary conditions 63
1.3.7 Common thermodynamic equations 64
1.4 Other kinds of balances 65
1.4.1 Balances on surfaces 65
1.4.2 Population balance models 68
1.4.3 Stochastic processes 72
1.5 Ad-hoc modelling 76

2. Model Simplifications and Approximations 85

2.1 Dimensionless models 85
2.2 Linearized models 90
2.3 Lumping and the quasi-steady state assumption 92
2.4 Geometric simplifications 106

3. Examining a Model 111

3.1 Before even writing any equations 111

3.2 While writing the model . . . 112
3.3 Dimensions and units . . . 112
3.4 Rudimentary approximate solutions . . . 113
3.4.1 Initial value problems . . . 113
3.4.2 Boundary value problems . . . 116
3.5 Examine the solution . . . 117

4. Some Ordinary Differential Equations 123

4.1 First order equations . . . 123
4.1.1 Separable equations . . . 124
4.1.2 Linear, first order equations . . . 127
4.1.3 Exact equations . . . 129
4.1.4 Homogeneous equations . . . 131
4.1.5 Clairaut's equation . . . 132
4.1.6 Bernoulli equation . . . 134
4.1.7 Riccati equation . . . 135
4.2 Second order equations . . . 136
4.2.1 Dependent variable does not occur explicitly . . . 137
4.2.2 Free variable does not occur explicitly . . . 137
4.2.3 Homogeneous equations . . . 138
4.2.4 Constant coefficients and Euler equations . . . 140
4.2.5 Common transformations of linear second order equations . 140
4.3 Higher order equations . . . 142

5. Finite Dimensional Vector Spaces 147

5.1 Basic concepts . . . 147
5.2 Examples . . . 149
5.3 Span, linear independence and basis . . . 151
5.3.1 Coordinates . . . 156
5.4 Isomorphisms . . . 157
5.4.1 Isomorphisms of vector spaces . . . 157
5.4.2 Subspaces . . . 160
5.4.3 Sums . . . 161
5.4.4 Representation of subspaces . . . 164
5.5 Matrices . . . 166
5.5.1 Matrix algebra . . . 167
5.5.2 Gauss elimination . . . 169
5.5.3 Determinants . . . 173
5.5.3.1 Basic properties of determinants . . . 175
5.5.3.2 Calculation of determinants . . . 177
5.5.3.3 The derivative of a determinant . . . 179
5.5.4 The classical adjoint matrix . . . 180

5.6 Systems of linear algebraic equations 181
5.6.1 Rank 181
5.6.2 Applications of rank 186
5.6.3 Solution structure 191
5.6.4 The null and range space of a matrix 198
5.6.5 Overdetermined systems 201
5.7 The algebraic eigenvalue problem 202
5.7.1 Finding eigenvalues and eigenvectors 203
5.7.2 Multiplicity 210
5.7.3 Similar matrices 211
5.7.3.1 Equivalence relations 213
5.7.4 Eigenspaces and eigenbases 214
5.7.4.1 Diagonalization of simple and semi-simple matrices 216
5.7.5 Generalized eigenspaces 217
5.7.5.1 Generalized eigenbases 221
5.7.6 Jordan canonical form 225
5.7.7 Jordan form of real matrices with complex eigenvalues . . . 229
5.7.8 Powers and exponentials of matrices 234
5.7.9 Location of eigenvalues 237
5.8 Geometry of vector spaces 238
5.8.1 Vector products 239
5.8.1.1 Inner product 239
5.8.1.2 Cross product 241
5.8.1.3 Triple scalar product 242
5.8.1.4 Dyad or outer product 245
5.8.2 Gram–Schmidt orthogonalization 245
5.8.3 Eigenrows 248
5.8.4 Real, symmetric matrices 250

6. Tensors 251

6.1 Definitions and basic concepts 252
6.2 Examples 253
6.2.1 Matrices as operators 254
6.2.2 Equivalence transformations 259
6.3 The adjoint operator 262
6.4 Tensors 263
6.4.1 Transformation rules 265
6.4.2 Invariants of tensors 270
6.5 Some tensors from physics and engineering 272
6.5.1 Fourier's law 272
6.5.2 Symmetry and group theory 279

6.5.3 The stress tensor . . . 282
6.6 Vectors and tensors in curvilinear coordinates . . . 290
6.6.1 Proper transformations . . . 291
6.6.2 Vectors and transformations at a point . . . 293
6.6.3 Covariance and contravariance . . . 295
6.6.4 The physical components . . . 301

7. Linear Difference Equations 303

7.1 Linear equations with constant coefficients . . . 314
7.1.1 Homogeneous solutions . . . 314
7.1.2 Particular solutions . . . 317
7.2 Single, first order equations . . . 321
7.3 Single, higher order equations . . . 322
7.3.1 Solution by variable transformation . . . 322
7.3.1.1 Euler's equation . . . 322
7.3.2 Reduction of order . . . 322
7.3.3 Particular solution by variation of parameters . . . 324
7.4 Systems of linear difference equations . . . 326
7.4.1 Basic theorems . . . 327
7.4.2 Particular solution by variation of parameters . . . 330
7.4.3 Equations with constant coefficients . . . 331
7.4.3.1 Homogeneous solutions . . . 332
7.4.3.2 Particular solutions for constant inhomogeneous term . . . 338
7.5 Nonlinear equations . . . 341
7.5.1 Riccati's equation . . . 342

8. Linear Differential Equations 343

8.1 Linear equations with constant coefficients . . . 344
8.1.1 Homogeneous solutions . . . 344
8.1.2 Method of matched coefficients . . . 347
8.2 Single, higher order equations . . . 353
8.2.1 Reduction of order . . . 353
8.2.2 Particular solution by variation of parameters . . . 355
8.3 Systems of linear differential equations . . . 357
8.3.1 Basic theorems . . . 359
8.3.2 Particular solution by variation of parameters . . . 360
8.3.3 Equations with constant coefficients . . . 361
8.3.3.1 Homogeneous solutions . . . 362
8.3.3.2 Particular solutions for constant inhomogeneous term . . . 367
8.3.3.3 Dealing with complex eigenvalues . . . 370

8.3.3.4 Classification of steady states 371
8.4 Stability of nonlinear ODEs 379
8.5 Series solutions 386
8.6 Some common functions defined by ODEs 394
8.6.1 Exponential and trigonometric functions 394
8.6.2 Bessel functions 395
8.6.3 Legendre functions 402

9. Hilbert Spaces 405

9.1 Infinite dimensional vector spaces 406
9.1.1 Countable and uncountable infinities 407
9.1.2 Normed spaces 408
9.1.3 Bases in infinite dimensional spaces 411
9.1.4 The function spaces $\mathcal{L}^p[0,1]$ 413
9.2 Hilbert spaces 414
9.2.1 Inner products 414
9.2.2 Examples 416
9.2.3 Orthogonality 417
9.2.4 Orthogonal projections 421
9.2.5 Orthogonal complements 423
9.3 Linear operators in Hilbert spaces 423
9.3.1 The adjoint operator 424
9.3.2 Examples 425
9.3.3 Sturm–Liouville operators 427
9.4 Eigenvalue problems 428
9.4.1 Sturm–Liouville problems 433
9.4.2 Conversion of linear equations to SLP 436
9.5 Fourier series 438
9.5.1 Fourier sine series 438
9.5.2 Fourier cosine series 438
9.5.3 Complete Fourier series 439
9.5.4 Gibb's phenomena 441
9.5.5 Generalized Fourier series 444

10. PDEs with Self-Adjoint Operators 449

10.1 Fourier series methods 449
10.1.1 Classification of second order PDEs 449
10.1.2 Inner product method 453
10.1.3 PDEs with Sturm–Liouville operators 457
10.1.3.1 Homogeneous problem 457
10.1.3.2 Homogeneous problem with transcendental equation for eigenvalues 460

10.1.3.3 Inhomogeneous PDE 472
10.1.3.4 Inhomogeneous, time varying boundary conditions 476
10.1.4 Other self-adjoint PDEs 479
10.2 Finite Fourier transform 490

11. First Order PDEs 493

11.1 Models that give rise to first order PDEs 493
11.2 First order PDE and Cauchy's method 498
11.2.1 Cauchy's method for linear equations 500
11.2.2 Cauchy's method for semi-linear equations 513
11.3 Shocks 516
11.4 Reduction of higher order PDEs 525

12. Similarity Transformations 529

12.1 Semi-infinite media 529

13. Problems 543

Appendix 613
A.1 Complex numbers 613

Index 617

Chapter 1

Model Formulation

Engineers write mathematical models for numerous reasons. To design processes, to scale them up or down, to optimize process performance and financial gain, to minimize formation and release of undesirable or hazardous compounds, to design control strategies and so on. There hardly seems a need to justify why engineers need to learn some mathematics and develop the basic skills needed for writing mathematical models.

Writing mathematical models falls into the class of dreaded "word problems" and is not an art by a skill that can be acquired by practising writing models for increasingly complex systems. Engineering models are not true or false in the sense that fundamental laws of nature are true or false. Engineering models are good or they are bad. Good engineering models accurately reflect, within measurement error, the process being modelled over the relevant range of operating conditions. Good models are mechanistic in the sense that they are based on established physics and chemistry, are as simple as possible in that they only incorporate mechanisms that are significant for the system being modelled and have only a small number of model parameters. Bad models on the other hand use ad-hoc[1] assumptions, are unnecessarily complex by incorporating a smattering of mechanisms without any regard for whether or not these are important and have "too many" model parameters. It is often possible to devise several different models of the same physical reality and one can pick and choose among these depending on the desired model accuracy, the purpose of the model and on the ease of model analysis.

Suppose we want to develop a model of the number of students in a classroom. It is a good idea to start by specifying the nomenclature that will be used when writing the model. Let $N(t)$ be the number of students in the classroom at time t, then

$$\begin{aligned} N(t) = N(t_0) + &\text{ Number of students that entered between time } t_0 \text{ and time } t \\ &- \text{ Number of students that left between time } t_0 \text{ and time } t. \end{aligned}$$

[1]Merriam Webster defines this as: for the particular end or case at hand without consideration of wider application.

This ridiculously simple model is an example of an initial value problem; an initial value, the value of $N(t_0)$, must be specified to solve the problem. In spite of its simplicity, the model encompasses three concepts that are common to all models we will consider in this book. The three concepts are the idea of a conserved property, a control volume and transport across the control volume surface.

The conserved property in this model is the number of students. The number of students in the classroom may change but only through known processes, by students entering and exiting the room. Students do not appear or vanish into thin air spontaneously. It is this property that allows us to write a *conservation balance* on the number of students. Conservation balances can be written only on conserved properties such as mass, moles, energy, number of particles, etc. These are called extensive properties, physical properties that are additive and depend on the size of the system. One cannot write conservation balances on properties such as temperature, concentration, pressure, density, etc. These are intensive properties or bulk properties that are independent of the size of the system and not additive when systems are joined together.

A conservation balance is always written over a control volume (CV), the classroom in this example. The CV over which a balance equation is written is defined by its surface and the conservation balance above states that the number of students in the room change depending on how many students cross the CV surface and whether they leave or enter the CV. The term is slightly misleading because the CV may not be a volume in physical space and the size of the CV may therefore not have units of length cubed. For instance, in the case of the classroom, it would be more reasonable to measure the size of the room as floor area. The size of a CV may also change with time. For instance, the combustion engine of an ordinary car can be modelled as a chemical reactor that repeatedly expands and contracts as fuel is added, combusted and expelled. The engine works by virtue of the fact the physical volume of the CV changes with time.

The students entering and exiting the room represent a rate of transport across the CV surface and to complete the model one must have some way of modelling this rate of transport. In most chemical engineering applications, the rates of transport are given by either the flow rates and compositions of process streams or by constitutive equations such as Fourier's law of conduction (for energy balances), Fick's law (for mass or mole balances) or Newton's law of viscosity (for momentum balances). Constitutive equations are not fundamental laws of nature, but rather expression or models based on empirical observations or more fundamental models such as models of molecular dynamics. The modeller may have available a large number of possible constitutive equations that can describe a given rate of transport. Unless there are substantial reasons for doing otherwise, the best choice is usually to pick a linear constitutive equation since this will give the simplest mathematical model. For instance, if one does not know a priori that a fluid is non-Newtonian, then one should use Newton's law of viscosity rather than the power law model in a momentum balance.

It is important to be mindful of the fact that the constitutive equations mentioned above, as well as most other constitutive equations, are not expressions for the rate of transport but expressions for the flux. Flux is amount (of the conserved property) per time per area, so the rate of transport across the control volume surface must be calculated as the product of the flux at a right angle to this surface and the surface area. For instance, Fick's law for diffusion is

$$N_x = -D\frac{dC}{dx}$$

where N_x is the molar or mass flux in the x-direction. The diffusion coefficient D has dimension of area per time and the concentration gradient in the x-direction has dimension of mole or mass per volume per length. The expression on the right hand side therefore has dimension

$$D\frac{dC}{dx}[=]\frac{\text{length}^2}{\text{time}}\frac{\text{moles}}{\text{length}^3\text{length}}[=]\frac{\text{moles}}{\text{time length}^2}$$

or moles per time per area, not moles per time.

Suppose now that we want to model, not the number of students in a classroom, but the number of people in a country. The size of the control volume, i.e. the size of the country, is huge compared to the size of a classroom and the number of people may run into millions. In this case, it makes sense to try and derive an approximate model by using a continuous variable, the population density, as the unknown. The error we make is minor if the number of people in the country is huge and is similar to the error we routinely make whenever we calculate the number of moles inside a CV as the product of a concentration and the physical volume of the CV. Strictly speaking, the value of this product should correspond to an integer number of molecules, the total number of molecules inside the CV, but it would be quite surprising if it in fact did. However, the difference between 10^{24} molecules and $10^{24}+1$ molecules is so insignificant we might as well accept the idea of $10^{24}+0.5$ molecules if it simplifies the modelling problem. With this in mind, we can define the population density $C(t)$ as the number of people per area at time t and the above conservation balance on the number of people in a country becomes

$$C(t)A(t) = C(t_0)A(t_0) + \int_{t_0}^{t} \text{Rate in } dt - \int_{t_0}^{t} \text{Rate out } dt\,.$$

$A(t)$ is the area of the country and it is written explicitly as a function of time to emphasize the fact that the size of a CV may change with time. Even sizes of countries may change through coastal erosion, wars and annexations. The balance above is correct under our assumptions but does not look like a familiar type of mathematical problem. We can obtain a more familiar form by differentiating with respect to time.

$$\frac{dCA}{dt} = \text{Rate in} - \text{Rate out}$$

which we recognize as a differential equation, a type of mathematical problem that most readers are bound to recognize. In fact, in all that follows, the differential equation form of the conservation equation will be our starting point.

Most people would probably object to the model above because emigration (rate of people leaving) and immigration (rate of people entering) are the only mechanisms through which the population density can change. While this may be a reasonable assumption for a classroom, the rates of birth and death must matter when we consider an entire country. Thus a better model is

$$\frac{dCA}{dt} = \text{Rate in} - \text{Rate out} + \text{Birth rate} - \text{Death rate}.$$

This equation is a special instance of our basic transient conservation equation which is

$$\text{Accumulation rate} = \text{rate in} - \text{rate out} + \text{production rate} - \text{consumption rate}.$$

The rates of formation and consumption that appear in most chemical engineering models are not birth or death rates, but rates of chemical reactions. The rates of chemical reactions are customarily given as specific rates, amount of moles of reactant formed in a given reaction per volume per time. So, just like fluxes must be multiplied by the area for transport in or out of a CV, specific chemical reaction rates must be multiplied by the volume of the CV.

We can state a general form of the conservation balance which is valid for most chemical engineering models.

$$\begin{aligned}\frac{dV(t)C(t)}{dt} =& \sum \text{Rate of transport in/out via process streams}\\ &+ \sum \text{Rate of transport in/out across surfaces } A(t)\\ &+ \sum \text{Rate of formation in chemical reactions } V(t)\end{aligned}$$

where $C(t)$ is the concentration of the extensive variable (mole per volume, energy per volume, number of particles per volume, etc.) The "accumulation" term in the balance is the transient or time dependent term. For steady state balances, this term is zero. Transport terms for transport across the control volume is split into two terms, one for process streams, another for transport across control volume surfaces such as heat loss to a cooling jack or evaporation from the surface of a liquid. These two kinds of transport phenomena are very different with only the latter requiring the use of constitutive equations for the flux. It is important that the sums are taken over the entire CV surface, that all fluxes are counted. Overlooking one or more surface fluxes is a frequent source of modelling error.

An assumption implicit in the above conservation balance is that the concentration inside the control volume, $C(t)$, is well defined and independent of what point inside the CV we consider. This is the "well mixed" assumption which is used frequently when modelling vessels and containers in chemical processes equipment. Since, in a well mixed system, the value of intensive variables do not depend on the position inside the CV, the properties of the exit streams are identical to those of the vessel content. This makes it particularly easy to state the expressions for the transport out of the CV via process streams. However, the well mixed assumption is exactly that, an assumption, and may not always be valid. For instance, a chemical reactor may not be well mixed if the reaction rate is much faster than the rate of mixing. In this case, the reactant concentrations close to the feed stream inlets will be different from the concentrations in the bulk of the vessel and close to the vessel outlets. This type of situation can create challenging problems for the modeller.

We can formally write a conservation balance on a control volume that is not well mixed by finding the total amount of the extensive variable inside the control volume by a volume integral. The accumulation terms would then take the form

$$\text{Accumulation rate} = \frac{d}{dt}\left(\int_{V(t)} C(x, y, z, t)dV\right).$$

This is a perfectly correct model but it is no longer a differential equation. It is something else, not immediately recognizable by most engineers as a standard mathematical problem. To avoid the difficulties that follow from this, we will never use a control volume that is not well mixed if there is any way to avoid it. The common way to avoid non-well mixed control volumes is to make the control volume so small that it becomes well mixed. This is so important that it should be stated as a rule of thumb:

Rule of Thumb. *Pick the control volume as large as possible but still so small that it is well mixed.*

In some situations, such as the abovementioned chemical reactor in which the rate of reaction is much faster than the rate of mixing, the way to accomplish this is to use a differential CV. One would pick a differential CV inside the reactor, a volume defined, for instance, by the differential distances dx, dy and dz in a classical rectangular coordinate system. One can write conservation balances on this CV such as momentum balances and molar balances. The solution to these balances will provide the fluid flow field inside the reactor and the reactant concentrations at each point inside the reactor. A simpler example of a differential balance is a model of the temperature profile in the pipe wall of a heat exchanger. The temperature in the pipe wall will be a function of the radial position and the pipe wall is therefore not well mixed. The appropriate CV to use will be a differentially thin section of the pipe wall defined by the radial distance dr. An energy balance on this CV gives rise to a differential equation that can be solved for the temperature profile. Balances that use differential CVs are called *differential balances* while balances that do not are called *macroscopic balances.* The two types of balances are discussed in separate sections. However, we will first briefly mention *point balances* which, as the name implies, are balances where the CV is a point.

Mass, energy and momentum balances are by far the most common and important chemical engineering models. We will consider these balances for three different types of control volumes: (1) point balances, (2) macroscopic balances and (3) differential balances. This is followed by a brief discussion of other types of balances that do not easily fit into this pattern.

1.1 Point balances

Point balances are those balances that are derived using a point of zero volume as the CV, such as splitters and mixing points. When the control volume has zero physical volume, then the accumulation and generation terms are necessarily zero and the balance equation is reduced to

$$In = Out\,.$$

In particular, the transient balances are identical to the steady state balances because the accumulation term in the balance is zero.

Example 1.1: Mass and energy balance on mixing point.

Two process streams with volumetric flow rates $Q_1(t)$ and $Q_2(t)$, temperatures $T_1(t)$ and $T_2(t)$ and with molar concentrations of a solute of $C_1(t)$ and $C_2(t)$ respectively, are merged at a point into a single stream, see Fig. 1.1. The physical properties of the two streams are functions of temperature. We want to write a model for the temperature and solute concentration of the exit stream.

In this case the CV is the mixing point and it will be modelled as having zero volume. This is strictly speaking not true but it is a common and usually valid modelling assumption. An energy balance requires that we know the flow rates of all the streams, and a model for the exit temperature therefore requires both a mass and an energy balance. The mass balance is

$$Q_1(t)\rho_1(T_1(t)) + Q_2(t)\rho_2(T_2(t)) = Q_e(t)\rho_e(T_e(t))$$

where ρ is the density of the streams. The energy, or more correctly the enthalpy balance, is

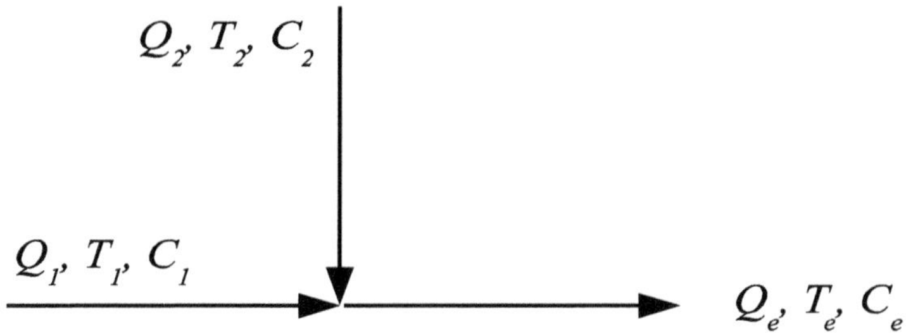

Fig. 1.1 Process streams with mixing point.

$$Q_1(t)\rho_1(T_1(t))\int_{T_{ref}}^{T_1(t)} c_{p,1}(T)dT + Q_2(t)\rho_2(T_2(t))\int_{T_{ref}}^{T_2(t)} c_{p,2}(T)dT$$
$$= Q_e(t)\rho_e(T_e(t))\int_{T_{ref}}^{T_e(t)} c_{p,e}(T)dT$$

where c_p is the heat capacities of the streams. T_{ref} is the reference temperature used when calculating enthalpies.

The molar balance on the solute is

$$Q_1(t)C_1(t) + Q_2(t)C_2(t) = Q_e(t)C_e(t)\,.$$

A frequently used simplifying assumption is that the physical properties of a substance are constant. If this can be invoked, then $\rho_1 = \rho_2 = \rho_e$ and the mass balance is reduced to

$$Q_1 + Q_2 = Q_e$$

where the explicit time dependence has been dropped to simplify the nomenclature. This result looks deceptively like a volume balance, but as volume is not a preserved property it does not make sense to talk about conservation of volume. The exception to this is of course systems in which density is constant and a volume balance becomes a stand-in for a mass balance.

If furthermore the heat capacity can be assumed constant, the balances simplifies greatly to

$$Q_1c_p(T_1 - T_{ref}) + Q_2c_p(T_2 - T_{ref}) = Q_ec_p(T_e - T_{ref}) \;\Rightarrow$$
$$Q_1T_1 + Q_2T_2 = Q_eT_e + T_{ref}(Q_e - Q_1 - Q_2)\,.$$

Using the mass balance, this is reduced further to

$$Q_1T_1 + Q_2T_2 = Q_eT_e\,.$$

Unless stated otherwise, we will assume constant physical properties in the rest of the book.

It would seem that a splitting point, a point where a process stream splits into two or more streams, is too simple to be worthy of modelling analysis. But this is not quite true. Any separation device that operates at steady state can be modelled as a splitting point because at steady state the accumulation term is 0 and a balance on the device simplifies to the same form as a balance on a splitting point. Balances on a splitting point therefore provide constraints that any separation device must satisfy when operating at steady state.

Example 1.2: Mass balances on a splitting point.

A feed stream with a volumetric flow rate of Q, consisting of a water-ethanol mixture with a volumetric ethanol concentration C_{feed} of 10%, must be separated into a product stream with a volumetric ethanol concentration C_p of 45% and a waste stream, see Fig. 1.2. It is of obvious interest to maximize the flow rate Q_p of the product stream while minimizing the flow rate Q_w of the waste stream. What is the largest volumetric flow rate one can hope to obtain for the product stream?

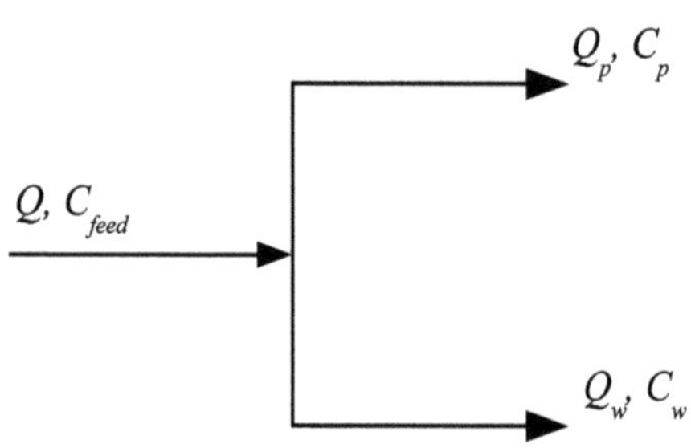

Fig. 1.2 Modelling a separator operating at steady state as a splitting point.

Assuming that there is no volume change associated with changes in composition, a questionable assumption in the case of water-ethanol mixtures as it turns out, the total mass balance and the ethanol mass balance over the splitting point are

$$Q = Q_p + Q_w$$

$$QC_{feed} = C_pQ_p + C_wQ_w\,.$$

The volumetric flow rate of the product stream will obviously be proportional to the volumetric flow rate of the feed stream so what we are really seeking is the maximum value of Q_p/Q. The two equations above are readily solved to give

$$\frac{Q_p}{Q} = \frac{C_w - 0.1}{C_w - 0.45}$$

and the maximum value of $0.2222... = 0.\bar{2}$ is attained when $C_w = 0$. This result should not come as a surprise; the maximum flow rate of the product stream is attained when all the ethanol is stripped out of the waste stream. It may not be possible or economically feasible to build a separator that can do this, but that is beside the point here. The point here is that only the balance equation was able to provide an actual number for the maximum possible flow rate of the product stream. Anticipating that this happens when all ethanol is striped from the waste stream does not suffice for obtaining this number.

1.2 Macroscopic balances

Macroscopic balances are balances on well mixed CVs that have an integral physical volume, usually process vessels and containers. It is frequently helpful to split the terms for transport across the CV surface into contributions from process streams and contributions from surface transport.

1.2.1 *Balances in which fluxes are process streams*

In simple cases involving only process streams, the *rates in* and *rates out* of a variable are the volumetric flow rates of the process streams multiplied by their concentrations, energy density or some other intensive variable.

Example 1.3: Mass and energy balance on mixing tank.

Several feed streams are combined in a tank with fixed volume V. The tank is assumed well mixed and has a single exit stream. The volumetric flow rates of the streams are indicated by Qs, their temperatures by Ts and the molar concentrations of a solute in the streams are indicated Cs. see Fig. 1.3.

Assuming no volume change in the mixing, the transient mass balance using the content of the tank as the CV is

$$\frac{dV}{dt} = \sum_{n=1}^{N} Q_n(t) - Q_e(t) \,.$$

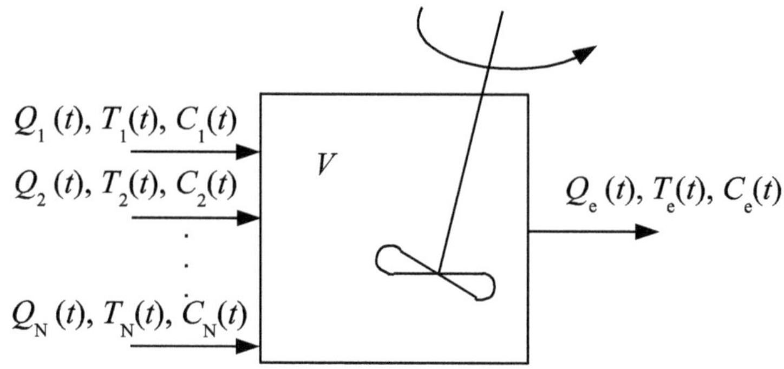

Fig. 1.3 Mixing of several process streams in a well mixed tank.

The balance on the solute is

$$\frac{dC_eV}{dt} = C_e\frac{dV}{dt} + V\frac{dC_e}{dt} = \sum_{n=1}^{N} C_n(t)Q_n(t) - C_e(t)Q_e(t)\,.$$

The energy balance is

$$\rho c_p\frac{dT_eV}{dt} = \rho c_pT_e\frac{dV}{dt} + \rho c_pV\frac{dT_e}{dt} = \sum_{n=1}^{N} \rho c_pT_n(t)Q_n(t) - \rho c_pT_e(t)Q_e(t)\,.$$

It is customary to write models in an agreed upon standard form. In this case as coupled ordinary differential equations, each of which contains a single derivative, usually isolated on the left side of the equation.

$$\frac{dV}{dt} = \sum_{n=1}^{N} Q_n(t) - Q_e(t)$$

$$\frac{dC_e}{dt} = \frac{1}{V}\left(\sum_{n=1}^{N} C_n(t)Q_n(t) - C_e(t)Q_e(t)\right) - \frac{C_e}{V}\left(\sum_{n=1}^{N} Q_n(t) - Q_e(t)\right)$$

$$\frac{dT_e}{dt} = \frac{1}{V}\left(\sum_{n=1}^{N} T_n(t)Q_n(t) - T_e(t)Q_e(t)\right) - \frac{T_e}{V}\left(\sum_{n=1}^{N} Q_n(t) - Q_e(t)\right)$$

The model derived in the example above is quite simple but models quickly become mathematically complicated if the number of process streams, vessels, mixing points and splitters becomes larger. In such situations, one can write balance equations for each piece of equipment and for each mixing point and splitter or one can write overall balances on several pieces of connected equipment. The balances or CVs that are chosen for writing the model of a process must be picked such that the balances are independent of one another, i.e. such that none of the balance equations are obtainable by manipulating the other balance equations. So, use common sense when picking CVs. If you use two separate vessels as CVs for two balances, you should not expect that a CV enclosing both vessels will give a balance equation that is independent of the first two. Additionally, a model must be complete. A model is said to be complete or closed if it has enough equations to allow one to solve, in principle at least, for all the unknowns. From a practical point, this means that a model must have at least as many independent equations as unknowns.

This is illustrated in the next example which is a steady state model of a recycle reactor. The example also illustrates the point that, in general, balance equations alone are insufficient for a complete model and that additional equations must be obtained from other sources.

Example 1.4: Steady State Balances on Recycle Reactor.

The reactor to be considered is sketched in Fig. 1.4. The reactor inlet stream is made by mixing the process feed stream and a recycle stream. The outlet stream from the reactor is separated into a product stream and a side stream, and the latter is split into a recycle stream and a purge. All flow rates are molar flow rates and are indicated by Fs. Composition is given in terms of mole fractions, indicated by ωs.

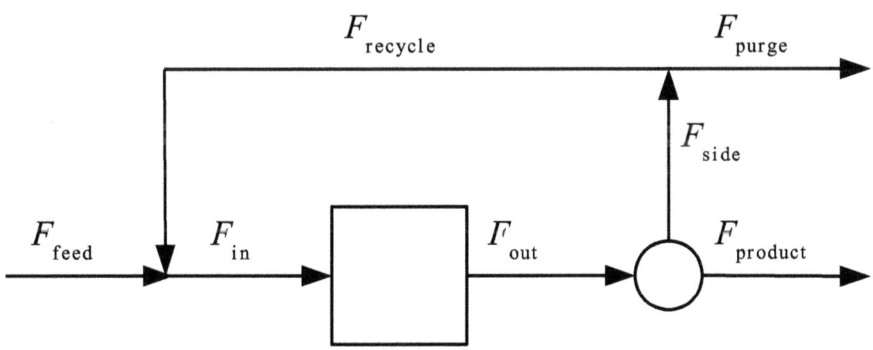

Fig. 1.4 Recycle reactor.

The feed to the process consists of H_2, N_2 and an inert called I. The mole fractions in the feed are $\omega_{H_2,\text{feed}} = 0.69$, $\omega_{N_2,\text{feed}} = 0.23$ and $\omega_{I,\text{feed}} = 0.08$. In the reactor the hydrogen and nitrogen are converted to ammonia by

$$3H_2 + 2N_2 \leftrightarrow 2NH_3$$

and we seek a model which will allow us to determine the composition and magnitude of the product stream as a function of the recycle ratio $\alpha = F_{\text{recycle}}/F_{\text{feed}}$.

The first problem one is faced with when writing a model is to make sure that there is enough information available to even make this possible. Clearly, this is not yet the case for this reactor. Obviously, one must specify something about how much ammonia is formed in the reactor and something about how the components are partitioned between the product and side stream in the separator. We will eschew complicated reactor models at this time and assume that 90% of the hydrogen that enters the reactor is converted to ammonia. Thus, the hydrogen balance over the reactor is

$$0.1F_{\text{in}}\omega_{H_2,\text{in}} = F_{\text{out}}\omega_{H_2,\text{out}}.$$

For the separator, we will assume that it separates the ammonia completely from the remaining compounds so that the product stream is pure ammonia, $\omega_{NH_3\text{products}} = 1$, and the side stream contains no ammonia.

There are a large number of control volumes that can now be used. Some of these are depicted by the ovals in Fig. 1.5. Number 1 is a CV around the mixing point of the feed stream and the recycle stream. These two streams represent the "in" terms in the balances over this CV, and the reactor inlet stream represents the "out" term in the balance. Number 2 is the reactor, number 3 the separator, number 4 the splitter and number 5 the entire process. Other CVs are possible such as the reactor plus the separator, the separator plus the splitter, etc.

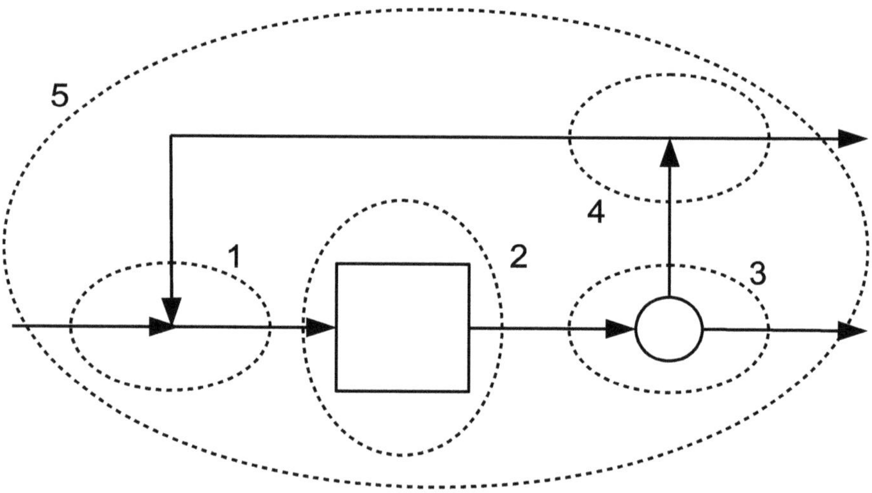

Fig. 1.5 Some of the possible control volumes that can be used to write balance equations on the recycle reactor.

Three types of balances can be written over these CVs: elemental balances, compound balances and, when there is no reaction inside the control volume, total mole balances. The elemental balances and the total mole balances are the easiest to write since they are independent of the reactions that take place but they can never suffice for a model precisely because they do not capture any information about the chemical reactions and at least one component balance over a CV enclosing the reactor is required. Similarly, at least one of the model equations must capture information about the performance of the separator, one equation must capture information about the feed stream and its composition and one equation must capture information about the stoichiometry of the reaction.

One can mindlessly start writing all possible balances over all possible control volumes but this wastes a lot of effort because many of the balances will be dependent upon one another. For instance, the total mole balances on the separator and the splitter are

$$F_{\text{out}} = F_{\text{side}} + F_{\text{products}}$$

$$F_{\text{side}} = F_{\text{purge}} + F_{\text{recycle}}$$

and summing these gives the total mole balance over the CV which encloses the separator and the splitter

$$F_{\text{out}} = F_{\text{purge}} + F_{\text{recycle}} + F_{\text{products}}.$$

This balance is therefore not independent of the balances over the separator and splitter and one of the three balances is therefore of no use.

It is better to start with the simplest balances and work backwards until one has a complete model. The challenge is to recognize how to get to that point in the easiest way possible.

Two equations of the model are essentially given already. That is the specified recycle ratio α, from which

$$F_{\text{recycle}} = \alpha F_{\text{feed}} \tag{1.1}$$

and the equation for conversion of hydrogen over the reactor

$$0.1 F_{\text{in}} \omega_{H_2,\text{in}} = F_{\text{out}} \omega_{H_2,\text{out}}. \tag{1.2}$$

According to the stoichiometry of the reaction, 3 moles of hydrogen are consumed for each mole of nitrogen consumed. Since the feed contains hydrogen and nitrogen in this ratio, this ratio will remain 3-to-1 at all points in the system and it follows that 90% of the nitrogen in the reactor inlet must also be converted in the reactor. Thus, a nitrogen balance over the reactor is

$$0.1 F_{\text{in}} \omega_{N_2,\text{in}} = F_{\text{out}} \omega_{N_2,\text{out}}. \tag{1.3}$$

Notice that the mole fractions appear in identical ways in the hydrogen and nitrogen balances. This is only possible because the feed to the process contains hydrogen and nitrogen in the proper stoichiometric ratio for formation of ammonia. A different feed composition would have made these two balances very different in appearance.

Since the unknown we want to find is the magnitude of the product stream (its composition has been specified as pure ammonia) it makes sense to let the next equation in the model be an equation for this variable. The simplest equation is a total mole balance over the separator.

$$F_{\text{out}} = F_{\text{side}} + F_{\text{product}}. \tag{1.4}$$

Two more easy equations are obtained from the total mole balances over the CVs that do not enclose chemical reactions, the splitter and the mixing point.

$$F_{\text{side}} = F_{\text{purge}} + F_{\text{recycle}} \tag{1.5}$$

$$F_{\text{feed}} + F_{\text{recycle}} = F_{\text{in}} \tag{1.6}$$

which gives 6 equations with 10 unknowns, F_{recycle}, F_{side}, F_{purge}, F_{out}, F_{in}, F_{product}, $\omega_{H_2,\text{in}}$, $\omega_{H_2,\text{out}}$, $\omega_{N_2,\text{in}}$ and $\omega_{N_2,\text{out}}$. The inert balance over the reactor provides another equation

$$F_{\text{in}}\omega_{I,\text{in}} = F_{\text{out}}\omega_{I,\text{out}} \tag{1.7}$$

as does the ammonia balance, which states that the moles of ammonia leaving the reactor is the amount entering plus the amount formed.

$$F_{\text{in}}\omega_{NH_3,\text{in}} + 2 \cdot 0.9 \cdot F_{\text{in}}\omega_{N_2,\text{in}} = F_{\text{out}}\omega_{NH_3,\text{out}}$$

However, as all the ammonia formed is removed in the separator, the recycle stream is ammonia free and thus $\omega_{NH_3,\text{in}} = 0$. The ammonia balance therefore simplifies to

$$2 \cdot 0.9 \cdot F_{\text{in}}\omega_{N_2,\text{in}} = F_{\text{out}}\omega_{NH_3,\text{out}}. \tag{1.8}$$

The model now has 8 equations and 13 unknowns. We have still not made use of the information about the composition of the feed stream. This information can be included by writing a balance over the mixing point. The inert balance is the simplest.

$$0.08F_{\text{feed}} + F_{\text{recycle}}\omega_{I,\text{recycle}} = F_{\text{in}}\omega_{I,\text{in}} \tag{1.9}$$

which add another unknown, $\omega_{I,\text{recycle}}$ for a total of 9 equations with 14 unknowns.

To keep the model as simple as possible, one should look for additional equations that do not introduce new unknowns. An obvious source, but for some reason often overlooked, comes from the fact that the mole fractions in a stream sum to 1. This gives three additional equations, one for the reactor inlet, one for the reactor outlet and one for the recycle stream (which has the same composition as the purge steam). Only the first two do not introduce new unknowns, while the equation for the recycle stream introduces two additional unknowns $\omega_{H_2,\text{recycle}}$ and $\omega_{N_2,\text{recycle}}$.

$$\omega_{H_2,\text{in}} + \omega_{N_2,\text{in}} + \omega_{I,\text{in}} = 1 \tag{1.10}$$

$$\omega_{H_2,\text{out}} + \omega_{N_2,\text{out}} + \omega_{NH_3,\text{out}} + \omega_{I,\text{out}} = 1 \tag{1.11}$$

$$\omega_{I,\text{recycle}} + \omega_{H_2,\text{recycle}} + \omega_{N_2,\text{recycle}} = 1. \tag{1.12}$$

These three equations must always be satisfied and must be included in the model to make it complete. They could have been stated at the very beginning.

The two new unknowns $\omega_{H_2,\text{recycle}}$ and $\omega_{N_2,\text{recycle}}$ are related by the fact that the hydrogen-to-nitrogen ratio equals 3 in all streams.

$$\omega_{H_2,\text{recycle}} = 3\omega_{N_2,\text{recycle}}. \tag{1.13}$$

Three possible additional equations that do not introduce new unknowns are the ammonia balance over the separator

$$F_{\text{product}} = F_{\text{out}}\omega_{NH_3,\text{out}}, \tag{1.14}$$

the inert balance over the separator

$$F_{\text{out}}\omega_{I,\text{out}} = F_{\text{side}}\omega_{I,\text{recycle}} \tag{1.15}$$

and a total hydrogen balance over the entire process

$$0.69F_{\text{feed}} = F_{\text{purge}}\omega_{H_2,\text{recycle}} + \frac{3}{2}F_{\text{product}}. \tag{1.16}$$

Equation (1.1) through (1.16) constitute the model, 16 nonlinear algebraic equations with 16 unknowns. Some of the equations are arbitrary in the sense that they can be replaced by other balances while others, such as the equations for the sum of the mole fractions, must be included in the model. The model certainly looks cumbersome but it can be solved with a bit of effort, thereby confirming that the model is complete or well posed. The solution is

$$F_{\text{product}} = \frac{146\alpha + 189\alpha^2 - 43 \pm (1+\alpha)\sqrt{50625\alpha^2 - 1350\alpha + 1849}}{182\alpha + 225\alpha^2 - 43 \pm (1+\alpha)\sqrt{50625\alpha^2 - 1350\alpha + 1849}}\frac{F_{\text{feed}}}{2}$$

$$F_{\text{recycle}} = \alpha F_{\text{feed}}$$

$$F_{\text{in}} = F_{\text{feed}}(1+\alpha)$$

$$F_{\text{purge}} = \frac{36}{225\alpha - 43 \pm \sqrt{50625\alpha^2 - 1350\alpha + 1849}}\alpha F_{\text{feed}}$$

$$F_{\text{side}} = \frac{225\,\alpha - 7 \pm \sqrt{50625\alpha^2 - 1350\alpha + 1849}}{225\alpha - 43 \pm \sqrt{50625\alpha^2 - 1350\alpha + 1849}}\alpha F_{\text{feed}}$$

$$F_{\text{out}} = \frac{450\alpha^2 + 175\alpha \pm (2\alpha+1)\sqrt{50625\alpha^2 - 1350\alpha + 1849} - 43}{225\alpha - 43 \pm \sqrt{50625\alpha^2 - 1350\alpha + 1849}}\frac{F_{\text{feed}}}{2}.$$

The result for the mole fraction can be expressed most compactly in terms of the molar flow rates

$$\omega_{I,\text{in}} = \frac{9F_{\text{out}} - 11F_{\text{product}}}{9F_{\text{in}}}$$

$$\omega_{I,\text{out}} = \frac{9F_{\text{out}} - 11F_{\text{product}}}{9F_{\text{out}}}$$

$$\omega_{I,\text{recycle}} = \frac{9F_{\text{out}} - 11F_{\text{product}}}{9F_{\text{side}}}$$

$$\omega_{N_2,\text{in}} = \frac{5F_{\text{product}}}{9F_{\text{in}}}$$

$$\omega_{N_2,\text{out}} = \frac{F_{\text{product}}}{18F_{\text{out}}}$$

$$\omega_{N_2,\text{recycle}} = \frac{9F_{\text{side}} + 11F_{\text{product}} - 9F_{\text{out}}}{36F_{\text{side}}}$$

$$\omega_{H_2,\text{in}} = \frac{2F_{\text{product}} - 3F_{\text{out}} + 3F_{\text{in}}}{3F_{\text{in}}}$$

$$\omega_{H_2,\text{out}} = \frac{F_{\text{product}}}{6F_{\text{out}}}$$

$$\omega_{H_2,\text{recycle}} = \frac{9F_{\text{side}} + 11F_{\text{product}} - 9F_{\text{out}}}{12F_{\text{side}}}$$

$$\omega_{NH_3,\text{out}} = \frac{F_{\text{product}}}{F_{\text{out}}}.$$

Notice that solution of the model equations at one step required finding the roots of a second order polynomial. Formally, there are therefore two solutions. Although multiple valid solutions are possible when solving nonlinear algebraic models, quite often some of these solutions must be discarded because they are not physically meaningful. In this case, evaluating the result for various values of the recycle ration α will show that one of the solutions has values of mole fractions that are not between zero and one. Obviously, this solution must be discarded.

Example 1.5: Transient balance on density gradient device.

The contraption shown in Fig. 1.6 is used to make so-called density gradients, which are used in separation of molecular species based on the density.

The device consists of two tanks. Tank 1 contains a less dense sucrose solution while Tank 2 contains a more dense sucrose solution. A stream with volumetric flow rate Q_e is continuously withdrawn from Tank 2. As the dense sucrose solution is removed from Tank 2, it is replenished by the less dense solution from Tank 1. The sucrose concentration in the exit stream therefore continues to drop and it can be layered carefully in a test tube to create the density gradient through the test tube. A thin layer of the mixture to be separated is then placed on top of the sucrose solution and the test tube is placed in a centrifuge. During centrifugation,

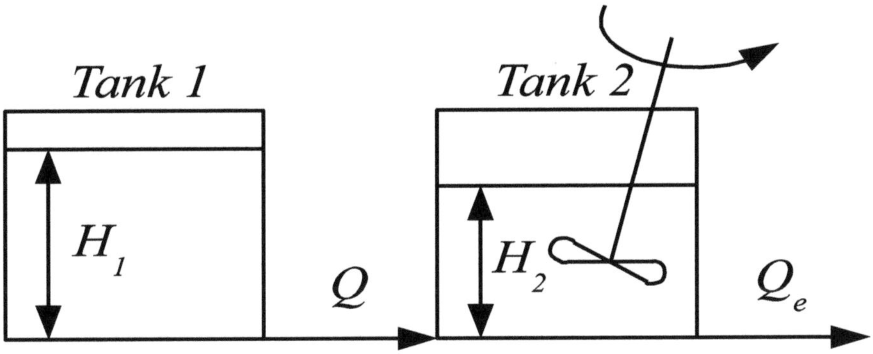

Fig. 1.6 Two connected mixing tanks used for making density gradients.

the molecules in the sample will sediment through the sucrose density gradient until they reach steady state at a position where the density of the sucrose solution is equal to that of the molecule. The molecules in the sample are thus separated based on their density.

We will model the device for making the sucrose gradient and determine how the density of the exit stream depends on how the device is operated. The following assumptions will be used in the model.

- Tank 1 and 2 are well mixed, i.e. the concentration in the exit stream equals the concentration in tank 2.
- The two tanks are geometrically identical and the liquid volume in the tanks can be found as AH_n where A is the cross sectional area of the tanks and H_n the liquid level in tank n.
- The volumetric flow rate between the tanks can be expressed as $Q = (H_1 - H_2)/R$, where R is a constant fluid resistance.

There are three unknowns in the problem, the two liquid levels, $H_1(t)$ and $H_2(t)$, and the sucrose concentration in Tank 2, $C_2(t)$, so three equations are needed. Liquid volume balances on Tank 1 and 2 gives

$$\frac{dH_1}{dt} = -(H_1 - H_2)/R$$

$$A\frac{dH_2}{dt} = (H_1 - H_2)/R - Q_e\,.$$

A sucrose balance on Tank 2 gives

$$A\frac{dH_2C_2}{dt} = C_1(H_1 - H_2)/R - C_2Q_e\,.$$

Notice that the accumulation terms, the time derivative, is the derivative of the total amount of sucrose inside Tank 2, i.e. concentration times volume, not just concentration. Since the cross sectional area A is constant, it can be taken outside the derivative, but as the liquid height, H_2, is a dependent variable, it cannot. The accumulation terms is therefore a derivative of a product. The model is easily simplified to the standard form for coupled ordinary differential equations with the derivatives of the dependent variables isolated on the left hand sides of the equations.

$$\frac{dH_1}{dt} = -\frac{H_1 - H_2}{AR}$$

$$\frac{dH_2}{dt} = \frac{H_1 - H_2}{AR} - \frac{Q_e}{A}$$

$$\frac{dC_2}{dt} = (C_1 - C_2)\frac{H_1 - H_2}{ARH_2}$$

and, to obtain a well posed problem, let the three initial conditions be

$$H_1(0) = H_2(0) = H \ , \ C_2(0) = C_0 \, .$$

1.2.2 *Balances with surface fluxes*

There are many types of chemical process equipment for which transport of energy and mass occur both via process streams and by transport across surfaces between two different phases. For instance, heat transfer between a vessel and a cooling/heating jacket, evaporation of a liquid from a free surface, diffusion of a compound from one phase to another, etc.

If a system contains several phases that exchange matter and energy across a surface and if there is *not* thermodynamic equilibrium between phases, then balances must be written for each phase. For instance, if a compound is transferred from an organic to an aqueous phase and an equilibrium is not present between the phases, then a mass balance must be written for each phase and the flux between the two phases must explicitly be included in each balance.

If, however, there is equilibrium between the two phases, then there is no transport of anything between the phases and a single mass balance on the entire process will suffice. In fact, thermodynamic equilibrium implies that the rate of transport between the phases is effectively infinitely large and since an infinite rate cannot be accommodated in a balance equation, balances on each phase are not even feasible.

We can therefore replace the previously given rule of thumb for picking a CV with the following improved rule of thumb.

Rule of Thumb. *Pick the CV as large as possible, but still so small that there is thermodynamic equilibrium inside the CV.*

Thermodynamic equilibrium usually means that there are no gradients in the system, but gradients may exist if they do not give rise to any transport phenomena. Thus a concentration gradient, a discontinuity in fact, can exist between two phases at thermodynamic equilibrium because equilibrium is determined by equality of chemical potential, not concentration. Similarly, a fluid flow field can have a discontinuity in velocity gradient at an interface between two liquids with different viscosity.

These two possibilities, whether there is thermodynamic equilibrium between phases or not, are illustrated in the next two examples.

Example 1.6: Equilibrium extraction.

A process stream contains a hazardous compound with the concentration C_0 which must be brought down to the concentration C_2 before the stream can be released into the environment. The hazardous compound is removed in a two-stage extraction process sketched below, Fig. 1.7, where the process stream is contacted with a solvent which takes up the hazardous compound. The volumetric flow rate of the process stream is Q_p and the volumetric flow rates of the two solvent streams are Q_1 and Q_2, respectively. The concentration of the hazardous compound in the inlet stream is C_0, between the stages it is C_1 and in the outlet stream it is C_2. The concentrations of the hazardous compound in the solvent exit streams, $C_{e,1}$ and $C_{e,2}$, are in equilibrium with the exit concentrations in the process stream and equilibrium is given by $C_e = KC$ where K is a constant partition coefficient.

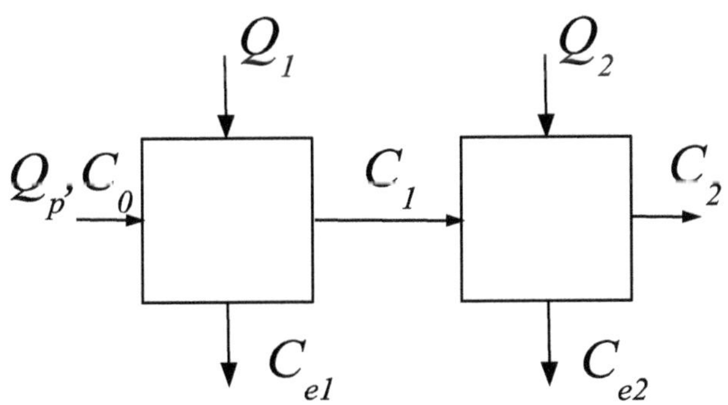

Fig. 1.7 Two stage extraction.

A steady state balance on the hazardous compound over the first extraction unit gives

$$Q_p(C_0 - C_1) = Q_1 C_{e,1}.$$

Using the equilibrium relationship $C_e = KC$, we can eliminate the concentration in the solvent exit stream

$$Q_p(C_0 - C_1) = Q_1 K C_1 .$$

Similarly, a balance on the second extraction unit gives,

$$Q_p(C_1 - C_2) = Q_2 K C_2 .$$

This model can now be used to determine the value of the concentration between the stages, C_1, which minimize the total volumetric flow rate of the solvent, Q_1+Q_2. For notational simplicity, let $f(C_1) = Q_1 + Q_2$.

$$f(C_1) = Q_p \left(\frac{C_0 - C_1}{KC_1} + \frac{C_1 - C_2}{KC_2} \right) = Q \left(\frac{C_0}{K} \frac{1}{C_1} - \frac{1}{K} + \frac{C_1}{KC_2} - \frac{1}{K} \right)$$

The minimum value is obtained when the derivative of f with respect to C_1 is zero

$$f' = Q \left(-\frac{C_0}{K} \frac{1}{C_1^2} + \frac{1}{KC_2} \right) = 0 \Rightarrow$$

$$C_1 = \sqrt{C_0 C_2} .$$

The next example, a continuous extractor that does not operate at equilibrium conditions, illustrates two things: writing balances on CVs that are not process vessels but individual phases in a multiphase system which change size with time and writing balances for which the **In** and **Out** terms are not just process streams. The example also briefly reviews how to model interfacial mass transfer.

Example 1.7: Continuous extraction.

A continuous extraction process in which compounds are transferred between two phases, referred to as I and II respectively, is sketched in Fig. 1.8. The two exit streams are assumed to come from separation of a single stream but the separator will not be included explicitly in the model. The volumetric feed rates of the two phases are called $Q_{I,f}$ and $Q_{II,f}$ and are assumed known. To write the most general model possible, the solvents and solutes of the two streams will not be specified. Instead, **all** the compounds in phase I, irrespective of whether or not they are transferred between the phases, are labeled 1,2,3,.. etc. and their respective concentrations in phase I are indicated as C_n. Since the two phases appear perfectly symmetrically in this problem, the component balances on components in phase II will be identical in form to the balances in phase I, so we will not state these explicitly and therefore need not, at this point, introduce a nomenclature for the concentrations in phase II. The feed concentrations in stream I are called $C_{n,f}$ and are also assumed known. Finally, the total extractor volume, V, is assumed constant and known and the extractor and both phases in the extractor are assumed to be well mixed.

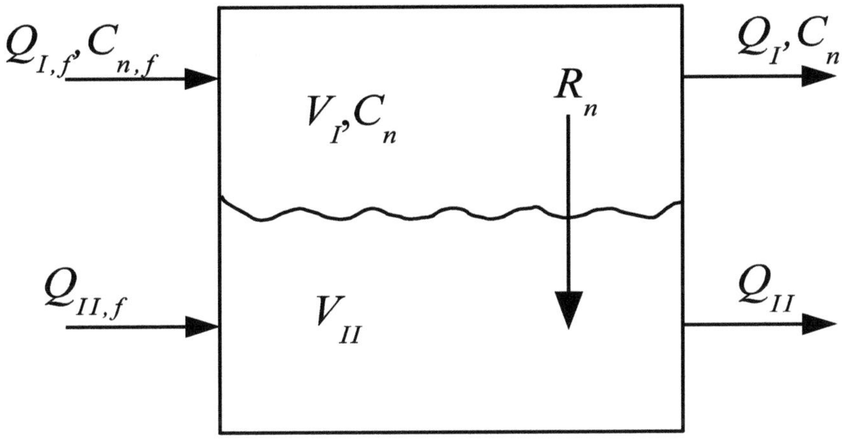

Fig. 1.8 Nomenclature for continuous extraction unit.

The unknowns in this problem are all the concentrations in both phases, the volumes of the two phases inside the extractor, V_I and V_{II} respectively, and the volumetric exit flow rates for both phases, Q_I and Q_{II} respectively.

The appropriate control volume for writing component balances is the volume of each phase. A mole balance on compound n using the volume of phase I as control volume is

$$\frac{dV_I C_n}{dt} = Q_{I,f} C_{n,f} - Q_I C_n - R_n \,.$$

Notice that the accumulation term on the left hand side is a derivative of the total number of moles of compound n, the product of the volume and the concentration. Since the volume of the phases may change with time, the physical volume of the control volume cannot be assumed constant and must appear inside the derivative.

The term R_n is the rate of transfer of compound n out of phase I to phase II. This transfer rate will be modeled using film theory. Simply put, in film theory, the bulk of the fluid is assumed well mixed, i.e. to have the same concentration, temperature and other physical properties throughout. However, close to an interphase the fluid is assumed stagnant, creating a film. The volume of the film is assumed negligible compared to the volume of the bulk fluid. Transfer across the film is by diffusion, conduction or some other non-convective mechanism. In the case of diffusion, the flux across the film, N, is described by means of a mass transfer coefficient, k_m, as $N = k_m \Delta C$ where ΔC is the concentration difference over the film, often referred to as *the driving force.* In the case of thermal conduction, the energy flux across the film is modelled by a similar equation using a heat transfer coefficient and a temperature difference as driving force.

The physical situation at the interface is shown in Fig. 1.9. In the well mixed bulk phases, the concentrations of the compound in question are C_I and C_{II}, respectively. Starting in phase I and moving towards the right, the concentration

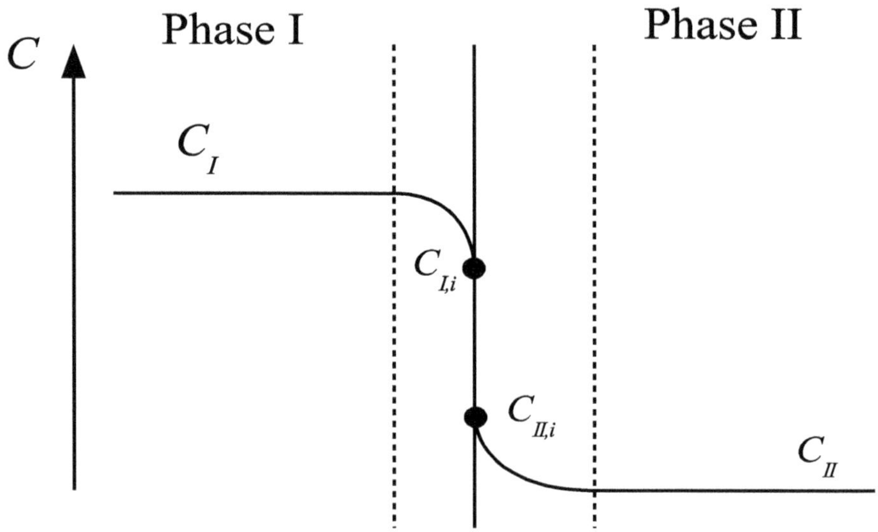

Fig. 1.9 Concentration of compound which is transferred between phase I and phase II. The interface between phase I and phase II is indicated by a solid line. The boundaries between the well mixed bulk fluids and the interfacial films are indicated by dotted lines.

remains constant until the film is reached. Passing through the film, the concentration changes continuously until it reaches $C_{I,i}$ at the interface. At the interface, the concentration changes discontinuously as required by the equilibrium which will be modelled here by a linear expression as $C_{I,i} = KC_{II,i}$, where K is a partition coefficient. More complex equilibrium expressions can of course be used. Finally, the concentration in phase II changes through the film until, at the top of the film, the bulk concentration C_{II} is attained. Notice that the two films do not necessarily have the same thickness. Neither do they support the same concentration difference.

With this image established, we can model the flux through the film by using the interface as a control volume with zero physical volume. Since the CV or interface has zero physical volume, there can be no accumulation taking place inside the CV and the balance simply states that the flux in equals the flux out. Modelling the fluxes through the two films by the respective single phase mass transfer coefficients, we get

$$N_n = k_{m,I}(C_I - C_{I,i}) = k_{m,II}(C_{II,i} - C_{II})$$

where N_n is the flux, mole per time per area of compound n, from phase I to phase II. It remains to eliminate the unknown interfacial concentrations by introducing an overall mass transfer coefficient.

The overall mass transfer coefficient can be defined two ways, based on either phase I or phase II as

$$N_n = k_m(C_I - KC_{II}) \text{ or } N_n = \hat{k}_m(C_I/K - C_{II}).$$

Continuing using k_m,

$$\frac{N_n}{k_m} = C_I - C_{I,i} + KC_{II,i} - KC_{II} = \frac{N_n}{k_{m,I}} + \frac{KN_n}{k_{m,II}} \Rightarrow$$

$$\frac{1}{k_m} = \frac{1}{k_{m,I}} + \frac{K}{k_{m,II}}.$$

The transfer rate, R_n, can therefore be written as

$$R_n = k_m(C_I - KC_{II})A = k_m(C_I - KC_{II})aV$$

where A is the total interfacial area in the extractor, usually expressed as the product of a, the specific interfacial area per volume, and V, the total volume of the vessel. This gives the final version of the component balance

$$\frac{dV_I C_n}{dt} = Q_{I,f}C_{n,f} - Q_I C_n - k_m(C_n - KC_{n,II})aV \Rightarrow$$

$$\frac{dC_n}{dt} = \frac{Q_{I,f}}{V_I}C_{n,f} - \frac{Q_I}{V_I}C_n - k_m(C_n - KC_{n,II})\frac{aV}{V_I} - \frac{C_n}{V_I}\frac{dV_I}{dt}.$$

A balance equation of this kind can be written for each compound. However, additional unknowns are the volumes of the two phases and the magnitude of the two exit streams. Four more equations are therefore needed for a complete model. Two of these are obvious. Since the total volume of the extractor is constant

$$V_I + V_{II} = V$$

and since the extractor is well mixed, the ratio of the exit volumetric flow rates must equal the ratio of the volumes of the two phases inside the extractor.

$$\frac{V_I}{V_{II}} = \frac{Q_I}{Q_{II}}$$

The remaining two equations are obtained from the mixing rules for the two phases. For ideal mixtures, the volume of the mixture is the sum over all components of the partial molar volumes multiplied by the number of moles, or

$$V_I = \sum v_n N_n = \sum v_n C_n V_I \,, \; V_{II} = \sum v_n N_n = \sum v_n C_n V_{II}$$

where v_n is the partial molar volume of component n and N_n is the number of moles of component n in the phase.

Mixing rules for non-ideal mixtures can give complex equations and simpler models are often sufficient. A simple way to close the model without using mixing rules is to assume that the densities of the two phases, ρ_I and ρ_{II} are constant.

Let the molecular weight of compound n be M_n, multiply each of the component balances found above my M_n and add them over all the compounds to get

$$\sum_{n=1}^{N} M_n \frac{dC_n}{dt} = \frac{Q_{I,f}}{V_I}\sum_{n=1}^{N} M_n C_{n,f} - \frac{Q_I}{V_I}\sum_{n=1}^{N} M_n C_n$$

$$- k_m \left(\sum_{n=1}^{N} M_n C_{n,I} - K\sum_{n=1}^{N} M_n C_{n,II}\right)\frac{aV}{V_I} - \frac{\sum_{n=1}^{N} M_n C_n}{V_I}\frac{dV_I}{dt} \Rightarrow$$

$$\frac{d}{dt}\left(\sum_{n=1}^{N} M_n C_n\right) = \frac{d\rho_I}{dt} = 0$$

$$= \frac{Q_{I,f}}{V_I}\rho_I - \frac{Q_I}{V_I}\rho_I - k_m\left(\rho_I - K\rho_{II}\right)\frac{aV}{V_I} - \frac{\rho_I}{V_I}\frac{dV_I}{dt} \Rightarrow$$

$$\frac{dV_I}{dt} = \left(Q_{I,f} - Q_I\right) - k_m\left(1 - K\frac{\rho_{II}}{\rho_I}\right)aV$$

with a similar result for the volume of phase II.

1.2.3 *Balances with chemical kinetics*

Chemical reactions are defined by their stoichiometry, usually given as an equation of the form

$$C_6H_{12}O_6 + 6O_2 \rightarrow 6CO_2 + 6H_2O$$

or

$$C_6H_{12}O_6 + 6O_2 - 6CO_2 - 6H_2O = 0$$

in which form products are recognized as the compounds with negative stoichiometric coefficients and reactants as compounds with positive coefficients. Reactions that proceed only one way are called irreversible while reactions that can proceed both ways are called reversible.

The specific rate of a reaction (number of moles per time per volume) will be indicated r_A where A is one of the compounds taking part in the reaction. r_A indicates the number of moles produced or consumed in the reaction in question and may be only one of several production/consumption terms in a balance equation. Each reaction will contribute one such term. The specific reaction rate can be specified using any of the compounds taking part in the reaction and the different ways of indicating the rate are related by the stoichiometry. For instance, for the reaction

$$\alpha A + \beta B \rightarrow \gamma C$$

it holds that

$$\frac{r_B}{\beta} = \frac{r_A}{\alpha} = -\frac{r_C}{\gamma}.$$

Reaction rates are concentration dependent and elementary reactions are often assumed to follow mass action kinetics. For mass action kinetics, the specific reaction rate is proportional to concentration of the reactants raised to the power of their stoichiometric coefficient. For instance, if the reaction above follows mass action kinetics, then

$$r_A = kC_A^{\alpha}C_B^{\beta}.$$

The reaction is said to be of order α in A and of order β in B. Of course, the same reaction is indicated if all stoichiometric coefficients are multiplied by the same factor, completely changing the order of the reaction, suggesting that the order of a reaction is not a well defined concept. However, mass action kinetics assumes that the reaction is written in a way that reflects the actual reaction mechanism and not as a combination of several reactions. The reaction rate constant k has whatever units are required to give r_A units of moles per time per volume and depends strongly on temperature. Other kinetic expressions are also seen. Particularly popular are expressions of the form

$$r_A = \frac{r_{\max}C_A}{K + C_A}.$$

In catalysis, this expression is known as Langmuir–Hinshelwood kinetics while in enzymology it is called Michaelis–Menten kinetics and in microbial growth kinetics it is known as the Monod model. $r_{\max}$ and K are constants, but not elemental reaction rate constants and therefore do not exhibit the same temperature dependence as these. For elemental reaction rate constants, Arrhenius' law has been found to represent temperature dependence well.

$$k(T) = k_0 e^{-E/RT}$$

where R is the gas constant, T absolute temperature, E is the so-called activation energy and k_0 is called the frequency factor.

Chemical reactions are associated with energy effects and reactions that evolve heat are called *exothermic* while reactions that consume heat are called *endothermic reactions*. The energy *consumed* in a reaction is quantified by the heat of reaction, ΔH_r in energy/mole. A chemical reaction therefore contributes a consumption term in an energy balance of the form $\Delta H_r rV$ where r is the specific reaction rate and V the volume of the CV. Clearly, ΔH_r is negative for exothermic reactions and positive for endothermic reactions.

There are several idealized reactor concepts that are commonly used in chemical engineering: the continuously stirred tank reactor (CSTR), the well mixed batch reactor, the well mixed fed batch reactor and the plug flow and packed bed reactor.

Models of the plug flow and packed bed reactor both require differential control volumes for their derivations and these reactors will be covered later, page 44 and page 53. The CSTR is simply a well mixed vessel with process streams going in and out of the vessel. If neither inlet and outlet steams are present, the reactor is a batch reactor and if there is no outlet stream, the reactor is a fed batch reactor. In other words, the batch and fed batch reactors are just special cases of the CSTR.

The CSTR can be modelled using the content of the reactor vessel as a control volume, Fig. 1.10.

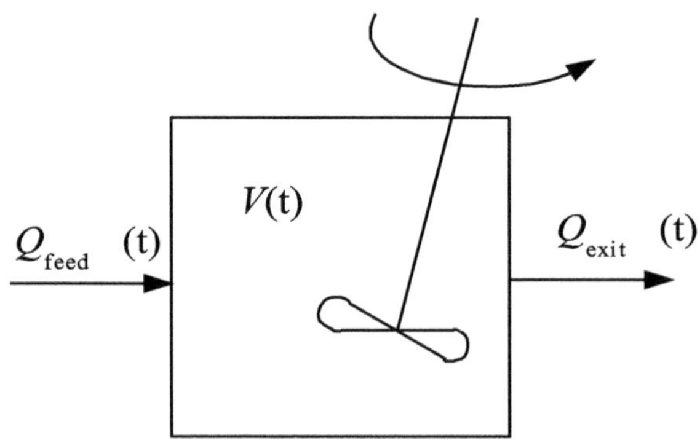

Fig. 1.10 CSTR diagram. Q represent the volumetric flow rates of the inlet and outlet streams, V is the volume of reacting solution inside the CSTR.

A general mass balance can be written over the CSTR. To identify all the reactants, let them be numbered 1 through N and let the reactions that occur inside the CSTR be numbered 1 through X. Further, let r_{xn} be the specific rate of formation of compound n in reaction x and assume kinetic expressions are known for all these rates and that there is no volume change associated with any of the reactions. A mole balance on compound n over the volume of the mixture inside the CSTR is

$$\frac{dVC_n}{dt} = Q_{\text{feed}}C_{n,feed} - Q_{\text{exit}}C_n + \sum_{x=1}^{X} r_{xn}V.$$

To make the model complete, an equation for the volume is needed. Obviously, since we are assuming no volume change in the reactions, this is

$$\frac{dV}{dt} = Q_{\text{feed}} - Q_{\text{exit}}\,.$$

Using this balance, the component balances can be rearranged to standard form.

$$\frac{dC_n}{dt} = \frac{Q_{\text{feed}}}{V}(C_{n,feed} - C_n) + \sum_{x=1}^{X} r_{xn}.$$

The next example considers a CSTR in which an exothermic reaction takes place. This example reviews how to handle thermal effects of reactions. It also an example of a model for which multiple solutions can exist, in spite of the fact that the model is complete. Multiple solutions may arise in nonlinear models and when multiple steady state solutions occur, it is reasonable to ask if all the solutions are stable. Only stable solutions can be observed empirically and unstable solutions are rarely of practical interest. The steady state CSTR with an exothermic reaction is a classical chemical engineering example in stability analysis. Two of the solutions, corresponding to an extinguished reactor and an ignited reactor, can be shown to be stable, while the remaining solution is unstable.

Example 1.8: Mass and energy balances for CSTR

Consider a CSTR of constant volume V which is equipped with temperature control through the use of a jacket with a constant coolant temperature T_c. In the reactor, a first order exothermic reaction with reaction rate constant which obey Arrhenius law and heat of reaction ΔH_r, consumes a reactant in the feed stream. The heat of reaction, ΔH_r, must be negative since we have assumed that the reaction is exothermic. The volumetric flow rate through the reactor is Q and the inlet reactant concentration and temperature are C_{feed} and T_{feed} respectively. Using the whole vessel as control volume, the reactant balance equation contains terms for two fluxes across the control volume surface, the inlet and outlet stream, a term for the rate of consumption of the reactant and the accumulation term

$$\frac{dVC}{dt} = QC_{\text{feed}} - QC - k_0 e^{-\frac{E}{RT}} CV\,.$$

The energy balance contains similar terms, but in addition it will have terms for loss of energy due to cooling and due to shaft work, the work done by the vessel on the surroundings (in thermodynamics, it is customary to consider work positive when done by the system on the surroundings). The loss term due to cooling equals the product of an overall heat transfer coefficient for the vessel, U, the area available for cooling, A, and the driving force or temperature for cooling, $T - T_c$. The shaft work is clearly negative, stirring of the vessel adds energy to the system, and will just be indicated by W. The energy balance becomes

$$\frac{dV\rho c_p T}{dt} = Q\rho c_p T_{\text{feed}} - Q\rho c_p T + (-\Delta H_r)CVk_0 e^{-\frac{E}{RT}} - UA(T - T_c) - W.$$

Assuming that there is no volume change associated with the reaction so that V is constant and that the density, ρ, and the heat capacity, c_p, are constant, i.e. temperature independent, the model becomes

$$V\frac{dC}{dt} = QC_{\text{feed}} - QC - k_0 e^{-\frac{E}{RT}} CV$$

$$V\rho c_p \frac{dT}{dt} = Q\rho c_p T_{\text{feed}} - Q\rho c_p T + (-\Delta H_r) CV k_0 e^{-\frac{E}{RT}} - UA(T - T_c) - W.$$

Later, page 87, we will simplify these two equations by introducing dimensionless variables. This will allow a simple analysis of the possible steady state solutions.

Example 1.9: Reaction with volume change.

Consider a CSTR of volume V in which the gas phase reaction $A+B \rightarrow P$ takes place as a first order reaction in each reactant. We will assume that the gas phase is an ideal gas, that the reactor is maintained at a constant pressure and that the pressure difference between the feed stream and exit stream is insignificant. The feed to the reactor consists of A, B and an inert with feed concentrations C_{Af}, C_{Bf} and C_{If} respectively. Finally, we assume that there are no significant heat effects of the reactions so that the reactor temperature is equal to that of the feed stream. Mass balances on each component, A, B, P and inert I are

$$V\frac{dC_A}{dt} = Q_f C_{Af} - QC_A - kC_A C_B V$$

$$V\frac{dC_B}{dt} = Q_f C_{Bf} - QC_B - kC_A C_B V$$

$$V\frac{dC_P}{dt} = -QC_P + kC_A C_B V$$

$$V\frac{dC_I}{dt} = Q_f C_{If} - QC_I$$

where Q_f and Q are the volumetric flow rates of the feed and exit stream. The model consists of four coupled ordinary differential equations but it is not complete because the volumetric flow rate of the exit stream, Q, is unknown. We need one more equation to close the model and it stands to reason that the fact that the gas is ideal must be included in the model at some point as must the fact of the stoichiometry of the reaction.

The ideal gas law tells us that since the reactor is operated at a constant pressure and temperature, the total molar concentration in the reactor is constant and equals p/RT. Hence $\frac{dC}{dt} = 0$ where $C = C_A + C_B + C_P + C_I$. Now take the sum of the four balance equations above.

The left hand sides add up to $V\frac{dC}{dt} = 0$ and the right hand sides can be rearranged to

$$Q(C_A + C_B + C_P + C_I) = Q_f(C_{Af} + C_{Bf} + C_{If}) - kC_AC_BV$$

or

$$QC = Q_fC - kC_AC_BV$$

which is the required equation for Q. This equation states that the molar flow rate of the exit stream equals the molar flow rate of the inlet stream minus the rate at which moles are consumed in the chemical reaction.

Example 1.10: Optimal feeding strategy for a fed batch.

Many bioreactors are operated as fed batch reactors because growth of the micro organism is inhibited by high concentrations of the nutrients required for growth. Therefore, charging the reactor with all the nutrients required for a desired yield results in slow growth initial growth rates and therefore slow overall production rates. In a fed batch reactor, a nutrient solution can be fed continuously at some rate that maintains the optimal or highest growth rate for optimal overall production rates. We will model this process here and determine what the optimal feed rate is.

Let X indicate the biomass concentration and C the nutrient concentration in the reactor. We will assume that a single reaction, conversion of nutrients to biomass, occurs at a specific rate of biomass formation r_X and that the specific rate of nutrient consumption equals $r_C = r_X/Y$. The constant parameter Y is called the yield. The biomass and nutrient balances on the fed batch reactor becomes

$$\frac{dVX}{dt} = Vr_X$$

$$\frac{dVC}{dt} = QC_f - \frac{V}{Y}r_X$$

where V is the volume of reaction mixture in the fed batch, Q the volumetric flow rate of the nutrient feed stream, C_f the nutrient feed concentration.

The precise kinetic expression for r_X does not concern us. All we need to know is that, since growth is inhibited by high concentration of nutrients and (obviously) also by a lack of nutrients, there is some optimal nutrient concentration at which r_X is maximal and we want to operate the fed batch such that this optimal reaction rate and optimal nutrient concentration is maintained. In other words, we can regard r_X and C as constant and equal to their optimal values. The only unknowns are therefore the biomass concentration $X(t)$, the volume of the reaction mixture, $V(t)$ and the volumetric feed rate $Q(t)$. The last two are obviously related through the volume balance

$$\frac{dV}{dt} = Q(t) .$$

Using the assumption of constant C and r_X, the nutrient balance and the volume balance can be rearranged to

$$\frac{dV}{dt} = \frac{r_X}{Y(C_f - C)} V.$$

The factor that multiplies V on the right side of this equation is a constant, so the differential equation is easily solved to give an expression for the volume in the fed batch

$$V(t) = V(0) e^{\frac{r_X}{Y(C_f - C)} t} .$$

The optimal volumetric feed rate is therefore

$$Q(t) = \frac{dV}{dt} = V(0) \frac{r_X}{Y(C_f - C)} e^{\frac{r_X}{Y(C_f - C)} t} .$$

1.3 Differential balances

In contrast to macroscopic balances, differential balances are used when the system being modelled is not well mixed or is not a thermodynamic equilibrium, i.e. when the dependent variables change with position. The balance equations are written over a differential CV which is defined using a coordinate system. Any coordinate system is a valid basis for writing a differential balance but the best choice is usually a coordinate system which reflects the geometry and symmetries of the system being modelled.

The models obtained from differential balances are at times called distributed models because the physical properties of the system are not lumped but distributed over space. A CV can even be differential in one spacial direction and macroscopic in another direction and a complete model can consist of multiple balance equations, some of which are macroscopic balances and others which are differential balances.

1.3.1 *Coordinate systems*

Three types of coordinate systems are routinely used in engineering: rectangular, cylindrical and spherical coordinates. Defining a coordinate system means defining both the origin of the system and the positive direction of each coordinate. In problems in fluid mechanics and problems that involve fluid flow, it is sometimes convenient to use a coordinate system that moves with the fluid flow. When a fixed coordinate system is used, we say that the model is *Eulerian* while a coordinate system that moves with the fluid flow is said to provide a *Lagrangian* description.

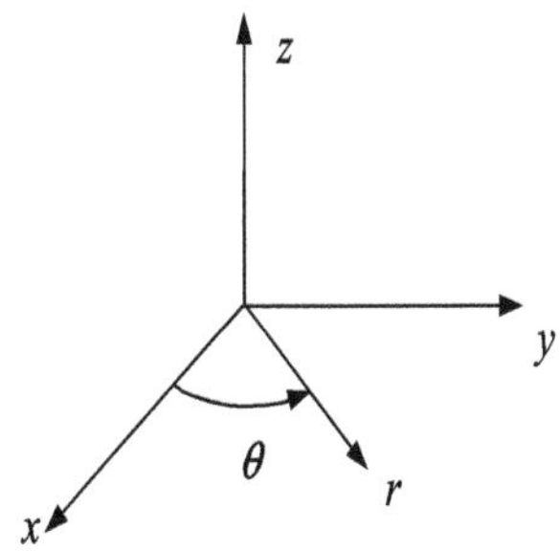

Fig. 1.11 Transformation between rectangular and cylindrical coordinate systems.

The type of coordinate system that is best suited for a given problem is often obvious and only in cases with fairly complex geometries is choosing a coordinate system a problem.

The simplest differential control volume is obtained in rectangular coordinates, (x, y, z). This CV is a differential cube with edges of lengths dx, dy and dz and the areas of the sides are $dxdy$, $dxdz$ and $dydz$. This CV is so familiar that we will not insult the reader's intelligence by spending any more time discussing it.

In cylindrical coordinate systems the coordinates are usually indicated by z, the axial coordinate, r, the radial coordinate and θ, the angular coordinate. The cylindrical coordinate system can be related to the rectangular coordinate system by letting the z-axis of the two coordinate systems coincide and placing a polar coordinate system in the xy-plane, Fig. 1.11. With this relative placement the transformation between the two coordinate systems can be written as

$$x = r\cos\theta$$

$$y = r\sin\theta$$

$$z = z.$$

The coordinate surfaces, the surfaces formed by constant values of r, θ and z, are concentric cylinders around the z-axis, the planes that pivot around the z-axis and the planes at a right angle to the z-axis.

A control volume is a volume swept out by the differentials dz, dr and $d\theta$ as shown in Fig. 1.12. Notice that the arc which spans the angle $d\theta$ has length $rd\theta$ and the area available for transport in the radial direction is therefore $rd\theta dz$ and the volume of the CV is $rd\theta dzdr$.

Spherical coordinates of a point are usually indicated by a radial coordinate r, a longitudinal coordinate ϕ and a latitudinal coordinate θ, which is the angle between the position vector of the point and the position vector of the north pole, Fig. 1.13.

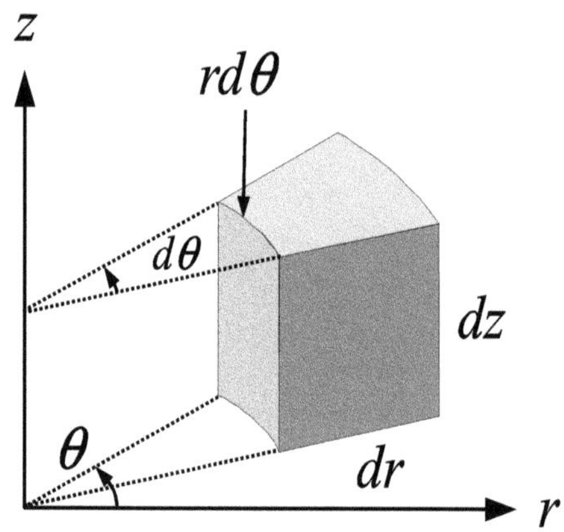

Fig. 1.12 The differential control volume in cylindrical coordinates.

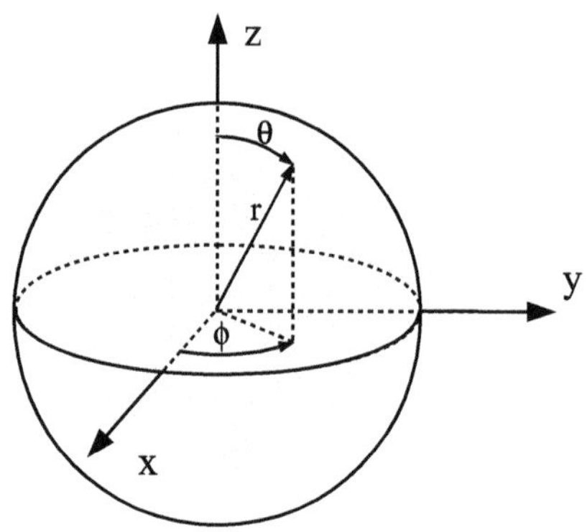

Fig. 1.13 Transformation between rectangular and spherical coordinate systems.

The transformation equations between rectangular and spherical coordinates can be written

$$
\begin{aligned}
x &= r\sin\theta\cos\phi \\
y &= r\sin\theta\sin\phi \\
z &= r\cos\theta
\end{aligned}
$$

where the origin of the two systems are superimposed, the $\theta = 0$-direction, the direction from the center to the north pole, is in the z-direction of the rectangular system and the $\phi = 0$-direction in the direction of the x axis.

The coordinate surfaces are concentric spheres (constant r), planes that pivot around the $\theta = 0$ axis (constant ϕ) and cones with their axis of symmetry on the $\theta = 0$-axis and their tips at the center of the r=constant spheres (constant θ).

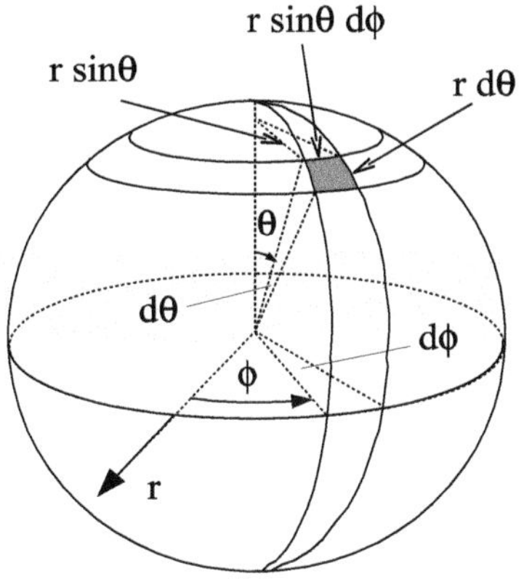

Fig. 1.14 The differential CV in spherical coordinates. The CV is not actually shown in the figure. It is a box of height dr sitting atop the gray square.

The differential control volume in spherical coordinates is shown in Fig. 1.14 and visualizing it well and figuring out the lengths of its sides does require some ability to visualize objects in three-dimension. The length of the arc spanned by $d\theta$ equals $rd\theta$, the length of the arc spanned by $d\phi$ equals $r \sin\theta d\phi$ so the differential area defined by $d\theta$ and $d\phi$ is $r^2 \sin\theta d\theta d\phi$. This is the area available for transport in the radial direction and is indicated by the gray square in Fig. 1.14. Similarly, the area for transport in the θ direction is $r \sin\theta dr d\phi$ and the area for transport in the ϕ direction is $r dr d\theta$. The physical volume of the differential element is $r^2 \sin\theta dr d\theta d\phi$.

The three simple coordinate systems mentioned here, rectangular, cylindrical and spherical, do not exhaust all possibilities. There are an infinity of possible coordinate systems and descriptions of the more useful ones can be found in the literature.[2]

All coordinate systems are created equal and are equally valid for setting up models. However, the aim of the modeler is to obtain the simplest possible mathematical problem and here the choice of coordinate system becomes important. Usually the boundary conditions dictate the choice of coordinate system such that the system that gives the simplest boundary conditions are used in a given problem. Boundary conditions are particularly nice if they can be specified along coordinate surfaces.

1.3.2 *Constitutive equations*

When writing balance equations on a differential CV, one must find expressions for the fluxes across the CV surfaces. This step typically requires the use of constitutive equations. Constitutive equations are equations that model the relationship

[2]See for example Morse, P.M. and Feshbach, H. [1953] *Methods of Theoretical Physics Vol. I* (McGraw-Hill, Boston MA) or Happel, J. and Brenner, H. [1983] *Low Reynolds Number Hydrodynamics* (Martinius Nijhoff, The Hague).

between physical quantities such as the heat flux and the temperature gradient or between a force applied and the amount of deformation it causes. Constitutive equations are not fundamental laws of nature and it does not make sense to talk about constitutive equations as being true or false; they are good or bad, depending on how well they describe reality. Picking the best constitutive equations is thus an important step in getting the best final model.

The four most important constitutive equations in chemical engineering can be stated in rectangular coordinates as follows: Fourier's law for conduction in an isotropic solid, used for calculating energy fluxes

$$q_x = -k\frac{\partial T}{\partial x}$$

where q_x is the energy flux in the x-direction (energy per time per area), k is the thermal conductivity of the substance (energy/(time length degree)) and T the temperature. Similar expressions hold for the flux in the y and z directions.

Fick's law for calculation of diffusive fluxes.

$$N_x = -D\frac{\partial C}{\partial x}$$

where N_x is the diffusive flux in the x-direction. Usually in mole/(area time) but other units can be used depending on the units used to measure concentration. D is the diffusion coefficient (area/time) and C concentration of the diffusion species. Similar expressions hold for the flux in the y and z directions.

Newton's law of viscosity which states that the stress in the fluid is proportional to a velocity gradient or a shear rate

$$\tau = \mu\frac{\partial v}{\partial x}$$

where τ is stress (pressure/area or mass/(length time^2)), μ viscosity (mass/(length time)) and v velocity.

Finally Hooke's law for deformation of an elastic solid, which states that the stress in the solid is proportional to the strain

$$\tau = E\frac{\Delta l}{l}$$

where τ is stress, E Young's modulus or shear modulus (Pressure or mass/(length time^2)), ΔL the change in length of a solid which has length L when unstressed. Thus $\Delta L/L$ is the strain.

Both Newton's law of viscosity and Hooke's law involve stress and specification of the stress in a continuum requires that all stress components be specified. Both laws are therefore considerably more complex than stated here and a proper statement of these laws require the use of tensors, a type of vector function that will be introduced in Section 6.4. Both laws can also be written with a minus sign on the right hand side. The sign depends on how the positive direction for stress is defined.

Note that all four constitutive equations are linear. Consequently, when they are used they (usually) give rise to linear models. Linear models are invariably easier to solve than nonlinear models and one should pick linear constitutive equations whenever this is possible.

Constitutive equations are specific to a given substance and thus make use of one or several material constants. Substances are often described by terms derived from the constitutive equations used to model them. For instance, if Newton's law of viscosity is used to model the relationship between stress and velocity gradients in a fluid, then the fluid is called a Newtonian fluid while a power law fluid is modeled using the power law.

1.3.3 *Balances in one coordinate direction*

The simplest differential balances are those for which the dependent variables are functions of only one coordinate direction. These balances are one-dimensional in the spacial coordinate. We will consider such balances in rectangular, cylindrical and spherical geometries and in situations with a high degree of symmetry such that dependent variables are functions of just a single coordinate direction.

Rectangular coordinates are used if the geometry of the system is almost like a flat plate. The problem becomes one-dimensional if the plate is thin compared to its height and width and if the boundary conditions at the surface of the plate do not change with position. In such a case, the "edge effects" at the edges of the plate can be ignored and one can use a single coordinate axis, orthogonal to the surface of the plate. The CV is a differentially thin slice orthogonal to this axis.

Example 1.11: Conduction in a wall.

Thermal conduction in a flat wall with a surface area of A and a thickness L is depicted in Fig. 1.15. The x-axis is oriented at a right angle to the surface of the wall and the CV is a semi infinite slab of differential thickness dx located between x and $x + dx$.

The external temperature to the left and the right of the wall is T_1 and T_2 respectively. Transport of energy in the wall is by conduction and a constitutive equation is required to model this process. We will use Fourier's law which states that the energy flux q (energy per time per area) in the positive coordinate direction is proportional to minus the temperature gradient. The proportionality constant is the thermal conductivity k.

$$q = -k\frac{\partial T}{\partial x}.$$

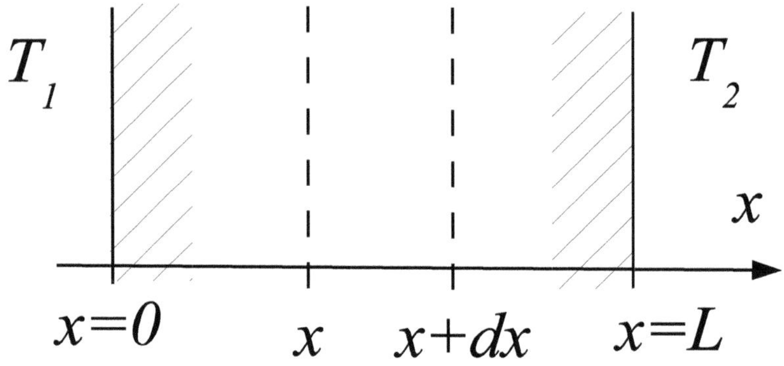

Fig. 1.15 Coordinate system and CV for energy balance on a flat wall. The CV is a semi infinite slab of differential thickness dx located between x and $x + dx$.

The transient energy balance can now be written

$$-kA\left.\frac{\partial T}{\partial x}\right|_x + kA\left.\frac{\partial T}{\partial x}\right|_{x+dx} = \frac{\partial \rho c_p T}{\partial t} A dx$$

where ρ and c_p is the density and heat capacity in the wall. The factor A is the surface area of the slab, i.e. the area available for conduction. The term which multiplies the time derivative, Adx, is the physical volume of the CV. A is a constant and can be divided out. This can be written

$$\frac{k\left.\frac{\partial T}{\partial x}\right|_{x+dx} - k\left.\frac{\partial T}{\partial x}\right|_x}{dx} = \frac{\partial \rho c_p T}{\partial t} .$$

The fraction on the left hand side is recognizable as the derivative of $\frac{\partial T}{\partial x}$ and the balance thus becomes

$$\frac{d}{dx}\left(k\frac{dT}{dx}\right) = \frac{\partial \rho c_p T}{\partial t} .$$

Assuming constant physical properties, this simplifies to

$$\frac{k}{\rho c_p}\frac{\partial^2 T}{\partial x^2} = \frac{\partial T}{\partial t} .$$

The factor $\frac{k}{\rho c_p}$ is called the thermal diffusivity, often indicated α (area/time).

To be complete, the model requires an initial condition and two boundary conditions. We will derive the boundary conditions for film resistance at the wall surface and solve for the steady state temperature profile. The steady state balance is obtained by setting the time derivative equal to zero and simplifying to get

$$\frac{\partial^2 T}{\partial x^2} = 0$$

which is readily solved to give

$$T = C_1 x + C_2$$

where C_1 and C_2 are arbitrary constants.

Film resistance at the surface is modelled by Newton's law of cooling. This model states that the flux through the film is proportional to the temperature difference across the film. The proportionality constant is the heat transfer coefficient h (Energy/(time length^2 degree)). The boundary condition at the wall surface is then obtained as an energy balance over the surface. Since the surface has zero volume the accumulation and generation terms are necessarily zero and the balance only contains the in and out fluxes. One of these is the conduction flux through the solid and the other is the flux through the film. This gives the boundary conditions

$$x = 0 \Rightarrow h(T_1 - T(0)) = -k\frac{dT}{dx}$$

$$x = L \Rightarrow -k \left.\frac{dT}{dx}\right|_L = h(T - T_2)\,.$$

Substituting the solution into these expressions and solving for the arbitrary constants give

$$T = \frac{h(T_2 - T_1)}{hL + 2k}x + \frac{hLT_1 + k(T_2 + T_1)}{hL + 2k}\,.$$

Cylindrical coordinates are used if the system is a cylinder or sort of looks like a cylinder. The problem is one-dimensional, depending only on the radial distance from the symmetry axis, if the radius of the cylinder is small compared to the length of the cylinder and if the boundary condition at the cylinder surface is the same everywhere. A single radial coordinate, pointing away from the axis of rotation, is used and the CV is a ring of differential thickness dr.

As an example of cylindrical symmetry, we will write a model of laminar flow in a cylindrical pipe. This example also illustrates how to write a momentum balance and it will demonstrate that different constitutive equations can lead to quite different results for the solution on the balance equation.

Example 1.12: Laminar flow in a cylindrical pipe. Consider a fluid inside a cylindrical pipe of radius R subjected to a pressure gradient of $\Delta P/L$. If we assume that the flow is laminar and that entrance and exit effects of the flow pattern can be ignored, then the flow is purely in the axial or z-direction, i.e. parallel to the walls of the pipe. The differential CV and the forces on it are shown in Fig. 1.16.

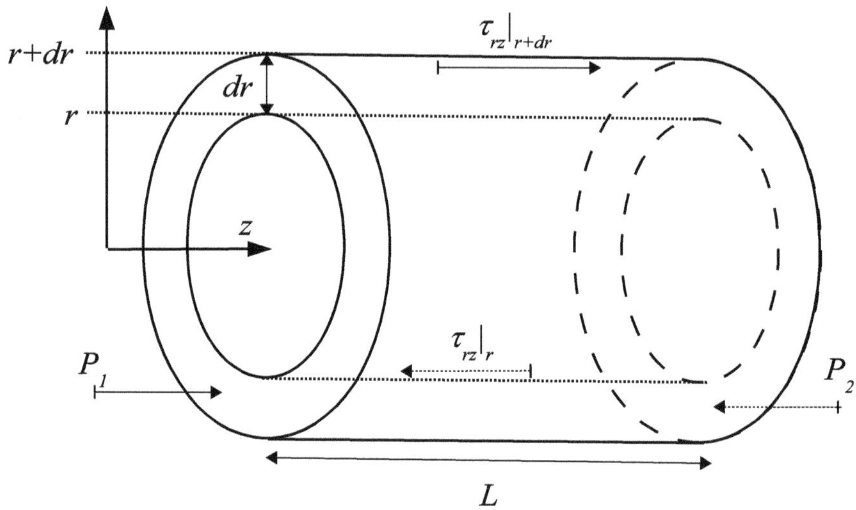

Fig. 1.16 Forces on a cylinder. Pressure or normal forces, P_1 and P_2, act on the ends of the CV while shear forces, τ, act on the cylindrical surfaces.

Since flow is only in the axial direction, parallel to the curved surfaces of the CV, there are no shear force on the ends of the CV, only normal forces, and there are no normal forces on the curved surfaces of the CV, only shear forces.

The normal forces at the ends of the CV are the pressures, P_1 and P_2 which act on the annular regions between r and $r + dr$.

$$\text{Pressure force in } z\text{-direction} = (P_2 - P_1)(\pi(r + dr)^2 - \pi r^2)$$

or

$$\text{Pressure force in } z\text{-direction} = \Delta P \pi (2rdr + dr^2)\,.$$

When calculating the surface areas of CVs in curvilinear coordinates it is common to encounter expressions, like the one above, containing terms with differentials to various powers. The higher order terms can always be ignored relative to the lower order terms and only the term of lowest order needs to be included. Thus

$$\text{Pressure force in } z\text{-direction} = \Delta P \pi 2rdr\,.$$

Notice that the factor that multiplies ΔP can be interpreted as the circumference of the annular region multiplied by its width. In the following, when writing balances over differential CVs, we will include only the lowest order differentials without any further comments.

The nomenclature for the shear forces requires a bit of explanation before the force balance can be stated. The cylindrical surfaces of the CV are called radial coordinate surfaces because the radial coordinate is constant along these surfaces. The outside radial coordinate surface is known as the positive radial coordinate surface because the outward normal to the CV on this surface points in the positive radial direction. Similarly, the inside radial coordinate surface is known as the negative radial coordinate surface. The shear force on a radial coordinate surface in the z-direction is indicated τ_{rz} and its positive direction is in the positive z-direction if the shear force acts on a positive radial coordinate surface. It is in the negative z-direction if the shear force acts on a negative radial coordinate surface.[3]

The shear forces on the CV can now be stated as

$$\text{Shear force in } z\text{-direction} = \tau_{rz}2\pi rL\big|_{r+dr} - \tau_{rz}2\pi rL\big|_{r} \,.$$

At steady flow, the forces on the CV must sum to zero. Thus

$$\Delta P\pi 2rdr + \tau_{rz}2\pi rL\big|_{r+dr} - \tau_{rz}2\pi rL\big|_{r} = 0\,.$$

Rearranging and recognizing the derivative, one obtains

$$-\frac{\Delta P}{L}r = \frac{\tau_{rz}2\pi rL\big|_{r+dr} - \tau_{rz}2\pi rL\big|_{r}}{dr} = \frac{d\tau_{rz}r}{dr}$$

which is integrated once to give

$$-\frac{\Delta P}{2L}r = \tau_{rz} + C1\,.$$

Symmetry requires that the shear stress is zero at the center of the pipe and therefore $C_1 = 0$.

$$-\frac{\Delta P}{2L}r = \tau_{rz}.$$

To proceed, we need a constitutive equation for the stress in the fluid. The most common constitutive equation, by far, is the already mentioned Newton's law of viscosity which, for the situation at hand, can be stated

$$\tau_{rz} = \mu\frac{dv_z}{dr}$$

where μ is the viscosity of the fluid and v_z the fluid velocity in the axial or z-direction. The following differential equation for the velocity profile is now obtained

$$\frac{dv_z}{dr} = -\frac{\Delta P}{2L\mu}r \;\Rightarrow$$

[3]It is important to point out that this convention for defining the positive direction of shear forces is not universally adopted. The positive direction can equally well be defined opposite to what is done here. The definition of positive direction is important because it determines the correct sign on the constitutive equation between shear stress and rate of strain. It is therefore always important to note what definition is used in a given text before proceeding.

$$v_z(r) = C_1 - \frac{\Delta P}{4L\mu} r^2 .$$

The boundary condition at the pipe wall is the no-slip condition stating the fluid velocity at the solid wall is equal to the velocity of the solid; zero in this case. I.e, $v_z(R) = 0$, giving the well known parabolic velocity profile.

$$C_1 = \frac{\Delta P}{4L\mu} R^2 \;\Rightarrow$$

$$v_z(r) = \frac{\Delta P R^2}{4L\mu}\left(1 - \left(\frac{r}{R}\right)^2\right)$$

Many fluids are non-Newtonian and are modelled using alternate constitutive equations for the fluid stress. For instance, many polymers exhibit shear thinning, meaning that the apparent viscosity decreases at higher flow rates. This behavior can be described with the power law model which in cylindrical coordinates takes the form

$$\tau_{rz} = -m\left|\frac{dv_z}{dr}\right|^{n-1}\frac{dv_z}{dr}$$

where the parameter m is called the modulus of viscosity and n is the power law index. The modulus is usually a strong function of temperature while the power law index is not. With this model, the differential equation for the velocity profile becomes

$$\frac{dv_z}{dr} = -\left(\frac{\Delta P}{2Lm}\right)^{1/n} r^{1/n} \;\Rightarrow$$

$$v_z(r) = \left(\frac{\Delta P R}{2Lm}\right)^{1/n}\frac{nR}{n+1}\left(1 - \left(\frac{r}{R}\right)^{\frac{n+1}{n}}\right).$$

Another frequently used rheological model is the Bingham model. This model is used for fluids such as ketchup, toothpaste and paints which exhibit a yield stress; a minimum stress required for the fluid to flow. The fluid may still move if the stress is below the yield stress, however, in this case it will move as a solid, without internal velocity gradients. When the stress in the substance exceeds the yield stress, flow is governed by

$$\tau_{rz} = -\mu_0\frac{dv_z}{dr} \pm \tau_0 \tag{1.17}$$

where τ_0 is the yield stress and μ_o the Bingham plastic constant. Both are empirical parameters and may be functions of temperature. For flow in a pipe, the stress is given by

$$\tau_{rz} = \frac{\Delta P r}{2L}$$

and there is a radius, r_0, below which the stress is less than the yield stress and the fluid therefore moves as a solid

$$r_0 = \frac{\tau_0 2L}{\Delta P} \Rightarrow v_z = C_1$$

where C_1 is an arbitrary constant, the velocity of the fluid which moves as a solid.

For $r > r_0$, the differential equation for the velocity profile is

$$\frac{dv_z}{dr} = \frac{\tau_0}{\mu_0} - \frac{\Delta P r}{\mu_0 2L} \Rightarrow$$

$$v_z(r) = \frac{\tau_0 r}{\mu_0} - \frac{\Delta P r^2}{\mu_0 4L} + C_2\,.$$

The arbitrary constants are found using the no-slip boundary condition at the pipe wall, $v_z(R) = 0$ and continuity of velocity at r_0.

$$v_z = \begin{cases} \dfrac{\tau_0 R}{\mu_0}\left(\dfrac{r}{R} - 1\right) + \dfrac{\Delta P R^2}{4\mu_0 L}\left(1 - \left(\dfrac{r}{R}\right)^2\right) & , r > r_0 \\[2ex] \dfrac{\tau_0 R}{\mu_0}\left(\dfrac{r_0}{R} - 1\right) + \dfrac{\Delta P R^2}{4\mu_0 L}\left(1 - \left(\dfrac{r_0}{R}\right)^2\right) & , r < r_0. \end{cases}$$

The flat velocity profile in the center of the Bingham fluid is often called a plug flow.

The final one-dimensional example is for spherical symmetry. The CV is here a spherical shell of thickness dr.

Example 1.13: Spherical catalytic pellet.

Consider a spherical catalytic particle of radius R in which a first order reaction takes place. Transport of the reactant through the pellet is given by Fick's law with an effective diffusion coefficient D. I.e. the diffusion flux in the positive radial direction, N_r, is

$$N_r = -D\frac{dC}{dr}\,.$$

We will assume that there is no mass transfer resistance between pellet surface and surroundings and that the concentration of the reactant in the surroundings is C_s. A steady state reactant balance on a spherical shell of radius r and thickness dr gives

$$-D4\pi r^2 \frac{dC}{dr}\bigg|_r + D4\pi r^2 \frac{dC}{dr}\bigg|_{r+dr} = kC4\pi r^2 dr\,.$$

After simplification, this becomes

$$\frac{D}{r^2}\frac{d}{dr}\left(r^2\frac{dC}{dr}\right) = kC\,.$$

Notice that when you divide through by dr you cannot divide out the r^2 factor at the same time because the r^2 factor that appears in the two terms are shorthand for two different factors, r^2 and $(r + dr)^2$, respectively.

The expression on the left hand side is common for problems in spherical coordinates and differential equations containing the expression can often be simplified using the following trick. Define a new variable as

$$y = Cr$$

which gives

$$\frac{dC}{dr} = \frac{d}{dr}\left(\frac{y}{r}\right) = \frac{1}{r}\frac{dy}{dr} - \frac{y}{r^2} \Rightarrow$$

$$r^2\frac{dC}{dr} = r\frac{dy}{dr} - y \Rightarrow$$

$$\frac{d}{dr}\left(r^2\frac{dC}{dr}\right) = \frac{dy}{dr} + r\frac{d^2y}{dr^2} - \frac{dy}{dr} = r\frac{d^2y}{dr^2}\,. \tag{1.18}$$

Using this variable transformation, the balance equation becomes

$$\frac{d^2y}{dr^2} = \frac{k}{D}y$$

which has the solution

$$y(r) = C_1\cosh\left(\sqrt{\frac{k}{D}}r\right) + C_2\sinh\left(\sqrt{\frac{k}{D}}r\right) \Rightarrow$$

$$C(r) = C_1\frac{\cosh\left(\sqrt{\frac{k}{D}}r\right)}{r} + C_2\frac{\sinh\left(\sqrt{\frac{k}{D}}r\right)}{r}\,.$$

In order to solve Eq. 1.18 for the steady state concentration profile, two boundary conditions are needed. However, it is possible to write more than just two conditions. For instance some possible boundary conditions in spherical coordinates are

$$r = R \Rightarrow C = C_s$$

$$r = 0 \Rightarrow \frac{dC}{dr} = 0$$

$$r = 0 \;\Rightarrow C \text{ finite}$$

$$\int_0^R kC(r)4\pi r^2\,dr = D4\pi R^2\left(\frac{dC}{dr}\right)_R\,.$$

The first boundary condition is a specification of the concentration at the pellet surface, the second follows from the assumption of spherical symmetry which requires the solution to be an even function. The third condition states that concentrations, since they are physical, measurable quantities, must be finite. The last is a total pellet balance equating the flux through the surface of the pellet to the total rate of conversion inside the pellet. All these boundary conditions must be satisfied by the solution. If they are not, the boundary conditions, the balance equation or the solution must be wrong. With a little experience, it is usually easy to pick the two boundary conditions that give the simplest solution procedure. For instance, in this case, the complete solution is

$$C(r) = C_1 \frac{\cosh\left(\sqrt{\frac{k}{D}}r\right)}{r} + C_2 \frac{\sinh\left(\sqrt{\frac{k}{D}}r\right)}{r}.$$

The first term in this solution has a singularity at $r = 0$. Since infinite concentrations are not physically possible, the arbitrary constant C_1 must equal zero. The boundary condition at the pellet surface, $C(R) = C_s$, then allows evaluation of C_2.

$$C(r) = C_s \frac{R}{r} \frac{\sinh\left(\sqrt{\frac{k}{D}}r\right)}{\sinh\left(\sqrt{\frac{k}{D}}R\right)}.$$

Notice that the result is in terms of a transcendental function sinh(). The argument of such a function must be a pure number, i.e. dimensionless and this can be used as a check on the solution. The dimension of the argument is

$$\sqrt{\frac{k}{D}}R[=]\sqrt{\frac{\text{time}}{\text{time length}^2}}\,\text{length}[=]\cdots$$

which checks out as dimensionless.

A more realistic model is obtained if mass transfer resistance is assumed at the surface of the pellet. In this case, the boundary condition at the pellet surface states that the flux through the film equals the flux into the pellet at the pellet surface

$$r = R \Rightarrow \quad -D\frac{\partial C}{\partial r} = k_m(C_s - C_\infty)$$

where k_m is a mass transfer coefficient (length/time), C_s the unknown concentration of the reactant at the surface of the pellet and C_∞ the concentration of the reactant outside the film. Of course, the previous boundary condition at the surface still holds and, therefore, so does the solution. However, the surface concentration, C_s is now an unknown which must be found from the condition above. One finds

$$C_s = \frac{k_m}{D} \frac{C_\infty}{\sqrt{\frac{k}{D}}\coth\left(\sqrt{\frac{k}{D}}R\right) - \frac{1}{R} + \frac{k_m}{D}}.$$

The analysis done so far is only valid for reactions for which temperature effects are not important. To model the effects of temperature, one must account for the amount of heat produced or consumed by the reaction and the effect of temperature on the rate of reaction. The mole balance remains the same when temperature effects are taken into account. However, the reaction rate constant is now temperature dependent and this dependence is modelled by Arrhenius' law. Thus, in spherical coordinates, the steady state mole balance is

$$\frac{1}{r^2}\frac{d}{dr}\left(r^2 D\frac{dC}{dr}\right) = k_0 e^{-\frac{E}{RT}} C.$$

The diffusion coefficient has not been taken outside the differentiation operation because diffusion coefficients are usually temperature sensitive and therefore, for this situation, dependent on position.

The energy produced or consumed by a reaction is accounted for by the heat of reaction ΔH_r. A steady state energy balance on a spherical shell of thickness dr gives

$$\left(-4\pi r^2 k_T\frac{dT}{dr}\right)_r - \left(-4\pi r^2 k_T\frac{dT}{dr}\right)_{r+dr} = k\Delta H_r C 4\pi r^2 dr$$

where k_T is the thermal conductivity of the pellet. The two flux terms on the left hand side are straightforward applications of Fourier's law. The right hand side represents the rate of energy consumption. From the definition of the heat of reaction, this term equals the volume of the spherical shell times the heat of reaction times the rate of conversion of the reactant. Dividing through by dr and rearranging gives

$$\frac{1}{r^2}\frac{d}{dr}\left(r^2 k_T\frac{dT}{dr}\right) = k_0 e^{-\frac{E}{RT}}\Delta H_r C$$

subject to boundary conditions similar to those for the concentration. The two coupled model equations are nonlinear and too complex to solve analytically. However, numerical solutions have been obtained and presented in the literature.[4]

The plug flow reactor (PFR) was mentioned earlier as an example of an idealized chemical reactor, and is a reactor that requires a differential control volume for modelling. A plug flow reactor is a flow through reactor where each lump of fluid that enters the reactor passes through the reactor without exchanging material or energy with adjacent lumps of fluid, a reactor without any mixing, the antithesis of a well mixed CSTR in other words. In addition, all lumps of fluid take the same time to pass through the reactor. The actual shape of the reactor is not the least bit relevant, but the PFR is often visualized as a pipe with a constant cross section

[4]Weisz, P.B. and Hicks, J.S. [1932] "The behavior of porous catalysts particles in view of internal mass and heat diffusion effects," *Chem. Eng. Sci.* **17**, 265.

in which fluid moves down the axial direction of the pipe with a flat velocity profile. To formulate a model for this reactor, place a z-coordinate pointing in the direction of the fluid flow and do a balance on a CV which is a slice of the reactor between z and $z + dz$, assume that the volumetric flow rate into the reactor is Q and that the pipe has a cross sectional area A, see Fig. 1.17.

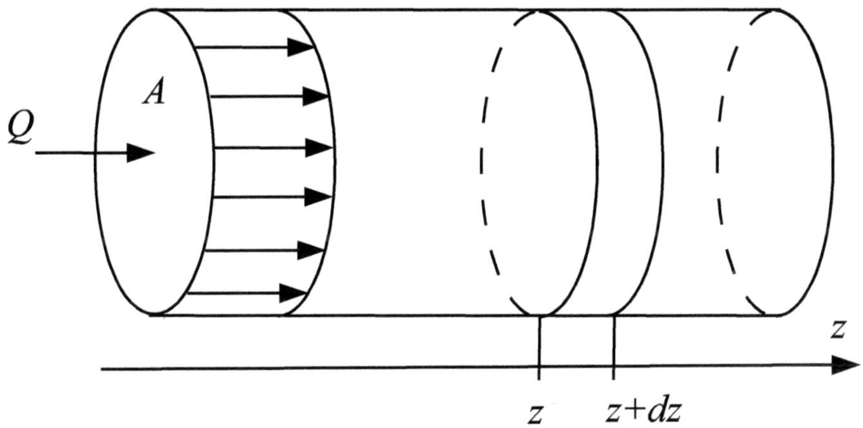

Fig. 1.17 Differential control volume for reactant balance on a PFR.

Let the reactant concentration be C and assume the reactant is consumed in a reaction with the specific rate $r(C)$, that there is no volume change associated with the reaction and that the fluid flow rate is v. The reactant balance is

$$vCA|_z - vCA|_{z+dz} = r(C)Adz\,.$$

Note the the fluid flow rate is $v = Q/A$. Upon simplification, this becomes

$$v\frac{dC}{dz} = -r(C).$$

This model is based on a coordinate system fixed in space, it is an Eulerian description. However, since all fluid elements move with the same velocity, one can just as well use a coordinate system that moves with the flow, a Lagrangian description. In this case, pick any differential element of fluid moving with the flow. Since there is no flow across the CV boundary and no exchange of material with adjacent fluid elements, the reactant balance reduces to "Rate of accumulation = Rate of consumption".

$$\frac{dC}{dt} = -r(C).$$

This is the same balance as one would have on a batch reactor. This is to be expected since, in the PFR model, each fluid element is essentially viewed as a batch reactor moving through the PFR.

The Eulerian and Lagrangian models are equally valid descriptions of the reactor and one can convert one to the other by noting that the position of a fluid element that enters the reactor at $z = 0$ at the time t_0 is $z = v(t - t_0)$. Hence

$$\frac{dz}{dt} = v \Rightarrow$$

$$\frac{dC}{dt} = \frac{dC}{dz}\frac{dz}{dt} = v\frac{dC}{dz}.$$

The next example considers a PFR in which the reaction is associated with a volume change.

Example 1.14: Plug flow reactor with volume change.

Consider a plug flow reactor in which the first order reaction

$$A \rightarrow 3B$$

with the reaction rate constant k takes place. All components in the reactor can be considered as ideal gases and the reactor operates at a constant pressure and temperature. The inlet mixture to the reactor consists of 75% inert and 25% A and the inlet volumetric flow rate, $Q(0)$, is given. What can be said about the size of reactor needed to obtain a given outlet concentration C_A?

Place a z-axis pointing in the direction of the fluid flow and do a balance on each component on a differential reactor slice of thickness dz. Notice that we cannot introduce a constant velocity v since the volume change associated with the reaction will cause the volumetric flow rate Q to change with position. The volumetric flow rate is therefore an unknown in the problem and the model is only complete if it includes an equation for $Q(z)$. The component balances are

$$\frac{dQC_A}{dz} = -kC_A A$$
$$\frac{dQC_B}{dz} = 3kC_A A$$
$$\frac{dQC_I}{dz} = 0$$

where C_A, C_B and C_I are the concentrations of A, B and inert, respectively. Summing these three equations to obtain

$$\frac{dQ(C_A + C_B + C_I)}{dz} = \frac{dQC}{dz} = 2kC_A A.$$

Now make use of the fact that the gas is an ideal gas. For an ideal gas at fixed pressure and temperature, the total molar concentration is constant (and equals p/RT) so C can be moved outside the derivative

$$\frac{dQ}{dz} = \frac{2kC_A A}{C} = \frac{2kC_A A}{C_A(0) + C_I(0)}.$$

Physically, this balance states that the rate of increase in volumetric flow equals the rate of increase in total number of moles.

The model can be restated in standard form by solving for the time derivatives.

$$\begin{aligned}
\frac{dC_A}{dz} &= -\frac{kC_A A}{Q}\left(1 + \frac{2C_A}{C_A(0) + C_I(0)}\right)\\
\frac{dC_B}{dz} &= \frac{kC_A A}{Q}\left(3 - \frac{2C_B}{C_A(0) + C_I(0)}\right)\\
\frac{dC_I}{dz} &= -\frac{2kC_A C_I A}{Q(C_A(0) + C_I(0))}\\
\frac{dQ}{dz} &= \frac{2kC_A A}{C_A(0) + C_I(0)}.
\end{aligned}$$

1.3.4 *Balances in several coordinate directions*

The next example illustrates a differential balance in which a reacting fluid flow with a known velocity profile passes through the control volume surface. When this occurs, fluxes are most easily calculated if the coordinate systems is chosen such that streamlines are at a right angle to the surface.

Example 1.15: Tubular reactor with laminar flow.

Write a model for the concentration of the reactant in a tubular reactor in which the fluid flow is laminar and diffusion occurs in both the radial and axial direction. Assume that the reactant is converted by an nth order reaction with reaction constant k.

The obvious choice of coordinate system is a cylindrical coordinate system with the origin at the center of the reactor inlet and the z-axis pointing in the direction of the flow. Let the radius of the reactor be R, the laminar profile then is given by

$$v_z(r) = v_m\left(1 - \left(\frac{r}{R}\right)^2\right).$$

The reactant concentration will be a function of r and z but not a function of the angle θ (unless of course the inlet concentration is a function of θ. The control volume will therefore be an annular ring of length dz and thickness dr. The area available for transport in the axial and radial directions are $2\pi r dr$ and $2\pi r dz$ respectively and the volume of the ring has the volume $2\pi r dr dz$. See Fig. 1.18.

The following transport terms will appear in the mass balance for the reactant. The term for convective transport in the axial direction:

$$2\pi r\, dr v(r)C(r,z))_z - (2\pi r\, dr v(r)C(r,z))_{z+dz}.$$

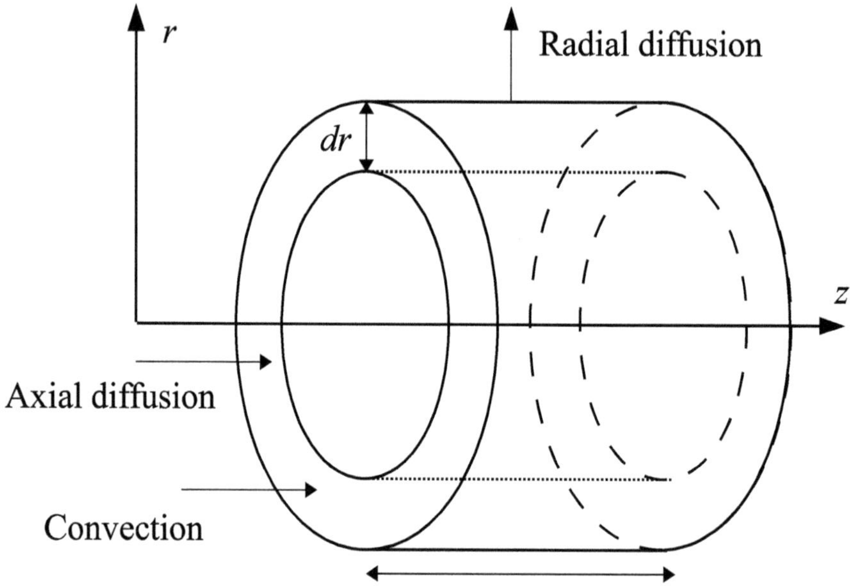

Fig. 1.18 Control volume and fluxes for the tubular reactor.

The term for diffusional transport in the axial direction:

$$\left(2\pi r\, dr\left(-D\frac{\partial C}{\partial z}\right)\right)_z - \left(2\pi r\, dr\left(-D\frac{\partial C}{\partial z}\right)\right)_{z+dz}.$$

The term for diffusional transport in the radial direction:

$$\left(2\pi r\, dz\left(-D\frac{\partial C}{\partial r}\right)\right)_r - \left(2\pi r\, dz\left(-D\frac{\partial C}{\partial r}\right)\right)_{r+dr}$$

where D is the diffusion coefficient and $C(r, z)$ the concentration of the reactant. Substituting all this into the conservation equation gives

$$(2\pi r\, dr v(r) C(r,z))_z - (2\pi r\, dr v(r) C(r,z))_{z+dz}$$

$$+ \left(2\pi r\, dr\left(-D\tfrac{\partial C}{\partial z}\right)\right)_z - \left(2\pi r\, dr\left(-D\tfrac{\partial C}{\partial z}\right)\right)_{z+dz}$$

$$+ \left(2\pi r\, dz\left(-D\tfrac{\partial C}{\partial r}\right)\right)_r - \left(2\pi r\, dz\left(-D\tfrac{\partial C}{\partial r}\right)\right)_{r+dr} = \tfrac{\partial}{\partial t}(C 2\pi r\, dr dz) + kC^n 2\pi r\, dr dz$$

dividing through by $dr dz$ and recognizing derivatives gives

$$D\frac{\partial^2 C}{\partial z^2} + \frac{D}{r}\frac{\partial}{\partial r}\left(r\frac{\partial C}{\partial r}\right) - v_m\left(1 - \left(\frac{r}{R}\right)^2\right)\frac{\partial C}{\partial z} = \frac{\partial C}{\partial t} + kC^n$$

where a constant diffusion coefficient has been assumed.

The boundary conditions in the radial direction are straightforward.

$$r = 0 \Rightarrow C \text{ finite and } \frac{\partial C}{\partial r} = 0$$

$$r = R \Rightarrow \frac{\partial C}{\partial r} = 0\,.$$

We will not attempt to write the boundary conditions in the z-direction here.

In the examples considered so far, the constitute equations that were used were simple enough that we did not worry about the fact that even simple constitutive equations take different forms in different coordinate systems. When writing balance equations with transport terms in several directions, this fact becomes important.

Consider, for instance, a transient energy balance in rectangular coordinates on a solid in which heat is generated at the specific rate $\dot{q}$ (Energy/(time volume)). The transient energy balance becomes

$$q_x|_x dydz - q_x|_{x+dx} dydz + q_y|_y dxdz - q_y|_{y+dy} dxdz + q_z|_z dxdy - q_z|_{z+dz} dxdy$$
$$= \rho c_p \frac{\partial T}{\partial t} dxdydz - \dot{q} dxdydz$$

where the term $q_x|_x dydz$ represents the rate of energy transport into the differential cube, through the surface parallel to the y, z-plane (or, equivalently, the surface which is at a right angle to the x-axis) and at position x. q_x is the flux and $dydz$ the area of the side of the differential cube. Similarly, the term $-q_x|_{x+dx} dydz$ represents the rate of transport out of the cube at position $x + dx$ and so forth. Density ρ and heat capacity c_p are assumed constant. This balance is rewritten using the standard method for rewriting differential balances. Divide through by the differential factors dx, dy and dz

$$-\frac{q_x|_{x+dx} - q_x|_x}{dx} - \frac{q_y|_{y+dy} - q_y|_y}{dy} - \frac{q_z|_{z+dz} - q_z|_z}{dz}$$
$$= \rho c_p \frac{\partial T}{\partial t} - \dot{q}$$

and recognize that the three fractions on the left hand side are derivatives with respect to x, y and z respectively, to get

$$-\frac{\partial q_x}{\partial x} - \frac{\partial q_y}{\partial y} - \frac{\partial q_z}{\partial z} = \rho c_p \frac{\partial T}{\partial t} - \dot{q}\,.$$

The flux q can now be modelled using Fourier's law. Substitution into the energy balance gives

$$\frac{\partial}{\partial x}\left(k\frac{\partial T}{\partial x}\right)+\frac{\partial}{\partial y}\left(k\frac{\partial T}{\partial y}\right)+\frac{\partial}{\partial z}\left(k\frac{\partial T}{\partial z}\right)=\rho c_p\frac{\partial T}{\partial t}-\dot{q}$$

and if the thermal conductivity k is independent of position, it can be taken outside the derivatives and divided out to get the simplified balance, usually written

$$\frac{k}{\rho C_p}\left(\frac{\partial^2 T}{\partial x^2}+\frac{\partial^2 T}{\partial y^2}+\frac{\partial^2 T}{\partial z^2}\right)+\frac{\dot{q}}{\rho c_p}=\frac{\partial T}{\partial t}\,.$$

If the same balance is now done in cylindrical coordinates, one must first content with the greater complexity of the differential CV. The size of the sides of the differential element are no longer simply products of the three differentials that define the CV but are given by the expression discussed earlier and the transient energy balance becomes

$$q_r r|_r d\theta dz - q_r r|_{r+dr} d\theta dz + q_\theta|_\theta dr dz - q_\theta|_{\theta+d\theta} dr dz + q_z|_z r d\theta dr - q_z|_{z+dz} r d\theta dr$$
$$= \rho c_p \frac{\partial T}{\partial t} r d\theta dr dz - \dot{q} r d\theta dr dz$$

which upon simplification becomes

$$-\frac{\partial r q_r}{\partial r}-\frac{\partial q_\theta}{\partial \theta}-r\frac{\partial q_z}{\partial z}=\rho c_p\frac{\partial T}{\partial t}r-\dot{q}r.$$

To continue, we need Fourier's law in cylindrical coordinates, and it is slightly more complicated looking than in rectangular coordinates.

$$q_r=-k\frac{\partial T}{\partial r}\,,\quad q_\theta=-\frac{k}{r}\frac{\partial T}{\partial \theta}\,,\quad q_z=-k\frac{\partial T}{\partial z}$$

giving

$$\frac{1}{r}\frac{\partial}{\partial r}\left(kr\frac{\partial T}{\partial r}\right)+\frac{1}{r^2}\frac{\partial}{\partial \theta}\left(k\frac{\partial T}{\partial \theta}\right)+\frac{\partial}{\partial z}\left(k\frac{\partial T}{\partial z}\right)=\rho c_p\frac{\partial T}{\partial t}-\dot{q}$$

and for constant thermal conductivity

$$\frac{k}{\rho C_p}\left(\frac{1}{r}\frac{\partial}{\partial r}\left(r\frac{\partial T}{\partial r}\right)+\frac{1}{r^2}\frac{\partial^2 T}{\partial \theta^2}+\frac{\partial^2 T}{\partial z^2}\right)+\frac{\dot{q}}{\rho c_p}=\frac{\partial T}{\partial t}\,.$$

The point to note here is that the form of both the balance equation and the constitutive equation change with the coordinate system. The different forms can be converted to one another, but doing so between general coordinate systems is not trivial and the appropriate form of most common constitutive equations can be looked up.

Example 1.16: Spherical catalytic pellet without spherical symmetry.

Consider again the case of a spherical catalytic pellet in which a first order reaction occurs but with a reactant surface concentration that varies with position and is given by $C_s(\theta, \phi)$. The boundary condition is not spherically symmetric and this causes symmetry breaking such that the reactant concentration inside the pellet is not spherically symmetric either. The concentration inside the pellet will be a function of all three spacial coordinates, r, θ and ϕ, when using a spherical coordinate system. The proper CV is therefore the differential volume defined by dr, $d\theta$ and $d\phi$ and the balance equation must account for the diffusive flux in all three coordinate directions. In all cases, the rate of transport in and out of the CV is the product of the flux given by Fick's law and the area of the surface of the CV that is at a right angle to the coordinate direction. In spherical coordinates, Fick's law takes the form

$$N_r = -D\frac{\partial C}{\partial r}\ ,\ N_\theta = -D\frac{1}{r}\frac{\partial C}{\partial \theta}\ ,\ N_\phi = -D\frac{1}{r\sin\theta}\frac{\partial C}{\partial \phi}$$

where the three Ns are the fluxes in the radial, θ and ϕ directions, respectively. For transport in the radial direction, the area available for the flux equals $rd\theta r\sin\theta d\phi$ at a radial distance of r. Thus

$$\text{In - out} = -D\,\frac{\partial C}{\partial r} rd\theta r\sin\theta d\phi\Big|_r - \left(-D\,\frac{\partial C}{\partial r} rd\theta r\sin\theta d\phi\Big|_{r+dr}\right)$$

where the subscript r indicates that the expression in parentheses is evaluated at r. For the θ direction, the area for transport equals $r\sin\theta d\phi dr$, giving

$$\text{In-out} = -D\,\frac{1}{r}\frac{\partial C}{\partial \theta}(r\sin\theta d\phi dr)\Big|_\theta - \left(-D\frac{1}{r}\,\frac{\partial C}{\partial \theta}(r\sin\theta d\phi dr)\Big|_{\theta+d\theta}\right)$$

and, finally, for transport in the ϕ direction, the area for transport is $rd\theta dr$.

$$\text{In-out} = -D\frac{1}{r\sin\theta}\,\frac{\partial C}{\partial \phi}(rd\theta dr)\Big|_\phi - \left(-D\frac{1}{r\sin\theta}\,\frac{\partial C}{\partial \phi}(rd\theta dr)\Big|_{\phi+d\phi}\right).$$

The accumulation term is the rate of change or time derivative of the concentration inside the control volume multiplied by the physical volume of the control volume, $r^2\sin\theta drd\phi d\theta$.

$$\text{Accumulation} = \frac{\partial}{\partial t}(Cr^2\sin\theta drd\phi d\theta) = r^2\sin\theta drd\phi d\theta\frac{\partial C}{\partial t}$$

Similarly, the generation term equals the specific reaction rate inside the control volume multiplied by the physical volume of the control volume.

$$\text{Generation} = -kCr^2 \sin\theta dr d\phi d\theta.$$

Putting everything together gives the following expression.

$$-D\,\frac{\partial C}{\partial r} r^2 \sin\theta d\phi d\theta\bigg|_r + D\,\frac{\partial C}{\partial r} r^2 \sin\theta d\phi d\theta\bigg|_{r+dr}$$

$$-D\,\frac{\partial C}{\partial \theta} \sin\theta d\phi dr\bigg|_\theta + D\,\frac{\partial C}{\partial \theta} \sin\theta d\phi dr\bigg|_{\theta+d\theta}$$

$$-D\frac{1}{\sin\theta}\,\frac{\partial C}{\partial \phi} d\theta dr\bigg|_\phi + D\frac{1}{\sin\theta}\,\frac{\partial C}{\partial \phi} d\theta dr\bigg|_{\phi+d\phi}$$

$$= r^2 \sin\theta dr d\phi d\theta \frac{\partial C}{\partial t} + kCr^2 \sin\theta dr d\phi d\theta.$$

Dividing through by $r^2 \sin\theta dr d\phi d\theta$ and recognizing derivatives, one obtains

$$\frac{1}{r^2}\frac{\partial}{\partial r}\left(Dr^2\frac{\partial C}{\partial r}\right) + \frac{1}{r^2\sin\theta}\frac{\partial}{\partial\theta}\left(D\sin\theta\frac{\partial C}{\partial\theta}\right) + \frac{1}{r^2\sin^2\theta}\frac{\partial}{\partial\phi}\left(D\frac{\partial C}{\partial\phi}\right)$$

$$= \frac{\partial C}{\partial t} + kC.$$

The boundary condition is

$$r = R \Rightarrow C = C_s(\theta, \phi).$$

It may strike some readers as odd, but all coordinate systems are equally valid and one can just as well write the balance equation in rectangular coordinates. In that case the governing differential equation will be (convince yourself of this)

$$\left(\frac{\partial}{\partial x}\left(D\frac{\partial C}{\partial x}\right) + \frac{\partial}{\partial y}\left(D\frac{\partial C}{\partial y}\right) + \frac{\partial}{\partial z}\left(D\frac{\partial C}{\partial z}\right)\right) = \frac{\partial C}{\partial t} + kC.$$

The big difference between models in the two coordinate systems become apparent when the boundary condition at the surface is written for the balance in rectangular coordinates. If the origin of the coordinate system is in the center of the pellet, then the boundary condition at the surface is

$$x^2 + y^2 + z^2 = R^2 \;\Rightarrow\; C = C_s(\theta(x, y, z), \phi(x, y, z)).$$

This condition is not specified on a coordinate surface and the specified surface condition is now in general a function of three coordinates, x, y, z. The difference between the two models becomes even greater when one considers the steady state problem with the simplifications that are possible if the surface concentration has spherical symmetry. In that case, the concentration is independent of θ and ϕ so all derivatives with respect to these variables vanish and the model in spherical coordinates becomes

$$\frac{1}{r^2}\frac{d}{dr}\left(Dr^2\frac{dC}{dr}\right) = kC.$$

Given the spherical symmetry, one could have derived this balance equation from a reactant balance using a control volume which is a spherical shell of thickness dr.

Similar simplifications are not possible in rectangular coordinates. Thus, in spherical coordinates the model is an ordinary differential equation while it is a partial differential equation in rectangular coordinates. There can be little disagreement that the model in spherical coordinates is far simpler and easier to solve than the model in rectangular coordinates even though both models are valid and give the same solution.

The main difficulty in the next example lies in the fact that the content of a packed bed reactor is not a single phase. The usual CV, an annular ring of thickness dr and length dz, will therefore not work because the concentration of the reactant is not constant inside this CV. The best CV includes matter from one phase only and is therefore the fluid content of a differential annular ring.

Example 1.17: Packed bed reactor with dispersion.

Consider a packed bed, tubular reactor in which a reactant in the fluid phase is converted instantaneously on the surface of catalytic pellets. The reactant must diffuse through a stagnant film surrounding the catalyst in order to react. The mass transfer coefficient across the film is k_m. The fluid flow is approximately plug flow and the mixing which occurs due to the deviation from plug flow can be modelled by assuming that back mixing of the reactant follows a law similar to Fick's law with a so-called dispersion coefficient taking the place of the diffusion coefficient. To account for the difference between radial and axial mixing, different dispersion coefficients are used in the radial and axial directions, D_r and D_z respectively. The volumetric flow rate through the reactor is Q, the cross sectional area of the reactor is A and the void fraction in the bed is ϵ. Write a model for this reactor assuming a negligible heat of reaction.

The best control volume is the fluid phase inside an annular ring of length dz and thickness dr. One cannot draw such a control volume but must think of it in more abstract terms. The surfaces of the control volume can be partitioned into two types: the surfaces of the annular ring which are covered by fluid and the interface between the fluid and the catalytic particles inside the control ring. Fluxes across all these surfaces must be accounted for when setting up the balance on the reactor.

Since the volumetric flow rate through the reactor is Q and the cross sectional area of the reactor is A, the fluid velocity based on an empty tube is Q/A. The interstitial fluid velocity, the fluid velocity between the catalyst particles, is the empty tube velocity divided by the void fraction of the bed, i.e. it equals $Q/(A\epsilon)$. This is the velocity needed in the convective term. The area available for flow in the axial direction is the area of the CV surface at a right angle to the axial direction, $2\pi rdr$, multiplied by the void fraction, ϵ. Thus, convective transport in the axial direction is given by

$$\left(2\pi r\, dr\epsilon\frac{Q}{A\epsilon}C(r,z)\right)_z - \left(2\pi r\, dr\epsilon\frac{Q}{A\epsilon}C(r,z)\right)_{z+dz}.$$

The term for transport by dispersion in the axial direction must similarly take into account the fact that fluid phase diffusion is restricted to the void volume.

$$\left(2\pi r\epsilon\, dr\left(-D_z\frac{\partial C}{\partial z}\right)\right)_z - \left(2\pi r\epsilon\, dr\left(-D_z\frac{\partial C}{\partial z}\right)\right)_{z+dz}.$$

Likewise for the term for transport by dispersion in the radial direction.

$$\left(2\pi r\epsilon\, dz\left(-D_r\frac{\partial C}{\partial r}\right)\right)_r - \left(2\pi r\epsilon\, dz\left(-D_r\frac{\partial C}{\partial r}\right)\right)_{r+dr}.$$

Finally, the flux across the film surrounding the catalytic pellets must be determined. The area available for transport across the film inside a control volume must be proportional to the physical volume of the control volume. The proportionality constant is the specific surface area of the bed, call it a. The rate of transport through the film now is

$$a2\pi rdrdzk_m(C - C_s)$$

where C_s is the concentration of the reactant at the solid surface. However, since the reaction is instantaneous, this concentration must be zero, i.e. the rate of transport through the film simply is

$$a2\pi rdrdzk_mC.$$

The accumulation term is

$$\frac{\partial}{\partial t}(C2\pi r\epsilon\, dr\, dz)$$

and since there is no reaction in the fluid phase, the generation term is zero. Putting the terms into the mass balance and carrying out the standard rearrangement of dividing through by $dr\, dz$ and recognizing derivatives gives

$$\frac{\partial}{\partial z}\left(\epsilon D_z\frac{\partial C}{\partial z}\right) + \frac{1}{r}\frac{\partial}{\partial r}\left(r\epsilon D_r\frac{\partial C}{\partial r}\right) - \frac{\partial}{\partial z}\left(\frac{Q}{A}C\right) = \epsilon\frac{\partial C}{\partial t} + ak_mC.$$

Assuming constant properties though the bed, this simplifies to

$$\epsilon D_z\frac{\partial^2 C}{\partial z^2} + \epsilon D_r\frac{1}{r}\frac{\partial}{\partial r}\left(r\frac{\partial C}{\partial r}\right) - \frac{Q}{A}\frac{\partial C}{\partial z} = \epsilon\frac{\partial C}{\partial t} + ak_mC.$$

The boundary conditions in the radial direction are straight forward

$$r = 0 \Rightarrow C \text{ finite and } \frac{\partial C}{\partial r} = 0.$$

The boundary condition at the wall is a no-flux condition indicating that the reactant does not diffuse through the reactor wall.

$$r = R \Rightarrow \frac{\partial C}{\partial r} = 0$$

where R is the radius of the reactor.

The boundary conditions in the axial or z-direction are difficult to state, if for no other reason that the problem is not well defined in the axial direction. Dispersion in the z-direction will cause mixing of fluid that has already reacted with unreacted fluid and the entrance concentration at $z = 0$ is therefore not the same as the feed concentration to the reactor. Similarly, the concentration at the very end of the bed will not be the same as the exit concentration from the reactor due to the dispersion and mixing at the exit. Neither the entrance nor the exit mixing effects can be modelled unless the geometry of the entrance and exit regions are known, yet such detailed modelling is usually excessively detailed and complex. In most cases, it is reasonable to assume that the reactor performance is not strongly dependent on how the entrance and exit regions are shaped as long as mixing of the reactants is fast and most of the reaction occurs inside the reactor rather than in the entrance and exit regions. So how does one write boundary conditions that reflect this? Different conditions have been proposed and a generally accepted solution has emerged. These generally accepted boundary conditions can be obtained as follows. (Interested readers are referred to the original paper by Wehner and Wilhelm[5] where different proposed boundary conditions are discussed and a more detailed derivation is given.)

Make the assumption that there is no dispersion for $z < 0$. A balance over the surface $z = 0$ gives

$$\left.\frac{Q}{A}C\right|_{-dz} = \left.\frac{Q}{A\epsilon}\epsilon C\right|_{dz} + \left(-D_z\epsilon \left.\frac{\partial C}{\partial z}\right|_{dz}\right).$$

If there is no dispersion for $z < 0$, i.e. no mixing of the feed with the reactor content, then $C(-dz)$ must equal the feed concentration C_{feed} and the boundary condition takes the form

$$z = 0 \Rightarrow C_{feed} = C - \frac{D_z A\epsilon}{Q}\frac{\partial C}{\partial z}.$$

The concentration in the entrance region before the reactive bed, $z < 0$, equals C_{feed}, so this boundary condition does not require that the concentration or the concentration gradient be continuous across $z = 0$. In a rigorous sense, this is

[5]Wehner, J.F. and Wilhelm, R.H. [1956] "Boundary conditions of flow reactor," *Chem. Eng. Sci.* **6**, 89–93.

obviously not possible, but what the boundary condition states is that at the bed entrance, concentrations change so fast or over such short distances that they can be adequately modelled assuming a discontinuous concentration and concentration gradient. In other words, this boundary condition is not dictated by some law of nature but is an approximate model of the bed entrance.

The commonly used boundary condition at the end of the bed is simply

$$z = L \Rightarrow \frac{\partial C}{\partial z} = 0.$$

1.3.5 *Operator notation*

Balance equations and constitutive equations can be stated in a coordinate system independent form using operator notation. An "operator" is just an expression that occurs frequently enough that it has been given a name.

In rectangular coordinates, the energy balance contain the expression $\frac{\partial^2 T}{\partial x^2} + \frac{\partial^2 T}{\partial y^2} + \frac{\partial^2 T}{\partial z^2}$. This expression arises frequently in balance equations when rectangular coordinates are used, and is written using operator notation as

$$\nabla^2 T = \frac{\partial^2 T}{\partial x^2} + \frac{\partial^2 T}{\partial y^2} + \frac{\partial^2 T}{\partial z^2}.$$

The operator ∇^2 is called the Laplacian operator and is, in rectangular coordinates, defined as

$$\nabla^2 \equiv \frac{\partial^2}{\partial x^2} + \frac{\partial^2}{\partial y^2} + \frac{\partial^2}{\partial z^2}.$$

With this operator, the transient energy balance on a solid with heat generation can be written

$$\frac{k}{\rho c_p} \nabla^2 T + \frac{\dot{q}}{\rho c_p} = \frac{\partial T}{\partial t}.$$

One can now look up the explicit expression for ∇^2 in any coordinate system one wants to use. For instance, in spherical coordinates, the ∇^2 operator is

$$\nabla^2 = \frac{1}{r^2} \frac{\partial}{\partial r} \left(r^2 \frac{\partial}{\partial r} \right) + \frac{1}{r^2 \sin\theta} \frac{\partial}{\partial \theta} \left(\sin\theta \frac{\partial}{\partial \theta} \right) + \frac{1}{r^2 \sin^2\theta} \frac{\partial^2}{\partial^2 \phi}$$

and the model result obtained in rectangular coordinates can be converted to spherical coordinates by substituting this expression into the operator form of the model giving

$$\frac{k}{\rho c_p} \left(\frac{1}{r^2} \frac{\partial}{\partial r} \left(r^2 \frac{\partial T}{\partial r} \right) + \frac{1}{r^2 \sin\theta} \frac{\partial}{\partial \theta} \left(\sin\theta \frac{\partial T}{\partial \theta} \right) + \frac{1}{r^2 \sin^2\theta} \frac{\partial^2 T}{\partial^2 \phi} \right) + \frac{\dot{q}}{\rho c_p} = \frac{\partial T}{\partial t}.$$

The symbol ∇ is known as "nabla" and is used in some form in almost all operators encountered in chemical engineering problems. It is called "the gradient" and is defined in rectangular coordinates as

$$\nabla = \frac{\partial}{\partial x}e_x + \frac{\partial}{\partial y}e_y + \frac{\partial}{\partial z}e_z$$

where e_x, e_y and e_z are the three basis vectors of the coordinate system. Thus ∇ is a vector and has both magnitude and direction. Taking the scalar product of ∇ with itself gives ∇^2, the Laplacian operator.

Many of the operators that make use of the ∇ symbol have common names such as gradient (**grad**), divergence (div) and curl (**curl**) or rotation (**rot**). Curl and rotation are synonymous. The **grad**, **div**, etc. nomenclature is found mostly in the German literature. Notice that the gradient is taken of a scalar function but is itself a vector while the divergence is taken of a vector but is itself a scalar. This is the reason for often writing **grad** in boldface, which in many texts is an indication of a vector. The different names, the different uses and the slightly different ways of writing different operators can initially be confusing, but keep in mind that the operator notation is simply a compact, coordinate free nomenclature. Working with operators only requires knowledge of what a given symbol stands for. The most commonly used operators, their definitions and names, are given in Tables 1.1 through 1.3 for the three basic coordinate systems.

Table 1.1 Common operators in rectangular coordinates.

Gradient of a scalar function

$$\mathbf{grad} f = \nabla f(x, y, z) = \frac{\partial f}{\partial x}e_x + \frac{\partial f}{\partial y}e_y + \frac{\partial f}{\partial z}e_z$$

Laplacian of a scalar function

$$\nabla^2 f(x, y, z) = \frac{\partial^2 f}{\partial x^2} + \frac{\partial^2 f}{\partial y^2} + \frac{\partial^2 f}{\partial z^2}$$

Divergence of a vector function

$$\text{div} f = \nabla \cdot v = \frac{\partial v_x}{\partial x} + \frac{\partial v_y}{\partial y} + \frac{\partial v_z}{\partial z}$$

Laplacian of a vector function

$$\nabla^2 v = \nabla^2 v_x e_x + \nabla^2 v_y e_y + \nabla^2 v_z e_z$$

Rotation or curl of a vector function

$$\mathbf{rot} v = \mathbf{curl} v = \nabla \times v = \left(\frac{\partial v_z}{\partial y} - \frac{\partial v_y}{\partial z}\right) e_x + \left(\frac{\partial v_x}{\partial z} - \frac{\partial v_z}{\partial x}\right) e_y + \left(\frac{\partial v_y}{\partial x} - \frac{\partial v_x}{\partial y}\right) e_z$$

Table 1.2 Common operators in cylindrical coordinates.

Gradient of a scalar function

$$\mathbf{grad} f = \nabla f(r,\theta,z) = \frac{\partial f}{\partial r} e_r + \frac{1}{r}\frac{\partial f}{\partial \theta} e_\theta + \frac{\partial f}{\partial z} e_z$$

Laplacian of a scalar function

$$\nabla^2 f(r,\theta,z) = \frac{1}{r}\frac{\partial}{\partial r}\left(r\frac{\partial f}{\partial r}\right) + \frac{1}{r^2}\frac{\partial^2 f}{\partial \theta^2} + \frac{\partial^2 f}{\partial z^2}$$

Divergence of a vector function

$$\text{div} f = \nabla \cdot v = \frac{1}{r}\frac{\partial r v_r}{\partial r} + \frac{1}{r}\frac{\partial v_\theta}{\partial \theta} + \frac{\partial v_z}{\partial z}$$

Laplacian of a vector function

$$\begin{aligned}\nabla^2 v = &\left(\frac{\partial}{\partial r}\left(\frac{1}{r}\frac{\partial r v_r}{\partial r}\right) + \frac{1}{r^2}\frac{\partial^2 v_r}{\partial \theta^2} - \frac{2}{r^2}\frac{\partial v_\theta}{\partial \theta} + \frac{\partial^2 v_r}{\partial z^2}\right) e_r \\ &+ \left(\frac{\partial}{\partial r}\left(\frac{1}{r}\frac{\partial r v_\theta}{\partial r}\right) + \frac{1}{r^2}\frac{\partial^2 v_\theta}{\partial \theta^2} + \frac{2}{r^2}\frac{\partial v_r}{\partial \theta} + \frac{\partial^2 v_\theta}{\partial z^2}\right) e_\theta \\ &+ \left(\frac{1}{r}\frac{\partial}{\partial r}\left(r\frac{\partial v_z}{\partial r}\right) + \frac{1}{r^2}\frac{\partial^2 v_z}{\partial \theta^2} + \frac{\partial^2 v_z}{\partial z^2}\right) e_z\end{aligned}$$

Rotation or curl of a vector function

$$\mathbf{rot} v = \mathbf{curl} v = \nabla \times v = \left(\frac{1}{r}\frac{\partial v_z}{\partial \theta} - \frac{\partial v_\theta}{\partial z}\right) e_r + \left(\frac{\partial v_r}{\partial z} - \frac{\partial v_z}{\partial r}\right) e_\theta + \left(\frac{1}{r}\frac{\partial r v_\theta}{\partial r} - \frac{1}{r}\frac{\partial v_r}{\partial \theta}\right) e_z$$

It is possible to find the explicit expression for an operator in any coordinate system if an explicit expression is known in a reference coordinate system and if the transformations which convert the reference coordinates to the new coordinates are known. The reference coordinates are usually rectangular coordinates. However, doing so is tedious and not straightforward. So unless one works in a very unusual coordinate system which is not described in the literature, expressions for operators are looked up in tables rather than derived.

Using operator notation, we can now restate Fourier's and Fick's laws. They are, Fourier's law

$$q = -k\nabla T$$

and Fick's law

$$N = -D\nabla C.$$

Notice that both fluxes are vectors and that Fourier's and Fick's laws are mathematically identical, which is at the root of the famous Chilton-Colburn analogy.[6]

[6]Chilton, T.H. and Colburn, A.P. [1934] "Mass transfer (absorption) coefficients, prediction from data on heat transfer and fluid friction," *Ind. Eng. Chem.* **26**, 1183–1187.

Table 1.3 Common operators in spherical coordinates.

Gradient of a scalar function

$$\mathbf{grad} f = \nabla f(r,\theta,\phi) = \frac{\partial f}{\partial r} e_r + \frac{1}{r}\frac{\partial f}{\partial \theta} e_\theta + \frac{1}{r\sin\theta}\frac{\partial f}{\partial \phi} e_\phi$$

Laplacian of a scalar function

$$\nabla^2 f(r,\theta,\phi) = \frac{1}{r^2}\frac{\partial}{\partial r}\left(r^2\frac{\partial f}{\partial r}\right) + \frac{1}{r^2\sin\theta}\frac{\partial}{\partial \theta}\left(\sin\theta\frac{\partial f}{\partial \theta}\right) + \frac{1}{r^2\sin^2\theta}\frac{\partial^2 f}{\partial \phi^2}$$

Divergence of a vector function

$$\mathrm{div} f = \nabla\cdot v = \frac{1}{r^2}\frac{\partial r^2 v_r}{\partial r} + \frac{1}{r\sin\theta}\frac{\partial v_\theta \sin\theta}{\partial \theta} + \frac{1}{r\sin\theta}\frac{\partial v_\phi}{\partial \phi}$$

Laplacian of a vector function

$$\begin{aligned}\nabla^2 v = &\left(\nabla^2 v_r - \frac{2v_r}{r^2} - \frac{2}{r^2}\frac{\partial v_\theta}{\partial \theta} - \frac{2v_\theta\cot\theta}{r^2} - \frac{2}{r^2\sin\theta}\frac{\partial v_\phi}{\partial \phi}\right) e_r \\ &+ \left(\nabla^2 v_\theta + \frac{2}{r^2}\frac{\partial v_r}{\partial \theta} - \frac{v_\theta}{r^2\sin^2\theta} - \frac{2\cos\theta}{r^2\sin^2\theta}\frac{\partial v_\phi}{\partial \phi}\right) e_\theta \\ &+ \left(\nabla^2 v_\phi - \frac{v_\phi}{r^2\sin^2\theta} + \frac{2}{r^2\sin\theta}\frac{\partial v_r}{\partial \phi} + \frac{2\cos\theta}{r^2\sin^2\theta}\frac{\partial v_\theta}{\partial \phi}\right) e_\phi\end{aligned}$$

Rotation or curl of a vector function

$$\begin{aligned}\mathbf{rot} v = \mathbf{curl} v = \nabla\times v = &\left(\frac{1}{r\sin\theta}\frac{\partial}{\partial \theta}(v_\phi\sin\theta) - \frac{1}{r\sin\theta}\frac{\partial v_\theta}{\partial \phi}\right) e_r \\ &+ \left(\frac{1}{r\sin\theta}\frac{\partial v_r}{\partial \phi} - \frac{1}{r}\frac{\partial r v_\phi}{\partial r}\right) e_\theta + \left(\frac{1}{r}\frac{\partial r v_\theta}{\partial r} - \frac{1}{r}\frac{\partial v_r}{\partial \theta}\right) e_\phi\end{aligned}$$

Strictly speaking, Fick's law is only valid for diffusion of a dilute solute through the solvent and one should be cautions about using it to model multicomponent diffusion or diffusion in concentrated solutions. In the form states here, both Fourier's law and Fick's law are for isotropic substances, i.e. substances in which properties are independent of direction. In non-isotropic substances such as crystals, either solid or liquid, properties change with direction and the thermal conductivity of or diffusion coefficient inside a crystal will depend on the direction of transport. Stating the relevant constitutive equations for such substances requires the use of tensors. Both equations must also be used with caution when modelling transport in porous media and the coefficients should more correctly be called effective conductivity and effective diffusivity. The value of the effective transport coefficients will not equal that of the coefficients in an unrestricted medium due to the convoluted path of the pores that serve as transport conduits.

Diffusion in porous media is particularly complex and may involve both Fickian diffusion and Knudsen and surface diffusion. Estimation of the effective diffusion coefficient in porous media is an important problem when modelling catalysts and an extensive literature exists on this topic.[7]

As demonstrated in the example of laminar flow in a pipe, page 37, momentum balances can be derived just as mass and energy balances by using differential control volumes. Doing so may be necessary for complex fluids and is a good exercise when first learning about fluid mechanics. But no person in their right mind would re-derive the momentum balance for a Newtonian fluid, the Navier–Stokes equation, each time it is used. The Navier–Stokes equation is not easy to derive but it is tabulated for all common coordinate systems. More important than being able to derive this equation is the ability to simplify it to the maximum extent possible for a given physical situation.

The fluid mechanical problems we will now consider are laminar flow problems of linear or Newtonian fluids governed by the Navier–Stokes equation, which, in spite of its complicated appearance, is nothing but a momentum balance or Newton's second law of motion

$$\rho \frac{Dv}{Dt} = -\nabla P + \mu \nabla^2 v + \rho g$$

and the continuity equation

$$\nabla \cdot v = 0.$$

These equations hold for fluids of constant density, ρ, and constant viscosity, μ. The operator D/Dt, called the substantial time derivative, is listed for the three basic coordinate systems in Table 1.4.

Boundary conditions for Newtonian fluids are often obtained from the requirement of continuity of velocity and continuity of stress. Making use of the latter requires the use of Newton's law of viscosity. In particular, these two continuity requirements give the no-slip condition at a fluid-solid interphase and the condition of zero fluid shear stress at a fluid-gas interphase, derived assuming that the gas viscosity is negligible compared to the fluid viscosity.

Like any other models, momentum balances can be simplified using symmetry considerations and additional simplifications are possible if one can conclude that some of the velocity components must be zero. This is usually concluded from physical insight but it can be difficult to do so if the geometry is complex and it is especially difficult for people who do not easily visualize objects in three-dimension.

Example 1.18: Simplification of Navier–Stoke's equation.

Consider steady, laminar flow of a Newtonian fluid in a cylindrical bearing for which the inner rod rotates with an angular velocity ω and the outer bearing is stationary, see Fig. 1.19. Assume that gravity can be ignored.

[7]The reader interested in learning more may want to consult the old but still valuable monograph: Jackson, R. [1977] *Transport in Porous Catalysts* (Elsevier, Amsterdam).

Table 1.4 The operator D/Dt in the three common coordinate systems.

Rectangular coordinates

$$\frac{Dv}{Dt} = \left(\frac{\partial v_x}{\partial t} + v_x\frac{\partial v_x}{\partial x} + v_y\frac{\partial v_x}{\partial y} + v_z\frac{\partial v_x}{\partial z}\right) e_x$$

$$+ \left(\frac{\partial v_y}{\partial t} + v_x\frac{\partial v_y}{\partial x} + v_y\frac{\partial v_y}{\partial y} + v_z\frac{\partial v_y}{\partial z}\right) e_y$$

$$+ \left(\frac{\partial v_z}{\partial t} + v_x\frac{\partial v_z}{\partial x} + v_y\frac{\partial v_z}{\partial y} + v_z\frac{\partial v_z}{\partial z}\right) e_z$$

Cylindrical coordinates

$$\frac{Dv}{Dt} = \left(\frac{\partial v_r}{\partial t} + v_r\frac{\partial v_r}{\partial r} + \frac{v_\theta}{r}\frac{\partial v_r}{\partial \theta} - \frac{v_\theta^2}{r} + v_z\frac{\partial v_r}{\partial z}\right) e_r$$

$$+ \left(\frac{\partial v_\theta}{\partial t} + v_r\frac{\partial v_\theta}{\partial r} + \frac{v_\theta}{r}\frac{\partial v_\theta}{\partial \theta} + \frac{v_r v_\theta}{r} + v_z\frac{\partial v_\theta}{\partial z}\right) e_\theta$$

$$+ \left(\frac{\partial v_z}{\partial t} + v_r\frac{\partial v_z}{\partial r} + \frac{v_\theta}{r}\frac{\partial v_z}{\partial \theta} + v_z\frac{\partial v_z}{\partial z}\right) e_z$$

Spherical coordinates

$$\frac{Dv}{Dt} = \left(\frac{\partial v_r}{\partial t} + v_r\frac{\partial v_r}{\partial r} + \frac{v_\theta}{r}\frac{\partial v_r}{\partial \theta} + \frac{v_\phi}{r\sin\theta}\frac{\partial v_r}{\partial \phi} - \frac{v_\theta^2 + v_\phi^2}{r}\right) e_r$$

$$+ \left(\frac{\partial v_\theta}{\partial t} + v_r\frac{\partial v_\theta}{\partial r} + \frac{v_\theta}{r}\frac{\partial v_\theta}{\partial \theta} + \frac{v_\phi}{r\sin\theta}\frac{\partial v_\theta}{\partial \phi} + \frac{v_r v_\theta}{r} - \frac{v_\phi^2\cot\theta}{r}\right) e_\theta$$

$$+ \left(\frac{\partial v_\phi}{\partial t} + v_r\frac{\partial v_\phi}{\partial r} + \frac{v_\theta}{r}\frac{\partial v_\phi}{\partial \theta} + \frac{v_\phi}{r\sin\theta}\frac{\partial v_\phi}{\partial \phi} + \frac{v_\phi v_r}{r} + \frac{v_\theta v_\phi}{r}\cot\theta\right) e_\phi$$

Place a cylindrical coordinate system with the z-axis located on the axis of rotation. The boundary conditions in this coordinate system are

$$v_z = 0 \text{ at } r = R_1\,,\ r = R_2$$

$$v_r = 0 \text{ at } r = R_1\,,\ r = R_2$$

$$v_\theta = \omega R_1 \text{ at } r = R_1 \text{ and } v_\theta = 0 \text{ at } r = R_2.$$

There are two symmetry operations one can perform, rotation around and translation along the center line. In other words: the physical situation is not changed by a rotation of the entire bearing around the z-axis nor is the situation changed

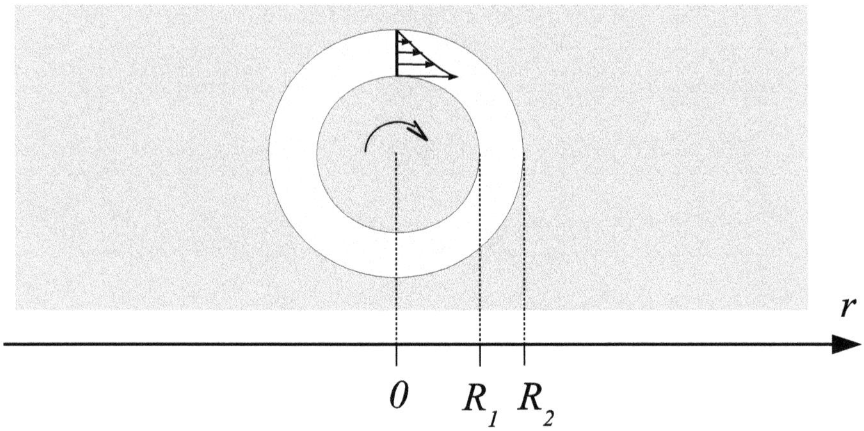

Fig. 1.19 Laminar flow in a cylindrical bearing.

by a translation along the z-axis. This is also shown by the fact that the origin of the z-axis is completely arbitrary and can be placed anywhere without affecting the problem and similarly for the $\theta = 0$ direction. For laminar flow one can therefore assume that the solution is independent of the angular coordinate θ and the axial coordinate z and all derivatives with respect to these two coordinates must be zero. Using this, the Navier–Stokes and continuity equations simplify to

$$\rho\left(v_r\frac{dv_r}{dr} - \frac{v_\theta^2}{r}\right) = -\frac{dp}{dr} + \mu\frac{d}{dr}\left(\frac{1}{r}\frac{d}{dr}(rv_r)\right)$$

$$\rho\left(v_r\frac{dv_\theta}{dr} + \frac{v_r v_\theta}{r}\right) = \mu\frac{d}{dr}\left(\frac{1}{r}\frac{d}{dr}(rv_\theta)\right)$$

$$\rho v_r\frac{dv_z}{dr} = \mu\frac{1}{r}\frac{d}{dr}\left(r\frac{dv_z}{dr}\right)$$

$$\frac{1}{r}\frac{d}{dr}(rv_r) = 0.$$

The last of these equations is the simplified continuity equation. It is solved in combination with the boundary conditions on v_r to yield $v_r = 0$, an expected result. Substituting this result into the remaining equations give the even simpler model

$$\rho\frac{v_\theta^2}{r} = \frac{dp}{dr}$$

$$0 = \mu\frac{d}{dr}\left(\frac{1}{r}\frac{d}{dr}(rv_\theta)\right)$$

$$0 = \mu \frac{1}{r}\frac{d}{dr}\left(r\frac{dv_z}{dr}\right).$$

The last two equations are solved to give

$$v_z = 0$$

as expected, and

$$v_\theta = \frac{R_1^2}{R_2^2 - R_1^2}\frac{\omega}{r}(R_2^2 - r^2).$$

Finally, the remaining equation, which is the simplified θ-component of the Navier–Stokes equation, can be solved for the pressure.

$$p(r) = \rho\omega^2 R_1^4 \frac{r^2 - R_2^4/r^2 - 4R_2^2\ln(r)}{2(R_1 - R_2)^2(R_1 + R_2)^2} + C_1$$

where C_1 is an arbitrary constant.

1.3.6 *Common boundary conditions*

Boundary conditions are required for balances on differential control volumes. Almost always, these conditions are derived from physical arguments using balances over a CV surface or from symmetry considerations. The most frequently encountered boundary conditions are summarized as follows.

Continuity: It is usually taken for granted that most physical variables such as temperature, concentration and velocity are continuous functions of position. Thus, if there is no film transport resistance at an interface, the value of the dependent variable at an interface is the same in the two phases. At a solid-fluid interface, the fluid velocity equals the solid velocity and at a fluid-fluid interface, the velocities of the two fluids and the stresses in the two fluids are equal.

Film resistance: The concept of a film at a solid-fluid surface implies that the magnitude of the quantity being transported changes rapidly across a stagnant, fluid phase film adjacent to the solid phase. The fluid film is assumed so thin that accumulation in the film can be neglected and a balance over the film states that the flux through the film towards the interphase equals the flux away from the interphase in the other phase. The film flux is usually modelled using a mass or heat transfer coefficient. Thus for a heat flux, the film condition becomes

$$-k\left.\frac{dT}{dx}\right|_{surface} = h(T_{surface} - T_{infinity})$$

where h is the heat transfer coefficient, $T_{infinity}$ is the fluid temperature "far away" from the solid surface and x represents the direction of the outward normal at the solid surface.

Boundedness: Physical quantities must remain bounded. This requirement can sometimes be used as a boundary condition because it allows one to discard solutions which have singularities, i.e. points at which the solution become infinite.

Symmetry: Many physical situations exhibit some sort of symmetry in the sense that the geometry of the object being modelled has a plane, axis or point of symmetry and the boundary conditions are such that they do not break this symmetry. When this is the case, the solution for the dependent variable may be assumed to exhibit the same symmetry. This requirement can be used as a boundary condition since it leads one to eliminate those solutions which do not exhibit the required symmetry. However, the assumption that the solution possesses the symmetry imposed by the physical situation can be wrong. Most importantly, symmetry breaking occurs at the onset of turbulent flow.

Approximate models: The example with the packed bed reactor showed that when the conditions at a boundary are very complex and difficult to account for, approximate boundary conditions, derived from models using simplifying assumptions, can be used.

1.3.7 *Common thermodynamic equations*

The thermodynamic equations that are most important to chemical engineers are equations of state, such as the ideal gas law

$$PN = nRT$$

where P is pressure, V volume, n number of moles, R the gas constant and T absolute temperature. The value of the gas constant in various different systems of units is given below

R	Units
1.987	cal/(g mole ^{o}K)
82.06	(atm cm^3)/(g mole ^{o}K)
1.314	(atm ft^3)/(lb mole ^{o}K)
0.08206	(atm liter)/(g mole ^{o}K)
8.3144	(Joule)/(g mole ^{o}K)

A step up in complexity from the ideal gas law is van der Waals equation.

$$P = \frac{RT}{V/n - b} - \frac{a}{(V/n)^2}$$

where a and b are constants with values that depend on the gas. More complicated equations of state exists but will not be covered here.

Another type of thermodynamic equation important when formulating mathematical models are equilibrium equations. We mention two here, Henry's law which gives the partial pressure of a solute over a solution

$$P_n = H_n \omega_n$$

where P_n is the partial pressure of the solute indicated by n, H_n the Henry's law constant for the solute and ω_n the mole fraction of the solute. And Raoult's law which gives the partial pressure of the solvent

$$P_n = \omega_n P_n^0$$

where P_n^0 is the vapour pressure of the pure solvent. The two equilibrium relationships are easily confused but are very different. Henry's law is for the solute, the compound present at a low mole fraction, while Raoult's law is for the solvent, the compound present at mole fraction close to 1. Both laws are valid for ideal dilute solutions, i.e. in the limit as the mole fraction of the solute goes to 0 and the mole fraction of the solvent goes to 1.

1.4 Other kinds of balances

Lest the reader should be left with the impression that differential balances in chemical engineering are only applicable to transport problems (mass, energy and momentum transport), this section provides some examples of differential balances from other areas. This often require the use of control volumes that are volumes in some abstract space as opposed to physical space. However, the general method for deriving and manipulating the balance remains the same. The main difficulty in writing models for many of these non-classical chemical engineering systems is the lack of familiarity with the expressions for the fluxes. Well known constitutive equations may not be available to describe the flux terms and expressions for the fluxes may therefore have to be modelled from first principles; a difficult task unless one is familiar with the system being modelled.

1.4.1 *Balances on surfaces*

We will consider a chemical reaction taking place on a surface. The CV is a unit area of the surface and it therefore has zero physical volume. This may seem like a paradox, but only serves to illustrate the fact that the term "control volume" must not be taken too literally.

Surface reactions play an important role in catalysis. Not just in industrial reactors, but also in environmental problems where airborne pollutants may adsorb to natural surfaces such as ice and water or on man-made surfaces such as the grimy film that covers most windows, and react in ways that alter the compositions of the pollutants in the air. Adsorbed molecules often react faster than molecules in suspension because reactant molecules only need to diffuse in two dimensions to encounter one another and, all other things being equal, two dimensional diffusion is faster than three dimensional diffusion. To convince yourself of this, just think of how much quicker it is to find someone's office if you know the floor on which the office is located as opposed to when you only know the building. Even greater rate enhancement results from restricting diffusion of reactants to one-dimension. This effect probably plays a role in biological systems where regulatory proteins may bind to a DNA molecule and diffuse along the DNA strand to rapidly find their binding site.

Example 1.19: Surface reaction.

The situation to be considered is sketched in Fig. 1.20. Two reactants in a gas stream, call them A and B, adsorb[8] to a surface and react to form a product P which then detach from the surface. The surface can at most be covered by a monolayer of reacting molecules and this will be modelled by assuming that there is a finite number of sites per surface area onto which reactants can adsorb. Adsorption is then modelled as a reaction between a gas phase molecule and a free site leading to an adsorbed molecule. The number of free sites must therefore be modelled by writing a conservation equation on the number of free sites, similar to the conservation equations on the reactants. The gas phase concentrations will be assumed constant while surface concentrations can change with time. Rates are assumed to follow mass action kinetics in both reactant concentrations and adsorption sites.

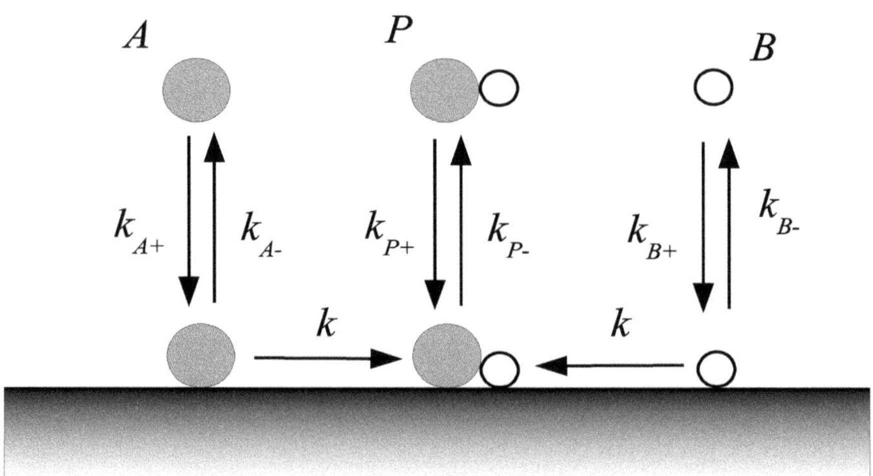

Fig. 1.20 Schematic of a surface catalyzed reaction between two molecules.

[8]The term adsorption is often confused with the similar sounding term absorption. Adsorption refers to retention of atoms and molecules on the surface of a substance while absorption refers to retention into the bulk of a phase.

Let the gas phase concentrations be $C_{A,g}$, $C_{B,g}$ and $C_{P,g}$ and the surface concentrations be $C_{A,s}$, $C_{B,s}$ and $C_{P,s}$. The number of free sites per area will be indicated as N and, assuming that N is large, it can be approximated as a continuous variable. Reaction rate constants are as indicated in the figure. Before we can write balance equations on the reactants, we must decide how to treat surface sites in a reaction. Does the product take up a single site on the surface or does it occupy two, as one might expect since it is formed from two reactants that occupy one site each? We will assume that the product occupies two sites so the number of free sites do not change in a reaction. (The reader may want to write the balances when it is assumed that the product occupies only one site and that a free site is therefore formed in each reaction). Mass balances on the two reactants and the product give

$$\frac{dC_{A,s}}{dt} = k_{A+}C_{A,g}N - k_{A-}C_{A,s} - kC_{A,s}C_{B,s}$$

$$\frac{dC_{B,s}}{dt} = k_{B+}C_{B,g}N - k_{B-}C_{B,s} - kC_{A,s}C_{B,s}$$

$$\frac{dC_{P,s}}{dt} = kC_{A,s}C_{B,s} + k_{P+}C_{P,g}N - k_{P-}C_{P,s}$$

and a balance on the number of free sites per area gives

$$\frac{dN}{dt} = k_{A-}C_{A,s} + k_{B-}C_{B,s} + 2k_{P-}C_{P,s} - k_{A+}C_{A,g}N - k_{B+}C_{B,g}N - 2k_{P+}C_{P,g}N.$$

The model consists of four coupled differential equations. But, as is often the case in kinetic problems of this kind, one of the differential equations can be replaced by an algebraic equation by taking some weighted sum of the balances. Multiply the balance for $C_{P,s}$ by two (since adsorbed product occupies two sites) and sum this equation and the remaining three balances to get the differential equation

$$\begin{aligned}
\frac{d}{dt}(C_{A,s} + C_{B,s} + 2C_{P,s} + N) &= k_{A+}C_{A,g}N - k_{A-}C_{A,s} - kC_{A,s}C_{B,s} \\
&+ k_{B+}C_{B,g}N - k_{B-}C_{B,s} - kC_{A,s}C_{B,s} \\
&+ 2kC_{A,s}C_{B,s} + 2k_{P+}C_{P,g}N - 2k_{P-}C_{P,s} \\
&+ k_{A-}C_{A,s} + k_{B-}C_{B,s} + 2k_{P-}C_{P,s} - k_{A+}C_{A,g}N - k_{B+}C_{B,g}N - 2k_{P+}C_{P,g}N
\end{aligned}$$

which upon simplification gives

$$\frac{d}{dt}(C_{A,s} + C_{B,s} + 2C_{P,s} + N) = 0 \Rightarrow C_{A,s} + C_{B,s} + 2C_{P,s} + N = \text{Constant}.$$

This constant is simply the total number of sites, free or filled, per area and it must be a constant since none of the reactions consume or produce sites. This result is of course so obvious that one could have written it down without the need for calculating the weighted sum of the balances. However, it is important to appreciate that this balance on the total number of sites is not independent of the four reactant balances and, when it is used in a model, one of the reactant balances can be ignored by eliminating N to get a model consisting of three coupled ordinary differential equations.

$$\frac{dC_{A,s}}{dt} = k_{A+}C_{A,g}(N_{\text{total}} - C_{A,s} - C_{B,s} - 2C_{P,s}) - k_{A-}C_{A,s} - kC_{A,s}C_{B,s}$$

$$\frac{dC_{B,s}}{dt} = k_{B+}C_{B,g}(N_{\text{total}} - C_{A,s} - C_{B,s} - 2C_{P,s}) - k_{B-}C_{B,s} - kC_{A,s}C_{B,s}$$

$$\frac{dC_{P,s}}{dt} = kC_{A,s}C_{B,s} + k_{P+}C_{P,g}(N_{\text{total}} - C_{A,s} - C_{B,s} - 2C_{P,s}) - k_{P-}C_{P,s}.$$

These equations are nonlinear and will have to be solved numerically.

1.4.2 *Population balance models*

Population balances are conservation equations on the number of particles in a system. The particles can be almost anything; cells in a microbial culture, droplets in an aerosol or emulsion, cars in traffic, etc. When modelling populations of particles, one is often interested in finding the so called distribution of states. This is a function that describes how a property such as size, mass, age, chemical composition or any other meaningful measure is distributed among the individuals in the population. Distributions of states can be multi-dimensional such as the age and mass distribution, etc. The balance equations that govern the distribution of states are the population balance equations. We will derive a population balance for the simple case of cells growing in a CSTR, using a differential CV over the abstract state space to formulate the model.

Example 1.20: Population balance equations.

The state space parameter will be indicated m to suggest mass, and we will assume that the parameter m increases with cell age so that divisions of cells in state m results in formation of cells with smaller m and such that the sum of the ms of the newborn cells equal the m of the dividing cell. However, derivation of the population balances for other state parameters is quite similar to the derivation shown here.

Let the distribution we seek be $C(m)$, i.e. $C(m)$ is the frequency in the population of cells in state m. This distribution can be scaled various ways and we will define $C(m)$ such that

$C(m)dm =$ Number concentration of cells in the population for which the property indicated by m takes a value between m and $m + dm$.

To anyone familiar with distributions it is obvious that one can divide $C(m)$ by the volume of the CSTR and obtain an equally valid distribution, the cell number concentration distribution. Or one can divide through by the total number of cells and obtain the frequency of cell states; a true distribution in the sense that the zeroth moment of the frequency is unity. In the latter case, $C(m)dm$ can be viewed as the probability that a randomly picked cell or individual has a value of the state parameter m that lies between m and $m + dm$.

To do a cell number balance on the differential control volume dm, Fig. 1.21, consider the processes through which cells can enter or leave the control volume. Cells can both enter and leave by growth, they can leave by dividing to become smaller cells and by being washed out of the CSTR and they can enter the control volume by birth or division of a larger cell.

The number of cells inside the control volume equals $C(m)dm$ and the cell number balance becomes

$$\frac{\partial C(m)dm}{\partial t} = \text{Flux in by growth} - \text{Flux out by growth}$$

$$\text{Flux in by birth} - \text{Flux out by division} - \text{Flux out by washout.}$$

These fluxes must be accounted for in order to finish the model but now there are no applicable constitutive equations. Each flux must be modelled.

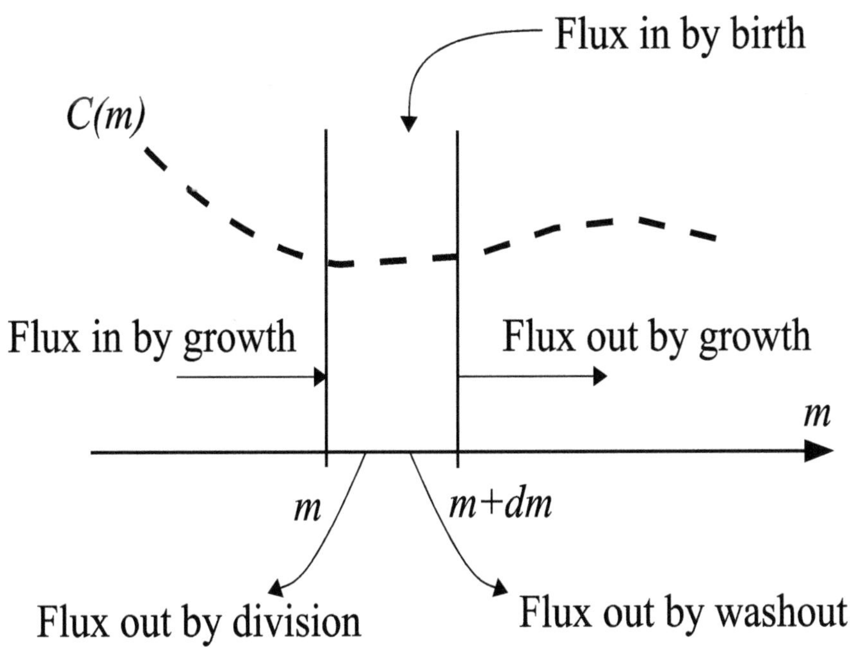

Fig. 1.21 Control volume and fluxes in m-space.

The flux out of the CV by washout is the easiest of the fluxes to calculate. Assuming that the dilution rate in the CSTR equals D (equal to the volumetric flow rate through the reactor divided by the reactor volume), then the rate of washout of cells from the CV simply equals the dilution rate times the number of cells inside the CV or $DC(m)dm$.

The growth flux into the CV at m is the value of the distribution at m, basically equal to the number of cells immediately smaller than the cells in the CV, multiplied by the rate at which cells grow into the CV. This is the single cell growth rate with respect to m and will be indicated $r(m)$ and it must be determined from some additional model of single cell growth such as a metabolic model. Modelling $r(m)$ requires some idea of how fast cells grow as a function of the cell state m and of the concentration of the nutrients in the medium. So the growth flux into the control volume equals $r(m)C(m)$ and, similarly, the growth flux out of the CV equals $r(m+dm)C(m+dm)$.

The flux out of the CV by cell division can be described by a function specifying the rate at which cells in a given state divide. This function is called the division intensity and will be indicated $\Gamma(m)$. It is defined as

$$\Gamma(m)dt = \text{Fraction of cells in state } m \text{ that divide in the next } dt \text{ time interval.}$$

The rate at which cells leave the CV is therefore equal to the division intensity multiplied by the number of cells inside the CV, $\Gamma(m)C(m)dm$.

Since cells split into two new cells in a division, the rate at which cells are born must equal twice the rate at which all cells divide. But the rate at which all cells divide is simply the rate of division discussed above, integrated over all cell states, $\int_0^\infty \Gamma(m)C(m)dm$. However, to find the flux of newborn cells into the CV, the rate of division must be multiplied by the probability that a division gives rise to a newborn cell with an m-value between m and $m+dm$. Let this probability be $p(m,\tilde{m})$

$$\begin{aligned} p(m,\tilde{m})dm =& \text{Probability that division of a cell in state } \tilde{m} \text{ results} \\ & \text{in a new cell with state between } m \text{ and } m+dm. \end{aligned}$$

The flux of new cells into the CV must therefore equal

$$2\int_0^\infty p(m,\tilde{m})dm\Gamma(\tilde{m})C(\tilde{m})d\tilde{m}\,.$$

Putting all these flux expressions into the balance equation gives

$$\begin{aligned} \frac{\partial C(m)dm}{\partial t} =& r(m)C(m) - r(m+dm)C(m+dm) \\ & 2\int_0^\infty p(m,\tilde{m})dm\Gamma(\tilde{m})C(\tilde{m})d\tilde{m} - \Gamma(m)C(m)dm - DC(m)dm. \end{aligned}$$

This result is then treated like any other differential balance, by dividing by the volume of the control volume, dm, and recognizing derivatives to get the population balance equation

$$\frac{\partial C}{\partial t} + \frac{\partial rC}{\partial m} = 2\int_0^\infty p(m,\tilde{m})\Gamma(\tilde{m})C(\tilde{m})d\tilde{m} - (D+\Gamma(m))C(m).$$

The next example shows an alternate method for deriving a population balance. The purpose of this is to demonstrate how it is sometimes possible to derive differential balances using macroscopic CVs. This is perhaps an odd approach, but the method does have its proponents who see the method as more rigorous than derivations that use questionable algebraic manipulations with differential quantities.

Example 1.21: Alternate derivation a population balance.

We will base the balance on the CV defined by the bracket from cell state b to state c where $c-b$ is not a differential quantity. We will assume without loss of generality that the cell state parameter m increases with cell age and that $b<c$. The number of cells in the CV is

$$\int_b^c C(m)\,dm.$$

The sum of the growth fluxes in and out of the control volume is

$$r(b)C(b) - r(c)C(c).$$

The division and washout fluxes of cells out of the control volume are

$$D\int_b^c C(m)dm + \int_b^c \Gamma(m)C(m)dm$$

and the birth flux into the control volume is

$$\int_b^c 2\int_0^\infty p(m,\tilde{m})\Gamma(\tilde{m})C(\tilde{m})d\tilde{m}dm.$$

Putting the flux and accumulation terms together, the balance becomes

$$\frac{\partial}{\partial t}\int_b^c C(m)\,dm = r(b)C(b) - r(c)C(c) + \int_b^c 2\int_0^\infty p(m,\tilde{m})\Gamma(\tilde{m})C(\tilde{m})d\tilde{m}dm$$

$$-\int_b^c (D+\Gamma(m))C(m)dm \Rightarrow$$

$$\int_b^c \frac{\partial C(m)}{\partial t}\,dm + \int_b^c \frac{\partial r(m)C(m)}{\partial m}dm = \int_b^c 2\int_0^\infty p(m,\tilde{m})\Gamma(\tilde{m})C(\tilde{m})d\tilde{m}dm$$

$$-\int_b^c (D+\Gamma(m))C(m)dm \Rightarrow$$

$$\int_b^c \left(\frac{\partial C}{\partial t} + \frac{\partial rC}{\partial m} - 2\int_0^\infty p(m,\tilde{m})\Gamma(\tilde{m})C(\tilde{m})d\tilde{m} - (D+\Gamma(m))C\right) dm = 0.$$

The key argument that allows one to obtain a differential balance from this macroscopic balance is as follows: as the limits on the integral are arbitrary, the integrand must be identically zero, giving us the population balance.

$$\frac{\partial C}{\partial t} + \frac{\partial rC}{\partial m} - 2\int_0^\infty p(m,\tilde{m})\Gamma(\tilde{m})C(\tilde{m})d\tilde{m} - (D+\Gamma(m))C = 0\,.$$

The population balance equation derived in the previous two examples is a simple case of the more general situation where the cell state is characterized by a large number of parameters and where the population balance is coupled to balances on the microbial substrates and products. Reader who would like to learn more on this should consult the paper by Fredrickson *et al.*[9] Population balance models are in fact a huge topic and they are important not just when modeling living cells, but when modeling any particulate system, such as aerosols, emulsions and crystallizers.[10]

The next example shows a situation in which the CV is a single point in state space and where the flux terms are stochastic and must be modelled using probabilistic arguments.

1.4.3 *Stochastic processes*

Chemical reactions are commonly modeled with the assumption that concentration can be treated as a continuous variable. This is a valid approximation in most cases but it fails when only a small number of molecules react. In this case, the state of the system must be indicated by specifying the numbers of each of the participating molecules as a function of time and reaction rates are no longer deterministic but stochastic. Similarly, the result is not deterministic, but is a probability function describing the likelihood of finding the system in a state with a given number of molecules of each of the reactants.

Example 1.22: Stochastic chemical kinetics.

To illustrate how to write a model for such a system, consider the very simple reaction

$$A \underset{c_{-1}}{\overset{c_1}{\rightleftarrows}} B$$

where c_1 and c_{-1} are the stochastic equivalents of reaction rate constants, they are known as reaction propensities.

[9]Fredrickson A., Ramkrishna, G.D. and Tsuchiya, H.M. [1967] "Statics and dynamics of procaryotic cell populations," *Math. Biosci.* **1**, 327–374.
[10]Ramkrishna, D. [2000] Population Balances (Academic Press, San Diego, CA, USA).

The state space is the set of point of the form (N_A, N_B) where N_A and N_B are the number of molecules of A and B respectively and the probability of finding the system in the state (N_A, N_B) at time t will be indicated $P(t, N_A, N_B)$. To write a model of this system, pick *a single point* (N_A, N_B) in the state space as CV and write a balance for the probability measure of that state.

For readers who have not studied probability calculus this is unfamiliar territory and it may be more convenient to consider instead a very large number of independent realizations, say $\mathcal{N}$, of this reaction system. Some number, $\mathcal{N}_{N_A,N_B}$, of these systems will be in the state specified by the control volume, and for a large enough number of realizations, the probability $P(t, N_A, N_B)$ equals $\mathcal{N}_{N_A,N_B}/\mathcal{N}$. We can therefore do a balance on the number of systems which are in the state (N_A, N_B) and divide the balance equation by $\mathcal{N}$ to get the model for the probability.

Systems that are in the state equivalent to the CV can leave the state through two possible reactions: A reacting to form B or $(N_A, N_B) \to (N_A - 1, N_B + 1)$ and B reacting to form A or $(N_A, N_B) \to (N_A + 1, N_B - 1)$.

Similarly, systems in the state $(N_A - 1, N_B + 1)$ can enter the CV if B reacts to form A and systems in the state $(N_A + 1, N_B - 1)$ can enter the CV if A reacts to form B.

The fluxes corresponding to each of these reactions are proportional to the number of systems in the required reactive state and to the number of molecules that can react. The proportionality constants are the reaction propensities c_1 and c_{-1} defined earlier. For the four reactions that affect our CV, we can therefore write

$$
\begin{aligned}
(N_A + 1, N_B - 1) \to (N_A, N_B) &: \text{ Flux in } = c_1 \mathcal{N}(t, N_A + 1, N_B - 1)(N_A + 1) \\
(N_A - 1, N_B + 1) \to (N_A, N_B) &: \text{ Flux in } = c_{-1} \mathcal{N}(t, N_A - 1, N_B + 1)(N_B + 1) \\
(N_A, N_B) \to (N_A - 1, N_B + 1) &: \text{ Flux out } = c_1 \mathcal{N}(t, N_A, N_B) N_A \\
(N_A, N_B) \to (N_A + 1, N_B - 1) &: \text{ Flux out } = c_{-1} \mathcal{N}(t, N_A, N_B) N_B.
\end{aligned}
$$

We hasten to add that the flux terms are only this simple because the reactions are all simple first order reactions. If several molecules are required for a reaction to proceed, then the rate is not proportional to the number of reactant molecules, but to the number of distinct molecular combinations for the reaction. For instance, had the reaction $A \to B$ been second order in A, then the factor N_A in the first flux term would have to be replaced by $N_A(N_A - 1)/2!$ since the first molecule can be picked N_A different ways, the second molecule $N_A - 1$ different ways and the product is divided by 2! since this particular sample of reactant molecules can be obtained 2! different ways. Substituting the flux terms into the conservation balance yield the equation

$$\begin{aligned}\frac{d\mathcal{N}(t,N_A,N_B)}{dt} &= c_1\mathcal{N}(t,N_A+1,N_B-1)(N_A+1)\\ &\quad + c_{-1}\mathcal{N}(t,N_A-1,N_B+1)(N_B+1)\\ &\quad - c_1\mathcal{N}(t,N_A,N_B)N_A\\ &\quad - c_{-1}\mathcal{N}(t,N_A,N_B)N_B\end{aligned}$$

and dividing through by $\mathcal{N}$ gives the model in terms of the fractions or probabilities of each state.

$$\begin{aligned}\frac{dP(t,N_A,N_B)}{dt} &= c_1P(t,N_A+1,N_B-1)(N_A+1)\\ &\quad + c_{-1}P(t,N_A-1,N_B+1)(N_B+1)\\ &\quad - c_1P(t,N_A,N_B)N_A\\ &\quad - c_{-1}P(t,N_A,N_B)N_B\end{aligned}$$

which is known as *the chemical master equation.*[11] For all but the simplest cases, the master equation must be solved numerically by Monte Carlo methods.[12] The few cases that are amenable to analytical methods are given in the literature.[13] Here, we will restrict ourselves to find the analytical solution for the probability distributions at steady state. Start by setting $N_A + N_B = N$, where N, the total number of molecules, is clearly constant and eliminate N_B from the master equation.

$$\begin{aligned}\frac{dP(t,N_A)}{dt} &= c_1P(t,N_A+1)(N_A+1) + c_{-1}P(t,N_A-1)(N+1-N_A)\\ &\quad - c_1P(t,N_A)N_A - c_{-1}P(t,N_A)(N-N_A)\end{aligned}$$

where $0 \le N_A \le N$. It is illustrative to write the equations for $N_A = 0$ and $N_A = N$ as these are special cases.

$$\frac{dP(t,0)}{dt} = c_1P(t,1) - c_{-1}P(t,0)N$$

$$\frac{dP(t,N)}{dt} = c_{-1}P(t,N-1) - c_1P(t,N)N$$

A quick investigation of these equations reveal that each flux term appears twice in the system, but with opposite sign. Thus, summing all the equations give

$$\frac{d\sum P}{dt} = 0 \Rightarrow \sum_{n=0}^{N} P(t,n) = 1$$

[11] For a rigorous derivation of the master equation for chemical kinetics, see Gillespie, D.T. [1992] "A rigorous derivation of the chemical master equation," *Phys. A* **188**, 404–425.
[12] Gillespie, D.T. [1976] "A general method for numerically simulating the stochastic time evolution of coupled chemical reactions," *J. Comp. Phys.* **22**, 403–434.
[13] McQuarrie, D.A. [1967] "Stochastic approach to chemical kinetics," *J. Appl. Prob.* **4**, 413–478.

where the sum of the probabilities necessarily must equal 1. Now, let $c = c_1/c_{-1}$ and set the derivatives equal to zero to get the set of equations

$$0 = cP(1) - P(0)N$$

$$0 = cP(N_A+1)(N_A+1) + P(N_A-1)(N+1-N_A) - cP(N_A)N_A - P(N_A)(N-N_A)$$

$$0 = cP(N-1) - P(N)N$$

$$\sum_{n=0}^{N} P(n) = 1.$$

The model has $N+2$ equations for $N+1$ unknowns, but the steady state master equations are not independent, any one of them can be obtained from the others. We therefore can discard the master equation for $N_A = N$ from the model. Now take the master equation for $N_A = 0$ and subtract it from the master equation for $N_A = 1$. Take this equation and subtract it from the master equation for $N_A = 3$, etc. The following set of equations is obtained

$$\begin{aligned}
0 =& cP(1) - P(0)N \\
0 =& cP(2)2 - P(1)(N-1) \\
&\vdots \\
0 =& cP(n)n - P(n-1)(N-(n-1)) \\
&\vdots \\
0 =& cP(N)N - P(N-1)
\end{aligned}$$

or as a recursive relationship, we can write

$$P(n) = P(n-1)\frac{N-(n-1)}{n}\frac{1}{c}$$

which is a very simple difference equation (Chap. 7) with the solution

$$P(n) = P(0)\frac{1}{c^n}\prod_{x=1}^{n}\frac{N-(x-1)}{x} = \frac{P(0)}{c^n}\frac{N!}{n!(N-n)!}$$

which can be substituted into the normalization equation to give

$$1 = P(0)\sum_{x=0}^{N}\frac{N!}{c^n n!(N-n)!} = P(0)\left(1+\frac{1}{c}\right)^N \Rightarrow$$

$$P(0) = \left(\frac{c}{c+1}\right)^N.$$

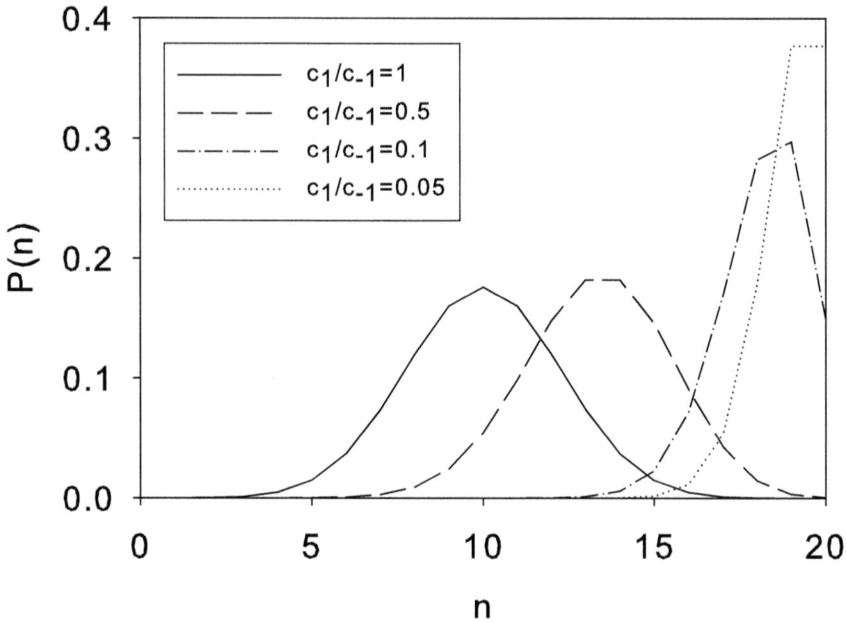

Fig. 1.22 Distributions of number of molecules of A for $N = 20$ and for different values of $c = c_1/c_{-1}$.

The problem is essentially solved since the remaining $P(n)$ values can easily be found recursively. A few distributions are plotted in Fig. 1.22 and they exhibit exactly the trends one would expect. For $c = 1$, the two reactions have the same rate and a symmetric distribution is obtained. As c decreases, the rate of the forward reaction, $A \to B$ decreases relative to the reverse reaction and the probability of finding a large number of molecules of A therefore increases and the distributions shift to the right.

1.5 Ad-hoc modelling

The expression *ad hoc* comes from Latin and means "for this". Ad hoc models are models that have no mechanistic justification or, at best, a very flimsy justification. Ad hoc models are usually justified by arguing that the model does replicate the qualitative trend one would expect and the model usually has one or more adjustable parameters that can be used to fit the model to some physical situation.

Ad hoc models lack a mechanistic justification and have no basis in terms of the underlying physics and chemistry, they are in a sense nothing but glorified curve fitting and the challenge when using these models is to pick a model that can reproduce the qualitative trends one expects. Clearly, one must be very careful about applying such a model to situations other than the specific ones for which the value of the model parameters were obtained.

A type of ad hoc model that may be familiar to many readers are the spring-dash pot models of viscoelastic fluids, see Fig. 1.23.

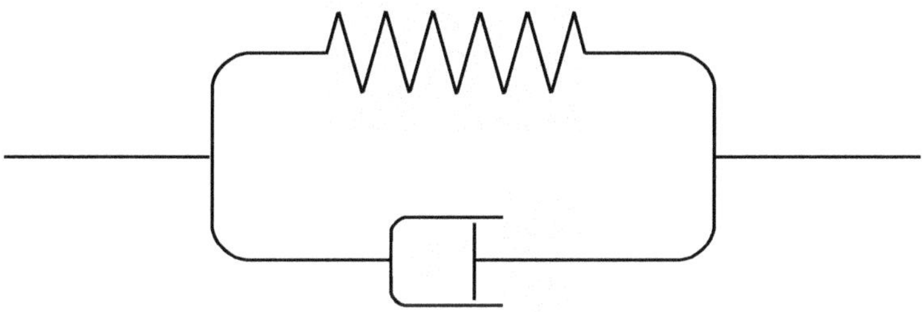

Fig. 1.23 A spring-dash pot model of a viscoelastic fluid.

The spring models the elastic properties of the substance while the dash pot, a shock absorber, models the viscous properties. The model therefore has two adjustable parameters, a spring constant and a viscosity. Evidently, a viscoelastic fluid is not made of tiny, molecular springs and dash pots, but the model still captures some of the essential character of a viscoeleastic fluid and more elaborate and presumably better models can be obtained by adding more springs and dash pots in series and parallel arrangements. Conversion of a spring-dash pot model to a full rheological model will not be attempted here. It is not a trivial task and is best left to experts.

In the remainder of this chapter, we present a few examples of ad hoc models that have proved successful in chemical engineering.

Example 1.23: Tanks in series model.

Ideal reactor models such as the CSTR or the plug flow reactor fail to adequately describe mixing in a reactor. Real reactors are neither perfectly well mixed nor fully free of mixing. A simple model that attempts to capture the reality of mixing in a flow through reactor is the tank in series model. In this model, a flow through reactor of volume V is modelled as a series of N identical CSTRs, each with the volume V/N, see Fig. 1.24.

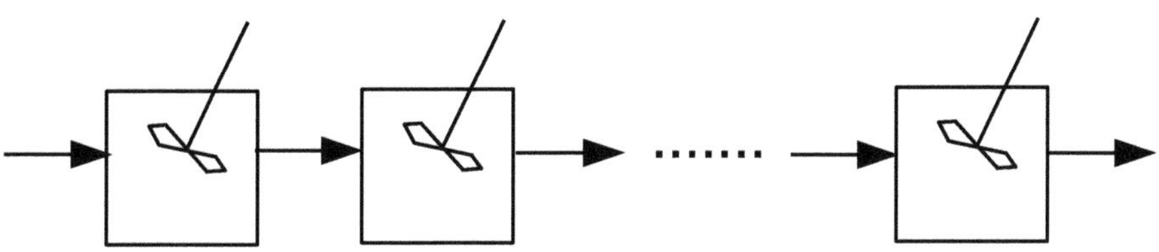

Fig. 1.24 The tank in series model: A sequence of N identical CSTRs.

We will argue here that this model does in fact describe a mixing that is between well mixed and plug flow.

Consider a flow through reactor of volume V with a feed stream with volumetric flow rate Q and inlet reactant concentration C_0 and assume that the reactant is consumed in a first order reaction with the specific rate constant k.

If the reactor is a CSTR, then the steady state outlet concentration is $C = \frac{C_0}{1+kV/Q}$. If the reactor is a plug flow reactor, then the outlet concentration is $C = C_0 e^{-k\frac{V}{Q}}$. Finding the outlet concentration from the tank in series model takes a bit more work. Number the vessels in the models consecutively from 1 to N and let the concentration in vessel n be C_n. A reactant balance on vessel n gives

$$QC_{n-1} - QC_n = \frac{V}{N}kC_n \Rightarrow$$

$$C_n = \frac{Q}{Q + kV/N}C_{n-1}.$$

Hence,

$$C_1 = \frac{Q}{Q + kV/N}C_0\ ,\ C_2 = \left(\frac{Q}{Q + kV/N}\right)^2 C_0 \cdots \Rightarrow$$

$$C = \left(\frac{1}{1 + \frac{kV}{QN}}\right)^N C_0.$$

When $N = 1$ we recover, as expected, the solution for a CSTR. Less obvious is the fact that we recover the solution for the plug flow reactor when $N \to \infty$. To see this, recall from calculus that

$$\lim_{N\to\infty}\left(1 + \frac{x}{N}\right)^N = e^x.$$

It easily follows that

$$\lim_{N\to\infty}\left(\frac{1}{1 + \frac{kV}{QN}}\right)^N C_0 = C_0 e^{-k\frac{V}{Q}}$$

which is the result for the plug flow reactor. The tank in series model therefore does appear to capture a state of mixing which lies between well mixed and totally unmixed. However, it is usually simpler to examine mixing by looking at systems without reactions and instead consider how the system responds in tracer studies. A tracer is an inert compound that is fed to the system and the mixing in the system is examined by looking at the concentration of the tracer in the exit stream. For instance, mixing in the tank in series model can be investigated by subjecting the system to a step change in the inlet concentration of a tracer. The simplest case is a step change in input, i.e. when the initial tracer concentration equals zero in all the stages and at time zero, the tracer concentration in the inlet is changed from zero to some fixed value, say C_0. The mass balance on the tracer for tank n is

$$\frac{dC_n}{dt} = \frac{QN}{V}(C_{n-1} - C_n)\ ,\ C_n(0) = 0\ ,\ C_0(t > 0) = 1.$$

To find the solution for this initial condition, solve for the tracer concentration in the first couple of vessels. Before doing so, it is convenient to make the model dimensionless. Let $\theta_n = \frac{C_n}{C_0}$ and $\tau = \frac{QN}{V}t$. This gives

$$\frac{d\theta_n}{d\tau} = \theta_{n-1} - \theta_n\ ,\ \theta_n(0) = 0\ ,\ \theta_0(t > 0) = 1$$

and the solutions for the first couple of tanks are

$$\theta_1(\tau) = 1 - e^{-\tau}$$

$$\theta_2(\tau) = 1 - e^{-\tau} - \tau e^{-\tau}$$

$$\theta_3(\tau) = 1 - e^{-\tau} - \tau e^{-\tau} - \frac{\tau^2}{2!}e^{-\tau}.$$

These results suggest the following general solution

$$\theta_n(\tau) = 1 - e^{-\tau}\sum_{k=0}^{n-1}\frac{\tau^k}{k!}.$$

Check that this is in fact the correct solution by substitution into the ODE.

$$1 - e^{-\tau}\sum_{k=0}^{n-2}\frac{\tau^k}{k!} - \left(1 - e^{-\tau}\sum_{k=0}^{n-1}\frac{\tau^k}{k!}\right) = e^{-\tau}\sum_{k=0}^{n-1}\frac{\tau^k}{k!} - e^{-\tau}\sum_{k=0}^{n-2}\frac{\tau^k}{k!} \Rightarrow$$

$$\frac{\tau^{n-1}}{(n-1)!} = \frac{\tau^{n-1}}{(n-1)!}.$$

Hence $\theta_N(\tau) = 1 - e^{-\tau}\sum_{k=0}^{N-1}\frac{\tau^k}{k!}$. This solution is plotted for 100 tanks for several values of τ in Fig. 1.25.

It is obvious from the graphs that the tank in series exhibit less mixing than a single well mixed tank. The step input passes through the tanks with only a slight broadening of the front as the time increases. A similar plot for an even larger number of tanks would show even less broadening of the front and it can be shown that in the limit as the number of tanks goes to infinity, the tank in series model

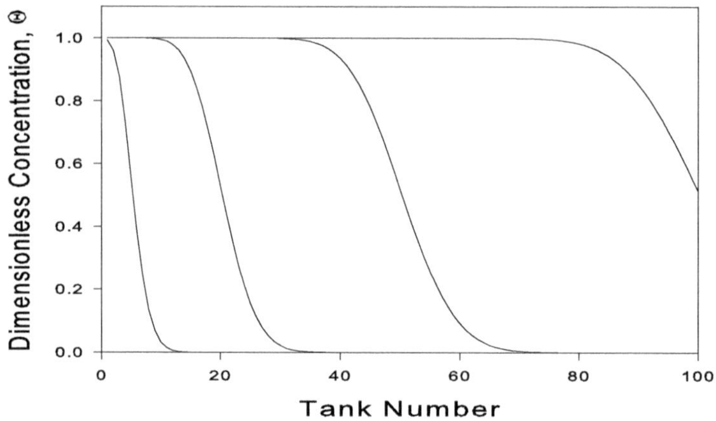

Fig. 1.25 Dimensionless concentration versus tank number for a step change input to the tank in series model. The graphs are, left to right, $\tau = 5, 20, 50, 100$.

behaves like a hard delay, changes in inlet concentration of a tracer are replicated in the exit stream but are delayed by V/Q, the residence time in the system.[14]

Example 1.24: Microbial growth models.

Mechanistic models of microbial growth requires detailed knowledge of the metabolic pathways and the manner in which the rates of the pathways are controlled. This is a daunting task and results in models that are so huge that they rarely can be analyzed or solved by anything but numerical means. Simpler ad hoc models that are easy to work with and that do a respectable job of modelling the qualitative and quantitative aspects of microbial growth may often suffice for reactor design, scale up and control, to state a few. The simplest such model use a single parameter to indicate the amount of biological matter, e.g. the concentration of biomass, and a single parameter, "the substrate", to indicate the amount of nutrients available for growth. These models are called unstructured (since they are devoid of chemical structure) and distributed (since the biological material is not modelled as segregated into individual cells but as distributed over a single well mixed bio phase). The conversion of nutrient to biomatter is usually assumed to be adequately modelled by a single constant stoichiometric constant called the yield, Y. It is defined as the specific rate of biomass increase, r_X, divided by the specific rate of substrate consumption, $-r_S$, or

$$Y = -\frac{r_X}{r_S}$$

---

[14]For readers familiar with Laplace transformation, there is an easy proof of this fact: the transfer function for each tank is $\frac{1}{\frac{V}{QN}s+1}$. The transfer function for N tanks is therefore $\frac{1}{(\frac{V}{QN}s+1)^N}$ and $\lim N \to \infty \frac{1}{(\frac{V}{QN}s+1)^N} = e^{-\frac{V}{Q}s}$, which is the Laplace transform of a delay of magnitude $\frac{V}{Q}$.

where X and S indicate biomass concentration and substrate concentration respectively. For a batch reactor, the biomass and substrate balances become

$$\frac{dX}{dt} = r_X \text{ and } \frac{dS}{dt} = -\frac{1}{Y} r_X.$$

The challenge when coming up with a good ad hoc model is to find a good expression for r_X. It is generally found (though not always) that the specific rate of biomass formation increases as the substrate concentration increases. More food means faster growth. But this trend abates at higher substrate concentrations when the specific growth rate approaches a constant maximum value. This qualitative aspect of microbial growth can be captured by any number of expressions and many have been suggested in the literature. The most celebrated model is the Monod model.

$$r_X = \frac{\mu_m S}{K_S + S} X$$

where μ_m = is known as the maximum specific growth rate of the organism and K_S as the Monod constant. Some other proposed models are:

Moser's equation:

$$r_X = \frac{\mu_m S^n}{K_S + S^n} X$$

Contoi's equation:

$$r_X = \frac{\mu_m S}{K_S X + S} X$$

Tessier's equation:

$$r_X = \mu_m (1 - e^{-K_S S}) X$$

Blackman's model:

$$r_X = \begin{cases} \mu_m X \ , \ S > K_S \\ \frac{\mu_m}{K_S} S X \ , \ S \leq K_S \end{cases}.$$

So how does one pick the best model for a given situation? When trying to discriminate between mechanistic models, one can legitimately ask which model fit gives parameter values that are closest to the true values. For instance, the best model can be picked as the one which, when fitted to data, gives a value of a diffusion coefficient which is closest to its true value. (Assuming of course that the true value can be determined with high accuracy in experiments designed to do just that).

However, this is not an option for ad hoc models because the model parameters do not have a true value, they are simply fitting parameters.

To help discriminate between the models, a least square fit of the five models to growth data from an *Eschericia coli* batch culture are shown in Fig. 1.26. The fits were done using the model parameters as well as the initial condition, the initial biomass and substrate concentrations as fitting parameters. The biomass concentrations were measured indirectly by measuring the optical density or OD of the cell suspension.

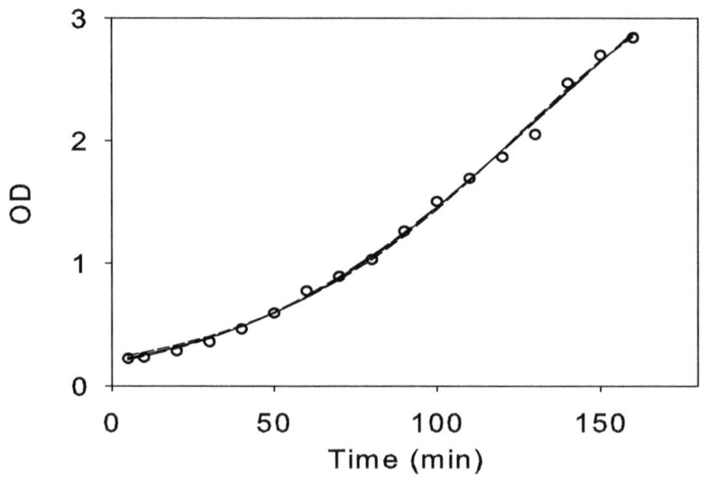

Fig. 1.26 Least square fits of five different kinetic models to *E. coli* batch growth data.

The five curve fits are so similar that attempting to discriminate between them by visual inspection is pointless. This is not surprising since the models were all chosen precisely because they were capable of reflecting the qualitative patterns in growth data.

Of course, one cannot expect to get good model discrimination without good data and a common reaction at this point is to collect more data. Optimally, the additional data should be collected under conditions where the models do differ, but this is difficult to do since the models are chosen to give the same qualitative behavior. A better approach may be to collect data under conditions that are quite different from those of the already acquired data. For instance, if the models are intended to be used to design or scale up a reactor, one may wish to collect data at a range of reactor sizes, fit the data to the candidate models and decide which model is best at fitting the entire range of reactor sizes. In essence, one wants to determine the model that can best be used to extrapolate to new operating conditions.

There may be any number of reasons why collecting more data is not an option; financial, technical or time constraints may apply. So one is often faced with having to decide on which model to use with only limited and imperfect data. One can try to discriminate between the models by comparing the sum of the square of the errors of each fit, S^2. This works well if the models have the same number of fitting parameters, and in this case, all but Moser's model have three model parameters (μ_m, K_s and Y). Moser's model has a fourth model parameter, the power n and the Moser model encompasses the Monod model as a special case, the case for $n = 1$. The Moser model will therefore always give a better fit, i.e. a smaller sum of squares of errors, than the Monod model. Indeed, one usually anticipates that models with more fitting parameters give better fits than models with fewer fitting parameters. In order to compare models with a different number of fitting parameters, one can estimate the variance in the data, σ^2. The estimate of the variance is calculated as the sum of the square of the errors divided by the degrees of freedom for each model fit and picking the model that gives the smallest value. The degrees of freedom of a least square fit is the number of data points minus the number of fitting parameters. If the number of fitting parameters equal the number of data points, the fit can usually be made to pass through all the points, but the degrees of freedom is zero and so the sum of the squares divided by the degrees of freedom is infinite; the fit of the model is essentially rubbish. In the table below, the sum of the squares of errors, S^2, and the estimate of the variance, σ^2, are tabulated for the five growth models fitted to the data in Fig. 1.26. The results indicate that the Moser model is the best model for these particular growth data.

Model	S^2	σ^2
Monod	0.0340	0.00681
Moser	0.0329	0.00548
Contois	0.0314	0.00628
Tessier	0.0420	0.00840
Blackmann	0.0339	0.00677

However, at this point, one should probably exercise some engineering judgment and common sense and ask if the better fit of the Moser model is so significant that it can justify using this model over models that are mathematically simpler and easier to work with. The Moser model is cumbersome to work with because it contains a dependent variable to a fractional power and most people would be content to use the Monod model which is easier to work with and which does an adequate job of modelling many kinds of microbial growth. It is probably due to this fact — doing an adequate job while being easy to work with — that the Monod model has become the most celebrated and used ad hoc microbial growth model.

Chapter 2

Model Simplifications and Approximations

In this chapter we will discuss some common tools for simplifying and editing models. The first topic, dimensionless models, is a tool for simplifying models without introducing any approximations. All the remaining topics are tools that give rise to a simpler but also approximate model.

After describing the tools and illustrating them with simple examples, we will revisit several models that were derived in the previous chapter and simplify and edit these.

2.1 Dimensionless models

Engineers make models dimensionless to reduce the parameter space thereby greatly simplifying model analysis as well as extending the use of a model. Through dimensionless models, situations that are seemingly very different can be compared. For instance, flow of a gas, of water and of an oil through a pipe can all be compared as long as all flows are at the same Reynolds number. This is the basis for doing physical model studies, e.g. using model air planes in a wind tunnel to investigate the properties of a specific air plane design.

Model solutions can be made dimensionless, but it is far more convenient to make the model dimensionless before attempting to solve it if, for no other reason, a dimensionless model can be solved using less writing that is required for solution of the full dimensioned model.

Example 2.1: Dimensionless Navier–Stokes equation.

An important example is the dimensionless version of the Navier–Stokes equation.The Navier–Stokes equation can be written in operator notations as

$$\rho \frac{Dv}{Dt} = -\nabla P + \mu \nabla^2 v + \rho g.$$

The dimensions of the parameters and variables in this equation are

$$\begin{aligned}
\text{Density: } &\rho[=]\text{mass/length}^3 \\
\text{Velocity: } &v[=]\text{length/time} \\
\text{Pressure: } &P[=]\text{mass/(time}^2\text{length)} \\
\text{Viscosity: } &\mu[=]\text{mass/(length time)} \\
\text{Gravity: } &g[=]\text{lenght/time}^2.
\end{aligned}$$

The Navier–Stokes equation does not contain an intrinsic characteristic length nor a characteristic velocity, so it is customary to introduce these from the boundary conditions, for instance the diameter of the pipe through which the fluid is flowing as the characteristic length and the average fluid velocity through the pipe (volumetric flow rate divided by pipe cross sectional area) as the characteristic velocity. Call these L and v_c. Now define a dimensionless length and velocity as $\mathbf{x} = x/L$ and $\mathbf{v} = v/v_c$. In this example, bold font will indicate the dimensionless variables. Consider the effect on the operator on the left hand side. In rectangular coordinates, we get

$$\begin{aligned}
\frac{Dv}{Dt} &= \frac{\partial v}{\partial t} + v_x\frac{\partial v}{\partial x} + v_y\frac{\partial v}{\partial y} + v_z\frac{\partial v}{\partial z} \\
&= \frac{v_c^2}{L}\left(\frac{L}{v_c}\frac{\partial \mathbf{v}}{\partial t} + \mathbf{v}_x\frac{\partial \mathbf{v}}{\partial \mathbf{x}} + \mathbf{v}_y\frac{\partial \mathbf{v}}{\partial \mathbf{y}} + \mathbf{v}_z\frac{\partial \mathbf{v}}{\partial \mathbf{z}}\right).
\end{aligned}$$

A dimensionless time suggests itself

$$\mathbf{t} = \frac{tv_c}{L},$$

giving

$$\rho\frac{Dv}{Dt} = \frac{\rho v_c^2}{L}\left(\frac{\partial \mathbf{v}}{\partial \mathbf{t}} + \mathbf{v}_x\frac{\partial \mathbf{v}}{\partial \mathbf{x}} + \mathbf{v}_y\frac{\partial \mathbf{v}}{\partial \mathbf{y}} + \mathbf{v}_z\frac{\partial \mathbf{v}}{\partial \mathbf{z}}\right) = -\nabla P + \mu\nabla^2 v + \rho g \Rightarrow$$

$$\begin{aligned}
\frac{\partial \mathbf{v}}{\partial \mathbf{t}} + \mathbf{v}_x\frac{\partial \mathbf{v}}{\partial \mathbf{x}} + \mathbf{v}_y\frac{\partial \mathbf{v}}{\partial \mathbf{y}} + \mathbf{v}_z\frac{\partial \mathbf{v}}{\partial \mathbf{z}} &= \frac{L}{\rho v_c^2}\left(-\nabla p + \mu\nabla^2 v + \rho g\right) \\
&= -\nabla^*\mathbf{P} + \frac{1}{Re}\nabla^{*2}\mathbf{v} + \mathbf{g}
\end{aligned}$$

where ∇^* indicates the dimensionless operator, and the additional dimensionless parameters are

$$\begin{aligned}
\text{Pressure: } \mathbf{P} &= \frac{P}{\rho v_c} \\
\text{Reynolds number: } Re &= \frac{v_c\rho L}{\mu} \\
\text{Gravity: } \mathbf{g} &= \frac{gL}{v_c^2}.
\end{aligned}$$

Making the problem dimensionless has reduced the number of model parameters from five to three.

There are always multiple different ways that a model can be made dimensionless and one should always try and make a model dimensionless in such a way that recognizable dimensionless groups appear. The only way to do this is to learn what the common dimensionless groups are. The most common groups in chemical engineering are listed in Table 2.1.

Table 2.1 Common dimensionless groups in chemical engineering.[a]

Name	Symbol	Definition	Name	Symbol	Definition
Biot number[b]	Bi	hl/k_s	Peclet number[e]	Pe	lv/D
Euler number	Eu	$P/(\rho v^2)$	Prandtl number	Pr	$\mu C_p/k$
Fourier number	Fo	$\alpha t/l^2$	Reynolds number	Re	$vl\rho/\mu$
Froude number	Fr	$v^2/(gl)$	Schmidt number	Sc	$\mu/(\rho D)$
Lewis number	Le	$k/(\rho C_p D)$	Sherwood number[f]	Sh	$k_m/(lD)$
Nusselt number[c]	Nu	hl/k_f	Stanton number	St	$h/(\rho v C_p)$
Peclet number[d]	Pe	$lv\rho C_p/k$	Weber number	We	$\rho v^2 l/\sigma$

[a] The characteristic length l is frequently calculated as the volume of the body divided by its surface area. [b] The thermal conductivity k_s is the conductivity of the solid body. [c] The thermal conductivity k_f is the conductivity of the fluid. [d] For thermal diffusion. [e] For mass diffusion. [f] Also known as the mass transfer Nusselt number.

If a model is just a little bit complex, there are usually numerous different ways it can be made dimensionless. However, for flow through systems with defined feed streams, it is recommended as a first step to use the properties of the feed streams to make dependent variables dimensionless. Once this is done, additional dimensionless variables will often suggest themselves. For initial value problems, it is recommended to use the initial values to make dependent variables dimensionless.

It is possible to find the dimensionless groups that are relevant to a system even without having an explicit model of the system. All that is required is that one knows what physical parameters are relevant. The analysis is called dimensional analysis and it requires some knowledge of linear algebra so it is described later, in the linear algebra chapter, page 188.

Example 2.2: Dimensionless mass and energy balances for CSTR.

Previously, (page 27), we derived the mass and energy balances for a CSTR of constant volume V, cooled by a coolant with a constant temperature T_c. In the reactor, a first order exothermic reaction with reaction rate constant which obey Arrhenius law and heat of reaction ΔH_r takes place. The volumetric flow rate through the reactor is Q and the inlet reactant concentration and temperature are C_{feed} and T_{feed} respectively. The two model equations, the mass and energy balances, are

$$V\frac{dC}{dt} = QC_{\text{feed}} - QC - k_0 e^{-\frac{E}{RT}} CV$$

$$V\rho c_p \frac{dT}{dt} = Q\rho c_p T_{\text{feed}} - Q\rho c_p T + (-\Delta H_r) CV k_0 e^{-\frac{E}{RT}} - UA(T - T_c) - W.$$

To make these model equations dimensionless, start by defining a dimensionless concentration, Θ, and a dimensionless temperature, Ψ, by dividing each variable by its feed value.

$$\Theta \equiv \frac{C}{C_{\text{feed}}}$$

$$\Psi \equiv \frac{T}{T_{\text{feed}}}.$$

In terms of these dimensionless variables, the balances take the form

$$\frac{V}{Q}\frac{d\Theta}{dt} = 1 - \Theta - \frac{k_0 V}{Q} e^{-\frac{E}{RT_f}\frac{1}{\Psi}} \Theta$$

$$\frac{V}{Q}\frac{d\Psi}{dt} = 1 - \Psi + \frac{(-\Delta H) C_f}{T_f \rho c_p} \cdot \frac{k_0 V}{Q} e^{-\frac{E}{RT_f}\frac{1}{\Psi}} \Theta - \frac{UA}{Q\rho c_p}\left(\Psi - \frac{T_c}{T_f}\right) - \frac{W}{T_f \rho c_p Q}.$$

Looking at the balances written this way, a bunch of dimensionless variables suggest themselves, namely

$$\alpha = \frac{k_0 V}{Q}.$$

This can be thought of as the ratio of the reaction rate and the dilution rate in the vessel. A large value of α indicates that the reaction is rapid relative to the residence time in the reactor.

$$h = \frac{-\Delta H_r C_f}{T_{\text{feed}} \rho c_p}$$

which is a dimensionless heat of reaction. For an exothermic reaction, h must be positive. And

$$u = \frac{AU}{Q\rho c_p}, \quad \Psi_c = \frac{T_c}{T_{\text{feed}}}, \quad w = \frac{W}{T_{\text{feed}} Q \rho c_p}, \quad \epsilon = \frac{E}{RT_{\text{feed}}}, \quad \tau = \frac{Qt}{V}$$

which gives us

$$\frac{d\Theta}{d\tau} = 1 - \Theta - \alpha e^{-\epsilon/\Psi}\Theta\,.$$

$$\frac{d\Psi}{d\tau} = 1 - \Psi + \alpha h e^{-\epsilon/\Psi}\Theta - u(\Psi - \Psi_c) - w.$$

The two balances are coupled, nonlinear and impossible to solve analytically. Here, we will only be concerned with solving for the steady states even though this simpler problem cannot be solved analytically. However, notice that the nonlinear term, $e^{-\frac{\epsilon}{\Psi}}\Theta$, is the same in both the mass and the energy balance. This strongly suggests eliminating this term from one of the steady state balances with the hope that a simpler problem will be obtained. Elimination of the term from the energy balance gives the steady state model

$$1 - \Theta = \Theta \alpha e^{-\frac{\epsilon}{\Psi}}$$

$$(1 - \Psi) + (1 - \Theta)h - u\Psi + (u\Psi_c - w) = 0.$$

The group $u\Psi_c - w$ can be thought of as a single dimensionless model parameter. The number of model parameters has thus been reduced from 13 to 5, a substantial reduction in the complexity of the problem.

Solving the mass and energy balances for Θ yields

$$\Theta = \frac{1}{1 + \alpha\exp(-\frac{\epsilon}{\Psi})}$$

$$\Theta = \left(1 + \frac{1}{h} + \frac{u\Psi_c - w}{h}\right) - \frac{1+u}{h}\Psi.$$

In the last result, we can again combine parameters to get the even simpler expression

$$\Theta = A - B\Psi$$

and for an exothermic reaction, B is positive so plotting Θ versus Ψ will give a family of straight lines with a negative slope. Examining the expression for the intercept A also shows that this must be greater than 1. The line for $A = 1$ therefore shows the lower limit on where these lines can occur. The steady states are found where these straight lines intersects the curve given by the equation for the mass balance. We cannot find these points analytically but can do so graphically by plotting the two equations for various parameter values. This is done in Fig. 2.1.

Notice that, depending on the value of A, there may be a unique steady state or there may be three steady states. For the latter case, there is an extinguished steady state for low values of A, an ignited steady state for large values of A and, for intermediate values of A, an intermediate steady state. This intermediate steady

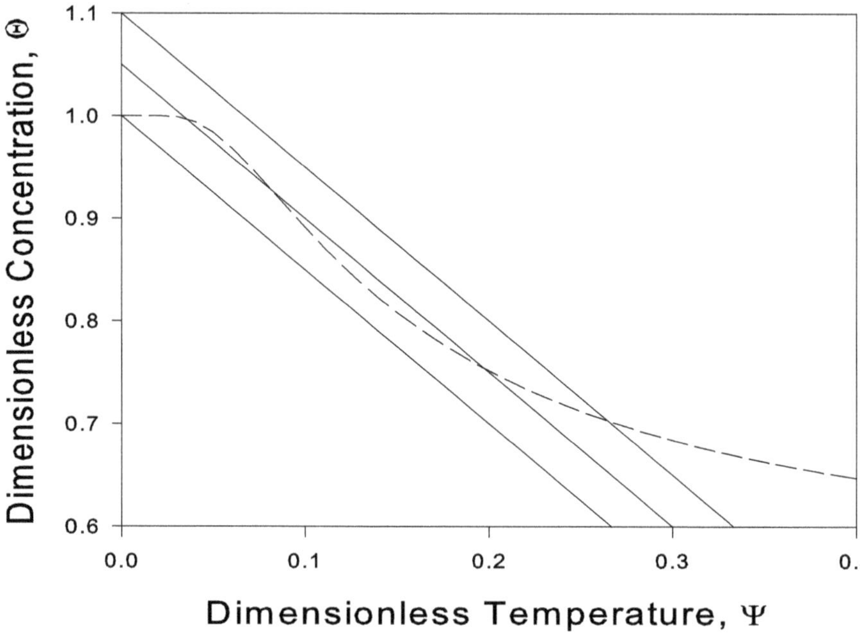

Fig. 2.1 Dimensionless mass and energy balances for a CSTR with an exothermic first order reaction. Parameter values are $\alpha = 0.9$, $\epsilon = 0.2$, $B = 1.5$. For the bottom straight line, $A = 1$, for the middle straight line $A = 1.05$ and for the top straight line $A = 1.1$.

state can be shown to be unstable while the ignited and extinguished steady states can be shown to be stable.[1]

2.2 Linearized models

It is always easier to solve linear models than nonlinear models, so the basic idea is to linearize nonlinear models around an operating point to obtain an approximate linear model. For many purposes, an approximate linear model will suffice. In particular, constitutive equations and reaction rate expressions can be linearized around an operating point. The mathematical technique is to truncate a Taylor series expansion around the point in question to create a linear expression. This of course assumes that the function in question has a Taylor expansion, i.e. that it is analytic, a valid assumption in most practical situations.

For instance, the temperature dependence of the heat capacity of many substances is often modelled with an expression of the form

[1]The problem of the stability of the steady state is clearly of great importance and it has been analyzed in significantly more depth than the presentation given here. Readers who would like to learn more should consult: Aris, R. and Amundson, N.R. [1958] "An analysis of chemical reactor stability and control - I," *Chem. Eng. Sci.* **7**(3), 121–131.

$$c_p = a + bT + cT^2 + d/T.$$

One can linearize this expression around any given operating temperature, not by simply discarding the nonlinear terms, but by expanding the expression around the operating temperature and then discarding higher order terms. If the operating temperature is T_0 then a Taylor expansion around this point gives

$$c_p = a + bT_0 + cT_0^2 + \frac{d}{T_0} + \left(b + 2cT_0 - \frac{d}{T_0^2}\right)(T - T_0) + \left(c + \frac{d}{T_0^3}\right)(T - T_0)^2 + \cdots$$

Discarding nonlinear terms gives the approximate linear model

$$c_p = a + bT_0 + cT_0^2 + \frac{d}{T_0} + \left(b + 2cT_0 - \frac{d}{T_0^2}\right)(T - T_0).$$

Example 2.3: CSTR with linearized reaction rates.

Consider a CSTR in which the reaction $A + B \to P$ takes place. The reaction is first order in each reactant and there are no heat effects or volume changes associated with the reaction. Derive a linearized model around the steady state operating point.

The two mass balances are

$$V\frac{dC_A}{dt} = Q(C_{Af} - C_A) - kC_AC_BV$$

$$V\frac{dC_B}{dt} = Q(C_{Bf} - C_B) - kC_AC_BV$$

where C_A and C_B are the two reactant concentrations and an f-subscript indicates the values in the feed stream. Q is the volumetric flow rate through the reactor, V the reactor volume and k a reaction rate constant. The two reactant concentrations appear fully symmetrically in the problem, so we will restrict ourselves to deriving the linearized model for A. The steady state operating concentration C_{Ae} is readily found as

$$C_{Ae} = \frac{2QC_{Bf}}{\sqrt{(C_{Af} - C_{Bf})^2V^2k^2 + (C_{Af} + C_{Bf})QVk + Q^2} - (C_{Af} - C_{Bf})Vk - Q} - \frac{Q}{Vk}.$$

Linearizing the A balance around this point gives

$$\begin{aligned} V\frac{dC_A}{dt} =& [Q(C_{Af} - C_A) - kC_AC_BV]_{A_{Ae},C_{Be}} \\ &+ \left[\frac{d(Q(C_{Af} - C_A) - kC_AC_BV)}{dC_A}\right]_{C_{Ae},C_{Be}} (C_A - C_{Ae}) \\ &+ \left[\frac{d(Q(C_{Af} - C_A) - kC_AC_BV)}{dC_B}\right]_{C_{Ae},C_{Be}} (C_B - C_{Be}). \end{aligned}$$

The first term on the right hand side must necessarily equal zero since it is the right hand side of the original mass balance evaluated at the steady state. Expanding the remaining two terms give

$$\begin{aligned} V\frac{dC_A}{dt} =& [-Q - kC_BV]_{C_{Ae},C_{Be}}(C_A - C_{Ae}) \\ &- [kC_AV]_{C_{Ae},C_{Be}}(C_B - C_{Be}) \end{aligned}$$

or

$$\frac{dC_A}{dt} = -\left(\frac{Q}{V} + kC_{Be}\right)(C_A - C_{Ae}) - kC_{Ae}(C_B - C_{Be}).$$

2.3 Lumping and the quasi-steady state assumption

By lumping we mean the process of taking a distributed variable and approximating it by a single average value. For instance, if a reactant diffuses through a catalytic pellet much faster than it reacts, then the reactant concentration will be approximately constant through the pellet and close to the steady state value it would have without a reaction taking place. We say that, in this case, the concentration is in a quasi-steady state and the concentration can be lumped into a single average value for the pellet.

The lumping approximation is thus closely aligned with the quasi-steady state assumption which states that a variable can be modelled as having its steady state value if the processes that drive this variable towards its steady state are much faster than the other processes in the system. For instance, if the rate of mixing in a stirred tank is much faster than the rates of the reactions that take place in the tank, then the tank can be assumed to be well mixed and the reactant concentrations can be assumed to have the same values at all points in the tank.

But how does one compare rates of processes to determine if some processes can be assumed faster than others? For instance, how does one compare rates of diffusion to rates of reaction? Diffusion rates increase with the value of the diffusion coefficient which has units of area per time and the rate of a first order reaction increase with the value of the reaction rate constant which has units of inverse time. Since the two parameters one would like to compare in magnitude do not even have the same dimension, how can one do a comparison? The answer is to make use of dimensionless groups. Many dimensionless groups are ratios of two rates or effects that influence a process. For instance, it is well known that the Reynolds number is the ratio of the inertial forces to the viscous forces. If the Reynolds number is large, then the flow is turbulent and dominated by inertial forces while a low value of the Reynolds number indicates that the flow is laminar and dominated by viscous forces.

Example 2.4: Diffusion and rate limited reaction in a catalytic pellet.

In a previous example, we derived a model for the reactant concentration profile in a spherical catalytic pellet with a first order reaction, page 41. The reactant balance on the pellet was

$$\frac{D}{r^2}\frac{d}{dr}\left(r^2\frac{dC}{dr}\right) = kC$$

subject to the boundary conditions

$$r = R \Rightarrow C = C_s$$
$$r = 0 \Rightarrow \frac{dC}{dr} = 0 \text{ or } C \text{ finite}.$$

To make this model dimensionless, define $\zeta = r/R$ and $\theta = C/C_s$, then

$$\frac{d}{dr} = \frac{d}{dR\zeta} = \frac{1}{R}\frac{d}{d\zeta} \text{ and } \frac{dC}{dr} = C_s\frac{d\theta}{dr}$$

giving

$$\frac{D}{R^2\zeta^2}\frac{d}{d\zeta}\left(\zeta^2\frac{d\theta}{d\zeta}\right) = k\theta$$

or

$$\frac{1}{\zeta^2}\frac{d}{d\zeta}\left(\zeta^2\frac{d\theta}{d\zeta}\right) = \frac{kR^2}{D}\theta.$$

Of course, in order to solve the dimensionless balance equation, one must also make the boundary conditions dimensionless

$$r = R \Rightarrow C = C_s \text{ becomes } \zeta = 1 \Rightarrow \theta = 1$$
$$r = 0 \Rightarrow \frac{dC}{dr} = 0 \text{ becomes } \frac{d\theta}{d\zeta} = 0 \text{ or } \theta \text{ finite}.$$

A dimensionless group that jumps out is

$$\Pi = \frac{kR^2}{D}.$$

Evidently, there are any number of possible choices such as $R\sqrt{\frac{k}{D}}$ or $\frac{D}{kR^2}$ but all combine the three model parameters in essentially the same manner. The choice of what number to actually use is somewhat arbitrary, but in most cases there is an established tradition for what number to use. For the situation at hand, the established choice is

$$\Phi = \frac{R}{3}\sqrt{\frac{k}{D}}.$$

This is the so-called Thiele modulus for a first order reaction in a spherical pellet.[2] The reason for defining the modulus this way, with the peculiar choice of $R/3$ as the characteristic length, is that this characteristic length equals the volume of the particle divided by its surface area. This definition of the characteristic length makes comparison between results for different shapes or geometries easier.

The Thiele modulus is basically a dimensionless measure of the rate of reaction relative to the rate of diffusion. The definition of the Thiele modulus depends on both the geometry of the catalytic pellet and on the expression for the reaction rate. However, it is possible to avoid the confusion caused by this situation dependent definition by defining a generalized Thiele modulus which can be applied to any type of kinetics in any catalytic pellet.[3]

Making the model dimensionless has reduced the number of parameters in the model from 3, k, D and R to 1, Φ, and the model is perhaps slightly easier to solve now. The solution for the dimensionless concentration profile is

$$\theta = \frac{1}{\zeta}\frac{\sinh(3\Phi\zeta)}{\sinh(3\Phi)}.$$

When the Thiele modulus is large, i.e. when the reaction rate constant is "large" and the diffusion coefficient is "small", then the reaction is diffusion limited and the reaction mainly occurs in a thin layer close to the catalyst surface. When the Thiele modulus is small, the reaction rate is limited and one may assume that the reactant concentration in the pellet will be approximately constant throughout. All this is corroborated by plotting the solution for different values of Φ as is done in Fig. 2.2.

The top graph, for $\Phi = 0.2$ is almost flat indicating an almost constant reactant concentration through the pellet. A faster diffusion rate would have almost no effect on the overall reaction rate since the reactant concentration through the pellet is close to the concentration at the pellet surface. For this value of Φ as well as for smaller values, one might consider using a lumped model. In a lumped model, the reactant concentration in the pellet equals the concentration at the pellet surface, so the rate of reactant consumption is simply

$$\text{Rate of reactant consumption in pellet} = kC_s \cdot \frac{4}{3}\pi R^3.$$

[2]Thiele, E.W. [1939] "Relation between catalytic activity and size of particle," *Ind. Eng. Chem.* **31**, 916–920.

[3]Bischoff, K.B. [1965] "Effectiveness factors for general reaction rate forms," *AIChE J.* **11**, 351–355.

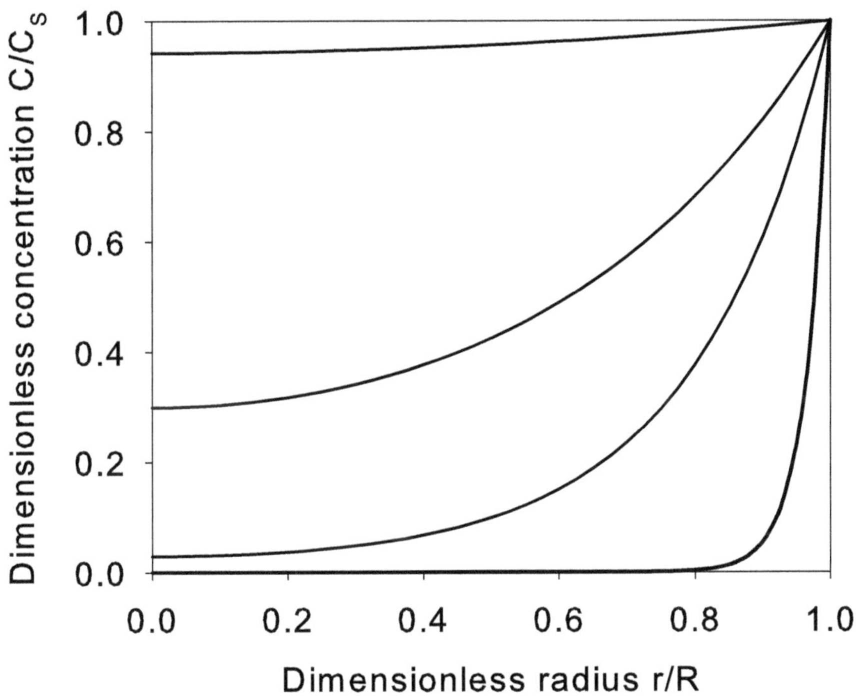

Fig. 2.2 Dimensionless reactant concentration in spherical catalytic pellet with a first order reaction. The graphs, top to bottom, are for $\Phi = 0.2$, $\Phi = 1$, $\Phi = 2$ and $\Phi = 10$.

The diffusion coefficient does not even occur in this model so the reaction is clearly rate limited.

The bottom graph, for $\Phi = 10$ shows that the reactant concentration in the pellet core is virtually zero. The diffusion rate is so slow that it cannot transport reactant into the pellet fast enough to keep up with the rate of consumption and the catalyst in the pellet center is essentially wasted as the reaction only occurs close to the pellet surface. The reaction is clearly diffusion limited.

Chemical reaction rates are notorious for their wide range of magnitudes and the quasi-steady state assumption is probably never invoked more than when modelling chemical kinetics. A famous case of this is the derivation of kinetics for enzyme catalyzed reactions.

Example 2.5: Enzyme kinetics.

Enzymes work by binding to a substrate (the reactant) to form an enzyme-substrate complex and this complex then reacts to form the free product. The reversibility of the reaction leading to formation of the enzyme-substrate complex and the irreversibility of the reaction leading to the product are both common assumptions in enzyme kinetics. The reaction can be written formally as

$$E + S \underset{k_{-1}}{\overset{k_1}{\rightleftarrows}} ES \overset{k_p}{\rightarrow} E + P$$

where E indicates the enzyme, S the substrate, ES the enzyme-substrate complex and P the product. The rate equations for a well mixed batch reactor can now be stated

$$\frac{dC_P}{dt} = k_p C_{ES}$$

$$\frac{dC_{ES}}{dt} = k_1 C_S C_E - (k_{-1} + k_p) C_{ES}$$

$$\frac{dC_E}{dt} = k_{-1} C_{ES} - k_1 C_E C_S + k_p C_{ES}$$

$$\frac{dC_S}{dt} = k_{-1} C_{ES} - k_1 C_E C_S.$$

Adding the second and third rate equations give

$$\frac{d}{dt}(C_E + C_{ES}) = 0 \Rightarrow C_E + C_{ES} = C_{E0}$$

where C_{E0} is the constant total enzyme concentration, free enzyme plus enzyme-substrate complex. This result is simply a total enzyme balance.

The rate equations above are nonlinear and cannot be solved analytically. In order to obtain a model which can be solved analytically, two simplifying assumptions have been suggested in the literature, the Michaelis–Menten[4] or equilibrium assumption and the Briggs–Haldane[5] or steady-state assumption. The latter assumption is less restrictive than the former and thus models the true kinetics more accurately, but it is slightly more cumbersome to work with.

The Michaelis–Menten assumption is that the product forming reaction is the rate limiting reaction in the system. In other words, the rate of formation of ES and the rate of breakdown of ES into free enzyme and substrate are fast compared to the rate at which the enzyme-substrate complex breaks down to form product. The two fast reactions can therefore be considered to be in a quasi-steady state at all times. Because the assumption implies that the enzyme-substrate complex is in equilibrium with the free substrate and enzyme, Michaelis–Menten analysis is also referred to as rapid equilibrium analysis. Using the assumption of equilibrium, we obtain,

[4]Michaelis, L. and Menten, M.L. [1913] "Die Kinetik der Invertinwirkung," *Biochem. Z.* **49**, 333.

[5]Briggs, G.E. and Haldane, J.S.B. [1925] "A note on the kinetics of enzyme action," *Biochem. J.* **19**, 338–339.

$$k_1 C_S C_E = k_{-1} C_{ES} \Rightarrow \frac{C_S C_E}{C_{ES}} = \frac{k_{-1}}{k_1} = K_m$$

and using this result to eliminate C_E from the total enzyme balance gives

$$C_{ES} = \frac{C_S C_{E0}}{K_m + C_S}.$$

Substituting this result into the rate equation for substrate formation gives the desired result

$$\frac{dC_P}{dt} = k_p \frac{C_S C_{E0}}{K_m + C_S} = \frac{r_{\max} C_s}{K_m + C_s}.$$

The Briggs–Haldane analysis is also called steady-state analysis because the basic assumption is that the concentration of the enzyme-substrate complex is at a quasi steady-state. This is **not** the same as the Michaelis–Menten assumption which stated that the complex was in equilibrium with the free species, but is slightly less restrictive. In fact, Michaelis–Menten results can be obtained from the Briggs–Haldane results as a special case. Mathematically, the Briggs–Haldane assumption can be stated as

$$\frac{dC_{ES}}{dt} = k_1 C_S C_E - k_{-1} C_{ES} - k_p C_{ES} = 0.$$

While in the Michaelis–Menten analysis we discarded the k_p-term, in this analysis we will keep it and the reaction leading to the product is therefore not necessarily the rate limiting reaction. Obviously, we should be able to obtain the Michaelis–Menten results from the Briggs–Haldane results by setting k_p equal to zero. To obtain the desired rate expression, we proceed essentially as before. Find C_E from the equation above and substitute the result into the total enzyme balance. Solve for C_{ES} and substitute this result into the equation for the rate of product formation.

$$\frac{dC_P}{dt} = k_p \frac{C_S C_{E0}}{\frac{k_{-1}+k_p}{k_1} + C_S}.$$

Notice, that if we invoke the Michaelis–Menten assumption, then $k_p << k_{-1}$ and this result simplifies to the Michaelis–Menten result.

Example 2.6: Lumped energy balance on catalytic slurry reactor.

Consider a flow through reactor in which catalytic pellets are kept in suspension by mechanical stirring of the fluid, see Fig. 2.3. The volumetric flow rate of the feed and exit stream is Q and their temperatures are T_f and T, respectively. The physical properties of the fluid, density and heat capacity are considered constant and the fluid is assumed well mixed.

Reactions occurring inside the catalytic pellets generate heat with a constant specific rate $\dot{q}$ (Energy per time per volume). It will be assumed that the rate of thermal conduction inside the pellets is fast compared to the rate of heat generation such that the temperature can be lumped to the value T_{pellet}. Since T_{pellet} is a lumped temperature, it will depend only on time, not on position in the pellets, and the energy balance will therefore be an ordinary differential equation. If the temperature was not lumped, one would have to write a differential energy balance on the pellets giving rise to a partial differential equation. (Exercise: Write an energy balance with a non-lumped temperature). For simplicity, we will also assume that all the pellets have the same shape and size. Heat transfer between pellets and fluid will be modelled using film theory with a heat transfer coefficient h.

The temperature in the reactor is controlled by a cooling jacket. The volumetric flow rate of the coolant stream is Q_{cool} and the inlet and outlet temperatures are T_{in} and T_{out} respectively. The liquid in the jacket is also assumed well mixed.

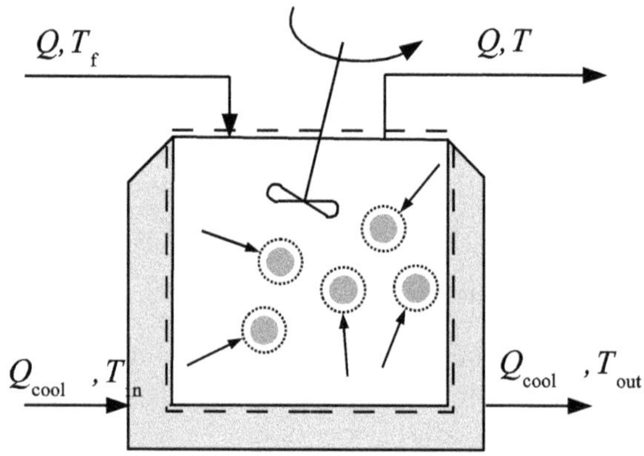

Fig. 2.3 Catalytic slurry reactor with cooling. Fluid inside the reactor is white, coolant is hatched. The dotted line indicates the CV boundaries; between the catalytic pellets and the fluid and between the fluid inside the reactor and the cooling jacket.

An energy balance can now be done on several possible CVs. The first will be the fluid phase inside the reactor, i.e. excluding the catalytic pellets themselves.

$$V\rho c_p \frac{dT}{dt} = Q\rho c_p(T_f - T) + hA_{pellet}(T_{pellet} - T) - UA_{cool}(T - T_{out}) \tag{2.1}$$

where V is the volume of the fluid inside the reactor, ρ and c_p the density and heat capacity of the fluid, A_{pellet} the total surface area of the pellets, A_{cool} the surface area between the reactor and the cooling jacket and U the overall heat transfer coefficient between reactor and cooling jacket. An energy balance on the pellets give

$$\rho_{pellet} c_{pellet} V_{pellet} \frac{dT_{pellet}}{dt} = \dot{q}V_{pellet} - hA_{pellet}(T_{pellet} - T) \tag{2.2}$$

where ρ_{pellet} and c_{pellet} is the density and heat capacity of the pellets and V_{pellet} the total volume of the pellets. The balance on the coolant becomes

$$V_{\text{cool}}\rho_{\text{cool}}c_{\text{cool}}\frac{dT_{\text{out}}}{dt} = Q_{\text{cool}}\rho_{\text{cool}}c_{\text{cool}}(T_{\text{in}} - T_{\text{out}}) + UA_{\text{cool}}(T - T_{\text{out}})\,. \qquad (2.3)$$

Equations (2.1) through (2.3) constitute the model. Three ordinary differential equations with three unknowns, the three temperatures.

The solution of this model depends on a huge number of model parameters, six material constants (densities and heat capacities), five geometric parameters (volumes and areas), the parameters h, U and $\dot{q}$ and the input feed steam properties, Q, Q_{cool}, T_f and T_{in}, which may vary with time. To simplify the model, reduce the effective number of model parameters by collecting the original parameters into a smaller number of groups. Taken to the extreme, this will give a dimensionless model with the smallest possible number of dimensionless parameters. However, in this case, we will only simplify the model equations to the point where they are in standard mathematical form, with the derivatives of the unknowns isolated.

$$\frac{dT}{dt} = D(T_f - T) + \left(\frac{hA_{\text{pellet}}}{V\rho c_p}\right)(T_{\text{pellet}} - T) - \left(\frac{UA_{\text{cool}}}{V\rho c_p}\right)(T - T_{\text{out}})$$

$$\frac{dT_{\text{pellet}}}{dt} = \left(\frac{\dot{q}}{\rho_{\text{pellet}}c_{\text{pellet}}}\right) - \left(\frac{h}{\rho_{\text{pellet}}c_{\text{pellet}}L_{\text{pellet}}}\right)(T_{\text{pellet}} - T)$$

$$\frac{dT_{\text{out}}}{dt} = D_{\text{cool}}(T_{\text{in}} - T_{\text{out}}) + \left(\frac{U}{L_{\text{cool}}\rho_{\text{cool}}c_{\text{cool}}}\right)(T - T_{\text{out}})$$

where we have introduced the dilution rates $D = Q/V$ and $D_{\text{cool}} = Q_{\text{cool}}/V_{\text{cool}}$ and the characteristic lengths of the pellets and cooling jacket, $L_{\text{pellet}} = V_{\text{pellet}}/A_{\text{pellet}}$ and $L_{\text{cool}} = V_{\text{cool}}/A_{\text{cool}}$. The fractions inside the brackets are grouped parameters with dimension of inverse time or temperature per time. The number of constant parameters is thus reduced from 14 to 5 making model analysis far simpler.

There are several ways one can proceed to make the model fully dimensionless. Both of the inlet temperatures, T_f and T_{in}, can be used to define a dimensionless temperature and time can similarly be made dimensionless two way using either the characteristic time of the reactor, $1/D$ or the characteristic time of the cooling jacket, $1/D_{\text{cool}}$.

Example 2.7: Return of the density gradient device.

In Example 1.5 on page 16, we looked at two connected, well mixed tanks for making density gradients. The relevant figure is repeated here, see Fig. 2.4.

We will now simplify the model, first by making it dimensionless, then by applying the quasi-steady state assumption to the liquid levels in the two tanks. The three balance equations, two on the liquid volume and one on the concentration of sucrose in tank 2, were

$$A\frac{dH_1}{dt} = -(H_1 - H_2)/R$$

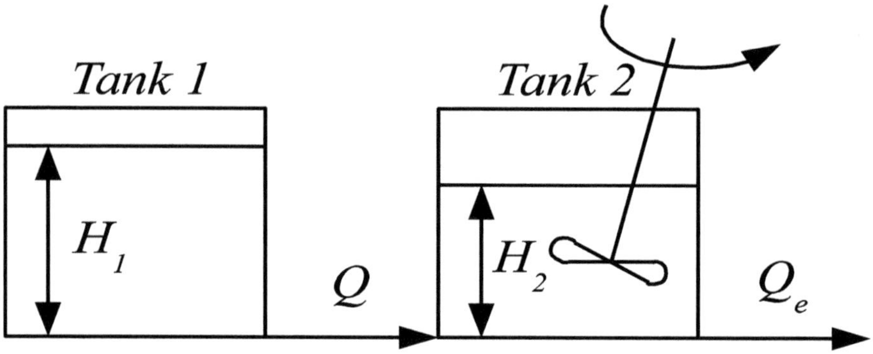

Fig. 2.4 Two connected mixing tanks used for making density gradients.

$$A\frac{dH_2}{dt} = (H_1 - H_2)/R - Q_e$$

$$A\frac{dH_2C_2}{dt} = C_1(H_1 - H_2)/R - C_2Q_e\,.$$

The model contains three parameters, A, R and Q_e, plus the values specified in the initial conditions. The product AR has dimension time and it therefore seems natural to define a dimensionless time, τ as

$$\tau = \frac{t}{AR}\,.$$

From the chain rule, it then follows that

$$\frac{dH_1}{dt} = \frac{dH_1}{d\tau}\frac{d\tau}{dt} = \frac{1}{AR}\frac{dH_1}{d\tau}\,.$$

Similarly for the other two derivatives. When substituted back into the model equations, these simplify to

$$\frac{dH_1}{d\tau} = H_2 - H_1$$

$$\frac{dH_2}{d\tau} = H_1 - H_2 - Q_eR$$

$$\frac{dC_2}{d\tau} = (C_1 - C_2)\frac{H_1 - H_2}{H_2}\,.$$

Using the initial values, the following dimensionless dependent variables can be defined

$$h_1 = \frac{H_1}{H}\,,\ h_2 = \frac{H_2}{H}\,,\ c_1 = \frac{C_1}{C_0}\,,\ c_2 = \frac{C_2}{C_0}$$

giving the fully dimensionless model

$$\frac{dh_1}{d\tau} = h_2 - h_1\,,\ h_1(0) = 1$$

$$\frac{dh_2}{d\tau} = h_1 - h_2 - \frac{Q_e R}{H}\ , \ h_2(0) = 1$$

$$\frac{dc_2}{d\tau} = (c_1 - c_2)\frac{h_1 - h_2}{h_2}\ , \ c_2(0) = 1\,.$$

In this version, the model has only one parameter, the dimensionless group $\frac{Q_e R}{H}$. With the number of parameters this low, it is easier to plot or tabulate a numerical solution versus parameter values. Such a numerical solution is shown in Fig. 2.5 as a family of curves for different values of $\frac{Q_e R}{H}$. The concentration in tank 1 is assumed to be zero in all cases.

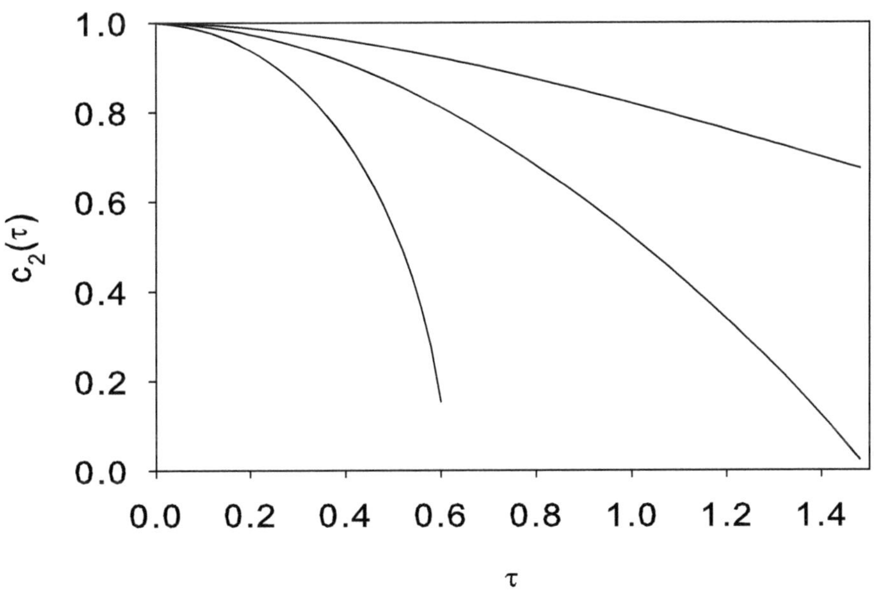

Fig. 2.5 Numerical solutions of the dimensionless sucrose concentration versus dimensionless time for $\frac{Q_e R}{H} = 0.5$ (top curve) 1 and 2 (bottom curve).

Although nonlinear, the model equations can in fact be partially solved. The two volume balances are linear and can be solved by methods that will be discussed later to give

$$h_1(\tau) = 1 + \frac{Q_e R}{H}\left(\frac{1}{4} - \frac{\tau}{2} - \frac{e^{-2\tau}}{4}\right)$$

$$h_2(\tau) = 1 + \frac{Q_e R}{H}\left(\frac{e^{-2\tau}}{4} - \frac{1}{4} - \frac{\tau}{2}\right).$$

Setting $c_1 = 0$ for simplicity and substituting the results for h_1 and h_2 into the dimensionless sucrose balance gives

$$\frac{1}{c_2}\frac{dc_2}{d\tau} = \frac{2\frac{Q_eR}{H}(e^{-2\tau}-1)}{\frac{Q_eR}{H}(e^{-2\tau}-2\tau-1)+4}$$

which is a separable equation that can be solved subject to the initial condition $c_2(0) = 1$

$$\int_1^{c_2}\frac{1}{c}\,dc = \int_0^{\tau}\frac{2\frac{Q_eR}{H}(e^{-2t}-1)}{\frac{Q_eR}{H}(e^{-2t}-2t-1)+4}\,dt \Rightarrow$$

$$c_2(\tau) = \exp\left(\int_0^{\tau}\frac{2\frac{Q_eR}{H}(e^{-2t}-1)}{\frac{Q_eR}{H}(e^{-2t}-2t-1)+4}\,dt\right).$$

The solution is finished in the sense that the unknown has been isolated on one side of the equality sign but is unfinished in the sense that the integral on the right hand side does not have an explicit solution. Solutions containing unfinished expressions of this type are often called *closed form solutions*, meaning that the unknown has been found in terms of some expression that is as simple as possible.

If we now assume that the sucrose stream, Q_e, is withdrawn slowly compared to how fast the liquid flows from tank 1 to tank 2, then it is reasonable to assume that the liquid levels in the two tanks are equal or in a quasi-steady state. This quasi-steady state assumption clearly implies that the volumetric flow rate from tank 1 to tank 2 equals half the flow rate out of the device, giving the model

$$A\frac{dH_1}{dt} = -\frac{Q_e}{2}$$

$$A\frac{dH_2}{dt} = -\frac{Q_e}{2}$$

$$A\frac{dH_2C_2}{dt} = C_1\frac{Q_e}{2} - C_2Q_e \Rightarrow H_2\frac{dC_2}{dt} = C_1\frac{Q_e}{2A} - C_2\frac{Q_e}{2A}.$$

Reintroducing the previously defined dimensionless variables

$$\tau = \frac{t}{AR}\,,\ h_1 = \frac{H_1}{H}\,,\ h_2 = \frac{H_2}{H}\,,\ c_1 = \frac{C_1}{C_0}\,,\ c_2 = \frac{C_2}{C_0}$$

the dimensionless model equations become

$$\frac{dh_1}{d\tau} = -\frac{Q_eR}{2H}$$

$$\frac{dh_2}{d\tau} = -\frac{Q_eR}{2H}$$

$$h_2\frac{dc_2}{d\tau} = c_1\frac{Q_eR}{2H} - c_2\frac{Q_eR}{2H}$$

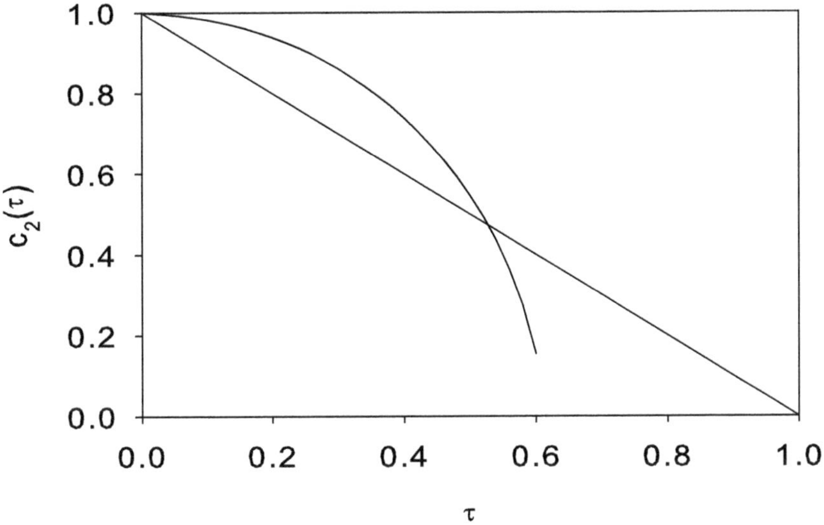

Fig. 2.6 Solutions of the dimensionless sucrose concentration versus dimensionless time for full model (curved solution) and for quasi-steady state model (straight line solution). Both plots are for $\frac{Q_e R}{H} = 2$ and $c_1 = 0$.

which can be solved much more easily than the original model to give

$$h_1(t) = 1 - \frac{Q_e R}{2H}\tau \ , \ 0 \leq \tau \leq \frac{2H}{Q_e R}$$

$$h_2(t) = 1 - \frac{Q_e R}{2H}\tau \ , \ 0 \leq \tau \leq \frac{2H}{Q_e R}$$

$$c_2(t) = c_1 + (1 - c_1)\left(1 - \frac{Q_e R}{2H}\tau\right) \ , \ 0 \leq \tau \leq \frac{2H}{Q_e R} \, .$$

The slow withdrawal of the solution therefore creates a linear concentration gradient in the test tube, with a bottom concentration of C_0 and a top concentration of C_1.

The quasi-steady state assumption used was that the flow out of the second vessel was slow compared to the flow between the vessels. This occurs if Q_E or if R are small and plotting the quasi-steady state solution against the solution of the full model confirms this. Figure 2.6 shows the two solutions for $\frac{Q_e R}{H} = 2$ and the quasi-steady state solution is obviously poor.

Figure 2.7 shows the sucrose concentration for $\frac{Q_e R}{H} = 0.1$ and the solution obtained using the quasi steady state assumption is seen to be quite good.

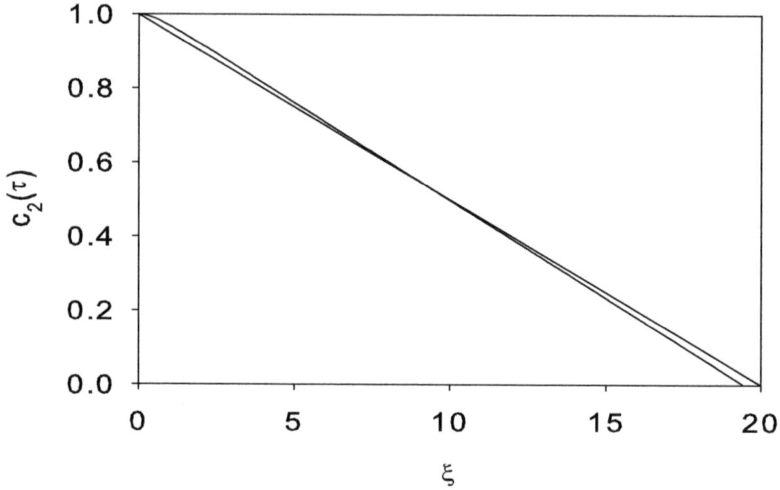

Fig. 2.7 Solutions of the dimensionless sucrose concentration versus dimensionless time for full model (curved solution) and for quasi-steady state model (straight line solution). Both plots are for $\frac{Q_e R}{H} = 0.1$ and $c_1 = 0$.

Example 2.8: Quasi steady state surface reaction.

On page 66, we modelled a surface catalyzed reaction in which two reactants, A and B, adsorb to a surface and react to form a product P which then detach from the surface. The sketch of the system is repeated in Fig. 2.8.

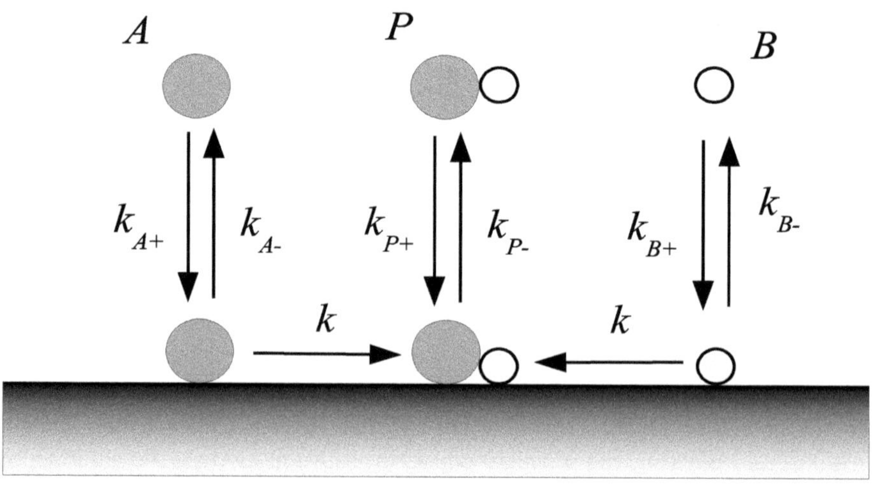

Fig. 2.8 Schematic of a surface catalyzed reaction between two molecules.

Assuming mass action kinetics, the reactant balances gave a model consisting of coupled, nonlinear differential equations that could only be solved numerically. We will now simplify this model using a quasi-steady state approximation and obtain an approximate model that is easily solved. If one can assume that the adsorption and desorption reactions are very fast compared to the chemical reaction, then the concentration of the adsorbed species can be assumed to be in a quasi-steady state. Steady state balances on adsorption and desorption of A, B and P gives three algebraic equations

$$k_{A+}C_{A,g}(N_{\text{total}} - C_{A,s} - C_{B,s} - 2C_{P,s}) = k_{A-}C_{A,s}$$

$$k_{B+}C_{B,g}(N_{\text{total}} - C_{A,s} - C_{B,s} - 2C_{P,s}) = k_{B-}C_{B,s}$$

$$k_{P+}C_{P,g}(N_{\text{total}} - C_{A,s} - C_{B,s} - 2C_{P,s}) = k_{P-}C_{P,s}$$

which can be solved for $C_{A,s}$, $C_{B,s}$ and $C_{P,s}$

$$C_{A,s} = \frac{k_{A+}k_{B-}k_{P-}C_{A,g}N_{\text{total}}}{k_{A+}k_{B-}k_{P-}C_{A,g} + k_{A-}k_{B+}k_{P-}C_{B,g} + 2k_{A-}k_{B-}k_{P+}C_{P,g} + k_{A-}k_{B-}k_{P-}}$$

$$C_{B,s} = \frac{k_{A-}k_{B+}k_{P-}C_{B,g}N_{\text{total}}}{k_{A+}k_{B-}k_{P-}C_{A,g} + k_{A-}k_{B+}k_{P-}C_{B,g} + 2k_{A-}k_{B-}k_{P+}C_{P,g} + k_{A-}k_{B-}k_{P-}}$$

$$C_{P,s} = \frac{k_{A-}k_{B-}k_{P+}C_{P,g}N_{\text{total}}}{k_{A+}k_{B-}k_{P-}C_{A,g} + k_{A-}k_{B+}k_{P-}C_{B,g} + 2k_{A-}k_{B-}k_{P+}C_{P,g} + k_{A-}k_{B-}k_{P-}}.$$

For most practical purposes, we are done because the interesting result, the steady rate of product formation, is easily calculated as a function of the gas phase concentrations.

$$\begin{aligned} r_P &= kC_{A,s}C_{B,s} \\ &= \frac{kk_{A+}k_{A-}k_{B+}k_{B-}k_{P-}^2C_{A,g}C_{B,g}N_{\text{total}}^2}{(k_{A+}k_{B-}k_{P-}C_{A,g} + k_{A-}k_{B+}k_{P-}C_{B,g} + 2k_{A-}k_{B-}k_{P+}C_{P,g} + k_{A-}k_{B-}k_{P-})^2}. \end{aligned}$$

Notice that although the surface reaction itself is irreversible and independent of the adsorbed product concentration, the overall reaction rate does depend on the gas phase concentration of the product. This is caused by the fact that adsorbed product limits the number of free sites available to the reactants and therefore slows down the overall reaction. The highest reaction rate is therefore obtained when the gas phase concentration of the product equals zero, in which case the expression above simplifies to

$$r_P = \frac{kk_{A+}k_{A-}k_{B+}k_{B-}C_{A,g}C_{B,g}N_{\text{total}}^2}{(k_{A+}k_{B-}C_{A,g} + k_{A-}k_{B+}C_{B,g} + k_{A-}k_{B-})^2}.$$

2.4 Geometric simplifications

By geometric simplifications, we mean ignoring the detailed geometry or shape of an object, using instead a simpler shape for modelling purposes. This is essentially what we do if we limit ourself to using rectangular, cylindrical or spherical coordinate systems. If the shape of an object is such that it is "small" in one direction compared to the other directions, then it is essentially "flat". This idealized flat object may not be planar but might be bent in some way. However, if the radius of curvature of the object is large compared to its thickness, then it can be approximated as a flat plane and rectangular coordinates are to describe the conditions inside the object. Similar idealizations can be used to justify cylindrical and spherical coordinates.

Another type of geometric simplification is the lubrication approximation used in fluid mechanics. Lubricating flow occurs between two solid bodies that move relative to one another. Flow is almost always assumed to be purely viscous and the lubrication approximation further makes the assumption that the flow is unidirectional. For instance, the lubricating flow between the inner and outer cylinder in a bearing can be approximated by flow that only has an angular velocity component. Or, consider the case of laminar flow through a cylindrical pipe with varying radius as in the following example.

Example 2.9: Lubricating flow in a pipe.

If a fluid passes through a pipe with a slowly varying radius, then the velocity profile can be approximated by the steady laminar flow velocity profile. For Newtonian flow this is

$$v(z) = v_m(z)\left(1 - \left(\frac{r}{R(z)}\right)^2\right)$$

where z is an axial coordinate in the direction of the flow. Assuming incompressible flow, the pipe radius and the center velocity can be related through a volume balance. Let the volumetric flow rate through the pipe be Q, then

$$Q = \int_0^{R(z)} v_m(z)\left(1 - \left(\frac{r}{R(z)}\right)^2\right) 2\pi r dr = \frac{\pi}{2} v_m(z) R(z)^2.$$

Hence, the velocity profile in the pipe is

$$v_m(z) = \frac{2Q}{\pi R(z)^2}\left(1 - \left(\frac{r}{R(z)}\right)^2\right).$$

The lubrication approximation assumes that it is acceptable to ignore small or slow changes in the shape of an object. A common implementation of this idea is to ignore the curvature of a surface. For instance, if a reaction in a catalytic pellet is diffusion limited and hence occurs primarily in a thin reaction layer close to the pellet surface, one may ignore the curvature of the pellet and do the model using rectangular coordinates. A more drastic implementation of this idea is to use rectangular coordinates to model a system, even when cylindrical or spherical coordinates would appear more reasonable. The point of doing this is obviously that models in rectangular coordinates are usually more easily solved than models in curvilinear coordinates. The next example illustrates this fact through calculation of the effectiveness factor of a catalytic pellet using the three basic coordinate systems. In rectangular and spherical coordinates, the model is solved using the well known hyperbolic functions while in cylindrical coordinates, the solution requires Bessel functions (which will be covered in section 8.6.2).

Example 2.10: Effectiveness factor for a catalytic pellet in the three basic geometries.

The reactant concentration profile for a first order reaction in a spherical catalytic pellet of radius R was found above to be

$$\frac{C(\zeta)}{C_s} = \frac{1}{\zeta}\frac{\sinh(3\Phi_s\zeta)}{\sinh(3\Phi_s)}$$

where $\zeta = r/R$ is the dimensionless radial coordinate. The Thiele modulus has been given the subscript s to indicate that this is the modulus in spherical coordinates, $\Phi_s = \frac{R}{3}\sqrt{\frac{k}{D}}$.

For a flat catalytic pellet of thickness $2L$, the steady state reactant balance takes the form

$$D\frac{d^2C}{dx^2} = kC \text{ subject to } C(L) = C_s \, , \; \left.\frac{dC}{dx}\right|_{x=0} = 0 \Rightarrow$$

$$\frac{C(\zeta)}{C_s} = \frac{\cosh(\Phi_r\zeta)}{\cosh(\Phi_r)}$$

where $\zeta = x/L$ and the Thiele modulus in rectangular coordinates is $\Phi_r = \sqrt{\frac{k}{D}}L$.

In cylindrical coordinates, the reactant balance is

$$\frac{D}{r}\frac{d}{dr}\left(r\frac{dC}{dr}\right) = kC \text{ subject to } C(R) = C_s \, , \; \left.\frac{dC}{dr}\right|_{r=0} = 0 \Rightarrow$$

$$\frac{C(\zeta)}{C_s} = \frac{I_0(2\Phi_c\zeta)}{I_0(2\Phi_c)}$$

where the $\zeta = r/R$ and the Thiele modulus in cylindrical coordinates is $\Phi_c = \frac{R}{2}\sqrt{\frac{k}{D}}$. The function I_0 is called the modified Bessel function of first kind of order zero.

The concentration profiles for the three different geometries are not that interesting to compare since they are really just a means to an end, to determine how good the catalytic particle is. The quality of a catalytic particle is usually measured by its effectiveness factor which is defined as the observed reaction rate over the particle divided by the rate maximum rate. The maximum rate is the rate observed without mass transfer resistance or equivalently, the rate observed at the particle surface. Thus, for a first order reaction

$$\eta = \frac{\int_V kC\,dV}{\int_V kC_s\,dV}$$

evaluation of the integral in the denominator is straight forward. It obviously equals the particle volume multiplied by the reaction rate at the surface, $kC_s\frac{4}{3}\pi R^3$. The integral in the numerator is more cumbersome to evaluate and it is easier to calculate the observed rate from the flux into the pellet. A reactant balance over the whole pellet states that the observed rate of reaction over the pellet equals the rate of diffusion from the pellet surface into the pellet. Thus, for a spherical pellet,

$$\int_V kC\,dV = -(-D4\pi R^2)\left(\frac{dC}{dr}\right)_{r=R} \Rightarrow \eta_s = \frac{D4\pi R^2\left(\frac{dC}{dr}\right)_{r=R}}{\int_V kC_s\,dV}\,.$$

The minus sign outside the parentheses is needed because the flux into the pellet is a flux in the negative radial direction. This expression is usually easier to evaluate because differentiation is simpler to carry out than integration. The result for the effectiveness factor in rectangular, cylindrical and spherical geometries become

$$\eta_r = \frac{1}{\Phi}\tanh\Phi$$

$$\eta_c = \frac{I_1(2\Phi)}{\Phi I_0(2\Phi)}$$

where I_1 and I_0 are the modified Bessel functions of first kind of order 1 and zero respectively.

$$\eta_s = \frac{1}{\Phi}\left(\frac{1}{\tanh(3\Phi)} - \frac{1}{3\Phi}\right).$$

The graphs for the three geometries are shown in Fig. 2.9 and they are seen to be very similar. Most importantly, they share the same asymptotes: at large values of the Thiele modulus, when the reaction is diffusion limited, $\eta \simeq 1/\Phi$ while at small values when the reaction rate is limiting, $\eta \simeq 1$. The Thiele modulus for different rate expressions is usually defined in such a way that this limiting behavior is maintained. The graphs also illustrate that, in this case at least, models based on different geometries but otherwise equivalent give close to the same results. This

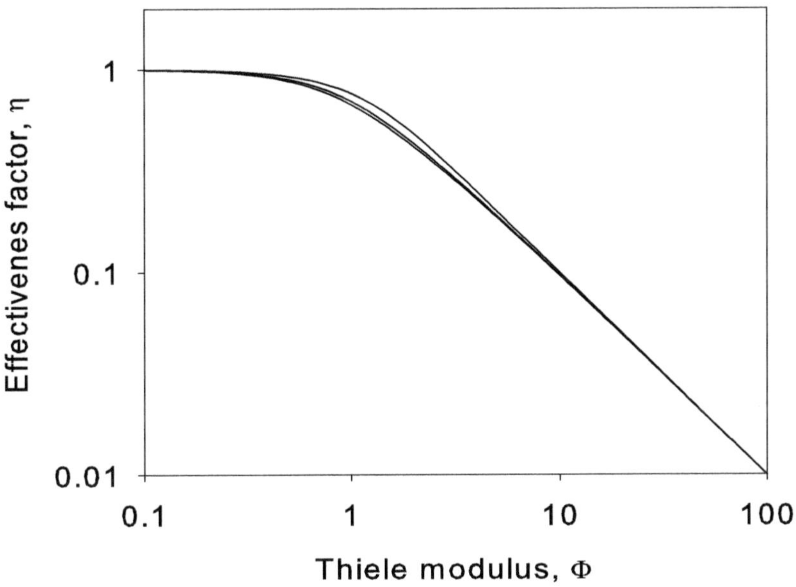

Fig. 2.9 The effectiveness factor, η, versus the Thiele modulus, Φ, for a first order reaction in rectangular, cylindrical and spherical geometries. The top curve is for rectangular geometry, the curve in the middle for cylindrical geometries and the curve at the bottom for spherical geometry.

suggest that approximate modelling can often be done in rectangular coordinates which are easier to handle than in curvilinear coordinates.

One more geometric simplification must be mentioned. This is the simplification obtained when assuming that the substance being modelled is semi-infinite in one direction. For instance, when energy is conducted or a compound diffuses into a flat wall, it is reasonable to assume that at short exposure times, transport of energy/compound has only penetrated a short distance from the exposed surface. At short exposure times, the distance to the center of the wall is therefore largely irrelevant since the transport process does not "see" the center yet. One might therefore just as well assume that the wall is semi-infinite. So, taking thermal conduction as an example, a transient energy balance on a wall of thickness $2L$ is Example 1.11 on page 35.

$$\frac{k}{\rho c_P}\frac{\partial^2 T}{\partial x^2} = \frac{\partial T}{\partial t}$$

where all symbols have their usual significance. The boundary conditions are

$$\left.\frac{\partial T}{\partial x}\right|_{x=0} = 0 \text{ and } T(L) = T_{\text{Surface}}\,.$$

If we now assume that the wall is semi-infinite, the boundary condition at $x = 0$ is discarded in favor of a boundary condition at $-\infty$ (or at ∞, depending on the orientation of the x-axis).

$$\lim_{x \to -\infty} T = T_0$$

where T_0 is the initial temperature of the wall. The boundary condition states that "far away" from the exposed surface, the temperature has not changed from its initial value. Obviously, this simplified model is only a good model at short exposure times, and it is far from obvious why it even represents a desirable simplification. The justification for the simplification becomes apparent when the two models, the finite domain model and the semi-infinite domain models are solved. The finite domain model can be solved, but the solution is in terms of an infinite series that converges very so slowly at short times that it is virtually useless as a solution. The semi-infinite model, however, has a solution in terms of a single, relatively simple function which is convenient to work with at short exposure times. The tools for solving the finite domain model are discussed in Chapter 10, the tools for the semi-infinite domain model are discussed in Chapter 12. In closing, we mention that the semi, which we hope is an obvious point infinite approximation of a substance only makes sense in rectangular coordinate.

Chapter 3

Examining a Model

Probably everyone who has ever taught mathematical modelling of physical systems will have amassed a collection of common errors that students make when writing models. Many of these common errors can be avoided by being mindful of common pitfalls while developing the model and by doing a brief preliminary model analysis to check the validity of the model. The purpose of this deliberately short chapter is to point out some of these pitfalls and some of the tools that can be used for preliminary model analysis. And, by presenting this material after the reader has had a chance to work through the modelling problems in the first two chapters, to convey an appreciation of these pitfalls while writing the model and to examine a model to detect errors and to develop a feel for the solution.

3.1 Before even writing any equations

Read the problem statement carefully and draw a sketch of the process while reading the problem. The sketch should indicate appropriate CVs that will be used, e.g. an entire vessel, each of the phases in a vessel and so on. The sketch should also indicate all transport processes with clearly marked arrows. Process streams are usually obvious, but other transport terms such as an evaporation flux or heat transfer across a vessel surface are less so. It is usually easy to mark all fluxes by arrows as one reads through the problem, possibly repeatedly, and indicating all fluxes on the figure makes it much more likely that flux terms will be included when writing the terms in the conservation balance. Once this is done, decide what CVs to use and what conserved substances to write balances on.

Homework assignments tend to be well posed (though not always) while real engineering problems are less so. To clarify the problem, it can be helpful to write a list of the unknowns that you are trying to find as well as a list of the known variables such as feed streams and their properties, the geometry and size of vessels and objects and the relevant physical properties, including equations of states and other thermodynamic properties. A list of unknowns, whether written down or just kept mentally, is particularly useful in determining if the model is complete, usually done by verifying that it has the same number of independent equations as the number of unknowns.

3.2 While writing the model

Write supporting prose explaining what the CV is for each conservation balance and what conserved quantity the balance is written on. State in words and in the same order as they appear in the balance equation, what the transport terms in and out of the CV are. This will not only help you to exercise care when writing the balance, it will help communicate your work to the people that you work with.

To help you clarify your work to yourself and the people you work with, use informative, sensible nomenclature. Do not indicate molar concentrations by Θ, use C. Be particularly careful when your model includes variables that are commonly referred to by the same symbol such as, for instance, thermal conductivity, mass transfer coefficients and reaction rate constants, that are all commonly indicated k. Use informative subscripts to indicate the difference between these.

As you write the terms for transport in and out of the control volume, check that they all have the correct dimension. In a conservation balance, this is usually straightforward, all terms must have dimension of amount (moles, mass, energy, etc.) per time.

Do not hesitate to write several models; a "complete" model as well as simplified models such as models that employ the quasi-steady state assumption or models that leave out terms you suspect may be insignificant. Even if the simplified models turn out to be inadequate for the purpose at hand, they are easier to solve than a complete model and can be used to develop a feel for the solution of the complete model as well as provide solutions that must match solutions to the complete model in the limit or for particular model parameter values.

3.3 Dimensions and units

Make the model dimensionless. This is not just an important step in simplifying the model, it is also a final check that the model is dimensionally correct. Do this check by confirming that all dimensionless groups are in fact dimensionless.

Dimensionless models may seem more abstract than dimensioned models but are in fact often more informative. For instance, being told what fluid flows through a pipe of given size and at what the velocity of the fluid is, is very concrete but being told just the Reynolds number of the flow is a much easier way of summarizing the character of the flow. Another way of stating this point is to note that the magnitude of model parameters are not that informative by themselves. For instance, is a reaction rate constant of 1 $\sec^{-1}$ indicative of a fast or slow reaction? The question does not even make sense unless we also know the size of the catalytic particle and the diffusion coefficient of the reactant. However, knowing that the Thiele modulus is 0.01 for example, immediately tells us that the reaction is in fact slow and therefore reaction limited.

It should be obvious, but it is sometimes forgotten, that all model parameters must be entered in the same system of units. The recommended system is the International System of Units or SI Units. This system uses meter for length, kilogram for mass, seconds for time, Kelvin for temperature and moles for amounts of matter.

3.4 Rudimentary approximate solutions

To develop a feel for the solution of a model and to check if the solution matches physical intuition, one can develop rudimentary approximate solutions. The two methods mentioned below are too rudimentary to be worthy of the name "numerical solutions" but they can be quickly and easily executed even without having access to a computer and the approximate solutions can be compared against expected solution properties to provide a check on the model and a tool for debugging a model for errors.

3.4.1 *Initial value problems*

Initial value problems can be analyzed by assuming that the solution has a Taylor expansion around the initial point in time, usually $t = 0$ and then discarding all but the lowest order terms. Keeping just the linear terms in the expansion is often sufficient to just convince yourself that the model does exhibit the correct/expected initial trend.

Example 3.1: Approximate solution of initial value problem.

Consider a well mixed batch reactor where the reaction $A + B \leftrightarrow P$ takes place. If the reaction is first order in each reactant and the initial concentration of A and B are equal, then a model of this system is

$$\frac{dC_A}{dt} = -k_1 C_A C_B + k_2 C_P \ , \ C_A(0) = C_0$$

$$\frac{dC_B}{dt} = -k_1 C_A C_B + k_2 C_P \ , \ C_B(0) = C_0$$

$$\frac{dC_P}{dt} = k_1 C_A C_B - k_2 C_P \ , \ C_P(0) = C_{P0} \, .$$

The nomenclature is hopefully self explanatory.

Since the initial concentration of A and B are identical, it follows from the stoichiometry that $C_A = C_B$ so we can simplify the model by considering only the balance equation for A. Furthermore, we can replace the balance equation in P with an algebraic equation derived from the stoichiometry

$$C_P = C_{P0} + (C_{A0} - C_A)$$

giving the model

$$\frac{dC_A}{dt} = -k_1 C_A^2 + k_2(C_{P0} + C_0 - C_A)\ ,\ C_A(0) = C_0.$$

This is a nonlinear model and consequently difficult to solve. However, one would anticipate that if the initial concentration of A and B is high compared to the initial concentration of P, then the solution at short times would show a consumption of A and B and a production of P. The reverse would be the case if the initial concentrations of A and B is low compared to the initial concentration of P. We will check that the model does in fact reflect this intuition by developing an approximate solution for short times.

Start by making the model dimensionless using the initial concentration to make the concentration of A dimensionless and define the dimensionless time as $\tau = tk_1C_0$. The dimensionless model is (convince yourself of this)

$$\frac{dx_A}{d\tau} = -x_A^2 + \frac{k_2}{k_1C_0}\left(\frac{C_{P0}}{C_0} + 1 - x_A\right)\ ,\ x_A(0) = 1.$$

Approximate the solution of this model by a second order polynomial in τ. Notice that the polynomial is written such that it already satisfies the initial condition

$$x_A(\tau) = 1 + a_1\tau + a_2\tau^2.$$

Substitution into the model gives

$$a_1 + 2a_2\tau = -(1 + a_1\tau + a_2\tau^2)^2 + \frac{k_2}{k_1C_0}\left(\frac{C_{P0}}{C_0} + 1 - 1 - a_1\tau - a_2\tau^2\right).$$

Dropping all terms of order 2 and higher on the right hand side gives

$$\begin{aligned}
a_1 + 2a_2\tau &= -(1 + a_1\tau + a_2\tau^2)^2 + \frac{k_2}{k_1C_0}\left(\frac{C_{P0}}{C_0} + 1 - 1 - a_1\tau - a_2\tau^2\right)\\
&= -1 - 2a_1\tau + \frac{k_2}{k_1C_0}\left(\frac{C_{P0}}{C_0} - a_1\tau\right)\\
&= -1 + \frac{k_2}{k_1C_0}\frac{C_{P0}}{C_0} - \left(2 + \frac{k_2}{k_1C_0}\right)a_1\tau.
\end{aligned}$$

Comparing coefficients of powers of τ gives

$$a_1 = \frac{k_2}{k_1 C_0}\frac{C_{P0}}{C_0} - 1$$

$$a_2 = -\left(1 + \frac{k_2}{2k_1C_0}\right) a_1 = \left(1 + \frac{k_2}{2k_1C_0}\right)\left(1 - \frac{k_2}{k_1C_0}\frac{C_{P0}}{C_0}\right).$$

Hence, the crude approximation for x_A at short times is

$$x_A = 1 + \left(\frac{k_2}{k_1C_0}\frac{C_{P0}}{C_0} - 1\right)\tau + \left(1 + \frac{k_2}{2k_1C_0}\right)\left(1 - \frac{k_2}{k_1C_0}\frac{C_{P0}}{C_0}\right)\tau^2. \quad (3.1)$$

This crude approximation is a function of two dimensionless groups, $\frac{k_2}{2k_1C_0}$ and $\frac{C_{P0}}{C_0}$ and it is plotted in Fig. 3.1 as a function of these two groups.

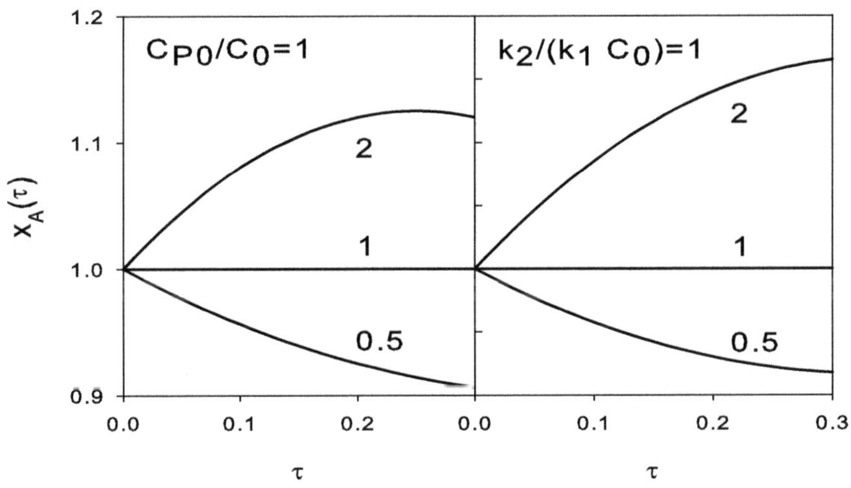

Fig. 3.1 Graphs of the approximate solution in Eq. (3.1). The parameter next to the curve on the left hand plot is the value of $\frac{k_2}{2k_1C_0}$, while the parameter on the right hand plot is the value of $\frac{C_{P0}}{C_0}$.

The initial dynamics exhibited by these graphs are exactly as expected. On the left hand graph, increasing the value of the reaction rate constant of the reverse reaction relative to the constant of the forward reaction drives the system towards a higher concentration of A at equilibrium. Hence A goes from being initially consumed, at $\frac{k_2}{2k_1C_0} = 0.5$ to being produced at $\frac{k_2}{2k_1C_0} = 2$. Similarly on the right hand graph as the initial concentration of P is increased relative to the initial concentration of A.

Of course, this very quick analysis does not prove anything. However, it does give us some confidence in the model that the crude solution exhibits the dynamics we would expect. Had this not been the case, it would cast doubt on the validity of the model or the crude solution and one would want to examine both before proceeding to develop the model further or to solve it.

3.4.2 *Boundary value problems*

Quick, approximate solutions of two point boundary value problems can be obtained by one point collocation. The basic idea is to represent the solution by a so called *trial function* which is designed to satisfy the boundary conditions but which has a single adjustable parameter. A polynomial is usually used. To obtain a value of the parameter, substitute the trial function into the differential equation and require that the equation is satisfied at a single point, the collocation point. Unless there are substantive reasons for doing otherwise, the midpoint of the interval is usually used as the collocation point.

Example 3.2: One point collocation.

Consider the model of a spherical catalyst pellet in which a second order reaction takes place. Convince yourself that a dimensionless model of this pellet is

$$\frac{1}{\zeta^2}\frac{d}{d\zeta}\left(\zeta^2\frac{dx}{d\zeta}\right) = \Phi^2 x^2 \ , \ x(1) = 1$$

where the Thiele modulus is defined as $\Phi = \sqrt{kR^2C_0/D}$. Instead of using a boundary condition at $\zeta = 0$, we will use the fact that the solution must be an even function and pick a trial function which is a polynomial in even powers of ζ

$$x(\zeta) = a + (1-a)\zeta^2$$

which does satisfy the boundary condition at the pellet surface. Substitution into the model gives

$$6(1-a) = \Phi^2(a + (1-a)\zeta^2)$$

Pick $\zeta = 0.5$ as the collocation point to obtain an equation for a.

$$6(1-a) = \Phi^2(a + (1-a)/4) \ \Rightarrow \ a = \frac{1}{3}\frac{24-\Phi^2}{\Phi^2+8}$$

Hence, the approximate solution, which is plotted in Fig. 3.2, is

$$x(\zeta) = \frac{1}{3}\frac{24-\Phi^2}{\Phi^2+8} + \left(1 - \frac{1}{3}\frac{24-\Phi^2}{\Phi^2+8}\right)\zeta^2 \qquad (3.2)$$

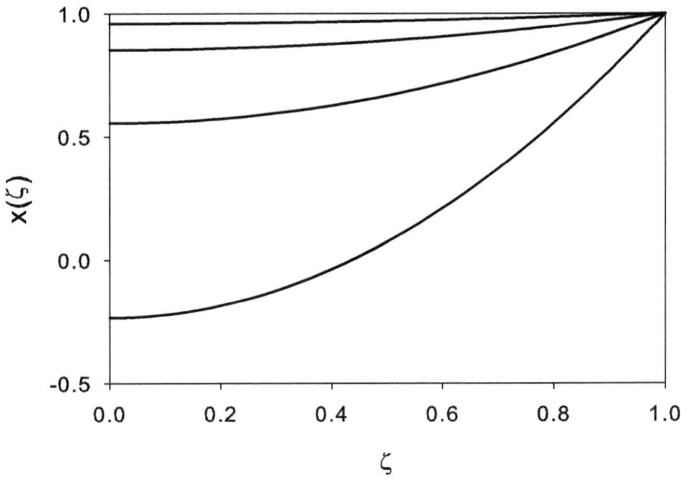

Fig. 3.2 Graphs of the approximate solution in Eq. (3.2). The solution is plotted for $\Phi = 0.5, 1, 2, 10$, top to bottom.

The approximate solution exhibits the expected trends. At low values of the Thiele modulus, the reactant concentration approaches that of the surface concentration throughout the pellet and the efficiency of the pellet is close to 1. As the Thiele modulus increases, the reaction occurs in an increasing narrow region close to the pellet surface. However, for large enough values of the Thiele modulus, the reactant concentration in the center of the pellet becomes negative. Negative concentrations are obviously not permitted, so there must be a problem with either the model or the approximate solution (or both). In this case the problem is with the approximate solution and the choice of trial function. A second order polynomial was used as a trial function and polynomials are notorious for being unable to fit functions that are approximately constant over an interval. In this case, the reactant concentration is close to zero in the core of the pellet and the trial function is unable to represent this, no matter what value is picked for the fitting parameter.

3.5 Examine the solution

After a solution is found for a model, one should always examine the solution to make sure it makes sense. Surprisingly often, this is not done as anyone who has reviewed manuscripts for scientific journals can attest to.

The last example discusses the case of a zero order reaction which can lead to model solutions with negative concentrations, an observation that unambiguously indicate an error in the model or in the solution. The correct way of resolving the problem of the negative concentrations is also discussed.

Example 3.3: Catalytic pellet with a zero order reaction.

Enzymes are the protein based catalysts that carry out and regulate cellular metabolism. Many enzymes are used industrially such as in production of syrups and flavors and in food processing or appear in consumer products such as in laundry detergents. Due to their cost, enzymes are most often used in immobilized form for easy recovery. Analysis of a simple enzyme reaction mechanism shows the reaction rate to be of so-called Michaelis-Menten form

$$r(C) = \frac{r_m C}{K + C}$$

where C is the concentration of the reactant and r the rate of consumption of the reactant. (In enzyme catalysis, the reactant is commonly called the substrate and the reaction rate is commonly called the velocity of the reaction. Thus, in the biochemical literature, the reactant concentration is often indicated S and the reaction rate is indicated by v). The two constant parameters, r_m and K, are the maximum reaction rate and the Michaelis-Menten constant respectively.

For obvious reasons, enzyme catalyzed reactions are carried out most efficiently at high reactant concentrations when the reaction rate is approximately equal to the maximum reaction rate, r_m, i.e. when the reactant is, to a good approximation, consumed in a zero order reaction. When this assumption is valid, a mass balance on the reactant in a flat immobilized enzyme pellet is

$$D\frac{d^2C}{dx^2} = r_m.$$

Placing the origin of the x-axis in the center of the particle, the boundary conditions are

$$x = 0 \Rightarrow \frac{dC}{dx} = 0$$

$$x = L \Rightarrow C = C_s$$

where L is half the pellet thickness and also the characteristic length of the pellet. C_s is the surface concentration of the reactant. Make this model dimensionless by introducing the following dimensionless variables and parameters.

$$\zeta = \frac{\xi}{L}\ ,\ \Theta = \frac{C}{C_s}\ ,\ \Phi = \sqrt{\frac{r_m L^2}{2C_s D}}$$

where ξ is the dimensionless spatial coordinate, Θ the dimensionless concentration and Φ the Thiele modulus for a zero order reaction. This gives the dimensionless model

$$\frac{d^2\Theta}{d\xi^2} = 2\Phi^2\ ,\ \Theta(1) = 1\ ,\ \left.\frac{d\Theta}{d\xi}\right|_0 = 0.$$

The reason for defining the Thiele modulus this way for a zero order reaction, creating the factor of 2 in the mass balance, is not obvious. However, we will not discuss the reasons for this other than to point out that this choice gives a result for the effectiveness factor with the desired limiting behavior at large values of the Thiele modulus, $\eta \approx 1/\Phi$. Solving for Θ

$$\Theta = \Phi^2\xi^2 + C_1\xi + C_2$$

and applying boundary conditions gives

$$\Theta(\xi) = \Phi^2(\xi^2 - 1) + 1.$$

It is now very important to note that the solution becomes negative at the center, a physical impossibility, if $\Phi > 1$. It is tempting to dismiss this problem by resetting the negative values of Θ to zero. However, this is not a valid solution to the problem of negative concentrations. To see this, note that at the distance at which the concentration becomes zero, call this distance x_c, one can write a reactant balance stating that the flux of reactant to this point equals the flux of reactant away from this point, i.e.

$$0 = -\left. D\frac{dC}{dx}\right|_{x_c}$$

but it is easily checked that the solution above does not have a zero derivative at the point where Θ becomes zero. The solution is still valid when $\Phi \leq 1$, but a different approach is needed for $\Phi \geq 1$.

The problem with the case $\Phi \geq 1$ is that the reactant concentration does become zero at some point inside the pellet and the boundary condition at $\xi = 0$ is therefore no longer applicable. In its place, one can use the reactant balance stated above, the balance at the point at which the concentration becomes zero. In dimensionless form, this is

$$\left.\frac{d\Theta}{d\xi}\right|_{\xi_c} = 0$$

which must be used in conjunction with a defining equation for ξ_c

$$\Theta(\xi_c) = 0\,.$$

The following solution is obtained

$$\Theta(\xi) = \begin{cases} \Phi^2(\xi^2 + (1 - \frac{1}{\Phi})^2 - 2\xi(1 - \frac{1}{\Phi}))\,, & \xi \geq 1 - \frac{1}{\Phi} \\ 0\,, & \xi \leq 1 - \frac{1}{\Phi} \end{cases}$$

and

$$\xi_c = 1 - \frac{1}{\Phi}\,.$$

As this solution is only valid when $\Phi \geq 1$, the complete solution can be written as

$$\Theta(\xi) = \begin{cases} \Phi^2(\xi^2 - 1) + 1\,, \Phi \leq 1 \\ \begin{cases} \Phi^2(\xi^2 + (1 - \frac{1}{\Phi})^2 - 2\xi(1 - \frac{1}{\Phi}))\,, & \xi \geq 1 - \frac{1}{\Phi} \\ 0\,, & \xi \leq 1 - \frac{1}{\Phi} \end{cases}, \Phi \geq 1 \end{cases}$$

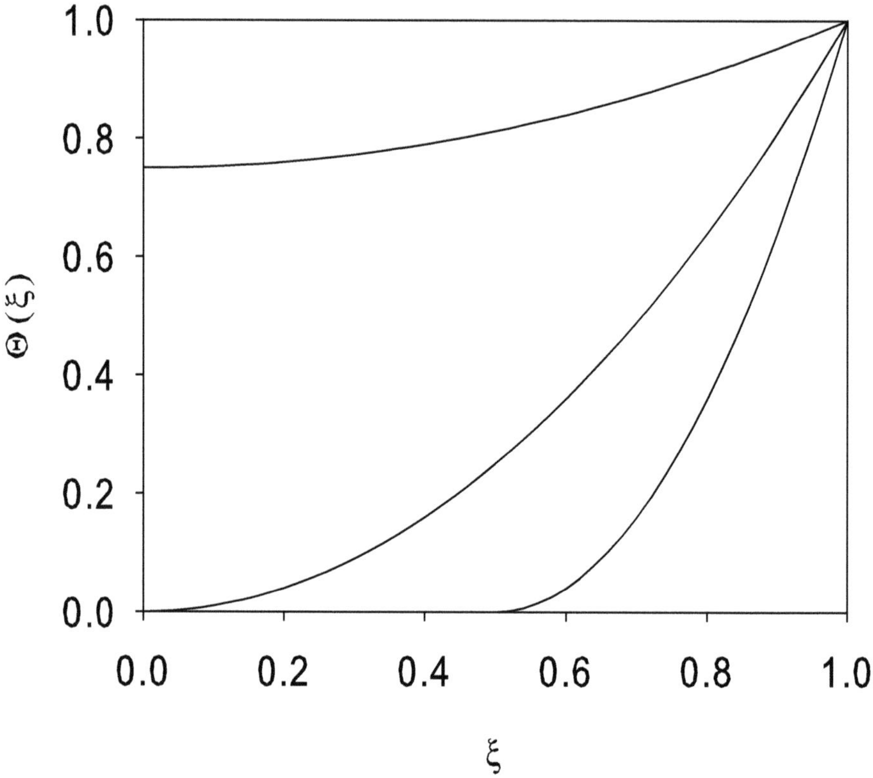

Fig. 3.3 Dimensionless concentration profiles for a zero order reaction with Φ =0.5, 1, 2, top graph to bottom graph.

and graphs of these solutions are shown in Fig. 3.3.

The effectiveness factor is found to be

$$\eta = \begin{cases} \int_0^L v_m \, dx/(v_m L) \, , \Phi \leq 1 \\ \int_{x_c}^L v_m \, dx/(v_m L) \, , \ \Phi \geq 1 \end{cases}$$

or

$$\eta = \begin{cases} 1 & , \Phi \leq 1 \\ \frac{1}{\Phi} & , \Phi \geq 1 \end{cases} .$$

Thus, the effectiveness factor curve for the zero order reaction consists of the two asymptotes of the curves of the effectiveness factors for a first order reaction, Fig. 2.9.

In the book *Mathematical Modelling Techniques* by Rutherford Aris,[1] the author provides a list of "maxims for mathematical modelling" as well as a list of tools that

[1]Aris, R. [1994] *Mathematical Modelling Techniques* (Dover Publications, New York).

are available to the modeller for getting a feel for the solution of the model. The list is more extensive than the short list of suggestions given here and many of the tools and maxims suggested by Aris require a broader knowledge of mathematical methods than can be expected of the intended readers of this book. Nevertheless, this book by Aris is highly recommended for anyone who wants to become more skilled at writing and studying mathematical modelling techniques.

Chapter 4

Some Ordinary Differential Equations

This chapter covers various ordinary differential equations that can be solved by quadrature or for which a theory exists for analyzing and possibly solving the equation. A particular differential equation may match several of the types listed below and either of the possible solution methods can then be used. Pick the method that is the easiest to use. The treatment below is sketchy and is intended as a summary and review of solution methods and not as a rigorous analysis of these methods.

4.1 First order equations

These are equations in which the highest order derivative appearing in the equation is of first order. Assuming that one can solve for the highest order derivative, the initial value problem for first order equations can be stated as

$$\frac{dx}{dt} = f(t,x)\ ,\ x(t_0) = x_0 \tag{4.1}$$

In order to state the existence and uniqueness theorem for this problem we will need the following definition.

Definition 4.1. A function, $f(t,x)$, is said to satisfy a Lipschitz condition in some domain D if there exists a constant $K > 0$ such that

$$|f(t,x_1) - f(t,x_2)| \leq K|x_1 - x_2|$$

for all (t,x) in D.

Unsurprisingly, the constant K is called a Lipschitz constant. The Lipschitz condition places an upper bound on how much the function can change as the argument of the function changes. If the function in question is differentiable, it is not hard to show that it satisfies a Lipschitz condition if the absolute value of the derivative of the function has an upper bound over the domain D.

We can now state the uniqueness and existence theorem.

Theorem 4.1. *Given the initial value problem 4.1. If $f(t,x)$ is continuous and satisfies a Lipschitz condition in the rectangular domain R*

$$R: \; |t-t_0| \le a \; , \; |x-x_0| \le b$$

and $|f(t,x)| \le M$ in R, then there exists a unique solution, $x(t)$, on the interval $|t-t_0| \le h$ where $h = min\{a, b/M\}$.

The proof of this theorem is not terribly hard but it is a bit long and will not be given here. It can be found in most advanced text on differential equations.[1] However, a few comments are in order: first of all, the theorem assumes that the function $f(t,x)$ is continuous and Lipschitz when $|t-t_0| < a$ but only assures that a unique solution exists for the interval $|t-t_0| < h$ where h may be smaller than a. The reason for this is hard to appreciate without studying the proof of the theorem. Secondly, the reader may recall from calculus that a continuous function over a closed interval is bounded and the stated inequality, $|f(t,x)| \le M$, is therefore always true and not an additional assumption. The inequality is introduced to define the constant M, the upper bound on $f(t,x)$, which is needed to determine the interval length h. Thirdly, the Lipschitz condition is not required for existence of the solution, but for the uniqueness. For instance, the initial value problem

$$\frac{dx}{dt} = 3x^{2/3} \; , \; x(0) = 0$$

is not Lipschitz (at $x = 0$) but has a solution, in fact infinitely many, all of the form

$$x(t) = \begin{cases} 0 \text{ for } t \le C \\ (t-C)^3 \text{ for } t \ge C \end{cases}$$

where C can be any positive constant. Finally, the theorem is true even if $x(t)$ and $f(t,x)$ are vectors, i.e. for coupled first order equations. This is important since many higher order equations can be recast as sets of coupled first order equations.

The next couple of sections describes several kinds of first order equations which have closed form solutions.

4.1.1 *Separable equations*

Equations which can be rearranged to the form

$$f(x)\,dx = g(t)\,dt$$

[1]See for example, Coddington, E.A. [1961] *An Introduction to Ordinary differential Equations* (Dover, New York). This book is also an excellent source of many other proofs of basic theorems of ODEs.

have the solution

$$\int f(x)\,dx = \int g(t)\,dt.$$

This is as simple as a differential equation can possibly get. However, it is not always immediately obvious that a differential equation is separable and if unable to recognize this, one can waste a lot of effort when solving the equations using alternate and less straightforward methods.

Example 4.1: Constant volume batch reactor.

Consider a well mixed batch reactor in which the irreversible reaction $A \rightarrow$Products takes place. The reaction proceeds at constant volume and the specific reaction rate is $r_A(C)$, where C is the concentration of the reactant A. A reactant balance over a unit volume of the vessel gives

$$\frac{dC}{dt} = -r_A(C)$$

with the initial condition

$$t = t_0 \;\Rightarrow\; C = C_0\,.$$

The right hand side is independent of t and the equation is separable

$$\int_{C_0}^{C} \frac{dC}{r_A(C)} = -\int_{t_0}^{t} dt.$$

Consider a first order reaction, $r_A = kC$. One easily obtains

$$\frac{1}{k}\ln\left(\frac{C}{C_0}\right) = t_0 - t \;\Rightarrow$$

$$C = C_0 e^{-k(t-t_0)}.$$

Or consider the more complex rate expression $r_A(C) = \frac{kC^n}{K+C}$. This gives

$$t_0 - t = \int_{C_0}^{C} \frac{K}{k} C^{-n} dc + \frac{1}{k}\int_{C_0}^{C} C^{1-n} dC$$

$$= \frac{K}{k(n-1)}(C_0^{1-n} - C^{1-n}) + \frac{1}{k(n-2)}(C_0^{2-n} - C^{2-n}).$$

The important thing to notice about this result is that division by zero results for $n = 1$ and $n = 2$. These are therefore special cases that must be handled individually. One should always check that solutions which are dependent on one or several parameters are valid for all possible values of these parameters and, if not, one should find the solutions for these special values. In this case, the solutions for $n = 1$ and $n = 2$ are

$n = 1$:

$$t_0 - t = \frac{K}{k}\ln\left(\frac{C}{C_0}\right) + \frac{1}{k}(C - C_0)$$

$n = 2$:

$$t - t_0 = \frac{K}{k}\left(\frac{1}{C} - \frac{1}{C_0}\right) + \frac{1}{k}\ln\left(\frac{C}{C_0}\right).$$

A special case of a separable equation is an equation for which the derivative is a function of only one of the variables, either the free or the dependent variable. If the derivative is a function of only the dependent variable, one must be careful not to miss the constant solutions, solutions of $f(x) = 0$. Consider for instance the logistic equation.

Example 4.2: The logistic equation.

The logistic equation is an ad hoc model of population growth and has the form

$$\frac{dx}{dt} = ax - bx^2$$

where t is time and x is a measure of population density, e.g. individuals or amount of biomass per volume, a is the specific growth rate of the population and b a death rate. Both a and b are positive and the population density x is obviously non-negative so we will only look for non-negative solutions to this equation. The equation can be rewritten as

$$\frac{dx}{(a-bx)x} = \frac{1}{ax} + \frac{b}{a(a-bx)} = dt \Rightarrow$$

$$\frac{1}{a}\ln\left(\frac{x}{a-bx}\right) = t + C \Rightarrow$$

$$x(t) = \frac{aC}{e^{-at} + Cb} \tag{4.2}$$

where C is an arbitrary constant. However, there are two more solutions, the constant solutions corresponding to $\frac{dx}{dt} = 0$. These constant solutions are often overlooked and are in this case

$$x_1 = \frac{a}{b} \text{ and } x_2 = 0\,.$$

Solutions to Eq. (4.2) are plotted in Fig. 4.1. The graphs strongly suggest that all unsteady solutions converge to the non-zero steady state solution. In other words, the zero solution is unstable while the non-zero solution is stable. We shall return to this issue later, in section 8.4, and show how stability and instability of steady states can be determined rigorously for equations such as the logistic equation.

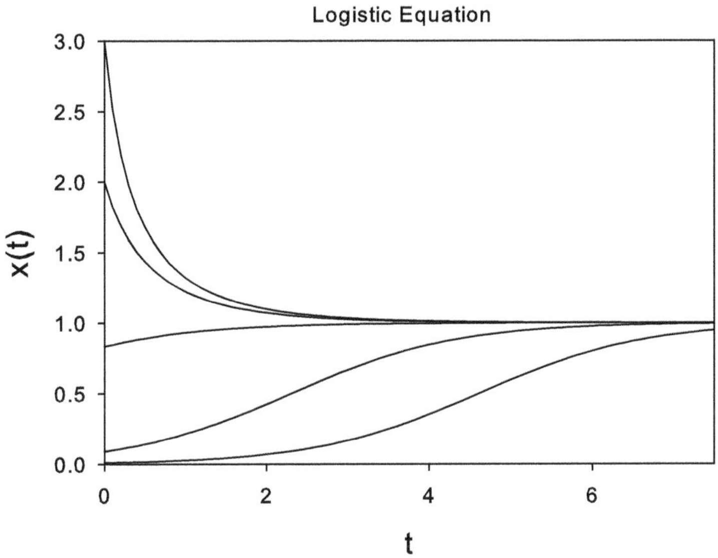

Fig. 4.1 Plots of Eq. (4.2) for $a = b = 1$ and for $c = -1.5, -2, 5, 0.1, 0.01$, top to bottom graphs.

4.1.2 *Linear, first order equations*

These are equations which can be written in the form

$$\frac{dx}{dt} + p(t)x = q(t)\,. \tag{4.3}$$

The equation can be solved by multiplying each side by a so-called integrating factor. An integrating factor is a function which, when multiplying the equation above, allow the terms on the left hand side to be collected into a single first order derivative. Let $P(t)$ be any function such that $P(t) = \int p(t)\,dt$. The integrating factor then is $e^{P(t)}$. To see this, multiply both sides of Eq. (4.3) by the integrating factor to get

$$e^{P(t)}\frac{dx}{dt} + e^{P(t)}p(t)x = e^{P(t)}q(t)$$

which is equivalent to

$$\frac{d}{dt}\left(e^{P(t)}x\right) = e^{P(t)}q(t)$$

in which all terms involving $x(t)$ have been collected into a single derivative. This equation is separable and solved to give

$$x(t) = e^{-P(t)}\int e^{P(t)}q(t)\,dt + Ce^{-P(t)}$$

where C is an arbitrary constant. Notice that the solution is simply the general solution to the homogeneous equation, $Ce^{-P(t)}$, (the equation obtained when $q(x) =$

0) plus a particular solution to the inhomogeneous equation, a pattern that will be repeated in linear higher order equations.

When an initial condition is provided, it often pays to use this during the solution procedure, rather than apply it later to find the arbitrary constant. For instance, if the initial condition is

$$t = t_0 \Rightarrow x = x_0$$

then the separable equation obtained after applying the integrating factor solves as

$$\int_{x_0}^{x} d\left(e^{P(t)}x\right) = \int_{t_0}^{t} e^{P(t)}q(t)\,dt \Rightarrow$$

$$x(t) = e^{-P(t)}\left(x_0 e^{P(t_0)} + \int_{t_0}^{t} e^{P(t)}q(t)\,dt\right).$$

Example 4.3: CSTR with first order reaction.

The reactant balance on a CSTR with a first order reaction, $r = kC$, is

$$V\frac{dC}{dt} = QC_f - QC - kCV \ , \ C(0) = C_0$$

where C is the reactant concentration, C_f the reactant concentration in the feed, V the reactor volume and Q the volumetric flow rate through the reactor. This balance can be written

$$\frac{dC}{dt} + (D+k)C = DC_f$$

where $D = Q/V$, known as the dilution rate. The integrating factor is $e^{(D+k)t}$ so one gets

$$e^{(D+k)t}\frac{dC}{dt} + e^{(D+k)t}(D+k)C = DC_f e^{(D+k)t} \Rightarrow$$

$$\frac{d}{dt}\left(e^{(D+k)t}C\right) = DC_f e^{(D+k)t}.$$

Now, multiply through by dt and integrate. If an initial condition is given, e.g. $C(0) = C_0$, this is the appropriate time to apply it.

$$\int_{C_0}^{C} d\left(e^{(D+k)t}C\right) = DC_f \int_0^t e^{(D+k)t}dt \Rightarrow$$

$$e^{(D+k)t}C - C_0 = \frac{DC_f}{D+k}\left(e^{(D+k)t} - 1\right) \Rightarrow$$

$$C = C_0 e^{-(D+k)t} + \frac{DC_f}{D+k}\left(1 - e^{-(D+k)t}\right).$$

4.1.3 *Exact equations*

The total differential of a function of two variables $f(x,t)$ is

$$df = \frac{\partial f}{\partial x}dx + \frac{\partial f}{\partial t}dt.$$

It follows that a differential equation of the form

$$\frac{\partial f}{\partial x}dx + \frac{\partial f}{\partial t}dt = 0$$

has the solutions

$$f(x,t) = C$$

where C is an arbitrary constant. This type of differential equation is called exact. The main difficulty in applying this result is to recognize that a given equation is exact. Assuming that a given differential equation can be rearranged to the form

$$M(x,t)dx + N(x,t)dt = 0$$

then it can be shown[2] that the equation is exact iff

$$\frac{\partial M}{\partial t} = \frac{\partial N}{\partial x}$$

and the solution is

$$\int_a^x M(\xi,t)\,d\xi + \int_b^t N(a,\zeta)\,d\zeta = C.$$

Example 4.4: Exact equation.

Given

$$\frac{dx}{dt} = \frac{3t^2 - x}{2x+t},$$

this equation can be rewritten as

$$(2x+t)\,dx + (x - 3t^2)\,dt = 0\,.$$

The criterion for exactness checks out

$$\frac{d}{dt}(2x+t) = 1\;,\;\frac{d}{dx}(x - 3t^2) = 1$$

so the solution is

$$\int_a^x (2\xi + t)\,d\xi + \int_b^t (a - 3\zeta^2)\,d\zeta = C \;\Rightarrow$$

$$x^2 + tx - t^3 = C + a^2 + ab - b^3 = C_1\,.$$

---

[2]Again, see Coddington, [1961].

Some equations that are not exact can be made exact by multiplication by an integrating factor. The method is illustrated in the next example.

Example 4.5: Exact equation from integrating factor.
Consider

$$(t + 2xt^2)dx + (2x + 3x^2t)dt = 0.$$

This equation is seen not to be exact as

$$\frac{d(t + 2xt^2)}{dt} = 1 + 4xt \neq \frac{d(2x + 3x^2t)}{dx} = 2 + 6xt.$$

To be able to find an integrating factor we will assume that it is a function of only one of the variables, t or x. In general, neither may work, both may work or only one of these choices will work. In this case, it turns out that an integrating factor which is a function of t only works. Call the integrating factor $u(t)$ and multiply the differential equation by this function

$$u(t)(t + 2xt^2)dx + u(t)(2x + 3x^2t)dt = 0.$$

In order for the equation to be exact, we require that

$$(t + 2xt^2)u'(t) + (1 + 4xt)u(t) = (2 + 6xt)u(t) \Rightarrow$$

$$\frac{du}{dt} = \frac{u}{t} \Rightarrow$$

$$u(t) = C_1 t.$$

Setting the constant $C_1 = 1$ and multiplying the original differential equation by t gives the problem

$$(t^2 + 2xt^3)dx + (2xt + 3x^2t^2)dt = 0$$

which is seen to be exact

$$\frac{d(t^2 + 2xt^3)}{dt} = \frac{d(2xt + 3x^2t^2)}{dx} = 2t + x6t^2$$

and which has the solution

$$\int_a^x (t^2 + 2\xi t^3)\, d\xi + \int_b^t (2a\zeta + 3a^2\zeta^2)\, d\zeta = C_1 \Rightarrow$$

$$t^2x + x^2t^3 = C_2.$$

4.1.4 *Homogeneous equations*

The use of the term homogeneous to indicate the type of differential equation described below is lamentable because the same term is used to indicate the completely different concept that the "right hand side" or inhomogeneous term of a linear differential equation of any order is identically zero. However, the terminology is universally accepted, so we are stuck with the ambiguity. A differential equation is homogeneous of degree n if it can be written in the form

$$\frac{dx}{dt} = \frac{N(x,t)}{M(x,t)}$$

and the functions M and N have the properties that

$$M(\lambda x, \lambda t) = \lambda^n M(x,t)\ ,\ N(\lambda x, \lambda t) = \lambda^n N(x,t)$$

where $\lambda > 0$. (It is important in the derivation which follows that the factor of λ^n can only be extracted when λ is positive.) Another way of stating this is that M and N both are of the same total degree in the variables x and t. Homogeneous equations are solved by a variable transformation. Define a new variable by $y = t/x$, then $dt = ydx + xdy$ giving

$$\frac{dx}{ydx + xdy} = \frac{N(x,yx)}{M(x,yx)}\,.$$

The free and dependent variables occur completely symmetrical in a first order homogeneous differential equation, so alternatively, one can define the new variable as $y = x/t$. For the region where $x > 0$ the homogeneity property is used to extract the factor x^n from both N and M giving

$$\frac{dx}{ydx + xdy} = \frac{N(1,y)}{M(1,y)}\,.$$

This equation is separable. One finds

$$\int \frac{1}{x}\,dx = \int \frac{\frac{N(1,y)}{M(1,y)}}{1 - \frac{N(1,y)}{M(1,y)}y}\,dy.$$

For the region where $x < 0$ the homogeneity property is used to extract the factor $(-x)^n$ from both N and M giving

$$\frac{dx}{ydx + xdy} = \frac{N(-1,-y)}{M(-1,-y)}$$

which is separable.

Example 4.6: Homogeneous equation.

Consider

$$\frac{dx}{dt} = \frac{x^2 + tx}{t^2 + tx}$$

and let $N = x^2 + tx$ and $M = t^2 + tx$. It is easily checked that both functions are homogeneous of degree 2. For positive values of x one obtains the equation

$$\int \frac{1}{x}\, dx = \int \frac{\frac{1+y}{y^2+y}}{1 - \frac{1+y}{y^2+y} y}\, dy.$$

Unfortunately, the integrand on the right hand side simplifies to a fraction with a denominator equal to 0, thus the integral on the right hand side is not well defined. We can avoid this problem by carrying out the variable substitution carefully instead of blindly plugging into the formula. The differential equation is first rewritten as

$$\frac{dx}{dt} = \frac{1 + t/x}{(t/x)^2 + t/x} = \frac{1}{t/x}$$

defining $y = t/x$ one obtains

$$\frac{dx}{ydx + xdy} = \frac{1}{y}$$

which is rearranged to

$$yx\frac{dy}{dx} = 0$$

and the solution is readily obtained as

$$y(x) = t/x = C \;\Rightarrow\; x = t/C.$$

4.1.5 *Clairaut's equation*

This equation is named after the French mathematician Alexis Claude Clairaut (1713–1765) and is an equation of the form

$$x = t\frac{dx}{dt} + f\left(\frac{dx}{dt}\right).$$

To solve this, differentiate the equation once with respect to t, giving

$$\frac{dx}{dt} = t\frac{d^2x}{dt^2} + \frac{dx}{dt} + f'\left(\frac{dx}{dt}\right)\frac{d^2x}{dt^2}$$

or

$$\frac{d^2x}{dt^2}\left(t + f'\left(\frac{dx}{dt}\right)\right) = 0.$$

A solution is obtained if either of the two factors is equal to zero. Taking the second derivative as equal to zero indicate that the first derivative is equal to an arbitrary constant C. Substituting this constant into the original differential equation gives the solution

$$x = Ct + f(C)$$

a straight line. This family of straight line solutions is known as the general solution. The alternative solution, which is called the singular solution, is given by

$$t = -f' \frac{dx}{dt} \, .$$

Substituting this into the original differential equation yields

$$x = -\frac{dx}{dt} f' \left(\frac{dx}{dt} \right) + f \left(\frac{dx}{dt} \right) .$$

Eliminating dx/dt from the last two equations give the desired equation between x and t, the singular solution. Quite often, it may not be possible or easy to do this elimination and it is simpler to regard dx/dt as a parameter and think of the last two equations as a parametric representation of the singular solution.

This analysis makes the solution structure appear simpler than it really is. The singular solution does not have an arbitrary constant so it is tempting to conclude that it is only valid in the special case when it passes through the point given by the initial condition. Most often this will not be the case and the solution therefore is the very simple straight line general solution. However, it can be shown that the lines of the general solution all are tangent to the singular solution. Thus a C^1 solution to the Clairaut equation can be pieced together by e.g. starting on the general solution which satisfies the initial condition. At the point where this solution is tangent to the singular solution, switch to the singular solution. At any other tangent point, one can again switch back and forth between the singular and general solutions. There can thus be an infinity of possible solutions for a given initial condition.

Example 4.7: Clairaut equation.

Find all the solutions to

$$x = t\frac{dx}{dt} - \exp\left(\frac{dx}{dt}\right) \, , \; x'(0) - 1 \, .$$

The general solution are the lines given by

$$x = Ct - e^C .$$

The singular solution is given by

$$t = \exp\left(\frac{dx}{dt}\right) \, , \; x = \frac{dx}{dt} \exp\left(\frac{dx}{dt}\right) - \exp\left(\frac{dx}{dt}\right) .$$

In this case, dx/dt can be eliminated to give an explicit expression for the singular solution.

$$x(t) = t \, \ln(t) - t$$

The initial condition is given at $t = 0$ at which point the singular solution is not even defined. The initial solution must therefore come from the general solution and is

$$x = t - e.$$

This solution is valid for all time but is only a possible solution to the problem. Another possible solution is obtained by switching to the singular solution at the point where the solution above is tangent to the singular solution. The tangent point is at $t = e$ giving the other possible solution as

$$x = \begin{cases} t - e\,,\ t \leq e \\ t\ \ln(t) - t\,,\ \geq e \end{cases}.$$

4.1.6 *Bernoulli equation*

The Bernoulli family is known for the large number of outstanding mathematicians and physicists it produced. The Bernoulli equation is named for Jakob Bernoulli (1654–1705) and is an equation of the form

$$\frac{dx}{dt} = p(t)x + q(t)x^n.$$

When $n = 0$ or $n = 1$ the equation is linear. If $n = 1$ it is homogeneous. For other values of n, the solution is obtained by the variable transformation $y = x^{1-n}$ which gives rise to a linear equation in y. Alternatively, instead of remembering the variable transformation, divide the equation through by x^n to get

$$\begin{aligned} x^{-n}\frac{dx}{dt} &= p(t)x^{1-n} + q(t) \ \Rightarrow \\ \frac{1}{1-n}\frac{d}{dt}(x^{1-n}) &= p(t)x^{1-n} + q(t) \ \Rightarrow \\ \frac{1}{1-n}\frac{dy}{dt} &= p(t)y + q(t) \end{aligned}$$

which is equivalent to the result obtained by the variable substitution.

Example 4.8: Bernoulli equation.

Solve

$$\frac{dx}{dt} + tx = \exp(t^2)x^3, \quad x(0) = 1.$$

For this equation, n equals 3 so the variable substitution is $y = x^{1-3} = x^{-2}$ or $x = y^{-1/2}$. Using the chain rule

$$\frac{dx}{dt} = -\frac{1}{2}y^{-3/2}\frac{dy}{dt}$$

which substituted into the differential equation gives

$$-\frac{1}{2}y^{-3/2}\frac{dy}{dt} + ty^{-1/2} = e^{t^2}y^{-3/2} \ \Rightarrow$$

$$\frac{dy}{dt} - 2ty = -\,2e^{t^2}.$$

This equation is solved using an integrating factor

$$\frac{d}{dt}\left(e^{-t^2}y\right) = -2 \Rightarrow$$

$$y = \frac{1}{x^2} = -2te^{t^2} + Ce^{t^2} \Rightarrow$$

$$x^{-1} = \pm\sqrt{-2te^{t^2} + Ce^{t^2}}.$$

Applying the initial condition gives $1 = \pm\sqrt{C}$ which is obviously only true for the plus sign and when $C = 1$. So the solution becomes

$$x = \frac{1}{\sqrt{e^{t^2} - 2te^{t^2}}}.$$

4.1.7 *Riccati equation*

The Riccati equation is an equation of the form

$$\frac{dx}{dt} = p(t)x^2 + q(t)x + r(t).$$

Notice that if $r(x) = 0$, then the Ricatti equation simplifies to the Bernoulli equation. If $p(t) = 0$ it simplifies to a first order linear equation.

The equation is named for the Italian mathematician Jacopo Francesco Riccati (1676–1754) who published a form of the Riccati equation but was unable to solve it. Another form of the Riccati equation was also considered by John Bernoulli (1667–1748) who tried in vain to solve it.

The Ricatti does not appear often in engineering models, but it is important because it can be shown, as will be done in Section 4.2.5, that it is equivalent to a linear second order equation and linear second order equations are very important in engineering applications.

It can be shown that the general Ricatti equation cannot be solved by elementary means, but if a single solution can be found, for instance by guessing, then the complete solution can be obtained by quadrature. Let the known solution be $x_1(t)$ and make use of the variable transformation $x(t) = x_1(t) + x_2(t)$ which transforms the Riccati equation into the first order equation

$$\frac{dx_2}{dt} = p(t)x_2^2 + (2p(t)x_1(t) + q(t))x_2.$$

This is a Bernoulli equation and it is solved by the variable transformation $x_1 = 1/y$. The complete solution to the Riccati equation then is

$$x(t) = x_1(t) + x_2(t).$$

Example 4.9: Riccati equation.

The equation

$$\frac{dx}{dt} = x^2 - (t^2 + 1)x + 2t + t^2$$

has the solution $x_1(t) = t^2$. Using the variable transformation $x(t) = t^2 + x_2(t)$ gives the equation

$$\frac{dx_2}{dt} = x_2^2 + (t^2 - 1)x_2.$$

Solve this Bernoulli equation by setting $x_2 = 1/y$.

$$\frac{dy}{dt} + (t^2 - 1)y(t) = -1$$

which has the solution

$$y(t) = e^{-\frac{1}{3}t^3+t}\left(C - \int e^{\frac{1}{3}t^3-t}\right).$$

Thus, the complete solution to the Riccati equation is

$$x(t) = t^2 + \frac{e^{\frac{1}{3}t^3-t}}{C - \int e^{\frac{1}{3}t^3-t}}.$$

4.2 Second order equations

Second order ordinary differential equations (ODEs) arise frequently in engineering models. They typically occur where a flux across the CV boundary is proportional to the gradient of the dependent variable, for instance models that employ Fick's law or Fourier's law. Unfortunately, many of the second order equations that do appear in engineering models cannot be solved in closed form but require the introduction of new transcendental functions such as Bessel or Legendre functions. (Of course, transcendental functions themselves are not "new". The exponential function and the trigonometric functions were introduced as solutions to simple ODEs that could not be solved by other means).

This section will first present various second order equations for which solutions can be found by quadrature, then describe some transformations that can be used as tools to analyze and understand second order ODEs.

4.2.1 *Dependent variable does not occur explicitly*

These equations should be recognized as first order equations in the first derivative. If the equation can be solved for the first derivative, another first order equation is obtained. This second equation will be separable and can therefore always be solved.

Example 4.10: Second order ODE where dependent variable does not occur explicitly.

Consider

$$\frac{d^2x}{dt^2} + \frac{dx}{dt} = t^2$$

which is simply a first order equation in $y = \frac{dx}{dt}$.

$$\frac{dy}{dt} + y = t^2 \;\Rightarrow$$

$$y = \frac{dx}{dt} = 2 - 2t + t^2 + C_1 e^{-t} \;\Rightarrow$$

$$x(t) = 2t - t^2 + \frac{t^3}{3} - C_1 e^{-t} + C_2\,.$$

4.2.2 *Free variable does not occur explicitly*

Equations of the form

$$\frac{d^2x}{dt^2} = f\left(x, \frac{dx}{dt}\right)$$

can be solved using the variable substitution

$$y = dx/dt$$

which gives

$$\frac{d^2x}{dt^2} = \frac{dy}{dt} = \frac{dy}{dx}\frac{dx}{dt} = y\frac{dy}{dx}\,.$$

This variable substitution changes the equation into a first order equation in which the new free variable is the former dependent variable.

Example 4.11: Second order ODE where free variable does not occur explicitly.

Consider a second order reaction taking place in a flat catalytic pellet of thickness $2L$. A balance on the reactant becomes

$$D\frac{d^2C}{dx^2} = kC^2$$

where D is the diffusion coefficient of the reactant through the pellet and k is the reaction rate constant. The boundary conditions are

$$C = C_0 \text{ at } x = \pm L$$

$$\frac{dC}{dx} = 0 \text{ at } x = 0\,.$$

Using the substitution $y = dC/dx$ gives

$$\frac{d}{dx}\left(\frac{dC}{dx}\right) = \frac{dy}{dC}\frac{dC}{dx} = yy'$$

where prime indicates differentiation with respect to C. The variable transformation changes the former dependent variable into the new free variable

$$y\frac{dy}{dC} = \frac{k}{D}C^2$$

which upon separation gives

$$\int_0^{dC/dx} y\,dy = \frac{k}{D}\int_{C(0)}^{C(x)} C^2\,dC$$

where the boundary condition at the center has been applied. One gets

$$\frac{dC}{dx} = \pm\sqrt{\frac{2k}{3D}(C^3 - C^3(0))}$$

and further

$$\int_x^L dx = \pm\int_{C(x)}^{C_0} \frac{dC}{\sqrt{\frac{2k}{3D}(C^3 - C^3(0))}}\,.$$

4.2.3 *Homogeneous equations*

Homogeneous second order equations are equations of the form

$$t\frac{d^2x}{dt^2} = f\left(\frac{x}{t}, \frac{dx}{dt}\right).$$

Notice that each of the terms, $td^2x/dt^2, dx/dt, x/t$ have the same dimension. These equations are solved using the same variable transformation as the one used with homogeneous first order equations. Let $y = x/t$, giving

$$\frac{dx}{dt} = t\frac{dy}{dt} + y$$

$$\frac{d^2x}{dt^2} = t\frac{d^2y}{dt^2} + 2\frac{dy}{dt}\,.$$

When substituted into the homogeneous second order equation, the following is obtained

$$t^2\frac{d^2y}{dt^2} + 2t\frac{dy}{dt} = f\left(y, t\frac{dy}{dt} + y\right) \Rightarrow$$

$$t^2\frac{d^2y}{dt^2} = f\left(y, t\frac{dy}{dt} + y\right) - 2t\frac{dy}{dt} = g\left(y, t\frac{dy}{dt}\right).$$

This problem is now solved using the variable substitution $\tau = \ln(|t|)$ to get

$$t\frac{dy}{dt} = \frac{dy}{d\tau}$$

$$t^2\frac{d^2y}{dt^2} = \frac{d^2y}{d\tau^2} - \frac{dy}{d\tau}.$$

Substitution into the last form of the differential equation yields

$$\frac{d^2y}{d\tau^2} - \frac{dy}{d\tau} = g\left(y, \frac{dy}{d\tau}\right)$$

but this is an equation in which the free variable does not occur explicitly so it can be solved by the method outlined above.

Example 4.12: Homogeneous, second order equation.

Consider

$$t\frac{d^2x}{dt^2} = 4\frac{x}{t} + \frac{dx}{dt}.$$

First carrying out the variable substitution $y = x/t$ gives

$$t^2\frac{d^2y}{dt^2} + 2t\frac{dy}{dt} = 4y + t\frac{dy}{dt} + y \Rightarrow$$

$$t^2\frac{d^2y}{dt^2} = 5y - t\frac{dy}{dt}$$

then using $\tau = \ln(|t|)$ gives

$$\frac{d^2y}{d\tau^2} - \frac{dy}{d\tau} = 5y - \frac{dy}{d\tau} \Rightarrow$$

$$\frac{d^2y}{d\tau^2} = 5y \Rightarrow$$

$$y(\tau) = C_1e^{\sqrt{5}\tau} + C_2e^{-\sqrt{5}\tau}$$

back substituting the variables and simplifying gives the solution in the final form

$$x(t) = C_1t^{1+\sqrt{5}} + C_2t^{1-\sqrt{5}}.$$

4.2.4 *Constant coefficients and Euler equations*

Two more second order equations must be mentioned, linear equations with constant coefficients, equations that can be written in the form

$$\frac{d^2x}{dt^2} + a_1\frac{dx}{dt} + a_0x = f(t)$$

where a_1 and a_0 are constants, and Euler equations which are equations that can be written in the form

$$t^2\frac{d^2x}{dt^2} + a_1t\frac{dx}{dt} + a_0x = f(t).$$

Both of these are special second order instances of their general N'th order form and are covered in Section 4.3.

4.2.5 *Common transformations of linear second order equations*

As mentioned already, any homogeneous (homogeneous in this context means that "the right hand side" of the equation equals zero) linear second order ODE can be transformed into a Riccati equation. Given

$$a_2(t)\frac{d^2x}{dt^2} + a_1(t)\frac{dx}{dt} + a_0(t)x = 0$$

the variable substitution

$$\frac{dx}{dt} = xye^{-\int_{t_0}^{t} a_1(s)/a_2(s)ds}$$

gives rise to the following Riccati equation

$$\frac{dy}{dt} + e^{-\int_{t_0}^{t} a_1(s)/a_2(s)ds}y^2 = -\frac{a_0(t)}{a_2(t)e^{-\int_{t_0}^{t} a_1(s)/a_2(s)ds}}.$$

Conversely, any Riccati equation can be transformed into a linear homogeneous second order differential equation. The variable substitution

$$x = -\frac{1}{py}\frac{dy}{dt}$$

when substituted into

$$\frac{dx}{dt} = p(t)x^2 + q(t)x + r(t)$$

gives the horrid expression

$$-\frac{1}{py}\frac{d^2y}{dt^2} + \frac{dy}{dt}\left(\frac{p}{yp^2} + \frac{1}{y^2p}\frac{dy}{dt}\right) = \frac{p}{p^2y^2}\left(\frac{dy}{dt}\right)^2 - \frac{q}{py}\frac{dy}{dt} + r$$

which simplifies to the linear second order equation

$$p(t)\frac{d^2y}{dt^2} - (p'(t) + q(t)p(t))\frac{dy}{dt} + r(t)p^2(t)y(t) = 0\,.$$

A linear second order equation of the form

$$a_2(t)\frac{d^2x}{dt^2} + a_1(t)\frac{dx}{dt} + a_0(t)x = f(t) \tag{4.4}$$

where $a_2(t) \neq 0$ can be transformed by dividing the equation by $a_2(t)$, multiplying by $\exp(-\int_{t_0}^{t} \frac{a_1(s)}{a_2(s)} ds)$ and simplifying to obtain

$$\frac{dy}{dt}\left(e^{-\int_{t_0}^{t}\frac{a_1(s)}{a_2(s)}ds}\frac{dx}{dt}\right) + e^{-\int_{t_0}^{t}\frac{a_1(s)}{a_2(s)}ds}\frac{a_0(t)}{a_2(t)}x = e^{-\int_{t_0}^{t}\frac{a_1(s)}{a_2(s)}ds}\frac{f(t)}{a_2(t)}\,.$$

We will simplify the nomenclature and write this as

$$(p(t)x')' + q(t)x = f(t).$$

It is this form of the second order ordinary differential operator that appears in the so-called Sturm–Liouville problems, Section 9.4.1 problems that has been studied extensively and for which many solution properties are known.

Assuming now that $p(t)$ is such that

$$\tau(t) = \int_{t_0}^{t} \frac{1}{p(s)} ds$$

has an inverse $t(\tau)$, then

$$\frac{dx}{d\tau} = \frac{dx}{dt}\frac{dt}{d\tau} = p(t)\frac{dx}{dt}$$

and

$$\frac{d^2x}{d\tau^2} = \frac{d}{d\tau}\left(p(t)\frac{dx}{dt}\right) = \frac{d}{dt}\left(p(t)\frac{dx}{dt}\right)\frac{dt}{d\tau} = p(t)\frac{d}{dt}\left(p(t)\frac{dx}{dt}\right).$$

Using this result we obtain

$$\frac{d^2x}{d\tau^2} + p(t(\tau))q(t(\tau))x = p(t(\tau))f(t(\tau))$$

in which there is no first order derivative present in the equation. In some situations, this is a convenient way of posing the problem.

4.3 Higher order equations

The only higher order ODEs that can be easily solved are linear equations with constant coefficients and Euler's equation. The former are equations of the form

$$a_N \frac{d^N x}{dt^N} + a_{N-1} \frac{d^{N-1} x}{dt^{N-1}} + \cdots + a_1 \frac{dx}{dt} + a_0 x = q(t) \tag{4.5}$$

where the a_ns are constant.

Linear ODEs with constant coefficient are hugely important in engineering models and most readers are probably somewhat familiar with these types of equations. The theory of these equations will be covered in detail in chapter 8. Here, we will summarize the basic facts about the solution of these equations.

We will assume that all the constants, a_n, are real and that the right hand side of the equation, the function $q(t)$ in the expressions above, is zero. The complete solution to Eq. (4.5) is found by guessing a solution of the form $x(t) = e^{\lambda t}$ and substituting this expression into the differential equation. Upon simplification, one obtains the characteristic polynomial

$$a_N \lambda^N + a_{N-1} \lambda^{N-1} + \cdots + a_1 \lambda + a_0 x = 0 .$$

If λ_n is a root of this polynomial, the $e^{\lambda_n t}$ solves the differential equation. If the N roots are distinct, the complete solution to the differential equation is

$$x(t) \sum_{n=1}^{N} C_n e^{\lambda_n t}$$

where the C_ns are arbitrary constants. If the roots are not all distinct and the root λ_n has multiplicity m, the complete solution can be shown to be

$$x(t) = \sum_{n=1}^{M} \left(\sum_{m=1}^{m_n} C_{n,m} \exp(\lambda_n t) t^{m-1} \right) .$$

The complete solution when the right hand side $q(t)$ does not equal zero will not be covered here.

Example 4.13: Linear ODE with constant coefficients.

Consider the equation

$$\frac{d^6 x}{dt^6} - 3 \frac{d^5 x}{dt^5} - \frac{d^4 x}{dt^4} + 11 \frac{d^3 x}{dt^3} - 12 \frac{d^2 x}{dt^2} + 4 \frac{dx}{dt} = 0 .$$

The characteristic polynomial is

$$x^6 - 3x^5 - x^4 + 11x^3 - 12x^2 + 4x = x(x-1)^3(x-2)(x+2).$$

The roots of the polynomial are $x = 0, 1, 2, -2$. The root $x = 1$ has multiplicity 3 while the remaining roots have multiplicity 1. The complete solution to the ODE thus is

$$x = C_1 + C_{2,1}e^t + C_{2,2}te^t + C_{2,3}t^2e^t + C_3e^{2t} + C_4e^{-2t}.$$

In most engineering models, only real solutions are of interest. This makes complex roots of the characteristic polynomial a nuisance because they only yield a real solution if some of the arbitrary constants are complex. However, since we are assuming that the constants a_n are real, any complex roots of the characteristic polynomial must appear as complex conjugate pairs, $\lambda = \alpha + \beta i$ and $\bar{\lambda} = \alpha - \beta i$ (where an overbar indicates complex conjugation). To avoid having to work with complex arithmetic, it can be shown that a real solution is obtained if terms in the solution of the form $C_n t^m e^{\lambda t} + C_{\bar{n}} t^m e^{\bar{\lambda} t}$ are replaced by terms of the form $C_k t^m e^{\alpha t}\cos(\beta t) + C_l t^m e^{\alpha t}\sin(\beta t)$.

Example 4.14: Linear ODE with complex roots.

Consider the equation

$$\frac{d^4x}{dt^4} - 4\frac{d^3x}{dt^3} + 14\frac{d^2x}{dt^2} - 20\frac{dx}{dt} + 25x = 0$$

with the characteristic polynomial

$$\lambda^4 - 4\lambda^3 + 14\lambda^2 - 20\lambda + 25 = (x - 1 - 2i)^2(x - 1 + 2i)^2.$$

The complete real solution is

$$x(t) = C_1e^t\cos(2t) + C_2e^t\sin(2t) + C_3te^t\cos(2t) + C_4te^t\sin(2t)$$

where the C_ns are real.

Euler's equation is named after the Swiss mathematician Leonard Euler (1707–1783), one of the greats among mathematicians as well as the most productive mathematician ever. Naming this rather unimportant equation after him is almost misleading because it seems to imply that this is his only important contribution to differential equations. That is far from the truth. In fact, many of the basic concepts and ideas that are used to solve differential equations are due to Euler. This being said, Euler or Cauchy-Euler equations are equations of the form

$$a_N t^N \frac{d^N x}{dt^N} + a_{N-1}t^{N-1}\frac{d^{N-1}x}{dt^{N-1}} + \cdots + a_1 t\frac{dx}{dt} + a_0 x = q(t) \tag{4.6}$$

which is a linear Nth order equation in which the coefficient of the nth order term is proportional to the free variable raised to the Nth power. These equations are solved through use of the variable transformation $\tau = \ln(|t|)$ which gives rise to a linear equation with constant coefficients.

Example 4.15: Euler equation.

Consider

$$t^2\frac{d^2x}{dt^2} + 2t\frac{dx}{dt} - 5x = 0\,.$$

The suggested variable transformation gives

$$\frac{dx}{dt} = \frac{dx}{d\tau}\frac{d\tau}{dt} = \frac{1}{t}\frac{dx}{d\tau}$$

$$\frac{d^2x}{dt^2} = \frac{d}{dt}\left(\frac{1}{t}\frac{dx}{d\tau}\right) = \frac{1}{t^2}\frac{d^2x}{d\tau^2} - \frac{1}{t^2}\frac{dx}{d\tau}$$

which, when substituted into the differential equation, yields

$$\frac{d^2x}{d\tau^2} + \frac{dx}{d\tau} - 5x = 0$$

a linear differential equation with constant coefficient which is readily solved by the methods already given

$$x(\tau) = C_1e^{(-\frac{1}{2}-\frac{\sqrt{21}}{2})\tau} + C_2e^{(-\frac{1}{2}+\frac{\sqrt{21}}{2})\tau}$$

or

$$x(t) = C_1t^{-\frac{1}{2}-\frac{\sqrt{21}}{2}} + C_2t^{-\frac{1}{2}+\frac{\sqrt{21}}{2}}\,.$$

An alternative way to solve Euler equations is to note that functions of the form $x = t^m$ are obviously solutions and valid values of m can be found by substituting these functions directly into the Euler equations. After simplifications, a polynomial equation in m, known as the indicial equation or polynomial, is obtained and the valid values of m are the roots of the indicial polynomial.

Example 4.16: Euler equation.

Consider

$$t^3\frac{d^3x}{dt^3} - 3t^2\frac{d^2x}{dt^2} + 2t\frac{dx}{dt} = 0\,.$$

Substitute $x = t^m$ into this equation to get

$$m(m-1)(m-2)t^m - 3m(m-1)t^m + 2mt^m = 0 \;\Rightarrow$$

$$m(m-1)(m-2) - 3m(m-1) + 2m = m(m-3+\sqrt{2})(m-3-\sqrt{2}).$$

Hence

$$x(t) = C_1 + C_2 t^{3-\sqrt{2}} + C_3 t^{3+\sqrt{2}}.$$

This second method may appear as a simpler method than going through the variable transformation to the linear equation with constant coefficient, but it only works well if the roots of the characteristic polynomial all distinct. It does not find all the solutions when this is not the case. Also, complex roots may occur and extracting the real part of t^m requires that it is rewritten as $\exp(m\ln(t))$ so that Euler's formula can be applied. It is therefore usually worth the trouble of going through the variable transformation to the linear equation with constant coefficients.

Chapter 5

Finite Dimensional Vector Spaces

One usually first encounters a *vector* depicted as a directed line segment in Euclidean space, or what amounts to the same thing, as an ordered n-tuple of numbers of the form $\{a_i\}_{i=1,\ldots,n}$. The former is a geometrical description of a vector, whereas the latter can be interpreted as an algebraic representation with respect to a given coordinate system. However, the interplay between geometry and algebra is not transparent in more general vector spaces that could be infinite dimensional.

Finite dimensional vector spaces are relatively easy to work with, and most people have a well developed intuition for them. Infinite dimensional vector spaces can be considerably more difficult to appreciate, but they arise naturally in applications as function spaces. A function space consists of all functions from a given set X to either the real or complex numbers, and is finite dimensional precisely when the set X is a finite set. Hence the Euclidean space mentioned above can be viewed as the function space with $X = \{1, \ldots, n\}$. Infinite dimensional spaces require additional concepts in order to use them effectively, as we shall see in later chapters. But there are similarities between finite and infinite spaces. For example, solving partial differential equations by Hilbert space methods is a close analogue to a simpler method employed in the study of finite dimensional spaces. The objective of this chapter is to explore the basic properties of finite dimensional vector spaces from an abstract point of view so as to draw similarities and distinctions between finite versus infinite dimensional spaces.

5.1 Basic concepts

A vector space V is a set of objects (called *vectors*) in association with two operations called addition and scalar multiplication. The addition property says two vectors can be added together to form another vector, and scalar multiplication is that a vector can be multiplied by a scalar to form a new vector. These operations must satisfy some rather transparent axioms that are listed below. By a *scalar*, we mean an element of a *scalar field* $\mathbb{F}$, which in itself is a set of objects with the operations of addition and multiplication that satisfy a set of axioms. The precise statement of the field axioms is rather long and tedious, and since we shall only consider scalar fields that are either the set of real numbers $\mathbb{R}$ or the complex numbers $\mathbb{C}$, we shall not explicitly state these. Suffice to say that scalars are closed under the usual rules of addition, multiplication and division (except by 0) that are familiar to all. Thus, the rational numbers form a field while integers do not because division of two integers may produce a result which is not an integer. We say that rational numbers are closed under addition, multiplication and division while the integers are not. The field of rational numbers is indicated $\mathbb{Q}$, but $\mathbb{Q}$ is not generally encountered in applications since it is not *complete* (this will be elaborated on later). It should be noted that by referring to a vector space V, it is not just the elements of V that are relevant, but also the field $\mathbb{F}$. The terminology that V is a *real* (respectively, *complex*) vector space is used if the underlying field $\mathbb{F} = \mathbb{R}$ (respectively $\mathbb{F} = \mathbb{C}$) is particularly important. The precise definition of a *vector space* follows.

Definition: *A vector space V over the scalar field $\mathbb{F}$ is a non-empty set on which addition (+) and scalar multiplication is defined. This means that for each $v_1 \in V$ and $v_2 \in V$, there is a vector denoted by $v_1 + v_2$ that also belongs to V; and for each $v \in V$ and $\alpha \in \mathbb{F}$, there is a vector denoted by αv that also belongs to V. There is also a special vector in V denoted by 0. These operations satisfy the following axioms, where $v_1, v_2, v_3 \in V$ and $\alpha_1, \alpha_2 \in \mathbb{F}$:*

(i)	$v_1 + v_2 = v_2 + v_1$	*(addition is cumulative)*
(ii)	$(v_1 + v_2) + v_3 = v_1 + (v_2 + v_3)$	*(addition is associative)*
(iii)	$v_1 + 0 = v_1$	*(0 is the additive identity)*
(iv)	*there exists* $w \in V$ *so that* $v + w = 0$	*(existence of additive inverse)*
(v)	$\alpha_1(v_1 + v_2) = \alpha_1 v_1 + \alpha_1 v_2$	*(distributive law)*
(vi)	$(\alpha_1 + \alpha_2)v_1 = \alpha_1 v_1 + \alpha_2 v_1$	*(distributive law)*
(vii)	$\alpha_1(\alpha_2 v_1) = (\alpha_1 \alpha_2)v_1$	*(scalar multiplication is associative)*
(viii)	$1v_1 = v_1$	*(scalar identity)*

One should note that the symbol 0 can be used in two different ways: one usage is the additive identity in (iii), and the other as the zero element in the field. This should not cause any confusion. It is useful, if for nothing else than for simplifying the nomenclature, to denote the element w that appears in (*iv*) as $-v$. One can deduce from the distributive property of scalar multiplication that $-v = (-1)v$.

It is important to note that a vector space structure does not allow two vectors to be multiplied. At some later point, product operations will be introduced, but a general vector space does not have an inherent product. Also note that abstract vectors do not resemble row or column vectors, but rather a row or column vector description is an example or a notational convenience to represent a vector. We shall see this and more in the following examples.

5.2 Examples

We now list several examples to illustrate some contexts in which vector spaces occur. In all these examples, the field $\mathbb{F}$ can be either the real numbers $\mathbb{R}$ or the complex numbers $\mathbb{C}$.

1) Let $V = \mathbb{F}$ with $\mathbb{F}$ also as the underlying field. Then V is a vector space over $\mathbb{F}$ where the vector addition and scalar multiplication are just the field operations of addition and multiplication.
2) More generally, let N be a positive integer and V equal the set of an ordered N-tuple of elements of $\mathbb{F}$ written as a row. An element $v \in V$ has the form
$$v = (\alpha_1, \alpha_2, \ldots, \alpha_N)$$
where each $\alpha_i \in \mathbb{F}$, $n = 1, \ldots, N$. Addition in V is defined by adding the corresponding components: if $v_1 = (\alpha_1, \alpha_2, \ldots, \alpha_N)$ and $v_2 = (\beta_1, \beta_2, \ldots, \beta_N)$ belong to V, then
$$v_1 + v_2 = (\alpha_1 + \beta_1, \alpha_2 + \beta_2, \ldots, \alpha_N + \beta_N).$$
Scalar multiplication is defined by multiplying each component by the scalar: if $v = (\alpha_1, \alpha_2, \ldots, \alpha_N) \in V$ and $\alpha \in \mathbb{F}$, then
$$\alpha v = (\alpha\alpha_1, \alpha\alpha_2, \ldots, \alpha\alpha_N).$$
In a similar manner, one can consider the vector space of N-tuples written as a column with the same operations. It should cause no confusion if we use the same notation $\mathbb{F}^N$ to denote either the row or column space, although usually it denotes the column space. When the scalar field is the set of real numbers, these vectors can be interpreted geometrically as points in the N-dimensional Euclidean space $\mathbb{R}^N$. Actually one can consider $V = \mathbb{C}^N$ as a vector space over either field $\mathbb{R}$ or $\mathbb{C}$.
3) Let $V = \mathcal{M}_{MN}$ be the set of all $M \times N$ matrices, where M and N are positive integers and the elements come from the same scalar field as the scalars. The operations of addition and scalar multiplication are defined component wise in a manner analogous to Example 2. Often, elements $A \in \mathcal{M}_{MN}$ are thought of as maps $A : \mathbb{F}^N \to \mathbb{F}^M$ that take a column vector $v \in \mathbb{R}^N$ to the column vector $Av \in \mathbb{R}^M$ that is defined by the usual matrix multiplication (see Section 5.5.1). Note also that $\mathcal{M}_{MN}$ reduces to the space of N-dimensional row vectors if $M = 1$, and the space of M-dimensional column vectors if $N = 1$.
4) Let $V = \mathcal{P}$ be the set of all polynomials P of arbitrary degree N, where N is non-negative integer. These are of the form
$$P(t) = \alpha_0 + \alpha_1 t + \cdots + \alpha_N t^N,$$
where the coefficients $\alpha_1, \ldots, \alpha_N$ belong to $\mathbb{F}$. For fixed N, the space $\mathcal{P}_N$ consisting of all polynomials of degree less than or equal to N is also a vector space. The vector space operations are the usual ones of elementary algebra.

5) Suppose $1 \leq p < \infty$, and let V be the set of (infinite) sequences
$$v = \{\alpha_1, \alpha_2, \cdots, \alpha_n, \cdots\}$$
where
$$\|v\|_p := \left[\sum_{n=1}^{\infty} |\alpha_n|^p\right]^{\frac{1}{p}} < \infty.$$
The addition of two such sequences is done coordinate-wise, and scalar multiplication is similarly defined. The verification that V is a vector space requires the fact that the sum of two sequences in V also belongs to V. The proof of this fact is a direct consequence of Minkowski's inequality, which we do not prove, but states that
$$\|v_1 + v_2\|_p \leq \|v_1\|_p + \|v_2\|_p.$$
The vector space in this example is denoted by l_p.

6) The space $V = l_\infty$ is defined as the set of bounded sequences. That is, l_∞ consists of those sequences $v = \{\alpha_n\}_n$ for which[1]
$$\|v\|_\infty := \sup_{n \in \mathbb{N}}\{|\alpha_n|\} < \infty.$$
The space c_0 (respectively c_{00}) consists of the subset of l_∞ for which $\alpha_n \to 0$ as $n \to \infty$ (respectively, $\alpha_n = 0$ for all large n). Clearly the sum of two sequences from, respectively, l_∞, c_0, and c_{00} again belongs to l_∞, c_0, and c_{00}.

7) Suppose $I \subseteq \mathbb{R}$ is an interval of the real line, and let $V = C[I]$ be the set of all continuous, $\mathbb{F}$-valued functions defined on I. This again is a vector space with the usual operations of addition and scalar multiplication. For $k \in \mathbb{N}$, the space $C^k(I)$ of k-times continuously differentiable functions defined on I is likewise a vector space. Further examples could be given where I is instead a subset of $\mathbb{R}^n$, and/or the range of the function belongs to $\mathbb{F}^m$.

8) Suppose V and W are both vector spaces over the same field $\mathbb{F}$, and let $\mathcal{L}(V, W)$ consist of all the *linear transformations* from V to W (the term *linear operator* is also frequently used). That is, $L \in \mathcal{L}(V, W)$ provided
$$L(\alpha_1 v_1 + \alpha_2 v_2) = \alpha_1 L(v_1) + \alpha_2 L(v_2) \quad \forall v_1,\, v_2 \in V \text{ and } \alpha_1, \alpha_2 \in \mathbb{F}.$$
Note that the vector addition and scalar multiplication on the left side of the equation takes place in V, and on the right hand side takes place in W. The set $\mathcal{L}(V, W)$ forms a vector space over $\mathbb{F}$ by defining vector addition $L_1 + L_2$ and scalar multiplication αL as
$$(L_1 + L_2)(v) = L_1(v) + L_2(v) \quad \text{and} \quad (\alpha L)(v) = \alpha L(v)$$
for all $v \in V$ and $\alpha \in \mathbb{F}$. Again, note the multiple uses of the addition symbol "+" and scalar multiplication. One can verify these operations satisfy the axioms for $\mathcal{L}(V, W)$ to form a vector space. We shall study operators in greater depth in the next chapter.

[1] Here "sup" is short for supremum. The supremum of a set of real numbers A is defined as the smallest number, $\sup(A)$, such that $a \leq \sup(A)$, where $a \in A$.

5.3 Span, linear independence and basis

Let V be a given vector space over the field $\mathbb{F}$. By definition, the sum of two vectors is another vector, but it is immediate that the sum of any *finite* number of vectors is again a vector. Such a multiple sum is called a linear combination, and is an expression of the form

$$v = \alpha_1 v_1 + \alpha_2 v_2 + \cdots + \alpha_N v_N$$

where the α_ns are scalars and the v_ns are vectors. It is clear that for any given set of vectors $\mathcal{B} \subseteq V$, the set W consisting of all possible linear combinations of $\mathcal{B}$ again forms a vector space in its own right. The space W is called the *span* of $\mathcal{B}$, and denoted by $W = \text{span}\, \mathcal{B}$. Any set of vectors with the property that their span equals W is called a *spanning set* of W.

Let $\mathcal{B} \subseteq V$ be a set of vectors. Then $\mathcal{B}$ is called *linearly independent* if for any finite subset of N vectors $\{v_n\}$ of $\mathcal{B}$, the only linear combination of these vectors that equals zero is the one where the coefficients are all zero. In other words, if N is any positive integer and $v_n \in \mathcal{B}$, $\alpha_n \in \mathbb{F}$, and $\sum_{n=1}^{N} \alpha_n v_n = 0$, then $\alpha_n = 0$ for all $n = 1, \ldots, N$. The negation of this property is that a set $\mathcal{B}$ is *linearly dependent*, which means for some N there exists N vectors $\{v_i\} \subseteq \mathcal{B}$, and there exists N scalars α_n not all zero, such that $\sum_{n=1}^{N} \alpha_n v_n = 0$. It is clear that any set containing the vector 0 is dependent. It is also immediate that any non-empty subset of an independent set is independent.

Example 5.1: Linear dependence of vectors. Consider the vectors

$$v_1 = \begin{pmatrix} 1 \\ 3 \\ 1 \end{pmatrix}, \quad v_2 = \begin{pmatrix} -1 \\ -1 \\ -1 \end{pmatrix}, \quad v_3 = \begin{pmatrix} 2 \\ 5 \\ 3 \end{pmatrix}.$$

To check if these are independent, suppose there are scalars α_1, α_2 and α_3 for which

$$\alpha_1 v_1 + \alpha_2 v_2 + \alpha_3 v_3 = 0.$$

Writing out the equations component-wise leads to the following system of equations

$$\begin{aligned} \alpha_1 - \alpha_2 + 2\alpha_3 &= 0 \\ 3\alpha_1 - \alpha_2 + 5\alpha_3 &= 0 \\ \alpha_1 - \alpha_2 + 3\alpha_3 &= 0. \end{aligned}$$

We shall soon review Gauss elimination which is a technique that systematically solves such a system. In this case, it yields the unique solution is $\alpha_1 = \alpha_2 = \alpha_3 = 0$, which shows that the set of vectors $\{v_1, v_2, v_3\}$ is linearly independent.

As another example, consider the vectors

$$v_1 = \begin{pmatrix} 1 \\ 2 \\ 1 \end{pmatrix}, \quad v_2 = \begin{pmatrix} 0 \\ 1 \\ 3 \end{pmatrix}, \quad v_3 = \begin{pmatrix} 1 \\ 0 \\ -5 \end{pmatrix}.$$

One can directly verify that $v_1 - 2v_2 - v_3 = 0$. Thus, there are α_ns that can be chosen not all zero for which the linear combination is zero, and consequently, these vectors are linearly dependent.

The problem of linear independence is more complicated for functions. For instance, everyone knows that the integer powers of x are linearly independent but a rigorous proof is of course required. Consider the linear combination

$$\sum_{n=0}^{N} C_n x^n = 0 .$$

Differentiate this N times to get

$$C_N N! = 0 \Rightarrow C_N = 0 .$$

So the last coefficient in the linear combination must equal 0 and the last term can be dropped from the sum. The last term in the new sum has the coefficient C_{N-1} and, by the same argument, this coefficient must equal 0. Consequently, all the coefficient must equal 0 and the integer powers of x are thus linearly independent.

Example 5.2: Linear independence of the trig-functions.

To show that $\cos(x)$ and $\sin(x)$ are linearly independent, consider the linear combination

$$C_1 \sin(x) + C_2 \cos(x) = 0 .$$

Differentiating with respect to x gives

$$C_1 \cos(x) - C_2 \sin(x) = 0 .$$

Eliminating $\sin(x)$ gives

$$C_1 \frac{C_1}{C_2} \cos(x) + C_2 \cos(x) = 0 \Rightarrow$$

$$C_1^2 + C_2^2 = 0 \Rightarrow C_1 = C_2 = 0 .$$

The concepts of span and independence are brought together in the notion of a basis, which also leads to the definition of the dimension of a vector space.

Definition: *A set $\mathcal{B} \subseteq V$ of vectors is a basis for V if (i) $V = span\mathcal{B}$ and (ii) $\mathcal{B}$ is linearly independent.*

Definition: *A vector space is said to have dimension N if there exists a basis consisting of N vectors. If there is no basis consisting of finite many elements, then the space is said to be infinite dimensional.*

It can be shown (although we shall not do it here) that every vector space has a basis, which is true even if it is not possible to find a finite set of basis vectors. Our definition of a basis is more accurately called a *Hamel basis* since the definition of the span of a set of vectors allows for only a linear combination of finite many elements. It turns out that there is no distinction in finite dimensional spaces, but it is very pertinent in infinite dimensions, as we shall see later. It turns out that Hamel bases (which are purely of an *algebraic* nature) in infinite dimensions are typically useless for applications. An adequate theory can only be developed there if a notion of convergence (which is a *topological* notion) of a sequence of vectors is introduced.

For the dimension of a vector space to be well-defined, we must know that every basis has the same number of elements. This is the content of the next theorem.

Theorem 5.1. *Suppose $\{v_1, v_2, \ldots, v_N\}$ and $\{w_1, w_2, \ldots, w_M\}$ are both bases for the vector space V. Then $N = M$.*

Proof: Assume the result is false, and without loss of generality, that $M < N$. Consider now the set $\{w_1, v_1, v_2, \ldots, v_N\}$. Since the v_ns form a basis, we can write w_1 as a linear combination of these:

$$w_1 = \sum_{n=1}^{N} \alpha_n v_n \;\Rightarrow\; 0 = w_1 - \sum_{n=1}^{N} \alpha_n v_n \,.$$

Since w_1 is not the vector 0, it is immediate that not all the α_ns are zero. Reindexing the v_ns if necessary, assume that $\alpha_1 \neq 0$, then

$$v_1 = \frac{1}{\alpha_1} w_1 - \sum_{n=2}^{N} \frac{\alpha_n}{\alpha_1} v_n.$$

We conclude that any vector that can be written as a linear combination of $\{v_1, v_2, \ldots, v_N\}$ can equally be written as a linear combination of $\{w_1, v_2, \ldots, v_N\}$ and thus the latter set spans V. But this set is also independent: suppose there are scalars $\beta_1, \ldots, \beta_N$ satisfying

$$0 = \beta_1 w_1 + \sum_{n=2}^{N} \beta_n v_n = \beta_1 \alpha_1 v_1 + \sum_{n=2}^{N} (\beta_1 \alpha_n + \beta_n) v_n.$$

Then, since the v_ns are independent, we must have

$$0 = \beta_1 \alpha_1 = \beta_1 \alpha_2 + \beta_2 = \cdots = \beta_1 \alpha_N + \beta_N = 0.$$

Now $\alpha_1 \neq 0$, so it follows that $\beta_1 = 0$, and then subsequently each $\beta_n = 0$ for $n = 2, \ldots, N$. In other words, $\{w_1, v_2, \ldots, v_N\}$ is linearly independent, and so forms a basis for V.

We now want to repeat the argument and replace one of $v_2, \ldots, v_n$ by w_2. There exist scalars $\alpha_1, \ldots, \alpha_N$ so that

$$w_2 = \alpha_1 w_1 + \sum_{n=2}^{N} \alpha_n v_n,$$

and not all of $\alpha_2, \ldots, \alpha_N$ can be zero, since otherwise we have $w_1 - \alpha_1 w_2 = 0$ which contradicts the fact that the set $\{w_1, w_2\}$ is independent. Again by reindexing if necessary, we can assume $\alpha_2 \neq 0$, and by rearranging terms, we can write

$$v_2 = \frac{1}{\alpha_2} w_2 - \frac{\alpha_1}{\alpha_2} w_1 - \sum_{n=3}^{N} \frac{\alpha_n}{\alpha_2} v_n.$$

Thus in a similar manner as the argument above, we can conclude that the set $\{w_1, w_2, v_3, \ldots, v_N\}$ is a basis for V. After repeating this argument M times, we have that $\{w_1, w_2, w_3, \ldots, w_M, v_{M+1}, \ldots, v_N\}$ is then a basis. However, we are also assuming that $\{w_1, w_2, w_3, \ldots, w_M\}$ is a basis, and so v_{M+1} is a nontrivial linear combination of $\{w_1, w_2, w_3, \ldots, w_M\}$, which is a contradiction. □

The above argument also can be used in the infinite dimensional case, and one can conclude that if there exists one basis that is infinite, then every basis is infinite. It also follows from the previous theorem that in an N dimensional space, any set of N linearly independent vectors forms a basis. It is often easy to find at least one basis since the very description of the vector space actually invokes a particular coordinate system (this was the case in Examples 1–4 above). In such a case, it is therefore immediate to determine the dimension. However, in practice, one often desires a particular basis that has desirable properties that may differ from the obvious one.

We return to the previous examples of vector spaces and determine their dimensions.

1) The vector space $V = \mathbb{F}$ is one dimensional over the field $\mathbb{F}$, where a basis consists of just the multiplicative identity $\{1\}$. In fact, any non-zero element of the field is a basis.
2) The space $\mathbb{F}^N$ of ordered N-tuples (either considered as rows or columns) is N-dimensional over the field $\mathbb{F}$, which can be easily seen by noting that the set $\{e_1, \ldots, e_N\}$ is a basis, where the vector e_n is the element with 0s in all the positions except the nth where there is a 1. This basis is called the *canonical basis* or *usual set of basis vectors*. Note that there are many other bases, for example $\{e_1, e_1 + e_2, e_1 + e_2 + e_3, \ldots, e_1 + e_2 + \cdots + e_N\}$ is another basis.

   The complex space $\mathbb{C}^N$ can also be considered as a real vector space, in which the dimension is $2N$. A basis consists of $\{e_1, \ldots, e_N\} \cup \{ie_1, \ldots, ie_N\}$, where i is the purely imaginary number satisfying $i^2 = -1$.

3) The vector space $\mathcal{M}_{MN}$ of $M \times N$ matrices has dimension MN. The set of matrices $\{E_{mn} : m = 1, \ldots, M, n = 1, \ldots N\}$ is a basis, where E_{mn} denotes the matrix with 1 in the mth row and n^{th} column and zeros elsewhere.
4) The set of polynomials $\mathcal{P}$ is infinite dimensional, since any finite set of polynomials would have a maximum degree. In fact, $\mathcal{P}$ has a basis given by $\{1, t, t^2, \ldots, t^n, \ldots\}$. The subcollection of polynomials $\mathcal{P}_N$ that have degree less than or equal to N has dimension $N+1$, since the set of polynomials $\{1, t, t^2, \ldots, t^N\}$ forms a basis.

5–6) The sequence spaces in these examples are all infinite dimensional, but unlike the polynomial example, it is impossible to explicitly give a (Hamel) basis. On the face of it, it seems that the set of canonical sequences $\{e_n\}_n$ would form a basis, where e_n is the sequence with 1 in the nth slot and zeros elsewhere. However, with the only exception of c_{00}, this cannot be a basis because a sequence with infinitely many non-zero coordinate entries cannot be a *finite* sum of the e_ns.

7) All of the function spaces are infinite dimensional, but unlike the polynomials $\mathcal{P}$, there is no Hamel basis.
8) The dimension of the set of linear transformations $\mathcal{L}(V, W)$ is equal to the product of the dimensions of V and W, as we shall see later.

The following result reveals some properties of a set of linear independent vectors if the dimension of the space is *a priori* known.

Theorem 5.2. *Suppose V has dimension N and $\mathcal{B} = \{w_1, w_2, \ldots, w_M\}$ is an independent set. Then we have the following.*

(1) We must have $M \leq N$.
(2) If $M = N$, then $\mathcal{B}$ is a basis.
(3) If $M < N$, then there exist vectors $v_{M+1}, \ldots, v_N$ in V so that $\mathcal{B} \cup \{v_{M+1}, \ldots, v_N\}$ is a basis.

Proof: We refer back to the steps of the proof of the previous theorem, and use the consequences of related arguments. Suppose $\{v_1, \ldots, v_N\}$ is a basis for V. If $M > N$, then the above shows that after possibly reindexing the v_ns, the set $\{w_1, v_2, \ldots, v_N\}$ is also a basis. After carrying out this replacement $N-1$ more times (which only used the property that $\mathcal{B}$ was an independent set), we conclude that $\{w_1, \ldots, w_N\}$ is a basis. This contradicts that w_{N+1} is in the span while at the same time being independent of $\{w_1, \ldots, w_N\}$, and hence $M \leq N$.

If $M = N$ and $\mathcal{B}$ was not a basis, then $\mathcal{B}$ must not span V, and so there exists a vector $v \notin \operatorname{span} \mathcal{B}$. We claim that $\mathcal{B} \cup \{v\}$ is linear independent. Indeed, suppose $\sum_{n=1}^{N} \alpha_n w_n + \alpha_{N+1} v = 0$. If $\alpha_{N+1} \neq 0$, then $v = \sum_{n=1}^{N} \frac{\alpha_n}{\alpha_{N+1}} w_n$ belongs to span $\mathcal{B}$, and so we must have $\alpha_{N+1} = 0$. However, since $\mathcal{B}$ is independent, it now follows that all the α_ns are zero, and the claim is proven. So we have that $\mathcal{B} \cup \{v\}$ is an independent set of size $N+1$, which contradicts the assertion of part (1).

Finally, if $M < N$, by the previous theorem $\mathcal{B}$ is not a basis. Thus since it is independent, it must not span V, and there exists a vector v_{M+1} in V that is not in span $\mathcal{B}$. As in the previous paragraph, we see that $\mathcal{B} \cup \{v_{M+1}\}$ is linearly independent. This argument can be repeated a total of $N - M$ times to select vectors $v_{M+1}, \ldots v_N$, and we arrive at a set $\mathcal{B} \cup \{v_{M+1}, \ldots, v_N\}$ of size N that is independent and therefore must form a basis. □

The previous result shows that any independent set can be expanded to a basis. The next result shows that any spanning set can be contracted to a basis.

Theorem 5.3. *Suppose V is finite dimensional and is spanned by a set of vectors $\mathcal{B}$. Then there exists a subset $\tilde{\mathcal{B}} \subseteq \mathcal{B}$ that is a basis of V.*

Proof: We build up $\tilde{\mathcal{B}}$ by adding vectors to this set as follows. Let v_1 be any non-zero vector in $\mathcal{B}$ and first let $\tilde{\mathcal{B}} = \{v_1\}$. If $V = \text{span}\, \tilde{\mathcal{B}}$, then we are done. Inductively, suppose $\tilde{\mathcal{B}} = \{v_1, \ldots, v_k\}$ is a subset of $\mathcal{B}$ that is independent. If $V = \text{span}\, \tilde{\mathcal{B}}$, then since $\tilde{\mathcal{B}}$ is independent, it is a basis. If not, then there exists another non-zero vector $v_{k+1} \in \mathcal{B}$ that does not belong to span $\tilde{\mathcal{B}}$. This vector is now added to $\tilde{\mathcal{B}}$ to create a new $\tilde{\mathcal{B}}$ that is still independent. Eventually this process must stop since V is finite dimensional, and at that point $\tilde{\mathcal{B}}$ is the desired basis. □

5.3.1 *Coordinates*

If $\mathcal{B} = \{v_1, \ldots, v_N\}$ is a basis for a vector space V, then any vector can be written as a linear combination of the elements of $\mathcal{B}$. Suppose now that the basis $\mathcal{B}$ is ordered. Then the ordered set of coefficients which multiply the basis vectors completely define the vector of the linear combination, which is the content of the next result.

Theorem 5.4. *Suppose $\mathcal{B} = \{v_1, \ldots, v_N\}$ is an ordered basis and $v \in V$. Then there exists a unique element $[v]_{\mathcal{B}} = (\alpha_1, \ldots, \alpha_N) \in \mathbb{F}^N$ so that $v = \alpha_1 v_1 + \cdots + \alpha_N v_N$.*

Proof: Let $v \in V$. Since $\mathcal{B}$ spans V, there exists $(\alpha_1, \ldots, \alpha_N) \in \mathbb{F}^N$ so that $v = \alpha_1 v_1 + \cdots + \alpha_N v_N$. We must prove the uniqueness assertion. Suppose $(\beta_1, \ldots, \beta_N) \in \mathbb{F}^N$ also satisfies $v = \beta_1 v_1 + \cdots + \beta_N v_N$. Then

$$0 = v - v = (\alpha_1 - \beta_1)v_1 + \cdots + (\alpha_N - \beta_N)v_N,$$

and since $\mathcal{B}$ is independent, it follows that $\alpha_n = \beta_n$ for each $n = 1, \ldots, N$. Thus the representation is unique. □

The vector $[v]_{\mathcal{B}}$ of ordered coefficients is referred to as the *coordinates* or the *coordinate representation* of the vector relative to the basis $\mathcal{B}$. The previous theorem helps to clarify the concept of dimension: the dimension of a vector space is the least number of parameters that are needed to describe all its elements.

Let us consider $\mathbb{R}^2$ with ordered basis $\{e_1, e_2\}$. If $(a, b) \in \mathbb{R}^2$, then the coordinate representation of (a, b) in this basis is actually just the same vector (a, b). However, if a different basis is chosen, for instance $\mathcal{B} = \{(2, 1), (1, 0)\}$, then the coordinate representation is different. One can find $[(a, b)]_{\mathcal{B}}$ by writing the vector as a linear combination $(a, b) = \alpha_1(2, 1) + \alpha_2(1, 0)$ and solving the system of equations

$$\begin{aligned} \alpha_1 + \alpha_2 &= a \\ \alpha_1 &= b \end{aligned}$$

that this generates, from which one finds $\alpha_1 = b$ and $\alpha_2 = a - 2b$. The vector itself is therefore different from its coordinate representation. In fact, one should think of the original vector as just its representation in the usual basis, and be mindful that other bases could be used and are often preferred.

Here is another example in the vector space $\mathcal{P}_2$. Find the coordinate representation of the polynomial $1 + t + t^2$ in each of the two bases

$$\mathcal{B}_1 = \{1, t - 1, (t - 2)(t - 1)\} \text{ and } \mathcal{B}_2 = \{1, t, t^2\}.$$

Writing the vector as a linear combination of the basis vectors in $\mathcal{B}_1$, one obtains the equation

$$1 + t + t^2 = \alpha_1 + \alpha_2(t - 1) + \alpha_3(t - 2)(t - 1),$$

and expanding the right hand side gives

$$1 + t + t^2 = \alpha_1 - \alpha_2 + 2\alpha_3 + t(\alpha_2 - 3\alpha_3) + \alpha_3 t^2.$$

Comparing coefficients of terms of equal power on the two sides of the equation results in the three equations

$$\alpha_3 = 1\ ,\ \alpha_2 - 3\alpha_3 = 1\ ,\ \alpha_1 - \alpha_2 + 2\alpha_3 = 1$$

which are solved for the coordinates

$$\alpha_1 = 3\ ,\ \alpha_2 = 4\ ,\ \alpha_3 = 1.$$

Therefore the coordinate representation can be written simply as $[1 + t + t^2]_{\mathcal{B}_1} = (3, 4, 1)$. While this basis required some algebraic calculations, the other basis, $\mathcal{B}_2$, is more convenient, and one can see immediately that $[1 + t + t^2]_{\mathcal{B}_2} = (1, 1, 1)$.

5.4 Isomorphisms

5.4.1 *Isomorphisms of vector spaces*

Mathematics is full of objects that can be represented in various different ways. For instance, the multiplicative identity of real numbers can be written as 1, $\sin^2\theta + \cos^2\theta$, $\frac{7}{7}$, e^0, etc. A considerable portion of the mathematics used in engineering is an exercise in finding the representation of an object that is usable and practical. For instance, a differential equation really contains the same information as its solution, but the differential equation is not a very convenient form to use for most purposes.

In other situations one is faced with the problem of determining if two expressions are different representations of the same object. We have seen above that the same vector will have different coordinates for different bases. Another familiar example is the complex numbers $\mathbb{C} = \{x + iy : x, y \in \mathbb{R}\}$, which can naturally be identified with $\mathbb{R}^2 = \{(x, y) : x, y \in \mathbb{R}\}$. These are not the same spaces, but they exhibit exactly the same vector space properties since there is a manner to transfer all such properties from one space into the other.

Suppose V and W are vector spaces over the same field $\mathbb{F}$. Recall from Example 8 that a mapping $\Psi : V \to W$ is called a linear transformation (or an operator) if $\Psi(v_1 + v_2) = \Psi(v_1) + \Psi(v_2)$ for all $v_1, v_2 \in V$ and $\Psi(\alpha v) = \alpha\Psi(v)$ for $v \in V$ and $\alpha \in \mathbb{F}$. A linear operator is thus a mapping that preserves the vector space operations. If in addition Ψ is one-to-one ($\Psi(v_1) = \Psi(v_2) \Rightarrow v_1 = v_2$) and onto (for all $w \in W$, there exists $v \in V$ so that $\Psi(v) = w$), then Ψ is called an *isomorphism* from V to W. If such an isomorphism exists, then the two vector spaces V and W are said to be *isomorphic*, a property denoted by $V \simeq W$. Informally, two vector spaces are isomorphic if the only difference between them is the nomenclature used to designate the elements, where the isomorphism is the rule needed to change from one type of nomenclature to another.

It is straightforward to show that: (1) the identity map from V to V is an isomorphism; (2) if Ψ is an isomorphism from V to W, then its inverse Ψ^{-1} is well defined and an isomorphism from W to V; and (3) if Ψ_1 is an isomorphism from V_1 to V_2, and Ψ_2 is an isomorphism from V_2 to V_3, then its composition $\Psi_2 \circ \Psi_1$ is an isomorphism from V_1 to V_3. Thus the property of two vector spaces being isomorphic is a so called equivalence relation, a concept that will be discussed further in Section 5.7.3.1.

A key property of an isomorphism, and in fact a defining one, is that it maps a basis to a basis. This result is valid for infinite dimensional spaces as well, but we shall prove it only in finite dimensions.

Theorem 5.5. *Suppose V and W are finite dimensional vector spaces and $\Psi : V \to W$ is a linear transformation. Then the following are equivalent:*

(1) Ψ is an isomorphism.
(2) For any basis $\mathcal{B} = \{v_1, \ldots, v_N\}$ of V, the image $\mathcal{B}' = \{\Psi(v_1), \ldots, \Psi(v_N)\}$ is a basis for W.
(3) There exists a basis $\mathcal{B} = \{v_1, \ldots, v_N\}$ of V so that $\mathcal{B}' = \{\Psi(v_1), \ldots, \Psi(v_N)\}$ is a basis for W.

Proof: ((1)⇒(2)): Suppose $\Psi : V \to W$ is an isomorphism and $\mathcal{B} = \{v_1, \ldots, v_N\}$ is a basis of V. We must show $\mathcal{B}'$ is a basis for W.

To show $\mathcal{B}'$ is independent, suppose $\{\alpha_1, \ldots, \alpha_N\}$ are scalars satisfying $\sum_{n=1}^N \alpha_n \Psi(v_n) = 0$. Then $0 = \sum_{n=1}^N \alpha_n \Psi(v_n) = \Psi\left(\sum_{n=1}^N \alpha_n v_n\right)$ since Ψ is linear, hence $\sum_{n=1}^N \alpha_n v_n = 0$ since Ψ is one-to-one. Consequently $\alpha_1 = \cdots = \alpha_N = 0$

since $\mathcal{B}$ is independent, which shows $\mathcal{B}'$ is independent. To show $\mathcal{B}'$ spans W, let $w \in W$. Since Ψ is onto, there exists $v \in V$ with $\Psi(v) = w$. Since $\mathcal{B}$ spans V, there exists scalars $\alpha_1, \ldots, \alpha_N$ with $v = \sum_{n=1}^{N} \alpha_n v_n$, and we have $w = \Psi(v) = \Psi\left(\sum_{n=1}^{N} \alpha_n v_n\right) = \sum_{n=1}^{N} \alpha_n \Psi(v_n)$, and hence $\mathcal{B}'$ spans W.
((2)⇒(3)): Since every vector space has a basis, this implication is immediate.
((3)⇒(1)): Suppose $\Psi : V \to W$ is a linear transformation that maps the basis $\mathcal{B} = \{v_1, \ldots, v_N\}$ of V to a basis of W. We must show Ψ is one-to-one and onto. Suppose $v \in V$ is such that $\Psi(v) = 0$. There exist scalars $\alpha_1, \ldots, \alpha_N$ so that $v = \sum_{n=1}^{N} \alpha_n v_n$, and so $0 = \Psi(v) = \Psi\left(\sum_{n=1}^{N} \alpha_n v_n\right) = \sum_{n=1}^{N} \alpha_n \Psi(v_n)$. Since $\mathcal{B}'$ is independent, we have $\alpha_1 = \cdots = \alpha_N = 0$, or that $v = 0$. Hence Ψ is one-to-one. Now suppose $w \in W$. By assumption $\{\Psi(v_1), \ldots, \Psi(v_N)\}$ is a basis of W, and so there exist scalars $\alpha_1, \ldots, \alpha_N$ so that $w = \sum_{n=1}^{N} \alpha_n \Psi(v_n) = \Psi\left(\sum_{n=1}^{N} \alpha_n v_n\right)$. Thus $\Psi(v) = w$ where $v = \sum_{n=1}^{N} \alpha_n v_n$, and we conclude that Ψ is onto. □

An important consequence of the previous theorem is that any two finite dimensional vector spaces of the same dimension are isomorphic to each other.

Theorem 5.6. *Two finite dimensional vector spaces over the same field have the same dimension if and only if they are isomorphic.*

Proof: Suppose $\{\mathcal{B}\} = \{v_1, \ldots, v_N\}$ is a basis of V and $\{\mathcal{B}'\} = \{w_1, \ldots, w_N\}$ a basis of W. Define $\Psi : V \to W$ by $\Psi\left(\sum_{n=1}^{N} \alpha_n v_n\right) = \sum_{n=1}^{N} \alpha_n w_n$. Then Ψ is linear and maps the basis $\{\mathcal{B}\}$ to the basis $\{\mathcal{B}'\}$, and is therefore an isomorphism. □

This theorem in essence says that all vector calculations in an N-dimensional vector space can be carried out in the most natural of N-dimensional spaces. This space is $\mathbb{R}^N$ or $\mathbb{C}^N$ depending upon the underlying field. Thus, there is no need to investigate the properties of unusual and uncommon finite dimensional vector spaces because their vectorial properties can all be determined by investigating $\mathbb{F}^N$. This is the reason for focusing on $\mathbb{F}^N$ in the following sections.

It should be remembered, however, that only the algebraic operations of addition and scalar multiplication are preserved by an isomorphism. Other natural operations and properties of a particular space do not generally carry over in any obvious way. For instance, the complex numbers as a real vector space is isomorphic to $\mathbb{R}^2$. There is a natural way to multiply two complex numbers that is not so natural in $\mathbb{R}^2$. Another instance is to consider the space of polynomials $\mathcal{P}_N$, in which one could consider the roots of its members. The space $\mathcal{P}_N$ is isomorphic to the space $\mathbb{F}^{N+1}$, but on the face of it, it makes no sense to speak of a root of an element of $\mathbb{F}^{N+1}$.

Another caveat is in order for interpreting the above theorem. If two vector spaces have the same dimension, why not just say they are equal and dispense with the notion of isomorphism altogether? The reason is that different vector spaces may have coordinates representing very different concepts, and referring to them as equal would cause confusion by suggesting that coordinates and vectors from different spaces could be interchanged. For instance, in problems in mechanics, vectors have coordinates that represent positions, velocities, angular momentums, forces and torques. The position coordinates are very different from the velocities, and both of these are different from the forces, etc.

5.4.2 *Subspaces*

Given a vector space V, there are often smaller spaces that naturally sit inside V but that are vector spaces in their own right. These are called subspaces. For example, the physical world we inhabit and move around is three-dimensional, but often we work on, say, a piece of paper, which is two-dimensional. The concept of a subspace is intuitively natural if one thinks of a line or plane through the origin sitting inside as the Euclidean space $\mathbb{R}^3$. This idea is formalized in the following definition.

Definition: *Let W be a subset of the vector space V. Then W is a subspace of V if W is also a vector space in its own right with the operations of addition and scalar multiplication inherited from V.*

Since the operations of V satisfy the the axioms of a vector space, it is immediate that W is a subspace if and only if the addition of two vectors from W again belongs to W, and that scalar multiplication of a vector from W again belongs to W. These two criteria can be combined in one expression by noting that W is a subspace of V if and only if

$$w_1,\ w_2 \in W,\ \alpha \in \mathbb{F} \quad \text{implies} \quad \alpha w_1 + w_2 \in W.$$

Let us first note a few facts about every vector space V: (1) There are two trivial subspaces, namely $\{0\}$ and the whole space V; (2) If W_1 and W_2 are subspaces, then $W_1 \cap W_2$ is also a subspace; (3) If $\mathcal{B}$ is any set of vectors in V, then $W = \text{span}\ \mathcal{B}$ is a subspace; (4) If $\{v_1, \dots\}$ is a basis for V, then for each $n = 1, \dots, N$, the subspace $W_n = \text{span}\ \{v_1, \dots, v_n\}$ is a subspace of dimension n, and thus every vector space has subspaces of each dimension less than equal to the dimension of V.

We next give specific examples.

Consider the two-dimensional Euclidean plane $\mathbb{R}^2$, for which we write the elements as (x, y). Since $\mathbb{R}^2$ has dimension 2, the only subspaces that are not $\{0\}$ and $\mathbb{R}^2$ itself must have dimension 1. Now every subspace of dimension one has one basis element, so every nontrivial subspace has the form $\{\alpha(x_0, y_0) : \alpha \in \mathbb{R}\}$ where $(x_0, y_0) \in \mathbb{R}^2$ is a non-zero vector. Two obvious such subspaces are those corresponding to the usual coordinate directions, i.e. the subspaces consisting of the x-axis where $y_0 = 0$, and the y-axis where $x_0 = 0$. In general, the subspaces are straight lines through the origin with a slope of y_0/x_0.

In three-dimensional Euclidean space $\mathbb{R}^3$, the nontrivial subspaces are either one (lines) or two (planes) dimensional. Since every subspace must contain the origin, the one dimensional subspaces have the form $\{s(x_0, y_0, z_0) : s \in \mathbb{R}\}$ for some non-zero $(x_0, y_0, z_0) \in \mathbb{R}^3$, and the two-dimensional ones can be written as $\{(x, y, z) : s_1x + s_2y + s_3z = 0\}$ for some non-zero $(s_1, s_2, s_3) \in \mathbb{R}^3$. More specifically, the vectors $(0, 7, 1)$ and $(1, 2, 3)$ span a subspace of $\mathbb{R}^3$ of dimension 2. The parametric representation of this plane is $(s_2, 7s_1 + 2s_2, s_1 + 3s_2)$, where $s_1, s_2 \in \mathbb{R}$. This set of points can equivalently be written as all those points (x, y, z) satisfying $19x+y-7z = 0$.

Examples of function subspaces include the following. On the real line, let $\mathcal{P}_N$ $(N \geq 0)$ be the set of polynomials of degree less than or equal to N, $\mathcal{P}$ the set of polynomials of arbitrary degree, and C^k $(k \geq 0)$ the k-times continuously differentiable functions, where C^0 is just the set of continuous functions. It is clear that for each non-negative integers $N_1 < N_2$ and $k_1 < k_2$, we have the subset inclusions

$$\mathcal{P}_{N_1} \subset \mathcal{P}_{N_2} \subset \mathcal{P} \subset C^{k_2} \subset C^{k_1}.$$

However, since all of these are vector spaces, it is also the case that each smaller space is a subspace of the larger one.

Evidently, all subspaces must contain the zero element, so any two subspaces must have at least this element in common. If the zero element is the only element that two subspaces have in common, then the two subspaces are said to be *disjoint*. This terminology is justified by the following consideration. Suppose $\mathcal{B}$ is a linear independent set in V, $\mathcal{B}_1$ and $\mathcal{B}_2$ are subsets of $\mathcal{B}$, and $W_1 = \text{span } \mathcal{B}_1$, $W_2 = \text{span } \mathcal{B}_2$. Then the subspaces W_1 and W_2 are disjoint subspaces if and only if $\mathcal{B}_1$ and $\mathcal{B}_2$ are disjoint subsets.

5.4.3 *Sums*

Subspaces can be used to break apart a vector space into lower dimensional spaces, which may be easier to work with or have desirable properties. This is achieved using the idea of a sum of subspaces.

Definition: *If W_1 and W_2 are both subspaces of the same vector space V, the sum $W_1 + W_2$ of W_1 and W_2 is the set of all vectors of the form $v = w_1 + w_2$, where $w_n \in W_n$, $n = 1, 2$. If W_1 and W_2 are also disjoint and $V = W_1 + W_2$, then W_1 and W_2 are said to be complement to each other in V.*

We have thus far taken subspaces sitting inside a given vector space, but there is a related and complementary approach of starting with two vector spaces over the same field and combining them to form a new vector space.

Definition: *Let V_1 and V_2 be two vector spaces over the same field. The direct sum $V_1 \oplus V_2$ of V_1 and V_2 is the vector space with elements v that have the form*

$v = (v_1, v_2)$ of an ordered pair, where $v_n \in V_n$, $n = 1, 2$. Vector addition and scalar multiplication are naturally defined component wise:

$$\alpha(v_1, v_2) + \beta(w_1, w_2) = (\alpha v_1 + \beta w_1, \alpha 2 v_2 + \beta w_2).$$

Both of these definitions can be extended in an obvious way to involve finite sums of n spaces $V_1, \ldots, V_n$ instead of just 2.

Note that if $V = V_1 \oplus V_2$, then $W_1 = \{(v_1, 0) : v_1 \in V_1\}$ and $W_2 = \{(0, v_2) : v_2 \in V_2\}$ are disjoint subspaces of V that are complements. If $\mathcal{B}_1$ and $\mathcal{B}_1$ are bases for V_1 and V_2 respectively, then $\{(v_1, 0) : v_1 \in \mathcal{B}_1\}$ and $\{(0, v_2) : v_2 \in \mathcal{B}_2\}$ are bases for W_1 and W_2 respectively, and their union is a basis for V. It is therefore immediate that the dimension of V is the sum of the dimensions of W_1 and W_2. On the other hand, if W_1 and W_2 are complements in a vector space V, then one can form the direct sum $W_1 \oplus W_2$, which turns out to be isomorphic to V. Indeed, let $\Psi : W_1 \oplus W_2 \to V$ be given by $\Psi(w_1, w_2) = w_1 + w_2$. Clearly Ψ is linear, and it is onto since $V = W_1 + W_2$. To show that it is one-to-one, it suffices to show that each $v \in V$ can be *uniquely* written as $v = w_1 + w_2$, where $w_n \in W_n$, $n = 1, 2$. Suppose that $v = w_1' + w_2'$ is another representation. Then $0 = v - v = (w_1 + w_2 - w_1' - w_2')$, which implies $w_1 - w_1' = w_2' - w_2$. Since this element belongs to $W_1 \cap W_2 = \{0\}$, we conclude that $w_1 = w_1'$ and $w_2 = w_2'$. Finally, we have shown that Ψ is an isomorphism and $V \simeq W_1 \oplus W_2$.

The direct sum concept is related with the earlier discussion on coordinates in the following way. Suppose V is N-dimensional with $\mathcal{B} = \{v_1, \ldots, v_N\}$ as basis. For each $n = 1, \ldots, N$, let $W_n = \text{span}\{v_n\}$. Since $\mathcal{B}$ spans V, one has $V = W_1 + \ldots W_N$, and we have seen that $v \in V$ has a unique representation as $v = \alpha_1 v_1 + \cdots + \alpha_N v_N$ where $(\alpha_1, \ldots, \alpha_N) = [v]_{\mathcal{B}}$. Alternatively, we could form the direct sum $W_1 \oplus \cdots \oplus W_N$, which is defined as the set of ordered N-tuples of the form $(\alpha_1 v_1, \ldots, \alpha_N v_N)$. Thus the difference between the coordinate representation $[v]_{\mathcal{B}}$ and its direct sum representation is only the presence of writing the basis vectors in the latter.

The sum $W_1 + W_2$ of two vector subspaces W_1 and W_2 is of course defined and a subspace even if they are not disjoint. If $W_1 = \text{span } \mathcal{B}_1$ and $W_2 = \text{span } \mathcal{B}_2$, then $W_1 + W_2 = \text{span } \{\mathcal{B}_1, \mathcal{B}_2\}$. However, if $\mathcal{B}_n$ is a basis for W_n $(n = 1, 2)$, it is not necessarily the case that $\{\mathcal{B}_1, \mathcal{B}_2\}$ will be a basis for $W_1 + W_2$ since it may not be independent. However, one can still calculate the dimension of $W_1 + W_2$ from the dimensions of W_1, W_2, and $W_1 \cap W_2$.

For instance, consider the space $\mathbb{R}^3$ and the two subspaces W_1 and W_2 defined by

$$W_1 = \text{span}\{e_1, e_2\} \quad \text{and} \quad W_2 = \text{span}\{e_2, e_3\}.$$

These two subspaces are two-dimensional but not disjoint since $(0, 1, 0)$ is an element of both. In fact the sum is the entire space of dimension 3 and the intersection is the one-dimensional subspace span $\{e_2\}$. Notice that the direct sum $W_1 \oplus W_2$ gives a different non-isomorphic object which has dimension 4.

We see in this simple example that the dimension of the sum equals the sum of the dimensions of each term minus the dimension of the intersection. This turns out to hold in general, which is the content of the next theorem.

Theorem 5.7. *If W_1 and W_2 are finite dimensional subspaces of a vector space V, then*

$$\dim(W_1 + W_2) = \dim W_1 + \dim W_2 - \dim(W_1 \cap W_2).$$

Proof: Set $N_i = \dim W_n$ for $i = 1, 2$, and $M = \dim(W_1 \cap W_2)$. Let $\mathcal{B}_0 = \{u_1, \ldots, u_M\}$ be a basis for $W_1 \cap W_2$. Theorem 5.2(3) says that $\mathcal{B}_0$ can be extended to a basis $\mathcal{B}_1 = \mathcal{B}_0 \cup \{v_{M+1}, \ldots, v_{N_1}\}$ of W_1 and to a basis $\mathcal{B}_2 = \mathcal{B}_0 \cup \{w_{m+1}, \ldots, w_{N_2}\}$ of W_2. We claim that $\mathcal{B} = \mathcal{B}_0 \cup \{v_{M+1}, \ldots, v_{N_1}\} \cup \{w_{m+1}, \ldots, w_{N_2}\}$ is a basis for $W_1 + W_2$, which would prove the result.

To show linear independence of $\mathcal{B}$, suppose

$$\sum_{m=1}^{M} \mu_m u_m + \sum_{n=M+1}^{N_1} \alpha_n v_n + \sum_{n=M+1}^{N_2} \beta_n w_n = 0.$$

Then $\sum_{m=1}^{M} \mu_m u_m + \sum_{n=M+1}^{N_1} \alpha_n v_n = -\sum_{n=M+1}^{N_2} \beta_n w_n$ which belongs to $W_1 \cap W_2$, and therefore can also be written as $\sum_{m=1}^{M} \nu_m u_m$ since $\mathcal{B}_0$ spans $W_1 \cap W_2$. Hence $\sum_{m=1}^{M} (\mu_m - \nu_m) u_m + \sum_{n=M+1}^{N_1} \alpha_n v_n = 0$. Since $\mathcal{B}_1$ is independent, it follows that $\mu_m = \nu_m$ for $m = 1, \ldots, M$ and $\alpha_n = 0$ for $n = M+1, \ldots, N_1$. In a similar manner, one can replace the roles of W_1 and W_2 and conclude that $\beta_n = 0$ for $n = M+1, \ldots, N_2$. We now have that $\sum_{m=1}^{M} \mu_m u_m$, and since the u_ms are independent, it follows that $\mu_m = 0$ for $m = 1, \ldots, M$, which finishes the proof of linear independence.

We now show that $\mathcal{B}$ spans $W_1 + W_2$. Indeed, let $v = w_1 + w_2$ where $w_i \in W_i$. Since $\mathcal{B}_i$ is a basis for W_i, we can write

$$w_1 = \sum_{m=1}^{M} \mu_m u_m + \sum_{n=M+1}^{N_1} \alpha_n v_n \quad \text{and} \quad w_2 = \sum_{m=1}^{M} \nu_m u_m + \sum_{n=M+1}^{N_2} \alpha_n w_n$$

for an appropiate choice of scalars. Then we have

$$v = w_1 + w_2 = \sum_{m=1}^{M} (\mu_m + \nu_m) u_m + \sum_{n=M+1}^{N_1} \alpha_n v_n + \sum_{n=M+1}^{N_2} \alpha_n w_n$$

which shows $\mathcal{B}$ spans $W_1 + W_2$. □

Notice if the subspaces W_1 and W_2 are disjoint, then $W_1 \cap W_2 = \{0\}$, and the previous theorem then agrees with an earlier observation that the dimension of a direct sum is the sum of the dimensions. In fact, one sees that if

$$V = V_1 \oplus V_2 \oplus \cdots \oplus V_n$$

then $\dim V = \sum_{n=1}^{n} \dim V_n$.

It is not obvious yet of the advantage to do this, but it turns out to be a convenient way of describing a vector space in the context of eigenvalue problems which will be introduced in Section 5.7.

5.4.4 *Representation of subspaces*

Subspaces are typically indicated two different ways, either by specifying a set of basis vectors for the subspace or by a parametric representation of the subspace. The two types of representation are easily converted to one another. For instance, if a basis for a subspace is the set of vectors $\{v_n\}_{n=1,N}$, then any vector in the subspace can be written

$$v = \sum_{n=1}^{N} s_n v_n$$

which is a parametric representation in the terms of the parameters $\{s_n\}_{n=1,N}$. Similarly, a set of basis vectors can easily be extracted from a standard (soon to be defined) parametric representation. E.g.

$$v = \begin{pmatrix} s_1 \\ s_2 \\ s_1 - 5s_2 \\ 3s_1 + s_2 \end{pmatrix} = s_1 \begin{pmatrix} 1 \\ 0 \\ 1 \\ 3 \end{pmatrix} + s_2 \begin{pmatrix} 0 \\ 1 \\ -5 \\ 1 \end{pmatrix}$$

which gives the basis vectors $\begin{pmatrix} 1 \\ 0 \\ 1 \\ 3 \end{pmatrix}$ and $\begin{pmatrix} 0 \\ 1 \\ -5 \\ 1 \end{pmatrix}$.

Neither the basis vector representation or the parametric representation of a subspace are unique. To determine if two subspaces are in fact the same subspace, one can systematically check if all elements of the first basis are in the space spanned by the second basis and vice versa. Doing so is computationally expensive and an easier method will be described later after the concept of rank has been introduced. To determine if two parametric representations are of the same subspace, it is convenient to write the representations in the same standard form. For instance, write the vector in such a way that individual parameters appear as the earliest possible elements of the vector and all other elements are linear combinations of these parameters. I.e. as

$$V = \begin{pmatrix} s_1 \\ s_2 \\ s_1 - 2s_2 \\ s_3 \\ 3s_1 + s_3 \end{pmatrix}$$

as opposed to

$$V = \begin{pmatrix} s_1 + 2s_2 \\ s_2 \\ s_1 \\ s_3 \\ 3s_1 + 6s_2 + s_3 \end{pmatrix}.$$

That these two parametric representations are in fact of the same vector space can be shown by converting the second representation to the standard form. This conversion is essentially nothing but a change in nomenclature and is carried out as follows: consider the parametric representation

$$V = \begin{pmatrix} s_1 + s_4 \\ -s_1 + s_2 \\ s_1 + 2s_3 + s_4 \\ 2s_1 - s_2 + s_4 \\ s_2 + 2s_3 + s_4 \end{pmatrix}$$

and define a new parameter equal to the first element of the vector, e.g. $\hat{s}_1 = s_1 + s_4$. Now, eliminate the old parameter s_1, giving

$$V = \begin{pmatrix} \hat{s}_1 \\ -\hat{s}_1 + s_2 + s_4 \\ \hat{s}_1 + 2s_3 \\ 2\hat{s}_1 - s_2 - s_4 \\ s_2 + 2s_3 + s_4 \end{pmatrix}.$$

Define $\hat{s}_2 = -\hat{s}_1 + s_2 + s_4$ and eliminate s_2 to get

$$V = \begin{pmatrix} \hat{s}_1 \\ \hat{s}_2 \\ \hat{s}_1 + 2s_3 \\ \hat{s}_1 - \hat{s}_2 \\ \hat{s}_2 + \hat{s}_1 + 2s_3 \end{pmatrix}.$$

Define $\hat{s}_3 = \hat{s}_1 + 2s_3$ and eliminate s_3

$$V = \begin{pmatrix} \hat{s}_1 \\ \hat{s}_2 \\ \hat{s}_3 \\ \hat{s}_1 - \hat{s}_2 \\ \hat{s}_2 + \hat{s}_3 \end{pmatrix} = \begin{pmatrix} s_1 \\ s_2 \\ s_3 \\ s_1 - s_2 \\ s_2 + s_3 \end{pmatrix}.$$

Notice that the number of parameters is also reduced from four to three.

When a vector space is given in the standard parametric form above, the dimension of the space equals the number of parameters.

Claim: *A vector space, represented in standard parametric form and using M parameters, $s_1, \cdots, s_M$, is M-dimensional.*

Proof: For notational convenience we can assume, without loss of generality, that the elements that equal the parameters $s_1, \cdots, s_M$ are the first M elements of the vector. The parametric representation therefore has the form

$$V = \begin{pmatrix} s_1 \\ \vdots \\ s_M \\ \sum_{n=1}^{M} a_{M+1,n} s_n \\ \vdots \\ \sum_{n=1}^{M} a_{N,n} s_n \end{pmatrix}.$$

To obtain M linearly independent vectors, let $s_n = 1$ for $n = 1, \cdots, M$, while all other s's are zero. All other vectors are readily seen to be linear combinations of these M vectors.

$$V = \begin{pmatrix} s_1 \\ 0 \\ \vdots \\ s_M \\ \sum_{n=1}^{M} a_{M+1,n} s_n \\ \vdots \\ \sum_{n=1}^{M} a_{N,n} s_n \end{pmatrix} = s_1 \begin{pmatrix} 1 \\ 0 \\ \vdots \\ 0 \\ a_{M+1,1} \\ \vdots \\ a_{N,1} \end{pmatrix} + s_2 \begin{pmatrix} 0 \\ 1 \\ \vdots \\ 0 \\ a_{M+1,2} \\ \vdots \\ a_{N,2} \end{pmatrix} + \cdots + s_M \begin{pmatrix} 0 \\ 0 \\ \vdots \\ 1 \\ a_{M+1,M} \\ \vdots \\ a_{N,M} \end{pmatrix}.$$

□

5.5 Matrices

Matrices are ordered sets of numbers of the form $\{a_{mn}\}$ where the indices range from $m = 1, \ldots, M$ and $n = 1, \ldots, N$. The first subscript of an element is the row number and the second subscript is the column number, and the matrix is said to have M rows and N columns. Matrices, if they are not too large, are usually written as tables of numbers

$$A = \begin{pmatrix} a_{11} & a_{12} & \cdots & a_{1N} \\ a_{21} & a_{22} & \cdots & a_{2N} \\ \vdots & \vdots & \ddots & \vdots \\ a_{M1} & a_{M2} & \cdots & a_{MN} \end{pmatrix}.$$

The *main diagonal* of a matrix are the elements with the same row and column index, a_{nn}. When the number of rows equals the number of columns a matrix is said to be *square*. A square matrix is called a *diagonal matrix* if the elements outside the main diagonal are zero. A square matrix is said to be *upper triangular* if all elements below the main diagonal (that is, those a_{mn} with $m < n$) are zero. Similarly a *lower triangular* matrix is a square matrix with zeros in the elements above the diagonal. The *trace* of a matrix, $\mathrm{tr}(A)$, is the sum of the diagonal terms.

The *transpose* of a matrix A is written A^T and is the matrix one obtains by interchanging the rows and columns of A, i.e. $\{a_{n,m}\}^T = \{a_{m,n}\}$ or

$$\begin{pmatrix} a_{11} & \cdots & a_{1N} \\ \vdots & \ddots & \vdots \\ a_{M1} & \cdots & a_{MN} \end{pmatrix}^T = \begin{pmatrix} a_{11} & \cdots & a_{M1} \\ \vdots & \ddots & \vdots \\ a_{1N} & \cdots & a_{MN} \end{pmatrix}.$$

A matrix is said to be *symmetric* if it is equal to its own transpose, which of course can happen only for square matrices. A matrix is *antisymmetric* or *skew-symmetric* if it equals minus its transpose. Only square matrices with zeros in the diagonal elements can be antisymmetric. The *identity matrix* is a square matrix with 1s on the main diagonal and zeroes elsewhere.

$$I = \begin{pmatrix} 1 & 0 & \cdots & 0 \\ 0 & 1 & \cdots & 0 \\ \vdots & \vdots & \ddots & \vdots \\ 0 & 0 & \cdots & 1 \end{pmatrix}.$$

If the dimension of the identity matrix is relevant, then we write I_N for the square matrix of size N.

5.5.1 *Matrix algebra*

As mentioned earlier, matrices with the same number of M rows and N columns can be added element by element

$$\{a_{mn}\} + \{b_{mn}\} = \{a_{mn} + b_{mn}\}.$$

Moreover, multiplication of a matrix by a scalar is defined by $\alpha\{a_{mn}\} = \{\alpha a_{mn}\}$. These operations satisfy the axioms of a vector space, and in fact one can easily produce an isomorphism between $\mathcal{M}_{MN}$ and $\mathbb{F}^{MN}$, thus $\mathcal{M}_{MN}$ has dimension MN.

In addition to the vector space structure of $\mathcal{M}_{MN}$, matrices can be multiplied if the first matrix has the same number of columns as the second has rows. Suppose $A = \{a_{mn}\} \in \mathcal{M}_{MN}$ and $B = \{b_{nk}\} \in \mathcal{M}_{NK}$ are two matrices. Their product $C = \{c_{mk}\} = AB$ belongs to $\mathcal{M}_{MK}$ and is defined by

$$c_{mk} = \sum_{n=1}^{N} a_{mn} b_{nk}.$$

One should note that even if $M = N = K$, the product AB will generally be different from BA, which means that multiplication is not commutative. Matrix-vector multiplication is defined by interpreting a vector in $\mathbb{F}^N$ as a $1 \times N$ or $N \times 1$ matrix (usually called row and column vectors respectively).

The following properties hold, where in each case we assume the dimensions of the matrix are such that the operations are defined.

(i)	$C(A+B) = CA + CB$	(left distributive law)
(ii)	$(A+B)C = AC + BC$	(right distributive law)
(iii)	$A(BC) = (AB)C$	(multiplication is associative)
(iv)	If $A \in \mathcal{M}_{MN}$ then $I_M A = A = A I_N$	(identity)
(v)	$(AB)^T = B^T A^T$	(transpose of product)

Suppose $A \in \mathcal{M}_{MN}$. Then A has a left inverse if there exists a matrix $A_l^{-1} \in \mathcal{M}_{MM}$ satisfying $A_l^{-1} A = I_M$, and has a right inverse if there exists $A_r^{-1} \in \mathcal{M}_{NN}$ satisfying $A A_r^{-1} = I_N$.

Let us note that the columns of an $M \times N$ matrix are linearly independent vectors in $\mathbb{F}^M$ if and only if whenever $Av = 0$ for $v \in \mathbb{F}^N$, then necessarily $v = 0$; this is just another way of writing the definition of linear independence in terms of matrix multiplication. If A has a left inverse A_l^{-1}, then $Av = 0$ implies $v = I_N v = A_l^{-1} A v = 0$, and hence the columns of A are independent. In particular, the existence of a left inverse implies $N \leq M$.

We can apply the same reasoning to the transpose. The transpose of product rule implies that if A has a right inverse A_r^{-1}, then A^T has $(A_r^{-1})^T$ as a left inverse. Hence in this case, the columns of A^T are independent, which is equivalent to saying the rows of A are independent, and in particular $M \leq N$.

If a matrix has both a left and a right inverse, then it follows from above that $M = N$. In fact the left and right inverses must be equal, as can be seen from the following short calculation

$$A_l^{-1} = A_l^{-1}(A A_r^{-1}) = (A_l^{-1} A) A_r^{-1} = A_r^{-1}.$$

A square matrix A is called invertible if it has a left and right inverse, and since these must be the same matrix, we write it simply as A^{-1}. We shall see later that for a square matrix, the existence of a right inverse implies the existence of a left one, and vice versa.

We emphasize that not every non-zero matrix has a multiplicative inverse, which is unlike ordinary numbers. For example, the matrix

$$A = \begin{pmatrix} 1 & 0 \\ 0 & 0 \end{pmatrix}$$

has no right inverse because the second row of the product AB will be all zeros for any matrix B. It also has no left inverse because the second column of the product BA will be all zeros for any matrix B.

If the inverse of a square matrix does exist, it is a useful tool for solving matrix problems. For instance, a set of coupled, linear equations with the same number of unknowns as equations can be written as $Ax = b$, so the solution is simply obtained by premultiplying each side by the inverse of A: $x = A^{-1}b$. Something more general is true as well for any $A \in \mathcal{M}_{MN}$ and $b \in \mathbb{F}^M$. Namely, x solves $Ax = b$ if and only if x solves the system $SAx = Sb$ for any $M \times M$ invertible square matrix S. The goal, then, is to find an invertible matrix S so that the solutions of $SAx = Sb$ are very simple to find.

5.5.2 *Gauss elimination*

We now explore systematically how to solve a linear system $Av = b$ of algebraic equations. Suppose A is an $M \times N$ matrix with entries (a_{mn}) and $b \in \mathbb{R}^M$ with components b_m. We seek to find $v \in \mathbb{F}^N$ with components x_n that will satisfy the system of equations $Av = b$, which written out in full looks like

$$\begin{aligned}
a_{11}x_1 + \ a_{12}x_2 + \cdots + \ a_{1n}x_n \ + \cdots + \ a_{1N}x_N &= b_1 \\
a_{21}x_1 + \ a_{22}x_2 + \cdots + \ a_{2n}x_n \ + \cdots + \ a_{2N}x_N &= b_2 \\
\vdots \qquad\qquad\qquad \ddots \qquad\qquad\qquad &\ \ \vdots \\
a_{m1}x_1 + a_{m2}x_2 + \cdots + a_{mn}x_n + \cdots + a_{mN}x_N &= b_m \\
\vdots \qquad\qquad\qquad \ddots \qquad\qquad\qquad &\ \ \vdots \\
a_{M1}x_1 + a_{M2}x_2 + \cdots + a_{Mn}x_n + \cdots + a_{MN}x_N &= b_M \,.
\end{aligned}$$

There may be no solution at all; for example, if all $a_{mn} = 0$ and one of the $b_m \neq 0$, there obviously cannot be a solution. Or there may be infinitely many solutions; for example, with $M = 1$, $N = 2$, $a_{11} = a_{12} = 1$, and $b_1 = 0$, then any $v \in \mathbb{R}^2$ with $x_1 = -x_2$ solves the system. *Gauss elimination* is a procedure that routinely converts a system of equations into an equivalent system (that is, it has exactly the same set of solutions) which is amenable to determining if solutions exist, and if so, to finding them.

It is usually convenient to save notation by identifying the system above with the *augmented* matrix

$$\left(\begin{array}{cccc|c} a_{11} & a_{12} & \cdots & a_{1N} & b_1 \\ a_{21} & a_{22} & \cdots & a_{2N} & b_2 \\ \vdots & & \ddots & & \vdots \\ a_{m1} & a_{m2} & \cdots & a_{mN} & b_m \\ \vdots & & \ddots & & \vdots \\ a_{M1} & a_{M2} & \cdots & a_{MN} & b_M \end{array}\right).$$

Notice the nth column of the augmented matrix corresponds to the coefficients of the x_n variable, and the mth row corresponds to the mth equation of the original system.We shall do matrix manipulations on the augmented matrix with the tacit understanding that it corresponds to a system of equations.

There are two main steps to Gauss elimination. The first step is called forward substitution and it converts the original system into an equivalent system with 0s and 1s strategically placed in the matrix. The second step, called backsubstitution or backsolving, reads off the solutions from the new system.

First, let us see with a few examples how the second step works. The system that is easiest to solve would have the form

$$\left(\begin{array}{ccc|c} 1 & 0 & 0 & 2 \\ 0 & 1 & 0 & 1 \\ 0 & 0 & 1 & -2 \end{array}\right)$$

since in this case the equation associated to each row gives the solution: $x_1 = 2$, $x_2 = 1$, $x_3 = -2$. In this case there is a unique solution to the system.

Consider next the augmented matrix

$$\left(\begin{array}{ccccc|c} 1 & -1 & 0 & 0 & 1 & 2 \\ 0 & 0 & 1 & 0 & 3 & 1 \\ 0 & 0 & 0 & 1 & -1 & -2 \end{array}\right)$$

that is associated with a system of three equations and five unknowns. There is a cascading feature to this matrix which has a so-called leading "1" in each row, and all the other entries in the columns with those 1s are 0. This is ideally suited to finding all solutions to the system of equations in the following way. Notice that the third equation says that $x_4 - x_5 = -2$, and so if x_5 is allowed to be an arbitrary value s_1, then the equation is satisfied when $x_4 = -2 + s_1$. Now the second equation is $x_3 + 3x_5 = 1$, and since we have already set $x_5 = s_1$, this equation will be satisfied when $x_3 = 1 - 3s_1$. Finally, the first equation is $x_1 - x_2 + x_5 = 2$. We already have set $x_5 = s_1$, but now we have a new choice for x_2, and so we set $x_2 = s_2$, another arbitrary parameter. Then the first equation holds when $x_1 = 2 + s_2 - s_1$. Hence for any choice of s_1 and s_2, any vector of the form

$$\begin{pmatrix} x_1 \\ x_2 \\ x_3 \\ x_4 \\ x_5 \end{pmatrix} = \begin{pmatrix} 2 + s_2 - s_1 \\ s_2 \\ 1 - 3s_1 \\ -2 + s_1 \\ s_1 \end{pmatrix} = \begin{pmatrix} 2 \\ 0 \\ 1 \\ -2 \\ 0 \end{pmatrix} + s_1 \begin{pmatrix} -1 \\ 0 \\ -3 \\ 1 \\ 1 \end{pmatrix} + s_2 \begin{pmatrix} 1 \\ 1 \\ 0 \\ 0 \\ 0 \end{pmatrix}$$

is a solution to the system, and every solution is of this form.

An augmented matrix is in *row-reduced echelon form* if it has the structure of this example. More specifically, the first non-zero entry of each row should be 1, and every other entry in that corresponding column should be 0; this last requirement is not always necessary for "backsolving" routines, but it simplifies the formula for representing solutions. Furthermore, these leading entries should appear in an increasing column as the row increases, and 0s should be in every entry to the bottom left of these leading 1s.

The goal of the first step in Gauss elimination is to transform the original system into one that is in row-echelon form. The transformation consists of repeatedly using one of the following three types of row operations:

1) Switching two rows.
2) Replacing a row by a non-zero scalar multiple of that row.
3) Replacing a row by adding it to the elements of a non-zero multiple of another row.

It is important to realize that these operations do not change the solution set, and thus the solutions of the original system can be found easily after it has been transformed into row-echelon form.

Again, it is perhaps easiest to illustrate the procedure with an example. Consider the augmented system

$$\left(\begin{array}{ccccc|c} 0 & 0 & 3 & -1 & 10 & 5 \\ 2 & -2 & 1 & 1 & 4 & 3 \\ 1 & -1 & 0 & 1 & 0 & 0 \end{array}\right).$$

The nth row of the matrix is written as R_n, and the arrow $\Rightarrow$ indicates that the indicated row operations are being performed to transform the matrix into another one that is equivalent. Here is Gauss elimination in action:

$$\left(\begin{array}{ccccc|c} 0 & 0 & 3 & -1 & 10 & 5 \\ 2 & -2 & 1 & 1 & 4 & 3 \\ 1 & -1 & 0 & 1 & 0 & 0 \end{array}\right) \overset{R_1 \leftrightarrow R_3}{\Rightarrow} \left(\begin{array}{ccccc|c} 1 & -1 & 0 & 1 & 0 & 0 \\ 2 & -2 & 1 & 1 & 4 & 3 \\ 0 & 0 & 3 & -1 & 10 & 5 \end{array}\right)$$

$$\overset{-2R_1+R_2 \to R_2}{\Rightarrow} \left(\begin{array}{ccccc|c} 1 & -1 & 0 & 1 & 0 & 0 \\ 0 & 0 & 1 & -1 & 4 & 3 \\ 0 & 0 & 3 & -1 & 10 & 5 \end{array}\right) \overset{-3R_2+R_3 \to R_3}{\Rightarrow} \left(\begin{array}{ccccc|c} 1 & -1 & 0 & 1 & 0 & 0 \\ 0 & 0 & 1 & -1 & 4 & 3 \\ 0 & 0 & 0 & 2 & -2 & -4 \end{array}\right)$$

$$\overset{\frac{1}{2}R_3 \to R_3}{\Rightarrow} \left(\begin{array}{ccccc|c} 1 & -1 & 0 & 1 & 0 & 0 \\ 0 & 0 & 1 & -1 & 4 & 3 \\ 0 & 0 & 0 & 1 & -1 & -2 \end{array}\right) \overset{\substack{-R_3+R_1 \to R_1 \\ R_3+R_2 \to R_2}}{\Rightarrow} \left(\begin{array}{ccccc|c} 1 & -1 & 0 & 0 & 1 & 2 \\ 0 & 0 & 1 & 0 & 3 & 1 \\ 0 & 0 & 0 & 1 & -1 & -2 \end{array}\right).$$

The final matrix is in row echelon form and can be solved by backsolving as was done earlier.

There are many instances where a system $Av = b$ needs to be solved with $b = 0$. In this case, the final column of the augmented matrix consists of all 0s, and each new augmented matrix obtained after performing a row operation does not change the final column. For notational simplicity, then, the final column of 0s is often omitted in the calculations.

Suppose Gauss elimination is performed on a system $Av = b$ where A is a square matrix. If the associated row-echelon matrix is of the form $[I|c]$ where I is the identity matrix and $c \in \mathbb{F}^N$, then the unique solution to the system is $v = c$. There are often situations where one wants to solve several systems of linear equations with a square system matrix but different right hand sides, say $Av^k = b^k$, $k = 1, \ldots, K$. This can be done most effectively by augmenting the system matrix with all the right hand sides, where the kth right hand side becomes the $N + k$th column of the augmented matrix. This would have the form $[A|b^1 \ldots b^K]$. In the same way then, the enlarged augmented matrix is transformed by Gauss elimination to a matrix of the form $[I|c^1 \ldots c^K]$, and the solution corresponding to the system $Av = b^k$ is $v = c^k$.

A particular implementation of the preceding paragraph leads to determining if the inverse of a square matrix A exists, and finding it if it does. Recall that the inverse A^{-1} satisfies $AA^{-1} = I$, and thus the kth column of A^{-1} is nothing but the solution of $Av^k = b^k$ where all components of b^k are to equal zero except the kth which equals 1. The considered augmented matrix has the the form $[A|I]$. When subjected to Gauss elimination, if the result is a transformed system of the form $[I|C]$, then the matrix C is the inverse A^{-1}. If Gauss elimination instead leads to a row of zeroes on the left side, then the conclusion is that the inverse of A does not exist since the right hand side will not be zero. The method is illustrated in the following example.

Example 5.3: Calculation of matrix inverse.

Consider the matrix

$$A = \begin{pmatrix} 1 & 2 & 0 \\ 1 & 2 & -1 \\ 0 & 4 & 8 \end{pmatrix},$$

for which we seek its inverse, if it exists. We perform Gauss elimination:

$$\left(\begin{array}{ccc|ccc} 1 & 2 & 0 & 1 & 0 & 0 \\ 1 & 2 & -1 & 0 & 1 & 0 \\ 0 & 4 & 8 & 0 & 0 & 1 \end{array}\right) \overset{-R_1+R_2\to R_2}{\Longrightarrow} \left(\begin{array}{ccc|ccc} 1 & 2 & 0 & 1 & 0 & 0 \\ 0 & 4 & 8 & 0 & 0 & 1 \\ 0 & 0 & -1 & -1 & 1 & 0 \end{array}\right) \overset{\substack{\frac{1}{4}R_2\to R_2 \\ -R_3\to R_3}}{\Longrightarrow} \left(\begin{array}{ccc|ccc} 1 & 2 & 0 & 1 & 0 & 0 \\ 0 & 1 & 2 & 0 & 0 & 1/4 \\ 0 & 0 & 1 & 1 & -1 & 0 \end{array}\right)$$

$$\overset{-2R_2+R_1\to R_1}{\Longrightarrow} \left(\begin{array}{ccc|ccc} 1 & 0 & -4 & 1 & 0 & -1/2 \\ 0 & 1 & 2 & 0 & 0 & 1/4 \\ 0 & 0 & 1 & 1 & -1 & 0 \end{array}\right) \overset{\substack{4R_3+R_1\to R_1 \\ -2R_3+R_2\to R_2}}{\Longrightarrow} \left(\begin{array}{ccc|ccc} 1 & 0 & 0 & 5 & -4 & -1/2 \\ 0 & 1 & 0 & -2 & 2 & 1/4 \\ 0 & 0 & 1 & 1 & -1 & 0 \end{array}\right).$$

The conclusion is that A^{-1} exists and

$$A^{-1} = \frac{1}{4}\begin{pmatrix} 20 & -16 & -2 \\ -8 & 8 & 1 \\ 4 & -4 & 0 \end{pmatrix}.$$

It is easy, and usually a good idea, to verify the result by calculating AA^{-1} or $A^{-1}A$, which if correct should equal I.

5.5.3 *Determinants*

The determinant is a scalar that is associated with a square matrix A, and is denoted by $\det(A)$, $D(A)$, or $|A|$. Its general definition is somewhat complicated and involves the notion of a permutation.

A map $\sigma : \{1, \ldots, N\} \to \{1, \ldots, N\}$ is called a permutation if it is one-to-one and onto. Every permutation can be obtained by first starting with the ordered set $\{1, \ldots, N\}$ and repeatedly "switching" the order of two elements. For example with $N = 3$, the permutation $\sigma(1) = 2$, $\sigma(2) = 3$, $\sigma(1) = 2$, which we can describe succinctly as $\{1, 2, 3\} \to \{2, 3, 1\}$, can be obtained by the chain of switches $\{1, 2, 3\} \to \{2, 1, 3\} \to \{2, 3, 1\}$. We denote by $|\sigma|$ the number of switches needed to form σ. Although the same permutation can be derived in many ways with varying numbers of switches, it turns out that the number of switches is always either even or odd (that is, the same permutation cannot be derived in one manner with an even number of switches and by another method with odd many). Thus the value $(-1)^{|\sigma|}$ is well-defined for every permutation, and is equal to one if an even number of switches are needed, and equal to -1 if it is an odd number.

Definition: *The determinant of an $N \times N$ matrix A is the scalar defined by*

$$\det(A) = \sum_{\sigma} (-1)^{|\sigma|} a_{1,\sigma(1)} a_{2,\sigma(2)} \cdots a_{N,\sigma(N)} \tag{5.1}$$

where the sum is taken over all possible permutations σ.

Let us work out the determinants for low dimensional matrices. With $N = 1$ and $A = [a]$, one has $\det(A) = a$. With $N = 2$ and $A = \begin{pmatrix} a_{11} & a_{12} \\ a_{21} & a_{22} \end{pmatrix}$, then there are only two permutations and $\det(A) = a_{11}a_{22} - a_{12}a_{21}$. For $N = 3$ and

$$A = \begin{pmatrix} a_{11} & a_{12} & a_{13} \\ a_{21} & a_{22} & a_{23} \\ a_{31} & a_{32} & a_{33} \end{pmatrix}$$

it starts getting more complicated:

$$\det(A) = a_{11}a_{22}a_{33} + a_{12}a_{23}a_{31} + a_{13}a_{21}a_{32} - a_{12}a_{21}a_{33} - a_{11}a_{23}a_{32} - a_{13}a_{22}a_{31}.$$

There is a mnemonic to help calculate determinants in 3×3 matrices. First, repeat the first and second columns as new fourth and fifth columns, which now looks like

$$\begin{pmatrix} a_{11} & a_{12} & a_{13} & a_{11} & a_{12} \\ a_{21} & a_{22} & a_{23} & a_{21} & a_{22} \\ a_{31} & a_{32} & a_{33} & a_{31} & a_{32} \end{pmatrix}.$$

Beginning at the top left, multiply diagonally down to the right and obtain the term $a_{11}a_{22}a_{33}$. Repeat twice more by starting at the next entry to the right, and obtain the terms $a_{12}a_{23}a_{31}$ and $a_{13}a_{21}a_{32}$. These all have positive signs. Then start at the top right and multiply diagonally down to the left and obtain the term $a_{12}a_{21}a_{33}$. Again repeat twice more by starting one entry to the left, and obtain the terms $a_{11}a_{23}a_{32}$ and $a_{13}a_{22}a_{31}$. The latter terms all have negative signs. Adding all these together gives the determinant.

It should be emphasized that no such simple devices are available to find determinants of matrices of size greater than three, however, determinants can be defined recursively as follows. Suppose A is an $N \times N$ matrix. For a double index (m, n), let $A(m|n)$ be the matrix constructed from A by deleting the mth row and n^{th} column of A. Each of these has dimension $(N-1) \times (N-1)$. Define the determinant of the 1×1 matrix $[a]$ as just a, and inductively assume that the determinant of an $(N-1) \times (N-1)$ matrix has been defined. The determinant of an $N \times N$ then equals

$$\det(A) = \sum_{n=1}^{N} (-1)^{1+n} a_{1n} \det(A(1|n)). \tag{5.2}$$

This is the so-called cofactor expansion along the first row, but in fact, any row or column could be used. That is, we also have

$$\det(A) = \sum_{n=1}^{N} (-1)^{m+n} a_{mn} \det(A(m|n)) \tag{5.3}$$

for each $m = 1, \dots, N$, which in turn equals

$$\sum_{m=1}^{N} (-1)^{m+n} a_{mn} \det(A(m|n))$$

for each $n = 1, \dots, N$. The scalar $(-1)^{m+n} \det(A(m|n))$ is called the (m, n) cofactor of A.

To see that all these formulas agree, note that each term of the sum in Eq. (5.1) contains exactly one factor from each row and column, and the entire sum consists of all possible ways these can appear. The formula (5.2) is just another way of writing the same thing, as are the other cofactor expansions. Often, by exploiting the particular structure of A, one of these formulas may be relatively easy to calculate. We will see examples of this below, but next we turn to deriving some properties of determinants.

5.5.3.1 *Basic properties of determinants*

The next theorem contains a list of properties that facilitate calculating determinants.

Theorem 5.8.

(1) A square matrix and its transpose have the same determinant, $A = A^T$. Thus the following assertions that are stated for rows apply equally to columns.
(2) Multiplication of the elements of any row by a scalar is equivalent to multiplication of the determinant by that scalar.
(3) If all the elements of any row of a matrix are zero, the determinant of the matrix equals zero.
(4) If two rows of a matrix are interchanged, the sign of the determinant is changed.
(5) The determinant of a matrix is equal to zero if two of its rows are identical. The same is true if a row is a multiple of another row.
(6) Suppose the mth row $(a_{m1}, \dots, a_{mN})$ of a matrix A is of the form $a_{mn} = b_{mn} + c_{mn}$, $n = 1, \dots, N$. Let A_1 (respectively A_2) be the matrix whose m^{th} row is $(b_{m1}, \dots, b_{mN})$ (respectively $(c_{m1}, \dots, c_{mN})$) and whose other entries are those of A. Then $\det(A) = \det(A_1) + \det(A_2)$.
(7) Suppose a matrix B is obtained from a matrix A by performing the third row operation that was introduced in Gauss elimination. Then $\det(B) = \det(A)$.
(8) The determinant of a product of two matrices equal the product of their determinants. In particular, if A is invertible, then $\det(A^{-1}) = \det(A)^{-1}$.
(9) The determinant of a diagonal matrix is the product of the diagonal entries. More generally, if A has a block structure of the form

$$A = \begin{pmatrix} A_1 & 0 \\ 0 & A_2 \end{pmatrix}, \tag{5.4}$$

where A_1 *and* A_2 *are square matrices whose dimensions sum to the dimension of* A*, then* $\det A = (\det A_1)(\det A_2)$.

Proof (1): This is clear if $N = 1$ since in this case $A = A^T$. Assume the result is true for dimension $N - 1$, and note that $A(1|n) = A(n|1)^T$ for each $n = 1, \ldots, N$. Recall that the rows of A are precisely the columns of A^T. The result follows for dimension N after observing that the cofactor expansion of A along the first row equals the cofactor expansion of A^T along the first column.
(2): Each term in the sum in Eq. (5.1) contains only one factor from each row and column. The scalar factor will therefore appear only once in each term and can be factored out of the sum as a multiplier of the determinant.
(3): By taking the scalar in part (2) to be zero, this follows immediately. It can also be verified directly from the definition, since one of the factors of every summand in Eq. (5.1) is 0.
(4): Interchanging two rows has the effect of adding one more switch to a permutation. Thus any even permutation becomes odd, and an odd one becomes even. This has the effect of multiplying Eq. (5.1) by -1, which is equivalent to changing its sign.
(5): If two rows are the same, then interchanging them does not change the matrix but changes the sign of its determinant by part (4). The only way this could happen is if the determinant is zero. If one row is a scalar multiple of another, then by part (1), the determinant equals that scalar times the determinant of a matrix with two identical rows, and thus is zero.
(6): This is clear from either the definition Eq. (5.1) or the cofactor expansion Eq. (5.3).
(7): Suppose R_m and R_n are the mth and n^{th} rows of A respectively (with $m \neq n$), α is a scalar, and B is the matrix with the same entries as A except for the mth row which is replaced by $R_m + \alpha R_n$. Let A_1 be the matrix A whose m^{th} row is replaced by R_n. Then $\det(B) = \det(A) + \alpha \det(A_1)$ by parts (6) and (2), and $\det(A_1) = 0$ by part (5). Thus $\det(B) = \det(A)$.
(8): The proof of this statement is considerably more involved than the previous ones. We only illustrate the idea with $N = 2$, although the same line of reasoning works in general. Suppose

$$A = \begin{pmatrix} a_{11} & a_{12} \\ a_{21} & a_{22} \end{pmatrix}; \quad B = \begin{pmatrix} b_{11} & b_{12} \\ b_{21} & b_{22} \end{pmatrix}; \text{ and}$$

$$C = AB = \begin{pmatrix} a_{11}b_{11} + a_{12}b_{21} & a_{11}b_{12} + a_{12}b_{22} \\ a_{21}b_{11} + a_{22}b_{21} & a_{21}b_{12} + a_{22}b_{22} \end{pmatrix}.$$

To calculate $\det(C)$, apply parts (2) and (6) to the first row to see that $\det(C) =$

$$a_{11} \begin{vmatrix} b_{11} & b_{12} \\ a_{21}b_{11} + a_{22}b_{21} & a_{21}b_{12} + a_{22}b_{22} \end{vmatrix} + a_{12} \begin{vmatrix} b_{21} & b_{22} \\ a_{21}b_{11} + a_{22}b_{21} & a_{21}b_{12} + a_{22}b_{22} \end{vmatrix}. \quad (5.5)$$

Now apply parts (2) and (6) on the second row of the first determinant of (5.5) to obtain

$$\begin{vmatrix} b_{11} & b_{12} \\ a_{21}b_{11} + a_{22}b_{21} & a_{21}b_{12} + a_{22}b_{22} \end{vmatrix} = a_{21} \begin{vmatrix} b_{11} & b_{12} \\ b_{11} & b_{12} \end{vmatrix} + a_{22} \begin{vmatrix} b_{11} & b_{12} \\ b_{21} & b_{22} \end{vmatrix} = a_{22} \det(B)$$

where the last equality follows from parts (1) and (5). Similarly, the second determinant in Eq. (5.5) can be found equal to $-a_{21} \det(B)$, where the minus sign comes from using part (4). Inserting all of this into Eq. (5.5) yields that $\det(C) = a_{11}a_{22} \det(B) - a_{12}a_{21} \det(B) = \det(A) \det(B)$.

The general case $N > 2$ is considerably more notationally complicated so we omit the details, but the idea is still the same.

The final statement involving invertible matrices follows because $I = A^{-1}A$, and so $1 = \det(I) = \det(A^{-1}) \det(A)$.

(9): It is immediate from the definition that the determinant of a diagonal matrix is the product of its diagonal entries. Suppose A has the block structure as in Eq. (5.4). Then if either A_1 or A_2 are the identity matrices, the result follows quickly from the cofactor expansion. The general result follows from this and part (8) by noting that

$$A = \begin{pmatrix} A_1 & 0 \\ 0 & A_2 \end{pmatrix} = \begin{pmatrix} A_1 & 0 \\ 0 & I_2 \end{pmatrix} \begin{pmatrix} I_1 & 0 \\ 0 & A_2 \end{pmatrix},$$

where I_1 and I_2 are identity matrices of the appropriate dimensions.

5.5.3.2 *Calculation of determinants*

We have seen that there are relatively simple formulas for calculating determinants for matrices of dimension $N \leq 3$. This section provides some examples that show how the properties in the previous theorem can be used to calculate determinants. These techniques can be quite useful if the matrix is of a larger dimension than 3.

First note that the determinant of an upper or lower diagonal matrix is just the product of its diagonal entries. For example, if

$$A = \begin{pmatrix} 2 & 3 & 1 & 2 \\ 0 & 5 & 3 & 1 \\ 0 & 0 & -1 & 3 \\ 0 & 0 & 0 & 2 \end{pmatrix}$$

then the cofactor expansion along the first column is simple because of the presence of so many zeros, and reduces to

$$\det(A) = \begin{vmatrix} 2 & 3 & 1 & 2 \\ 0 & 5 & 3 & 1 \\ 0 & 0 & -1 & 3 \\ 0 & 0 & 0 & 2 \end{vmatrix} = 2 \begin{vmatrix} 5 & 3 & 1 \\ 0 & -1 & 3 \\ 0 & 0 & 2 \end{vmatrix} = (2)(5) \begin{vmatrix} -1 & 3 \\ 0 & 2 \end{vmatrix} = -20,$$

which is seen to be just the product of the diagonal entries.

Notice that the properties (2), (4) and (7) describe how determinants change under the row operations of Gauss elimination. The next example illustrates this by attempting to keep computation to a minimum while reducing the matrix to one that is upper triangular.

Example 5.4: Calculation of determinant by Gauss elimination.

$$\begin{vmatrix} 2 & -3 & 3 & 1 \\ -2 & 5 & 2 & 3 \\ 4 & -10 & -1 & 3 \\ 1 & -3 & 1 & 2 \end{vmatrix} \overset{R_1 \leftrightarrow R_4}{=} (-1) \begin{vmatrix} 1 & -3 & 1 & 2 \\ -2 & 5 & 2 & 3 \\ 4 & -10 & -1 & 3 \\ 2 & -3 & 3 & 1 \end{vmatrix} \overset{\substack{2R_1+R_2\to R_2 \\ -4R_1+R_3\to R_3 \\ -2R_1+R_4\to R_4}}{=} (-1) \begin{vmatrix} 1 & -3 & 1 & 2 \\ 0 & -1 & 4 & 7 \\ 0 & 2 & -5 & -5 \\ 0 & 3 & 1 & -3 \end{vmatrix}$$

$$\overset{\substack{2R_2+R_3\to R_3 \\ 3R_2+R_4\to R_4}}{=} (-1) \begin{vmatrix} 1 & -3 & 1 & 2 \\ 0 & -1 & 4 & 7 \\ 0 & 0 & 3 & 9 \\ 0 & 0 & 13 & 18 \end{vmatrix} \overset{\frac{1}{3}R_3\to R_3}{=} (-3) \begin{vmatrix} 1 & -3 & 1 & 2 \\ 0 & -1 & 4 & 7 \\ 0 & 0 & 1 & 3 \\ 0 & 0 & 13 & 18 \end{vmatrix} \overset{-13R_3+R_4\to R_4}{=} (-3) \begin{vmatrix} 1 & -3 & 1 & 2 \\ 0 & -1 & 4 & 7 \\ 0 & 0 & 1 & 3 \\ 0 & 0 & 0 & -21 \end{vmatrix}$$

which equals $(-3)(21) = -63$.

Here is one final example that finds the determinant by always expanding along the row or column that has the most zeros.

Example 5.5: Calculation of determinant by expansion.

Consider

$$A = \begin{vmatrix} 0 & \frac{1}{2} & 7 & 8 & 0 \\ \frac{3}{2} & 0 & 0 & -1 & -\frac{1}{2} \\ 1 & 0 & 0 & 4 & -\frac{5}{3} \\ 0 & 2 & 3 & 0 & 0 \\ 0 & \frac{1}{2} & 0 & -\frac{1}{2} & \frac{3}{2} \end{vmatrix}.$$

In order to make as many of the terms in the expansion zero, start the expansion with the row or column with the largest number of zeros. This could be either row 4 or column 1. Expanding by column 1 gives,

$$\det(A) = -\frac{3}{2} \begin{vmatrix} \frac{1}{2} & 7 & 8 & 0 \\ 0 & 0 & 4 & -\frac{5}{3} \\ 2 & 3 & 0 & 0 \\ \frac{1}{2} & 0 & -\frac{1}{2} & \frac{3}{2} \end{vmatrix} + \begin{vmatrix} \frac{1}{2} & 7 & 8 & 0 \\ 0 & 0 & -1 & -\frac{1}{2} \\ 2 & 3 & 0 & 0 \\ \frac{1}{2} & 0 & -\frac{1}{2} & \frac{3}{2} \end{vmatrix}$$

and each of the two 3-by-3 matrices can now be expanded further. Using row 2 of both gives,

$$\det(A) = -\frac{3}{2}\left(-4\begin{vmatrix} \frac{1}{2} & 7 & 0 \\ 2 & 3 & 0 \\ \frac{1}{2} & 0 & \frac{3}{2} \end{vmatrix} - \frac{5}{3}\begin{vmatrix} \frac{1}{2} & 7 & 8 \\ 2 & 3 & 0 \\ \frac{1}{2} & 0 & -\frac{1}{2} \end{vmatrix}\right)$$

$$+\left(\begin{vmatrix} \frac{1}{2} & 7 & 0 \\ 2 & 3 & 0 \\ \frac{1}{2} & 0 & \frac{3}{2} \end{vmatrix} - \frac{1}{2}\begin{vmatrix} \frac{1}{2} & 7 & 8 \\ 2 & 3 & 0 \\ \frac{1}{2} & 0 & -\frac{1}{2} \end{vmatrix}\right).$$

Continuing this process,

$$\det(A) = 6\left(\frac{3}{2}\begin{vmatrix} \frac{1}{2} & 7 \\ 2 & 3 \end{vmatrix}\right) + \frac{5}{2}\left(\frac{1}{2}\begin{vmatrix} 7 & 8 \\ 3 & 0 \end{vmatrix} - \frac{1}{2}\begin{vmatrix} \frac{1}{2} & 7 \\ 2 & 3 \end{vmatrix}\right)$$

$$+\frac{3}{2}\begin{vmatrix} \frac{1}{2} & 7 \\ 2 & 3 \end{vmatrix} - \frac{1}{2}\left(\frac{1}{2}\begin{vmatrix} 7 & 8 \\ 3 & 0 \end{vmatrix} - \frac{1}{2}\begin{vmatrix} \frac{1}{2} & 7 \\ 2 & 3 \end{vmatrix}\right)$$

$$= 9\left(\frac{3}{2} - 14\right) + \frac{5}{4}(-24) - \frac{5}{4}\left(\frac{3}{2} - 14\right) + \frac{3}{2}\left(\frac{3}{2} - 14\right) - \frac{1}{4}(-24) + \frac{1}{4}\left(\frac{3}{2} - 14\right)$$

$$= -\frac{571}{4} = -142.75\,.$$

5.5.3.3 *The derivative of a determinant*

One occasionally needs to differentiate a determinant $D(x)$ of an $N \times N$ matrix $A(x)$ whose entries $a_{mn}(x)$ are differentiable functions of a free variable x. From the definition in Eq. (5.1) of $D(x)$, one has by the product rule that

$$\begin{aligned}\frac{dD}{dx} &= \sum_{\sigma}(-1)^{|\sigma|}\left(\frac{da_{1,\sigma(1)}}{dx}\right)a_{2,\sigma(2)}\cdots a_{N,\sigma(N)} \\ &+ \sum_{\sigma}(-1)^{|\sigma|}a_{1,\sigma(1)}\left(\frac{da_{2,\sigma(2)}}{dx}\right)\cdots a_{N,\sigma(N)} \\ &\vdots \\ &+ \sum_{\sigma}(-1)^{|\sigma|}a_{1,\sigma(1)}a_{2,\sigma(2)}\cdots\left(\frac{da_{N,\sigma(N)}}{dx}\right).\end{aligned}$$

Let $\dot{D}_n(x)$ denote the determinant of the matrix whose entries are the same as $A(x)$ except for the nth row which is replaced by the derivative of that row. The previous calculation shows that

$$\frac{dD}{dx}(x) = \sum_{n=1}^{N}\dot{D}_n(x)$$

where $\dot{D}_m$ is the determinant obtained by differentiating the m'th row of D and the sum is taken over all such determinants.

5.5.4 *The classical adjoint matrix*

The adjoint of a matrix, sometimes called the classical adjoint to avoid confusion with the adjoint of an operator, a concept which will be defined later, is defined for square matrices only and is the transpose of the matrix obtained by replacing each element by its cofactor. That is, if we let $A_{mn} = (-1)^{m+n}A(m|n)$ be the (m,n) cofactor, the classical adjoint is the matrix

$$\operatorname{adj}(A) = \begin{pmatrix} A_{11} & A_{21} & \cdots & A_{N1} \\ A_{12} & A_{22} & & A_{N2} \\ \vdots & \vdots & \ddots & \vdots \\ A_{1N} & A_{2N} & \cdots & A_{NN} \end{pmatrix}.$$

The product of the matrix and its adjoint is particularly simple:

$$A\operatorname{adj}(A) = \begin{pmatrix} a_{11} & a_{12} & \cdots & a_{1N} \\ a_{21} & a_{22} & \cdots & a_{2N} \\ \vdots & \vdots & \ddots & \vdots \\ a_{N1} & a_{N2} & \cdots & a_{NN} \end{pmatrix} \begin{pmatrix} A_{11} & A_{21} & \cdots & A_{N1} \\ A_{12} & A_{22} & \cdots & A_{N2} \\ \vdots & \vdots & \ddots & \vdots \\ A_{1N} & A_{2N} & \cdots & A_{NN} \end{pmatrix}$$

$$= \begin{pmatrix} \sum a_{1n}A_{1n} & \sum a_{1n}A_{2n} & \cdots & \sum a_{1n}A_{Nn} \\ \sum a_{2n}A_{1n} & \sum a_{2n}A_{2n} & \cdots & \sum a_{2n}A_{Nn} \\ \vdots & \vdots & \ddots & \vdots \\ \sum a_{Nn}A_{1n} & \sum a_{Nn}A_{2n} & \cdots & \sum a_{Nn}A_{Nn} \end{pmatrix}$$

where each of the sums are over the index n ranging from 1 to N. Each of the diagonal elements is of the form $\sum a_{mn}A_{mn}$, $m = 1, \ldots, N$, and is the cofactor expansion along the mth row. Hence each diagonal entry equals $\det(A)$. The off-diagonal elements are all zero for the following reason. Suppose $m_1 \neq m_2$ are indices between 1 and N. Then $\sum a_{m_1n}A_{m_2n}$ is the cofactor expansion along the m_1^{st} row of the matrix that has the same entries as A except with the m_1^{nd} row is replaced by the m_1^{st} row. Since this matrix has two identical rows, its determinant is zero by Theorem 5.8(5). We have shown that the product of a matrix and its adjoint equals a diagonal matrix with the value of the determinant in the diagonal elements.

$$A\operatorname{adj}(A) = \det(A)I$$

where I is the $N \times N$ identity matrix. The same result holds if the matrix post-multiplies its adjoint. It is now easy to see that if $\det(A) \neq 0$, then

$$\left(\frac{1}{\det(A)}\operatorname{adj}(A)\right) A = A\left(\frac{1}{\det(A)}\operatorname{adj}(A)\right) = I_N,$$

and so A^{-1} exists with

$$A^{-1} = \frac{1}{\det(A)}\operatorname{adj}(A). \tag{5.6}$$

Conversely, if the determinant of a square matrix A has an inverse, then $1 = \det(I) = \det(A^{-1}A) = \det(A^{-1})\det(A)$, and so $\det(A) \neq 0$. We have shown

Theorem 5.9. *A square matrix A has an inverse if and only if* $\det(A) \neq 0$.

When a square matrix has an inverse, it is often called nonsingular, and when it does not, it is called singular.

5.6 Systems of linear algebraic equations

We now study a system of linear equations in further detail. Suppose A is an $M \times N$ matrix and b is a column vector in $\mathbb{F}^M$. A system of linear algebraic equations can be written in vector-matrix notation as $Av = b$, where x is the vector of unknowns. Formally, the nth row of A in combination with the nth element of b comprises the nth equation. If A has M rows and N columns, the system of equations has M equations and N unknowns. In this section, we will determine the solution structure of coupled linear equations by the simple method of simplifying $Av = b$ by Gauss elimination and investigating the resulting simpler system. However, before doing so, we will define and discuss the concept of rank which plays an important role in systems of linear equations.

5.6.1 *Rank*

The *row rank* of an $M \times N$ matrix A is defined as the dimension of the subspace that is spanned by the rows of A. Similarly, the *column rank* of A is the dimension of the subspace that is spanned by the columns of A. Obviously, the row rank is less than or equal to M and the column rank less than or equal to N.

Theorem 5.10. *Suppose A and B are two matrices in which B is obtained from A by an elementary row operation. Then the rows of A and B span the same subspace, and in particular, A and B have the same row rank.*

Proof: Let W_1 be the subspace spanned by the rows of A, and W_2 be the subspace spanned by the rows of B. The first row operation is switching two rows, and the second operation is multiplication of a row by a non-zero scalar; neither of these have any effect on the span, and so $W_1 = W_2$ in these cases. As for the third row operation, let R_j denote the jth row of A and $\tilde{R}_j$ denote the jth row of B. The third operation says that one of the rows (say, the mth row $\tilde{R}_m$) of B is of the form $\tilde{R}_m = \alpha R_j + R_m$, where $\alpha \neq 0$ and $m \neq j$. Since the rest of the rows of B are the same as A, it is immediate that $W_2 \subseteq W_1$. The opposite inclusion holds as well, since $R_m = \tilde{R}_m - \alpha R_j$ and $R_j = \tilde{R}_j$ and so R_m is a linear combination of the rows of B. Hence $W_1 \subseteq W_2$, and so the subspaces spanned by the rows of each matrix is the same. It follows immediately that the row ranks of A and B coincide. □

If a matrix is in row-echelon form, then its row rank can be found very easily, since it equals the number of non-zero rows. Theorem 5.10 applied repeatedly says the row rank of a matrix is the same as the row rank of any matrix obtained from elementary row operations, and thus one can find the row rank by Gauss elimination. We illustrate by an example.

Example 5.6: Rank by Gauss elimination.

$$\begin{pmatrix} 1 & 3 & 4 \\ 1 & -2 & -1 \\ 3 & -1 & 2 \end{pmatrix} \Rightarrow \begin{pmatrix} 1 & 3 & 4 \\ 0 & -5 & -5 \\ 0 & -10 & -10 \end{pmatrix} \Rightarrow \begin{pmatrix} 1 & 3 & 4 \\ 0 & 1 & 1 \\ 0 & 0 & 0 \end{pmatrix} \Rightarrow \begin{pmatrix} 1 & 0 & 1 \\ 0 & 1 & 1 \\ 0 & 0 & 0 \end{pmatrix}.$$

The last matrix has two independent rows, and so the row rank of the original matrix is two.

As revealed in this example, and which is true generally as an immediate consequence of Theorem 5.10, the row rank equals the number of non-zero rows in a row-echelon form of the matrix. Obviously the number of these non-zero rows is the same as the number of "leading ones" in those rows. If we now consider the column rank of a row reduced matrix, the number of columns that contain these leading ones is the same as the number of non-zero rows, and moreover one can easily show that these columns form a basis (in fact, these columns are elements of the canonical basis) of the subspace spanned by the columns. Thus for a matrix in row-echelon form, one sees immediately that the row rank is equal to the column rank, and one may think this should hold for all matrices. This is indeed the case, which is the content of the next theorem.

Theorem 5.11. *Suppose A and B are two matrices in which B is obtained from A by an elementary row operation. Then A and B have the same column rank. In fact, if a set of columns of A are independent, then the corresponding columns in B are independent.*

Proof: Suppose the matrices are $M \times N$ dimensional. Let $C_1 = (c_{m1}), \ldots, C_n = (c_{mn})$ denote the columns of A that form a basis for the subspace spanned by all the columns of A, and let $\tilde{C}_1 = (\tilde{c}_{m1}), \ldots, \tilde{C}_n = (\tilde{c}_{mn})$ be the corresponding columns of B. We will show that $\{\tilde{C}_1, \ldots, \tilde{C}_n\}$ is a basis for the subspace spanned by the columns of B, which implies the result. If the first or second elementary row operation is used, then the result is easy to prove, so assume B is obtained from A by the third elementary operation. This means there exists $j \neq i$ and a scalar

$\alpha \neq 0$ so that $\tilde{c}_{jk} = c_{jk} + \alpha c_{ik}$ for all $k = 1, \ldots, n$, and $\tilde{c}_{mk} = c_{mk}$ for all $m \neq j$ and $k = 1, \ldots, n$.

It is routine to show that $\{\tilde{C}_1, \ldots, \tilde{C}_n\}$ spans the subspace, so we direct our attention to showing they are independent. Now suppose $\alpha_1, \ldots, \alpha_n$ are scalars with $\sum_{k=1}^n \alpha_k \tilde{C}_k = 0$, which written componentwise says that $\sum_{k=1}^n \alpha_k \tilde{c}_{mk} = 0$ for all $m = 1, \ldots, M$. Since $\tilde{c}_{mk} = c_{mk}$ for all $m \neq j$, we have $\sum_{k=1}^n \alpha_k c_{mk} = 0$ for all $m \neq j$. We also have

$$0 = \sum_{k=1}^n \alpha_k \tilde{c}_{jk} = \sum_{k=1}^n \alpha_k c_{jk} + \alpha \sum_{k=1}^n \alpha_k c_{ik} = \sum_{k=1}^n \alpha_k c_{jk},$$

where in the last equality we used that $0 = \sum_{k=1}^n \alpha_k \tilde{c}_{ik} = \sum_{k=1}^n \alpha_k c_{ik}$. Therefore $\sum_{k=1}^n \alpha_k c_{mk} = 0$ for all $m = 1 \ldots, M$, which is the componentwise way of saying that $\sum_{k=1}^n \alpha_k C_k = 0$. Since the C_ks are independent, it follows that $\alpha_k = 0$ for all $k = 1, \ldots, n$, or that the the $\tilde{C}_k$s are independent. □

The previous theorem implies that the column rank of a matrix is the same as the column rank of the row-reduced matrix that is in row-echelon form. As observed above, the row and column ranks of a row-echelon matrix coincide, and so we have the following corollary.

Corollary. *The row rank of a matrix is the same as its column rank, and equals the maximum number of linearly independent rows, or equivalently, the maximum number of linearly independent columns.*

It makes sense now to just refer to the rank of a matrix without specifying if it is the row or column rank since these are the same. A matrix is said to have *full rank* if its rank is as large as possible, which of course is the lesser number of the number of rows and columns.

As seen in Theorem 5.10, row operations do not change the subspace spanned by the rows, and thus in the previous example above, the row vectors (1 0 1) and (1 1 0) are a basis of the subspace spanned by the rows. To obtain a basis for the subspace spanned by the columns, the row-echelon matrix can also be used, but only as a means to indicate which of the original columns should be used. The columns of the orginal matrix that correspond to the columns of the associated row-echelon matrix that contain the leading ones form a basis. In the previous example, the first two columns of the row-echelon matrix contain the leading ones, and so the column vectors

$$\begin{pmatrix} 1 \\ 1 \\ 3 \end{pmatrix} \quad \begin{pmatrix} 3 \\ -2 \\ -1 \end{pmatrix}$$

form a basis of the subspace spanned by all the column vectors.

There is another equivalent definition of rank which is practical to use in relatively small dimensions. Consider all square matrices obtained from A by removing rows and/or columns of A; such submatrices are called *minors*. The rank of A is the size of the largest (square) minor that is nonsingular. According to this definition, the rank of a matrix can be determined by calculating the determinants of all square matrices (starting with largest and including the determinant of the matrix itself if it is square), and continuing until one finds a minor with a non-zero determinant. This is another way to see that the row and column ranks of a matrix are always equal, since the determinant of the transpose equals the determinant of the original matrix.

Example 5.7: Calculation of the rank by using determinants.

Consider the matrix

$$A = \begin{pmatrix} 2 & 1 & 3 & 4 \\ -1 & 1 & -2 & -1 \\ 0 & 3 & -1 & 2 \end{pmatrix}.$$

We calculate its rank by considering the determinants of its minors. The largest possible minors are of size 3-by-3 obtained by deleting a single column, and these determinants are

$$\begin{vmatrix} 1 & 3 & 4 \\ 1 & -2 & -1 \\ 3 & -1 & 2 \end{vmatrix} = \begin{vmatrix} 2 & 3 & 4 \\ -1 & -2 & -1 \\ 0 & -1 & 2 \end{vmatrix} = \begin{vmatrix} 2 & 1 & 4 \\ -1 & 1 & -1 \\ 0 & 3 & 2 \end{vmatrix} = \begin{vmatrix} 2 & 1 & 3 \\ -1 & 1 & -2 \\ 0 & 3 & -1 \end{vmatrix} = 0\,.$$

Thus the rank must be less than three. It is now easy to see that the rank is two since

$$\begin{vmatrix} 2 & 1 \\ -1 & 1 \end{vmatrix} = 3 \neq 0,$$

which is the determinant of the 2×2 minor obtained be deleting the third row and the third and fourth columns.

The rank of a matrix equals the number of linearly independent rows or columns, but the term could also be used to describe a set of vectors. The rank of a set of vectors is the largest number of linearly independent vectors in the set.

Example 5.8: Rank and linear independence.

This example finds the rank of the set

$$v_1 = \begin{pmatrix} 1 \\ 3 \\ 2 \\ 5 \\ 2 \end{pmatrix}, \; v_2 = \begin{pmatrix} 2 \\ 4 \\ 0 \\ 8 \\ -1 \end{pmatrix}, \; v_3 = \begin{pmatrix} 3 \\ 7 \\ 2 \\ 13 \\ 1 \end{pmatrix}, \; v_4 = \begin{pmatrix} 1 \\ 1 \\ -2 \\ 3 \\ -3 \end{pmatrix},$$

and will in fact find the maximal subset of linear independent vectors. We write the vectors as the columns of a matrix and reduce to row-echelon form by Gauss elimination.

$$\begin{pmatrix} 1 & 2 & 3 & 1 \\ 3 & 4 & 7 & 1 \\ 2 & 0 & 2 & -2 \\ 5 & 8 & 13 & 3 \\ 2 & -1 & 1 & -3 \end{pmatrix} \implies \begin{pmatrix} 1 & 0 & 1 & -1 \\ 0 & 1 & 1 & 1 \\ 0 & 0 & 0 & 0 \\ 0 & 0 & 0 & 0 \\ 0 & 0 & 0 & 0 \end{pmatrix}$$

which has rank 2, and the first two columns of the latter has the leading ones. Thus the set of vectors has rank 2, therefore contains only two linearly independent vectors. Furthermore, we conclude that v_1 and v_2 are linearly independent, and one can verify $v_3 = v_1 + v_2$ and $v_4 = v_2 - v_1$. In other words, v_1 and v_2 is a basis for the vector space spanned by v_1 through v_4.

We now consider the problem of determining if two linear independent sets span the same vector space. Suppose the sets $\{v_1, v_2, \ldots, v_N\}$ and $\{w_1, w_2, \ldots, w_N\}$ are both linearly independent subsets of $\mathbb{R}^M$. If they do span the same space, then all the w_ns are linear combinations of the v_ns and vice versa. Thus the rank of $\{v_1, \ldots, v_N, w_1, \ldots, w_N\}$ must be N. If the rank is greater then N, the two bases do not span the same space.

Example 5.9: Determining if two bases define the same vector space.

Let

$$\{v_1, v_2, v_3, v_4\} = \left\{ \begin{pmatrix} 1 \\ 5 \\ 0 \\ -1 \\ 4 \\ 0 \end{pmatrix}, \begin{pmatrix} 0 \\ -2 \\ 3 \\ 1 \\ 5 \\ 0 \end{pmatrix}, \begin{pmatrix} 2 \\ 2 \\ 0 \\ 3 \\ -3 \\ 7 \end{pmatrix}, \begin{pmatrix} 1 \\ 5 \\ 5 \\ -1 \\ 4 \\ 0 \end{pmatrix} \right\}$$

$$\{w_1, w_2, w_3, w_4\} = \left\{ \begin{pmatrix} 0 \\ -3 \\ 7 \\ -1 \\ 3 \\ 10 \end{pmatrix}, \begin{pmatrix} 1 \\ 0 \\ 2 \\ -3 \\ 9 \\ -8 \end{pmatrix}, \begin{pmatrix} -4 \\ 7 \\ 3 \\ -9 \\ 10 \\ 1 \end{pmatrix}, \begin{pmatrix} 3 \\ -5 \\ 5 \\ 4 \\ -1 \\ -1 \end{pmatrix} \right\}.$$

Both sets are independent and span four-dimensional subspaces of $\mathbb{R}^6$. To determine if they span the same subspaces, consider the matrix formed by letting the columns consist of all the vectors and reduce by Gauss elimination

$$\begin{pmatrix} 1 & 0 & 2 & 1 & 0 & 1 & -4 & 3 \\ 5 & -2 & 2 & 5 & -3 & 0 & 7 & -5 \\ 0 & 3 & 0 & 5 & 7 & 2 & 3 & 5 \\ -1 & 1 & 3 & -1 & -1 & -3 & -9 & 4 \\ 4 & 5 & -3 & 4 & 3 & 9 & 10 & -1 \\ 0 & 0 & 7 & 0 & 10 & -8 & 1 & -1 \end{pmatrix} \Longrightarrow \begin{pmatrix} 1 & 0 & 2 & 1 & 0 & 1 & -4 & 3 \\ 0 & -2 & -8 & 0 & -3 & -5 & 27 & -20 \\ 0 & 0 & 2 & 0 & -5 & -9 & 1 & -6 \\ 0 & 0 & 0 & 10 & -55 & -119 & 99 & -122 \\ 0 & 0 & 0 & 0 & -82 & -147 & 109 & -156 \\ 0 & 0 & 0 & 0 & 0 & -\frac{4231}{164} & \frac{5585}{164} & -\frac{1325}{41} \end{pmatrix}.$$

There is no need to reduce it all the way to row-echelon form, for we can readily see that the rank is 6, and consequently, the two bases do not span the same subspace.

5.6.2 *Applications of rank*

The link between rank and linear independence can be used to cast some engineering problems in terms of matrix rank. Consider for instance the problem of finding the equilibrium composition of a single phase reaction mixture. How many concentrations, and of which compounds, must be known or specified before the remaining concentrations can be calculated from equilibrium constants? One cannot pick component concentrations and reactions arbitrarily and be sure of being able to do these calculations. The reason for this becomes clear if we consider a simple case, such as a mixture of oxygen, hydrogen, water and hydrogen peroxide. One can specify equilibrium constants for many possible reactions. For instance, $H_2 + \frac{1}{2}O_2 \leftrightarrow H_2O$, $2H_2 + O_2 \leftrightarrow 2H_2O$ and $3H_2 + \frac{3}{2}O_2 \leftrightarrow 3H_2O$, but it is obvious to everyone that these are really the same reaction, written three different ways and this reactions do not provide any information from which the concentration of hydrogen peroxide can be calculated. Also, one cannot specify the concentration of H_2O, H_2 and O_2 because these concentrations are not independent, but linked through the equilibrium constant of the above reaction. The point is that as one specifies known concentration and reactions, one must make certain that enough concentrations are known, that these known concentrations do not overspecify the system and that the reactions are not linear combinations of one another.

In mixtures with a large number of compounds and reactions it is not possible, just by inspection, to pick a correct set of known concentrations and a full set of linearly independent reactions. One needs to proceed systematically. To do so, consecutively number all the compounds in the mixture and all the elements in the system, starting with 1. Define the elemental matrix such that element n, m of this matrix is the number of atoms of element n in compound m. Determine the rank of this matrix and find a set of linearly independent columns with this rank. The rank equals the number of concentrations that must be known; the set of linearly independent columns indicates which concentrations must be known;

and the remaining columns are used to derive the reactions for which equilibrium constants must be known. An example will make the method clearer.

Example 5.10: Maximum number of independent chemical reactions.

Consider a mixture containing the compounds H_2O, C_2H_4, C_2H_5OH, $C_4H_{10}O$ and CH_4. Number the compounds in the order given here and the elements such that hydrogen is number 1, carbon number 2 and oxygen number 3. The elemental matrix becomes

	H_2O	C_2H_4	C_2H_5OH	$C_4H_{10}O$	CH_4
H	2	4	6	10	4
C	0	2	2	4	1
O	1	0	1	1	0

Specifying a chemical reaction between any of the components in the system is equivalent to specifying a linear combination between columns of this matrix. For instance, the reaction $H_2O + C_2H_4 \leftrightarrow C_2H_5OH$, is equivalent to the linear combination $v_1 + v_2 - v_3 = 0$, where v_n is the vector defined by the nth column of the elemental matrix. One might say that H_2O, C_2H_4 and C_2H_5OH are linearly dependent compounds and the problem is to find a set of linearly independent compounds from which all other compounds are formed through chemical reactions. The maximum number of linearly independent compounds or columns is the rank of the elemental matrix. Doing standard row operations, the elemental matrix is reduced to the from

	H_2O	C_2H_4	C_2H_5OH	$C_4H_{10}O$	CH_4
H	1	2	3	5	2
C	0	1	1	2	$\frac{1}{2}$
O	0	0	0	0	1

The rank is 3. Thus, one can specify three compounds that can form all remaining compounds in the system through chemical reactions. Since there are five compounds in the system or equivalently five columns in the elemental matrix, $5-3=2$ reactions must be specified. One cannot randomly pick the three compounds from which the remaining compounds are formed, as the three compounds must be linearly independent. Inspection of the reduced matrix shows that columns 1, 2 and 5 in the non-reduced matrix are linearly independent and the linearly independent compounds can therefore be picked as H_2O, C_2H_4 and CH_4. Through chemical reactions, these three compounds can form all other compounds in the mixture.

To specify the reactions needed to form the remaining compounds, write the columns of the remaining compounds (taken from the elemental matrix in non-reduced form as linear combinations of the columns that correspond to the three linearly independent compounds. Thus $v_3 = v_1 + v_2$ represents the reaction

$$C_2H_5OH \leftrightarrow H_2O + C_2H_4$$

and $v_4 = v_1 + 2v_2$ represents

$$C_4H_{10}O \leftrightarrow H_2O + 2C_2H_4 .$$

Thus, given the concentrations of CH_4, H_2O and C_2H_4, all other equilibrium concentrations can be found. To be a bit more concrete: assume we mix M_{0,CH_4} moles of CH_4, M_{0,H_2O} moles of H_2O and M_{0,C_2H_4} in a vessel with volume V. What are the concentrations after equilibrium has been reached?

The concentration of the inert is trivial

$$C_{CH_4} = \frac{M_{0,CH_4}}{V} .$$

Four equations are needed to find the remaining four concentrations. The equilibrium constants of the two reactions we determined above provide two equations

$$\frac{C_{H_2O}C_{C_2H_4}}{C_{C_2H_5OH}} = K_1; \qquad \frac{C_{H_2O}C^2_{C_2H_4}}{C_{C_4H_{10}O}} = K_2 .$$

The stoichiometry of the reactions provide two more

$$M_{H_2O} = M_{0,H_2O} - M_{C_2H_5OH} - M_{C_4H_{10}O} \Rightarrow$$

$$C_{H_2O} = C_{0,H_2O} - C_{C_2H_5OH} - C_{C_4H_{10}O}$$

$$M_{C_2H_4} = M_{0,C_2H_4} - M_{C_2H_5OH} - 2M_{C_4H_{10}O} \Rightarrow$$

$$C_{C_2H_4} = C_{0,C_2H_4} - C_{C_2H_5OH} - 2C_{C_4H_{10}O} .$$

All that remains is to solve the last four equations for the four unknown concentrations.

Another engineering problem in which the concept of rank is of central importance is dimensional analysis. Mathematical models of physical systems are frequently made dimensionless, resulting in formation of a model with a set of dimensionless groups of parameters. This process greatly simplifies a solution because the dimensionless solution has to correlate fewer parameters than the dimensioned solution. For instance, it is much simpler to correlate the friction factor or dimensionless pressure drop in a pipe with the Reynolds number than to correlate the dimensioned pressure drop with fluid viscosity, density, pipe diameter and fluid velocity.

In some situations it may be very difficult to write an explicit model that can be made dimensionless but one may still want to obtain a set of dimensionless groups that are dependent on one another for this situation. This is useful for model studies or for correlating experimental results. If one knows all the physical parameters that enter a problem, one can find such sets even if one cannot write an explicit model. This type of analysis is called dimensional analysis.

One wants to find a set of dimensionless groups that make use of all the physical parameters of the problem while at the same time avoiding groups that are dependent on one another. For instance, if Re (the Reynolds number) and Pr (the Prandtl number) are valid dimensionless groups, then so is the group $Re^N Pr^M$ where N and M are real numbers. Finding such a set of independent groups is closely related to the problem of finding a basis of a vector space, i.e. of finding a set of linearly independent vectors.

To see this, note that the physical parameters of the problem have dimensions that are products of the fundamental dimensions (length, mass, time, etc.) raised to some power. Let the parameters of the problem be P_n and let the fundamental dimensions be m_m. We can then write

$$\text{Dimension of } P_n = m_1^{\alpha_{1n}} \cdot m_2^{\alpha_{2n}} \cdots m_M^{\alpha_{nM}}$$

and we can arrange this information in a vector

$$\begin{pmatrix} \alpha_{1n} \\ \alpha_{2n} \\ \vdots \\ \alpha_{nM} \end{pmatrix}.$$

The vectors of all the parameters are now arranged into a table and the entries form the dimensional matrix of the problem.

$$\begin{array}{ccccc} & P_1 & P_2 & \dots & P_N \\ m_1 & \alpha_{11} & \alpha_{12} & \dots & \alpha_{1N} \\ m_2 & \alpha_{21} & \alpha_{22} & \dots & \alpha_{2N} \\ \vdots & \vdots & \vdots & & \vdots \\ m_M & \alpha_{M1} & \alpha_{M2} & \dots & \alpha_{MN} \end{array} \quad \text{or simply} \quad \begin{pmatrix} \alpha_{11} & \alpha_{12} & \dots & \alpha_{1N} \\ \alpha_{21} & \alpha_{22} & \dots & \alpha_{2N} \\ \vdots & \vdots & & \vdots \\ \alpha_{M1} & \alpha_{M2} & \dots & \alpha_{MN} \end{pmatrix}.$$

The rank of the matrix determines the number of linearly independent rows and the number of fundamental dimensions. Let the rank be r, and pick r linearly independent rows. Without any loss of generality, we can assume, by renumbering if nothing else, that these r linearly independent rows are the first r rows. The remaining $N-r$ rows can now be written as linear combinations of the first r rows, e.g. as

$$P_n = \sum_{k=1}^{r} C_{nk} P_k \ , \ n = r+1, r+2, \dots, N.$$

Clearly, the dimension of P_n equals the dimension of $\prod_{k=1}^{r} P_k^{C_{nk}}$ so from each such linear combination, one can form a dimensionless group Π_n as

$$\Pi_n = \frac{\prod_{k=1}^{r} P_k^{C_{nk}}}{P_n}.$$

The $N - r$ dimensionless groups formed this way are all independent because they each contain a parameter, P_n which does not appear in the other groups. Furthermore, the number of independent groups cannot exceed $N - r$ so we have obtained our desired set of dimensionless groups.

We have shown that if a physical situation requires N physical parameters for its description and these parameters require r fundamental dimensions, then the physical situation can be modelled or described as an equation (or a set of equations) between $N - r$ dimensionless groups. This result is usually known as Buckinghams Π theorem after Edgar Buckingham who first published it.[2]

Example 5.11: Dimensional analysis.

To illustrate the use of the theorem, consider a model study of the dynamics of a boat. The situation is described by the liquid viscosity μ, density ρ and surface tension σ and by the boat velocity v, mass m and length l and by the acceleration of gravity g. This gives the dimensional matrix below

$$\begin{array}{cccccccc} & \mu & \rho & \sigma & v & m & l & g \\ L & -1 & -3 & 0 & 1 & 0 & 1 & 1 \\ M & 1 & 1 & 1 & 0 & 1 & 0 & 0 \\ T & -1 & 0 & -2 & -1 & 0 & 0 & -2 \end{array}$$

where L, M and T indicate the fundamental dimensions of length, mass and time. It is easily seen that columns 4, 5 and 6 are linearly independent, so write the remaining columns as linear combinations of these three. E.g.

$$\begin{pmatrix} -1 \\ 1 \\ -1 \end{pmatrix}_{\mu} = \begin{pmatrix} 1 \\ 0 \\ -1 \end{pmatrix}_{v} + \begin{pmatrix} 0 \\ 1 \\ 0 \end{pmatrix}_{m} - 2 \begin{pmatrix} 1 \\ 0 \\ 0 \end{pmatrix}_{l}$$

which gives the dimensionless group

$$\Pi_1 = \frac{\mu l^2}{vm}.$$

Similarly, one obtains

$$\Pi_2 = \frac{\rho l^3}{m}$$

$$\Pi_3 = \frac{\sigma l^2}{mv^2}$$

[2]Buckingham, E. [1914] "On physically similar systems; illustrations of the use of dimensional equations," *Phys. Rev.* **4**, 345376.

$$\Pi_4 = \frac{v^2}{gl} .$$

The last group, Π_4 is recognizable as the Froude number, Fr, while the remaining dimensionless groups look unfamiliar. However, it is easy to obtain the Reynolds number as

$$Re = \frac{vl\rho}{\mu} = \frac{\rho l^3}{m}\frac{vm}{\mu l^2} = \frac{\Pi_2}{\Pi_1}$$

and the Weber number as

$$We = \frac{\rho v^2 l}{\sigma} = \frac{\rho l^3}{m}\frac{mv^2}{\sigma l^2} = \frac{\Pi_2}{\Pi_3}$$

and one can therefore use the dimensionless groups Re, Fr, We and Π_2.

5.6.3 *Solution structure*

We now consider more fully the structure of the solutions to a general system of M coupled linear equations with N unknowns. Suppose we have the system

$$\begin{pmatrix} a_{11} & a_{12} & \cdots & a_{1N} \\ a_{21} & a_{22} & \cdots & a_{2N} \\ \vdots & \vdots & \ddots & \vdots \\ a_{n1} & a_{n2} & \cdots & a_{nN} \\ \vdots & \vdots & \ddots & \vdots \\ a_{M1} & a_{M2} & \cdots & a_{MN} \end{pmatrix} \begin{pmatrix} x_1 \\ x_2 \\ \vdots \\ x_n \\ \vdots \\ x_N \end{pmatrix} = \begin{pmatrix} b_1 \\ b_2 \\ \vdots \\ b_n \\ \vdots \\ b_M \end{pmatrix},$$

where $v = (x_1, \ldots, x_N)^T$ is the unknown. Assume after Gauss elimination the following augmented matrix in row-echelon form is obtained

$$\left(\begin{array}{cccccccc|c} 1 & 0 & \cdots & 0 & \gamma_{1,r+1} & \cdots & \gamma_{1,N} & & \beta_1 \\ 0 & 1 & \cdots & 0 & \gamma_{2,r+1} & \cdots & \gamma_{2,N} & & \beta_2 \\ \vdots & \vdots & \ddots & \vdots & \vdots & \ddots & \vdots & & \vdots \\ 0 & 0 & \cdots & 1 & \gamma_{r,r+1} & \cdots & \gamma_{r,N} & & \beta_r \\ 0 & 0 & \cdots & 0 & 0 & \cdots & 0 & & \beta_{r+1} \\ \vdots & \vdots & \ddots & \vdots & \vdots & \ddots & \vdots & & \vdots \\ 0 & 0 & \cdots & 0 & 0 & \cdots & 0 & & \beta_M \end{array}\right).$$

As we have seen in several earlier examples, this is not the most general form a row-echelon matrix can take after Gauss elimination, since some of the rows or columns may not appear. Nonetheless, we will get an idea how to find and represent all its solutions. We break our analysis into several cases.

Case I: $r < M$. This means there are rows in the row-echelon matrix that consists of all zeros.

(a) If there exists an index n with $r+1 \leq n \leq M$ with $\beta_n \neq 0$, then the system has no solution. Such a system is said to be *inconsistent* or *overdetermined*, and will be further considered below.

(b) Suppose $\beta_{r+1} = \cdots = \beta_M = 0$. Then there are $M - r$ redundant equations that give no relevant information regarding the set of solutions. That is, there are $M - r$ equations that are multiples of the other r equations, and removing them from the system does not change the set of solutions. The solutions can be found as in the ensuing Cases III and IV.

Case II: $M > N$. This means there are more equations than unknowns, and therefore $r < M$ and we are back to Case I.

Case III: $r = M$ and $M = N$. In this case, the system matrix has full rank and the γs do not appear. There is a unique solution $v = (x_1, \ldots, x_N)^T$ given by $x_n = \beta_n$ for each $n = 1, \ldots, N$.

Case IV: $r = M$ and $M < N$. The values of the first r unknown variables can be determined from the remaining unknowns, since

$$x_n = \beta_n - \sum_{m=r+1}^{N} \gamma_{nm} x_m \ , \ n = 1, 2, \ldots, r. \tag{5.7}$$

For any choice of scalars for the unknowns x_m, $m = r+1, \ldots, N$, then letting the other x_n be given by Eq. (5.7) provides a solution, and in fact, all solutions have this form. Each solution is different for a different choice of scalars, and thus there are infinitely many solutions. From a practical point of view this situation often implies that the model represented by the equations is not complete. For instance, more mass balances may have to be written before one can solve for the unknowns.

It may happen (as in some of the previous examples) that the leading 1s in the row-echelon matrix may not cascade downwards along consecutive rows. As a simple example with $M = 2$ and $N = 4$, suppose Gauss elimination produces a matrix of the form

$$\left(\begin{array}{cccc|c} 1 & \gamma_1 & \gamma_2 & 0 & \beta_1 \\ 0 & 0 & 0 & 1 & \beta_2 \end{array}\right).$$

In this case, the solutions satisfy $x_4 = \beta_2$, x_2 and x_3 could be any scalars, and $x_1 = \beta_1 - \gamma_1 x_2 - \gamma_2 x_3$. Again, it is easy to see that all solutions have this form.

In general, the row-echelon matrix may have many rows that do not have leading ones. Assume the system is not overdetermined and has r non-zero rows. Let I be the column indices that contain a leading one, and J the remaining indices. (In the previous example, for instance, $I = \{1, 4\}$ and $J = \{2, 3\}$). We observe from our analysis that if arbitrary scalars x_j are assigned to those coordinates with $j \in J$, then a solution to the system can be obtained by solving for the remaining coordinates x_i, $i \in I$, in terms of those x_j with $j \in J$ and $j > i$. More specifically, the set of solutions can be described in the following way. We denote the entries of the row-echelon matrix by a_{mn}, and note that the a_{mn}s that are not necessarily 0 or a leading 1 is when $1 \le m \le r$ and $j > m$ with $j \in J$. Now let x_j, $j \in J$, be any collection of scalars. For $i \in I$, let m be the row index that contains the leading one that is in column i. Finally, for the remaining indices $i \in I$, define

$$x_i = \beta_m - \sum_{j>i,\ j\in J} a_{mj} x_j. \tag{5.8}$$

Then $v \in \mathbb{F}^N$ whose n^{th} coordinate is x_n is a solution of the system. Conversely, if v is a solution then Eq. (5.8) must hold for all $i \in I$, so it is the case that every solution to the system has such a representation.

Example 5.12: Metabolic flux analysis.

In systems where a large number of coupled reactions take place, it is reasonable to ask how many reaction rates (at steady state) one must measure in order to be able to calculate them all. This problem is particularly acute in biological systems where some of the reactions are partially extracellular and their rates therefore easy to measure while other reactions are entirely intracellular and their rates therefore difficult to measure. In systems biology, this analysis of reaction rates is called metabolic flux analysis and this example illustrates the technique for the reaction network shown in Fig. 5.1. The reader who is even a little bit familiar with biochemistry will recognize it as a simplified depiction of the Embden-Meyerhof pathway and Tricarboxylic (TCA) or Krebs or citric acid cycle, the primary energy yielding pathway in higher organisms.

A few comments about the reaction network is in order. The Embden-Meyerhof pathway and *TCA* cycle are presented slightly differently in different texts depending on the amount of detail used and the assumptions made. However, these issues will not be addressed here, only the metabolic flux analysis is of interest. Reaction rates that are easy to measure are extracellular reactions such as glucose consumption, reaction 1, or carbon dioxide evolution, reaction 3. Reaction rates are also easy to measure for certain classes of compounds in the biophase such as carbohydrates or lipids, reactions 2 and 4 respectively. These rates are simply measured by measuring the biomass concentration and its composition in the exit stream from the reactor. All these reactions are indicated by broken arrows and their rates by r_n. Notice that carbon dioxide is both an intracellular component and extracellular component. The extracellular carbon dioxide is that released in reaction 3 only. All remaining reactions are intracellular and are indicated by unbroken arrows and boldface rates $\mathbf{r}_n$. The only purpose of this nomenclature is to make the figure easier to read. Finally, reaction 7 does not have the stoichiometry indicated in the figure, instead one molecule of G6P gives two molecules of Pyr. This stoichiometry was not used in the figure to avoid cluttering an already cluttered figure but must be used when writing mass balances.

We can now write steady state mass balances on each compound.

$$\text{G6P: } r_1 - r_2 - r_7/2 = 0$$

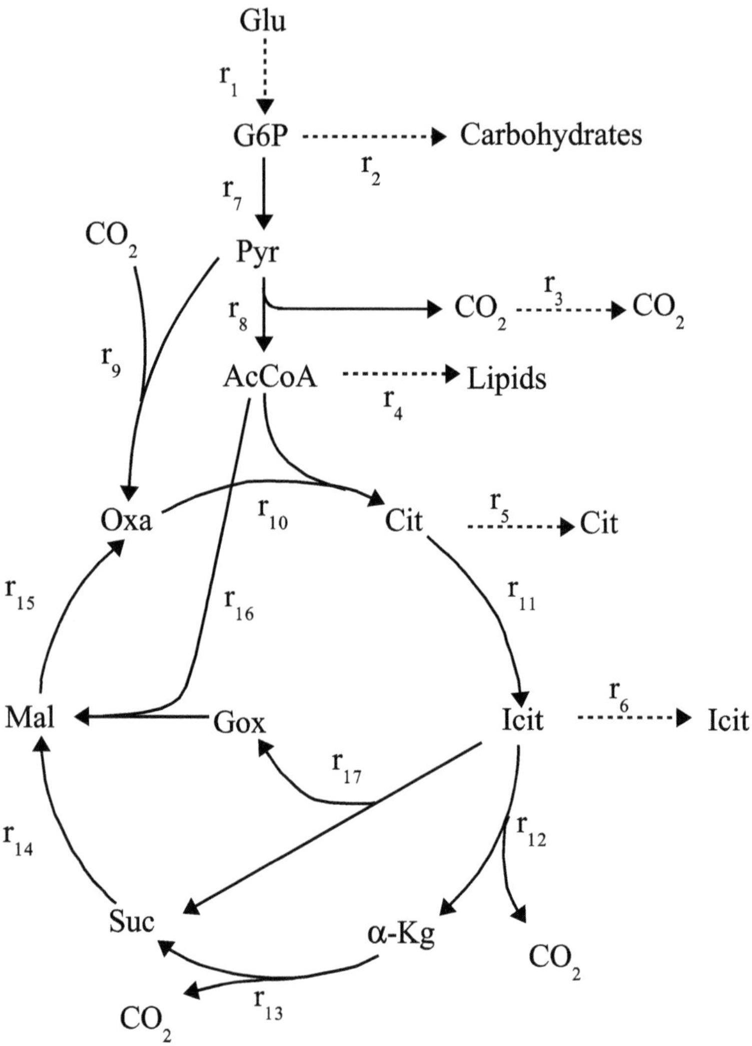

Fig. 5.1 The reactions of the Embden-Meyerhof pathway and the *TCA* cycle. The compounds are abreviated as follows: Glu, glucose; G6P, glucose-6-phosphate; Pyr, pyruvate; AcCoA, acetyl co-enzyme A; Cit, citrate; Icit, isocitrate; α-Kg, α-ketoglutarate; Suc, succinate; Mal, malate; Oxa, oxaloacdetate; Gox, glyoxylate. Reactions for which the rate can easily be measured are indicated by broken arrows and normal r_n, while hard to measure intracellular reactions are indicated by unbroken arrows and bold font $\mathbf{r}_n$.

$$\text{Pyr:}\ \ r_7 - r_8 - r_9 = 0$$

$$\text{AcCoA:}\ r_8 - r_4 - r_{10} - r_{16}$$

$$\text{Cit:}\ r_{10} - r_5 - r_{11} = 0$$

$$\text{Icit: } r_{11} - r_6 - r_{12} - r_{17} = 0$$

$$\alpha\text{-Kg: } r_{12} - r_{13} = 0$$

$$\text{Suc: } r_{13} + r_{17} - r_{14} = 0$$

$$\text{Mal: } r_{14} + r_{16} - r_{15} = 0$$

$$\text{Oxa: } r_{15} + r_9 - r_{10} = 0$$

$$\text{Gox: } r_{17} - r_{16} = 0$$

$$CO_2 : \ r_8 + r_{12} + r_{13} - r_9 - r_3 = 0 \,.$$

In vector-matrix form

$$\begin{pmatrix} 1 & -1 & 0 & 0 & 0 & 0 & -\frac{1}{2} & 0 & 0 & 0 & 0 & 0 & 0 & 0 & 0 & 0 & 0 \\ 0 & 0 & 0 & 0 & 0 & 0 & 1 & -1 & -1 & 0 & 0 & 0 & 0 & 0 & 0 & 0 & 0 \\ 0 & 0 & 0 & -1 & 0 & 0 & 0 & 1 & 0 & -1 & 0 & 0 & 0 & 0 & 0 & -1 & 0 \\ 0 & 0 & 0 & 0 & -1 & 0 & 0 & 0 & 0 & 1 & -1 & 0 & 0 & 0 & 0 & 0 & 0 \\ 0 & 0 & 0 & 0 & 0 & -1 & 0 & 0 & 0 & 0 & 1 & -1 & 0 & 0 & 0 & 0 & -1 \\ 0 & 0 & 0 & 0 & 0 & 0 & 0 & 0 & 0 & 0 & 0 & 1 & -1 & 0 & 0 & 0 & 0 \\ 0 & 0 & 0 & 0 & 0 & 0 & 0 & 0 & 0 & 0 & 0 & 0 & 1 & -1 & 0 & 0 & 1 \\ 0 & 0 & 0 & 0 & 0 & 0 & 0 & 0 & 0 & 0 & 0 & 0 & 0 & 1 & -1 & 1 & 0 \\ 0 & 0 & 0 & 0 & 0 & 0 & 0 & 0 & 1 & -1 & 0 & 0 & 0 & 0 & 1 & 0 & 0 \\ 0 & 0 & 0 & 0 & 0 & 0 & 0 & 0 & 0 & 0 & 0 & 0 & 0 & 0 & 0 & -1 & 1 \\ 0 & 0 & -1 & 0 & 0 & 0 & 0 & 1 & -1 & 0 & 0 & 1 & 1 & 0 & 0 & 0 & 0 \end{pmatrix} \begin{pmatrix} r_1 \\ r_2 \\ r_3 \\ r_4 \\ r_5 \\ r_6 \\ r_7 \\ r_8 \\ r_9 \\ r_{10} \\ r_{11} \\ r_{12} \\ r_{13} \\ r_{14} \\ r_{15} \\ r_{16} \\ r_{17} \end{pmatrix} = 0 \,.$$

The rank of this matrix is 11. Thus, one can solve for eleven variables in terms of six others. This is a nice result because the first six reaction rates are extracellular and easy to measure. However, the analysis also makes clear that all these six reaction rates must be measured in order to determine the intracellular rates. The intracellular rates are found by solving

$$\begin{pmatrix} -\frac{1}{2} & 0 & 0 & 0 & 0 & 0 & 0 & 0 & 0 & 0 & 0 \\ 1 & -1 & -1 & 0 & 0 & 0 & 0 & 0 & 0 & 0 & 0 \\ 0 & 1 & 0 & -1 & 0 & 0 & 0 & 0 & 0 & -1 & 0 \\ 0 & 0 & 0 & 1 & -1 & 0 & 0 & 0 & 0 & 0 & 0 \\ 0 & 0 & 0 & 0 & 1 & -1 & 0 & 0 & 0 & 0 & -1 \\ 0 & 0 & 0 & 0 & 0 & 1 & -1 & 0 & 0 & 0 & 0 \\ 0 & 0 & 0 & 0 & 0 & 0 & 1 & -1 & 0 & 0 & 1 \\ 0 & 0 & 0 & 0 & 0 & 0 & 0 & 1 & -1 & 1 & 0 \\ 0 & 0 & 1 & -1 & 0 & 0 & 0 & 0 & 1 & 0 & 0 \\ 0 & 0 & 0 & 0 & 0 & 0 & 0 & 0 & 0 & -1 & 1 \\ 0 & 1 & -1 & 0 & 0 & 1 & 1 & 0 & 0 & 0 & 0 \end{pmatrix} \begin{pmatrix} r_7 \\ r_8 \\ r_9 \\ r_{10} \\ r_{11} \\ r_{12} \\ r_{13} \\ r_{14} \\ r_{15} \\ r_{16} \\ r_{17} \end{pmatrix} = \begin{pmatrix} -1 & 1 & 0 & 0 & 0 & 0 \\ 0 & 0 & 0 & 0 & 0 & 0 \\ 0 & 0 & 0 & 1 & 0 & 0 \\ 0 & 0 & 0 & 0 & 1 & 0 \\ 0 & 0 & 0 & 0 & 0 & 1 \\ 0 & 0 & 0 & 0 & 0 & 0 \\ 0 & 0 & 0 & 0 & 0 & 0 \\ 0 & 0 & 0 & 0 & 0 & 0 \\ 0 & 0 & 0 & 0 & 0 & 0 \\ 0 & 0 & 0 & 0 & 0 & 0 \\ 0 & 0 & 1 & 0 & 0 & 0 \end{pmatrix} \begin{pmatrix} r_1 \\ r_2 \\ r_3 \\ r_4 \\ r_5 \\ r_6 \end{pmatrix}.$$

The example shows how experiment informs theory and vice versa. The theory or model specifies the number and type of experiments that must be done to identify all reaction rates while the number of experiments that can be done places a natural limit on the complexity of the model. For instance, if it is possible to easily measure more reaction rates than the six mentioned above, does that allow one to construct a more detailed model? In this case, one can find an additional extracellular reaction with a rate that can easily be measured and that is the rate of ammonia uptake from the medium. Ammonia is used to synthesize amino acid for protein production in the reaction

$$\alpha\text{-Kg} + NH_3 \ \rightarrow\ \text{Protein}\,.$$

This reaction can therefore be added to the model in Fig. 5.1 to create a more realistic model. Doing so will of course alter the α-Kg balance.

Similarly, one may ask if a simpler model can be found when it is not possible to measure all of the six rates mentioned above. For instance, if the rate of reaction 6, excretion of isocitrate, cannot be measured, how can the model be simplified to allow for this in such a way that all the intracellular reaction rates can still be found from the known rates? An obvious possibility is to drop glyoxylate (Gox) from the model which implies that reaction 16 and 17 must also be dropped. This gives the following set of balance equations

$$\begin{pmatrix} 1 & -1 & 0 & 0 & 0 & 0 & -\frac{1}{2} & 0 & 0 & 0 & 0 & 0 & 0 & 0 & 0 \\ 0 & 0 & 0 & 0 & 0 & 0 & 1 & -1 & -1 & 0 & 0 & 0 & 0 & 0 & 0 \\ 0 & 0 & 0 & -1 & 0 & 0 & 0 & 1 & 0 & -1 & 0 & 0 & 0 & 0 & 0 \\ 0 & 0 & 0 & 0 & -1 & 0 & 0 & 0 & 0 & 1 & -1 & 0 & 0 & 0 & 0 \\ 0 & 0 & 0 & 0 & 0 & -1 & 0 & 0 & 0 & 0 & 1 & -1 & 0 & 0 & 0 \\ 0 & 0 & 0 & 0 & 0 & 0 & 0 & 0 & 0 & 0 & 0 & 1 & -1 & 0 & 0 \\ 0 & 0 & 0 & 0 & 0 & 0 & 0 & 0 & 0 & 0 & 0 & 0 & 1 & -1 & 0 \\ 0 & 0 & 0 & 0 & 0 & 0 & 0 & 0 & 0 & 0 & 0 & 0 & 0 & 1 & -1 \\ 0 & 0 & 0 & 0 & 0 & 0 & 0 & 0 & 1 & -1 & 0 & 0 & 0 & 0 & 1 \\ 0 & 0 & -1 & 0 & 0 & 0 & 0 & 1 & -1 & 0 & 0 & 1 & 1 & 0 & 0 \end{pmatrix} \begin{pmatrix} r_1 \\ r_2 \\ r_3 \\ \vdots \\ r_{13} \\ r_{14} \\ r_{15} \end{pmatrix} = 0\,.$$

The rank of this matrix is 10, so the model simplification done in response to an inability to measure isocitrate production does have the desired effect. One can solve for the intracellular rates as well as the excretion rate of isocitrate in terms of the five known rates. (Whether or not the simplification is biologically reasonable is an altogether different question).

$$\begin{pmatrix} 0 & -\frac{1}{2} & 0 & 0 & 0 & 0 & 0 & 0 & 0 & 0 \\ 0 & 1 & -1 & -1 & 0 & 0 & 0 & 0 & 0 & 0 \\ 0 & 0 & 1 & 0 & -1 & 0 & 0 & 0 & 0 & 0 \\ 0 & 0 & 0 & 0 & 1 & -1 & 0 & 0 & 0 & 0 \\ -1 & 0 & 0 & 0 & 0 & 1 & -1 & 0 & 0 & 0 \\ 0 & 0 & 0 & 0 & 0 & 0 & 1 & -1 & 0 & 0 \\ 0 & 0 & 0 & 0 & 0 & 0 & 0 & 1 & -1 & 0 \\ 0 & 0 & 0 & 0 & 0 & 0 & 0 & 0 & 1 & -1 \\ 0 & 0 & 0 & 1 & -1 & 0 & 0 & 0 & 0 & 1 \\ 0 & 0 & 1 & -1 & 0 & 0 & 1 & 1 & 0 & 0 \end{pmatrix} \begin{pmatrix} r_6 \\ r_7 \\ r_8 \\ r_9 \\ r_{10} \\ r_{11} \\ r_{12} \\ r_{13} \\ r_{14} \\ r_{15} \end{pmatrix} = \begin{pmatrix} -1 & 1 & 0 & 0 & 0 \\ 0 & 0 & 0 & 0 & 0 \\ 0 & 0 & 0 & 1 & 0 \\ 0 & 0 & 0 & 0 & 1 \\ 0 & 0 & 0 & 0 & 0 \\ 0 & 0 & 0 & 0 & 0 \\ 0 & 0 & 0 & 0 & 0 \\ 0 & 0 & 0 & 0 & 0 \\ 0 & 0 & 0 & 0 & 0 \\ 0 & 0 & 1 & 0 & 0 \end{pmatrix} \begin{pmatrix} r_1 \\ r_2 \\ r_3 \\ r_4 \\ r_5 \end{pmatrix}.$$

5.6.4 *The null and range space of a matrix*

The vector space $\mathcal{M}_{MN}$ of all $M \times N$ dimensional matrices with entries belonging to $\mathbb{F}$ has dimension MN. As a vector space, then, $\mathcal{M}_{MN}$ is isomorphic to $\mathbb{F}^{MN}$ by Theorem 5.6. In addition to the vector space structure, an element $A \in \mathcal{M}_{MN}$ can also be thought of as a map or function that takes an element v from the domain space $\mathbb{F}^N$ by matrix multiplication into the vector Av in the image space $\mathbb{F}^M$. The concept of a root of a function is generalized to matrices through the definition of the null space of a matrix. The null space (also referred to as the kernel) of A is defined as $\mathcal{N}(A) = \{v \in \mathbb{F}^N : Av = 0\}$, i.e. the set of all vectors in the domain space that map to the zero vector in the image space. Similarly, the range of a matrix A is defined as $\mathcal{R}(A) = \{w \in \mathbb{F}^M :$ there exists $v \in \mathbb{F}^N$ with $Av = w\}$, i.e. the set of all vectors in the domain space that have a pre-image in the domain space.

It is not difficult to see that both $\mathcal{N}(A)$ and $\mathcal{R}(A)$ are subspaces of $\mathbb{F}^N$ and $\mathbb{F}^M$, respectively. Indeed, to see that $\mathcal{N}(A)$ is a subspace of $\mathbb{F}^N$, let v, w belong to $\mathcal{N}(A)$ and $\alpha \in \mathbb{F}$. Then by the properties of matrix algebra, $A(\alpha v + w) = \alpha A v + Aw = \alpha 0 + 0 = 0$, and so $\alpha v + w \in \mathcal{N}(A)$. One can also show directly that $\mathcal{R}(A)$ is a subspace, but alternatively, this can be deduced by observing that $\mathcal{R}(A) = \{Av : v \in \mathbb{F}^N\}$ is equal to the span of the columns of A, and thus must be a subspace. Moreover, by definition, the dimension of $\mathcal{R}(A)$ is the column rank of A, or as we have seen, the rank of A.

Example 5.13: Null and range spaces. The null space of a matrix A is simply the solution to $Av = 0$. For instance, for

$$A = \begin{pmatrix} 1 & -2 & 1 & 3 & 1 \\ 2 & -4 & 1 & 4 & 1 \\ 0 & 0 & -2 & -4 & 1 \end{pmatrix} v = 0 \Rightarrow \begin{pmatrix} 1 & -2 & 1 & 3 & 1 \\ 0 & 0 & 1 & 2 & 1 \\ 0 & 0 & 0 & 0 & 1 \end{pmatrix} v = 0 \Rightarrow v = \begin{pmatrix} 2s_2 - s_1 \\ s_2 \\ -2s_1 \\ s_1 \\ 0 \end{pmatrix}.$$

The rank is three, and a basis for $\mathcal{R}(A)$ consists of the first, third and fifth columns of A

$$\left\{ \begin{pmatrix} 1 \\ 2 \\ 0 \end{pmatrix} \begin{pmatrix} 1 \\ 1 \\ -2 \end{pmatrix} \begin{pmatrix} 1 \\ 1 \\ 1 \end{pmatrix} \right\}.$$

Note that the basis vectors are found as the columns of the original matrix A, not the matrix after reduction to row-echelon form. As it turns out, in this example, it does not matter because the first, third and fifth column of the row-echelon matrix gives a basis that spans the same vector space as the basis above. To see that this is not always the case, consider the matrix

$$B = \begin{pmatrix} 1 & -2 & 1 & 3 & 1 \\ 2 & -4 & 1 & 4 & 1 \\ 1 & -2 & 0 & 1 & 0 \end{pmatrix} \Rightarrow \begin{pmatrix} 1 & -2 & 1 & 3 & 1 \\ 0 & 0 & 1 & 2 & 1 \\ 0 & 0 & 0 & 0 & 0 \end{pmatrix}.$$

The basis for the range is the first and third column of B.

$$\left\{ \begin{pmatrix} 1 \\ 2 \\ 1 \end{pmatrix} \begin{pmatrix} 1 \\ 1 \\ 0 \end{pmatrix} \right\}$$

which (obviously) does not span the same space as the first and third column on the row-echelon matrix

$$\left\{ \begin{pmatrix} 1 \\ 0 \\ 0 \end{pmatrix} \begin{pmatrix} 1 \\ 1 \\ 0 \end{pmatrix} \right\}.$$

Another way to think about the range is to solve for the vectors in the domain space for an arbitrary vector in the image space. A solution is only possible if the vector in the image space is also in the range. For instance, consider again the matrix B from above and the set of equations

$$Bv = w \Rightarrow \begin{pmatrix} 1 & -2 & 1 & 3 & 1 & w_1 \\ 2 & -4 & 1 & 4 & 1 & w_2 \\ 1 & -2 & 0 & 1 & 0 & w_3 \end{pmatrix} \Rightarrow \begin{pmatrix} 1 & -2 & 1 & 3 & 1 & w_1 \\ 0 & 0 & 1 & 2 & 1 & 2w_1 - w_2 \\ 0 & 0 & 0 & 0 & 0 & w_1 - w_2 + w_3 \end{pmatrix}.$$

A solution is only possible if $w_1 - w_2 + w_3 = 0$ or if $w_3 = w_2 - w_1$. The range can thus be represented parametrically as

$$\mathcal{R}(B) = \begin{pmatrix} s_1 \\ s_2 \\ s_2 - s_1 \end{pmatrix}.$$

Convince yourself that this is in fact the same as the vector space spanned by the basis found previously.

It is not necessary to solve a set of equations to find the null space of a matrix. Reduction to row-echelon form is sufficient. We will need the following theorem concerning the dimension of $\mathcal{N}(A)$.

Theorem 5.12. *Suppose $A \in \mathcal{M}_{MN}$. Then the dimension of the null space of A equals N minus the rank of A. That is,* $\dim \mathcal{N}(A) = N - \dim \mathcal{R}(A)$.

Proof: The set of solutions to any system of equations is the same as the set of solutions to any other system obtained by Gauss elimination, and thus we may assume that A is in row-echelon form with entries $\{a_{mn}\}$. Let I be the column indices that contain a leading one, and J the remaining indices. Now, the number of elements in I is the rank of A, so it suffices to show that the number of elements in J is the dimension of $\mathcal{N}(A)$. We construct a basis of $\mathcal{N}(A)$ that is indexed by J, which will prove the result.

The idea is for each $j \in J$ to let the jth arbitrary scalar equal to 1 and the other arbitrary scalars equal to 0, and use Eq. (5.8) to produce a solution. Precisely, for each $j \in J$, let $v_j = (x_1^j, \ldots, x_N^j)^T$ be the solution of the system $Av = 0$ where

$$x_n^j = \begin{cases} 1 & \text{if } n = j \\ 0 & \text{if } n > j \text{ or } n \in J,\ n \neq j \\ -a_{m_n j} & \text{if } n \in I,\ n < j, \end{cases}$$

where m_n is equal to the number of indices in I that are less than or equal to n. One can check that each v_j has the form of a solution as described in Eq. (5.8), and so each belongs to $\mathcal{N}(A)$. We will show the set $\mathcal{B} := \{v_j\}_{j \in J}$ is a basis for $\mathcal{N}(A)$.

To see that $\mathcal{B}$ is linearly independent, suppose $\sum_{j \in J} \alpha_j v_j = 0$, which says that

$$\sum_{j \in J} \alpha_j x_n^j = 0 \tag{5.9}$$

for each $n = 1, \ldots, N$. For each $j \in J$, one has $x_n^j = 0$ if $n \neq j$ and $x_j^j = 1$. Thus the sum in Eq. (5.9) reduces to α_n for each $n \in J$, and so $\alpha_n = 0$ for all n and $\mathcal{B}$ is linearly independent.

We now show $\mathcal{B}$ spans $\mathcal{N}(A)$. Let $v = (x_1, \ldots, x_N)^T \in \mathcal{N}(A)$. Since $Av = 0$, v has the form described in Eq. (5.8) where the β's are all 0. Using this, one can verify that $v = \sum_{n=1}^{N} x_n v_n$, and so $\mathcal{B}$ spans $\mathcal{N}(A)$. □

We mentioned earlier that it is sometimes convenient to allow for non-zero entries in the column of a leading one. If this was the case here, however, the formula for x_n^j would be considerably more complicated.

Example 5.14: Null and range spaces. We illustrate the above theorem with a specific example. Let

$$A = \begin{pmatrix} 1 & -2 & 1 & 3 & 1 \\ 2 & -4 & 1 & 4 & 1 \\ 0 & 0 & -2 & -4 & 1 \end{pmatrix}$$

which can be reduced to the row-echelon form

$$\begin{pmatrix} 1 & -2 & 0 & 1 & 0 \\ 0 & 0 & 1 & 2 & 0 \\ 0 & 0 & 0 & 0 & 1 \end{pmatrix}.$$

The rank is three, $I = \{1, 3, 5\}$ and a basis for $\mathcal{R}(A)$ consists of the first, third and fifth columns of A (and *not* those columns of the echelon matrix). The Null space $\mathcal{N}(A)$ therefore has dimension 2 and $J = \{2, 4\}$. The basis for $\mathcal{N}(A)$ constructed in the proof of the theorem consists of $\{v_2, v_4\}$ where $v_2 = (-a_{12}\ 1\ 0\ 0\ 0)^T = (2\ 1\ 0\ 0\ 0)^T$ and $v_4 = (-a_{14}\ 0\ -a_{24}\ 1\ 0)^T = (-1\ 0\ -2\ 1\ 0)^T$.

The special case where $M = N$ is recorded in the following corollary.

Corollary. *For a square $N \times N$ matrix, the sum of the dimensions of the range and null space equals N.*

The last result may cause some readers to conjecture that the range and null space of a matrix are disjoint, but this does not hold in general. For instance, the matrix $A = \begin{pmatrix} 0 & 1 \\ 0 & 0 \end{pmatrix}$ has the same null and range space, which is the one-dimensional space spanned by $\begin{pmatrix} 1 \\ 0 \end{pmatrix}$.

5.6.5 *Overdetermined systems*

Overdetermined systems were defined above as those systems that have no solution. Such systems may arise in applications when repeated and inaccurate measurements of the state of a system is used to determine the values of model parameters. In these cases, although a solution may not exist, one can seek and find a best approximation $\bar{v}$ to a solution that makes the length of the vector $Av - b$ as small as possible.

To be specific, suppose $A = \{a_{mn}\} \in \mathcal{M}_{MN}$, $b \in \mathbb{F}^M$, and the system $Av = b$ is overdetermined. Now we are assuming $Av - b$ is not equal to 0 for all v, and the goal is to find a vector $v = (x_1, \ldots, x_N)^T$ such that the error $Av - b$ is minimized. In most cases, this minimization is carried out as a least square fit. In other words, one wants to minimize the function

$$f((x_1, \ldots, x_N)^T) = \sum_{m=1}^{M} \left(\sum_{n=1}^{N} a_{mn} x_n - b_m \right)^2. \tag{5.10}$$

A necessary and sufficient condition for the minimum of Eq. (5.10) to be attained at the vector v is that the partial derivatives with respect to the x_ns are all equal to zero. For fixed n, we calculate

$$\begin{aligned} 0 = \frac{\partial f}{\partial x_n} &= \sum_{m=1}^{M} a_{mn} \left(\sum_{k=1}^{N} a_{mk} x_k - b_m \right) \\ &= \sum_{m=1}^{M} \sum_{k=1}^{N} a_{mn} a_{mk} x_k - \sum_{m=1}^{M} a_{mn} b_m \end{aligned}$$

and notice that the last quantity is the nth coordinate of $A^TAx - A^Tb$. We conclude that the optimal solution v satisfies $A^TAv = A^Tb$.

Notice that if v is an optimal solution, then so is $v + w$ for any $w \in \mathcal{N}(A)$, since $A^TA(v+w) = A^TAv = A^Tb$. Therefore, a unique solution exists when $\mathcal{N}(A) = \{0\}$. In this case, the columns of A are independent, $N < M$, and A has full rank. In fact, one can then show that the square matrix A^TA has an inverse, and the solution v has the explicit representation

$$v = (A^TA)^{-1}A^Tb.$$

The matrix $(A^TA)^{-1}A^T$ is known as the *pseudoinverse* or Moore-Penrose inverse of A. If A is square and has an inverse, then the Moore-Penrose inverse agrees with the inverse.

5.7 The algebraic eigenvalue problem

It has already been mentioned that it often is a good idea to think of matrices as linear operators, simple functions that map one vector into another vector. If it is an $N \times N$ square matrix, then it maps a vector in $\mathbb{R}^N$ into another vector in $\mathbb{R}^N$ and in doing so effectively alters both the direction and the magnitude of the argument vector. For a given matrix, one may ask if there are vectors with the property that, when multiplied by the matrix, they change in magnitude but not in direction. For instance, this problem may arise when calculating forces on a surface in a fluid. The force at a point is calculated by Cauchy's law which states that the force equals the product of the so-called stress tensor, represented by a matrix, and the outward normal to the point. So asking when the force on the surface is purely normal is equivalent to asking when multiplication of the outward normal by the stress tensor matrix gives a force with the same direction as the outward normal. Mathematically, the question is if there are non-zero vectors that satisfy the equation

$$Av = \lambda v$$

alternatively written as

$$(A - \lambda I)v = 0 \tag{5.11}$$

where I is the N dimensional identity matrix and λ is a scalar which specifies the change in the magnitude of the vector v. Finding the non-zero vectors that satisfy this equation is known as the algebraic eigenvalue problem. The scalars λ are known as *eigenvalues* of the matrix A and the set of eigenvalues is called the *spectrum* of A. For a fixed value of λ, a non-zero solution v to Eq. (5.11) is called an *eigenvector*[3] of the eigenvalue λ. The system (Eq. (5.11)) always has the trivial solution $v = 0$,

[3]The term "eigen" must be a top candidate for the mathematical term with the most synonyms. Other commonly used terms include proper, latent, characteristic and singular.

and only when $v \neq 0$ is v considered an eigenvector. The set of all eigenvectors of a given eigenvalue plus the zero vector is called the *eigenspace* of that eigenvalue. The eigenspace is a subspace, as can be seen directly from the definitions, or by noticing that it equals the null space of the matrix $A - \lambda I$.

The key feature of the eigenvalue problem is that it decomposes the domain $\mathbb{F}^N$ into a direct sum of subspaces so that A is a very simple map when restricted to each of the subspaces.

5.7.1 *Finding eigenvalues and eigenvectors*

The key observation to finding an eigenvalue is to note that Eq. (5.11) has a non-trivial solution v precisely when $(A - \lambda I)$ does not have an inverse, and this is true if and only if

$$\det(A - \lambda I) = 0. \tag{5.12}$$

This equation is called the *characteristic equation*, and it is this equation which must be solved to find the eigenvalues. Notice that $p(\lambda) = \det(A - \lambda I)$ is an Nth degree polynomial in λ and is called the *characteristic polynomial*, and thus finding eigenvalues is the same as finding roots of the characteristic polynomial. It follows that the spectrum will consist of at most N elements, and will have less than N if at least one of the roots is repeated. Once an eigenvalue λ has been found, the associated eigenspace is then $\mathcal{N}(A - \lambda)$, and can be found in the manner described in the previous section. To make explicit that a vector is an eigenvector of the eigenvalue λ_n, the vector will be indicated as $\tilde{v}_n$.

Thus far, the underlying scalar field $\mathbb{F}$ has been allowed to be either $\mathbb{R}$ or $\mathbb{C}$, but now the difference between the two fields becomes important. The theory is more orderly if $\mathbb{F} = \mathbb{C}$ and A with entries in $\mathbb{C}$ maps $\mathbb{C}^N$ into $\mathbb{C}^N$. In this case, the existence of a non-zero solution $v \in \mathbb{C}^N$ to Eq. (5.11) is equivalent to λ satisfying the characteristic Eq. (5.12). If A has real entries and maps $\mathbb{R}^N$ into $\mathbb{R}^N$, then, as before, the existence of $v \neq 0$ that belongs to $\mathbb{R}^N$ and satisfies Eq. (5.11) is equivalent to λ satisfying Eq. (5.12). This can only happen if the eigenvalue λ is real, however, there are matrices with real entries but no real eigenvalues; e.g.

$$\text{if } A = \begin{pmatrix} 0 & -1 \\ 1 & 0 \end{pmatrix},$$

then $p(\lambda) = \lambda^2 + 1$ has no real roots. Nonetheless, any real or complex root of Eq. (5.12) is still called an eigenvalue. If the root is not a real number, then there can be no $\tilde{v} \neq 0$ belonging to $\mathbb{R}^N$ that satisfies Eq. (5.11), although there is such a $\tilde{v}$ belonging to $\mathbb{C}^N$ (which is still called an eigenvector) and has the form $\tilde{v} = v_1 + iv_2$ where $v_1, v_2 \in \mathbb{R}^N$. Thus even though the eigenvalue problem was originally stated in $\mathbb{R}^N$, we may have to consider complex eigenvalues with associated eigenvectors in $\mathbb{C}^N$. This unavoidable feature complicates some of the statements below when dealing with real matrices having complex eigenvalues.

The Fundamental Theorem of Algebra asserts that every N^{th} degree polynomial $p(\lambda)$ with complex coefficients has N roots in $\mathbb{C}$ if multiplicities are counted. This property of $\mathbb{C}$ is generally referred to as saying the field $\mathbb{C}$ is algebraically closed. If $\lambda_1, \ldots, \lambda_j$ are the distinct complex roots of the characteristic equation, then $p(\lambda)$ can be factored as

$$p(\lambda) = (-1)^N (\lambda - \lambda_1)^{k_1} \ldots (\lambda - \lambda_j)^{k_j}, \tag{5.13}$$

where the ks are the multiplicities and are positive integers satisfying $k_1 + \cdots + k_j = N$. k_n is called the *algebraic multiplicity* of λ_n.

We observe that if $p(\lambda)$ has real coefficients and $\lambda_k = \alpha + i\beta$ is a truly complex root (although real numbers are also complex numbers, we refer to a scalar as truly complex if $\beta \neq 0$), then its conjugate $\overline{\lambda} = \alpha - i\beta$ is also a root. That is, complex roots of real-coefficient polynomials appear in conjugate pairs. This happens because if p has real coefficients, then the conjugate of $p(\lambda)$ is equal to $p(\overline{\lambda})$, and so λ_k is a root if and only if $\overline{\lambda_k}$ is. The factors in Eq. (5.13) that contain λ_k and $\overline{\lambda}_k$ can be multiplied as $(\lambda - \lambda_k)(\lambda - \overline{\lambda_k}) = (\lambda - \alpha)^2 + \beta^2$ into a quadratic polynomial with real coefficients that cannot be linearly factored in $\mathbb{R}$.

Suppose A has real entries and so p has real coefficients. If λ is a truly complex eigenvalue with an associated eigenvector $\tilde{v} = v_1 + iv_2$, then, as we have seen, $\overline{\lambda}$ is also an eigenvalue. Let $\overline{\tilde{v}} = v_1 - iv_2$. We have $A\overline{\lambda} = \overline{A\lambda} = \overline{\lambda\tilde{v}} = \overline{\lambda}\overline{\tilde{v}}$, and hence $\overline{\tilde{v}}$ is an associated eigenvector to $\overline{\lambda}$. Thus not only do truly complex eigenvalues occur in conjugate pairs, but so do their associated eigenvectors.

We illustrate the method of solving the eigenvalue problem with an example.

Example 5.15: Calculation of eigenvalues and eigenvectors.

We find the eigenvalues and eigenvectors of the matrix

$$A = \begin{pmatrix} -1 & 0 & 3 \\ 1 & 1 & 3 \\ 0 & -1 & -2 \end{pmatrix}.$$

The characteristic polynomial is

$$\begin{vmatrix} -1-\lambda & 0 & 3 \\ 1 & 1-\lambda & 3 \\ 0 & -1 & -2-\lambda \end{vmatrix} = -\lambda^3 - 2\lambda^2 - 2\lambda - 4 = 0,$$

and its roots are

$$\lambda_1 = -2\,,\ \lambda_2 = i\sqrt{2}\,,\ \lambda_3 = -i\sqrt{2}\,.$$

Notice that the two complex eigenvalues are conjugates of each other. A first eigenvector $\tilde{v}_1$ is found by solving $(A + 2I)\tilde{v}_1 = 0$, which is equivalent to belonging to the null space of

$$\begin{pmatrix} 1 & 0 & 3 \\ 1 & 3 & 3 \\ 0 & -1 & 0 \end{pmatrix} \Rightarrow \begin{pmatrix} 1 & 0 & 3 \\ 0 & 1 & 0 \\ 0 & 0 & 0 \end{pmatrix}.$$

The latter matrix was obtained from the former through row operations, and so the null space consists of vectors of the form

$$\begin{pmatrix} -3s \\ 0 \\ s \end{pmatrix}$$

where s is any real scalar. Any choice of $s \neq 0$ produces an eigenvector $\tilde{v}_1$ associated to λ_1.

A second eigenvector is found in the null space of $A - i\sqrt{2}I$; this matrix equals and then reduces to row-echelon form as

$$\begin{pmatrix} -1-i\sqrt{2} & 0 & 3 \\ 1 & 1-i\sqrt{2} & 3 \\ 0 & -1 & -2-i\sqrt{2} \end{pmatrix} \Rightarrow \begin{pmatrix} 1 & 0 & -1+i\sqrt{2} \\ 0 & 1 & 2+i\sqrt{2} \\ 0 & 0 & 0 \end{pmatrix}.$$

The null space consists of vectors of form

$$\begin{pmatrix} (1-i\sqrt{2})s \\ (-2-i\sqrt{2})s \\ s \end{pmatrix} = s\left[\begin{pmatrix} 1 \\ -2 \\ 1 \end{pmatrix} + i\begin{pmatrix} -\sqrt{2} \\ -\sqrt{2} \\ 0 \end{pmatrix}\right] = \tilde{v}_2 \,.$$

Note the second eigenvector has complex entries, which will always be the case for a complex eigenvalue if A has real entries. As shown above, complex eigenvalues and associated eigenvectors are paired as conjugates, so we can immediately assert that the third eigenvector associated to λ_3 is of the form

$$\tilde{v}_3 = s\left[\begin{pmatrix} 1 \\ -2 \\ 1 \end{pmatrix} - i\begin{pmatrix} -\sqrt{2} \\ -\sqrt{2} \\ 0 \end{pmatrix}\right].$$

To get an idea of the general solution structure of the eigenvalue problem, we describe the case of $N = 2$ and $\mathbb{F} = \mathbb{R}$ in complete detail. Let

$$A = \begin{pmatrix} a_{11} & a_{12} \\ a_{21} & a_{22} \end{pmatrix}$$

with all of the entries being real numbers. The characteristic polynomial is $p(\lambda) = \det(A - \lambda I) = \lambda^2 - (a_{11} + a_{22})\lambda + a_{11}a_{22} - a_{12}a_{21} = \lambda^2 - \operatorname{tr}(A)\lambda + \det(A)$, where $\operatorname{tr}(A) = a_{11} + a_{22}$ is the trace of A and $\det(A)$ is the determinant of A. The roots of p are found by the quadratic formula, and equal

$$\frac{1}{2}\left(\operatorname{tr}(A) \pm \sqrt{d}\right),$$

where $d = \operatorname{tr}(A)^2 - 4\det(A)$ is the discriminant. There are three main cases to consider determined by the sign of d.

Case I: $d > 0$. This means there are two distinct real eigenvalues $\lambda_1 = \frac{1}{2}(\text{tr}A + \sqrt{d})$ and $\lambda_2 = \frac{1}{2}(\text{tr}(A) - \sqrt{d})$ that both have algebraic multiplicity 1. It turns out that the associated eigenvectors $\tilde{v}_1, \tilde{v}_2$ are linearly independent, and therefore form a basis for $\mathbb{R}^2$.
Case II: $d = 0$ and $a_{11} = a_{22}$. There is one eigenvalue only, $\lambda = a_{11} = a_{22}$. It has an algebraic multiplicity of 1, and since we are also assuming $d = 0$, it follows that $a_{12} = a_{21} = 0$. A is therefore the matrix λI, and any two independent vectors $\tilde{v}_1$ and $\tilde{v}_2$ are eigenvectors and form a basis for $\mathbb{R}^2$. Thus, the eigenspace of λ is two-dimensional and we say that the geometric multiplicity of the eigenvalue equals two.
Case III: $d = 0$ and $a_{11} \neq a_{22}$. In this case, again there is only one eigenvalue $\lambda = \frac{a_{11}+a_{22}}{2}$, and it has algebraic multiplicity equal to two. However, there is only one linearly independent eigenvector $\tilde{v}_1$, and so the geometric multiplicity of the eigenvalue is 1. One can find a basis for $\mathbb{R}^2$ that includes $\tilde{v}_1$ by solving the system $(A - \lambda I)\tilde{v}_2 = \tilde{v}_1$ for $\tilde{v}_2$. The second vector $\tilde{v}_2$ is called a *generalized eigenvector*, and $\{\tilde{v}_1, \tilde{v}_2\}$ forms a basis for $\mathbb{R}^2$.
Case IV: $d < 0$. The eigenvalues are now truly complex, and are the complex conjugates $\frac{1}{2}(\text{tr}A \pm i\sqrt{-d})$. The eigenvectors must be sought in $\mathbb{C}^2$, and are of the form $v_1 \pm iv_2$ where $v_n \in \mathbb{R}^2$, $n = 1, 2$. We will see below that $\{v_1, v_2\}$ are independent and so forms a basis of $\mathbb{R}^2$.

Let K be the 2×2 matrix whose columns are the basis vectors $\tilde{v}_1$ and $\tilde{v}_2$ as given in each of the cases. A clear understanding of how A behaves as an operator from $\mathbb{R}^2$ to $\mathbb{R}^2$ can be described in matrix terms. In case I, we have $A\tilde{v}_n = \lambda\tilde{v}_n$, $n = 1, 2$, and so

$$AK = (A\tilde{v}_1 \; A\tilde{v}_2) = (\lambda_1\tilde{v}_1 \; \lambda_2\tilde{v}_2) = K\begin{pmatrix} \lambda_1 & 0 \\ 0 & \lambda_2 \end{pmatrix}.$$

Recall that a square matrix K is invertible if and only if its columns form a basis, and hence we can also write

$$A = K\begin{pmatrix} \lambda_1 & 0 \\ 0 & \lambda_2 \end{pmatrix}K^{-1}. \tag{5.14}$$

Case II has the exact same form with the additional property that $\lambda = \lambda_1 = \lambda_2$. Case III has $A\tilde{v}_1 = \lambda\tilde{v}_1$ and $A\tilde{v}_2 = \tilde{v}_1 + \lambda\tilde{v}_2$, and thus

$$AK = K\begin{pmatrix} \lambda & 1 \\ 0 & \lambda \end{pmatrix} \quad \text{or, equivalently,} \quad A = K\begin{pmatrix} \lambda & 1 \\ 0 & \lambda \end{pmatrix}K^{-1}. \tag{5.15}$$

Case IV has the truly complex eigenvalues $\lambda = \alpha + i\beta$ and $\lambda = \alpha - i\beta$, where $\alpha = \frac{\text{tr}A}{2}$ and $\beta = i\frac{-d}{2}$. There are no eigenvectors in $\mathbb{R}^2$. Nonetheless, a useful description of A can still be obtained. Let $\lambda = \alpha + i\beta$ with $\alpha, \beta \in \mathbb{R}$, $\beta \neq 0$, be the eigenvalue with an associated (complex) eigenvector $\tilde{v} = v_1 + iv_2$. We have $A\tilde{v} = \lambda\tilde{v}$ in $\mathbb{C}^2$, and the real and imaginary parts must coincide. On the one hand, $A\tilde{v} = Av_1 + iAv_2$ since A has all real entries, and the other hand, $\lambda\tilde{v} = (\alpha + i\beta)(v_1 + iv_2) = (\alpha v_1 - \beta v_2) + i(\beta v_1 + \alpha v_2)$. Thus $Av_1 = \alpha v_1 - \beta v_2$ and $Av_2 = \beta v_1 + \alpha v_2$. In matrix terms, this says

$$AK = K \begin{pmatrix} \alpha & \beta \\ -\beta & \alpha \end{pmatrix} \quad \text{or equivalently,} \quad A = K \begin{pmatrix} \alpha & \beta \\ -\beta & \alpha \end{pmatrix} K^{-1}. \tag{5.16}$$

Example 5.16: Structure of eigenvalues and eigenvectors.
Case I: Consider

$$A = \begin{pmatrix} 4 & 1 \\ 8 & 6 \end{pmatrix} \Rightarrow p(\lambda) = \lambda^2 - 10\lambda + 16$$

giving

$$\lambda_1 = 2\ ,\ \tilde{v}_1 = \begin{pmatrix} s \\ 2s \end{pmatrix} \quad \text{and} \quad \lambda_2 = 8\ ,\ \tilde{v}_2 = \begin{pmatrix} 2 \\ 4s \end{pmatrix}.$$

Forming the matrix of eigenvectors, K, gives

$$K = \begin{pmatrix} 1 & 1 \\ -2 & 4 \end{pmatrix} \Rightarrow A = K \begin{pmatrix} 2 & 0 \\ 0 & 8 \end{pmatrix} K^{-1}.$$

Case II: Consider

$$A = \begin{pmatrix} 4 & 0 \\ 0 & 4 \end{pmatrix} \Rightarrow p(\lambda) = (4 - \lambda)^2$$

giving

$$\lambda_1 = 4\ ,\ \tilde{v}_1 = \begin{pmatrix} s_1 \\ s_2 \end{pmatrix} = s_1 \begin{pmatrix} 1 \\ 0 \end{pmatrix} + s_2 \begin{pmatrix} 0 \\ 1 \end{pmatrix}.$$

Forming the matrix of eigenvectors, K, gives

$$K = \begin{pmatrix} 1 & 0 \\ 0 & 1 \end{pmatrix} \Rightarrow A = K \begin{pmatrix} 4 & 0 \\ 0 & 4 \end{pmatrix} K^{-1}.$$

Case III: Consider

$$A = \begin{pmatrix} 3 & 1 \\ -1 & 1 \end{pmatrix} \Rightarrow p(\lambda) = \lambda^2 - 4\lambda + 4$$

giving

$$\lambda_1 = 2\ ,\ \tilde{v}_1 = \begin{pmatrix} s_1 \\ -s_1 \end{pmatrix}.$$

Now find $\tilde{v}_2$ as

$$(A - \lambda I)\tilde{v}_2 = \begin{pmatrix} 1 & 1 \\ -1 & -1 \end{pmatrix} \tilde{v}_2 = \tilde{v}_1 \;\Rightarrow\; \tilde{v}_2 = \begin{pmatrix} s_1 - s_2 \\ s_2 \end{pmatrix}.$$

Setting $s_1 = s_2 = 1$ and forming the matrix of eigenvectors, K, gives

$$K = \begin{pmatrix} 1 & 0 \\ -1 & 1 \end{pmatrix} \;\Rightarrow\; A = K \begin{pmatrix} 2 & 1 \\ 0 & 2 \end{pmatrix} K^{-1}.$$

Case IV: Consider

$$A = \begin{pmatrix} 4 & -1 \\ 8 & 8 \end{pmatrix} \;\Rightarrow\; p(\lambda) = \lambda^2 - 12\lambda + 40$$

giving

$$\lambda_{1,2} = 6 \pm 2i\ ,\ \tilde{v}_{1,2} = s \begin{pmatrix} 1 \\ -2 \mp 2i \end{pmatrix}.$$

Notice that the minus sign in the eigenvector corresponds to the plus sign in the eigenvalue. Forming the matrix of eigenvectors, K, gives

$$K = \begin{pmatrix} 1 & 1 \\ -2-2i & -2+2i \end{pmatrix} \;\Rightarrow\; A = K \begin{pmatrix} 6+2i & 0 \\ 0 & 6-2i \end{pmatrix} K^{-1}.$$

To avoid complex numbers, the K-matrix can be replaced by a matrix of the real and complex parts of $\tilde{v}_1$. Thus

$$K = \begin{pmatrix} 1 & 0 \\ -2 & 2 \end{pmatrix} \;\Rightarrow\; A = K \begin{pmatrix} 6 & 2 \\ -2 & 6 \end{pmatrix} K^{-1}.$$

We point out again that in the complex case $\mathbb{F} = \mathbb{C}$ where A has complex entries and maps $\mathbb{C}^2$ to $\mathbb{C}^2$, the characteristic polynomial has two complex roots (counting multiplicity). All possibilities are covered in exactly the same manner as above in Cases I–III, and as before, the columns of the matrix K form a basis for $\mathbb{C}^2$. There is no analogue for Case IV, which is why the decomposition we seek is more complicated if the underlying field is $\mathbb{R}$.

To make the following discussion less abstract we will first look at two examples that show the value of representing a matrix in the form $A = KJK^{-1}$. Suppose, for instance, that one needs to find the matrix raised to a power, then one can write

$$A^n = (KJK^{-1})^n = KJ^NK^{-1}.$$

But it is easily seen that a diagonal matrix raised to a power equals another diagonal matrix in which the diagonal terms have been raised to this power. Thus, taking a matrix from the example above

$$\begin{pmatrix} 4 & 1 \\ 8 & 6 \end{pmatrix}^n = \begin{pmatrix} 1 & 1 \\ -2 & 4 \end{pmatrix} \begin{pmatrix} 2 & 0 \\ 0 & 8 \end{pmatrix}^n \begin{pmatrix} \frac{2}{3} & -\frac{1}{6} \\ \frac{1}{3} & \frac{1}{6} \end{pmatrix} = \begin{pmatrix} 1 & 1 \\ -2 & 4 \end{pmatrix} \begin{pmatrix} 2^n & 0 \\ 0 & 8^n \end{pmatrix} \begin{pmatrix} \frac{2}{3} & -\frac{1}{6} \\ \frac{1}{3} & \frac{1}{6} \end{pmatrix} \Rightarrow$$

$$\begin{pmatrix} 4 & 1 \\ 8 & 6 \end{pmatrix}^n = \begin{pmatrix} \frac{2}{3}2^n + \frac{1}{3}8^n & -\frac{1}{6}2^n + \frac{1}{6}8^n \\ -\frac{4}{3}2^n + \frac{4}{3}8^n & \frac{1}{3}2^n + \frac{2}{3}8^n \end{pmatrix}.$$

Consider the two coupled ODEs

$$\begin{pmatrix} \frac{dx}{dt} \\ \frac{dy}{dt} \end{pmatrix} = \begin{pmatrix} 4 & 1 \\ 8 & 6 \end{pmatrix} \begin{pmatrix} x \\ y \end{pmatrix}.$$

The previous example shows that this can equally well be written as

$$\begin{pmatrix} \frac{dx}{dt} \\ \frac{dy}{dt} \end{pmatrix} = K \begin{pmatrix} 2 & 0 \\ 0 & 8 \end{pmatrix} K^{-1} \begin{pmatrix} x \\ y \end{pmatrix} \Rightarrow K^{-1} \begin{pmatrix} \frac{dx}{dt} \\ \frac{dy}{dt} \end{pmatrix} = \begin{pmatrix} 2 & 0 \\ 0 & 8 \end{pmatrix} K^{-1} \begin{pmatrix} x \\ y \end{pmatrix}$$

or

$$\begin{pmatrix} \frac{d(\frac{2}{3}x - \frac{1}{6}y)}{dt} \\ \frac{d(\frac{1}{3}x + \frac{1}{6}y)}{dt} \end{pmatrix} = \begin{pmatrix} 2 & 0 \\ 0 & 8 \end{pmatrix} \begin{pmatrix} (\frac{2}{3}x - \frac{1}{6}y) \\ (\frac{1}{3}x + \frac{1}{6}y) \end{pmatrix}.$$

The problem now takes a form with two decoupled equations which are trivial to solve

$$\begin{matrix} \frac{2}{3}x - \frac{1}{6}y = C_1e^{2t} \\ \frac{1}{3}x + \frac{1}{6}y = C_2e^{8t} \end{matrix} \Rightarrow \begin{matrix} x = & C_1e^{2t} + C_2e^{8t} \\ y = & -2C_1e^{2t} + 4C_2e^{8t}. \end{matrix}$$

The mapping properties of diagonal matrices are particularly easy to understand, since they simply multiply the kth canonical basis element by the kth diagonal element. What is the effect of the representation $A = KJK^{-1}$ that appears on the right hand sides of Eqs. (5.14)–(5.16)? Since K is invertible, its columns form a basis $\mathcal{B}$. If a vector v is viewed as the coordinates of a vector with respect to $\mathcal{B}$, then multiplying v by K^{-1} produces the same vector written in its canonical coordinates. Mapping by K has the opposite effect. Thus one can view the composed map KJK^{-1} in three stages: firstly, it takes a vector in its $\mathcal{B}$-coordinates to its canonical coordinates; secondly, the canonical coordinates are multiplied by J; and thirdly, this vector is mapped back to its coordinates of the canonical basis. The right hand sides of Eqs. (5.14)–(5.16) thus give a method of representing the map A in a particularly structured manner. We seek a similar structural description of a matrix A for the general situation where the dimension is $N > 2$. The goal is to find a basis for $\mathbb{F}^N$ (whose columns will define the matrix K) that reveals the basic behavior of A as a map from $\mathbb{F}^N$ to $\mathbb{F}^N$.

5.7.2 *Multiplicity*

We assume throughout this subsection that either $\mathbb{F} = \mathbb{C}$, *or that* $\mathbb{F} = \mathbb{R}$ *and the eigenvalue* λ_0 *is real.* The algebraic multiplicity of an eigenvalue has already been defined as the multiplicity of the eigenvalue as a root in the characteristic polynomial. The *geometric multiplicity* of such a λ_0 is defined only in these cases, and is equal to the dimension of the associated eigenspace. The spectrum of a matrix is said to be *degenerate* if any of the roots of the characteristic polynomial have a multiplicity greater than 1, and *nondegenerate* if there are distinct eigenvalues.

We can see the difference of the two types of multiplicities in the above description where $\mathbb{F} = \mathbb{R}$ in dimension $N = 2$. In Case I, where the eigenvalues are real and distinct, both the algebraic and geometric multiplicities are one for both eigenvalues. There is a repeated eigenvalue in Cases II and III. In Case II, the algebraic and geometric multiplicity is 2, and in Case III, the algebraic multiplicity is 2 and the geometric multiplicity is 1. Notice that in each case, the geometric multiplicity is less than or equal to the algebraic multiplicity, and the next result says this holds in general.

Theorem 5.13. *The geometric multiplicity of an eigenvalue* λ_0 *is less than or equal to its algebraic multiplicity.*

Proof: Suppose A is $N \times N$ and the geometric multiplicity of an eigenvalue λ_0 is m. The eigenspace associated to λ_0 is the null space of $(A - \lambda_0 I)$, and we are assuming $\dim \mathcal{N}(A - \lambda_0 I) = m$. From Theorem 5.12, the size of a square matrix equals the sum of the rank and the dimension of the null space, and so the rank of $(A - \lambda_0 I)$ equals $N - m$. We will use this result shortly.

The algebraic multiplicity of λ_0 is the multiplicity of λ_0 as a root in the characteristic polynomial $p(\lambda)$. If $m = 1$, then $(A - \lambda_0 I)$ has no inverse and λ_0 is a root of $p(\lambda)$, and therefore the algebraic multiplicity is at least one and the result is true in this case. Now in general, a root λ of a polynomial has multiplicity n if and only if all the derivatives evaluated at λ up to order $n - 1$ of the polynomial is equal to zero. Thus it suffices to show the jth derivative $\frac{d^j p}{d\lambda^j}(\lambda_0)$ is zero for each $j = 1, \ldots, m - 1$.

The derivative of $p(\lambda) = \det(A - \lambda I)$ was calculated in Subsection 5.5.3.3, and it greatly simplifies here since all of the entries of $(A - \lambda I)$ are constant except for the diagonal entries. Let $\dot{D}_n(\lambda)$ be the determinant of the matrix whose entries are the same as $(A - \lambda I)$ except for the nth row, which is replaced by the derivative of the nth row of $(A - \lambda I)$. The derivative of the nth row contains all 0s except for the nth column which is -1. Hence $\dot{D}_n(\lambda) = (-1)^n \det(A_n(\lambda))$, where $A_n(\lambda)$ is the $(N-1) \times (N-1)$ dimensional matrix obtained by deleting the nth row and column of $(A - \lambda I)$. The formula derived in Subsection 5.5.3.3 says that

$$\left.\frac{dp}{d\lambda}\right|_{\lambda_0} = \sum_{n=1}^{N} \dot{D}_n(\lambda_0) = \sum_{n=1}^{N} (-1)^n \det(A_n(\lambda_0)). \tag{5.17}$$

We have noted above that the rank of $(A - \lambda_0 I)$ is $N - m$, and this is the maximum number of independent rows. For each $n = 1, \ldots, N$, the structure of $A_n(\lambda_0)$ is such that the number of independent rows cannot be larger and so is at most $N - m$. Again by Theorem 5.12, we have $N - 1 = \dim \mathcal{N}(A(\lambda_0)) +$ the rank of $A(\lambda_0)$, and so $\dim \mathcal{N}(A(\lambda_0))$ is at least $m - 1$. Hence $\det(A_n(\lambda_0)) = 0$, and it follows from Eq. (5.17) that $\left.\frac{dp}{d\lambda}\right|_{\lambda_0} = 0$. This proves the theorem if $m = 2$. For larger m, the same process can be repeated on each of the smaller matrices $A_n(\lambda)$, and after $m-1$ times, the conclusion is that $\left.\frac{d^j p}{d\lambda^j}\right|_{\lambda_0} = 0$ for $j = 1, \ldots, m - 1$. □

5.7.3 *Similar matrices*

Before proceeding to the general decomposition results, we introduce formally the notion of similarity that was previously hinted at in Eqs. (5.14)–(5.16).

Definition: *Two matrices A and B are similar, which is written as $A \sim B$, if and only if there exists an invertible matrix K such that*

$$K^{-1} A K = B. \tag{5.18}$$

The transformation between similar matrices is called a similarity transformation. The interpretation of similar matrices viewed as maps on $\mathbb{F}^2$ was given in the paragraph following Eq. (5.16), and is equally valid for arbitrary dimension N. The crucial insight is that similar matrices can be viewed as representing the same map from $\mathbb{F}^N$ to $\mathbb{F}^N$ but with respect to different bases in $\mathbb{F}^N$. One may then suspect that similar matrices have a range of properties in common. That this is indeed the case is born out in the following theorem.

Theorem 5.14.

(1) *Similar matrices have the same determinant.*
(2) *Similar matrices have the same characteristic polynomial, and therefore have the same eigenvalues with the same algebraic multiplicity.*

(3) *Suppose A and B satisfy Eq. (5.18). Then $\tilde{v}_B$ is an eigenvector of B associated to λ if and only if $\tilde{v}_A = K\tilde{v}_B$ is an eigenvector of A associated λ.*

(4) *Suppose $A \sim B$. The geometric multiplicity of an eigenvalue with respect to A is the same as that with respect to B.*

Proof (1): Suppose Eq. (5.18) holds. Then by Theorem 5.8(8), we have $\det(B) = \det(K^{-1}AK) = \det(K^{-1})\det(A)\det(K) = \det(A)$.
(2): Suppose Eq. (5.18) holds. The characteristic polynomials p_A and p_B are calculated by

$$\begin{aligned} p_B(\lambda) &= \det(B - \lambda I) = \det(K^{-1}AK - \lambda I) \\ &= \det(K^{-1}(A - \lambda I)K) = \det(K^{-1})\det(A - \lambda I)\det(K) \\ &= \det(A - \lambda I) = p_A(\lambda). \end{aligned}$$

It immediately follows that the eigenvalues and their algebraic multiplicites are the same for both matrices.
(3): Suppose $K^{-1}AK = B$ and $B\tilde{v}_B = \lambda\tilde{v}_B$. Then $AK\tilde{v}_B = KB\tilde{v}_B = \lambda K\tilde{v}_B$, and hence $\tilde{v}_A = K\tilde{v}_B$ is an eigenvector of A associated to λ. Conversely, if $\tilde{v}_A = K\tilde{v}_B$ is an eigenvector of A, then $B\tilde{v}_B = BK^{-1}\tilde{v}_A = K^{-1}AKK^{-1}\tilde{v}_A = K^{-1}A\tilde{v}_A = \lambda K^{-1}\tilde{v}_A = \lambda\tilde{v}_B$, and thus $\tilde{v}_B$ is an eigenvector of B.
(4): The proof is based on the fact that invertible matrices map linearly independent sets to linearly independent sets. Indeed, suppose K is invertible, and $\{v_1, \dots, v_M\}$ is a linearly independent set of vectors. For scalars $\alpha_1, \dots, \alpha_M$ such that

$$0 = \sum_{m=1}^{M} \alpha_m K v_m = K\left(\sum_{m=1}^{M} \alpha_m v_m\right),$$

we must have $\sum_{m=1}^{M} \alpha_m v_m = 0$ since K is invertible, and therefore $\alpha_1 = \cdots = \alpha_M = 0$ since the vs are linearly independent.

Now suppose the geometric multiplicity with respect to B of an eigenvalue λ is M. That is, there exists a basis $\{v_1, \dots, v_M\}$ of $\mathcal{N}(B - \lambda I)$ consisting of M eigenvectors of B. The fact just mentioned says that $\{Kv_1, \dots, Kv_M\}$ is linearly independent, and Part (4) of this theorem implies each of the Kv_ms are eigenvectors of A. Thus $\mathcal{N}(A - \lambda I)$ contains at least M linearly independent vectors, and so the geometric multiplicity with respect to A must be at least M. That it equals M is deduced by using the same argument with the roles of A and B reversed. □

The properties in the previous theorem are not sufficient for matrices to be similar. That is, if A and B have the same eigenvalues with the same algebraic and geometric multiplicities, it does not follow that A and B are similar. As we see in the following, there are missing ingredients to characterize the similarity property that involve the dimension and structure of *generalized eigenspaces.*

We close this subsection with two simple observations. Suppose $K^{-1}AK = B$. Then

$$K^{-1}A^nK = B^n \quad \text{for all} \quad n = 1, \ldots. \tag{5.19}$$

Indeed, we have

$$\begin{aligned} K^{-1}A^nK &= K^{-1}\underbrace{A(KK^{-1})A(KK^{-1})\ldots(KK^{-1})A}_{n \text{ times}}K \\ &= (K^{-1}AK)(K^{-1}AK)\ldots\ (K^{-1}AK)(K^{-1}AK) \\ &= B^n. \end{aligned}$$

We secondly observe that

$$K^{-1}(A - \lambda I)K = (B - \lambda I) \quad \text{for all} \quad \lambda \in \mathbb{F}. \tag{5.20}$$

The proof consists of noting that

$$K^{-1}(A - \lambda I)K = K^{-1}AK - \lambda K^{-1}K = B - \lambda I.$$

The goal of the next several subsections is to find a matrix J that is similar to a given matrix A and that has a particularly nice structure. However, before proceeding to this, we will briefly introduce the concept of equivalence relations.

5.7.3.1 *Equivalence relations*

The property of two matrices being similar is an example of an equivalence relation, which is a general mathematical concept that is important enough to justify a brief digression.

An equivalence relation is relationship defined on a set of objects which generalizes the concept of equality. That is, elements of a set are considered equivalent if they share some property. For instance, two natural numbers may be considered equivalent if their difference is divisible by 2. A subset consisting of all equivalent objects is called an *equivalence class.* Note that in the equivalence relation just stated, the set of natural numbers is partitioned into two equivalence classes, one being the set of all even numbers and the other the set of all odd numbers. This is a general property of equivalence relations. It is often the case that one is primarily interested in the properties of equivalence classes instead of the properties of individual elements of a set. The precise definition of an equivalence relation follows.

Definition: *An equivalence relation on a set X is a relation $\sim$ between pairs of elements (x, y) of X that satisfies the axioms:*

(1) Reflexivity: $x \sim x$.
(2) Symmetry: $x \sim y \Rightarrow y \sim x$.
(3) Transitivity: $x \sim y$ and $y \sim z \Rightarrow x \sim z$.

The set $\{x \in X | x \sim y\}$ is called the equivalence class of X containing y, and is indicated by $[y]$.

It is easy to verify that matrix similarity is an equivalence relation. Clearly (1) $A \sim A$ since we can take $K = I$. (2) If $A \sim B$, then $B \sim A$ is seen to hold by replacing K by K^{-1} in Eq. (5.18). (3) If $A \sim B$ and $B \sim C$, then there exists invertible matrices K_1 and K_2 so that $K_1^{-1}AK_1 = B$ and $K_2^{-1}BK_2 = C$. Now $K = K_1K_2$ is invertible, and it satisfies

$$K^{-1}AK = (K_1K_2)^{-1}AK_1K_2 = K_2^{-1}K_1^{-1}AK_1K_2 = K_2^{-1}BK_2 = C,$$

and so $A \sim C$.

A partition of a set is a collection of nonempty and disjoint subsets whose union equals the set itself. There is a direct correspondence between all the equivalent relations that can be defined on a set and all the partitions of the set. From a partition, an equivalence relation can be obtained by defining that two elements are equivalent if they are both in the same subset. It is easy to see that this does in fact satisfy the three requirements of an equivalence relation. Conversely, suppose an equivalence relation $\sim$ is given. Then the collection of equivalence classes form a partition. To see this, first note that each x belongs to an equivalence class. We only need to show that different equivalent classes are disjoint. Let $[x]$ and $[y]$ be two different equivalence classes, and assume that z is an element belonging to both. If u is an element of $[x]$, then $u \sim x$ and $x \sim z$, and by transitivity we have $u \sim z$. But since also $z \sim y$, we therefore have $u \sim y$, or that $u \in [y]$. Thus $[x] \subseteq [y]$, and with an identical argument reversing the roles of x and y, one can see that $[y] \subseteq [x]$. Hence $[x] = [y]$, which contradicts the assumption that $[x]$ and $[y]$ are different equivalence classes. □

5.7.4 *Eigenspaces and eigenbases*

The eigenvectors found for different eigenvalues in the examples earlier were linearly independent, and the next result reveals that this is a general fact.

Theorem 5.15. *A set of eigenvectors in $\mathbb{C}^N$ that are associated to distinct eigenvalues is linearly independent over $\mathbb{C}$.*

Proof: Suppose $\tilde{v}_n$ is an eigenvector associated to an eigenvalue λ_n for $n = 1, \ldots, N$ where the λ_ns are distinct. By Theorem 5.3, there exists a subset $\mathcal{B}$ of $\{\tilde{v}_n\}_{n=1,N}$ that is linearly independent and spans the same subspace that is spanned by all of the $\tilde{v}_n$s. By reindexing if necessary, we can assume $\mathcal{B} = \{\tilde{v}_1, \ldots, \tilde{v}_M\}$ where $M \leq N$. The theorem states that all the eigenvectors are linearly independent, i.e. that $M = N$, so for the purpose of obtaining a contradiction, suppose $M < N$. Since $\tilde{v}_{M+1} \in \text{span}\,\mathcal{B}$, there exist scalars $\alpha_1, \ldots, \alpha_M$ so that

$$\tilde{v}_{M+1} = \alpha_1\tilde{v}_1 + \cdots + \alpha_M\tilde{v}_M. \tag{5.21}$$

The property of an eigenvector is that $A\tilde{v}_n = \lambda_n$, and so if A is applied to both sides of Eq. (5.21), we have

$$\lambda_{M+1}\tilde{v}_{M+1} = A\tilde{v}_{M+1} = \alpha_1 A\tilde{v}_1 + \cdots + \alpha_M A\tilde{v}_M = \alpha_1\lambda_1\tilde{v}_1 + \cdots + \alpha_M\lambda_M\tilde{v}_M. \quad (5.22)$$

On the other hand, we can multiply Eq. (5.21) by λ_{M+1} and see that

$$\lambda_{M+1}\tilde{v}_{M+1} = \alpha_1\lambda_{M+1}\tilde{v}_1 + \cdots + \alpha_M\lambda_{M+1}\tilde{v}_M. \quad (5.23)$$

When Eq. (5.23) is subtracted from Eq. (5.22), we obtain

$$0 = \alpha_1(\lambda_1 - \lambda_{M+1})\tilde{v}_1 + \cdots + \alpha_M(\lambda_M - \lambda_{M+1})\tilde{v}_M.$$

Since $\mathcal{B}$ is linearly independent, therefore $\alpha_1(\lambda_1 - \lambda_{M+1}) = \cdots = \alpha_M(\lambda_1 - \lambda_{M+1}) = 0$, and since we are assuming the eigenvalues are distinct, $\lambda_{M+1} \neq \lambda_n$ for each $n = 1, \ldots, M$. It now follows that $\alpha_1 = \cdots = \alpha_M = 0$, which inserted into Eq. (5.21) implies that $\tilde{v}_{M+1} = 0$. This contradicts that $\tilde{v}_{M+1}$ is an eigenvector (and in particular is not the zero vector), and finishes the proof. □

Corollary. *Eigenspaces that belong to different eigenvalues are disjoint.*

A basis whose elements are eigenvectors is called an *eigenbasis* for $\mathbb{F}^N$. The ideal situation to understand an operator A is when there exists an eigenbasis, but one may not exist. We next explore scenarios for any number of dimensions. We consider six main cases, but these by no means exhaust all possibilities. The first three are only pertinent when $\mathbb{F} = \mathbb{C}$, or if $\mathbb{F} = \mathbb{R}$ and the eigenvalues are real. Cases IV–VI are special to $\mathbb{F} = \mathbb{R}$ with truly complex values.

Case I: The simplest eigenvalue problems are those for which all the eigenvalues have algebraic multiplicity equal to one, which means the eigenvalues are distinct. If the matrix is viewed as a map from $\mathbb{R}^N$ to $\mathbb{R}^N$, then we also assume here the eigenvalues are all real. Such eigenvalues are called *simple*, and matrices that have only simple eigenvalues are called simple matrices. Since there are N eigenvalues (counting multiplicity) associated with any $N \times N$ matrix, if the matrix is simple, then by Theorem 5.15 the associated N eigenvectors form a basis for $\mathbb{F}^N$. The one-dimensional eigenspaces that are generated by the eigenvectors are disjoint subspaces, and $\mathbb{F}^N$ can be written as the direct sum of the eigenspaces.

Case II: The next simplest case is when the algebraic multiplicity is equal to the geometric multiplicity for each eigenvalue. The previous corollary says that eigenspaces associated to distinct eigenvalues are disjoint. If the dimension of each eigenspace is the same as that eigenvalue's algebraic multiplicity, then the sum of the dimensions of all the eigenspaces is N. This means there exists an eigenbasis, even though the eigenvalues may not be distinct. These eigenvalues and their matrices are called semi-simple.

Case III: Suppose $N > 1$ and A has only one eigenvalue λ with algebraic multiplicity N and geometric multiplicity equal to one. A *generalized eigenvector* is a non-zero vector v so that $(A-\lambda I)^j v = 0$ for some power j. Although there is only one independent eigenvector, it is a fact that there is in this case a basis of generalized eigenvectors, and these may in some cases be found by iteratively solving the systems $(A-\lambda I)v_{j+1} = v_j$ for $j = 0, \ldots, N-1$ where $v_0 = 0$. (The case $j = 0$ produces the eigenvector v_1). We will return later to the issue of how generalized eigenvectors can be calculated in general. A complete proof of this fact is somewhat involved, and we devote the next subsection to proving this.

A general matrix A may not be of any of the forms above, but the above three cases are the building blocks to cover the possible scenarios when $\mathbb{F} = \mathbb{C}$ or A is a map from $\mathbb{R}^N$ to $\mathbb{R}^N$ with real eigenvalues. As we shall see in the next section, we can partition $\mathbb{F}^N$ into a direct sum so that A acts on each summand in one of the three possible ways described in Cases I–III.

The situation where A has real entries with truly complex eigenvalues needs special attention and is briefly described in Cases IV–VI. The goal is the same, but we need to find a basis of $\mathbb{R}^N$ rather than $\mathbb{C}^N$, and that sheds light on the behavior of the mapping A from $\mathbb{R}^N$ to $\mathbb{R}^N$.

Case IV: Suppose first that $\lambda = \lambda^{\text{Re}} + i\lambda^{\text{Im}}$ is an eigenvalue with $\lambda^{\text{Im}} \neq 0$, and that $\tilde{v} = v^{\text{Re}} + iv^{\text{Im}}$ is an associated eigenvector. We call v^{Re} the real part and v^{Im} the imaginary part of the eigenvector. Since A has only real entries, $\overline{\tilde{v}} = v^{\text{Re}} - iv^{\text{Im}}$ is an eigenvector associated with $\overline{\lambda} = \lambda^{\text{Re}} + i\lambda^{\text{Im}}$. Of course $\tilde{v}$ and $\overline{\tilde{v}}$ belong to $\mathbb{C}^N$ and not $\mathbb{R}^N$. The linear independence asserted in Theorem 5.15 takes place in $\mathbb{C}^N$, and applies here for $\{\tilde{v}, \overline{\tilde{v}}\}$ since $\lambda \neq \overline{\lambda}$. We claim that v^{Re} and v^{Im} are independent in $\mathbb{R}^N$ over $\mathbb{R}$. Indeed, suppose $\alpha_1 v^{\text{Re}} + \alpha_2 v^{\text{Re}} = 0$ with $\alpha_1, \alpha_2 \in \mathbb{R}$. We have

$$\frac{\alpha_1 - i\alpha_2}{2}\left(v^{\text{Re}} + iv^{\text{Im}}\right) + \frac{\alpha_1 + i\alpha_2}{2}\left(v^{\text{Re}} - iv^{\text{Im}}\right) = \alpha_1 v^{\text{Re}} + \alpha_2 v^{\text{Im}} = 0,$$

and thus $\alpha_1 - i\alpha_2 = 0$ and $\alpha_1 + i\alpha_2 = 0$. This implies $\alpha_1 = \alpha_2 = 0$, or that v^{Re} and v^{Im} are independent over $\mathbb{R}$.

Cases V and VI: It could also happen that for real matrices A, a complex eigenvalue has algebraic multiplicity greater than one. Suppose λ is a truly complex eigenvalue with algebraic multiplicity equal to $m > 1$. The analogue to Case II is when the real and imaginary parts of all the eigenvectors span a subspace of dimension $2m$. A generalized eigenvector $v \in \mathbb{C}^N$ as introduced in Case III can be defined here in the same way, and the collection of the real and imaginary parts of each of these vectors are independent in $\mathbb{R}^N$.

5.7.4.1 *Diagonalization of simple and semi-simple matrices*

The map A from $\mathbb{F}^N$ to $\mathbb{F}^N$ is easy to understand in Cases I and II since there exists an eigenbasis $\mathcal{B} = \{\tilde{v}_1, \ldots, \tilde{v}_N\}$. Suppose $A\tilde{v}_n = \lambda_n \tilde{v}_n$ for $n = 1, \ldots, N$ (some of the λ_ns may be repeated in the semi-simple case), and a vector $v \in \mathbb{F}^N$ is represented as $v = \sum_{n=1}^{N} \alpha_n \tilde{v}_n$. Then

$$Av = \sum_{n=1}^{N} \alpha_n A\tilde{v}_n = \sum_{n=1}^{N} \alpha_n \lambda_n \tilde{v}_n \,.$$

In other words, once a given vector v is decomposed as a linear combination of the elements of an eigenbasis $\mathcal{B}$, pre-multiplication of the vector by a matrix A works on this linear combination simply by multiplying the $\tilde{v}_n^{\text{th}}$ component of the eigenbasis by λ_n. This can be succinctly described in terms of matrices by letting K be the $N \times N$ matrix whose n^{th} column is $\tilde{v}_n$. Let Λ be the diagonal matrix whose n^{th} diagonal entry is λ_n; that is

$$\Lambda = \begin{pmatrix} \lambda_1 & 0 & \cdots & 0 \\ 0 & \lambda_2 & \cdots & 0 \\ \vdots & \vdots & \ddots & \vdots \\ 0 & 0 & \cdots & \lambda_N \end{pmatrix}.$$

The number of times each eigenvalue appears in the diagonal equals its algebraic multiplicity, and the matrix Λ is unique except for the order in which the eigenvalues appear. Then $AK = K\Lambda$, and since the eigenvectors are linearly independent, K has full rank and its inverse exists. Therefore A is similar to a diagonal matrix, and is transformed by

$$K^{-1}AK = \Lambda. \tag{5.24}$$

Because of the diagonal nature of Λ, we say that a semi-simple matrix is *diagonalizable*.

Another approach to the diagonalization of a semi-simple matrix A uses direct sums. Recall that the eigenspace an eigenvalue λ is precisely the null space $\mathcal{N}(A - \lambda I)$ of $A - \lambda I$, and different eigenvalues have disjoint eigenspaces. Let $\lambda_1, \ldots, \lambda_n$ be the distinct eigenvalues of A with respective algebraic multiplicities $k_1, \ldots, k_n$. Thus $\sum_{j=1}^{n} k_j = N$, the size of the matrix and the semi-simple property is equivalent to saying $k_j = \dim(A - \lambda_j I)$ for each j. Hence the direct sum decomposition

$$\mathbb{F}^N = \mathcal{N}(A - \lambda_1 I) \oplus \cdots \oplus \mathcal{N}(A - \lambda_n I) \tag{5.25}$$

holds. Conversely, if a matrix A can be decomposed as a direct sum of its eigenspaces as in Eq. (5.25), then A is semi-simple.

5.7.5 *Generalized eigenspaces*

Not all matrices are semi-simple, and such a matrix cannot be transformed into a diagonal matrix as in Eq. (5.24) and its domain cannot be decomposed as in Eq. (5.25). Nonetheless, related transformations and decompositions can be given in terms of generalized eigenvectors and generalized eigenspaces, which is the topic of this section. *We assume throughout this section that either $\mathbb{F} = \mathbb{C}$ or that $\mathbb{F} = \mathbb{R}$*

and all the eigenvalues are real. As mentioned above in other contexts, the case where $\mathbb{F} = \mathbb{R}$ with truly complex eigenvalues needs special consideration and will be considered separately.

We make and will use the two following claims that are valid for any matrix B.

Claim 1: The null space $\mathcal{N}(B)$ of B belongs to the null space $\mathcal{N}(B^n)$ for every n^{th} power of B, $n > 1$.
Proof: If $v \in \mathcal{N}(B)$ and $n > 1$, then $B^n v = B^{n-1}(Bv) = B^{n-1}(0) = 0$, and so $v \in \mathcal{N}(B^n)$. $\square$

Claim 2: If $\mathcal{N}(B^M) = \mathcal{N}(B^{M+1})$, then $\mathcal{N}(B^M) = \mathcal{N}(B^{M+n})$ for every $n > 1$.
Proof: Let $M > 1$. Assume that $\mathcal{N}(B^M) = \mathcal{N}(B^{M+1})$. Claim 1 says that $\mathcal{N}(B^{M+n}) \subseteq \mathcal{N}(B^{M+n+1})$, and to prove the reverse inclusion, let $v \in \mathcal{N}(B^{M+n+1})$. Then $0 = B^{M+n+1}v = B^{M+1}(B^n v)$, which says that $B^n v \in \mathcal{N}(B^{M+1}) = \mathcal{N}(B^M)$, and so implies $B^M(B^n v) = B^{M+n}v = 0$, or that $v \in \mathcal{N}(B^{M+n})$. Thus, any vector in $\mathcal{N}(B^{M+n+1})$ is also in $\mathcal{N}(B^{M+n+})$ or $\mathcal{N}(B^{M+n+1}) \subseteq \mathcal{N}(B^{M+n})$. This proves the reverse inclusion and therefore $\mathcal{N}(B^{M+n+1}) = \mathcal{N}(B^{M+n})$. $\square$

Recall that the eigenspace of λ is the null space of the matrix $A - \lambda I$. By Claim 1 with $B = A - \lambda I$, we have

$$\mathcal{N}(A - \lambda I) \subseteq \mathcal{N}((A - \lambda I)^2) \subseteq \cdots \subseteq \mathcal{N}((A - \lambda I)^n) \tag{5.26}$$

for every $n > 1$. The inclusions in Eq. (5.26) cannot be strict for arbitrarily large n since the dimension of the null space cannot exceed the domain's dimension N. Hence there exists n so that

$$\mathcal{N}((A - \lambda I)^{n-1}) \subsetneq \mathcal{N}((A - \lambda I)^n) = \mathcal{N}((A - \lambda I)^{n+1}). \tag{5.27}$$

We shall soon see that $n > 1$ if and only if λ is not simple or semi-simple. Claim 2 implies n is unique, and it is called the index (sometimes called the Riesz index) of λ. The index of the eigenvalues λ_n will be indicated γ_n.

Definition: *Suppose λ is an eigenvalue of A that is not simple or semi-simple. The generalized eigenspace associated to λ is the null space of $(A - \lambda I)^j$, where j is the index of λ. The non-zero elements of the generalized eigenspace are called generalized eigenvectors.*

A matrix A is said to be invariant on a subspace W provided $Aw \in W$ for all $w \in W$. We shall require the following.

Claim 3: *For each $k \geq 1$, A is invariant on both the null space and range of $(A - \lambda I)^k$.*
Proof: Let $k \geq 1$, and set $\mathcal{N} = \mathcal{N}(A - \lambda I)^k$ and $\mathcal{R} = \mathcal{R}(A - \lambda I)^k$. If $v \in \mathcal{N}$, then

$$0 = (A - \lambda I)^k v = (A - \lambda I)^{k-1}(A - \lambda I)v,$$

or that $(A - \lambda I)v \in \mathcal{N}(A - \lambda I)^{k-1}$. But the latter null space is contained in $\mathcal{N}$ by Eq. (5.26), and therefore

$$Av = (A - \lambda I)v + \lambda v \in \mathcal{N},$$

or that A leaves the subspace $\mathcal{N}$ invariant.

Now suppose $v \in \mathcal{R}$, and so $v = (A - \lambda I)^k w \in \mathcal{R}$ for some w. Then $(A - \lambda I)v = (A - \lambda I)^k (A - \lambda I)w \in \mathcal{R}$, and so $Av = (A - \lambda I)v + \lambda v \in \mathcal{R}$ since $\mathcal{R}$ is a subspace. Thus A leaves $\mathcal{R}$ invariant. □

Properties of generalized eigenspaces are collected in the following theorem.

Theorem 5.16. *Suppose λ_0 is an eigenvalue of a matrix A.*

(1) The generalized eigenspace associated to λ_0 contains all the eigenvectors of λ_0 but no other eigenvectors.
(2) Generalized eigenspaces of different eigenvalues are disjoint.
(3) The dimension of the generalized eigenspace of λ_0 equals the algebraic multiplicity of λ_0.
(4) Suppose $K^{-1}AK = B$ and W is the generalized eigenspace of B associated to λ_0. Then $KW = \{Kw : w \in W\}$ is the generalized eigenspace of A associated to λ_0.

Proof (1): The eigenvectors of λ_0 are the non-zero vectors contained in $\mathcal{N}(A - \lambda_0 I)$, and these belong to the generalized eigenspace by Eq. (5.26). Now let $\tilde{v}_1$ be any eigenvector associated to an eigenvalue λ_1. If $\tilde{v}_1 \in \mathcal{N}(A - \lambda_0 I)^{\gamma_0}$, then

$$0 = (A - \lambda_0 I)^{\gamma_0}\tilde{v}_1 = (A - \lambda_0 I)^{\gamma_0 - 1}(A\tilde{v}_1 - \lambda_0 \tilde{v}_1) = (\lambda_1 - \lambda_0)(A - \lambda_0 I)^{\gamma_0 - 1}\tilde{v}_1.$$

Repeated use of these steps eventually give

$$0 = (\lambda_1 - \lambda_0)^{\gamma_0}\tilde{v}_1,$$

and since $\tilde{v}_1 \neq 0$, we must have $\lambda_1 = \lambda_0$. This shows that the generalized eigenspace cannot contain eigenvectors of any other eigenvalue.
(2): Let W be the intersection of two distinct generalized eigenspaces. Claim 3 says that W is invariant with respect to A, and so one can consider the restriction map $A|_W$ of A to W as a map in its own right. As such, if $W \neq \{0\}$, then $A|_W$ has at least one eigenvalue with an associated eigenvector. This eigenvector belongs to W and is also an eigenvector of A by definition. However, an eigenvector cannot belong to two distinct generalized eigenspaces by Part (1), a contradiction. Therefore $W = \{0\}$.
(3): Note that this result requires a logical connection between two disparate concepts, the geometric concept of dimension and the algebraic concept of root multiplicity. The idea is to transform A to a similar matrix where the determinant can be easily calculated.

Let $\mathcal{N} = \mathcal{N}(A - \lambda_0 I)^{\gamma_0}$ denote the generalized eigenspace of λ_0, and let the range of $(A - \lambda_0 I)^{\gamma_0}$ be denoted by $\mathcal{R}$. The previous Claim 3 says that $\mathcal{N}$ and $\mathcal{R}$ are both invariant with respect to A, and we claim that $\mathcal{R}$ and $\mathcal{N}$ are disjoint. To see that this is indeed the case, consider a vector v in the intersection of $\mathcal{R}$ and $\mathcal{N}$ and show that v must be the zero vector. Thus, let $v \in \mathcal{R} \cap \mathcal{N}$. Then $v \in \mathcal{R}$ says there exists w so that $v = (A - \lambda_0 I)^{\gamma_0} w$, and $v \in \mathcal{N}$ says $0 = (A - \lambda_0 I)^{\gamma_0} v = (A - \lambda_0 I)^{2\gamma_0} w$. Therefore $w \in \mathcal{N}(A - \lambda_0 I)^{2\gamma_0}$, but recall Eq. (5.27) and Claim 2, which imply that $\mathcal{N}(A - \lambda_0 I)^{\gamma_0} = \mathcal{N}(A - \lambda_0 I)^{2\gamma_0}$. We conclude $w \in \mathcal{N}$, or that $v = (A - \lambda_0 I)^{\gamma_0} w = 0$ and $\mathcal{N}$ and $\mathcal{R}$ are disjoint.

Next, note that $N = \dim \mathcal{N} + \dim \mathcal{R}$ by the Corollary to Theorem 5.12, and since $\mathcal{N}$ and $\mathcal{R}$ are disjoint, we must have $V = \mathcal{R} \oplus \mathcal{N}$. Let $m = \dim \mathcal{N}$, and choose a basis $\{v_1, \ldots, v_N\}$ of $\mathbb{F}^N$ by having the first m elements $\{v_1, \cdots v_m\}$ be a basis for $\mathcal{N}$ and the remaining $\{v_{m+1}, \ldots, v_N\}$ be a basis for $\mathcal{R}$. Let K be the invertible matrix whose columns are the vs. Since A is invariant on both $\mathcal{N}$ and $\mathcal{R}$, there exists square matrices $A_{\mathcal{N}}$ and $A_{\mathcal{R}}$ of size equal to the respective dimensions of $\mathcal{N}$ and $\mathcal{R}$ so that

$$AK = K \begin{pmatrix} A_{\mathcal{N}} & 0 \\ 0 & A_{\mathcal{R}} \end{pmatrix} \Rightarrow K^{-1}AK = \begin{pmatrix} A_{\mathcal{N}} & 0 \\ 0 & A_{\mathcal{R}} \end{pmatrix},$$

which shows A is similar to a block matrix. Similar matrices have the same determinant (Theorem 5.14(2)), and the determinant of a block matrix is the product of determinants of the blocks (Theorem 5.8(9)). Therefore we have

$$\det(A - \lambda I) = \det(A_{\mathcal{N}} - \lambda I)\det(A_{\mathcal{R}} - \lambda I) = p_{\mathcal{N}}(\lambda) p_{\mathcal{R}}(\lambda),$$

where $p_{\mathcal{N}}$ and $p_{\mathcal{R}}$ are the characteristic polynomials of $\mathcal{N}$ and $\mathcal{R}$ respectively. Recall from part (1) that $\mathcal{N}$ contains all of the eigenvectors associated with λ_0 and no others, so we must have $p_{\mathcal{N}}(\lambda) = (\lambda - \lambda_0)^m$ and λ_0 is not a root of $p_{\mathcal{R}}$. It follows that the algebraic multiplicity of λ_0 is m, the same as the dimension of $\mathcal{N}$, and the proof is complete.

(4): Suppose $K^{-1}AK = B$ and W is the generalized eigenspace of B associated to λ_0. By Eq. (5.19) and (5.20), we have $K^{-1}(A - \lambda_0 I)^j K = (B - \lambda_0 I)^j$ for all j. Recall that $W = \mathcal{N}(B - \lambda_0 I)^{\gamma_0}$ where γ_0 is the index of λ_0. Therefore $K^{-1}(A - \lambda_0 I)^{\gamma_0} K w = 0$ if and only if $w \in W$, and multiplying by K says that $v \in \mathcal{N}(A - \lambda_0)^{\gamma_0}$ if and only if $v \in KW$. □

Corollary *The index γ of an eigenvalue λ is strictly larger than one if and only if λ is not simple or semi-simple.*

Proof: An eigenvalue is simple or semi-simple if and only if its algebraic multiplicity is equal to its geometric multiplicity. By definition, the geometric multiplicity is the dimension of $\mathcal{N}(A - \lambda I)$, and Part (3) of the previous theorem says the algebraic multiplicity is equal to the dimension of $\mathcal{N}(A - \lambda I)^\gamma$. It follows that λ is simple or semi-simple if and only if $\mathcal{N}(A - \lambda I) = \mathcal{N}(A - \lambda I)^\gamma$, or that $\gamma = 1$. □

5.7.5.1 *Generalized eigenbases*

We see in this subsection how to find a particularly nice structure for the basis for the generalized eigenspace. Suppose $\gamma > 1$ is the index of an eigenvalue λ, and let $\mathcal{N}$ be its associated generalized eigenspace. A *chain of generalized eigenvectors* of rank r is a sequence $\{\tilde{v}_r, \ldots, \tilde{v}_1\}$ of non-zero vectors in $\mathcal{N}$ that satisfy

$$(A - \lambda I)\tilde{v}_k = \tilde{v}_{k-1} \quad \text{for } k = 2, \ldots, r, \tag{5.28}$$

and $(A - \lambda I)\tilde{v}_1 = 0$. $\tilde{v}_k$ is called the generalized eigenvector of rank k. Clearly a chain cannot be greater than the index γ, since $(A - \lambda I)^\gamma v = 0$ for all $v \in \mathcal{N}$. It is clear that $\tilde{v}_1$ is the only eigenvector for A that appears in the chain. The next result demonstrates that chains of vectors can be part of a basis.

Theorem 5.17.

(1) A chain of generalized eigenvectors $\{\tilde{v}_r, \ldots, \tilde{v}_1\}$ is linearly independent.
(2) For each $j = 1, \ldots, m$, suppose $\{\tilde{v}^j_{r_j}, \ldots, \tilde{v}^j_1\}$ is a chain of rank r_j. If $\tilde{v}^1_1, \ldots, \tilde{v}^m_1$ are linearly independent, then the entire collection $\{\tilde{v}^1_{r_1}, \ldots, \tilde{v}^1_1, \ldots, \tilde{v}^m_{r_m}, \ldots, \tilde{v}^m_1\}$ is linearly independent.
(3) Suppose $K^{-1}AK = B$ and $\{\tilde{v}_r, \ldots, \tilde{v}_1\}$ is a chain for B. Then $\{K\tilde{v}_r, \ldots, K\tilde{v}_1\}$ is a chain for A.

Proof (1): Suppose $\alpha_r, \ldots, \alpha_1$ are scalars that satisfy

$$\sum_{k=1}^{r} \alpha_k \tilde{v}_k = 0. \tag{5.29}$$

Note that

$$(A - \lambda I)^o \tilde{v}_k = \begin{cases} \tilde{v}_1 & \text{if } s = k - 1 \\ 0 & \text{if } s \geq k \end{cases} \tag{5.30}$$

and recall $\tilde{v}_1 \neq 0$. Multiplying Eq. (5.29) by $(A - \lambda I)^{r-1}$ gives $\alpha_r v_1 = 0$, and so $\alpha_r = 0$. Given that, we now multiply Eq. (5.29) by $(A - \lambda I)^{r-2}$ to obtain $\alpha_{r-1} = 0$. After continuing this process a total of $r - 1$ times, we conclude all of the αs are 0, and the chain is linearly independent.
(2) A similar argument used in part (1) works here as well. Suppose α^j_k ($1 \leq j \leq m$ and $1 \leq k \leq r_j$) are scalars that satisfy

$$\sum_{j=1}^{m} \sum_{k=1}^{r_j} \alpha^j_k \tilde{v}^j_k = 0. \tag{5.31}$$

Let r equal the largest among $r_1, \ldots, r_m$. Multiplying Eq. (5.31) by $(A - \lambda I)^{r-1}$ and using Eq. (5.30), we have

$$\sum_{j, r_j = r} \alpha^j_{r_j} v^j_1 = 0,$$

which by the linear independence hypothesis implies $\alpha_r^j = 0$ for all j with $r_j = r$. Again we can repeat the process by reducing the power $r-1$ times and conclude all of the αs are zero.
(3) Suppose $K^{-1}AK = B$ and $\{\tilde{v}_r, \ldots, \tilde{v}_1\}$ is a chain for B, and so $(B-\lambda I)\tilde{v}_{k+1} = \tilde{v}_k$ for $k = 1, \ldots, r-1$. By Eq. (5.20),

$$(A-\lambda I)K\tilde{v}_{k+1} = KK^{-1}(A-\lambda I)K\tilde{v}_{k+1} = K(B-\lambda I)\tilde{v}_{k+1} = K\tilde{v}_k,$$

for $k = 1, \ldots, r-1$, and the result follows. □

Chains of generalized eigenvectors are the building blocks of finding a basis for the generalized eigenspace, however in practical situations of large dimension, finding them can be a considerable challenge if not impossible. If the length of a chain is equal to the dimension of the generalized eigenspace, which happens precisely when the geometric multiplicity is 1, then the chain is a basis. If a chain is shorter than the dimension of the generalized eigenspace, then either the chain can be extended to a higher rank, or another chain must be found. What is worse, if one is attempting to add a chain to an existing independent set to build a basis, it may have to be discarded due to lack of linear independence between the vectors in the chain even though the vectors at the top of the chain are linearly independent.

An attractive method that often works in low dimensions is to go in the opposite direction by starting with an eigenvector and using Eq. (5.28) to calculate generalized eigenvectors of increasing rank. A serious problem is that it may not produce the entire generalized eigenspace. It is quite complicated to explain a method that works in all situations, and we illustrate some of these effects in the following examples.

Example 5.17: Calculation of generalized eigenvectors I.

Consider the matrix

$$A = \begin{pmatrix} 2 & 0 & -1 \\ 2 & -1 & 2 \\ 1 & -1 & 2 \end{pmatrix}$$

which has the characteristic polynomial $(1-\lambda)^3$, and therefore there is only eigenvalue $\lambda = 1$ with an algebraic multiplicity of 3. Theorem 5.16 implies that the generalized eigenspace is all of $\mathbb{R}^3$. The eigenspace is found by calculating the null space $\mathcal{N}(A-I)$, which can be done by row reduction as

$$A - I = \begin{pmatrix} 1 & 0 & -1 \\ 2 & -2 & 2 \\ 1 & -1 & 1 \end{pmatrix} \Rightarrow \mathcal{N}(A-I) = \left\{ \begin{pmatrix} s \\ 2s \\ s \end{pmatrix} : s \in \mathbb{R} \right\}.$$

Hence the dimension of the eigenspace (which is the geometric multiplicity) is one. We specify the eigenvector $\tilde{v}_1$ by letting the parameter $s = 1$. The index is three and a chain can be produced by using Eq. (5.28) twice recursively. The first time

$$(A-I)v_2 = v_1 \quad \Rightarrow \quad \tilde{v}_2 = \begin{pmatrix} 1+s \\ 2s \\ s \end{pmatrix},$$

and we specify $\tilde{v}_2$ by letting $s = 0$; the second time is

$$(A-I)\tilde{v}_3 = \tilde{v}_2 \quad \Rightarrow \quad \tilde{v}_3 = \begin{pmatrix} 1+s \\ 1+2s \\ s \end{pmatrix},$$

and again we can specify $\tilde{v}_3$ by letting $s = 0$. Hence $\{\tilde{v}_3, \tilde{v}_2, \tilde{v}_1\}$ is a chain, and so it is linearly independent and must be a basis for the generalized eigenspace.

We shall soon see an example where one cannot work backwards in any obvious way up the chain beginning from an eigenvector, and instead are compelled to consider chains generated by starting with a generalized eigenvector that has maximal rank. In the current example, the index is three, and so the generalized eigenspace is all of $\mathbb{R}^3$. Any vector that does not belong to the null space of $(A-I)^2$ could initiate a chain, since we would then easily calculate $\tilde{v}_2$ and $\tilde{v}_1$ from Eq. (5.28) and produce three linearly independent vectors. One notes that

$$(A-I)^2 = \begin{pmatrix} 0 & 1 & -2 \\ 0 & 2 & -4 \\ 0 & 1 & -2 \end{pmatrix} \quad \Rightarrow \quad \mathcal{N}(A-I)^2 = \left\{ \begin{pmatrix} s_1 \\ 2s_2 \\ s_2 \end{pmatrix} : s_1, s_2 \in \mathbb{R} \right\}.$$

The vector $\tilde{v}_3$ chosen above does not belong to $\mathcal{N}(A-I)^2$ and so successfully initiates a chain, but any other vector $\tilde{v}_3 = (x_1, x_2, x_3)^T$ with $2x_3 \neq x_2$ would do as well. No matter how the generalized basis $\{\tilde{v}_1, \tilde{v}_2, \tilde{v}_3\}$ is chosen, if K is the matrix whose columns are the $\tilde{v}$'s, then

$$AK = K \begin{pmatrix} 1 & 1 & 0 \\ 0 & 1 & 1 \\ 0 & 0 & 1 \end{pmatrix} \quad \Rightarrow \quad K^{-1}AK = \begin{pmatrix} 1 & 1 & 0 \\ 0 & 1 & 1 \\ 0 & 0 & 1 \end{pmatrix}.$$

Example 5.18: Calculation of generalized eigenvectors II.

Now let

$$A = \begin{pmatrix} 3 & 1 & -1 \\ 1 & 3 & -1 \\ 2 & 2 & 0 \end{pmatrix}.$$

The characteristic polynomial is $(2-\lambda)^3$, and so A has only one eigenvalue $\lambda = 2$ with an algebraic multiplicity of three. The eigenspace is found as before,

$$A - 2I = \begin{pmatrix} 1 & 1 & -1 \\ 1 & 1 & -1 \\ 2 & 2 & -2 \end{pmatrix} \quad \Rightarrow \quad \mathcal{N}(A-2I) = \left\{ \begin{pmatrix} -s_1 + s_2 \\ s_1 \\ s_2 \end{pmatrix} : s_1, s_2 \in \mathbb{R} \right\}$$

and has dimension two, which is the geometric multiplicity. Thus there are two independent eigenvectors, but the problem that confronts us is to know how to find an eigenvector that will belong to a chain of rank two. It is not immediately clear, since for example, if we choose the eigenvector $\tilde{v}$ by setting $s_1 = 0$ and $s_2 = 1$, then the system $(A - 2I)w = \tilde{v}$ is inconsistent, and so is the related system obtained by setting $s_1 = 1$ and $s_2 = 0$. Thus unlike the previous example, it is not obvious how to find the generalized eigenbasis by going up a chain from an eigenvector, and so we are compelled into finding a generalized eigenvector that will generate a chain of rank two by going downwards. Any element that does not belong to the null space of $(A - 2I)$ can be used as a generalized eigenvector, so for example, we can take $\tilde{v}_2 = (0\ 1\ 0)^T$, and define $\tilde{v}_1 = (A - 2I)\tilde{v}_2 = (1\ 1\ 2)^T$. We know $\tilde{v}_1$ must be an eigenvector since $(A - 2I)^2$ is the zero matrix, but it can also be seen belonging to $\mathcal{N}(A - 2I)$ by noting $\tilde{v}_1$ is parameterized with the choice of $s_2 = 2$ and $s_1 = 1$. The final basis vector must be an eigenvector $\tilde{v}_3$, and can be chosen as any eigenvector that is independent of $\tilde{v}_2$. For example, we can take $s_1 = 1$ and $s_2 = 1$ and use $\tilde{v}_3 = (0\ 1\ 1)^T$. If K is the invertible matrix whose columns are the $\tilde{v}$s, then

$$AK = K \begin{pmatrix} 2 & 1 & 0 \\ 0 & 2 & 0 \\ 0 & 0 & 2 \end{pmatrix} \quad \Rightarrow \quad K^{-1}AK = \begin{pmatrix} 2 & 1 & 0 \\ 0 & 2 & 0 \\ 0 & 0 & 2 \end{pmatrix}.$$

Example 5.19: Calculation of generalized eigenvectors III.

Finally, consider the 4×4 matrix

$$A = \begin{pmatrix} 3 & -1 & 0 & 2 \\ 1 & 1 & 0 & 2 \\ 1 & -1 & 2 & 3 \\ 0 & 0 & 0 & 2 \end{pmatrix},$$

which has characteristic polynomial $(2 - \lambda)^4$. The only eigenvalue is $\lambda = 2$ with algebraic multiplicity four. One has

$$(A - 2I) = \begin{pmatrix} 1 & -1 & 0 & 2 \\ 1 & -1 & 0 & 2 \\ 1 & -1 & 0 & 3 \\ 0 & 0 & 0 & 0 \end{pmatrix} \quad \Rightarrow \quad \mathcal{N}(A - 2I) = \left\{ \begin{pmatrix} s_1 \\ s_1 \\ s_2 \\ 0 \end{pmatrix} : s_1, s_2 \in \mathbb{R} \right\},$$

and the geometric multiplicity is two. Since $(A - 2I)^2 = 0$, the index is two. We need to find two chains of rank two, and again it is not clear how to do this by starting with eigenvectors and going upwards. Instead, we find two linearly independent generalized eigenvectors that do not belong to $\mathcal{N}(A - 2I)$, say $\tilde{v}_2 = (0\ 1\ 0\ 1)^T$ and

$\tilde{v}_4 = (2\ 1\ 0\ 0)^T$, and calculate

$$\tilde{v}_1 = (A - 2I)\tilde{v}_2 = \begin{pmatrix} 1 \\ 1 \\ 2 \\ 0 \end{pmatrix} \quad \text{and} \quad \tilde{v}_3 = (A - 2I)\tilde{v}_4 = \begin{pmatrix} 1 \\ 1 \\ 1 \\ 0 \end{pmatrix}.$$

As before, if K is the invertible matrix whose columns are the $\tilde{v}$s, then

$$AK = K \begin{pmatrix} 2 & 1 & 0 & 0 \\ 0 & 2 & 0 & 0 \\ 0 & 0 & 2 & 1 \\ 0 & 0 & 0 & 2 \end{pmatrix} \quad \Rightarrow \quad K^{-1}AK = \begin{pmatrix} 2 & 1 & 0 & 0 \\ 0 & 2 & 0 & 0 \\ 0 & 0 & 2 & 1 \\ 0 & 0 & 0 & 2 \end{pmatrix}.$$

We caution the reader that not every choice of independent vectors $\tilde{v}_2$ and $\tilde{v}_4$ will give rise to two different chains. For example $\tilde{v}_2 = (0\ 1\ 0\ 1)^T$ and $\tilde{v}_4 = (1\ 2\ 2\ 1)^T$ are independent and not in $\mathcal{N}(A - 2I)$, but they both map to $\tilde{v}_1$ by $A - 2I$. The way to choose $\tilde{v}_4$ is that it should not belong to $\mathcal{N}(A - 2I)$ and be independent of the entire chain containing $\tilde{v}_2$.

5.7.6 *Jordan canonical form*

We have seen several examples where a given matrix is similar to a matrix with a diagonal-like structure. We now formalize these notions.

Definition: *An* elementary Jordan block *of size n is an $n \times n$ square matrix $J_n(\lambda)$ of the form*

$$J_n(\lambda) = \begin{pmatrix} \lambda & 1 & 0 & 0 & \dots & 0 & 0 & 0 \\ 0 & \lambda & 1 & 0 & \dots & 0 & 0 & 0 \\ & \vdots & & & \ddots & & \vdots & \\ 0 & 0 & 0 & 0 & \dots & 0 & \lambda & 1 \\ 0 & 0 & 0 & 0 & \dots & 0 & 0 & \lambda \end{pmatrix}.$$

That is, $J_n(\lambda) = \lambda I_n + B$, where I_n is the n dimensional identity matrix and B is the matrix with all zeros except for ones along the superdiagonal.

The structural properties of an elementary Jordan block matrix include the following.

(1) There is only one eigenvalue λ with algebraic multiplicity equal to n. This is obvious since the determinant of an upper triangular matrix is the product of its diagonal entries.

(2) Observe that $J_n(\lambda) - \lambda I_n = B$, $B^k \neq 0$ for $k = 1, \ldots, n-1$, and $B^n = 0$. This shows the index of λ is n.
(3) The geometric multiplicity of λ is equal to one.
(4) The eigenspace equals the subspace spanned by first canonical basis element e_1.
(5) All of the canonical basis elements $\{e_n, \ldots, e_1\}$ constitute a chain of rank n, and are related by $(J_n(\lambda) - \lambda I)e_{k+1} = e_k$ for each $k = 1, \ldots, n-1$.

The matrix A from Example **5.16**, Case III is shown there to be similar to an elementary Jordan block matrix.

Definition: *A* Jordan block *with eigenvalue λ is of the form*

$$J(\lambda) = \begin{pmatrix} J_{n_1}(\lambda) & 0 & \cdots & 0 & 0 \\ 0 & J_{n_2}(\lambda) & \cdots & 0 & 0 \\ & \vdots & \ddots & & \vdots \\ 0 & 0 & \cdots & J_{n_{m-1}}(\lambda) & 0 \\ 0 & 0 & \cdots & 0 & J_{n_m}(\lambda) \end{pmatrix}$$

where each of the $J_k(\lambda)$'s are elementary Jordan blocks.

The structural properties of a Jordan block matrix include the following:

(1) There is only one eigenvalue λ with algebraic multiplicity equal to $n(\lambda) = \sum_{k=1}^{m} n_k$. The generalized eigenspace is $\mathbb{F}^n$.
(2) The index of λ is the largest value of n_k among $k = 1, \ldots, m$.
(3) The geometric multiplicity of λ is equal to m, the number of elementary Jordan blocks. In particular, λ is semi-simple if and only if each of the elementary blocks is one-dimensional.
(4) The eigenspace is spanned by the canonical basis elements that correspond to the first columns of the elementary Jordan blocks. Explicitly, set $s_0 = 0$ and $s_k = n_k + s_{k-1}$ for $k = 1, \ldots, m$. The vectors $\{e_1, e_{s_1+1}, \ldots, e_{s_{m-1}+1}\}$ is a basis of the eigenspace.
(5) For each $k = 1, \ldots, m$, the canonical basis vectors $\{e_{s_k}, e_{s_k-1}, \ldots, e_{s_{k-1}+1}\}$ constitute a chain of rank n_k that spans a subspace W_k that is invariant under $J(\lambda)$.

The matrices from Examples 5.17 and 5.18 were shown to be similar to Jordan blocks.

The general situation is where a matrix has more than one eigenvalue. A matrix will be said to have Jordan canonical form if it consists of Jordan blocks along its diagonal.

Definition: *A matrix J is said to be in Jordan canonical form if it has the structure*

$$J = \begin{pmatrix} J(\lambda_1) & 0 & \cdots & 0 \\ 0 & J(\lambda_2) & \cdots & 0 \\ \vdots & \vdots & \ddots & \vdots \\ 0 & 0 & \cdots & J(\lambda_l) \end{pmatrix}$$

where $\lambda_, \dots, \lambda_l$ are distinct and each of the diagonal block matrices $J(\lambda_k)$ are Jordan blocks.

If we detail the structural properties of a Jordan matrix as we did for the Jordan blocks, then the indexing becomes unduly complicated. Instead, we simply point out that the generalized eigenspace W_{λ_k} associated to an eigenvalue λ_k is spanned by the columns of J that contain $J(\lambda_k)$. All of the structural properties of Jordan blocks apply to the J- invariant subspace W_{λ_k}.

If A is similar to a Jordan matrix J, then there exists an invertible matrix K so that $K^{-1}AK = J$. Let $\mathcal{B} \subseteq \mathbb{F}^N$ be the columns of K, which is called a *Jordan basis* for A. One can infer the structural properties of A through the basis $\mathcal{B} = \{v_1, \dots, v_N\}$ by referring to the above structural properties stated for the canonical basis, and replacing each e_k by v_k. See Theorem 5.14(4). In particular, we note that a Jordan basis $\mathcal{B}$ contains a basis of the eigenspace, and every element of $\mathcal{B}$ belongs to some chain that ends in the eigenspace.

The fundamental structural theorem can now be given.

Theorem 5.18. *Suppose A is a square matrix, and suppose $\mathbb{F} = \mathbb{C}$, or if $\mathbb{F} = \mathbb{R}$ then A has all real eigenvalues. Then there exists a Jordan basis for A that gives rise to a similar matrix in Jordan canonical form. Except for the ordering of the Jordan blocks and the ordering of the elementary Jordan blocks within the Jordan blocks, the Jordan canonical form is unique.*

Proof: Consider an N dimensional square matrix A, and suppose first that A has only one eigenvalue λ. Our proof uses induction on the dimension N. The case $N = 1$ is trivial, so let $N > 1$ and assume the result is true for dimensions strictly less than N.

Let $\mathcal{N}$ and $\mathcal{R}$ be the null and range spaces of $(A - \lambda I)$, respectively. The assumption $\mathbb{F} = \mathbb{C}$, or if $\mathbb{F} = \mathbb{R}$ and A has all real eigenvalues implies that $r = \dim \mathcal{N} > 0$, and hence $\dim \mathcal{R} < N$ since $N = \dim \mathcal{N} + \dim \mathcal{R}$ (corollary to Theorem 5.12). By Claim 3 (that precedes Theorem 5.16), both $\mathcal{N}$ and $\mathcal{R}$ are invariant under A. Consider the restriction of A to $\mathcal{N}$, which is a map we denote by $A|_{\mathcal{N}}$. This is a map defined on lower dimensional space and to which the induction hypothesis can be invoked. Hence there exists a Jordan basis $\mathcal{B}_{\mathcal{N}}$ for $\mathcal{N}$ associated to $A|_{\mathcal{N}}$. The elements of $\mathcal{B}_{\mathcal{N}}$ can be denoted by $v^1_{n_1}, \dots, v^1_1, \dots, v^m_{n_m}, \dots, v^m_1$, where $v^k_{n_k}, \dots, v^k_1$ is a chain for each $k = 1, \dots, m$, and $\{v^1_1, \dots, v^m_1\}$ is a basis for the null space of $A|_{\mathcal{N}}$. Let $W = \mathcal{N} \cap \mathcal{R}$, which is the eigenspace of $A|_{\mathcal{N}}$, and hence has $\mathcal{B}_W = \{v_1^1, \ldots, v_1^m\}$ for a basis. Let $s = \dim \mathcal{N} - m$, and there are vectors $\{u_1, \ldots, u_s\}$ so that $\mathcal{B}_W \cup \{u_1, \ldots, u_s\}$ is a basis for $\mathcal{N}$. Now, since $v_{n_1}^1, \ldots, v_{n_m}^m$ belong to $\mathcal{R}$, for each $k = 1, \ldots, m$, there exists $w^k \in \mathbb{F}^N$ so that $(A-\lambda I)w^k = v_{n_k}^k$, and therefore $w^k, v_{n_k}^k, \ldots, v_1^k$ is a chain of length n_k+1. The collection $\mathcal{B}_\mathcal{R} \cup \{w^1, \ldots, w^m\}$ of all these chains is independent by Theorem 5.17(2), and there are $r + m$ many such vectors. The collection $\mathcal{B}_\mathcal{R} \cup \{w^1, \ldots, w^m\} \cup \{u_1, \ldots, u_s\}$ is independent, and there are $r + m + s = \dim \mathcal{R} + \dim N = N$ of these, and therefore it constitutes a Jordan basis associated to A.

Finally, if $\lambda_1, \ldots, \lambda_l$ are the distinct eigenvalues of A, then the first part of the proof can be applied l times to the restriction of A to the generalized eigenspace W_{λ_k} for each $k = 1, \ldots, l$. A Jordan basis $\mathcal{B}_k$ is obtained for W_{λ_k}, and the collection of all of these form a Jordan basis for A.

As for the uniqueness assertion, suppose $A \sim J_1$ and $A \sim J_2$ where J_1 and J_2 are both in Jordan form. It follows from Theorem 5.14(1) that $J_1 \sim J_2$. By Theorem 5.16(4) the generalized eigenspaces are isomorphic, and by Theorem 5.17(3) each of the chains inside the generalized eigenspaces are isomorphic. The only way this can happen is that the structure of the Jordan matrices are the same except for the order where the subspaces are located. □

Another way to interpret the Jordan canonical form is with a direct sum decomposition as was done in the semi-simple case in Eq. (5.25). As in the proof, if $\lambda_1, \ldots, \lambda_l$ are the distinct eigenvalues and W_{λ_k} are the associated generalized eigenspaces, then

$$\mathbb{F}^N = W_{\lambda_1} \oplus \cdots \oplus W_{\lambda_l}. \tag{5.32}$$

Moreover, each W_{λ_k} is the direct sum of m_k chains that contain an eigenvector associated to λ_k, where m_k is the geometric multiplicity of λ_k. The index of λ_k is the rank of the largest chain contained in W_{λ_k}.

The proof of Theorem 5.18 used induction and so does not indicate how a Jordan form can be found in practice. Finding a Jordan form can be very difficult if the dimension of the space is large and some eigenvalues have large multiplicities. There are two major difficulties. The first step is to calculate the characteristic polynomial $p(\lambda) = \det(A-\lambda I)$, and then factor it into linear factors. The fundamental theorem of algebra says this is always theoretically possible with $\mathbb{F} = \mathbb{C}$, but practically can be impossible. If $\mathbb{F} = \mathbb{R}$ and p has truly complex roots, then it cannot be linearly factored at all, but let assume for the moment that can be done and all the eigenvalues are real. Thus $p(\lambda) = (\lambda - \lambda_1)^{n_1} \ldots (\lambda - \lambda_l)^{n_l}$, where the λ_ks are distinct. The powers $n_1, \ldots, n_l$ are the algebraic multiplicities, and each factor $(\lambda - \lambda_k)^{n_k}$ corresponds to the generalized eigenspace W_{n_k} that appears in the direct sum Eq. (5.32). We next assume further that there is only one eigenvalue λ, for in general, the procedure is repeated by considering each eigenvalue separately. Let

W denote the generalized eigenspace, m the geometric multiplicity, and j the index of λ. Recall that $\mathcal{N}(A-\lambda I)^{j-1} \subsetneq \mathcal{N}(A-\lambda I)^j \subseteq \mathcal{N}(A-\lambda I)^j$, and so there exists $\tilde{v}_j^1 \in W$ so that $\tilde{v}_j^1 \notin \mathcal{N}(A-\lambda I)^{j-1}$. A chain of rank j can be created by setting $\tilde{v}_{k-1}^1 = (A-\lambda I)\tilde{v}_k^1$ for $k = j, j-1, \ldots, 2$. The vectors $\tilde{v}_j^1, \ldots, v_1^1$ are independent by Theorem 5.17, and so can be extended to a basis of W by Theorem 5.2.

The second difficult step is to find such an extension that spans an invariant subspace W_1 under A. This can typically be done practically only in very special cases, and the theoretical proof of its existence would involve considerably more development and would take us too far afield. But assuming we have found a subspace W_1 that it is invariant under A, then one can repeat the procedure on $A = A|_{W_1}$, which is the restriction of A to W_1. We are done if $W_1 = \{0\}$, so assuming $W_1 \neq \{0\}$, the only eigenvalue for A_1 is λ, and its index j_1 is no more than j. We can choose $\tilde{v}_{j_1}^2 \in W_1$ so that $\tilde{v}_{j_1}^2 \notin \mathcal{N}(A-\lambda I)^{j_1-1}$ and produce a chain $C_2 = \{\tilde{v}_{j_1}^2, \ldots, \tilde{v}_1^2\}$. We now have to find a complementary subspace W_2 that is invariant under A_1, etc. Each step produces a basis K_k for the chain subspace C_k whose columns consist of $\tilde{v}_1^k, \ldots, \tilde{v}_{j_k}^k$, and that satisfies $AK = KJ_{j_k}(\lambda)$ where $J_{j_k}(\lambda)$ is an elementary Jordan block. The procedure terminates after m steps. That is, the geometric multiplicity is equal to the number of elementary Jordan blocks that comprise the generalized eigenspace.

5.7.7 *Jordan form of real matrices with complex eigenvalues*

In this subsection, we study in further detail the case where A has real entries with possibly complex roots. First, suppose A has *only* truly complex roots, and recall that the complex eigenvalues appear in conjugate pairs. Thus A can be viewed as a map from $\mathbb{R}^N$ to $\mathbb{R}^N$ since the dimension of the space is even. The number of conjugate pairs of eigenvalues with multiplicity is N. The following development is based on the fact that $\mathbb{R}^{2N}$ is isomorphic to $\mathbb{C}^N$, where the underlying field is $\mathbb{R}$, and the isomorphism $\Psi : \mathbb{R}^{2N} \to \mathbb{C}^N$ implicit in the discussion is described as follows. Let $v \in \mathbb{R}^{2N}$ with components α_k, and define $\Psi(v)$ to be the element in $\mathbb{C}^N$ whose k^{th} component is $\alpha_{2k-1} + i\alpha_{2k}$ for $k = 1, \ldots, N$. That is, the components of a vector in $\mathbb{R}^{2N}$ are paired consecutively together as the real and imaginary parts of a vector in $\mathbb{C}^N$.

Suppose $\lambda_1 = \alpha_1 + i\beta_1, \ldots, \lambda_n = \alpha_n + i\beta_n$ is a list of the distinct eigenvalues that does not contain a conjugate pair. The characteristic polynomial $p(\lambda)$ factors over $\mathbb{R}$ as

$$p(\lambda) = \left((\lambda-\alpha_1)^2 + \beta_1^2\right)^{d_1} \cdots \left((\lambda-\alpha_n)^2 + \beta_n^2\right)^{d_n},$$

where the exponents d_k are the *algebraic multiplicities* of λ_k, and $d_1 + \cdots + d_n = N$. This definition of algebraic multiplicity agrees with the previous one since $(\lambda-\alpha_k)^2 + \beta_k^2 = (\lambda-\lambda_k)(\lambda-\overline{\lambda}_k)$, but the notion of geometric multiplicity was not defined earlier in the present context. We now extend that and subsequent notions, where the main change from the previous development is to replace the linear term $(A - \lambda I)$ by the quadratic term $\big((A - \alpha I)^2 + \beta^2 I\big)$.

The *geometric multiplicity* of an eigenvalue $\lambda = \alpha + i\beta$ is one half of the dimension of $\mathcal{N}\big((A - \alpha I)^2 + \beta^2 I\big)$, and is the same for λ as it is for $\overline{\lambda}$. Assume now in addition A has only one distinct eigenvalue $\lambda = \alpha + i\beta$ (and so $d = N$). If the geometric and algebraic multiplicity associated to λ coincide (and equal d), then λ is called a semi-simple eigenvalue, and there exist real vectors v_k^{Re} and v_k^{Im}, $k = 1, \ldots, d$, that are the real and imaginary parts of complex eigenvectors $\tilde{v}_k = v_k^{\mathrm{Re}} + iv_k^{\mathrm{Im}}$ of λ which form a basis of $\mathcal{N}\big((A - \alpha I)^2 + \beta^2 I\big)$. We have $Av_k^{\mathrm{Re}} = \alpha v_k^{\mathrm{Re}} - \beta v_k^{\mathrm{Im}}$ and $Av_k^{\mathrm{Im}} = \beta v_k^{\mathrm{R}} + \alpha v_k^{\mathrm{Im}}$. Let K be the $2N \times 2N$ matrix whose columns are in order given by v_1^{Re}, v_1^{Im}, v_2^{Re}, $v_2^{\mathrm{Im}}, \ldots, v_d^{\mathrm{Re}}$, v_d^{Im}. Then we have

$$AK = K \begin{pmatrix} \Lambda_{\alpha,\beta} & 0_2 & \cdots & 0_2 \\ 0_2 & \Lambda_{\alpha,\beta} & \cdots & 0_2 \\ \vdots & & \ddots & \vdots \\ 0_2 & 0_2 & \cdots & \Lambda_{\alpha,\beta} \end{pmatrix},$$

where

$$\Lambda_{\alpha,\beta} = \begin{pmatrix} \alpha & \beta \\ -\beta & \alpha \end{pmatrix} \quad \text{and} \quad 0_2 = \begin{pmatrix} 0 & 0 \\ 0 & 0 \end{pmatrix}.$$

This shows A is similar to a diagonal-like matrix consisting of the 2×2 block matrix $\Lambda_{\alpha,\beta}$ along the diagonal and zeroes elsewhere.

The matrix A is semi-simple if all its eigenvalues are semi-simple, in which case A is similar to a block diagonal matrix of the form above with the matrices $\Lambda_{\alpha_k,\beta_k}$ along the diagonal.

As before, not all matrices are semi-simple and admit an eigenbasis. The notions of generalized eigenspaces and bases are extended as follows. The index, γ, of A associated to an eigenvalue $\lambda = \alpha + i\beta$ is the least n so that

$$\mathcal{N}\big((A - \alpha I)^2 + \beta^2 I\big)^n = \mathcal{N}\big((A - \alpha I)^2 + \beta^2 I\big)^{n+1}.$$

Given the index γ, the generalized eigenspace of A associated to λ is the null space of $(A - \alpha I)^2 + \beta^2 I)^\gamma$, and its non-zero vectors are called generalized eigenvectors.

The definition that $\{\tilde{v}_r, \ldots, \tilde{v}_1\}$ is a chain of vectors was given in Eq. (5.28), and remains the same here with the $\tilde{v}$s as complex vectors. However we now want real-valued vectors, and the way to get them is to simply take the real and imaginary parts of the complex vectors. The chain property is that $\tilde{v}_{k-1} = (A - \lambda I)\tilde{v}_k$, and so the real and imaginary parts satisfy

$$\begin{aligned} Av_k^{\mathrm{Re}} &= v_{k-1}^{\mathrm{Re}} + \alpha v_k^{\mathrm{Re}} - \beta v_k^{\mathrm{Im}} \\ Av_k^{\mathrm{Im}} &= v_{k-1}^{\mathrm{Im}} + \alpha v_k^{\mathrm{Im}} + \beta v_k^{\mathrm{Re}}. \end{aligned}$$

An elementary Jordan block was defined in the last subsection as the representation of a chain of vectors, and the appropriate extension to real matrices with complex roots is given next.

Definition: *Let $\lambda \in \mathbb{C}$ be given by $\lambda = \alpha + i\beta$ with $\beta \neq 0$. A (real)* elementary Jordan block *of size $2n$ with eigenvalue λ is an $2n \times 2n$ square matrix $J_{2n}(\lambda)$ of the block form*

$$J_{2n}(\lambda) = \begin{pmatrix} \Lambda_{\alpha,\beta} & I_2 & 0_2 & \dots & 0_2 & 0_2 \\ 0_2 & \Lambda_{\alpha,\beta} & I_2 & \dots & 0_2 & 0_2 \\ & \vdots & & \ddots & & \vdots \\ 0_2 & 0_2 & 0_2 & \dots & \Lambda_{\alpha,\beta} & I_2 \\ 0_2 & 0_2 & 0_2 & \dots & 0_2 & \Lambda_{\alpha,\beta} \end{pmatrix},$$

where

$$I_2 = \begin{pmatrix} 1 & 0 \\ 0 & 1 \end{pmatrix}.$$

With the definition of an elementary Jordan block in hand, we can now proceed as before and define a Jordan block associated to an eigenvalue λ as having a diagonal block structure with elementary Jordan blocks on the diagonal. The Jordan canonical form is defined as a matrix in diagonal block form with Jordan blocks along the diagonal, and the analogue of Theorem 5.18 carries over to the present case as well.

For a general $N \times N$ matrix A with real entries, there may be both real and complex eigenvalues. The Jordan canonical form of A is thus a matrix that combines the Jordan form of the previous subsection with the one developed in this subsection. We illustrate with two examples.

Example 5.20: Jordan canonical form of a real matrix with complex eigenvalues. Consider the matrix

$$A = \begin{pmatrix} 4 & -1 & -1 & 1 \\ 4 & 0 & -1 & 0 \\ 2 & -1 & 2 & 0 \\ -1 & 1 & -1 & 2 \end{pmatrix}$$

which has characteristic polynomial $p(\lambda) = \lambda^4 - 8\lambda^3 + 26\,\lambda^2 - 40\lambda + 25 = (\lambda - 2 - i)^2(\lambda - 2 + i)^2 = \left((\lambda - 2)^2 + 1\right)^2$. Using Gauss elimination, we get

$$(A - (2+i)I) = \begin{pmatrix} 2-i & -1 & -1 & 1 \\ 4 & -2-i & -1 & 0 \\ 2 & -1 & -i & 0 \\ -1 & 1 & -1 & -i \end{pmatrix} \implies \begin{pmatrix} 1 & 0 & -1-i & 0 \\ 0 & 1 & -2-i & 0 \\ 0 & 0 & 0 & 1 \\ 0 & 0 & 0 & 0 \end{pmatrix},$$

and thus the eigenvalue $\lambda = 2 + i$ has geometric multiplicity equal to 1 with an associated eigenvector

$$\tilde{v}_1 = \begin{pmatrix} 1+i \\ 2+i \\ 1 \\ 0 \end{pmatrix} = \begin{pmatrix} 1 \\ 2 \\ 1 \\ 0 \end{pmatrix} + i \begin{pmatrix} 1 \\ 1 \\ 0 \\ 0 \end{pmatrix} = v_1^{\text{Re}} + i v_1^{\text{Im}}.$$

To find a generalized eigenvector, we solve $(A - (2+i)I)\tilde{v}_2 = \tilde{v}_1$ for $\tilde{v}_2$, and we see that

$$\left(\begin{array}{cccc|c} 2-i & -1 & -1 & 1 & 1+i \\ 4 & -2-i & -1 & 0 & 2+i \\ 2 & -1 & -i & 0 & 1 \\ -1 & 1 & -1 & -i & 0 \end{array}\right) \Longrightarrow \left(\begin{array}{cccc|c} 1 & 0 & -1-i & 0 & 0 \\ 0 & 1 & -2-i & 0 & -1 \\ 0 & 0 & 0 & 1 & i \\ 0 & 0 & 0 & 0 & 0 \end{array}\right)$$

and a generalized eigenvector $\tilde{v}_2$ is found as

$$\tilde{v}_2 = \begin{pmatrix} 1+i \\ 1+i \\ 1 \\ i \end{pmatrix} = \begin{pmatrix} 1 \\ 1 \\ 1 \\ 0 \end{pmatrix} + i \begin{pmatrix} 1 \\ 1 \\ 0 \\ 1 \end{pmatrix} = v_2^{\text{Re}} + i v_2^{\text{Im}}.$$

There are actually two Jordan canonical forms for A depending on whether A is viewed (i) as a map on $\mathbb{C}^4$ under the field $\mathbb{C}$ or (ii) as a map on $\mathbb{R}^4$ under the field $\mathbb{R}$. For (i), let K_1 be the matrix whose ordered columns are $\tilde{v}_1$, $\tilde{v}_2$, $\overline{\tilde{v}}_1$, $\overline{\tilde{v}}_2$, and then we have

$$AK_1 = K_1 \begin{pmatrix} \lambda & 1 & 0 & 0 \\ 0 & \lambda & 0 & 0 \\ 0 & 0 & \bar{\lambda} & 1 \\ 0 & 0 & 0 & \bar{\lambda} \end{pmatrix} \Longrightarrow K_1^{-1}AK_1 = \begin{pmatrix} \lambda & 1 & 0 & 0 \\ 0 & \lambda & 0 & 0 \\ 0 & 0 & \bar{\lambda} & 1 \\ 0 & 0 & 0 & \bar{\lambda} \end{pmatrix}.$$

For (ii), let K_2 have the columns v_2^{Re}, v_1^{Im}, v_2^{Re}, v_2^{Im}, and then we have

$$AK_2 = K_2 \begin{pmatrix} 2 & 1 & 1 & 0 \\ -1 & 2 & 0 & 1 \\ 0 & 0 & 2 & 1 \\ 0 & 0 & -1 & 2 \end{pmatrix} \Longrightarrow K_2^{-1}AK_2 = \begin{pmatrix} 2 & 1 & 1 & 0 \\ -1 & 2 & 0 & 1 \\ 0 & 0 & 2 & 1 \\ 0 & 0 & -1 & 2 \end{pmatrix}.$$

Example 5.21: Jordan canonical form of a real matrix with both real and complex eigenvalues.

As another example, consider

$$A = \begin{pmatrix} 5/2 & -1/2 & -1 & 3 \\ 1/2 & 3/2 & -1 & 3 \\ 1/2 & -1/2 & -1 & 4 \\ -1/2 & 1/2 & -2 & 3 \end{pmatrix},$$

where the characteristic polynomial is $\lambda^4 - 6\lambda^3 + 17\lambda^2 - 28\lambda + 20 = (\lambda - 2)^2((\lambda - 1)^2 + 4)$. The eigenvalues are 2 (with algebraic multiplicity 2) and $1 \pm 2i$. We calculate and reduce

$$(A - 2I) = \begin{pmatrix} 1/2 & -1/2 & -1 & 3 \\ 1/2 & -1/2 & -1 & 3 \\ 1/2 & -1/2 & -3 & 4 \\ -1/2 & 1/2 & -2 & 1 \end{pmatrix} \implies \begin{pmatrix} 1 & -1 & 0 & 0 \\ 0 & 0 & 1 & 0 \\ 0 & 0 & 0 & 1 \\ 0 & 0 & 0 & 0 \end{pmatrix},$$

which shows geometric multiplicity of 2 is 1 and the eigenspace is spanned by $\tilde{v}_1 = (1\,1\,0\,0)^T$. To find the generalized eigenspace, we solve $(A - 2I)\tilde{v}_2 = \tilde{v}_1$ for $\tilde{v}_2$, and find $\tilde{v}_2 = (1\,1\,-1\,-6)^T$. The complex eigenvalue $1 + 2i$ is analyzed by

$$(A - (1 + 2i)I) = \begin{pmatrix} -1/2 - 2i & -1/2 & -1 & 3 \\ 1/2 & -3/2 - 2i & -1 & 3 \\ 1/2 & -1/2 & -4 - 2i & 4 \\ -1/2 & 1/2 & -2 & -2 - 2i \end{pmatrix}$$

which reduces to

$$\begin{pmatrix} 1 & 0 & 0 & i \\ 0 & 1 & 0 & i \\ 0 & 0 & 1 & -1 + i \\ 0 & 0 & 0 & 0 \end{pmatrix}, \text{ and has eigenvector } \tilde{v}_3 = \begin{pmatrix} -i \\ -i \\ 1 - i \\ 1 \end{pmatrix} = \begin{pmatrix} 0 \\ 0 \\ 1 \\ 1 \end{pmatrix} - i \begin{pmatrix} 1 \\ 1 \\ 1 \\ 0 \end{pmatrix}.$$

Let K be the matrix with columns $\tilde{v}_1$, $\tilde{v}_2$, v_3^{Re}, v_3^{Im}, and one has

$$AK = K \begin{pmatrix} 2 & 1 & 0 & 0 \\ 0 & 2 & 0 & 0 \\ 0 & 0 & 1 & 2 \\ 0 & 0 & -2 & 1 \end{pmatrix} \implies K^{-1}AK = \begin{pmatrix} 2 & 1 & 0 & 0 \\ 0 & 2 & 0 & 0 \\ 0 & 0 & 1 & 2 \\ 0 & 0 & -2 & 1 \end{pmatrix}.$$

5.7.8 *Powers and exponentials of matrices*

Only vector space operations have been featured thus far in this chapter, but in the space of square matrices, the multiplication of these matrices is also defined and gives another element in the space. In particular, if A is a square matrix, then each of its powers A^n is defined for $n = 1, 2, \dots$. Moreover, if $p(\lambda) = a_n\lambda^n + \cdots + a_1\lambda + a_0$ is a polynomial whose coefficients belong to the scalar field $\mathbb{F}$, then the matrix polynomial can be defined as

$$p(A) = a_n A^n + \cdots + a_1 A + a_0 I.$$

If $f(\cdot)$ is any function with a convergent power series, then one may define $f(A)$ as the matrix that equals the infinite sum. We explore this in greater detail for the exponential function $f(x) = e^x = 1 + x + \frac{x^2}{2!} + \frac{x^3}{3!} + \cdots = \sum_{n=0}^{\infty} \frac{x^n}{n!}$.

The reader should appreciate that raising a matrix to some integer power is a much richer proposition than raising a real number to a power. When real numbers other than ± 1 are raised to increasingly higher powers, their absolute values approach either infinity or zero. This is not true for matrices. For example, the matrix $A = \begin{pmatrix} 0 & 1 & 0 \\ -1 & 0 & 0 \\ 0 & 0 & -1 \end{pmatrix}$ has the powers

$$A^2 = \begin{pmatrix} -1 & 0 & 0 \\ 0 & -1 & 0 \\ 0 & 0 & 1 \end{pmatrix}, \quad A^3 = \begin{pmatrix} 0 & -1 & 0 \\ 1 & 0 & 0 \\ 0 & 0 & -1 \end{pmatrix}, \quad A^4 = \begin{pmatrix} 1 & 0 & 0 \\ 0 & 1 & 0 \\ 0 & 0 & 1 \end{pmatrix},$$

after which the higher powers obviously begin to repeat: $A^5 = A$, $A^6 = A^2$, etc.

Another property of numbers is that 0 is the only number that may have a power to equal zero. This is not true for matrices, as can be seen by considering $A = \begin{pmatrix} 0 & 1 \\ 0 & 0 \end{pmatrix}$; it is easy to check that A^2 equals the zero matrix. Matrices with the property that a power equals zero are important enough to deserve a name.

Definition: *A square matrix B is said to be nilpotent of order k if $B^{k-1} \neq 0$, but $B^k = 0$.*

If two matrices A and J are similar, say $A = KJK^{-1}$, then their powers are also similar since $A^k = KJ^kK^{-1}$ for each $k = 1, 2, \dots$.

The Jordan form (Theorem 5.18) provides an effective tool for calculating arbitrary powers of a matrix A, at least if $\mathbb{F} = \mathbb{C}$ or $\mathbb{F} = \mathbb{R}$ and the eigenvalues are real. Suppose this is the case, and $A = KJK^{-1}$ where J is a matrix in Jordan form. Note that J is of the form $J = D + B$, where D is a diagonal matrix and B has zeroes everywhere except possibly along the superdiagonal. If the dimension of A is $N \times N$, then B is nilpotent of order $k \leq N$ (more precisely, the order of nilpotency is the largest geometric multiplicity of any of its eigenvalues). The matrix D^k is particularly easy to calculate since D is diagonal. Indeed, if D is diagonal with entries λ_n, then D^k is also a diagonal matrix with corresponding entries λ_n^k. The sum $I + B + B^2 + \cdots + B^n$ is also easy to calculate since it eventually becomes the zero matrix. For example, a matrix B for $N = 3$ with nilpotent order $k = 3$ satisfies

$$B = \begin{pmatrix} 0\,1\,0 \\ 0\,0\,1 \\ 0\,0\,0 \end{pmatrix}, \quad B^2 = \begin{pmatrix} 0\,0\,1 \\ 0\,0\,0 \\ 0\,0\,0 \end{pmatrix}, \quad \text{and} \quad B^3 = \begin{pmatrix} 0\,0\,0 \\ 0\,0\,0 \\ 0\,0\,0 \end{pmatrix}.$$

The powers $(D+B)^n$ can be calculated by using the binomial theorem, and so they have a somewhat complicated form. We shall not go into all the details here, but rather sketch an effective method to compute the exponential e^A of a general matrix A.

We define e^A as the matrix

$$e^A = I + A + \frac{A^2}{2!} + \frac{A^3}{3!} + \ldots,$$

but we still have to give meaning to the series converging. If A is a diagonal matrix D of the form

$$D = \begin{pmatrix} \lambda_1 & 0 & \cdots & 0 \\ 0 & \lambda_2 & \cdots & 0 \\ 0 & \vdots & \cdots & 0 \\ 0 & \cdots & 0 & \lambda_n \end{pmatrix},$$

then e^D can be computed directly as

$$e^D = \begin{pmatrix} e^{\lambda_1} & 0 & \cdots & 0 \\ 0 & e^{\lambda_2} & \cdots & 0 \\ 0 & \vdots & \cdots & 0 \\ 0 & \cdots & 0 & e^{\lambda_n} \end{pmatrix}.$$

On the other hand, if A is the nilpotent matrix B of order N with zeroes everywhere except with 1's along the super diagonal, and so has the form

$$B = \begin{pmatrix} 0\,1\,0 & 0 & \cdots & 0 \\ 0\,0\,1 & 0 & \cdots & 0 \\ \vdots & \vdots & & \vdots \\ 0\,0\,0 & \cdots & 1 & 0 \\ 0\,0\,0 & \cdots & 0 & 1 \\ 0\,0\,0 & \cdots & 0 & 0 \end{pmatrix},$$

then the terms B^n for $n \geq N$ all are zero, and the infinite sum reduces to the finite sum

$$e^B = \begin{pmatrix} 1 & 1 & \frac{1}{2!} & \frac{1}{3!} & \cdots & \frac{1}{(N-1)!} \\ 0 & 1 & 1 & \frac{1}{2!} & \cdots & \frac{1}{(N-2)!} \\ \vdots & & & \vdots & & \vdots \\ 0 & 0 & 0 & \cdots & 1 & 1 \\ 0 & 0 & 0 & \cdots & 0 & 1 \end{pmatrix},$$

Now suppose A is an elementary Jordan block $A = J = \lambda I + B$. Then it turns out (only because $\lambda IB = B\lambda I$) that $e^J = e^\lambda e^B$, which specifically says

$$e^J = e^\lambda \begin{pmatrix} 1 & 1 & \frac{1}{2!} & \frac{1}{3!} & \cdots & \frac{1}{(N-1)!} \\ 0 & 1 & 1 & \frac{1}{2!} & \cdots & \frac{1}{(N-2)!} \\ 0 & \vdots & \cdots & \vdots & & \\ 0 & 0 & 0 & \cdots & 1 & 1 \\ 0 & 0 & 0 & \cdots & 0 & 1 \end{pmatrix}.$$

Now note that powers of a matrix preserve the block structure. If A is any Jordan block matrix, its exponential will consist of a block matrix where the elementary Jordan blocks will become the exponentials of the blocks. Similarly if A is any Jordan matrix. For example,

$$A = \begin{pmatrix} 2 & 1 & 0 & 0 & 0 & 0 \\ 0 & 2 & 1 & 0 & 0 & 0 \\ 0 & 0 & 2 & 1 & 0 & 0 \\ 0 & 0 & 0 & 2 & 0 & 0 \\ 0 & 0 & 0 & 0 & 3 & 1 \\ 0 & 0 & 0 & 0 & 0 & 3 \end{pmatrix}$$

has exponential

$$e^A = \begin{pmatrix} e^2 & e^2 & \frac{e^2}{2!} & \frac{e^2}{3!} & 0 & 0 \\ 0 & e^2 & e^2 & \frac{e^2}{2!} & 0 & 0 \\ 0 & 0 & e^2 & e^2 & 0 & 0 \\ 0 & 0 & 0 & e^2 & 0 & 0 \\ 0 & 0 & 0 & 0 & e^3 & e^3 \\ 0 & 0 & 0 & 0 & 0 & e^3 \end{pmatrix}.$$

Finally, if A is any matrix that is similar to a Jordan matrix, say $A = KJK^{-1}$, then $e^A = Ke^JK^{-1}$.

One must be very cautious when applying scalar functions to matrix arguments. The matrix functions may still be well defined through their Taylor series, but the functions may no longer have the properties that are so familiar when they have scalar arguments. The exponential function occurs particularly often and when applied to a matrix, it does not in general hold that $e^{A+B} = e^A e^B$, though it does hold if A and B commute.

5.7.9 *Location of eigenvalues*

The precise values of the eigenvalues of a matrix are often not as important as simply knowing their approximate location in the complex plane. In stability analysis of continuous time models one may want to know if the eigenvalues are all in the negative half-plane, i.e. if they all have negative real parts or, for discrete time models, one may want to know if the eigenvalues are all located inside the unit circle. This type of problem is probably familiar to most readers as the similar problem for the location of the roots of a polynomial. For polynomial roots, various simple tests, such as the Routh test, are available that allows one to infer something about the location of the roots and these tests can *in principal* be applied to the characteristic polynomial. However, except for small or sparse matrices, finding the characteristic polynomial is usually a lot of work and it is better to try and conclude something about the location of the eigenvalues directly from the matrix.

For a diagonal matrix, the eigenvalues equal the diagonal elements and it seems reasonable to assume that if the off-diagonal elements are numerically much smaller than the diagonal elements, then the eigenvalues are not much different from the diagonal values. Thus, one can conjecture that the eigenvalues are within some distance from the diagonal elements, and that this distance is determined by the values of the off-diagonal elements. This is the idea behind the following theorem, known either as Gerschgorin's theorem or Brauer's theorem.

Theorem 5.19 (Gerschgorin). *Every eigenvalue of the matrix $A_{N\times N}$ lies in at least one of the discs centered at $a_{n,n}$ and with radius $\sum_{k=1,k\neq n}^{N} |a_{n,k}|$. Stated in mathematical nomenclature*

$$\forall\lambda\ \exists n \quad such\ that\ |\lambda - a_{n,n}| \leq \sum_{k=1,k\neq n}^{N} |a_{n,k}|.$$

Proof: Let λ and $\tilde{v}$ be an eigenvalue and an associated eigenvector and let v_n be the numerically largest element of $\tilde{v}$. Clearly, $v_n \neq 0$. Then

$$A\tilde{v} = \lambda\tilde{v} \;\Rightarrow$$

$$\sum_{k=1}^{N} a_{n,k}v_k = \lambda v_n \;\Leftrightarrow$$

$$\lambda - a_{n,n} = \sum_{k=1, k\neq n}^{N} a_{n,k} \frac{v_k}{v_n} \Rightarrow$$

$$|\lambda - a_{n,n}| \leq \sum_{k=1, k\neq n}^{N} |a_{n,k} \frac{v_k}{v_n}| \leq \sum_{k=1, k\neq n}^{N} |a_{n,k}| \qquad \square$$

The disks that limit the location of the eigenvalues in the complex plane are called Gerschgorin or Brauer disks. It can further be shown that if some number, x, of the Gerschgorin disks overlap and form a connected domain, isolated from the other disks, then this domain contains exactly x eigenvalues. It follows that if all the disks are isolated from one another, then each disk contains exactly 1 eigenvalue.[4]

The next theorem is due to Oskar Perron and concerns matrices with strictly positive entries. Such matrices are often just called positive matrices. We will not prove this theorem here.

Theorem 5.20 (Perron). *Let $A = \{a_{n,m}\}_{n,m=1..N}$ be a positive matrix, i.e. such that $a_{n,m} > 0 \, \forall \, n, m$, then the following holds.*

The matrix has a real simple eigenvalue r with the property that all other eigenvalues λ, whether real or complex, are numerically smaller than r, i.e. $|\lambda| < r$. Furthermore, r has an eigenvector for which all components are positive and r is bounded by the inequalities

$$min \sum_{m=1}^{N} a_{n,m} \leq r \leq max \sum_{m=1}^{N} a_{n,m}$$

where the minimum and maximum are over all possible values of n.

5.8 Geometry of vector spaces

Up till now, we have made no use of geometric concepts such as length, distance or direction. The reason for this is that the vector space definition does not include any geometric concepts. A geometry is usually imposed on a vector space by defining a so called inner product, a generalization of the well known scalar product. In this last section, we define this product as well as other vector products and explore some of the geometric aspects of vectors spaces that appear with the definition of this inner product.

[4]The proof of this result can be found in Wilkinson, J. H. [1965] *The Algebraic Eigenvalue Problem* (Clarendon Press, Oxford).

5.8.1 *Vector products*

There are several possible types of products between vectors, all of which have different names and are indicated by different types of multiplication signs. The inner product is particularly important.

5.8.1.1 *Inner product*

The classical scalar or dot product is indicated by a fat dot: $\bullet$ and is defined for vectors in $\mathbb{R}^N$ and $\mathbb{C}^N$ as

$$v \bullet w = \sum_{n=1}^{N} v_n \overline{w_n} \tag{5.33}$$

where the overbar indicates complex conjugation. A generalization of this concept is the inner product which is defined for vectors in $\mathbb{R}^N$ or $\mathbb{C}^N$. It is indicated by triangular brackets, $\langle \cdot, \cdot \rangle$, and is a mapping of an ordered pair of vectors into the scalar field $\mathbb{F}$.

Definition: *An inner product is a binary operation which maps two elements of a vector space over $\mathbb{C}$ into the scalar field and which has the following properties:*

(1) Additivity:

$$\langle v + w, q \rangle = \langle v, q \rangle + \langle w, q \rangle .$$

(2) Hermitian Symmetry:

$$\langle v, w \rangle = \overline{\langle w, v \rangle} .$$

(3) Homogeneity:

$$\langle \alpha v, w \rangle = \alpha \langle v, w \rangle$$

where α is a scalar.

(4) Positive definiteness:

$$\langle v, v \rangle > 0 \ , \ \text{if } v \neq 0 \, .$$

Notice that for real vectors, the inner product is commutative and that it follows from the symmetry property that the inner product of a vector with itself is real, even if the associated field is the complex numbers. The requirement of positive definiteness is therefore meaningful even when the vector contains complex elements. A vector space on which an inner product is defined is called an *inner product space.* Convince yourself that the scalar product is an inner product.

The scalar product is not the only possible inner product defined in $\mathbb{F}^N$. For example, let $\langle v, w \rangle$ denote the classical scalar product, and suppose A is any $N \times N$

matrix that satisfies $\langle Av, v\rangle > 0$ for every $v \neq 0$ (such a matrix is called *positive definite*). Then a new inner product is defined via

$$\langle v, w\rangle_A = \langle Av, w\rangle.$$

The inner product provides a convenient way of defining the geometric properties of a vector space. The length of a vector is defined as

$$|v| = \sqrt{\langle v, v\rangle}$$

and the angle θ between two vectors is

$$\langle v, w\rangle = |v||w| \cos\theta.$$

Once length and direction is defined, it becomes meaningful to visualize vectors in $\mathbb{R}^N$ as directed line segments. For non-zero vectors, the inner product can be interpreted geometrically as the length, calculated with sign, of the projection of the first factor onto the oriented line defined by the second factor, see Fig. 5.2.

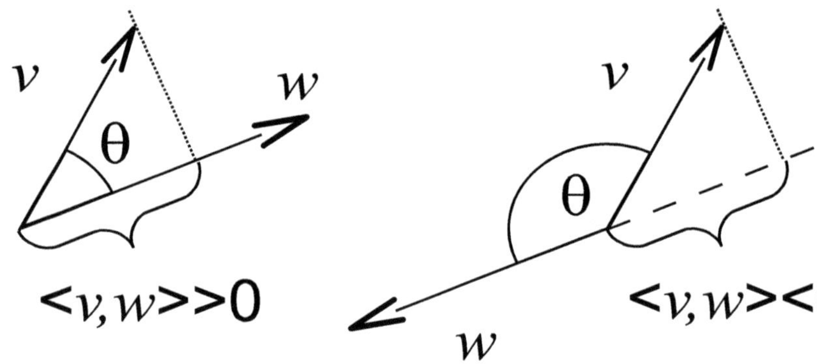

Fig. 5.2 Geometric interpretation of the inner product.

A vector is said to be *normalized* if it has length 1. Any non-zero vector can be normalized by dividing it by its length. Two vectors, different from zero, are said to be *orthogonal* if their inner product is zero.

It is convenient to pick the basis vectors of an inner product space such that they are pairwise orthogonal. When this holds, the basis is referred to as an *orthogonal basis* and the coordinate system as orthogonal. A vector, written in terms of an orthogonal basis, $\{e_1, e_2, \cdots, e_j, \cdots, e_n\}$, is

$$v = \sum_{n=1}^{N} v_n e_n$$

and the coordinates, v_n, are found by taking the inner product of both sides of this expression with an arbitrary basis vector e_m

$$\langle v, e_m\rangle = \sum_{n=1}^{N} v_n \langle e_n, e_m\rangle = v_m \langle e_m, e_m\rangle.$$

Thus, the expression for the coordinates of a vector in terms of an orthogonal basis is

$$v_m = \frac{\langle v, e_m \rangle}{\langle e_m, e_m \rangle}$$

and if the basis vectors have unit length, this result simplifies further to

$$v_m = \langle v, e_m \rangle.$$

A basis for which the basis vector are pairwise orthogonal and all have unit length is called an *orthonormal basis.* A similarly simple expression does not hold for non-orthogonal or oblique coordinates and oblique coordinates systems are therefore almost never used.

5.8.1.2 *Cross product*

The cross product is defined for vectors in $\mathbb{R}^3$ and is indicated by "$\times$". It can be written several ways. For instance as

$$v \times w = |v||w| \sin\theta \, n_{vw}$$

where θ is the angle between the two vectors and n_{vw} is the unit normal to the v, w plane, pointing in the direction indicated by the right hand rule (in a right handed coordinate system) applied to v, w or by the left hand rule (in a left handed coordinate system). Geometrically, $v \times w$ is a vector of magnitude equal to the area of the parallelogram spanned by v and w, orthogonal to the plane of this parallelogram and pointing in the direction indicated by the right hand rule, see Fig 5.3.

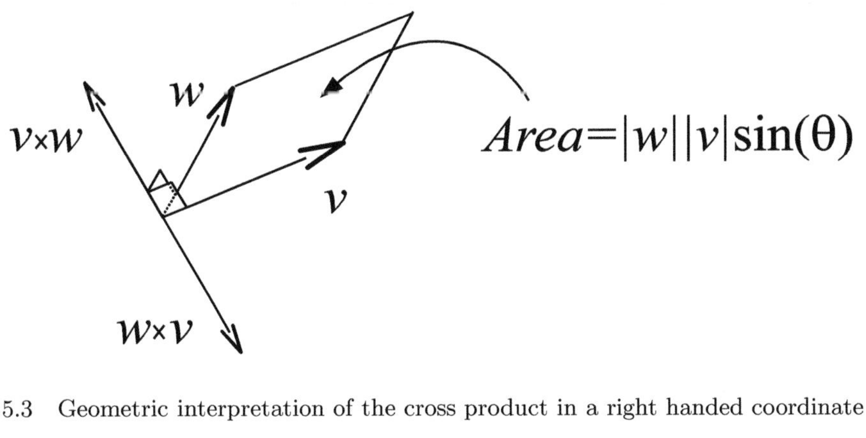

Fig. 5.3 Geometric interpretation of the cross product in a right handed coordinate system.

In terms of coordinate representation in a rectangular coordinate system, the cross product can be found as

$$v \times w = (v_1 e_1 + v_2 e_2 + v_3 e_3) \times (w_1 e_1 + w_2 e_2 + w_3 e_3)$$

where the e_ns are the basis vectors $e_1 = (1, 0, 0)$, $e_2 = (0, 1, 0)$ and $e_3 = (0, 0, 1)$. The cross product obeys the distributive law, so this becomes

$$\begin{aligned} v \times w = & v_1 w_1 e_1 \times e_1 + v_1 w_2 e_1 \times e_2 + v_1 w_3 e_1 \times e_3 \\ & + v_2 w_1 e_2 \times e_1 + v_2 w_2 e_2 \times e_2 + v_2 w_3 e_2 \times e_3 \\ & + v_3 w_1 e_3 \times e_1 + v_3 w_2 e_3 \times e_2 + v_3 w_3 e_3 \times e_3. \end{aligned}$$

Noting that $e_1 \times e_1 = 0$, $e_1 \times e_2 = e_3$ etc., this simplifies to

$$v \times w = (v_2 w_3 - v_3 w_2)e_1 + (v_3 w_1 - v_1 w_3)e_2 + (v_1 w_2 - v_2 w_1)e_3.$$

The cross product can also be written as

$$v \times w = \begin{vmatrix} e_1 & e_2 & e_3 \\ v_1 & v_2 & v_3 \\ w_1 & w_2 & w_3 \end{vmatrix}$$

where the vertical lines indicates a determinant. The cross product is not commutative. In fact

$$v \times w = -w \times v.$$

The vector that is found as the result of a cross product has the peculiar property that when the two factors in the cross product are reflected in a mirror, the cross product of these reflected vectors is a vector with the opposite direction of that of the original cross product, see Fig. 5.4. Vectors with this property are sometimes referred to as *axial* vectors as opposed to *polar* vectors which do not change directions under reflection.

Polar and axial vectors may be visualized differently: polar vectors as the usual directed line segment or arrow and axial vectors as a line segment associated with an oriented loop which indicates the sense of the vector, see Fig. 5.5.

Using this loop to visualize the cross product, the direction of the loop should be such that it indicates the direction that the first factor must turn in order to align with the second factor. Doing so does result in a visualization of the cross product that does work under reflection, see Fig. 5.6.[5]

5.8.1.3 *Triple scalar product*

The triple scalar product of three vectors is

$$[v, w, u] = v \bullet (w \times u).$$

[5]This short description of two ways of visualizing vectors only scratches the surface of possibilities. To learn more, everyone should consult the wonderful little book *Geometrical Vectors* by Weinreich, G., published in 1998 by The University of Chicago Press.

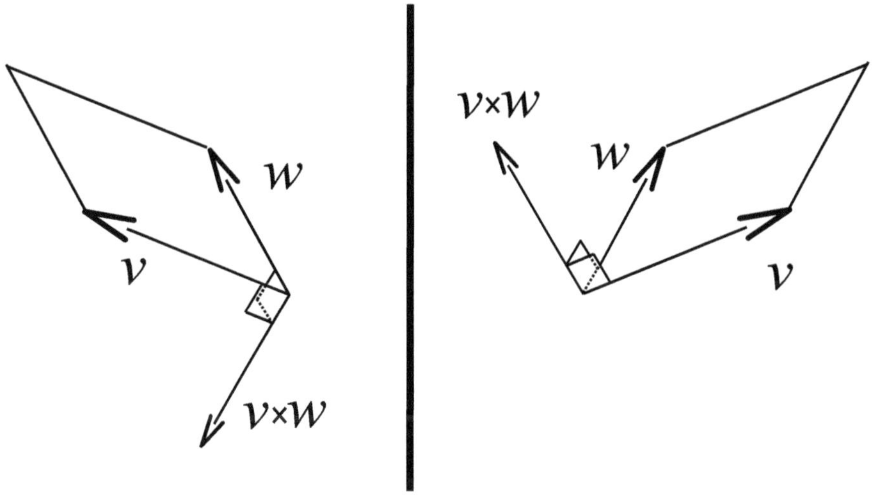

Fig. 5.4 Reflection in a mirror, the fat vertical line, of the two factors in a cross product gives a result that is not a reflection of the original cross product result.

Fig. 5.5 Common ways of visualizing polar (left) and axial (right) vectors.

It equals the volume of the parallelepiped formed by the three factors, as in Fig. 5.7. In terms of coordinates

$$[v, w, u] = v_1(w_2u_3 - w_3u_2) + v_2(w_3u_1 - w_1u_3) + v_3(w_1u_2 - w_2u_1)$$

$$= \begin{vmatrix} v_1 & v_2 & v_3 \\ w_1 & w_2 & w_3 \\ u_1 & u_2 & u_3 \end{vmatrix} .$$

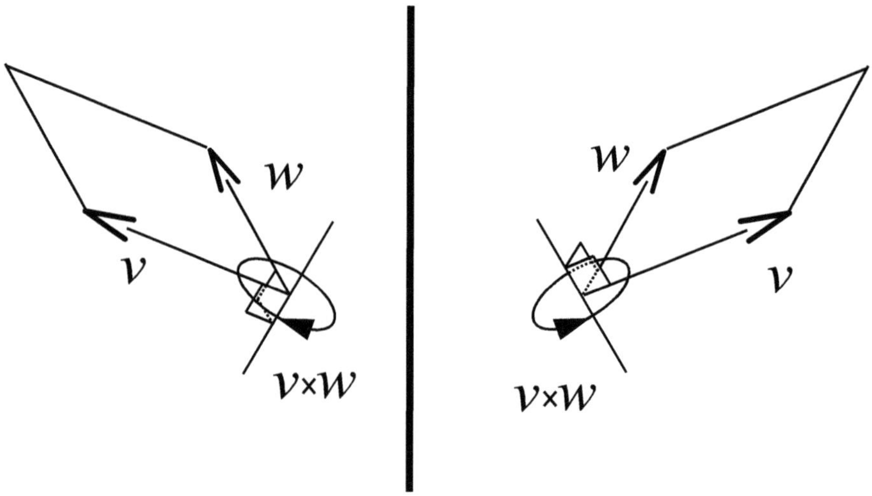

Fig. 5.6 Visualizing the cross product using a loop to indicate the sense or direction of the product provides a visualization that works correctly under reflection.

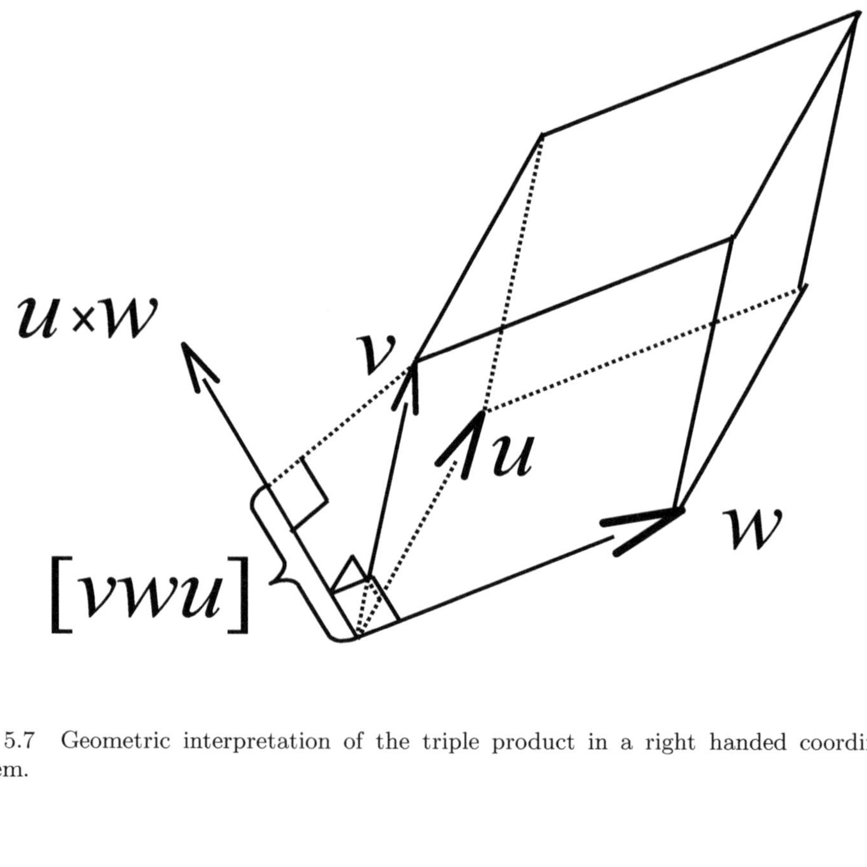

Fig. 5.7 Geometric interpretation of the triple product in a right handed coordinate system.

5.8.1.4 *Dyad or outer product*

The dyad or outer product of two vectors that are ordered sets of complex numbers is indicated by placing the two factors immediately next to one another, without a binary operator or spacial separation, or by the triangular brackets in the order opposite to that used to indicate the inner product, i.e. by $\rangle, \langle$. It is defined as follows.

Definition: *A dyad or outer product is a binary operation which maps two elements, v and w, of a vector space into the set of ordered numbers of the form $q_{n,m}$ where*

$$vw = \rangle v, w\langle = v_n w_m = q_{n,m}.$$

The outer product of two vectors in $\mathbb{R}^N$ can obviously be viewed as an array with N columns and N rows such that the element in row n and column m equals the product of the nth element of the first factor and the mth element of the second factor. Such an array is an N-by-N matrix. The outer product is not commutative.

Example 5.22: Outer product.

The two possible outer products of

$$v = (1, 2, 3) \quad \text{and} \quad w = (6, -2, 4)$$

are

$$\rangle v, w\langle = \begin{pmatrix} 6 & -2 & 4 \\ 12 & -4 & 8 \\ 18 & -6 & 12 \end{pmatrix}$$

and

$$\rangle w, v\langle = \begin{pmatrix} 6 & 12 & 18 \\ -2 & -4 & -6 \\ 4 & 8 & 12 \end{pmatrix}.$$

5.8.2 *Gram–Schmidt orthogonalization*

Nothing in the definition of a basis demands that the basis vectors are orthogonal to one another. However, orthogonal bases are more convenient to work with than oblique bases and they are preferentially used. Luckily, there is a simple algorithm, known as Gram–Schmidt orthogonalization, by which one can find an orthogonal basis of a vector space given any basis of the space. Gram–Schmidt orthogonalization is often presented with the assumption that the known basis is normalized. This is not a requirement for the algorithm to work, though it does simplify the result a bit. Gram–Schmidt orthogonalization works as follows.

Let $\{v_n\}_{n=1,N}$ be a given, non-orthogonal basis and let $\{w_n\}_{n=1,N}$ indicate the unknown, orthogonal basis. As the first basis vector of the orthogonal basis, pick

$$w_1 = v_1$$

and pick the second basis vector as the following linear combination of w_1 and v_2.

$$w_2 = v_2 - cw_1 \,.$$

The value of c must now be determined such that w_1 and w_2 are orthogonal. Thus

$$\langle w_2, w_1 \rangle = 0 \Rightarrow$$

$$\langle v_2, v_1 \rangle - c\langle v_1, v_1 \rangle = 0 \Rightarrow$$

$$c = \frac{\langle v_2, v_1 \rangle}{\langle v_1, v_1 \rangle} \Rightarrow$$

$$w_2 = v_2 - v_1 \frac{\langle v_2, v_1 \rangle}{\langle v_1, v_1 \rangle} \,.$$

Now pick the third vector in the orthogonal basis, w_3, as the following linear combination of w_1, w_2 and v_3.

$$w_3 = v_3 - c_1 w_1 - c_2 w_2 \,.$$

Orthogonality demands that

$$\langle w_3, w_1 \rangle = 0 \Rightarrow \langle v_3, w_1 \rangle - c_1\langle w_1, w_1 \rangle - c_2\langle w_2, w_1 \rangle = 0$$

$$\langle w_3, w_2 \rangle = 0 \Rightarrow \langle v_3, w_2 \rangle - c_1\langle w_1, w_2 \rangle - c_2\langle w_2, w_2 \rangle = 0$$

and using the fact that w_1 and w_2 are orthogonal, this readily simplifies and the two constants are found as

$$c_1 = \frac{\langle v_3, w_1 \rangle}{\langle v_1, v_1 \rangle}$$

$$c_2 = \frac{\langle v_3, w_2 \rangle}{\langle w_2, w_2 \rangle}$$

a result that simplifies further if the w-vectors are normalized.

In general, w_n is found recursively as

$$w_n = v_n - \sum_{k=1}^{n-1} c_k w_k$$

and the constant c_m is found by taking the inner product of this equation with w_m.

$$\langle w_n, w_m \rangle = \langle v_n, w_m \rangle - c_m \langle w_m, w_m \rangle = 0 \;\Rightarrow$$

$$c_m = \frac{\langle v_n, w_m \rangle}{\langle w_m, w_m \rangle} \,.$$

Example 5.23: Gram–Schmidt orthogonalization.

Given the basis v below, find an orthogonal basis for the vector space spanned by v.

$$v = \left\{ \begin{pmatrix} 0 \\ 1 \\ 2 \\ 3 \end{pmatrix}, \begin{pmatrix} 1 \\ 1 \\ 0 \\ 1 \end{pmatrix}, \begin{pmatrix} 1 \\ 0 \\ 2 \\ 2 \end{pmatrix}, \begin{pmatrix} 2 \\ 1 \\ 2 \\ 1 \end{pmatrix} \right\}.$$

The first vector in the orthogonal basis is

$$w_1 = \begin{pmatrix} 0 \\ 1 \\ 2 \\ 3 \end{pmatrix}$$

and the second vector is given by

$$w_2 = \begin{pmatrix} 1 \\ 1 \\ 0 \\ 1 \end{pmatrix} - c \begin{pmatrix} 0 \\ 1 \\ 2 \\ 3 \end{pmatrix}.$$

Orthogonality of w_1 and w_2 now requires that

$$\langle w_1, w_2 \rangle = 4 - 14c = 0 \Rightarrow c = \frac{2}{7} \Rightarrow w_2 = \begin{pmatrix} 1 \\ 5/7 \\ -4/7 \\ 1/7 \end{pmatrix}.$$

The third basis vector in the orthogonal basis, w_3, is

$$w_3 = \begin{pmatrix} 1 \\ 0 \\ 2 \\ 2 \end{pmatrix} - c_1 \begin{pmatrix} 0 \\ 1 \\ 2 \\ 3 \end{pmatrix} - c_2 \begin{pmatrix} 1 \\ 5/7 \\ -4/7 \\ 1/7 \end{pmatrix}$$

giving

$$\langle w_1, w_3 \rangle = 10 - 14c_1 = 0 \Rightarrow c_1 = \frac{5}{7}$$

$$\langle w_2, w_3 \rangle = \frac{1}{7} - c_2 \frac{13}{7} = 0 \Rightarrow c_2 = \frac{1}{13}$$

and thus

$$w_3 = \begin{pmatrix} 1 \\ 0 \\ 2 \\ 2 \end{pmatrix} - \frac{5}{7} \begin{pmatrix} 0 \\ 1 \\ 2 \\ 3 \end{pmatrix} - \frac{1}{13} \begin{pmatrix} 1 \\ 5/7 \\ -4/7 \\ 1/7 \end{pmatrix} = \begin{pmatrix} 12/13 \\ -10/13 \\ 8/13 \\ -2/13 \end{pmatrix}.$$

Finally

$$w_4 = \begin{pmatrix} 2 \\ 1 \\ 2 \\ 1 \end{pmatrix} - c_1 \begin{pmatrix} 0 \\ 1 \\ 2 \\ 3 \end{pmatrix} - c_2 \begin{pmatrix} 1 \\ 5/7 \\ -4/7 \\ 1/7 \end{pmatrix} - c_3 \begin{pmatrix} 12/13 \\ -10/13 \\ 8/13 \\ -2/13 \end{pmatrix}$$

gives

$$\langle w_1, w_4 \rangle = 8 - 14c_1 = 0 \quad \Rightarrow \ c_1 = \frac{4}{7}$$
$$\langle w_2, w_4 \rangle = \frac{12}{7} - \frac{13}{7}c_2 = 0 \Rightarrow c_2 = \tfrac{12}{13}$$
$$\langle w_3, w_4 \rangle = \frac{28}{13} - \frac{24}{13}c_3 = 0 \Rightarrow \ c_3 = \tfrac{7}{6}$$

and thus

$$w_4 = \begin{pmatrix} 2 \\ 1 \\ 2 \\ 1 \end{pmatrix} - \frac{4}{7} \begin{pmatrix} 0 \\ 1 \\ 2 \\ 3 \end{pmatrix} - \frac{12}{13} \begin{pmatrix} 1 \\ 5/7 \\ -4/7 \\ 1/7 \end{pmatrix} - \frac{7}{6} \begin{pmatrix} 12/13 \\ -10/13 \\ 8/13 \\ -2/13 \end{pmatrix} = \begin{pmatrix} 0 \\ 2/3 \\ 2/3 \\ -2/3 \end{pmatrix}.$$

5.8.3 *Eigenrows*

For a given matrix, a related eigenvalue problem is obtained by considering the complex conjugate, transpose matrix

$$\overline{A}^T w = \eta w.$$

For reasons which will become clearer after linear operators have been discussed, this problem is called the adjoint eigenvalue problem. Using the reversal rule for matrix products, $(AB)^T = B^T A^T$, this problem can also be written

$$w\overline{A} = \eta w.$$

The characteristic equation is

$$|\overline{A}^T - \eta I| = 0$$

but the determinant does not change when the rows and columns of the matrix are interchanged, so this is equivalent to

$$|\overline{A} - \eta I| = 0 \ \Rightarrow \ |A - \overline{\eta} I| = 0\,.$$

The η eigenvalues are therefore the complex conjugates of the λ eigenvalues. If all the eigenvalues are real, the two eigenvalue sets are identical. The eigenvectors of the complex conjugate, transpose matrix, the ws, are called the eigenrows of the original matrix. The set of eigenvectors and eigenrows have some nice properties.

Theorem 5.21. *Each eigenvector is orthogonal to each eigenrow, except for the one with which it shares a complex conjugate eigenvalue pair, or*

$$\langle \tilde{w}_m, \tilde{v}_n \rangle = 0 \quad \textit{iff } m \neq n\,.$$

Proof: Take the eigenvalue problem in the original form

$$A\tilde{v}_n = \lambda_n \tilde{v}_n$$

and premultiply by an eigenrow, $\tilde{w}_m$, to obtain an inner product

$$\langle \tilde{w}_m, A\tilde{v}_n \rangle = \langle \tilde{w}_m, \lambda_n \tilde{v}_n \rangle = \overline{\lambda_n} \langle \tilde{w}_m, \tilde{v}_n \rangle\,. \tag{5.34}$$

Then take the eigenrow problem in the form

$$\overline{A}^T \tilde{w}_m = \lambda_m \tilde{w}_m$$

and postmultiply this by an eigenvector, $\tilde{v}_n$, to obtain the inner product

$$\langle \overline{A}^T \tilde{w}_m, \tilde{v}_n \rangle = \lambda_m \langle \tilde{w}_m, \tilde{v}_n \rangle\,. \tag{5.35}$$

Taking the difference of Eq. (5.34) and Eq. (5.35) gives

$$(\lambda_m - \overline{\lambda_n}) \langle \tilde{w}_m, \tilde{v}_n \rangle = \langle \overline{A}^T \tilde{w}_m, \tilde{v}_n \rangle - \langle \tilde{w}_m, A\tilde{v}_n \rangle.$$

The last term on the right hand side is rearranged by writing it out in full and changing the order of summation.

$$\begin{aligned} \langle \tilde{w}_m, A\tilde{v}_n \rangle &= \sum_{k=1}^{N} \tilde{w}_{m,k} \overline{\left(\sum_{l=1}^{N} a_{kl} \tilde{v}_{n,l} \right)} = \sum_{k=1}^{N} \tilde{w}_{m,k} \left(\sum_{l=1}^{N} \overline{a_{kl}}\, \overline{\tilde{v}_{n,l}} \right) \\ &= \sum_{l=1}^{N} \left(\sum_{k=1}^{N} \overline{a_{kl}} \tilde{w}_{m,k} \right) \overline{\tilde{v}_{n,l}} = \langle \overline{A}^T \tilde{w}_m, \tilde{v}_n \rangle\,. \end{aligned} \tag{5.36}$$

Thus

$$(\lambda_m - \overline{\lambda_n}) \langle \tilde{w}_m, \tilde{v}_n \rangle = 0$$

if $n \neq m$ or equivalently, if $\overline{\lambda_n} \neq \lambda_m$ this can only be true if

$$\langle \tilde{w}_m, \tilde{v}_n \rangle = 0 \quad \text{iff } m \neq n$$

i.e. if the eigenvectors and eigenrows are orthogonal. □

Because eigenvectors and eigenrows are mutually orthogonal, they are said to be *biorthogonal sets*.

5.8.4 *Real, symmetric matrices*

When a matrix is real and symmetric, it equals its own complex conjugate transpose and the associated eigenvalue and adjoint eigenvalue problems are therefore identical. Real, symmetric matrices appear frequently in applications and they have a nice eigenvalue structure; their eigenvalues are real and their eigenvectors are orthogonal.

Theorem 5.22. *The eigenvectors of different eigenvalues of a real, symmetric matrix are orthogonal.*

Proof: Since the eigenvalue and adjoint eigenvalue problem are identical, the eigenvectors equal the eigenrows, or $\tilde{v}_n = \tilde{w}_n$. Thus

$$\langle \tilde{v}_m, \tilde{v}_n \rangle = \langle \tilde{v}_m, \tilde{w}_n \rangle = 0$$

where the biorthogonality property of the eigenvector and eigenrows was used. □

Theorem 5.23. *The eigenvalues of a real, symmetric matrix are real.*

Proof: Starting with Eq. (5.36)

$$\langle \tilde{w}_m, A\tilde{v}_n \rangle = \langle \overline{A}^T \tilde{w}_m, \tilde{v}_n \rangle .$$

Setting $\tilde{w}_m = \tilde{v}_n$ and using the fact that $A = \overline{A}^T$, one obtains

$$\langle \tilde{v}_n, A\tilde{v}_n \rangle = \langle A\tilde{v}_n, \tilde{v}_n \rangle \;\Rightarrow$$

$$\langle \tilde{v}_n, \lambda_n \tilde{v}_n \rangle = \langle \lambda_n \tilde{v}_n, \tilde{v}_n \rangle \;\Rightarrow$$

$$\overline{\lambda_n} \langle \tilde{v}_n, \tilde{v}_n \rangle = \lambda_n \langle \tilde{v}_n, \tilde{v}_n \rangle \;\Rightarrow$$

$$\overline{\lambda_n} = \lambda_n$$

which can only be true if the eigenvalues are real. □

Notice that if the eigenvectors and the eigenrows of a real, symmetric matrix have all been normalized such that they have length 1, then

$$\langle w_m, v_n \rangle = \begin{cases} 0 \,, & n \neq m \\ 1 \,, & n = m \end{cases}$$

and it follows that if K is a matrix with the nth eigenvector in column n and M a matrix with the nth eigenrow in row n, then

$$MK = I \;\Leftrightarrow\; KM = I .$$

Thus, when the eigenvectors and eigenrows of a real, symmetric matrix are normalized, the matrices of eigenvectors and eigenrows are inverses of one another.

Chapter 6

Tensors

It the previous chapter, matrices were viewed as fixed objects, sets of numbers indexed by two counters. The main new idea in this chapter is that matrices, just like vectors, can be viewed as coordinate representations of some underlying object. This underlying object is a linear operator, a concept that was briefly alluded to in Example 8 on page 150.

Linear operators are functions that map vectors into other vectors and the important aspect of linear operators is that they can be represented by a matrix and conversely, that a matrix can be thought of as a coordinate representation of a linear operator. Not surprisingly, since the matrix that represents a linear operator is a coordinate representation of this operator, it is not a pre-set object but has entries that depend on the choice of bases for the vector spaces involved. We say that the matrix representation of the linear operator is not natural. We will start by investigating how the matrix representation of a linear operator can be found and how the entries in the matrix change with a change of bases.

Tensors are a particular type of linear operator that occurs frequently in physics and engineering. For instance, in a nonisotropic[1] medium, thermal conduction of energy is not necessarily in the same direction as the temperature gradient, but in the direction associated with the highest thermal conductivity. Thus Fourier's law for a nonisotropic solid is more complex than for an isotropic solid and it requires a tensor to describe how the temperature gradient is mapped to the energy flux. While general linear operators can map between vector spaces of different dimensions, the property that tensors share is that tensors are always associated with vector spaces of the same dimension. Unfortunately, mathematicians, physicists and engineers have numerous different ways of defining tensors and it is not obvious that these definitions are in fact equivalent. We will give the most common tensor definitions here and present a casual argument for why they are equivalent.

[1]i.e. a medium in which properties depend on direction such as pure crystals or laminated materials.

6.1 Definitions and basic concepts

Linear operators are maps or functions, L, which map elements of a vector space V into elements of a vector space W, $L : V \to W$, such that

$$L(v_1 + v_2) = L(v_1) + L(v_2)$$
$$L(\alpha v) = \alpha L(v)$$

or, equivalently, such that

$$L(\alpha v_1 + v_2) = \alpha L(v_1) + L(v_2)$$

where L is a linear operator, v_1 and v_2 are vectors in V and α is a scalar. Obviously, for this definition to make sense, V and W must be over the same scalar field. By letting α equal 0, one can see that linear operators always map the zero element of V into the zero element of W. The space V, the vector space on which the linear operator is defined, is called the *domain* of L and is written $\mathcal{D}(L)$. The space W into which the operator maps is called the image space while the image of all the vectors in the domain of L is called the *range* of L and is indicated $\mathcal{R}(L)$. The range of an operator does not necessarily equal the entire image space. The set of vectors in the domain of an operator that are mapped to the 0 element is the *null space* or *kernel* of the operator.

In order to become comfortable with the concept of linear operators, it may be helpful to keep in mind that they are simply linear functions that have vectors as arguments and produce vectors as images. The simplest possible linear operator maps elements of $\mathbb{R}$ into elements of $\mathbb{R}$ and is scalar multiplication of the real numbers. Thus, this linear operator just changes the magnitude of a vector in its domain. However, vectors in $\mathbb{R}^N$, where N is greater than 1, are more complex than the real numbers, since they have both magnitude and direction, and the effect of a linear operator is to change both. Many of the concepts that are applied to scalar functions can also be applied to linear operators. Unfortunately, the concept of a graph, which is so useful for visualizing scalar functions, cannot be used with linear operators. Although linear operators have a graph in the abstract sense, this graph is generally a plane in some higher dimensional space that humans cannot envision.

Linear operators that map between the same vector space have a natural algebraic structure. Addition and scalar multiplication are defined as follows

$$L_1 + L_2 : V \to W \text{ such that } (L_1 + L_2)(v) \equiv L_1(v) + L_2(v)$$

$$\alpha L : V \to W \text{ such that } (\alpha L)(v) \equiv \alpha L(v).$$

It is easy to check that the sum of linear operators defined this way is still a linear operator and that the same holds for scalar multiplication of a linear operator.

Therefore, the set of linear operators that map from the vector space V to the vector space W form a vector space itself which we will indicate as $\mathcal{L}(V, W)$. A special linear operator space is the so-called dual vector space of V, often indicated V^*. It is the space of linear operators that map elements of V into the real numbers.

Definition: *The vector space of linear operators of the form $L : V \to \mathbb{R}$ is called the dual space of V and is indicated V^*.*

6.2 Examples

1) The operator $L_\theta : \mathbb{R}^2 \to \mathbb{R}^2$ defined by

$$L_\theta(x = (x_1, x_2)) = (x_1 \cos(\theta) - x_2 \sin(\theta), x_1 \sin(\theta) + x_2 \cos(\theta))$$

is a linear operator for all values of θ. To see this, first check addition

$$\begin{aligned} L_\theta(v + w) &= ((v_1 + w_1)\cos(\theta) - (v_2 + w_2)\sin(\theta), (v_1 + w_1)\sin(\theta) \\ &\quad +(v_2 + w_2)\cos(\theta)) \\ &= (v_1 \cos(\theta) - v_2 \sin(\theta), v_1 \sin(\theta) + v_2 \cos(\theta)) \\ &\quad +(w_1 \cos(\theta) - w_2 \sin(\theta), w_1 \sin(\theta) + w_2 \cos(\theta)) \\ &= L_\theta(v) + L_\theta(w) \end{aligned}$$

then scalar multiplication

$$\begin{aligned} L_\theta(\alpha v) &= (\alpha v_1 \cos(\theta) - \alpha v_2 \sin(\theta), \alpha v_1 \sin(\theta) + \alpha v_2 \cos(\theta)) \\ &= \alpha(v_1 \cos(\theta) - v_2 \sin(\theta), v_1 \sin(\theta) + v_2 \cos(\theta)) = \alpha L_\theta(v). \end{aligned}$$

This operator can be interpreted geometrically as a rotation through the angle θ of vectors in the plane.

2) The linear operator $L_p : \mathbb{R}^3 \to \mathbb{R}^2$ defined by

$$L_p(x_1, x_2, x_3) = (x_1, x_2)$$

can be interpreted geometrically as a projection of points in three-dimensional (x_1, x_2, x_3) space into the (x_1, x_2)-plane.

3) The linear operator $L : \mathbb{R}^2 \to \mathbb{R}^3$ defined by

$$L(x_1, x_2) = (x_1 + x_2, 2x_1 + 2x_2, x_1 + x_2)$$

maps vectors in two-dimensional Euclidean space into vectors in three-dimensional Euclidian space. However, the vectors in the range of the operator only span a one-dimensional vector space. Geometrically, the range is a straight line in three-dimensional space, which passes through the origin. In other words, the range of this operator is not the entire image space.

4) Differentiation is a linear operator which maps n-times differentiable functions into $n-1$-times differentiable functions, i.e.

$$\frac{d}{dt} : C^n[a,b] \to C^{n-1}[a,b]\,.$$

5) Small and therefore linear deformations of elastic solids can be described by specifying the linear operator that maps position vectors of physical points in the undeformed solid into the position vectors of the physical points in the deformed solid. This linear operator is known as the deformation tensor.
6) Cauchy's law provides an equation for calculation of the force per area on a given surface in a solid. It can be stated as

$$F = \tau(n)$$

where F is the force per area at a given point, n is the outward normal to the surface at the point and τ is a linear operator that maps the outward normal into the force. τ is known as the stress tensor and once a coordinate system has been chosen, e.g. the usual Cartesian system, Cauchy's law can be stated as

$$\begin{pmatrix} F_x \\ F_y \\ F_z \end{pmatrix} = \begin{pmatrix} \tau_{xx} & \tau_{xy} & \tau_{xz} \\ \tau_{yx} & \tau_{yy} & \tau_{yz} \\ \tau_{zx} & \tau_{zy} & \tau_{zz} \end{pmatrix} \begin{pmatrix} n_x \\ n_y \\ n_z \end{pmatrix}.$$

6.2.1 *Matrices as operators*

Suppose the matrix $A \in \mathcal{M}_{MN}$ has entries in the field $\mathbb{F}$, and consider the map $L_A : \mathbb{F}^N \to \mathbb{F}^M$ defined by $L_A(x) = Ax$, where $x \in \mathbb{F}^N$ and Ax are the usual matrix multiplication. Properties of matrix multiplication show that $L_A(\alpha_1 x_1 + \alpha_2 x_2) = \alpha_1 L_A(x_1) + \alpha_2 L_A(x_2)$, and so L_A is a linear operator.

It turns out that the converse is true as well. That is, every linear operator $L \in \mathcal{L}(\mathbb{F}^N, \mathbb{F}^M)$ is of the form $L = L_A$ for some $A \in \mathcal{M}_{MN}$. To see this, let $B_V = \{e_1, \ldots, e_N\}$ and $B_W = \{f_1, \ldots, f_M\}$ denote the bases of $V = \mathbb{F}^N$ and $W = \mathbb{F}^M$, respectively. Each $v \in V$ can be written as $v = \alpha_1 e_1 + \cdots + \alpha_N e_N$, and since L is linear, we have $L(v) = \alpha_1 L(e_1) + \cdots + \alpha_N L(e_N)$. For the image of the basis vectors, $L(e_n)$, we have

$$L(e_1) = a_{1,1} f_1 + a_{2,1} f_2 + \ldots + a_{M,1} f_M$$
$$\vdots$$
$$L(e_N) = a_{1,N} f_1 + a_{2,N} + \ldots + a_{M,N} f_M\,.$$

Notice that the indices on the $a_{m,n}$ are such that the first index matches the index of the f_m basis vector. We define the matrix of transition (from basis B_V to basis B_W) as

$$A = \begin{pmatrix} a_{1,1} & a_{1,2} & \dots & a_{1,N} \\ \vdots & \vdots & & \vdots \\ a_{M,1} & a_{M,2} & \dots & a_{M,N} \end{pmatrix}.$$

Column n of A is the coordinate representation of $L(e_n)$ in W using the basis $B_W = \{f_1, \dots, f_M\}$, or

$$[L(e_n)]_{W,B_W} = \text{Column } n \text{ of } A.$$

Consider now the effect of L on any $v \in V$. Let $[v]_{V,B_V} = (\alpha_1, \dots, \alpha_N)$.

$$v = \alpha_1 e_1 + \alpha_2 e_2 + \dots + \alpha_N e_N = \sum_{n=1}^{N} \alpha_n e_n$$

then

$$\begin{aligned} w = L(v) = L \sum_{n=1}^{N} \alpha_n e_n &= \sum_{n=1}^{N} \alpha_n L(e_n) \\ &= \sum_{n=1}^{N} \alpha_n \sum_{m=1}^{M} a_{m,n} f_m \\ &= \sum_{m=1}^{M} \left(\sum_{n=1}^{N} a_{m,n} \alpha_n \right) f_m . \end{aligned}$$

Clearly, the term in the parenthesis is the coordinate representation of w in the basis B_W.

$$[w]_{W,B_W} = \begin{pmatrix} \sum_{n=1}^{N} a_{1,n}\alpha_n \\ \sum_{n=1}^{N} a_{2,n}\alpha_n \\ \vdots \\ \sum_{n=1}^{N} a_{M,n}\alpha_n \end{pmatrix}$$

But we also have that

$$A[v]_{V,B_V} = \begin{pmatrix} a_{1,1} & a_{1,2} & \dots & a_{1,N} \\ \vdots & \vdots & & \vdots \\ a_{M,1} & a_{M,2} & \dots & a_{M,N} \end{pmatrix} \begin{pmatrix} \alpha_1 \\ \vdots \\ \alpha_N \end{pmatrix} = \begin{pmatrix} \sum_{n=1}^{N} a_{1,n}\alpha_n \\ \sum_{n=1}^{N} a_{2,n}\alpha_n \\ \vdots \\ \sum_{n=1}^{N} a_{M,n}\alpha_n \end{pmatrix}.$$

Hence, we have

$$[w]_{W,B_W} = [Lv]_{W,B_W} = A[v]_{V,B_V} = [L]_{V.B_V}^{W,B_W}[v]_{V,B_V} \,.$$

In other words, the action of a linear operator on any vector can be found by premultiplying the coordinate representation of the vector by the matrix of transition. There are three important points to remember here.

- Every linear operator mapping between finite dimensional vector spaces has a matrix representation.
- The matrix representation is basis dependent.
- The action of the linear operator is completely determined by its action on the basis vectors in the domain space and the columns of the matrix of transition is the coordinate representation of the images of these basis vectors.

In the case where $M = N$ and A is a square matrix, then it is easy to see L_A is an isomorphism if and only if the inverse of A exists.

Example 6.1: Consider the vector space of all third order polynomials, i.e. polynomials which can be written in the form

$$P(x) = a_0 + a_1x + a_2x^2 + a_3x^3$$

and let the basis for this vector space be

$$B_3 = \{1, 1+x, 1+x+x^2, 1+x+x^2+x^3\} \,.$$

Consider the linear operator L which maps $P(x)$ into $\mathbb{R}^4$ such that

$$L : P(x) \to (a_0, a_1, a_2, a_3)$$

and let the basis in $\mathbb{R}^4$ be the canonical basis, $B_{\mathbb{R}^4} = \{e_1, e_2, e_3, e_4\}$, where e_n is a vector of the form

$$(e_n)_m = \begin{cases} 0 \,, & n \neq m \\ 1 \,, & n = m \end{cases} \,.$$

Derive the matrix representation for this operator. The matrix is determined by considering its action on the basis vectors. This is seen to be

$$\begin{gathered} L : 1 \to (1,0,0,0) \\ L : 1+x \to (1,1,0,0) \\ L : 1+x+x^2 \to (1,1,1,0) \\ L : 1+x+x^2+x^3 \to (1,1,1,1) \end{gathered}$$

or, writing this result in terms of the coordinate representations of the vectors involved

$$\begin{aligned} L &: (1,0,0,0))_{B_3} \to (1,0,0,0)_{B_{\mathbb{R}^4}} \\ L &: (0,1,0,0)_{B_3} \to (1,1,0,0))_{B_{\mathbb{R}^4}} \\ L &: (0,0,1,0)_{B_3} \to (1,1,1,0))_{B_{\mathbb{R}^4}} \\ L &: (0,0,0,1)_{B_3} \to (1,1,1,1))_{B_{\mathbb{R}^4}} \end{aligned}.$$

Recall that the columns of the matrix of the operator are the coordinates of images of the basis vectors. Thus

$$L = \begin{pmatrix} 1\,1\,1\,1 \\ 0\,1\,1\,1 \\ 0\,0\,1\,1 \\ 0\,0\,0\,1 \end{pmatrix}.$$

This result can be checked by finding the coordinate representation of an arbitrary vector/polynomial, premultiplying this by the matrix of the operator and confirming that the result is the vector (a_0, a_1, a_2, a_3). Let the coordinates of an arbitrary polynomial be (P_1, P_2, P_3, P_4), then

$$\begin{aligned} P(x) &= P_1 \cdot 1 + P_2(1+x) + P_3(1+x+x^2) + P_4(1+x+x^2+x^3) \\ &= (P_1+P_2+P_3+P_4) + (P_2+P_3+P_4)x + (P_3+P_4)x^2 + P_4x^3. \end{aligned}$$

Comparing coefficients of terms of equal power gives the equations

$$\left.\begin{aligned} a_0 &= P_1+P_2+P_3+P_4 \\ a_1 &= P_2+P_3+P_4 \\ a_2 &= P_3+P_4 \\ a_3 &= P_4 \end{aligned}\right\} \Rightarrow \begin{aligned} P_1 &= a_0 - a_1 \\ P_2 &= a_1 - a_2 \\ P_3 &= a_2 - a_3 \\ P_4 &= a_3 \end{aligned}.$$

So one can write a vector by its coordinate representation as $P(x) = (a_0 - a_1, a_1 - a_2, a_2 - a_3, a_3)$ and the check on the matrix of the operator is

$$LP = \begin{pmatrix} 1\,1\,1\,1 \\ 0\,1\,1\,1 \\ 0\,0\,1\,1 \\ 0\,0\,0\,1 \end{pmatrix} \begin{pmatrix} a_0 - a_1 \\ a_1 - a_2 \\ a_2 - a_3 \\ a_3 \end{pmatrix} \begin{pmatrix} a_0 \\ a_1 \\ a_2 \\ a_3 \end{pmatrix}$$

which is the correct result.

Example 6.2: Consider the differentiation operator D which maps third order polynomials into the space of second order polynomial. Use the basis B_3

$$B_3 = \{1, 1+x, 1+x+x^2, 1+x+x^2+x^3\}$$

in the space of third order polynomials and the basis

$$B_2 = \{1, 1+x, 1+x+x^2\}$$

in the space of second order polynomials and find the matrix representation of D. Again, the matrix is determined from its action on the basis vectors.

$$D: 1 \to 0$$
$$D: 1+x \to 1$$
$$D: 1+x+x^2 \to 1+2x$$
$$D: 1+x+x^2+x^3 \to 1+2x+3x^2.$$

The images of the basis vectors in B_3 must all be represented by their coordinates in B_2. Clearly, the coordinate representation of 0 is $(0,0,0)$ and the coordinate representation of 1 is $(1,0,0)$. Rewriting $1+2x$ as $2(1+x)-1$ its coordinate representation is seen to be $(-1,2,0)$ and similarly $1+2x+3x^2 = 3(1+x+x^2) - (1+x) - 1$ so its coordinate representation is $(-1,-1,3)$. The action of D on the basis vectors of B_3 can now be written fully in coordinate representation as

$$D: (1,0,0,0)_{B_3} \to (0,0,0)_{B_2}$$
$$D: (0,1,0,0)_{B_3} \to (1,0,0)_{B_2}$$
$$D: (0,0,1,0)_{B_3} \to (-1,2,0)_{B_2}$$
$$D: (0,0,0,0)_{B_3} \to (-1,-1,3)_{B_2}$$

and the matrix representation of D is

$$D = \begin{pmatrix} 0 & 1 & -1 & -1 \\ 0 & 0 & 2 & -1 \\ 0 & 0 & 0 & 3 \end{pmatrix}.$$

Again, the result can be checked easily

$$DP = \begin{pmatrix} 0 & 1 & -1 & -1 \\ 0 & 0 & 2 & -1 \\ 0 & 0 & 0 & 3 \end{pmatrix} \begin{pmatrix} a_0 - a_1 \\ a_1 - a_2 \\ a_2 - a_3 \\ a_3 \end{pmatrix} = \begin{pmatrix} a_1 - 2a_2 \\ 2a_2 - 3a_3 \\ 3a_3 \end{pmatrix}.$$

This result is of course the coordinate representation of DP. To find the vector/polynomial itself, multiply the coordinates by the basis vectors of B_2 and simplify.

$$\begin{aligned} DP &= (a_1 - 2a_2)\cdot 1 + (2a_2 - 3a_3)(1+x) + 3a_3(1+x+x^2) \\ &= a_1 + 2a_2 x + 3a_3 x^2 \\ &= \frac{d}{dx}(a_0 + a_1 x + a_2 x^2 + a_3 x^3). \end{aligned}$$

Example 6.3: Dual space of $\mathbb{R}^N$.

The dual space of the vector space V was defined as the space of linear operators that maps elements of V into the real numbers. The dual space is particularly simple when the vector space is $\mathbb{R}^N$. Let the basis of $\mathbb{R}^N$ be the canonical basis $B = \{e_1, \cdots, e_N\}$, and let the basis of $\mathbb{R}$ be 1. If L is in the dual space of $\mathbb{R}^N$, then L maps the basis vectors of $\mathbb{R}^N$ into real numbers, i.e.

$$Le_1 = a_1$$
$$\vdots$$
$$Le_N = a_N$$

where $a_n \in \mathbb{R}$ and the matrix of transition A becomes

$$A = (a_1, \cdots, a_N)$$

which obviously can be viewed as a vector in $\mathbb{R}^N$. The space $\mathbb{R}^N$ and its dual space are therefore isomorphic or self-dual. In general, the dimension of a finite dimensional vector space and its dual space are equal. (It turns out that a similar result does not hold for infinite dimensional vector spaces.)

6.2.2 *Equivalence transformations*

Since the matrix representation of a linear operator depends on the choice of bases in the domain and image spaces, it is natural to ask how the matrix representation of the operator changes when these bases are changed. A change of basis can be viewed as a linear operator that maps a vector from the space V into the space $\hat{V}$, where V and $\hat{V}$ are isomorphic vector spaces but with different bases. Linear operators that map a vector space V into itself are important in mathematics and are called *endomorphisms* of V.

Let the basis vectors of V be $\{e_n\}$ and the basis vectors of $\hat{V}$ be $\{\hat{e}_m\}$. Since a change in basis does not change the vector itself, the matrix of transition for a basis change must be the identity map, i.e. $Ke_n = e_n$. Writing e_n in terms of the $\hat{e}_m$ basis vectors as

$$e_n = \sum_{m=1}^{N} a_{m,n}\hat{e}_m$$

and for notational convenience, define the matrices $B_V = (e_1, \ldots, e_N)$ and $B_{\hat{V}} = (\hat{e}_1, \ldots, \hat{e}_N)$ and let the matrix of transition be indicated K

$$K = \begin{pmatrix} a_{1,1} & a_{1,2} & \cdots & a_{1,N} \\ \vdots & \vdots & & \vdots \\ a_{N,1} & a_{N,2} & \cdots & a_{N,N} \end{pmatrix}$$

then

$$(e_1, \ldots, e_N) = (\hat{e}_1, \ldots, \hat{e}_N)\begin{pmatrix} a_{1,1} & a_{1,2} & \cdots & a_{1,N} \\ \vdots & \vdots & & \vdots \\ a_{N,1} & a_{N,2} & \cdots & a_{N,N} \end{pmatrix} \Rightarrow$$

$$B_V = B_{\hat{V}} K \;\Rightarrow\; K = B_{\hat{V}}^{-1} B_V.$$

The matrix of transition for a change in basis is thus easily calculated once the two sets of basis vectors are given. Calculation of this matrix is illustrated in the example below.

Example 6.4: Change of basis.

Let V and $\hat{V}$ both be $\mathbb{R}^3$ and let their bases be

$$B_V = \{e_1, e_2, e_3\} = \left\{ \begin{pmatrix} 1 \\ 0 \\ 2 \end{pmatrix}, \begin{pmatrix} -2 \\ 1 \\ -2 \end{pmatrix}, \begin{pmatrix} -1 \\ 0 \\ 1 \end{pmatrix} \right\}$$

$$B_{\hat{V}} = \{\hat{e}_1, \hat{e}_2, \hat{e}_3\} = \left\{ \begin{pmatrix} 1 \\ 1 \\ 1 \end{pmatrix}, \begin{pmatrix} 0 \\ 2 \\ 3 \end{pmatrix}, \begin{pmatrix} 0 \\ 2 \\ -1 \end{pmatrix} \right\}.$$

To find the matrix of transition, we need to find the coordinate representation of the e_ns in the basis $B_{\hat{V}}$.

$$Le_1 = e_1 = a_{1,1}\hat{e}_1 + a_{2,1}\hat{e}_2 + a_{3,1}\hat{e}_3 \;\Rightarrow$$

$$\begin{pmatrix} 1 \\ 0 \\ 2 \end{pmatrix} = a_{1,1}\begin{pmatrix} 1 \\ 1 \\ 1 \end{pmatrix} + a_{2,1}\begin{pmatrix} 0 \\ 2 \\ 3 \end{pmatrix} + a_{3,1}\begin{pmatrix} 0 \\ 2 \\ -1 \end{pmatrix}.$$

Solving for the $a_{n,m}$s give the coordinate representation we seek.

$$[e_1]_{\hat{e},\hat{V}} = \begin{pmatrix} a_{1,1} \\ a_{2,1} \\ a_{3,1} \end{pmatrix} = \begin{pmatrix} 1 \\ \frac{1}{8} \\ -\frac{5}{8} \end{pmatrix}$$

Similarly,

$$[e_2]_{\hat{e},\hat{V}} = \begin{pmatrix} -2 \\ \frac{3}{8} \\ \frac{9}{8} \end{pmatrix}$$

$$[e_3]_{\hat{e},\hat{V}} = \begin{pmatrix} -1 \\ \frac{5}{8} \\ -\frac{1}{8} \end{pmatrix}$$

giving the matrix representation of the change of basis

$$K = \begin{pmatrix} 1 & -2 & -1 \\ \frac{1}{8} & \frac{3}{8} & \frac{5}{8} \\ -\frac{5}{8} & \frac{9}{8} & -\frac{1}{8} \end{pmatrix}.$$

It is readily confirmed that

$$K = B_{\hat{V}}^{-1} B_V = \begin{pmatrix} 1 & 0 & 0 \\ 1 & 2 & 2 \\ 1 & 3 & -1 \end{pmatrix}^{-1} \begin{pmatrix} 1 & -2 & -1 \\ 0 & 1 & 0 \\ 2 & -2 & 1 \end{pmatrix}.$$

Because a change in basis can be reversed, there must be another matrix S such that $v = S\hat{v}$. But, $\hat{v} = K(S\hat{v})$ so $KS = I$ or $S = K^{-1}$. Not surprisingly, a basis change and the inverse basis change have matrix representations that are inverses of one another.

Consider now any linear operator $L : V \to W$. In some choice of bases, this operator is represented by the matrix A.

$$w = Av.$$

Do a basis change in both V and W, i.e. let

$$\hat{v} = K_V v \quad \text{and} \quad \hat{w} = K_W w$$

from which we obtain

$$w = Av \Rightarrow$$

$$K_W^{-1}\hat{w} = AK_V^{-1}\hat{v} \Rightarrow$$

$$\hat{w} = (K_W A K_V^{-1})\hat{v} = \hat{A}\hat{v}$$

where

$$\hat{A} = K_W A K_V^{-1}. \tag{6.1}$$

The matrices A and $\hat{A}$ are said to be *equivalent* and Eq. (6.1) is called an *equivalence transformation.*

In most practical applications, the same basis is used for V and W and one is primarily interested in knowing how a matrix representation of a linear operator changes when V and W are subject to the same change of basis. When that is the case, the matrices representing the change of basis, K_V and K_W, are identical and the equivalence transformation becomes

$$\hat{A} = KAK^{-1}. \tag{6.2}$$

Thus, the two matrices, A and $\hat{A}$, are similar and all matrix representation of the linear operators $L : V \to V$ are similar.

6.3 The adjoint operator

In an inner product space, one can define the adjoint of an operator L as the operator L^* which satisfies

$$\langle Lv, w\rangle = \langle v, L^* w\rangle.$$

If $L = L^*$ the operator is said to be self-adjoint. We shall see that self-adjoint operators have several very nice properties.

For finite dimensional vector spaces on which the inner product is defined by Eq. (5.33) the matrix of the adjoint operator can be found once the matrix of the operator itself is known. Interpreting this scalar product as the matrix product of a $1 \times N$ and an $N \times 1$ matrix, we get

$$\begin{aligned}\langle Av, w\rangle &= (Av)^T \overline{w} \\ &= v^T \overline{(\overline{A}^T w)} \\ &= \langle v, \overline{A}^T w\rangle\end{aligned}$$

so the adjoint of a matrix is the transpose, complex conjugate of this matrix. This is the reason for calling the eigenvalue problem of this matrix the adjoint eigenvalue problem. A real matrix is self-adjoint if it is square and equals its own transpose, i.e. if it is symmetric.

The proofs of the two theorems that follow do not rely on the operator being finite dimensional. They are valid even for operators on infinite dimensional vector spaces such as function spaces and the results will be used extensively in the next chapter.

Theorem 6.1. *All the eigenvalues of a self-adjoint operator are real.*

Proof: Let $\tilde{v}_n$ be an eigenvector of the self-adjoint, linear operator L and let λ_n be the associated eigenvalue. Then

$$\langle L\tilde{v}_n, \tilde{v}_n\rangle = \langle \lambda_n \tilde{v}_n, \tilde{v}_n\rangle = \lambda_n \langle \tilde{v}_n, \tilde{v}_n\rangle$$

but since L is self-adjoint, we also must have that

$$\langle L\tilde{v}_n, \tilde{v}_n\rangle = \langle \tilde{v}_n, L\tilde{v}_n\rangle = \langle \tilde{v}_n, \lambda_n \tilde{v}_n\rangle = \overline{\lambda_n} \langle \tilde{v}_n, \tilde{v}_n\rangle$$

and consequently

$$\begin{aligned}\lambda_n \langle \tilde{v}_n, \tilde{v}_n\rangle &= \overline{\lambda_n} \langle \tilde{v}_n, \tilde{v}_n\rangle \Rightarrow \\ \lambda_n &= \overline{\lambda_n}\end{aligned}$$

and since the eigenvalues equal their own complex conjugates, they must be real. □

Theorem 6.2. *Eigenvectors of a self-adjoint operator which belong to different eigenvalues are orthogonal.*

Proof: Let $\tilde{v}_n$ and $\tilde{v}_m$ be eigenvectors of the self-adjoint operator L and let their eigenvalues be λ_n and λ_m respectively. Then

$$\begin{aligned} \langle L\tilde{v}_n, \tilde{v}_m\rangle &= \langle \tilde{v}_n, L\tilde{v}_m\rangle \Rightarrow \\ \lambda_n\langle \tilde{v}_n, \tilde{v}_m\rangle &= \lambda_m\langle \tilde{v}_n, \tilde{v}_m\rangle \Rightarrow \\ (\lambda_n - \lambda_m)\langle \tilde{v}_n, \tilde{v}_m\rangle &= 0 \end{aligned}$$

and when the two eigenvalues are different, the factor in front of the inner product is different from zero and can be divided out given the desired result

$$\langle \tilde{v}_n, \tilde{v}_m\rangle = 0 . \qquad \square$$

6.4 Tensors

The linear transformations most often encountered in engineering and physics are called tensors. There is great confusion in the engineering literature as to what the proper definition of a tensor is. This is probably due to several factors. For instance, tensors have a so-called order and in engineering texts, zero order tensors are defined differently from first order tensors and both are defined differently from second order tensors. Tensors of higher orders are defined differently still, but these higher order tensors occur less frequently in engineering applications. In spite of the fact that tensors of different orders exist, the term "tensor" is often used in engineering literature to indicate only second order tensors. In the more rigorous mathematical literature, tensors of arbitrary order are given a definition, but this more abstract presentation of tensors is a less easy introduction to the concept. The confusion about the proper tensor definition is further exacerbated by the fact that the rigorously defined tensor in mathematics is seemingly a different object from the tensor used in engineering. In fact, the different ways of defining tensors are not in conflict, and the difference stems from the fact that the tensor concept is very rich and allows for multiple interpretations.

A quick but incomplete explanation of tensors are that they are a generalization of the classical vector concept and, once a coordinate system has been specified, they can therefore be written in terms of their coordinates. Central to the tensor concept is the idea that, although the coordinate representation of a tensor will change when the coordinate system is changed, the tensor itself is an object that is independent of the choice of coordinate system. This independence is only possible if the coordinate representation of tensors satisfy certain transformation laws which we will derive in this chapter. Tensors can and often are defined as objects that satisfy given transformation laws.

Tensors of different orders will be defined as follows.

- A zero order tensor is a linear operator that maps elements from $\mathbb{R}$ to $\mathbb{R}$, scalar multiplication in other words. The matrix representation of a zero order tensor is a scalar. Scalar quantities such as temperature, concentration and viscosity are zero order tensors.

- First order tensors are linear operators that map elements of $\mathbb{R}^N$ into $\mathbb{R}$, and first order tensors are therefore elements of the dual space of $\mathbb{R}^N$. However, this space is isomorphic to $\mathbb{R}^N$ and first order tensors are therefore vectors in the usual engineering sense, objects with magnitude and direction. Force, velocity, stress, the acceleration of gravity are all first order tensors.
- Second order tensors are linear operators that map from $\mathbb{R}^N$ to $\mathbb{R}^N$. For instance, the stress tensor which maps the outward normal at a point on a surface into the force on the surface at that point is a second order tensor. In a given coordinate system, a second order tensor can be represented by an $N \times N$ matrix.

We will next give a rigorous mathematical definition of a tensor and an argument will be made that, although the mathematical and the engineering definition of tensors appear to be different, the two types of tensors are in some sense the same. We start by defining a multi-linear function, which is an extension of the linear operator concept.

Definition: *A multi-linear function is a function which maps elements of the vector space V_1, V_2, ...,V_N into the vector space W such that F it is linear in each of its arguments.*

Being linear in each argument means that

$$L(\cdots, \alpha_1 v_1 + \alpha_2 v_2, \cdots) = \alpha_1 L(\cdots, v_1, \cdots) + \alpha_2 L(\cdots, v_2, \cdots).$$

Finally, a tensor can be defined.

Definition: *A tensor is a multi-linear function that maps elements of a vector space V and elements of its dual space V^* into the real numbers $\mathbb{R}$. The number of arguments of the tensor is known as the degree of the tensor. The number of arguments taken from V is the covariant degree, the number of arguments taken from V^* is the contravariant degree.*

The difference between covariant and contravariant degree is not important for Cartesian coordinates and can be ignored for now. However, the concepts of covariance and contravariance do become important for curvilinear coordinates.

Consider, as a simple example, a tensor that maps two vectors v and w into $\mathbb{R}$. For the sake of this argument, it does not matter whether these two vectors come from V or V^*. The only thing that matters is that the vector spaces of which they are elements have the same dimension N. The mapping of the two vectors into $\mathbb{R}$ can be represented algebraically as $(Av)w$, where A is an $N \times N$ matrix. It is easy to see that this map is linear in each argument and thus is a tensor. This suggests that a tensor of degree two can be represented as a square matrix. But a square matrix, when premultiplying a vector, maps that vector into another vector of the same size. Thus, given a square matrix, one can think of it either as a rule for mapping two vectors of the same dimension into $\mathbb{R}$ or as a rule for mapping one vector into another. It is this reinterpretation of a tensor that is the basis for tensor definitions in physics and engineering. This reinterpretation can be put on a rigorous basis by showing that tensor spaces, as defined in engineering, are isomorphic to tensor spaces as defined in mathematics.

The important thing to note is that a tensor, in the rigorous mathematical sense given above, only operates on vectors of the same dimension. Therefore, a linear map from a space with dimension N into a space with dimension M can only be interpreted as a tensor if either $N = M$ or if $M = 1$.

6.4.1 *Transformation rules*

In engineering texts, tensors are at times defined as i) objects that are invariant under coordinate transformations or as ii) objects that obey given transformation laws under coordinate transformations. These two definitions appear contradictory, but they are not. The first definition refers to the tensor itself. For instance, the acceleration of gravity is the same physical object and always points towards the center of the earth, no matter what coordinate system is used. Similarly, the stress tensor is always the same function, the function that maps the outward normal on a surface into the force on that surface, irrespective of the coordinate system used. The acceleration of gravity and the stress tensor are therefore invariant under coordinate transformations. The second definition refers to the matrix representation of the tensor. The matrix representation of a linear operator, and therefore of a tensor, is a function of the choice of basis. In order for a tensor to remain the same object in all coordinate systems, i.e. to be invariant under coordinate transformations, the matrix representation of the tensor must change in a systematic way as the coordinate system changes. For instance, the acceleration of gravity has the coordinate representation $(g, 0, 0)$ in a coordinate system where the first basis vector points down, but has the representation $(0, g, 0)$ in a coordinate system where the second basis vector points down. Yet, it remains the same physical object. This distinction between the tensor itself and its matrix representation must be kept in mind in order for the tensor concept to make sense.

The transformation rules that the matrix representation of tensors must obey are different for tensors of different order and transformation rules for first and second order tensors in Cartesian coordinate systems will be derived in this section.

A change of basis between Cartesian coordinate systems combine four types of changes: rotations, reflections, dilations/magnifications and translations. In Cartesian coordinates, the coordinate representation of a vector is independent of the origin of the coordinate system. Translations, which are relocations of the coordinate system origin, do not therefore have an effect on the coordinate representation of vectors and tensor and need not be considered here. Dilations and magnifications correspond to stretching and compression of the base vectors, which changes the lengths of vectors. Evidently, dilations and magnifications of the coordinate basis will change the coordinate representation of vectors and tensors, but this change corresponds physically to a change in units which is already well understood and will therefore not be considered here either. This leaves reflections and rotations to be considered and transformation rules of vectors and tensors only refer to these two types of transformations.

Reflections change a right handed coordinate system into a left handed coordinate system and vice versa, while rotations preserve the handedness of the system. Both preserve lengths. Consider a rotation/reflection that moves a set of canonical basis vectors e_n into another set $\hat{e}_n$. Each of the basis vectors in the rotated coordinate system can be written in terms of the original basis vectors as

$$\hat{e}_n = \sum_{m=1}^{N} \theta_{nm} e_m$$

which can be written in vector-matrix notation as

$$\hat{e} = \begin{pmatrix} \hat{e}_1 \\ \vdots \\ \hat{e}_N \end{pmatrix} = \begin{pmatrix} \theta_{11} & \theta_{12} & \cdots & \theta_{1N} \\ \vdots & \vdots & \ddots & \vdots \\ \theta_{N1} & \theta_{N2} & \cdots & \theta_{NN} \end{pmatrix} \begin{pmatrix} e_1 \\ \vdots \\ e_N \end{pmatrix} = \Theta e \tag{6.3}$$

where the matrix, Θ is known as *the rotation matrix.* Taking the inner product $\langle \hat{e}_n, e_m \rangle$ and using the orthogonality of the e_m-base vectors, one sees that $\theta_{nm} = \langle \hat{e}_n, e_m \rangle = \cos(\angle(\hat{e}_n, e_m))$. Thus, θ_{nm} is the cosine of the angle between the base vector e_m and the rotated base vector $\hat{e}_n$. The θs are therefore called *direction cosines.*

Obviously, the direction cosines must take values between 1 and -1 and must be related in some way. This places restrictions on what form Θ can take. To determine what these restrictions are, make use of the fact that rotations/reflections preserve lengths and angles between vectors. Therefore, since the basis vectors e_n are mutually orthogonal unit vectors, the $\hat{e}_n$ basis vectors must also be mutually orthogonal unit vectors. The inner product of two basis vectors in the new coordinate system is therefore either 1 or 0, depending on whether or not the two vectors are the same basis vector or not.

$$\langle \hat{e}_n, \hat{e}_m \rangle = \begin{cases} 1 \ , \ n = m \\ 0 \ , \ n \neq m \end{cases}.$$

This inner product can be written in terms of the coordinate representation of the vectors in the e-basis as

$$\begin{aligned} \langle \hat{e}_n, \hat{e}_m \rangle &= \langle \theta_{n1} e_1 + \cdots + \theta_{nN} e_N, \theta_{m1} e_1 + \cdots + \theta_{mN} e_N \rangle \\ &= \theta_{n1}\theta_{m1} + \cdots + \theta_{nN}\theta_{mN} \\ &= \sum_{k=1}^{N} \Theta_{nk}\Theta_{mk} = (\Theta\Theta^T)_{nm} = \begin{cases} 1 \ , \ n = m \\ 0 \ , \ n \neq m \end{cases}. \end{aligned}$$

It follows from the last equality that

$$\Theta\Theta^T = I \Leftrightarrow \Theta^T = \Theta^{-1} \tag{6.4}$$

which is the condition that matrix must satisfy in order to represent a rotation/reflection. Matrices that do satisfy Eq. (6.4) are said to be *orthogonal.* Each row or column of an orthogonal matrix has length unity and is orthogonal to all other rows or columns. Because the determinant of a product of matrices equal the product of the determinants, it follows that the determinant of the rotation matrix must equal plus or minus 1. If the determinant of the rotation matrix equals 1, the rotation is said to be proper, while it is said to be improper when the determinant equals minus 1. Physically, an improper rotation corresponds to a rotation combined with a reflection. It therefore changes a right handed coordinate system into a left handed system, and vice versa.

The transformation rule for vectors or first order tensors can now be found. Let the coordinates of v in the basis e_n be $(\alpha_1, \alpha_2, \cdots, \alpha_N)$, i.e.

$$v = \sum_{n=1}^{N} \alpha_n e_n .$$

Premultiply this equation by the rotation matrix Θ to get

$$\Theta v = \Theta \sum_{n=1}^{N} \alpha_n e_n = \sum_{n=1}^{N} \alpha_n \Theta e_n = \sum_{n=1}^{N} \alpha_n \hat{e}_n$$

but the last sum is the coordinate expansion in the basis $\hat{e}$, or simply $\hat{v}$. Therefore

$$\hat{v} = \Theta v.$$

The transformation rule for a second order tensor follows from Eq. (6.2).

$$\hat{A} = \Theta A \Theta^T .$$

These results are sufficiently important to be summarized in a theorem.

Theorem 6.3. *In Cartesian coordinate systems, changing the basis vectors from* e *to* $\hat{e} = \Theta e$*, changes the coordinate representation of tensors as follows.*

Zero order tensors or scalars: $\hat{\alpha} = \alpha$

First order tensors or vectors: $\hat{v} = \Theta v$

Second order tensors: $\hat{A} = \Theta A \Theta^T$

WARNING! POSSIBLE NAMING CONFUSION.

The rotation matrix is not always defined as in Eq. (6.3). In this definition, the direction cosine, α_{nm}, is the cosine of the angle between $\hat{e}_n$ and e_m. In some texts, element (n, m) of the rotation matrix is defined as the cosine of the angle between e_n and $\hat{e}_m$. The rotation matrix defined this way is the transpose of the rotation matrix defined in Eq. (6.3) and in texts where this definition is used, the transformation rules take the form

$$\hat{v} = \Theta^T v$$

$$\hat{A} = \Theta^T A \Theta.$$

In many books, the theorems above are not derived from the different definitions of tensors of various orders. Instead, tensors of different order are defined as objects that satisfy these transformation rules.

It is illustrative to expand the transformation rules above and write them in terms of their components. The first and second order tensor laws become respectively

$$\hat{v}_n = \sum_{k=1}^{N} \theta_{n,k} v_k$$

$$\hat{a}_{n,m} = \sum_{k=1}^{N} \sum_{l=1}^{N} \theta_{n,k} \theta_{m,l} a_{k,l}\,.$$

This suggests defining tensors of any order as objects that satisfy the transformation rule

$$\hat{a}_{n,m,\ldots,k} = \sum_{q=1}^{N} \sum_{r=1}^{N} \cdots \sum_{s=1}^{N} \theta_{n,q} \theta_{m,r} \ldots \theta_{k,s} a_{q,r,\ldots,s}\,.$$

Using this definition, a tensor of order N is an object that satisfies the transformation rule above and for which the number of subscripts and sums equal N.

Evidently, tensor expressions can become large and in order to economize the nomenclature many texts make use of a version of the following range and summation convention.

Range and Summation Convention *An index variable that occurs unrepeated in a term is understood to take all its possible values,* $1, 2, \ldots, N$*, where* N *usually is the dimension of the vector space in question.*

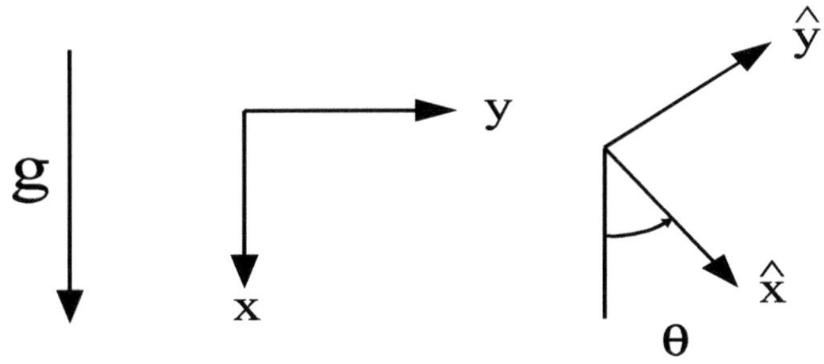

Fig. 6.1 Two coordinate systems for representing the acceleration of gravity.

An index variable that occurs repeatedly in a term is understood to be a summation variable over the range 1 *through* N.

Using this convention, the equation for transformation of an $\mathcal{N}$th order tensor is written

$$\hat{a}_{n,m,\ldots,k} = \theta_{n,q}\theta_{m,r}\ldots\theta_{k,s}a_{q,r,\ldots,s}\,.$$

Example 6.5: Rotation in the plane.

Consider the coordinate representation of the acceleration of gravity in two (x, y) coordinate systems. One system in which the x-axis points down, another rotated the angle θ relative to the first system, see Fig. 6.1. Straightforward geometric observations shows that the acceleration of gravity in the two systems have the coordinate representations

$$g = (g, 0)\ ,\ \hat{g} = (g\cos(\theta), -g\sin(\theta)).$$

We shall see that this is the result obtained with the rotation matrix. The direction cosines are easily found and the rotation matrix is

$$\Theta = \begin{pmatrix} \cos(\theta) & \cos(\pi/2-\theta) \\ \cos(\theta+\pi/2) & \cos(\theta) \end{pmatrix} = \begin{pmatrix} \cos(\theta) & \sin(\theta) \\ -\sin(\theta) & \cos(\theta) \end{pmatrix}$$

so

$$\hat{g} = \begin{pmatrix} \cos(\theta) & \sin(\theta) \\ -\sin(\theta) & \cos(\theta) \end{pmatrix} \begin{pmatrix} g \\ 0 \end{pmatrix} = \begin{pmatrix} g\cos(\theta) \\ -g\sin(\theta) \end{pmatrix}.$$

We end by making explicit the assumption used in the derivation of the transformation law of second order tensors: that the coordinate representation of the two vectors, v and Av, are given with respect to the same basis. It is perhaps difficult to see why anyone would use different bases to represent the two vectors, but it turns out that this allows a simpler matrix representation of the linear operator than is possible with one basis. When the same basis is used to represent both the vector in the domain and in the range, the simplest matrix representation of the operator is a Jordan canonical form matrix. Using two different bases to represent these vectors, it can be shown that any linear operator that maps elements of a vector space onto itself can be represented by a diagonal matrix that have 1s in the first k diagonal elements and zeros everywhere else. The number k, the number of non-zero diagonal elements, is the rank of the operator.

6.4.2 *Invariants of tensors*

The transformation rule for second order tensors show that all the matrix representations of a tensor are all similar. This means that properties that are shared by similar matrices are properties of second order tensors that do not change under basis rotations/reflections. These properties are called the invariants of the tensor and quite often have an important physical interpretation. Some of the important tensor invariants have already been found in the section on similar matrices. They are restated here and some other invariants of second order tensors are introduced.

Eigenvalues, their algebraic and geometric multiplicity, eigenvectors and generalized eigenvectors are tensor invariants.

This is a restatement of the result derived for similar matrices. It only needs to be noted that the result for similar matrices stated that if A and B are similar according to $A = K^{-1}BK$, then their eigenvectors are related by $\tilde{x}_A = K^{-1}\tilde{x}_B$. But the relationship between eigenvectors is just the transformation rule for first order tensors, so the eigenvectors must be invariants. This is a very important result because it makes it clear that one can talk in a meaningful way about the eigenvalue problem of a tensor, irrespective of the choice of coordinate system used.

The determinant is a tensor invariant.

It was shown previously that similar matrices have the same determinant.

A tensor is said to be *symmetric* if it can be represented by a symmetric matrix. *Symmetry is an invariant tensor property.*

Proof: We need to show that if A is symmetric, i.e. if $a_{n,m} = a_{m,n}$, then so is $\Theta A \Theta^T$. An element of $\Theta A \Theta^T$ is found as

$$(\Theta A \Theta^T)_{n,m} = \sum_{k=1}^{N}\sum_{l=1}^{N} \theta_{n,k} a_{k,l} \theta^T_{l,m} = \sum_{k=1}^{N}\sum_{l=1}^{N} \theta_{n,k} a_{k,l} \theta_{m,l}$$

and similarly

$$(\Theta A \Theta^T)_{m,n} = \sum_{k=1}^{N} \sum_{l=1}^{N} \theta_{n,k} a_{l,k} \theta_{m,l}$$

but, since $a_{k,l} = a_{l,k}$, the right hand sides of these two equations are equal, so $(\Theta A \Theta^T)_{n,m} = (\Theta A \Theta^T)_{m,n}$ and symmetry is preserved. □

Many of the tensors that occur in applications can be shown from physical arguments to be symmetric. A tensor is said to be *anti-symmetric* or *skew-symmetric* if it can be represented by an anti-symmetric or skew-symmetric matrix. By a proof similar to the proof for symmetry, anti-symmetry is seen to be an invariant tensor property.

The trace of a matrix representation of a tensor is a tensor invariant.

Proof: The trace of a matrix, $tr(A)$, is defined as the sum of the diagonal terms of the matrix. Let

$$B = \Theta A \Theta^T$$

then

$$b_{nm} = \sum_{k=1}^{N} \sum_{l=1}^{N} \theta_{nk} a_{kl} \theta_{lm}^T \Rightarrow$$

$$tr(B) = \sum_{n=1}^{N} \sum_{k=1}^{N} \sum_{l=1}^{N} \theta_{nk} a_{kl} \theta_{ln}^T = \sum_{k=1}^{N} \sum_{l=1}^{N} a_{kl} \left(\sum_{n=1}^{N} \theta_{ln} \theta_{nk}^T \right).$$

The sum in the parenthesis is element (l, k) of $\Theta \Theta^T$, but, since Θ is an orthogonal matrix, this is element (l, k) of the identity matrix. Therefore, the term in parenthesis is 1 when $k = l$ and zero otherwise, so

$$tr(B) = \sum_{n=1}^{N} a_{nn} = tr(A). \qquad \square$$

Notice that the orthogonality property of Θ is not used to its fullest in this proof. The only thing that is required is that the matrix that premultiplies A is the inverse of the matrix that postmultiplies A. The proof is therefore valid for any pair of similar matrices. All similar matrices have the same trace, in particular, matrices have the same trace as any similar Jordan Canonical matrix. In a Jordan Canonical matrix, the trace obviously equals the sum of the eigenvalues, counting their multiplicity and, since similar matrices also have the same eigenvalues, the trace of any matrix must equal the sum of the eigenvalues.

Finally, the range and the null space of a linear operator were defined previously in a basis independent way. Both are therefore tensor invariants.

6.5 Some tensors from physics and engineering

Tensors that appear in physics and engineering are most often second order tensors and in many contexts the term "tensor" is synonymous with second order tensors, i.e. linear operators that can be represented by a square matrix. Some such tensors are briefly introduced and discussed in this section.

Tensors that appear in applications are almost invariably much simpler than their size imply. At first glance, a tensor that is represented by an $N \times N$ matrix has N^2 entries and may seem therefore to require the determination of N^2 parameter values. However, using physical arguments one can often derive equations between the entries such that the number of parameter values that all must be determined from experimental data is much less than N^2. The method for deriving equations between the parameters of the tensor is based on a symmetry argument: if the physical situation is symmetric in some way, then the tensor must reflect this symmetry. This means that the matrix representation must be invariant under those rotations that do not alter the physical problem. The method for doing this type of simplification is illustrated in the first example of a tensor, the conduction tensor in Fourier's law.

6.5.1 *Fourier's law*

Tensors arise naturally in models of non-isotropic materials. For instance, in transport by diffusion or conduction in non-isotropic substances. In these situations, the direction of flux is not necessarily the same as the direction of the gradient vector. The flux will predominantly be in the direction of least resistance, which may not be the same as the direction of the gradient. In the extreme case when the substance is made of alternating layers of a conductor and a perfect insulator, the flux in the direction perpendicular to the layers is always zero, irrespective of the gradient across the layers. This is where tensors are needed to model the relationship between gradient and flux. For instance, in the rectangular coordinate system, (x, y, z), Fourier's law for conduction in a non-isotropic solid can be written

$$\begin{pmatrix} q_1 \\ q_2 \\ q_3 \end{pmatrix} = -\begin{pmatrix} k_{1,1} & k_{1,2} & k_{1,3} \\ k_{2,1} & k_{2,2} & k_{2,3} \\ k_{3,1} & k_{3,2} & k_{3,3} \end{pmatrix} \begin{pmatrix} \frac{\partial T}{\partial x} \\ \frac{\partial T}{\partial y} \\ \frac{\partial T}{\partial z} \end{pmatrix} = -k(\nabla T)$$

where k is a thermal conductivity tensor that maps the temperature gradient vector into the thermal flux vector. In three dimensions, the matrix of this tensor has nine elements implying that nine material constants must be determined to characterize the solid. However, even most non-isotropic substances have some symmetry property which can be used to reduce the number of material constants. For instance, if the material is isotropic, the tensor version of Fourier's law should simplify down to the familiar scalar law that contains a single material constant, the conductivity. Isotropy indicates the highest degree of symmetry, a situation in which properties are independent of direction, and for which the matrix representation, of say a conductivity tensor, must be the same in all coordinate systems. Consider therefore the effect of a coordinate basis rotation given by the rotation matrix

$$\Theta = \begin{pmatrix} 1 & 0 & 0 \\ 0 & -1 & 0 \\ 0 & 0 & 1 \end{pmatrix}$$

which geometrically correspond to a change in the direction of the second basis vector. In the new coordinate system, the conductivity tensor becomes

$$\hat{k} = \Theta k \Theta^T = \begin{pmatrix} k_{1,1} & -k_{1,2} & k_{1,3} \\ -k_{2,1} & k_{2,2} & -k_{2,3} \\ k_{3,1} & -k_{3,2} & k_{3,3} \end{pmatrix}$$

and, since the conductivity tensor being considered is independent of the choice of basis, it must hold that $k_{1,2} = -k_{1,2}$ which implies that $k_{1,2} = 0$. Similarly for $k_{2,1}$, $k_{2,3}$ and $k_{3,1}$. Another change of basis corresponding to a change in the direction of the first or third basis vector shows that $k_{1,3} = k_{3,1} = 0$. The conductivity tensor therefore has the simpler form

$$k = \begin{pmatrix} k_{1,1} & 0 & 0 \\ 0 & k_{2,2} & 0 \\ 0 & 0 & k_{3,3} \end{pmatrix} .$$

Consider now what happens to this tensor representation when the first and second basis vectors are switched. This switch is represented by the rotation matrix

$$\Theta = \begin{pmatrix} 0 & 1 & 0 \\ 1 & 0 & 0 \\ 0 & 0 & 1 \end{pmatrix}$$

and in the new coordinate system, the matrix representation of the conductivity tensor becomes

$$k = \begin{pmatrix} k_{2,2} & 0 & 0 \\ 0 & k_{1,1} & 0 \\ 0 & 0 & k_{3,3} \end{pmatrix}$$

showing that $k_{1,1} = k_{2,2}$. A switch of a different pair of basis vectors similarly shows that all the diagonal elements of k are equal. In an isotropic material, the conductivity tensor is therefore represented by a diagonal matrix and all the diagonal elements equal the single, well defined, conductivity of the material. Fourier's law can therfore be written in the more familiar form

$$\begin{pmatrix} q_1 \\ q_2 \\ q_3 \end{pmatrix} = - \begin{pmatrix} k & 0 & 0 \\ 0 & k & 0 \\ 0 & 0 & k \end{pmatrix} \begin{pmatrix} \frac{\partial T}{\partial x} \\ \frac{\partial T}{\partial y} \\ \frac{\partial T}{\partial z} \end{pmatrix}.$$

Example 6.6: Conduction tensor for a laminated solid.

Consider a solid which is made of alternating layers of equal thickness of two types of materials with different thermal conductivities. We want to determine the form of the conductivity tensor in a (x, y, z)-coordinate system for which the x-axis is orthogonal to the interphase between the layers, see Fig. 6.2.

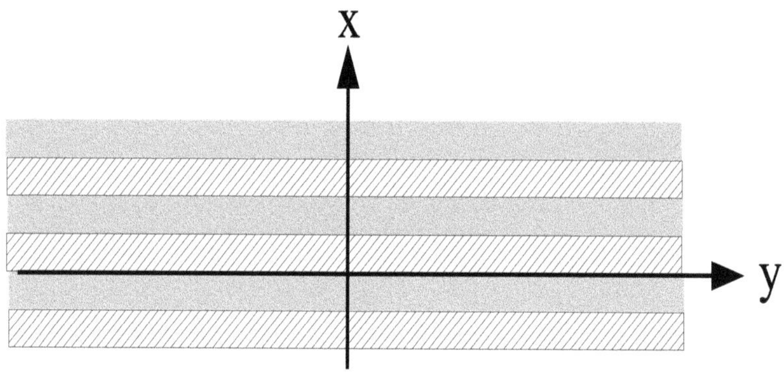

Fig. 6.2 Coordinate system used in writing the matrix representation of the conductivity tensor in a laminated material.

We will assume that the thickness of the layers is small compared to the size of the entire object being considered. There are thus many thin layers and it is therefore reasonable to model thermal conduction in the object by a single conductivity tensor, rather than to model the conduction in individual lamina. Let the conductivity tensor in the chosen coordinate system be

$$k = \begin{pmatrix} k_{1,1} & k_{1,2} & k_{1,3} \\ k_{2,1} & k_{2,2} & k_{2,3} \\ k_{3,1} & k_{3,2} & k_{3,3} \end{pmatrix}.$$

The physical situation is invariant with respect to a 180-degree rotation of the solid in the (x, y)-plane and this is equivalent to inverting the direction of the x-axis. Inversion of the x-axis is represented by the rotation matrix

$$\Theta = \begin{pmatrix} -1 & 0 & 0 \\ 0 & 1 & 0 \\ 0 & 0 & 1 \end{pmatrix}.$$

Call the conductivity tensor in the new coordinate system $\hat{k}$, then

$$\hat{k} = \Theta k \Theta^T = \begin{pmatrix} k_{1,1} & -k_{1,2} & -k_{1,3} \\ -k_{2,1} & k_{2,2} & k_{2,3} \\ -k_{3,1} & k_{3,2} & k_{3,3} \end{pmatrix}$$

and since $k = \hat{k}$, one can conclude that $k_{1,2} = k_{1,3} = k_{2,1} = k_{3,1} = 0$. Similarly, inversion of the y-axis, given by

$$\Theta = \begin{pmatrix} 1 & 0 & 0 \\ 0 & -1 & 0 \\ 0 & 0 & 1 \end{pmatrix}$$

does not alter the physical picture either and the conductivity tensor will therefore not change under this coordinate transformation. One finds

$$\hat{k} = \Theta k \Theta^T = \begin{pmatrix} k_{1,1} & -k_{1,2} & k_{1,3} \\ -k_{2,1} & k_{2,2} & -k_{2,3} \\ k_{3,1} & -k_{3,2} & k_{3,3} \end{pmatrix}$$

and therefore $k_{2,3} = k_{3,2} = 0$. At this point, before examining any other rotations, it pays to take advantage of the simplifications we have discovered and note that, in our coordinate system, the conductivity tensor can be represented by a matrix of the form,

$$k = \begin{pmatrix} k_{1,1} & 0 & 0 \\ 0 & k_{2,2} & 0 \\ 0 & 0 & k_{3,3} \end{pmatrix}.$$

Finally, consider rotations of θ-degrees around the x-axis, given by

$$\Theta = \begin{pmatrix} 1 & 0 & 0 \\ 0 & \cos\theta & \sin\theta \\ 0 & -\sin\theta & \cos\theta \end{pmatrix}$$

which preserve the physical picture and therefore should have no effect on the matrix representation of k. We find

$$\hat{k} = \Theta k \Theta^T = \begin{pmatrix} k_{1,1} & 0 & 0 \\ 0 & k_{2,2}\cos^2\theta + k_{3,3}\sin^2\theta & \cos\theta\sin\theta(-k_{2,2} + k_{3,3}) \\ 0 & \cos\theta\sin\theta(-k_{2,2} + k_{3,3}) & k_{2,2}\sin^2\theta + k_{3,3}\cos^2\theta \end{pmatrix}$$

and picking $\theta = \pi/2$ makes it obvious that $k_{2,2} = k_{3,3}$. The matrix representation of the tensor thus has only two independent parameters and can be written

$$k = \begin{pmatrix} k_{1,1} & 0 & 0 \\ 0 & k_{2,2} & 0 \\ 0 & 0 & k_{2,2} \end{pmatrix}.$$

The two material parameters that appear in the tensor can be found if the conductivities of the two lamina materials are known. Let these conductivities be k_1 and k_2 and consider the two cases with a temperature gradient imposed across the lamina and parallel with the lamina, as in Fig. 6.3.

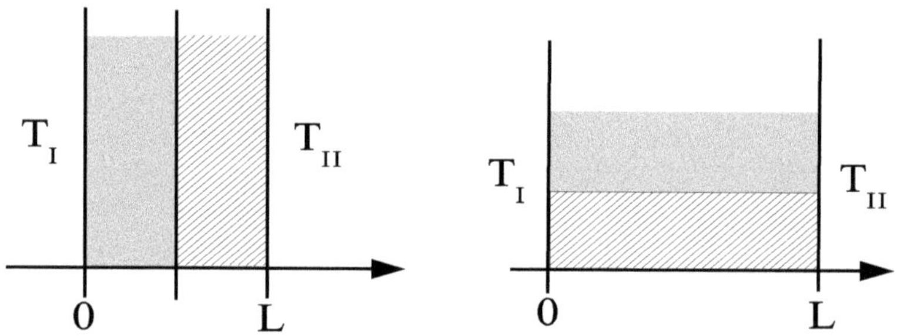

Fig. 6.3 Conduction at a right angle to the lamina and parallel with the lamina.

It is a simple matter to find the steady state energy flux, q, for both cases and determine the average conductivities as $q/((T_I - T_{II})/L)$. The average conductivity in the case of conduction across the lamina must equal $k_{1,1}$, while the average conductivity in the case of conduction parallel to the lamina must equal $k_{2,2}$. One finds

$$k_{1,1} = \frac{2k_1k_2}{k_1 + k_2}$$

and

$$k_{2,2} = \frac{k_1 + k_2}{2}$$

and thus

$$k = \begin{pmatrix} \frac{2k_1k_2}{k_1+k_2} & 0 & 0 \\ 0 & \frac{k_1+k_2}{2} & 0 \\ 0 & 0 & \frac{k_1+k_2}{2} \end{pmatrix}.$$

Example 6.7: Conduction in laminated wall.

Consider now conduction across a wall made of the laminated solid which was analyzed in the previous example. The wall will be a constant thickness with a fixed temperature gradient, $T_I - T_{II}$, imposed across it. The surface of the wall is not

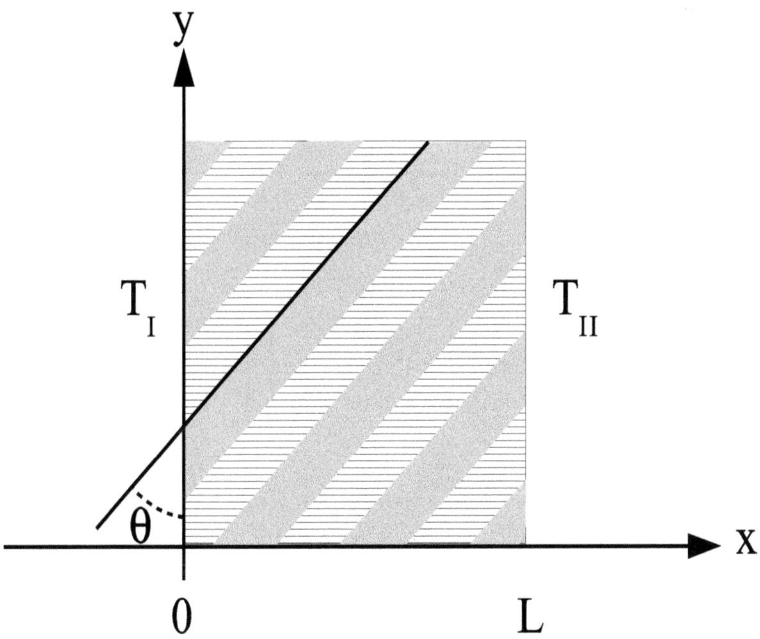

Fig. 6.4 Laminated wall in which the lamina are at an angle θ with the wall surface.

parallel to the lamina but forms an angle θ with a direction parallel to the lamina, see Fig. 6.4. The objective is to determine the heat flux through the wall, both its magnitude and direction.

We will first formulate the steady state energy balance for the wall. Let q_x, q_y and q_z be the conduction fluxes in the x, y and z-directions respectively and do a balance on a differential box with sides dx, dy and dz. This gives

$$q_x dydz|_x - q_x dydz|_{x+dx} + q_y dxdz|_y - q_y dxdz|_{y+dy} + q_z dxdy|_z - q_z dxdy|_{z+dz} = 0$$

or

$$\frac{\partial q_x}{\partial x} + \frac{\partial q_y}{\partial y} + \frac{\partial q_z}{\partial z} = \nabla \cdot q = 0$$

so the energy balance becomes

$$\nabla \cdot (k(\nabla T)) = 0\,.$$

The next step is to determine the matrix representation of the conductivity tensor in the coordinate system in which we choose to work. The best idea is usually to pick a coordinate system in which the boundary conditions are simple, rather than a system in which the tensor is simple. Therefore, we will use the coordinate system shown in Fig. 6.4 in which the wall surfaces are in the y, z-plane. This coordinate system is obtained from the coordinate system used in the previous example, the system in which the x-axis was at a right angle to the lamina, by rotating the former coordinate system an angle θ around the z-axis. This corresponds to the rotation matrix

$$\Theta = \begin{pmatrix} \cos\theta & \sin\theta & 0 \\ -\sin\theta & \cos\theta & 0 \\ 0 & 0 & 1 \end{pmatrix}$$

and, in the coordinate system shown in Fig. 6.4, the matrix of the conductivity tensor therefore takes the form

$$k = \begin{pmatrix} \cos\theta & \sin\theta & 0 \\ -\sin\theta & \cos\theta & 0 \\ 0 & 0 & 1 \end{pmatrix} \begin{pmatrix} k_{1,1} & 0 & 0 \\ 0 & k_{2,2} & 0 \\ 0 & 0 & k_{2,2} \end{pmatrix} \begin{pmatrix} \cos\theta & -\sin\theta & 0 \\ \sin\theta & \cos\theta & 0 \\ 0 & 0 & 1 \end{pmatrix}$$

or

$$k = \begin{pmatrix} k_{1,1}\cos^2\theta + k_{2,2}\sin^2\theta & \cos\theta\sin\theta(k_{2,2} - k_{1,1}) & 0 \\ \cos\theta\sin\theta(k_{2,2} - k_{1,1}) & k_{1,1}\sin^2\theta + k_{2,2}\cos^2\theta & 0 \\ 0 & 0 & k_{2,2} \end{pmatrix}.$$

In our coordinate system and with this conduction tensor, the energy balance thus becomes

$$\frac{\partial}{\partial x}\left((k_{1,1}\cos^2\theta + k_{2,2}\sin^2\theta)\frac{\partial T}{\partial x} + \cos\theta\sin\theta(k_{2,2} - k_{1,1})\frac{\partial T}{\partial y}\right) +$$

$$\frac{\partial}{\partial y}\left(\cos\theta\sin\theta(k_{2,2} - k_{1,1})\frac{\partial T}{\partial x} + (k_{1,1}\sin^2\theta + k_{2,2}\cos^2\theta)\frac{\partial T}{\partial y}\right) + k_{2,2}\frac{\partial^2 T}{\partial z^2} = 0\,.$$

The boundary conditions for the wall are

$$x = 0 \Rightarrow T = T_I \text{ all } y,\, z$$

$$x = L \Rightarrow T = T_{II} \text{ all } y,\, z\,.$$

The boundary conditions imply that the temperature is not a function of y and z, so the energy balance simplifies to

$$\frac{\partial^2 T}{\partial x^2} = 0$$

and the solution is

$$T(x) = (T_{II} - T_I)\frac{x}{L} + T_I$$

which is not a surprising result although the amount of work needed to find the result perhaps was. Putting everything back together, the heat flux becomes

$$q = \begin{pmatrix} q_x \\ q_y \\ q_z \end{pmatrix} = -\begin{pmatrix} (k_{1,1}\cos^2\theta + k_{2,2}\sin^2\theta)\frac{T_{II}-T_I}{L} \\ \cos\theta\sin\theta(k_{2,2} - k_{1,1})\frac{T_{II}-T_I}{L} \\ 0 \end{pmatrix}.$$

This, not the temperature profile, is of course the interesting result. The direction of the temperature gradient is not the same as the direction of the flux. The flux will tend to preferentially follow the better conductor, resulting in a flux component in the y-direction. In dimensionless form, using $k_{1,1}$ as the characteristic conductivity, and setting $\alpha = k_{2,2}/k_{1,1}$, the result becomes

$$\begin{pmatrix} \frac{q_x L}{k_{1,1}(T_{II}-T_I)} \\ \frac{q_y L}{k_{1,1}(T_{II}-T_I)} \\ \frac{q_z L}{k_{1,1}(T_{II}-T_I)} \end{pmatrix} = -\begin{pmatrix} (\cos^2\theta + \alpha\sin^2\theta) \\ \cos\theta\sin\theta(\alpha - 1) \\ 0 \end{pmatrix}. \tag{6.5}$$

A plot of these equations are shown in Fig. 6.5. Notice that when the difference in conductivity of the two materials becomes greater, the effect of tilting the lamina relative to the wall surface becomes more pronounced.

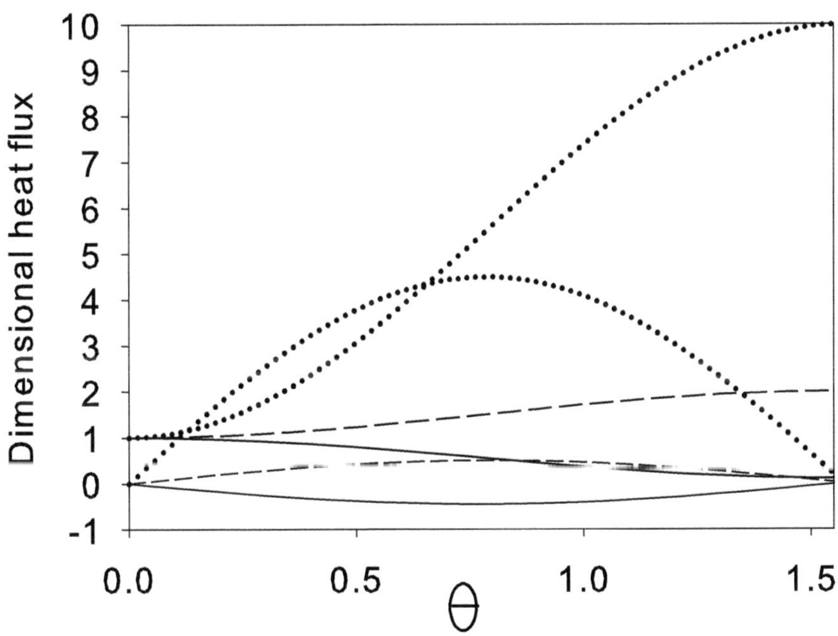

Fig. 6.5 Graphs of dimensionless fluxes in a laminar wall, Eq. (6.5). The three monotone graphs are fluxes in the x-direction, the three unimodal (symmetric around $\theta = \pi/4$) graphs are fluxes in the y-direction. Solid line: $\alpha = 0.1$, broken line: $\alpha = 2$, dotted line: $\alpha = 10$.

6.5.2 *Symmetry and group theory*

The previous example, conduction in a laminated material, may appear complicated compared to conduction in an isotropic material. However, the symmetry properties of a laminated material are simple compared to some of the non-isotropic material that one can encounter. Luckily, symmetry is well understood and the mathematical abstraction of symmetry is the concept of a group. Thus there is a well established body of mathematics for dealing with non-isotropic materials in a systematic way.

When we say that an object is symmetric, we mean that it is possible to perform so-called symmetry operations with the object. A symmetry operation is an act that moves an object away from its location in space and puts it back such that it fits perfectly in the space it occupied previously. For instance, a circle can be rotated any amount around its center or reflected in any diagonal and still fit perfectly in its original location. These operations, rotations around the center and reflection in a diagonal, are therefore symmetry operations on a circle. A plane can be translated any amount in a direction parallel to the plane, so these translations are symmetry operations on the plane. As another example, consider a rectangle with height h and width $w \neq h$. The symmetry of the rectangle is embodied by all the different ways one can move it around and return it to a position that covers the original rectangle perfectly. There are four ways this can be done for the rectangle: One can rotate it 180^0 in the plane (call this operation R_{180}), one can reflect it in either of the two lines of symmetry passing through the mid points of opposite sides (call these operations R_w and R_h) or one can simply do nothing, call this operation R_0), Fig. 6.6.

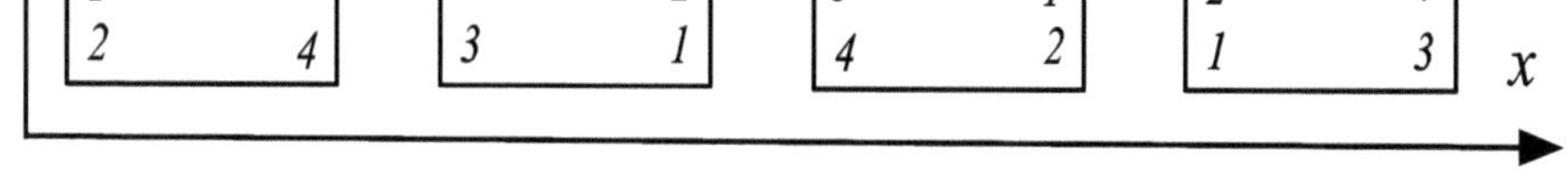

Fig. 6.6 Symmetry operations on a rectangle.

Symmetry operations can be combined. For instance, two applications of R_{180} is equivalent to R_0 and so forth. The set of all possible symmetry operations together with the rule that specifies how to combine operations form a symmetry group. For finite groups, the combinations of operations are often summarized in a *Cayley table.* The Cayley table for the rectangle is given in Table 6.1.

A group is defined rigorously as a set (the set of symmetry operations) and a binary operations between the elements of the set such that

i) The operation is associative, that is, $R_x(R_yR_z) = (R_xR_y)R_z$.

Table 6.1 Cayley table of the symmetry operations of a rectangle. Columns indicate the first factor, rows the second factor.

	R_0	R_w	R_h	R_{180}
R_0	R_0	R_w	R_h	R_{180}
R_w	R_w	R_0	R_{180}	R_h
R_h	R_h	R_{180}	R_0	R_w
R_{180}	R_{180}	R_h	R_w	R_0

ii) The set contains an identity element, R_0, such that $R_0 R_x = R_x R_0 = R_x$ for all R_x. (This simply states that one can always do nothing).

iii) Each element has an inverse R^{-1} such that $RR^{-1} = R^{-1}R = R_0$. (This states that one can always undo a change in the orientation of an object).

Using this abstract definition, one can readily find groups that do not correspond to any physical objects that normal, sane people can visualize.

At the risk of departing too far from the main topic of this chapter, it deserves to be mentioned that polynomials have symmetry properties and each polynomial can be identified with a unique group. For instance, the two roots of the polynomial x^2+1, i and $-i$, are interchangeable in the sense that there is no way to distinguish i from $-i$ based on their algebraic properties, as the difference between them is one of nomenclature, not of substance. This reflects a symmetry of the polynomial x^2+1 that is not possessed by x^2-1 for which the roots, -1 and 1, are not exchangeable. The group of a polynomial has great theoretical significance and it can be shown that the group determines whether or not a solution for the roots of the polynomial can be found in terms of algebraic operations and radicals.

The group of symmetries of the rectangle is commutative, i.e. for any two elements it always holds that $R_x R_y = R_y R_x$. This is not true in general. For instance the group that represents the symmetries of an equilateral triangle is easy to construct and it is seen not to be commutative. The group of symmetries of the rectangle is also finite, it has only four elements, but not all groups are finite. A circle can be rotated around its center any number of radians between 0 and 2π and reflected in any diagonal. The group of the circle is therefore infinite.

With regard to boundary conditions, consider as an example steady thermal conduction in the rectangle in Fig. 6.6, assume a constant thermal conductivity and let the temperature at the top equal the temperature at the bottom and the temperature on the left hand side equal the temperature on the right hand side. The specified surface temperatures give enough boundary conditions for a well posed problem, but sometimes it is more convenient to use boundary conditions based on the symmetry of the problem. Assuming rectangular symmetry of the solution, the solution at any point must equal the solution at other points that are mapped to this point by the symmetry operation. The temperature is therefore an even function relative to the planes of reflection. Using the coordinate system shown in Fig. 6.6, the temperature, T, at a point (x, y) must therefore satisfy

$$T(x,y) = T(2x_r - x, y) = T(x, 2y_r - y) = T(2x_r - x, 2y_r - y)$$

where x_r and y_r are the x and y coordinates of the vertical and horizontal planes of reflection, respectively. The equations above are particularly nice if x_r and y_r equal zero, stating that temperature is an even function with respect to x and y. This is the reason for placing coordinate axis in the symmetry planes whenever possible. Since the temperature must be finite at the planes of reflection, it can only be an even function if its derivatives with respect to x and y equal zero. An equivalent statement of the symmetry boundary conditions is therefore

$$\left.\frac{\partial T}{\partial x}\right|_{x=x_r} = 0 \ , \quad \left.\frac{\partial T}{\partial y}\right|_{y=y_r} = 0$$

which is often the most convenient boundary condition to use when solving the balance equation.

Arguably, the most important symmetry groups encountered in the study of non-isotropic materials are the symmetry groups associated with the seven crystal classes. Working with models of pure crystals may require a large number of material constants, how large depending on the specific class. The most complex (or least symmetric) crystal class is the class of triclinic crystals which may in fact require 36 material constants for their modeling. Luckily, the pertinent tensor expressions for transport and mechanical properties of all crystal class have been derived and tabulated and can be looked up.[2]

6.5.3 *The stress tensor*

In Cartesian coordinates, the components of the stress tensor, τ, are usually defined as the stress on the sides of a differentially small cube with edges parallel to the coordinate directions. These stresses are arranged in a 3×3 matrix as

$$\tau = \begin{pmatrix} \tau_{1,1} & \tau_{1,2} & \tau_{1,3} \\ \tau_{2,1} & \tau_{2,2} & \tau_{2,3} \\ \tau_{3,1} & \tau_{3,2} & \tau_{3,3} \end{pmatrix}$$

where $\tau_{n,m}$ is the stress on the n plane in the mth coordinate direction, see Fig. 6.7. In many texts, the normal stresses, $\tau_{n,n}$, are indicated by $\sigma_{n,n}$.

The n planes are those sides of the cube that are orthogonal to the nth coordinate direction and the positive n plane is the side for which the outward normal to the cube points in the positive n direction. The negative n plane is the side for which the outward normal points in the negative n direction. There is no agreement in

[2]Nye, J.F. [1985] *Physical Properties of Crystals* (Clarendon Press, Oxford).

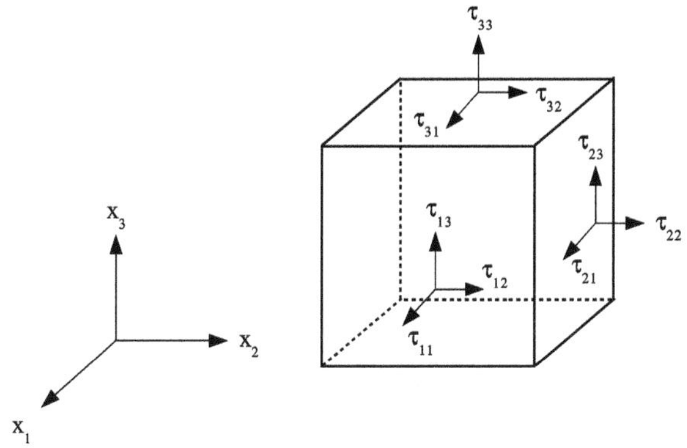

Fig. 6.7 Nomenclature used for indicating the components of stress in Cartesian coordinates. To avoid clutter, the stresses on the negative planes are not shown.

the literature on the choice of positive direction for stress. In this text, the positive direction of $\tau_{n,m}$ on the positive n plane is the positive direction of the m axis. On the negative planes, the opposite hold. Thus, normal stresses are positive when they point out of the cube.

Using physical arguments, one can show that the stress tensor is symmetric, i.e. that $\tau_{n,m} = \tau_{m,n}$. The argument is based on writing a torque balance on a differential element and showing that, unless the stress tensor is symmetric, the angular acceleration of the element is infinite, which is physically unacceptable.

The stress tensor is used in Cauchy's law which states that the stress t, the force per unit area, on a point of a surface equals the stress tensor operating on the outward normal to the surface at that point, n. (Cauchy's law is not a fundamental law of nature but follows from elementary considerations of the forces on a differential volume element. The derivation can be found in most books on continuum mechanics.) In vector matrix notation, Cauchy's law is simply

$$t = \tau n\,.$$

A coordinate rotation, given by the rotation matrix Θ, brings the coordinate representation of t and n into $\hat{t}$ and $\hat{n}$, Cauchy's law transforms as

$$\Theta t = \Theta \tau n = \Theta \tau \Theta^T \Theta n \;\Rightarrow\; \hat{t} = \Theta \tau \Theta^T \hat{n}$$

and we recognize that $\hat{\tau} = \Theta \tau \Theta^T$, so Cauchy's law does indeed require that τ transforms as a second order tensor.

Because the stress tensor is symmetric, it is a self-adjoint operator and the eigenvalues of the stress tensor are real and the eigenvectors are mutually orthogonal. The eigenvectors are those that are parallel to the outward normal and are thus the directions in which the stress is pure normal stress and with a magnitude equal to the eigenvalue. The eigendirections and eigenvalues of the stress tensor are known as the *principal stress directions* and the *principal stresses.* The principal stresses are usually indicated σ_1, σ_2 and σ_3. On transforming to a coordinate system that uses the principal directions as axis, the stress tensor becomes a diagonal matrix

$$\mathcal{T} = \begin{pmatrix} \sigma_1 & 0 & 0 \\ 0 & \sigma_2 & 0 \\ 0 & 0 & \sigma_3 \end{pmatrix}$$

and the three components of the stress on a surface with outward normal $n = (n_1, n_2, n_3)$ are

$$t_1 = \sigma_1 n_1 \ , \ t_2 = \sigma_2 n_2 \ , \ t_3 = \sigma_3 n_3$$

from which we obtain

$$\left(\frac{t_1}{\sigma_1}\right)^2 + \left(\frac{t_2}{\sigma_2}\right)^2 + \left(\frac{t_3}{\sigma_3}\right)^2 = n_1^2 + n_2^2 + n_3^2 = 1$$

which, in the coordinate system defined by the direction of the principal stresses, is an equation for an ellipsoid, the *stress ellipsoid*, as in Fig. 6.8. The stress ellipsoid can be a convenient way of visualizing the stress conditions inside a substance.

However, the definition of the stress tensor given above is entirely physical and there is no a priori reason to think that the stress tensor defined this way is a tensor in the mathematical sense. To see that this is so, we will need to consider the stress tensor as defined by the stresses on two differential cubes, rotated relative to one another, and determine if the two stress tensors transform as a mathematical tensor, see Fig. 6.9.

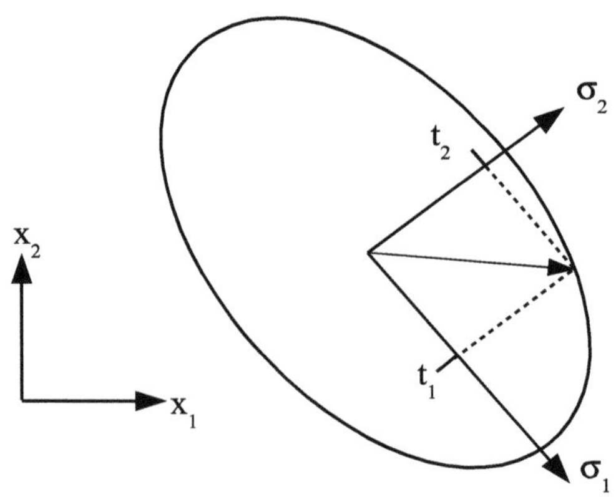

Fig. 6.8 The stress ellipsoid in two dimensions.

First use Cauchy's law to find the stress on the rotated cube in the coordinates of the unrotated (x_1, x_2, x_3)-system. To this end we will need the coordinate representation of the outward normals, the $\hat{n}$s, on the rotated cube in the (x_1, x_2, x_3)-system. In the rotated coordinate system the coordinates of the outward normals are simply

$$\hat{n}_1 = \begin{pmatrix} 1 \\ 0 \\ 0 \end{pmatrix}_{(\hat{x}_1,\hat{x}_2,\hat{x}_3)}, \quad \hat{n}_2 = \begin{pmatrix} 0 \\ 1 \\ 0 \end{pmatrix}_{(\hat{x}_1,\hat{x}_2,\hat{x}_3)}, \quad \hat{n}_3 = \begin{pmatrix} 0 \\ 0 \\ 1 \end{pmatrix}_{(\hat{x}_1,\hat{x}_2,\hat{x}_3)}$$

or minus this if we are considering the negative planes. The coordinates in the unrotated system are therefore found by premultiplication with Θ^T, the rotation matrix that brings the rotated basis vectors back into the unrotated basis vectors. In the unrotated coordinate system, the coordinate representation of the outward normal to plane $\hat{n}$ of the rotated cube therefore is $(\theta_{n,1}, \theta_{n,2}, \theta_{n,3})$. So by Cauchy's law, the stress on plane $\hat{n}$ of the rotated cube is

$$\tau|_{\text{plane } n} = \begin{pmatrix} \tau_{1,1} & \tau_{1,2} & \tau_{1,3} \\ \tau_{2,1} & \tau_{2,2} & \tau_{2,3} \\ \tau_{3,1} & \tau_{3,2} & \tau_{3,3} \end{pmatrix} \begin{pmatrix} \theta_{n,1} \\ \theta_{n,2} \\ \theta_{n,3} \end{pmatrix} = \begin{pmatrix} \sum_{m=1}^{3} \tau_{1,m}\theta_{n,m} \\ \sum_{m=1}^{3} \tau_{2,m}\theta_{n,m} \\ \sum_{m=1}^{3} \tau_{3,m}\theta_{n,m} \end{pmatrix}_{(x_1,x_2,x_3)}$$

where the subscript indicates that this is the coordinate representation in the unrotated system. The coordinates in the rotated system are obtained by projecting this vector onto the basis vectors of the rotated system, which is equivalent to taking the inner product of this vector with the basis vectors in the rotated system. Thus, the stress on plane $\hat{n}$ of the rotated cube, in the kth direction, in the rotated system,

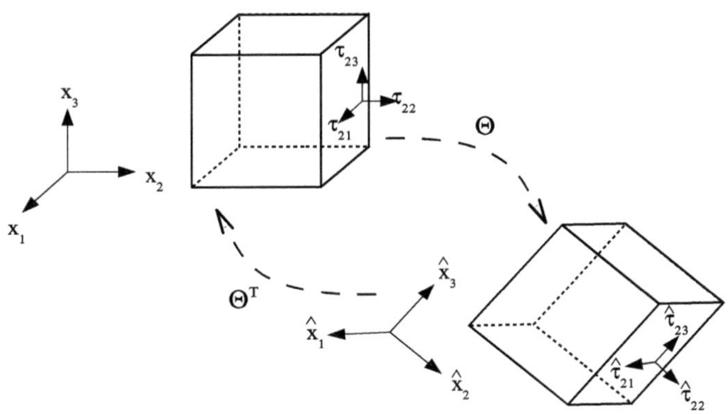

Fig. 6.9 Rotation of a differential cube and the associated coordinate system.

$\hat{\tau}_{n,k}$ is

$$\hat{\tau}_{n,k} = \left\langle \begin{pmatrix} \sum_{m=1}^{3} \tau_{1,m}\theta_{n,m} \\ \sum_{m=1}^{3} \tau_{2,m}\theta_{n,m} \\ \sum_{m=1}^{3} \tau_{3,m}\theta_{n,m} \end{pmatrix}, \begin{pmatrix} \theta_{k,1} \\ \theta_{k,2} \\ \theta_{k,3} \end{pmatrix} \right\rangle \Rightarrow$$

$$\hat{\tau}_{n,k} = \sum_{l=1}^{3}\sum_{m=1}^{3} \theta_{k,l}\tau_{l,m}\theta_{n,m} = \sum_{l=1}^{3}\sum_{m=1}^{3} \theta_{n,m}\theta_{k,l}\tau_{m,l} = \Theta\tau\Theta^T\Big|_{n,k}$$

which is precisely the mathematical transformation rule defining a second order tensor.

The stress tensor is a central object in two constitutive equations, Hooke's law for a linear elastic solid and Newton's law of viscosity for a linear fluid.

Hooke's law: Hooke's law relates the stresses in an elastic solid to the deformation of the solid and a brief introduction to a mathematical description of deformation is therefore in order.

We will be concerned with small deformations of a solid substance and can describe the deformation in terms of what happens to a differential element inside the solid.

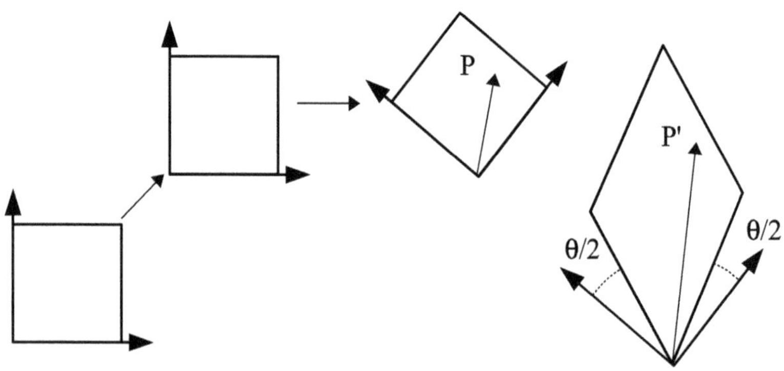

Fig. 6.10 The three steps associated with deformation of a solid: translation of a differential element as a solid body, rotation as a solid body and deformation.

The differential element can translate as a solid body, rotate as a solid body and finally deform. The deformation itself can be decomposed into two steps, a change in length of either or both sides of the square, taking the square into a rectangle, and a change in the angles of the rectangle, taking it into a parallelogram, as in Fig. 6.10. When describing the angle change, it is important that this is done in such a way that the sides of the parallelogram end up making the same angle with the undeformed sides. Unless this is the case, the deformation will contain a component of pure, solid body rotation.

The deformation will be described by what happens to a position vector P of a point inside the differential element. The deformation maps this to a point with the position vector P' and the displacement vector u is defined as

$$u = P' - P.$$

Consider a one-dimensional stretching process in the x-direction. The stretching increases the length of the object in the x-direction from e.g. l to $l + \Delta l$ and the displacement of a point initially at x is

$$u_x = \frac{\Delta l}{l} x = \epsilon_{xx} x$$

where we have introduced the strain ϵ_{xx}, which is the relative change in x-position when the object is stretched in the x-direction. Stretching may not be uniform but can vary from point to point and we therefore generalize the definition of strain to

$$\epsilon_{xx} = \frac{\partial u_x}{\partial x}$$

with similar definitions for stretching in the other coordinate directions.

The strain associated with the angle change deformation is a bit more complex, but it can be found in most texts on continuum mechanics and is given by

$$\epsilon_{nm} = \frac{1}{2}\left(\frac{\partial u_n}{\partial x_m} + \frac{\partial u_m}{\partial x_n}\right).$$

These strains form a second order tensor, the *strain tensor*, which evidently is symmetric

$$\epsilon = \begin{pmatrix} \epsilon_{11} & \epsilon_{12} & \epsilon_{13} \\ \epsilon_{21} & \epsilon_{22} & \epsilon_{23} \\ \epsilon_{31} & \epsilon_{32} & \epsilon_{33} \end{pmatrix}$$

and the diagonal terms of the tensor are called the *tensile strains* while the off-diagonal elements are called *shear strains*. Hooke's law can now be stated. It is

$$\epsilon = S\tau$$

or

$$\epsilon_{nm} = \sum_k \sum_l S_{nmkl}\tau_{kl}.$$

In words: each strain component is a linear function of all nine stress components. S is a tensor of elastic coefficients. It is a fourth order tensor and maps one second order tensor, the stress tensor, into another second order tensor, the strain tensor. Hooke's law is sometimes written as

$$\tau_{nm} = \sum_k \sum_l C_{nmkl}\epsilon_{kl}$$

where C is the stiffness tensor, also fourth order.

Writing out fourth order tensors requires non-existent four dimensional paper, but one can get around this problem by writing the stress and strain tensors as vectors. This is a perfectly legitimate thing to do, given the isomorphism of vectors spaces of the same dimension. Taking advantage of the fact that both tensors are symmetric, one can write Hooke's law as

$$\begin{pmatrix} \frac{\partial u_1}{\partial x_1} \\ \frac{\partial u_2}{\partial x_2} \\ \frac{\partial u_3}{\partial x_3} \\ \frac{\partial u_1}{\partial x_2} + \frac{\partial u_2}{\partial x_1} \\ \frac{\partial u_1}{\partial x_3} + \frac{\partial u_3}{\partial x_1} \\ \frac{\partial u_2}{\partial x_3} + \frac{\partial u_3}{\partial x_2} \end{pmatrix} = \begin{pmatrix} S_{11} & S_{12} & S_{13} & S_{14} & S_{15} & S_{16} \\ S_{21} & S_{22} & S_{23} & S_{24} & S_{25} & S_{26} \\ S_{31} & S_{32} & S_{33} & S_{34} & S_{35} & S_{36} \\ S_{41} & S_{42} & S_{43} & S_{44} & S_{45} & S_{46} \\ S_{51} & S_{52} & S_{53} & S_{54} & S_{55} & S_{56} \\ S_{61} & S_{62} & S_{63} & S_{64} & S_{65} & S_{66} \end{pmatrix} \begin{pmatrix} \tau_{11} \\ \tau_{22} \\ \tau_{33} \\ \tau_{12} = \tau_{21} \\ \tau_{13} = \tau_{31} \\ \tau_{23} = \tau_{32} \end{pmatrix}. \tag{6.6}$$

Notice that in this formulation, the factor of $\frac{1}{2}$ on the shear strain terms has been drooped and, in this formulation, the matrix S_{nm} is **not** a second order tensor. It does not even make sense to think of it as such since it has size 6-by-6 and cannot be multiplied by a 3-by-3 rotation matrix. The equation above is simply a convenient vector-matrix equation for applying Hooke's law. Any coordinate rotation requires that Hooke's law is expressed in the fourth order tensor form. The equation above does make it obvious, however, that at most 36 elasticity coefficients are required. 36 material constants is still a very large number to determine experimentally, and one should use any trick in the book to reduce this number.

One avenue for simplification is to use thermodynamics to argue that the strain state of an elastic solid is independent of the path taken to attain this state.[3] From this, it follows that the matrix S_{nm} in Eq. (6.6) is symmetric, $S_{nm} = S_{mn}$. Consequently, there can be at most 21 material constants in the equation. This number of material constants is in fact required for the most anisotropic of all solid materials, pure crystals of triclinic type. However, for solids that exhibit some kind of symmetry, such as the crystals from the other crystal classes or non-crystalline materials, symmetry considerations can be used to reduce the number of material constants. Isotropic materials are particularly important and for these, Hooke's law

[3]See Nye, [1985].

reduces to

$$\begin{pmatrix} \frac{\partial u_1}{\partial x_1} \\ \frac{\partial u_2}{\partial x_2} \\ \frac{\partial u_3}{\partial x_3} \\ \frac{\partial u_1}{\partial x_2} + \frac{\partial u_2}{\partial x_1} \\ \frac{\partial u_1}{\partial x_3} + \frac{\partial u_3}{\partial x_1} \\ \frac{\partial u_2}{\partial x_3} + \frac{\partial u_3}{\partial x_2} \end{pmatrix} = \begin{pmatrix} \frac{1}{E} & -\frac{\nu}{E} & -\frac{\nu}{E} & 0 & 0 & 0 \\ -\frac{\nu}{E} & \frac{1}{E} & -\frac{\nu}{E} & 0 & 0 & 0 \\ -\frac{\nu}{E} & -\frac{\nu}{E} & \frac{1}{E} & 0 & 0 & 0 \\ 0 & 0 & 0 & \frac{2(1+\nu)}{E} & 0 & 0 \\ 0 & 0 & 0 & 0 & \frac{2(1+\nu)}{E} & 0 \\ 0 & 0 & 0 & 0 & 0 & \frac{2(1+\nu)}{E} \end{pmatrix} \begin{pmatrix} \tau_{11} \\ \tau_{22} \\ \tau_{33} \\ \tau_{12} = \tau_{21} \\ \tau_{13} = \tau_{31} \\ \tau_{23} = \tau_{32} \end{pmatrix}$$

where E is called Young's modulus and ν is Poisson's ratio. Physically, Young's modulus is the ratio of the tensile tension and the relative extension produced by stretching while Poisson's ratio is the amount of contraction orthogonal to the direction of stretching and the extension in the direction of stretching. Hooke's law for an isotropic material is often written in an alternate form where the stiffness constants are used to calculate the stresses as functions of the deformations

$$\begin{pmatrix} \tau_{11} \\ \tau_{22} \\ \tau_{33} \\ \tau_{12} = \tau_{21} \\ \tau_{13} = \tau_{31} \\ \tau_{23} = \tau_{32} \end{pmatrix} = \begin{pmatrix} 2\mu+\lambda & \lambda & \lambda & 0 & 0 & 0 \\ \lambda & 2\mu+\lambda & \lambda & 0 & 0 & 0 \\ \lambda & \lambda & 2\mu+\lambda & 0 & 0 & 0 \\ 0 & 0 & 0 & \mu & 0 & 0 \\ 0 & 0 & 0 & 0 & \mu & 0 \\ 0 & 0 & 0 & 0 & 0 & \mu \end{pmatrix} \begin{pmatrix} \frac{\partial u_1}{\partial x_1} \\ \frac{\partial u_2}{\partial x_2} \\ \frac{\partial u_3}{\partial x_3} \\ \frac{\partial u_1}{\partial x_2} + \frac{\partial u_2}{\partial x_1} \\ \frac{\partial u_1}{\partial x_3} + \frac{\partial u_3}{\partial x_1} \\ \frac{\partial u_2}{\partial x_3} + \frac{\partial u_3}{\partial x_2} \end{pmatrix}.$$

The two stiffness constants, λ and μ, are called the Lamé constants and they are related to Young's modulus and Poisson's ratio by

$$\mu = \frac{E}{2(1+\nu)}, \quad \lambda = \frac{E\nu}{(1-2\nu)(1+\nu)}.$$

It is perhaps surprising that the number of material constants does not reduce to 1 for an isotropic material. But, consider what happens when pulling a rubber band: it deforms in two directions. It both stretches and becomes thinner. Thus two constants are needed. One to describe the resistance to stretching and another to describe the thinning of the rubber band in a plane orthogonal to the direction of stretching.

Newton's law of viscosity: A Newtonian or linear fluid is a fluid that exhibits a linear relationship between stress and rate of strain. It is usually given as

$$\tau_{nm} = \pm\mu\frac{dv_n}{dx_m}$$

where v_n is the velocity in the nth coordinate direction and the sign will depend on the chosen convention for the positive direction of stress. It is plus when the positive direction of the stress is pointing out of the differential element and minus when the positive stress direction is pointing towards the interior of the differential element. For an isotropic fluid, an analysis similar to that for an isotropic solid gives

$$\begin{pmatrix} \tau_{11} \\ \tau_{22} \\ \tau_{33} \\ \tau_{12} = \tau_{21} \\ \tau_{13} = \tau_{31} \\ \tau_{23} = \tau_{32} \end{pmatrix} = \begin{pmatrix} 2\mu + (\kappa - \frac{2}{3}\mu) & (\kappa - \frac{2}{3}\mu) & (\kappa - \frac{2}{3}\mu) & 0\ 0\ 0 \\ (\kappa - \frac{2}{3}\mu) & 2\mu + (\kappa - \frac{2}{3}\mu) & (\kappa - \frac{2}{3}\mu) & 0\ 0\ 0 \\ (\kappa - \frac{2}{3}\mu) & (\kappa - \frac{2}{3}\mu) & 2\mu + (\kappa - \frac{2}{3}\mu) & 0\ 0\ 0 \\ 0 & 0 & 0 & \mu\ 0\ 0 \\ 0 & 0 & 0 & 0\ \mu\ 0 \\ 0 & 0 & 0 & 0\ 0\ \mu \end{pmatrix} \begin{pmatrix} \frac{\partial v_1}{\partial x_1} \\ \frac{\partial v_2}{\partial x_2} \\ \frac{\partial v_3}{\partial x_3} \\ \frac{\partial v_1}{\partial x_2} + \frac{\partial v_2}{\partial x_1} \\ \frac{\partial v_1}{\partial x_3} + \frac{\partial v_3}{\partial x_1} \\ \frac{\partial v_2}{\partial x_3} + \frac{\partial v_3}{\partial x_2} \end{pmatrix}$$

where μ is the viscosity and κ the *bulk viscosity*. The bulk viscosity is close to zero for most practical problems and is therefore usually ignored. The similarity to Newton's law of viscosity is obvious.

6.6 Vectors and tensors in curvilinear coordinates

It was stressed in Chapter 1 that differential balance models should always be formulated in such a way that the boundary conditions are the simplest possible. Simplicity of the partial differential equation is less important. Curvilinear coordinates are indispensable for this purpose but their use is complicated by the fact that transformation rules for vectors and tensors are more complex than for rotations of Cartesian coordinates. In this section a method is described for transformation of vectors and important concepts such as contravariance and covariance are introduced. The problem of transforming differential operators to arbitrary curvilinear coordinates requires introduction of a good bit of additional material and is not covered here.[4]

Discussion of transformation rules for vectors will here be restricted to situations for which

- Space is Euclidean.
- Distances in space has dimension length.
- Space is three-dimensional.
- The curvilinear coordinates are orthogonal.

We have not defined what is meant by a Euclidean space, but most readers probably have a gut feeling for the concept. Euclidean spaces are flat spaces with the property that we can define a *global* Cartesian coordinate system that can be referred to at any point in the space and in which the distance between

[4]Suggested readings for readers who wants to learn more about this issue are
Aris, R. [1989] *Vectors Tensors, and the Basic Equations of Fluid Mechanics* (Dover Publications, Inc. New York); Morse and Feshbach [1953].

two points, (x_1, y_1, z_1) and (x_2, y_2, z_2), is given by the usual Euclidean distance, $\sqrt{(x_2 - x_1)^2 + (y_2 - y_1)^2 + (z_2 - z_1)^2}$.

Contrast this with a non-Euclidean space, such as the two-dimensional space on the surface of a sphere, for instance the surface of the earth. In this space one can define a local Cartesian coordinate system, as is done every day when calculating distances that are small compared to the radius of the earth. But this Cartesian system is local because the direction of the axis changes as one moves around the surface of the earth. For instance, the "North" direction is parallel to the North–South axis when one is at the equator but almost orthogonal to this axis when one is close to one of the poles. Furthermore, if one wants to calculate the distance between two points on the surface that are far apart, say Paris and Sydney, then a calculation based on the Euclidean distance gives the straight line distance right through the earth when, in fact, what one wants is the distance along the surface of the earth. This surface distance is a good bit harder to calculate than the simple Euclidean distance.

Another example of a non-Euclidean space is the space-time of relativity theory in which a point has both space and time coordinates, (x, y, z, t) and in which distances are given by $\sqrt{(x_2 - x_1)^2 + (y_2 - y_1)^2 + (z_2 - z_1)^2 - c^2(t_2 - t_1)^2}$. (This distance measure in space-time is called space-like and only applies when the distance thus calculated are real. When the space-like distance is complex, a different distance measure which is called time-like is used.)

Although non-Euclidean or curved spaces do appear in practical problems, for instance in terrestrial navigation, surface driven flow or, most famously, general relativity, none of these situations are of great importance to chemical engineers.

The restriction to three-dimensional spaces is made simply for the sake of notational convenience. Extension to higher dimensions is straight forward.

Orthogonal curvilinear coordinates are those for which the coordinate directions are mutually orthogonal at all points in space. It is not obvious at this point how to determine when this is the case, but it will become clear as we proceed. As a practical matter, orthogonal coordinate systems are almost always used since they are so much simpler to work with than non-orthogonal systems.

6.6.1 *Proper transformations*

In the most general case, transformation equations between coordinate systems are given implicitly as

$$\begin{aligned} f_1(x_1, x_2, x_3; \xi_1, \xi_2, \xi_3) &= 0 \\ f_2(x_1, x_2, x_3; \xi_1, \xi_2, \xi_3) &= 0 \\ f_3(x_1, x_2, x_3; \xi_1, \xi_2, \xi_3) &= 0 \end{aligned}$$

where the Cartesian coordinate system will be referred to by (x_1, x_2, x_3) and the curvilinear coordinate system as (ξ_1, ξ_2, ξ_3). Exceptions will be made for cylindrical and spherical coordinates which already have their well established nomenclature.

A coordinate transformation must by necessity be invertible, except perhaps at a few singular points such as the origin. Transformations that are invertible are called *proper*. The implicit function theorem can be used to determine if a transformation is proper. Various versions of the theorem exists and it is discussed and proven in many advanced calculus texts. For our purpose, we can state it as follows:

Theorem 6.4. *Implicit function theorem: If* (x_1, x_2, x_3) *is a solution of* $f_n(x_1, x_2, \xi_1, \xi_2, \xi_3) = 0$ *and if there is a neighborhood around* $(x_1, x_2, \xi_1, \xi_2, \xi_3)$ *where all the* f_n*s are continuous with continuous first partial derivatives and where the Jacobian matrix (defined page 380),* $\partial(f_1, f_2, f_3)/\partial(x_1, x_2, x_3)$*, has a non-zero determinant, then there exists a unique solution*

$$\begin{aligned} x_1 &= g_1(\xi_1, \xi_2, \xi_3) \\ x_2 &= g_2(\xi_1, \xi_2, \xi_3) \\ x_3 &= g_3(\xi_1, \xi_2, \xi_3). \end{aligned}$$

In particular, if a coordinate transformation is given in the form

$$\begin{aligned} x_1 &= x_1(\xi_1, \xi_2, \xi_3) \\ x_2 &= x_2(\xi_1, \xi_2, \xi_3) \\ x_3 &= x_3(\xi_1, \xi_2, \xi_3) \end{aligned}$$

then the transformation can be uniquely inverted when all the right hand sides are continuous with continuous partial derivatives and if the Jacobian, $\partial(x_1, x_2, x_3)/\partial(\xi_1, \xi_2, \xi_3)$, has a non-zero determinant.

The two most commonly used curvilinear coordinates, cylindrical and spherical coordinates, have already been introduced. Since they will be used used for illustrative purposes in this section, the transformations between them and Cartesian coordinates are given again. Between rectangular and cylindrical coordinates

$$\begin{aligned} x &= r\cos\theta \\ y &= r\sin\theta \\ z &= z \end{aligned} \quad \Leftrightarrow \quad \begin{aligned} r &= \sqrt{x^2 + y^2} \\ \theta &= \arctan\left(\frac{y}{x}\right) \\ z &= z \end{aligned}$$

and between rectangular and spherical coordinates

$$\begin{aligned} x &= r\sin\theta\cos\phi \\ y &= r\sin\theta\sin\phi \\ z &= r\cos\theta \end{aligned} \quad \Leftrightarrow \quad \begin{aligned} r &= \sqrt{x^2 + y^2 + z^2} \\ \theta &= \arctan\left(\frac{\sqrt{x^2 + y^2}}{z}\right) \\ \phi &= \arctan\left(\frac{y}{x}\right) \end{aligned}.$$

These two transformations are easily inverted and the transformations are shown in both directions.

6.6.2 *Vectors and transformations at a point*

The physical coordinates of a vector, in all but Cartesian coordinate systems, are associated with a specified point and the coordinates change when the point changes. Consider, for instance, the physical coordinates of the force of gravity in rectangular and polar coordinates, Fig. 6.11.

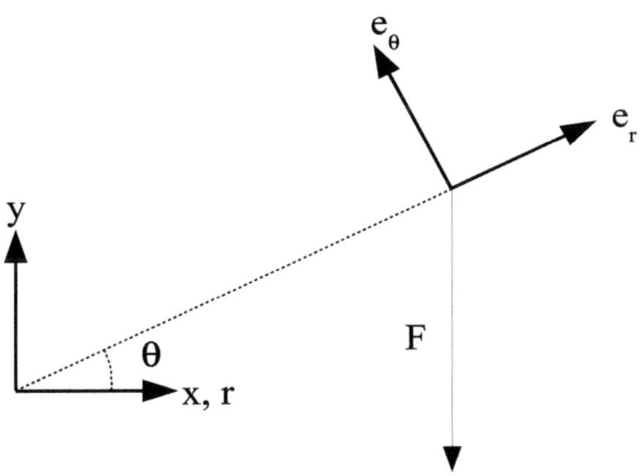

Fig. 6.11 Coordinate representation of the force of gravity in polar coordinates. The origins of the rectangular and polar coordinates are located at the same point and the r-axis at $\theta = 0$ is identical to the positive x-axis. The positive θ-direction is as indicated.

In the rectangular coordinate system indicated, the physical components of the force have the coordinate representation $(0, -|F|)$ in all points. To find the physical components in the polar coordinate system, place unit vectors in the positive r and θ directions at the point considered and project the acceleration onto these base vectors to obtain its components. Clearly, the coordinate representation is $(-|F|\sin\theta, -|F|\cos\theta)$, assuming the r-component is the first coordinate. Thus, the polar coordinates depend on the point considered and, in curvilinear coordinates, the coordinates of vectors are always coordinates at a specific point.

Thus, we are interested in transformation rules for vectors and tensors at fixed points in space. Because the physical components of a vector as well as the coordinate transformation vary from point to point, a vector will be defined as an ordered set of three numbers, at a point, that satisfy certain transformation rules to be determined. At a fixed point, one can use the standard engineering trick of linearizing the relevant equations around the point, thereby obtaining expressions that are much simpler to work with. Let the point in question be given as $(\xi_{10}, \xi_{20}, \xi_{30})$. A

Taylor expansion of x_n around this point gives

$$\begin{aligned} x_n(\xi_1,\xi_2,\xi_3) =& x_n(\xi_{10},\xi_{20},\xi_{30}) + \left.\frac{\partial x_n}{\partial \xi_1}\right|_{(\xi_{10},\xi_{20},\xi_{30})}(\xi_1-\xi_{10}) \\ &+ \left.\frac{\partial x_n}{\partial \xi_2}\right|_{(\xi_{10},\xi_{20},\xi_{30})}(\xi_2-\xi_{20}) \\ &+ \left.\frac{\partial x_n}{\partial \xi_3}\right|_{(\xi_{10},\xi_{20},\xi_{30})}(\xi_3-\xi_{30}) + \text{Higher order terms}\,. \end{aligned}$$

Discarding the higher order terms and rearranging, this can be written

$$\begin{pmatrix} dx_1 \\ dx_2 \\ dx_3 \end{pmatrix} = \begin{pmatrix} \frac{\partial x_1}{\partial \xi_1} & \frac{\partial x_1}{\partial \xi_2} & \frac{\partial x_1}{\partial \xi_3} \\ \frac{\partial x_2}{\partial \xi_1} & \frac{\partial x_2}{\partial \xi_2} & \frac{\partial x_2}{\partial \xi_3} \\ \frac{\partial x_3}{\partial \xi_1} & \frac{\partial x_3}{\partial \xi_2} & \frac{\partial x_3}{\partial \xi_3} \end{pmatrix}_{(\xi_{10},\xi_{20},\xi_{30})} \begin{pmatrix} d\xi_1 \\ d\xi_2 \\ d\xi_3 \end{pmatrix}$$

which relates small changes in the curvilinear coordinates to the equivalent changes in rectangular coordinates through the Jacobian of the transformation evaluated at the point in question, $(\xi_{10},\xi_{20},\xi_{30})$. In the following, we will drop the subscript on the Jacobian as it is always understood to be evaluated at the given point. The Jacobian determinant can be interpreted geometrically as the volume of the differential volume in the (ξ_1,ξ_2,ξ_3) coordinates. To see this, note that the differential volume in the Cartesian coordinates simply equals

$$dV = dx_1 dx_2 dx_3$$

but that this volume can also be calculated as the triple scalar product of the vectors $dx_1 = \frac{\partial x_1}{\partial \xi_1}d\xi_1 + \frac{\partial x_1}{\partial \xi_2}d\xi_2 + \frac{\partial x_1}{\partial \xi_3}d\xi_3$, $dx_2 = \frac{\partial x_2}{\partial \xi_1}d\xi_1 + \cdots$ and $dx_3 = \frac{\partial x_3}{\partial \xi_1}d\xi_1 + \cdots$, i.e.

$$dV = [dx_1 dx_2 dx_3] = \begin{vmatrix} \frac{\partial x_1}{\partial \xi_1}d\xi_1 & \frac{\partial x_1}{\partial \xi_2}d\xi_2 & \frac{\partial x_1}{\partial \xi_3}d\xi_3 \\ \frac{\partial x_2}{\partial \xi_1}d\xi_1 & \frac{\partial x_2}{\partial v}d\xi_2 & \frac{\partial x_2}{\partial \xi_3}d\xi_3 \\ \frac{\partial x_3}{\partial \xi_1}d\xi_1 & \frac{\partial x_3}{\partial v}d\xi_2 & \frac{\partial x_3}{\partial \xi_3}d\xi_3 \end{vmatrix} = \begin{vmatrix} \frac{\partial x_1}{\partial \xi_1} & \frac{\partial x_1}{\partial \xi_2} & \frac{\partial x_1}{\partial \xi_3} \\ \frac{\partial x_2}{\partial \xi_1} & \frac{\partial x_2}{\partial \xi_2} & \frac{\partial x_2}{\partial \xi_3} \\ \frac{\partial x_3}{\partial \xi_1} & \frac{\partial x_3}{\partial \xi_2} & \frac{\partial x_3}{\partial \xi_3} \end{vmatrix} d\xi_1 d\xi_2 d\xi_3\,.$$

For instance, for spherical coordinates, one finds

$$J = \begin{vmatrix} \sin\theta\cos\phi & r\cos\theta\cos\phi & -r\sin\theta\sin\phi \\ \sin\theta\sin\phi & r\cos\theta\sin\phi & r\sin\theta\cos\phi \\ \cos\theta & -r\sin\theta & 0 \end{vmatrix} = r^2\sin\theta$$

or

$$dV = r^2\sin\theta dr d\theta d\phi$$

which was found previously from geometrical considerations (page 33).

6.6.3 *Covariance and contravariance*

The transformation rules under rotation of a Cartesian coordinate system were derived by figuring out the coordinates of the rotated basis vectors. After these were determined, it was straightforward to find the coordinates of any vector in the rotated system. It is not possible to do a similar type of analysis for arbitrary coordinate transformations because one does not know what the set of new basis vectors corresponding to the variable transformation is. The emphasis on the *physical* coordinates of a vector in the previous section hints that it may be necessary to consider several different types of coordinates of a vector under a given variable transformation. In fact, we will need to consider three different kinds of vector coordinates; the well known physical coordinates as well as the covariant and contravariant coordinates which will be defined shortly. This complexity can be illustrated by a simple example which shows that the physical coordinates of different vectors may transform differently under the same variable transformation.

Consider a wall with different surface temperatures at the left and right surface and place an x-axis orthogonal to the wall surface. With this wall, one can associate a position vector v going from the left surface to the right surface and a temperature gradient ΔT across the wall. Let the basis unit vector be 1m, the wall thickness 1m and the temperature difference across the wall be 1^0C. Then

$$v = (1) \text{ and } \Delta T = (1)\,.$$

Consider next the simplest possible variable transformation possible in this system, e.g. change in the length of the basis vector from 1m to 1 cm. Using this basis vector, the other two vectors become

$$v = (100) \text{ and } \Delta T = (0.01)\,.$$

In other words, v becomes longer and ΔT becomes shorter so their physical coordinates must satisfy different transformation laws. At this point the reader may protest and point out, quite correctly, that the coordinate change is really a change in units from centimeters to meters and once the units are reported, the strangeness of the transformation behavior vanishes. While true, this is also irrelevant because we do not carry units when doing abstract math and the observation that different vectors may transform differently is valid. v is an example of a contravariant vector while ΔT is an example of a covariant vector.

Contravariant vectors: linearizing the transformation from Cartesian to curvilinear coordinates at the point in question gives the following relationship between the differentials

$$\begin{pmatrix} d\xi_1 \\ \\ d\xi_2 \\ \\ d\xi_3 \end{pmatrix} = \begin{pmatrix} \frac{\partial \xi_1}{\partial x_1} & \frac{\partial \xi_1}{\partial x_2} & \frac{\partial \xi_1}{\partial x_3} \\ \\ \frac{\partial \xi_2}{\partial x_1} & \frac{\partial \xi_2}{\partial x_2} & \frac{\partial \xi_2}{\partial x_3} \\ \\ \frac{\partial \xi_3}{\partial x_1} & \frac{\partial \xi_3}{\partial x_2} & \frac{\partial \xi_3}{\partial x_3} \end{pmatrix} \begin{pmatrix} dx_1 \\ \\ dx_2 \\ \\ dx_3 \end{pmatrix} = J_\xi \begin{pmatrix} dx_1 \\ \\ dx_2 \\ \\ dx_3 \end{pmatrix}$$

where J_ξ is the Jacobian of the transformation to curvilinear coordinates.

A contravariant vector is an ordered set of three numbers, the coordinates or components of the vector, which transform as these differentials, i.e. as

$$v^n = \begin{pmatrix} v^1 \\ \\ v^2 \\ \\ v^3 \end{pmatrix}_{\text{contra}} = \begin{pmatrix} \frac{\partial \xi_1}{\partial x_1} & \frac{\partial \xi_1}{\partial x_2} & \frac{\partial \xi_1}{\partial x_3} \\ \\ \frac{\partial \xi_2}{\partial x_1} & \frac{\partial \xi_2}{\partial x_2} & \frac{\partial \xi_2}{\partial x_3} \\ \\ \frac{\partial \xi_3}{\partial x_1} & \frac{\partial \xi_3}{\partial x_2} & \frac{\partial \xi_3}{\partial x_3} \end{pmatrix} \begin{pmatrix} v_{x_1} \\ \\ v_{x_2} \\ \\ v_{x_3} \end{pmatrix} = J_\xi v$$

where we have introduced the common usage of superscripts to indicate the index of contravariant components and of using a superscript on the vector itself to indicate its contravariant nature.

Scalar multiplication, addition and subtraction of contravariant vectors is defined the obvious way.

Covariant vectors: the gradient of a scalar function, such as temperature, is transformed from a dependence on the x coordinates to a dependence on the ξ coordinates using the chain rule.

$$\begin{pmatrix} \frac{\partial T}{\partial \xi_1} \\ \\ \frac{\partial T}{\partial \xi_2} \\ \\ \frac{\partial T}{\partial \xi_3} \end{pmatrix} = \begin{pmatrix} \frac{\partial x_1}{\partial \xi_1} & \frac{\partial x_2}{\partial \xi_1} & \frac{\partial x_3}{\partial \xi_1} \\ \\ \frac{\partial x_1}{\partial \xi_2} & \frac{\partial x_2}{\partial \xi_2} & \frac{\partial x_3}{\partial \xi_2} \\ \\ \frac{\partial x_1}{\partial \xi_3} & \frac{\partial x_2}{\partial \xi_3} & \frac{\partial x_3}{\partial \xi_3} \end{pmatrix} \begin{pmatrix} \frac{\partial T}{\partial x_1} \\ \\ \frac{\partial T}{\partial x_2} \\ \\ \frac{\partial T}{\partial x_3} \end{pmatrix}$$

Covariant vectors are vectors that transform this way. Specifically, covariant vector is an ordered set of three numbers which transform as

$$v_n = \begin{pmatrix} v_1 \\ \\ v_2 \\ \\ v_3 \end{pmatrix}_{\text{co}} = \begin{pmatrix} \frac{\partial x_1}{\partial \xi_1} & \frac{\partial x_2}{\partial \xi_1} & \frac{\partial x_3}{\partial \xi_1} \\ \\ \frac{\partial x_1}{\partial \xi_2} & \frac{\partial x_2}{\partial \xi_2} & \frac{\partial x_3}{\partial \xi_2} \\ \\ \frac{\partial x_1}{\partial \xi_3} & \frac{\partial x_2}{\partial \xi_3} & \frac{\partial x_3}{\partial \xi_3} \end{pmatrix} \begin{pmatrix} v_{x_1} \\ \\ v_{x_2} \\ \\ v_{x_3} \end{pmatrix} = J_x^T v$$

where J_x is the Jacobian of the transformation to Cartesian coordinates. Clearly, $J_x = J_\xi^{-1}$. Notice that the matrix used in covariant transformations is **not** the Jacobian matrix of the transformation to Cartesian coordinates, but the transpose of this matrix. The common usage of subscripts to indicate the covariant vector and its components has been used here.

Scalar multiplication, addition and subtraction of covariant vectors is defined the obvious way.

The transformation laws can be written compactly as

$$v^n = J_\xi v \quad \text{and} \quad v_n = J_x^T v = (J_\xi^{-1})^T v$$

and for a rotation of Cartesian coordinates, $J_\xi = \Theta$ and $(\Theta^{-1})^T = \Theta$, so they become

$$v^n = \Theta v \quad \text{and} \quad v_n = \Theta v$$

i.e. contravariant and covariant vectors are identical in Cartesian coordinates. This is the reason we can write simply v for the Cartesian vector in the transformation laws above.

The transformation laws are easily extended to transformations between two different curvilinear coordinate systems, e.g. transformations from cylindrical to spherical coordinates. The v on the right hand side in the transformation laws above is simply replaced by the contra or covariant vector in the originating coordinate system. Written in terms of sums rather than in matrix notation, contravariant vectors are defined as vectors that transform as

$$v^n = \sum_{m=1}^{3} \frac{\partial \xi_n}{\partial x_m} v^m$$

while covariant vectors are those that transform as

$$v_n = \sum_{m=1}^{3} \frac{\partial x_m}{\partial \xi_n} v_m \,.$$

Incidentally, this definition also holds in non-Euclidean spaces where it is not possible to refer back to the Cartesian coordinates of a vector.

Tensors: tensors are now defined in a completely analogous way. A tensor of rank N is an ordered set of 3^N numbers, the coordinates or components, which transform with respect to each index as a vector. For instance, a tensor with components that transform as

$$t_m^n = \sum_{k=1}^{3}\sum_{l=1}^{3} \frac{\partial \xi_n}{\partial x_k}\frac{\partial x_l}{\partial \xi_m} t_l^k$$

is a (mixed) tensor of rank 2 of contravariant degree 1 and covariant degree 1.

Addition and subtraction of tensors with the same number of indices of the same type is defined the obvious way. In curvilinear coordinates, tensor relationships can become just as unwieldy as in Cartesian systems and the summation convention is modified as follows.

Summation convention: any index that appears once as a superscript and once as a subscript in a product is a dummy index and is summed over its entire range (1 to 3 in three-dimensional systems).

Using this convention, the formula above is written more economically as

$$t_m^n = \frac{\partial \xi_n}{\partial x_k}\frac{\partial x_l}{\partial \xi_m} t_l^k \,.$$

Consider once more the representation of the force of gravity in rectangular and polar coordinates as depicted in Fig. 6.11. In the rectangular coordinate system, the physical components are given by $(0, -|F|)$, while in polar coordinates they are $(-|F|\sin\theta, -|F|\cos\theta)$. Contravariant transformation give

$$g^2 = \begin{pmatrix} \frac{\partial r}{\partial x} & \frac{\partial r}{\partial y} \\ \frac{\partial \theta}{\partial x} & \frac{\partial \theta}{\partial y} \end{pmatrix} \begin{pmatrix} 0 \\ -|F| \end{pmatrix} = \begin{pmatrix} \frac{x}{\sqrt{x^2+y^2}} & \frac{y}{\sqrt{x^2+y^2}} \\ -\frac{y}{x^2+y^2} & \frac{x}{x^2+y^2} \end{pmatrix} \begin{pmatrix} 0 \\ -|F| \end{pmatrix}$$

$$= \begin{pmatrix} \cos\theta & \sin\theta \\ -\frac{\sin\theta}{r} & \frac{\cos\theta}{r} \end{pmatrix} \begin{pmatrix} 0 \\ -|F| \end{pmatrix} = \begin{pmatrix} -|F|\sin\theta \\ -\frac{|F|\cos\theta}{r} \end{pmatrix}$$

which does not equal the physical components of F. The second coordinate does not even have the correct dimension. The same comment can be made about the covariant vector

$$g_2 = \begin{pmatrix} \frac{\partial x}{\partial r} & \frac{\partial y}{\partial r} \\ \frac{\partial x}{\partial \theta} & \frac{\partial y}{\partial \theta} \end{pmatrix} \begin{pmatrix} 0 \\ -|F| \end{pmatrix} = \begin{pmatrix} \cos\theta & \sin\theta \\ -r\sin\theta & r\cos\theta \end{pmatrix} \begin{pmatrix} 0 \\ -|F| \end{pmatrix} = \begin{pmatrix} -|F|\sin\theta \\ -r|F|\cos\theta \end{pmatrix}.$$

Because the two components of both the contravariant and the covariant vectors have different dimensions, it does not make sense to find the magnitude of F by taking the square root of the usual inner product of g^2 or g_2 with them self. However, the usual inner product of the two vectors gives the length squared of the vector.

$$\langle g^2, g_2\rangle = |F|^2\sin^2\theta + |F|^2\cos^2\theta = |F|^2.$$

This result suggests that the space of covariant vectors is the dual of the space of contravariant vectors and vice versa. This is in fact the case and while an inner product is usually not defined between two contravariant vectors or two covariant vectors, it is defined in the obvious way between a contravariant and a covariant vector.

The two vectors that are obtained by contravariant and covariant transformation of a vector ought not be thought of as different geometrical or physical objects but rather as equivalent representations of the same object and one representation can easily be converted into the other. Taking again the transformation laws in their matrix form and eliminating v, one finds

$$\left.\begin{aligned} v^m &= J_\xi v \;\Rightarrow\; v = J_\xi^{-1} v^m \\ v_n &= J_x^T v \end{aligned}\right\} \Rightarrow v_n = J_x^T J_\xi^{-1} v^m = J_x^T J_x v^m.$$

The factor $J_x^T J_x$ is of central importance here. It can be shown to be a second order tensor of covariant degree two and is called the (covariant) *metric tensor.* It is usually indicated g_{nm}. Thus

$$v_n = g_{nm} v^m$$

where the summation convention has been used.

Contravariant and covariant vectors that are related this way are said to be *associated* vectors. Calculation of the covariant vector from its associated contravariant vector is called *lowering of the index*. Carrying out the matrix multiplication in the expression previous, one obtains

$$g_{nm} = \sum_{k=1}^{3} \frac{\partial x_k}{\partial \xi_n} \frac{\partial x_k}{\partial \xi_m} \,.$$

Notice that one cannot use the summation convention here because the repeated index, k, appears as a subscript in both occurrences. An element of the metric tensor is the inner product of two rows of J_x, but in an orthogonal system of coordinates, these rows must be orthogonal and g_{nm} must therefore be diagonal. For cylindrical coordinates, the metric tensor becomes

$$g_{\text{cylindrical}} = \begin{pmatrix} 1 & 0 & 0 \\ 0 & r^2 & 0 \\ 0 & 0 & 1 \end{pmatrix}$$

while for spherical coordinates it is

$$g_{\text{spherical}} = \begin{pmatrix} 1 & 0 & 0 \\ 0 & r^2 & 0 \\ 0 & 0 & r^2 \sin^2(\theta) \end{pmatrix} .$$

The square roots of the diagonal terms of g_{nm} are called *scale factors* and are usually indicated h_n. Just like the Jacobian determinant of a coordinate transformation relates differential volumes in the two coordinate system, the scale factors provide expressions for the distance that correspond to differential changes. To see this, consider the differential vector ds which, in a Cartesian system has the coordinates dx_1, dx_2 and dx_3. Its length is given by

$$\begin{aligned}
|ds|^2 = dx_1^2 + dx_2^2 + dx_3^2 &= \left(\sum_{m=1}^{3} \frac{\partial x_1}{\partial \xi_m} d\xi_m\right)^2 + \left(\sum_{m=1}^{3} \frac{\partial x_2}{\partial \xi_m} d\xi_m\right)^2 + \left(\sum_{m=1}^{3} \frac{\partial x_3}{\partial \xi_m} d\xi_m\right)^2 \\
&= \sum_{m=1}^{3}\sum_{n=1}^{3} \frac{\partial x_1}{\partial \xi_n}\frac{\partial x_1}{\partial \xi_m} d\xi_n d\xi_m + \sum_{m=1}^{3}\sum_{n=1}^{3} \frac{\partial x_2}{\partial \xi_n}\frac{\partial x_2}{\partial \xi_m} d\xi_n d\xi_m + \sum_{m=1}^{3}\sum_{n=1}^{3} \frac{\partial x_3}{\partial \xi_n}\frac{\partial x_3}{\partial \xi_m} d\xi_n d\xi_m \\
&= \sum_{m=1}^{3}\sum_{n=1}^{3}\sum_{k=1}^{3} \frac{\partial x_k}{\partial \xi_n}\frac{\partial x_k}{\partial \xi_m} d\xi_n d\xi_m = \sum_{m=1}^{3}\sum_{n=1}^{3} g_{nm} d\xi_n d\xi_m = \sum_{n=1}^{3} g_{nn} d\xi_n d\xi_n \,.
\end{aligned}$$

So if one considers an infinitesimal change in only one of the ξ-coordinates, the associated distance is

$$|ds| = \sqrt{g_{nn}} d\xi_n = h_n d\xi_n \,.$$

Referring back to the metric tensors for cylindrical and spherical coordinates given above, the distance that correspond to a change of dr or dz in cylindrical coordinates is simple dr and dz, respectively (since g_{rr} and g_{zz} both equal 1).

However, the distance that correspond to a change $d\theta$ is $rd\theta$. Similarly, in spherical coordinates, a change of $d\theta$ correspond to a distance of $rd\theta$ while a change of $d\phi$ correspond to the distance $r\sin(\theta)d\phi$. Again, this is exactly what was found previously from geometrical arguments, page 33. Notice also that a differential volume element has a volume given by

$$dV = h_1h_2h_3d\xi_1d\xi_2d\xi_3 .$$

Returning to the force of gravity example with the contra and covariant representations in polar coordinates given by

$$g^2 = \begin{pmatrix} -|F|\sin\theta \\ -\frac{|F|\cos\theta}{r} \end{pmatrix} , \; g_2 = \begin{pmatrix} -|F|\sin\theta \\ -r|F|\cos\theta \end{pmatrix}$$

and a metric tensor

$$g_{nm} = \begin{pmatrix} 1 & 0 \\ 0 & r^2 \end{pmatrix} ,$$

one can easily check that $g_2 = g_{nm}g^2$ as

$$\begin{pmatrix} -|F|\sin\theta \\ -r|F|\cos\theta \end{pmatrix} = \begin{pmatrix} 1 & 0 \\ 0 & r^2 \end{pmatrix} \begin{pmatrix} -|F|\sin\theta \\ -\frac{|F|\cos\theta}{r} \end{pmatrix} .$$

The inverse operator of g_{nm} is indicated g^{nm} and is called the contravariant metric tensor. Not surprisingly,

$$v^n = g^{nm}v_m .$$

This calculation of a contravariant vector from its associated covariant vector is called raising of the index.

As noted already, the component of contra and covariant vectors may not all have the same dimensions or have the dimensions of the physical vector. Special care must therefore be taken when defining an inner product. Usually, an inner product is defined between a contravariant and a covariant vector as

$$\langle v^n, w_n\rangle = v^n w_n \quad \text{or} \quad \langle v_n, w^n\rangle = v_n w^n .$$

The inner product defined this way can be reinterpreted as a linear operator, the first factor, mapping the second factor into the real numbers. In other words, the space of contravariant vectors and covariant vectors are dual spaces of one another.

The length of a vector is defined as the square root of the inner product between the vector and its associated vector

$$|v|^2 = \langle v^n, v_n\rangle = g^{nm}v_m v_n .$$

The angle, θ, between two vectors, both either contravariant or covariant, is defined as

$$\cos\theta = \frac{g_{nm}v^m w^n}{|v||w|} \quad \text{or} \quad \cos\theta = \frac{g^{nm}v_m w_n}{|v||w|} .$$

6.6.4 *The physical components*

The physical components of a vector are the components relative to a basis of unit vectors in the ξ-directions. This basis can be obtained by rotation of the Cartesian basis vectors in the usual way by specifying the direction cosines. In the special case when the scale factors of a transformation are all unity, the variable transformation from Cartesian coordinates to the new coordinates is a pure rotation. The transformation is

$$\begin{pmatrix} \xi_1 \\ \xi_2 \\ \xi_3 \end{pmatrix} = \Theta \begin{pmatrix} x_1 \\ x_2 \\ x_3 \end{pmatrix}$$

and the direction cosines are

$$\theta_{nm} = \frac{\partial \xi_n}{\partial x_m}\,.$$

If the scale factors are not unity, then a differential change $d\xi_n$, associated with the differential change dx, correspond to a length of $h_n d\xi_n$ and this must therefore be the direction cosine giving the rotation matrix

$$\Theta = \begin{pmatrix} h_1 \frac{\partial \xi_1}{\partial x_1} & h_1 \frac{\partial \xi_1}{\partial x_2} & h_1 \frac{\partial \xi_1}{\partial x_3} \\ h_2 \frac{\partial \xi_2}{\partial x_1} & h_2 \frac{\partial \xi_2}{\partial x_2} & h_2 \frac{\partial \xi_2}{\partial x_3} \\ h_3 \frac{\partial \xi_3}{\partial x_3} & h_3 \frac{\partial \xi_3}{\partial x_2} & h_3 \frac{\partial \xi_3}{\partial x_3} \end{pmatrix}.$$

It can be shown that this can also be written

$$\Theta = \begin{pmatrix} \frac{1}{h_1}\frac{\partial x_1}{\partial \xi_1} & \frac{1}{h_1}\frac{\partial x_2}{\partial \xi_1} & \frac{1}{h_1}\frac{\partial x_3}{\partial \xi_1} \\ \frac{1}{h_2}\frac{\partial x_1}{\partial \xi_2} & \frac{1}{h_2}\frac{\partial x_2}{\partial \xi_2} & \frac{1}{h_2}\frac{\partial x_3}{\partial \xi_2} \\ \frac{1}{h_3}\frac{\partial x_1}{\partial \xi_3} & \frac{1}{h_3}\frac{\partial x_2}{\partial \xi_3} & \frac{1}{h_3}\frac{\partial x_3}{\partial \xi_3} \end{pmatrix}.$$

Each row in a rotation matrix is orthogonal to any other row so it follows from the matrix above that

$$\frac{1}{h_n h_m}\left\langle \left(\frac{\partial x_1}{\partial \xi_n}, \frac{\partial x_2}{\partial \xi_n}, \frac{\partial x_3}{\partial \xi_n}\right), \left(\frac{\partial x_1}{\partial \xi_m}, \frac{\partial x_2}{\partial \xi_m}, \frac{\partial x_3}{\partial \xi_m}\right)\right\rangle = 0\,.$$

In other words, the **columns** of the Jacobian matrix, J_x are mutually orthogonal, a fact used to conclude that the metric tensor is diagonal. This is only the case for orthogonal systems, however.

Once the rotation matrix at a point is known for a coordinate transformation, it is a simple matter to calculate the coordinates of any vector relative to the transformed basis, i.e. the physical coordinates, if the Cartesian coordinates of the vector are known. The component in the nth transformed coordinate is simply the inner product of the vector with the nth basis vector in the transformed coordinates. Let the Cartesian coordinates of the vector be (v_1, v_2, v_3) and the coordinates in the transformed system be (w_1, w_2, w_3). Note that in this case the use of subscripts does not imply a covariant vector. As for rotations of Cartesian systems, let e_1, e_2

and e_3 be the unit basis vectors in the Cartesian system and let unit basis vectors in the transformed system be $\hat{e}_1$, $\hat{e}_2$ and $\hat{e}_3$. Then

$$\begin{aligned} w_n &= \langle v_1 e_1 + v_2 e_2 + v_3 e_3, \hat{e}_n \rangle \\ &= \langle v_1 e_1 + v_2 e_2 + v_3 e_3, \theta_{n1} e_1 + \theta_{n2} e_2 + \theta_{n3} e_3 \rangle \\ &= v_1 \theta_{n1} + v_2 \theta_{n2} + v_3 \theta_{n3} \,. \end{aligned}$$

For the cylindrical coordinate system, one finds the scale factors $h_1 = 1$, $h_2 = r$ and $h_3 = 1$ and

$$\Theta_{\text{cylindrical}} = \begin{pmatrix} \cos\theta & \sin\theta & 0 \\ -\sin\theta & \cos\theta & 0 \\ 0 & 0 & 1 \end{pmatrix}$$

giving the following rule for transformation of vectors between Cartesian and cylindrical coordinates

$$\begin{aligned} w_1 &= v_1 \cos\theta + v_2 \sin\theta \\ w_2 &= -\, v_1 \sin\theta + v_2 \cos\theta \\ w_3 &= v_3 \end{aligned}$$

Returning one final time to the example of the force of gravity in polar coordinates, we started with the Cartesian coordinates $(0, -|F|)$. The components in polar coordinates therefore are

$$F_r = 0 \cdot \cos\theta - |F| \sin\theta \quad \text{and} \quad F_\theta = 0 \cdot (-\sin\theta) - |F| \cos\theta \;\Rightarrow$$

$$F_{\text{polar}} = \begin{pmatrix} -|F| \sin\theta \\ -|F| \cos\theta \end{pmatrix}$$

which was derived previously from geometric considerations.

Similarly, in spherical coordinates, one finds the scale factors $h_1 = 1$, $h_2 = r$ and $h_3 = r \sin\theta$ and

$$\Theta_{\text{spherical}} = \begin{pmatrix} \sin\theta\cos\phi & \sin\theta\sin\phi & \cos\theta \\ \cos\theta\cos\phi & \cos\theta\sin\phi & -\sin\theta \\ -\sin\phi & \cos\phi & 0 \end{pmatrix}$$

or written as a transformation rule between Cartesian coordinates, (v_1, v_2, v_3) and spherical coordinates, (w_1, w_2, w_3)

$$\begin{aligned} w_1 &= v_1 \sin\theta\cos\phi + v_2 \sin\theta\sin\phi + v_3 \cos\theta \\ w_2 &= v_1 \cos\theta\cos\phi + v_2 \cos\theta\sin\phi - v_3 \sin\theta \\ w_3 &= -\, v_1 \sin\phi + v_2 \cos\phi \,. \end{aligned}$$

Some readers may want to spend some time with this result and convince themselves that it agrees with the results one obtains from geometric analysis.

Chapter 7

Linear Difference Equations

Difference equations are recurrence relationships that relate the values in a sequence of numbers to one another. Most readers are probably familiar with these equations from the study of numerical integration of ODEs. For instance, the ODE initial value problem

$$\frac{dx}{dt} = f(t, x) \ , \ x(t_0) = x_0$$

can be integrated numerically with Euler's method as

$$x_{n+1} = x_n + f(t_n, x_n)h = x_n + f_n h$$

where h is a step size of the integration and x_n is the numerical solution for $x(t)$ after n time steps, i.e. at the time $t_n = nh + t_0$. The numerical solution is a sequence of values $\{x_0, x_1, \ldots, x_n\}$ that satisfy the difference equation and initial condition. Euler's method gives rise to a first order difference equation, first order because the difference between the highest value of the counter, $n+1$, and the lowest value, n, equals 1. Higher order difference equations are obtained with higher order methods of numerical integration. For instance, Simpson's rules gives a second order difference equation of the form

$$x_{n+2} = x_n + \frac{h}{3}(f_{n+2} + 4f_{n+1} + f_n).$$

In this equation, the unknown value x_{n+2} appears both explicitly on the left hand side and as an argument of f_{n+2}. Depending on the form of f, a unique solution for x_{n+2} may or may not exist. Notice that use of Euler's method requires a single value as initial condition, x_0, while Simpson's method requires two values, x_0 and x_1.

The counter will henceforth be named τ instead of n since most of the cases we consider are initial value problems where the counter indicates discrete time. A difference equation of Nth order is an equation of the form

$$F(\tau, x_\tau, x_{\tau+1}, \ldots, x_{\tau+N}) = 0 \tag{7.1}$$

where F is a given function and τ and N are integers. An Nth order difference equation requires a sequence of N consecutive values of the unknown as initial condition. An initial condition to this equation thus consists of a sequence of N values, usually written either of two ways, $\{x_1, \ldots, x_N\}$ or $\{x_0, \ldots, x_{N-1}\}$.

The linear operator that maps an element of the sequence of numbers into the next element of the sequence, i.e. $Lx_\tau = x_{\tau+1}$, is often called the *shift operator*. If the operator L_N is such that $L_N x_\tau = L^N x_\tau = x_{\tau+N}$, we say it is a shift operator of order N. One may also encounter problems in which several equations of this type are coupled to create a set of difference equations. If F, in Eq. (7.1) is not an explicit function of τ, then the equation(s) is said to be *autonomous*. Any non-autonomous system of difference equations can be rendered autonomous by adding an equation of the form

$$y_{\tau+1} = y_\tau + 1 \;, \; y_0 = 0$$

which clearly has the solution $y_\tau = \tau$, and substituting y_τ for τ in the remaining equations. Formally, a solution to Eq. (7.1) is a sequence of numbers $\{x_\tau\}_{\tau=-\infty}^{\tau=\infty}$ such that any N consecutive values satisfy the difference equation. However, most often, one is only concerned with the solution for increasing values of τ, i.e. sequences of the form $\{x_\tau\}_{\tau=0}^{\tau=\infty}$, and this is the only solution that will be considered in the following. It is a simple matter to extend the theory to include solutions for decreasing values of the counter.

Just as the origin of the time axis is arbitrary in ODE problems, so the initial value of τ is arbitrary in difference equations and it can be shifted any number of integers, say m, to produce the equivalent formulation

$$F(\tau - m, x_{\tau-m}, x_{\tau+1-m}, \ldots, x_{\tau+N-m}) = 0$$

which of course also induces a shift in the initial condition.

Difference equation models may occur in situations that have a build-in time step, such as calculation of compound interest or the dynamics of annual or daily changes of some variable. They can occur in models of periodic processes such as cell cycle models and are used in process control problems where the discrete sampling time of measuring devices make difference equations the natural choice of model. They also arise in models of staged processes in which case the counter no longer refers to time steps but to stage number. Some examples of difference equation models follow.

A further rationale for studying difference equations is that the theory of linear difference equations is a good preamble to the study of linear ordinary differential equations. Linear ODEs are of course widespread in models of chemical engineering systems and it turns out that the theory of linear ODEs is strikingly similar to the theory of linear difference equations. Furthermore, it is easier to develop and comprehend the theory for difference equations than for differential equations, so an understanding of the simpler difference equations is a good basis for developing an understanding of the more difficult differential equations.

The difference equations given are all examples of ODEs. They are called ordinary because the dependent variable is a function of only one counter. Problems in which the dependent variable is a function of several counters can also be found and these equations are known as partial difference equations. Partial difference equations will not be covered here.

Example 7.1: High finance.

After graduation your are left with a student loan that you have to pay off within some given period of time. Each month, interest accrues and each month you make a fixed payment on the loan. Determine the magnitude of this payment.

For this problem, the counter τ has an obvious meaning; it is the number of months since you graduated. The amount left in the principal in month $\tau + 1$ is $x_{\tau+1}$ and is readily found from the amount left in month τ as

$$x_{\tau+1} = \alpha x_\tau - q$$

where α is the factor with which the principal increases when interest accrues and q is the amount of the monthly payment. The initial condition for the problem is the given value of x_0, the initial amount in the principal. This problem is easily solved using the methods covered in this chapter. The solution is

$$x_\tau = \alpha^\tau \left(x_0 - \frac{q}{\alpha - 1} \right) + \frac{q}{\alpha - 1} .$$

If the loan must be paid off in N months, then the required monthly payment is found by setting $x_N = 0$ and solving for q

$$q = x_0 \frac{\alpha^N (\alpha - 1)}{\alpha^N - 1} .$$

Example 7.2: Growth and cell cycle.

Living cells pass through a so-called cell cycle, the process of cell growth followed by cell division. Under many conditions, the length of the cell cycle is approximately the same for all cells in a population. Thus, if the duration of the cell cycle is t_{cc}, then, assuming all cells live and that cell division gives rise to two new cells, a cell number balance between the time t and $t + t_{cc}$ gives

$$N(t + t_{cc}) = 2N(t)$$

where N is the number of cells. This balance can be written as a difference equation by letting the counter τ be the number of cell cycles that have occurred since time zero. Let the number of cells at time zero be N_0 and let N_τ indicate the number of cells at time τt_{cc}, then the number of cells is modeled by the initial value problem

$$N_{\tau+1} - 2N_\tau = 0\,.$$

The solution obviously is $N_\tau = 2^\tau N_0$.

The effect of cell death can easily be added to the model. If the probability that a cell dies before dividing equals θ, then the cell number balance is

$$N(t + t_{cc}) = 2N(t)(1 - \theta)$$

or, as a difference equation

$$N_{\tau+1} - 2N_\tau(1 - \theta) = 0 \;\Rightarrow\; N_\tau = (2(1 - \theta))^\tau N_0\,.$$

This model is obviously very simplistic. Cells do not all divide at exactly the same age and there is therefore some variation in the length of the cell cycle among different cells. We can try and take this variation into account by assuming that some fraction f of the cells divide at an age a_1 while the remaining cells divide at the greater age a_2. However, with this assumption, we no longer have a single cell cycle time and therefore no obvious time step that can be used in a difference equation. The solution is to make the time step of the difference equation so small that both division ages are integer multiples of this time step. This can always be done if the ratio of two division ages is a rational number and can in fact be done no matter how many division ages are used in the model as long as all the ages are rational multiples of one another.

For instance, let $a_1/a_2 = K/L$ where K and L are integers with a greatest common divisor equal to 1. Then, surely we can write $a_1 = K\Delta t$ and $a_2 = L\Delta t$ where the magnitude of Δt can be found from either of the two equations. This Δt will be the time step in the difference equation. To set up a model, let $N_{a,\tau}$ be the number of cells with ages between $a\Delta t$ and $(a+1)\Delta t$ at time $t = \tau\Delta t$. Over a time step Δt, most cells simply grow older by Δt, so their numbers are governed by the difference equation

$$N_{a+1,\tau+1} = N_{a,\tau}\,.$$

The only exceptions to this are for cells with ages between a_1 and $a_1 + \Delta t$ and for cells with ages between zero and Δt. For the former, the difference equation must reflect the fact that a fraction f of the cells have divided at a_1 and the difference equation therefore becomes

$$N_{K,\tau+1} = (1 - f)N_{K-1,\tau}\,.$$

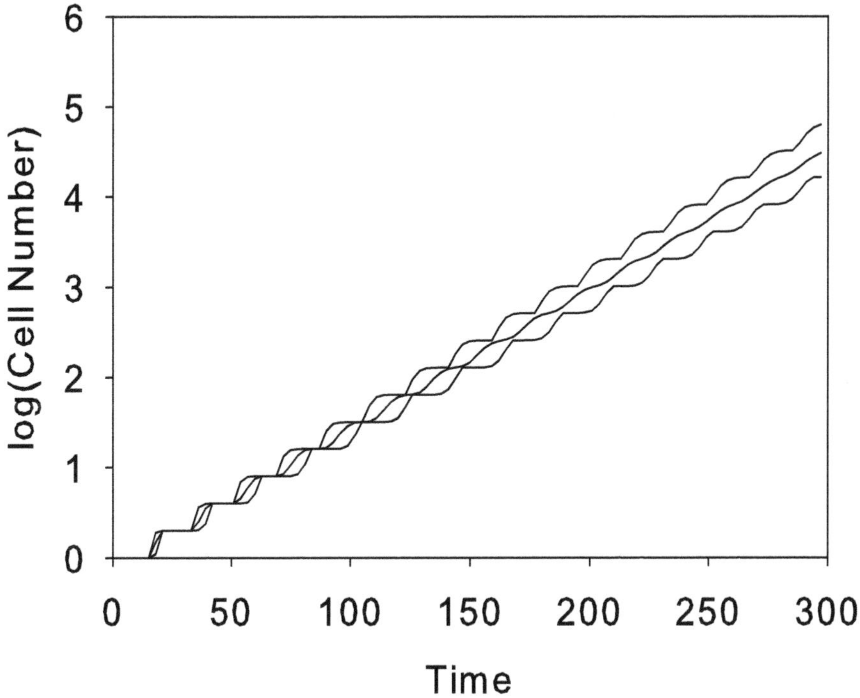

Fig. 7.1 Semi-log plot of the total cell number versus time obtained by simulation of Eq. (7.2). The parameters are $a_1 = 18$, $a_2 = 21$ and $f = 0.1$, bottom curve, 0.5, middle curve, and 0.9, top curve.

For the latter, the difference equation must reflect the fact that cells of age zero arise from all the dividing cells, i.e.

$$N_{0,\tau+1} = 2fN_{K-1,\tau} + 2N_{L-1,\tau}\,.$$

It is natural to write these equations in vector-matrix notation as

$$\begin{pmatrix} N_0 \\ N_1 \\ \vdots \\ N_{K-1} \\ N_K \\ N_{K+1} \\ \vdots \\ N_{L-2} \\ N_{L-1} \end{pmatrix}_{\tau+1} = \begin{pmatrix} 0\,0\ldots & 2f & 0\,0\ldots 0\,2 \\ 1\,0\ldots & 0 & 0\,0\ldots 0\,0 \\ \vdots\ \vdots & \vdots & \vdots\ \vdots\quad \vdots\ \vdots \\ 0\,0\ldots & 0 & 0\,0\ldots 0\,0 \\ 0\,0\ldots & (1-f) & 0\,0\ldots 0\,0 \\ 0\,0\ldots & 0 & 1\,0\ldots 0\,0 \\ \vdots\ \vdots & \vdots & \vdots\ \vdots\quad \vdots\ \vdots \\ 0\,0\ldots & 0 & 0\,0\ldots 0\,0 \\ 0\,0\ldots & 0 & 0\,0\ldots 1\,0 \end{pmatrix} \begin{pmatrix} N_0 \\ N_1 \\ \vdots \\ N_{K-1} \\ N_K \\ N_{K+1} \\ \vdots \\ N_{L-2} \\ N_{L-1} \end{pmatrix}_{\tau} \tag{7.2}$$

and simulating this equation on a computer is a trivial exercise. Results of such a simulation is shown in Fig. 7.1. The initial condition used in the simulations were a single cell of zero age, i.e. $N_0 = (1, 0, 0, \ldots, 0)$. The three curves are for the same values of a_1 and a_2, but for different values of f, the fraction of cells that divide at the younger age. As expected, the cell numbers increase faster as f increases because a larger fraction of the cells then divide at a younger age. Notice how the growth curves start off with the appearance of step functions but become increasingly smooth as time goes on. The well defined steps at early times are a result of the initial condition, a single cell or a population of cells all of the same age. When all cells are in the same state, age in this case, the population is said to be synchronous but this synchrony is slowly lost as different cells divide at different ages. Evidently, synchrony will never be lost in the model that was developed at the beginning of this example because this model only had a single division age and all cells therefore always divide synchronously. However, even in the model with two division ages, synchrony will not be completely lost and the growth curves will continue to maintain some of the character of step functions. This is unavoidable because the discretization on which the model is based is still quite coarse. Only when division ages are modeled as a continuous distribution will the model be able to predict full loss of synchrony.

Of course, one can improve the model by forcing a finer discretization. For instance, the simulations shown in Fig. 7.1 used $a_1 = 18$ and $a_2 = 21$, giving $\Delta t = 3$. A slight change in the model parameters to $a_1 = 18.1$ and $a_2 = 21.1$ forces a change in Δt from 3 to 0.1 thus giving a much finer discretization of age and time. However, there is a substantial computational penalty for such a model improvement. The size of the system matrix increases from 7-by-7 to 211-by-211, thus increasing the computational time for calculation of each step. Simultaneously, the smaller value of Δt necessitates calculations of more steps to arrive at the same final time.

Example 7.3: Population dynamics.

The most famous difference equation of all times is arguably the Fibonacci equation[1] The equation was derived as a model of a population of rabbits using a time step of one month. Initially, only one pair of rabbits is present. After one month, the pair is old enough to breed and give rise to a second pair. A newborn pair requires two months to mature before they can produce a new pair while older pairs produce a new pair each month. Thus, the following difference equations for the number of rabbit pairs in the τ month is

$$N_{\tau+2} = N_{\tau+1} + N_\tau \,.$$

With the initial condition $N_1 = N_2 = 1$, the Fibonacci numbers are obtained; a series in which each number is the sum of the two previous numbers;

[1]Named after the Italian mathematician Leonardo of Pisano, 1175–1250, who usually goes by the name Fibonacci or son of Bonaccio.

1,1,2,3,5,8,13,21, ...The numbers are given for any value of τ by the well known Binnet's formula

$$\frac{1}{\sqrt{5}}\left(\left(\frac{1+\sqrt{5}}{2}\right)^{\tau}-\left(\frac{1-\sqrt{5}}{2}\right)^{\tau}\right).$$

This population model is obviously only valid for short times. Rabbits eventually die and as the population becomes larger, it will deplete food sources faster, usually resulting in a decrease in fecundity and increase in mortality. We can model these effects by demanding that $N_{\tau+1} \propto N_\tau$ when N is small and $N_{\tau+1} \propto (1 - N_\tau)$ when $N \approx 1$ (in some suitably scaled measure of the population size). A simple equation that reflects both these demands is

$$N_{\tau+1} = \lambda N_\tau(1 - N_\tau). \tag{7.3}$$

Eq. (7.3) is clearly a discrete version of the logistic equation and is at times referred to as the logistic difference equation. It is justifiably famous because, although simple, it exhibits many of the kinds of complicated dynamics that are studied in nonlinear analysis. Numerically found solutions for different values of $\lambda =$ 1.8 are plotted in Fig. 7.2. As a nonlinear difference equation it is is a departure from the main topic of this text. However, a short digression to examine the dynamics of this equation is worthwhile if for no other reason than to illustrate various types of dynamics that that are **not** seen in linear models and that therefore indicate that linear models are insufficient. In addition, looking at the dynamics of a simple nonlinear model will hopefully make the reader appreciate how relatively simple linear models are.

When $\lambda = 2.5$, the solution rapidly approaches the value 0.6. The system is at a stable steady state. This simple solution structure is maintained up to a λ-value of 3, after which the structure of the solution is radically changed. The system is said to undergo a *bifurcation* and the λ-values at which this occurs is a *bifurcation point*. However, the rigorous definition of a bifurcation or a bifurcation point is beyond the scope of this text. When $\lambda = 3.2$, the solution never reaches a fixed value but instead rapidly approach a state where values alternate between 0.513045 and 0.799455. This state is known as a two cycle and is a type of dynamics only seen in nonlinear models. Superficially, a two cycle may look like the kind of oscillations that can be seen in linear difference equations of the form

$$x_{\tau+1} = -x_\tau$$

for which the solution is $\{x_0, -x_0, x_0, -x_0, \ldots\}$, however, there is a fundamental difference between the two states. In the linear case, the amplitude of the oscillations are determined by the initial condition, but in the nonlinear case, the two cycle, this is not so. All initial conditions (between 0 and 1) will approach the same two cycle as τ goes to infinity. The two cycle is an example of a so called *attractor*, a dynamic

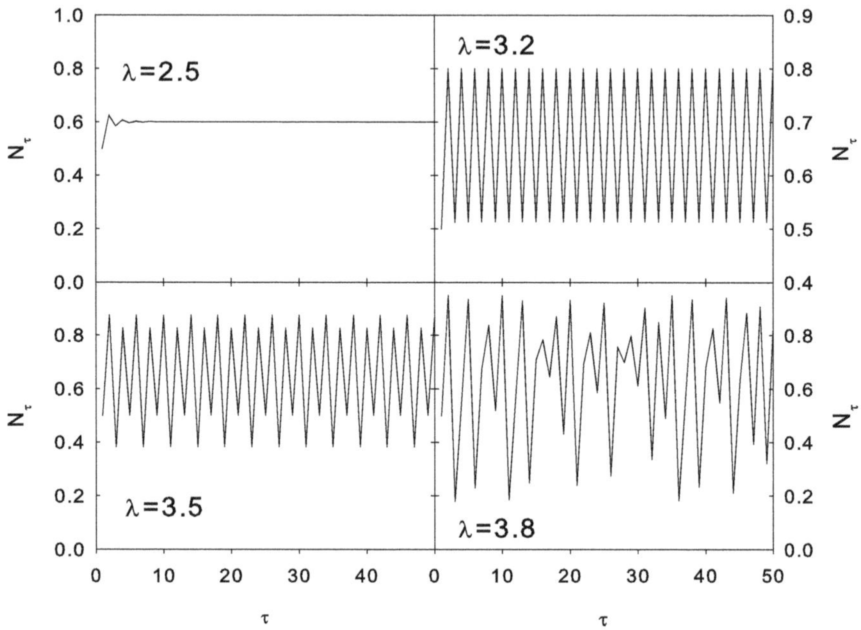

Fig. 7.2 Numerical solution of Eq. (7.3) for $\lambda = 2.5, 3.2, 3.5$ and 3.8 and $N_0 = 0.5$.

state that is approached arbitrarily close by initial conditions in the neighborhood of the attractor. Clearly, the oscillatory state of the linear model is not an attractor because different initial conditions attain different final states.

As λ is increased to 3.5, the system undergoes another bifurcation and the solution reaches a state where it alternates between four different values, a four cycle. When the solution changes from a two cycle to a four cycle, it is said to undergo a period doubling bifurcation. It turns out that as λ keeps increasing in value, the system will keep undergoing period doubling, passing through an eight cycle, a sixteen cycle and so forth. However, the interval of λ-values over which a given cycle is found becomes smaller and smaller such that this period doubling scenario terminates around $\lambda = 3.5699456...$ Between this value and $\lambda = 4$ the solution can be either periodic with a period different from 2^n or it can be chaotic such as the solution shown for $\lambda = 3.8$. Chaotic solutions or attractors are strictly nonlinear phenomena. They have no apparent structure and have the peculiar property that initial conditions that are close together diverge to give solutions that are far apart. For instance, the solutions for x_{100} with $\lambda = 3.8$ and initial conditions $x_1 = 0.5$ and $x_1 = 0.501$ are 0.221944 and 0.631368 respectively.[2]

[2]To learn more about the dynamics of Eq. (7.3), the original paper by Mitchell Feigenbaum in which many of the properties of the solutions were first described, is strongly recommended. Feigenbaum, M.J. [1980] "Universal Behavior in Nonlinear Systems," *Los Alamos Science*, 4–27.

Example 7.4: Digital process control.

Modern process control is mostly digital and the sampling interval of the controller/data acquisition system provides a natural time step for the counter in difference equation models of process control. Consider the liquid level control loop sketched in Fig. 7.3. Volumetric flow rates are indicated by Q and the exit flow rate is assumed given by $Q_{\text{exit}} = RV$, where R is a resistance and V the volume of liquid inside the tank. A balance on the liquid volume in the tank gives

$$\frac{dV}{dt} = Q_{\text{feed}} - RV + U$$

where U is a disturbance variable, e.g. other intermittent feeds or drains of the tank. Introducing the deviation variables $e = V - V_{\text{steady state}}$ and $q = Q_{\text{feed}} - Q_{\text{steady state}}$, the balance takes the form

$$\frac{de}{dt} = q - Re + U.$$

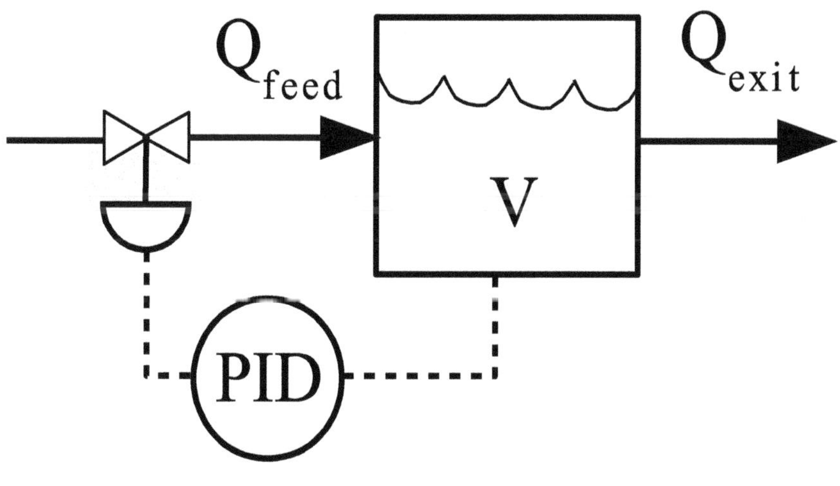

Fig. 7.3 PID control of liquid level in a tank.

The deviation variable q is determined by the controller; in this example a digital PID controller that uses a measurement of V to determine q. The output from a digital PID controller can be written as

$$q_\tau = q_{\tau-1} - K_c\left[(e_\tau - e_{\tau-1}) + \frac{\Delta t}{\tau_I}e_\tau + \frac{\tau_D}{\Delta t}(e_\tau - 2e_{\tau-1} + e_{\tau-2})\right] \tag{7.4}$$

where τ_I and τ_D are the integral and derivative times, respectively. Δt is the data acquisition time interval and e_τ is the measurement of $e(t)$ at time $\Delta t\tau$. To avoid a model that contains both a differential equation and a difference equation, the volume balance will be converted to a difference equation. This can be done several ways, all of which introduce some amount of error or approximation, usually insignificant though. Approximating the time derivative in the volume balance by a backwards difference, the balance becomes

$$\frac{e_\tau - e_{\tau-1}}{\Delta t} = q_\tau - Re_\tau + U_\tau\,. \tag{7.5}$$

Eqs. (7.4) and (7.5) constitute the difference equation model of the system. The model is linear and can be solved with the methods covered in this chapter. Often one only wants to simulate the model for selected parameter values and, to this end, the model must be put in a more convenient form; a form with explicit expressions for e_τ and q_τ in terms of earlier values. Solving Eqs. (7.4) and (7.5) for e_τ and q_τ gives

$$e_\tau = \frac{\tau_I((2\tau_D K_c + K_c\Delta t + 1)e_{\tau-1} - \tau_D K_c e_{\tau-2} + \Delta t q_{\tau-1} + U_\tau \Delta t)}{K_c\tau_I\Delta t + K_c\Delta t^2 + K_c\tau_D\tau_I + \tau_I + \tau_I R\Delta t}\,. \tag{7.6}$$

$$\begin{aligned} q_\tau =& \frac{(\tau_D\tau_I/\Delta t + 2\tau_I R\tau_D + \tau_I R\Delta t - \Delta t)K_c e_{\tau-1} - (1/\Delta t + R)\tau_D\tau_I K_c e_{\tau-2}}{K_c\tau_I\Delta t + K_c\Delta t^2 + K_c\tau_D\tau_I + \tau_I + \tau_I R\Delta t} \\ &+ \frac{(R\Delta t + 1)\tau_I q_{\tau-1} - (\Delta t\tau_I + \Delta t^2 + \tau_D\tau_I)K_c U_\tau}{K_c\tau_I\Delta t + K_c\Delta t^2 + K_c\tau_D\tau_I + \tau_I + \tau_I R\Delta t}\,. \end{aligned} \tag{7.7}$$

In this form, it is easy to specify an initial condition and use a spreadsheet to calculate values of q and e for increasingly large values of τ. Results of such calculations are shown in Fig. 7.4. The displayed trends will be obvious to any reader already familiar with process control.

In all cases, the system was assumed initially at steady state, i.e. $e_{-1} = e_0 = q_0 = 0$ and the disturbance was set equal to 1 for times between zero and 1. A shows the effect of the sampling time Δt. The continuous curve represents results for $\Delta t = 1$, the dashed curve for $\Delta t = 0.5$ and the dotted curve for $\Delta t = 0.05$. The remaining parameter values are $\tau_I = 1$, $\tau_D = 5$, $K_c = 5$ and $R = 0.5$. The slowest sampling time is of the same magnitude as the duration of the disturbance, i.e. only U_0 and $U_1 = 1$, the rest are zero, and the model almost completely fails to capture the disturbance and the resulting dynamics. Clearly, the shorter sampling time will give a more representative description of the dynamics unless the sampling time becomes so small that truncation errors in the calculations become significant. B shows the effect of the integration time τ_I, equal to 10 (continuous curve), 1 (dashed curve) and 0.1 (dotted curve). $\Delta t = 0.05$ and all other parameter values are as for A. Notice how the decrease in integration time makes the control respond more rapidly but also induces increasingly severe oscillations in the response. C shows the effect of different derivative times. τ_D, equal to 10 (continuous curve), 5 (dashed curve) and 1 (dotted curve). $\tau_I = 1$ and all other parameter values are as

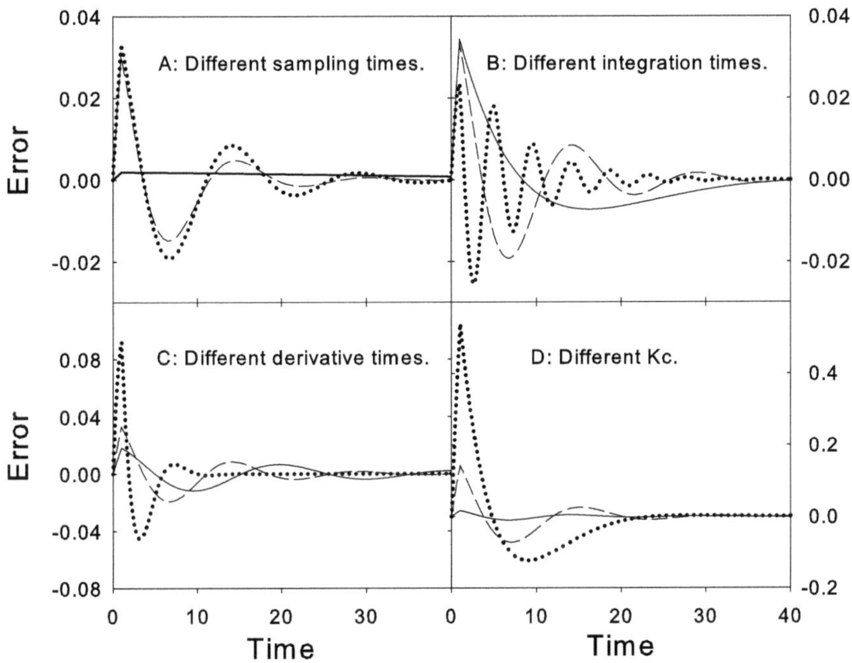

Fig. 7.4 Results of spreadsheet calculations of the digital control model in Eqs. (7.6) and (7.7). See text for detailed explanation.

for B. D shows the effect of different values of the controller gain, K_c. K_c, equal to 10 (continuous curve), 1 (dashed curve) and 0.1 (dotted curve). $\tau_D = 5$ and all other parameter values are as for C. Notice how the increase in controller gain makes the system respond more rapidly to the disturbance.

Example 7.5: Flow through reactor with back mixing.

A flow through reactor with back mixing can be modeled as a series of interacting CSTRs. Assume that all the vessels in the series have the same volume V, and let the volumetric feed flow rate be Q_f with a reactant concentration C_f, the forward flow rate between the vessels be Q and number the vessels consecutively, starting with the feed vessel as number 1, Fig. 7.5.

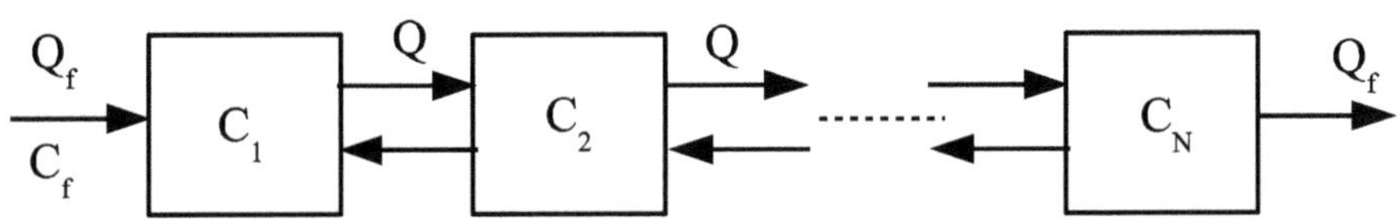

Fig. 7.5 A series of CSTRs as a model of a flow through reactor with back mixing.

A steady state balance on vessel τ gives

$$QC_{\tau-1} + (Q - Q_f)C_{\tau+1} - (Q - Q_f)C_\tau - QC_\tau = Vr(C_\tau) \ , \ \tau = 2, 3, \ldots, N-1$$

where $r(C_\tau)$ is the specific rate of consumption of the reactant. The balances on the first and last vessels are special cases and do not match the general form above. For vessel 1, the balance equation becomes

$$Q_f C_f + (Q - Q_f)C_2 - QC_1 = Vr(C_1)$$

and for the last vessel it becomes

$$QC_{N-1} - (Q - Q_f)C_N - Q_f C_N = Vr(C_N) \, .$$

The model is seen to consist of N coupled difference equations in the unknown concentrations.

7.1 Linear equations with constant coefficients

Equations with constant coefficients are a simple, special case for which complete solutions can be found. Let the difference equation be

$$a_N x_{\tau+N} + a_{N-1} x_{\tau+N-1} + \cdots + a_0 x_\tau = q_\tau \tag{7.8}$$

where the coefficients a_n are constants. Notice that the left hand side is simply a linear combination of shift operators operating on the sequence x_τ. The problem is said to be homogeneous if $q_\tau = 0$. We will later state and prove a theorem regarding the solution structure of sets of linear difference equations. A single equation is covered by this theorem which states that the complete solution to the equation above is the sum of a particular solution and the complete solution to the homogeneous problem. Furthermore, the complete solution to the homogeneous problem is a linear combination of N, linearly independent solutions called *basis solutions*.

7.1.1 *Homogeneous solutions*

To find a set of basis solutions for the homogeneous problem, guess a solution of the form $x_{h,\tau} = \lambda^\tau$. Substituting this into the homogeneous difference equation gives

$$\lambda^N + a_{N-1}\lambda^{N-1} + \cdots + a_1\lambda + a_0 = 0$$

which is known as the *characteristic polynomial* of the difference equation. For reasons that will become obvious once we start looking at difference equations in vector-matrix notation, the roots of this polynomium are often referred to as eigenvalues. If all the roots of this polynomial, say λ_1 through λ_N, have multiplicity one, then they provide the N linearly independent basis solutions needed and the complete homogeneous solution becomes

$$x_{h,\tau} = C_1\lambda_1^\tau + C_2\lambda_2^\tau + \cdots + C_N\lambda_N^\tau$$

where the C_ns are arbitrary constants which can be determined from the initial conditions.

If the characteristic polynomial has roots with multiplicity greater than 1, additional solutions must be found to obtain a complete set of basis solutions. To find these, suppose that λ is a double root of the characteristic polynomial and claim that $\tau\lambda^\tau$ is a solution. This solution is certainly linearly independent of the other solutions and substituting it into the homogeneous difference equation gives

$$(\tau + N)\lambda^{\tau+N} + a_{N-1}(\tau + N - 1)\lambda^{\tau+N-1} + \cdots + a_1(\tau + 1)\lambda^{\tau+1} + a_0\tau\lambda^\tau = 0 \Rightarrow$$

$$\tau\{\lambda^N + a_{N-1}\lambda^{N-1} + \cdots + a_0\} + \lambda\{N\lambda^{N-1} + a_{N-1}(N-1)\lambda^{N-2} + \cdots + a_1\} = 0\,.$$

The expression inside the first pair of brackets is the characteristic polynomial evaluated at the root, so this equals zero. The expression inside the second pair of brackets can be seen to equal the first derivative of the characteristic polynomial, evaluated at the root. But for a double root, the derivative of a polynomial at the root also equals zero, so $\tau\lambda^\tau$ is indeed a solution. This argument can be extended to roots with higher multiplicity and if a root, λ, has multiplicity M, a set of basis solutions that correspond to the root is λ^τ, $\tau\lambda^\tau$,$\tau^2\lambda^\tau$, ..., $\tau^{M-1}\lambda^\tau$. We can summarize these results in the closed form solution

$$x_{h,\tau} = \sum_{n=1}^{\hat{N}} \left\{ \sum_{m=1}^{M_n} C_{n,m}\tau^{m-1}\lambda_n^\tau \right\} \tag{7.9}$$

where $\hat{N}$ is the number of discrete roots and M_n is the multiplicity of the nth root, λ_n.

An important observation can be made at this point. As τ goes to ∞, λ^τ goes to zero or $\pm\infty$ depending on whether it is inside or outside the unit circle in the complex plane. Thus, the stability of the solutions of these equations depend exclusively on the location of the eigenvalues. The initial conditions are irrelevant in this context.

Theorem 7.1. *The solution of a linear, homogeneous, constant coefficient difference equation is stable if the eigenvalues are all inside the unit circle. Or equivalently*

$$\textit{Solution is stable if } |\lambda| < 1\,,\ \forall\,\lambda\,.$$

A solution is unstable if there is a single eigenvalue which is outside the unit circle. Or equivalently

$$\textit{Solution is unstable if } \exists\,\lambda \ni |\lambda| > 1\,.$$

Most readers are probably familiar with the equivalent result for ODEs which states that the system is stable if the eigenvalues are all in the negative half-plane and unstable if even a single eigenvalue is in the positive half-plane. It is important to remember the difference between the stability results for discrete and continuous systems and not confuse them.

One often knows from physical insight that the solution to a difference equation must be real but if any of the eigenvalues are complex, then so is the solution unless the arbitrary constant is complex and such that the imaginary parts cancel. Rather than fight with the complex algebra of determining the arbitrary constant in each case, one can extract the real part of the solution in Eq. (7.9) once and for all.

Assuming that the coefficients of the difference equation are real, then the complex eigenvalues must appear as complex conjugate pairs. Each such pair, $\lambda = \alpha \pm i\beta$, give rise to two basis solutions of the solution subspace

$$x_\tau = C_1 \tau^m (\alpha + i\beta)^\tau + C_2 \tau^m (\alpha - i\beta)^\tau .$$

The two powers are best calculated if the complex numbers are expressed in polar notation, i.e.

$$\begin{aligned} x_\tau &= C_1 \tau^m (\alpha^2 + \beta^2)^{\tau/2} e^{i\tau\theta} + C_2 \tau^m (\alpha^2 + \beta^2)^{\tau/2} e^{-i\tau\theta} \\ &= \tau^m (\alpha^2 + \beta^2)^{\tau/2} (C_1 e^{i\tau\theta} + C_2 e^{-i\tau\theta}) \end{aligned}$$

where $\theta = \arctan(\beta/\alpha)$. Expanding the exponential function using Euler's formula,[3] this becomes

$$\begin{aligned} x_\tau &= \tau^m (\alpha^2 + \beta^2)^{\tau/2} (C_1 (\cos(\tau\theta) + i \sin(\tau\theta)) + C_2 (\cos(\tau\theta) - i \sin(\tau\theta))) \\ &= \tau^m (\alpha^2 + \beta^2)^{\tau/2} ((C_1 + C_2) \cos(\tau\theta) + i(C_1 - C_2) \sin(\tau\theta)) \\ &= \tau^m (\alpha^2 + \beta^2)^{\tau/2} (C_3 \cos(\tau\theta) + C_4 \sin(\tau\theta)) . \end{aligned}$$

This expression is real if C_3 and C_4 are real and this, in turn, is accomplished if C_1 and C_2 are complex conjugates of one another. We conclude that the purely real basis solution for a complex conjugate pair of eigenvalues, $\lambda = \alpha \pm i\beta$, are $\tau^m (\alpha^2 + \beta^2)^{\tau/2} \cos(\tau\theta)$ and $\tau^m (\alpha^2 + \beta^2)^{\tau/2} \sin(\tau\theta)$. These basis solutions can be used in liu of λ^τ when the solution is known to be real.

Example 7.6: Homogeneous difference equation.

Solve the initial value problem

$$x_{\tau+4} - 2x_{\tau+3} + 9x_{\tau+2} + 2x_{\tau+1} - 10x_\tau = 0$$

subject to the initial condition $x_0 = 1$, $x_1 = 2$, $x_2 = 3$ and $x_3 = 4$. The characteristic polynomial is

$$\lambda^4 - 2\lambda^3 + 9\lambda^2 + 2\lambda - 10 = (\lambda - 1)(\lambda + 1)(\lambda - 1 + 3i)(\lambda - 1 - 3i) = 0$$

[3]Euler's formula: $e^{i\theta} = \cos(\theta) + i \sin(\theta)$

so the solution takes the form

$$x_\tau = C_1 1^\tau + C_2(-1)^\tau + C_3(1-3i)^\tau + C_4(1+3i)^\tau.$$

Substituting the initial conditions into this equation gives four equations for the four arbitrary constants.

$$\begin{aligned}
1 &= C_1 + C_2 + C_3 + C_4 \\
2 &= C_1 - C_2 + C_3(1-3i) + C_4(1+3i) \\
3 &= C_1 + C_2 - C_3(8+6i) + C_4(-8+6i) \\
4 &= C_1 - C_2 + C_3(-26+18i) - C_4(26+18i)
\end{aligned}$$

These are solved to give

$$\begin{aligned}
C_1 &= \frac{203}{146} + \frac{82}{219}i \\
C_2 &= -\frac{225}{1898} - \frac{738}{949}i \\
C_3 &= \frac{9}{73} + \frac{24}{73}i \\
C_4 &= -\frac{375}{949} + \frac{212}{2847}i
\end{aligned}$$

so the solution becomes

$$y_\tau = \frac{203}{146} + \frac{82}{219}i + \left(-\frac{225}{1898} - \frac{738}{949}i\right)(-1)^\tau +$$
$$\left(\frac{9}{73} + \frac{24}{73}\right)(1-3i)^\tau + \left(-\frac{375}{949} + \frac{212}{2847}i\right)(1+3i)^\tau.$$

7.1.2 *Particular solutions*

Particular solutions can always be found by variation of parameters, a method that will be covered later, but for simple right hand sides, *the method of matched coefficients* or *the method of undetermined coefficients* is easier to work with. In this method, a guessed solution, containing a number of free coefficients, is substituted into the difference equation and a set of equations for the coefficient values are obtained as will be illustrated in the following examples. The method only works for linear equations with constant coefficients and only for certain inhomogeneous terms, q_T. The q_τs for which the method works are the most commonly encountered cases, however. The form of the particular solution for several right hand sides is given in Table 7.1.

Example 7.7: Non-homogeneous difference equation.

Solve

$$x_{T+2} + x_{T+1} - 2x_\tau = 3^\tau \sin(\tau).$$

Table 7.1 Particular solutions corresponding to different inhomogeneous terms. The parameters to be matched are given by capital Latin letters.

q_τ	$x_{p,\tau}$
a^τ	Aa^τ
$\sin(b\tau)$ or $\cos(b\tau)$	$A\sin(b\tau) + B\cos(b\tau)$
$a^\tau \sin(b\tau)$ or $a^\tau \cos(b\tau)$	$a^\tau(A\sin(b\tau) + B\cos(b\tau))$
τ^n	$A_n\tau^n + A_{n-1}\tau^{n-1} + \cdots + A_1\tau + A_0$
$\tau^n a^\tau$	$a^\tau(A_n\tau^n + A_{n-1}\tau^{n-1} + \cdots + A_1\tau + A_0)$

The characteristic polynomial is

$$\lambda^2 + \lambda - 2 = 0 \Rightarrow \lambda_1 = 1 \ , \ \lambda_2 = -2$$

so the homogeneous solution is

$$x_{h,\tau} = C_1 1^\tau + C_2(-2)^\tau = C_1 + C_2(-2)^\tau.$$

From the table, the particular solution has the form

$$x_{p,\tau} = 3^\tau(A\sin(\tau) + B\cos(\tau))$$

which, when substituted into the difference equation, gives

$$3^{\tau+2}(A\sin(\tau+2) + B\cos(\tau+2)) + 3^{\tau+1}(A\sin(\tau+1) + B\cos(\tau+1))$$
$$-2 \cdot 3^\tau(A\sin(\tau) + B\cos(\tau)) = 3^\tau \sin(\tau).$$

Dividing through by 3^τ and using well known trigonometric identities, this can be written as

$$9A\sin(\tau)\cos(2) + 9A\cos(\tau)\sin(2) + 9B\cos(\tau)\cos(2) - 9B\sin(\tau)\sin(2)$$

$$+3A\sin(\tau)\cos(1) + 3A\cos(\tau)\sin(1) + 3B\cos(\tau)\cos(1) - 3B\sin(k)\sin(1)$$

$$-2A\sin(\tau) - 2B\cos(\tau)) = \sin(\tau).$$

Matching coefficients of the trigonometric terms on both sides of this equation gives

$$9A\cos(2) - 9B\sin(2) + 3A\cos(1) - 3B\sin(1) - 2A = 1$$

$$9A\sin(2) + 9B\cos(2) + 3A\sin(1) + 3B\cos(1) - 2B = 0$$

which are solved for A and B.

$$A = \frac{9\cos(2) + 3\cos(1) - 2}{\gamma}$$

$$B = -\frac{9\sin(2) + 3\sin(1)}{\gamma}$$

where

$$\gamma = 81\cos^2(2) + 54\cos(2)\cos(1) - 36\cos(2) + 9\cos^2(1)$$
$$-12\cos(1) + 4 + 54\sin(1)\sin(2) + 9\sin^2(1) + 81\sin^2(2).$$

Unfortunately, Table 7.1 cannot be expanded to include more inhomogeneous terms because the method of matched coefficients only works for these very special terms. Understanding why this is so makes the entire table so logical that you will not need to memorize the table at all. Note first that Eq. (7.8) can be written as

$$Lx_\tau = q_\tau$$

where the operator L is a linear combination of shift operators of various orders and the higher order shift operators are simply repeated applications of the first order shift operator. When the shift operator is repeatedly applied to the inhomogeneous term q_τ, a sequence of functions, q_τ, $q_{\tau+1}\cdots q_{\tau+n}$ is generated. All the functions listed in Table 7.1 have the property that, at some point, all the next elements in the sequence are linear combinations of the previous elements. Thus, there exists a smallest number N such that

$$q_{\tau+N} = \sum_{n=0}^{N-1} C_n q_{\tau+n}$$

and the sequence of functions q_τ through $q_{\tau+N-1}$ therefore span a space that contain all $q_{\tau+n}$. Call this space V_R. It is obviously a subspace of the entire space of functions of τ. For instance, if $q_\tau = \tau^3$ then the following sequence is generated

$$q_\tau = \tau^3$$

$$q_{\tau+1} = (\tau+1)^3 = \tau^3 + 3\tau^2 + 3\tau + 1$$

$$q_{\tau+2} = (\tau+2)^3 = \tau^3 + 6\tau^2 + 12\tau + 8$$

$$\vdots$$

$$q_{\tau+n} = \text{3'rd order polynomial in } \tau.$$

Evidently, in this case, the subspace V_R is the space of 3rd order polynomial in τ.

We now investigate if a particular solution is a vector in V_R. An arbitrary vector in V_R can be written as a linear combination of the vectors in the spanning set, so

$$v_\tau = \sum_{n=0}^{N-1} C_n q_{\tau+n} \,.$$

Substitute this into the difference equation

$$Lv_\tau = \sum_{n=0}^{N-1} C_n L q_{\tau+n} = q_\tau$$

which is trivially true as long as there is an n for which $Lq_{\tau+n} = q_\tau$. Assuming for the moment that this is the case, we conclude that a particular solution can be found as an element of V_R, i.e. as a linear combination of whatever basis vector one picks for this space. But this is precisely what the entries for the particular solutions in Table 7.1 represent!

We must still consider the possibility that there is no n for which $Lq_{\tau+n} = q_\tau$. When this occurs, the operator L of the difference equation maps the elements of V_R into a true subset of V_R and if this subset does not contain the inhomogeneous term q_τ there is no element in V_R that maps to q_τ and V_R therefore does not contain a particular solution. This will happen if q_τ is a basis solution to the homogeneous problem or if terms in q_τ are basis solutions. A careful discussion of this problem is a bit involved and will be skipped here. The solution of this type of problem (by clever extension of V_R with additional elements) will instead be illustrated in the following example.

Example 7.8: Another non-homogeneous difference equation.

Solve

$$x_{\tau+2} - 4x_{\tau+1} + 4x_\tau = 2^\tau \,.$$

The homogeneous solution is found to be

$$x_{h,\tau} = (C_1 + C_2\tau)2^\tau .$$

Judging from the table, the particular solution has the form $A2^\tau$, but this is not linearly independent of the homogeneous solution. Taking a clue from the solutions that are generated when a root of the characteristic equation has a multiplicity greater than 1, we might try a particular solution of the form $A\tau 2^\tau$. However, this is also a homogeneous solution, so we can try $A\tau^2 2^\tau$. Substituting this into the difference equation results in

$$A(\tau+2)^2 2^{\tau+2} - 4A(\tau+1)^2 2^{\tau+1} + 4\tau^2 2^\tau = 2^\tau \;\Rightarrow\; A = \frac{1}{8}$$

so the complete solution is

$$x_\tau = (C_1 + C_2\tau)2^\tau + \frac{1}{8}\tau^2 2^\tau .$$

7.2 Single, first order equations

We will now drop the assumption of constant coefficients and consider a single linear, first order equation with a variable coefficient

$$x_{\tau+1} + p_{0,\tau} x_\tau = q_z \ , \ x_0 \text{ given} .$$

For this equation, it also holds that the complete solution is the sum of the homogeneous and a particular solution. The homogeneous solution is obvious

$$x_{h,\tau} = x_0 \prod_{n=0}^{\tau-1} (-p_{0,n}).$$

To find a particular solution, we will exploit the similarity, touted at the opening of this chapter, between linear difference and differential equations. The first order, linear differential equation is

$$\frac{dx}{dt} + p_0(t)x(t) = q(t)$$

and it has the solution

$$x(t) = C_1 e^{-\int p_0(t)\,dt} + e^{-\int p_0(t)\,dt} \int q(t) e^{\int p_0(t)\,dt}\,dt .$$

Note that the general solution of the differential equation is the sum of the homogeneous solution and a particular solution. Furthermore, the particular solution equals the homogeneous solution multiplied by a factor dependent on the inhomogeneous term, $q(t)$. To find a particular solution to the difference equation, we will therefore assume that it has the same solution structure as the differential equation. In other words, that the particular solution can be written as the product of a homogeneous solution and something else. This assumption, that a particular solution is the product of a homogeneous solution and something else, turns out to be a powerful idea and it will be used to find particular solutions for higher order equations. The method based on this assumption is called *variation of parameters.* Thus, a particular solution to the difference equation will be assumed to have the form

$$x_{p,\tau} = f_\tau \prod_{n=0}^{\tau-1} (-p_{0,n}).$$

Substituting this expression into the difference equation gives

$$f_{\tau+1} \prod_{n=0}^{\tau} (-p_{0,n}) + p_{0,\tau} f_\tau \prod_{n=0}^{\tau-1} (-p_{0,n}) = q_\tau \ \Rightarrow$$

$$f_{\tau+1} \prod_{n=0}^{\tau} (-p_{0,n}) - f_\tau \prod_{n=0}^{\tau} (-p_{0,n}) = q_\tau \ \Rightarrow$$

$$f_\tau = f_0 + \sum_{m=0}^{\tau-1} \frac{q_m}{\prod_{n=0}^{m} = (-p_{0,n})}\,.$$

Any choice of initial condition for f_τ will provide an acceptable solution so one would usually pick the simplest conditions, $f_0 = 0$. With this choice, the complete solution becomes

$$x_\tau = \prod_{n=0}^{\tau-1}(-p_{0,n}) \left(\sum_{m=0}^{\tau-1} x_0 + \frac{q_m}{\prod_{n=0}^{m}(-p_{0,n})} \right).$$

7.3 Single, higher order equations

These are equations of the form

$$p_{N,\tau}x_{\tau+N} + p_{N-1,\tau}x_{\tau+N-1} + \cdots + p_{1,\tau}x_{\tau+1} + p_{0,\tau}x_\tau = q_\tau \tag{7.10}$$

where we will assume that the leading coefficient is different from zero for all values of τ. Again, the complete solution is the sum of a particular solution, $x_{p,\tau}$ and a linear combination of N linearly independent basis solutions, $x_{h,\tau}$, to the homogeneous problem. Alas, no method exists for finding the basis solutions to the general problem of order two or higher.

7.3.1 *Solution by variable transformation*

A few special cases are known for which a variable transformation brings the difference equation into an equation with constant coefficients which can then be solved by the methods already covered.

7.3.1.1 *Euler's equation*

Euler's difference equation is an equation of the form

$$(\tau + N)!x_{\tau+N} + a_{N-1}(\tau + N - 1)!x_{\tau+N-1} + \cdots + a_0\tau!x_\tau = 0\,.$$

To solve this, let $x_\tau = y_\tau/\tau!$, which gives

$$y_{\tau+N} + a_{N-1}y_{\tau+N-1} + \cdots + a_0 y_\tau = 0$$

a linear equation with constant coefficients that can be solved with the method described above.

7.3.2 *Reduction of order*

It is sometimes possible to find a basis solution to a linear difference equation by guessing or by trial and error. This known basis solution can be used to reduce the order of the problem for the remaining solutions, much like a known root of a polynomial can be divided out to obtain a lower order polynomial for the remaining roots. The method used is a straight forward application of the variation of parameters concept.

Let x_τ be a known basis solution of an Nth order, linear difference equation, Eq. (7.10). Construct a solution, linearly independent of x_τ, as $f_\tau x_\tau$ and substitute this expression in to the difference equation.

$$p_N f_{\tau+N} x_{\tau+N} + p_{N-1} f_{\tau+N-1} x_{\tau+N-1} + \cdots + p_1 f_{\tau+1} x_{\tau+1} + p_0 f_\tau x_\tau$$
$$= \sum_{n=0}^{N} p_n f_{\tau+n} x_{\tau+n} = 0$$

where, in order to simplify the nomenclature, the τ-dependence of the p_ns has been left out. This equation is now subjected to some non-obvious manipulations and simplifications which the reader can be forgiven for skipping by going straight to the result in Eq. (7.11)

$$\begin{aligned}
\sum_{n=0}^{N} p_n f_{\tau+n} x_{\tau+n} &= \sum_{n=0}^{N} f_{\tau+n} \left(\sum_{m=0}^{n} p_m x_{\tau+m} - \sum_{m=0}^{n-1} p_m x_{\tau+m} \right) \\
&= \sum_{n=0}^{N} (f_{\tau+n} - f_{\tau+n+1}) \left(\sum_{m=0}^{n} p_m x_{\tau+m} - \sum_{m=0}^{n-1} p_m x_{\tau+m} \right) \\
&\quad + \sum_{n=0}^{N} f_{\tau+n+1} \left(\sum_{m=0}^{n} p_m x_{\tau+m} - \sum_{m=0}^{n-1} p_m x_{\tau+m} \right) \\
&= \sum_{n=0}^{N} (f_{\tau+n} - f_{\tau+n+1}) \sum_{m=0}^{n} p_m x_{\tau+m} - \sum_{n=0}^{N} f_{\tau+n} \sum_{m=0}^{n-1} p_m x_{\tau+m} \\
&\quad + \sum_{n=0}^{N} f_{\tau+n+1} \sum_{m=0}^{n} p_m x_{\tau+m} \,.
\end{aligned}$$

Recognize that the Nth term in the first and last of the sums of the last line equal zero because they contain the factor $\sum_{m=0}^{N} p_x y_{\tau+m}$ which is the difference equation evaluated for the homogeneous solution. Also, the first term in the middle sum must equal zero since it contains the factor $\sum_{m=0}^{-1} p_m y_{\tau+m}$ which equals zero as the upper limit is less than the lower limit. The last two sums can thus be written

$$\begin{aligned}
&- \sum_{n=1}^{N} f_{\tau+n} \sum_{m=0}^{n-1} p_m x_{\tau+m} + \sum_{n=0}^{N-1} f_{\tau+n+1} \sum_{m=0}^{n} p_m x_{\tau+m} \\
&= - \sum_{n=1}^{N} f_{\tau+n} \sum_{m=0}^{n-1} p_m x_{\tau+m} + \sum_{n=1}^{N} f_{\tau+n} \sum_{m=0}^{n-1} p_m x_{\tau+m} \\
&= 0
\end{aligned}$$

and one obtains an $(N-1)$th order difference equation in $f_{\tau+1} - f_\tau$ which, for notational convenience, we will call y_τ.

$$\sum_{n=0}^{N-1} y_{\tau+n} \sum_{m=0}^{n} p_m x_{\tau+m} = 0 \text{ where } y_\tau = f_{\tau+1} - f_\tau \tag{7.11}$$

which is a linear difference equation in y_τ of order $N-1$. Assuming this equation can be solved and a complete set of basis solutions obtained, each of these basis solutions provides a first order difference equation for f_τ. Once these are solved, the basis solutions of the original equation in x_τ can be reconstructed.

Example 7.9: Reduction of order.

Consider

$$\tau(\tau+1)x_{\tau+2} - 2\tau(\tau+2)x_{\tau+1} + (\tau+1)(\tau+2)x_\tau = 0\,.$$

One possible basis solution is $x_{1,\tau} = \tau$ so the problem that needs to be solved to obtain the other basis solution is

$$y_\tau(\tau+1)(\tau+2)\tau + y_{\tau+1}((\tau+1)(\tau+2)\tau - 2\tau(\tau+2)(\tau+1)) = 0 \;\Rightarrow$$

$$y_{\tau+1} - y_\tau = 0\,.$$

The solution is y_τ equal to a constant different from zero (to avoid the trivial solution). For simplicity, pick the constant equal to 1, giving

$$f_{\tau+1} - f_\tau = 1 \;\Rightarrow\; f_\tau = \tau$$

using the initial condition $f_1 = 1$. The second basis solution becomes

$$x_{2,\tau} = \tau x_{1,\tau} = \tau^2.$$

7.3.3 *Particular solution by variation of parameters*

Assume that a complete set of basis solutions to Eq. (7.10) is known and call these solutions $x_{h,1,\tau},\ldots,x_{h,N,\tau}$. Let a particular, $x_{p,\tau}$, solution be

$$x_{p,\tau} = \sum_{n=1}^{N} f_{n,\tau}x_{h,n,\tau} \tag{7.12}$$

where the functions, $f_{n,\tau}$, are to be determined. The only requirement on these functions are that they are not constants (because that would make $x_{p,\tau}$ a basis solution) and that $x_{p,\tau}$ satisfies the difference equation. To determine expressions for the $f_{n,\tau}$s we will impose additional requirements that seem arbitrary, but that result in equations that are easy to work with. Start by demanding that the $N-1$ equations below hold

$$\begin{aligned} x_{p,\tau+1} &= \sum_{n=1}^{N} f_{n,\tau}x_{h,n,\tau+1} \\ &\;\;\vdots \\ x_{p,\tau+N-1} &= \sum_{n=1}^{N} f_{n,\tau}x_{h,n,\tau+N-1}\,. \end{aligned} \tag{7.13}$$

This provides N equations, all of the general form

$$x_{p,\tau+m} = \sum_{n=1}^{N} f_{n,\tau} x_{h,n,\tau+m}$$

where $0 \leq m \leq N-1$. Now, shift the index for τ to $\tau+1$ in the equation for $x_{p,\tau+x}$ to get

$$x_{p,\tau+m+1} = \sum_{n=1}^{N} f_{n,\tau+1} x_{h,n,\tau+m+1}$$

and subtract the equation for $x_{p,\tau+m+1}$. This gives $N-1$ equations of the form

$$0 = \sum_{n=1}^{N} (f_{n,\tau+1} - f_{n,\tau}) x_{h,n,\tau+m+1} \,. \tag{7.14}$$

One more equation is needed to determine the $f_{n,\tau}$'s. let this be

$$x_{p,\tau+N} = \sum_{n=1}^{N} f_{n,\tau} x_{h,n,\tau+N} + \sum_{n=1}^{N} (f_{n,\tau+1} - f_{n,\tau}) x_{h,n,\tau+N} \,. \tag{7.15}$$

Substitute the expressions for the particular solution, Eqs. (7.12), (7.13) and (7.15), into the difference equation

$$p_{N,\tau} \left(\sum_{n=1}^{N} f_{n,\tau} x_{h,n,\tau+N} + \sum_{n=1}^{N} (f_{n,\tau+1} - f_{n,\tau}) x_{h,n,\tau+N} \right) + \sum_{m=0}^{N-1} p_{m,\tau} \sum_{n=1}^{N} f_{n,\tau} x_{h,n,\tau+m}$$
$$= q_\tau \,.$$

Rearranging gives

$$p_{N,\tau} \sum_{n=1}^{N} (f_{n,\tau+1} - f_{n,\tau}) x_{h,n,\tau+N} + \sum_{n=1}^{N} f_{n,\tau} \sum_{m=0}^{N} p_{m,\tau} x_{h,n,\tau+m} = q_\tau \,.$$

The last, inner sum is simply the difference equation evaluated at the basis solutions and is therefore zero. Thus

$$\sum_{n=1}^{N} (f_{n,\tau+1} - f_{n,\tau}) x_{h,n,\tau+N} = \frac{q_\tau}{p_{N,\tau}} \,. \tag{7.16}$$

Eqs. (7.14) and (7.16) are the N equations from which the $f_{n,\tau}$ functions can be determined. In matrix-vector notation, these equations are

$$\begin{pmatrix} x_{h,1,\tau+1} & \cdots & x_{h,N,\tau+1} \\ \vdots & \ddots & \vdots \\ x_{h,1,\tau+n} & \cdots & x_{h,N,\tau+n} \\ \vdots & \ddots & \vdots \\ x_{h,1,\tau+N} & \cdots & x_{h,N,\tau+N} \end{pmatrix} \begin{pmatrix} f_{1,\tau+1} - f_{1,\tau} \\ \vdots \\ f_{n,\tau+1} - f_{n,\tau} \\ \vdots \\ f_{N,\tau+1} - f_{N,\tau} \end{pmatrix} = \begin{pmatrix} 0 \\ \vdots \\ 0 \\ \vdots \\ q_\tau / p_{N,\tau} \end{pmatrix} .$$

These equations can now be solved for $f_{n,\tau+1} - f_{n,\tau}$ and these solutions in turn solved for $f_{n,\tau}$ and the particular solution can be reconstructed.

Example 7.10: Particular solution by variation of parameters.

Consider

$$\tau(\tau+1)x_{\tau+2} - 2\tau(\tau+2)x_{\tau+1} + (\tau+1)(\tau+2)x_\tau = \tau(\tau+1)(\tau+2).$$

A complete set of basis solutions is $x_{h,1,\tau} = \tau$ and $x_{h,2,\tau} = \tau^2$. Therefore, to find a particular solution, we must solve

$$\begin{pmatrix} \tau+1 & (\tau+1)^2 \\ \tau+2 & (\tau+2)^2 \end{pmatrix} \begin{pmatrix} f_{1,\tau+1} - f_{1,\tau} \\ f_{2,\tau+1} - f_{2,\tau} \end{pmatrix} = \begin{pmatrix} 0 \\ (\tau+2) \end{pmatrix} \Rightarrow$$

$$f_{1,\tau+1} - f_{1,\tau} = -\tau - 1 \Rightarrow f_{1,\tau} = -\frac{\tau^2+\tau}{2}$$

$$f_{2,\tau+1} - f_{2,\tau} = 1 \Rightarrow f_{2,\tau} = \tau$$

where we have used the initial conditions $f_{1,0} = 0$ and $f_{2,0} = 0$. Reassembling everything and simplifying gives the complete solution

$$x_\tau = C_1\tau + C_2\tau^2 + \frac{\tau^3 - \tau^2}{2}.$$

7.4 Systems of linear difference equations

A difference equation like Eq. (7.10)

$$p_{N,\tau}x_{\tau+N} + p_{N-1,\tau}x_{\tau+N-1} + \cdots + p_{0,\tau}x_\tau = q_\tau$$

where the leading coefficient $p_{N,\tau}$ is different from zero for all values of τ, can be solved for $x_{\tau+N}$

$$y_{\tau+N} = \frac{q_\tau - p_{N-1,\tau}y_{\tau+N-1} - \cdots - p_{0,\tau}y_\tau}{p_{N,\tau}}$$

so Eq. (7.10) can be restated as

$$\begin{pmatrix} x_{\tau+1} \\ x_{\tau+2} \\ \vdots \\ x_{\tau+N-1} \\ x_{\tau+N} \end{pmatrix} = \begin{pmatrix} 0 & 1 & 0 & \cdots & 0 \\ 0 & 0 & 1 & \cdots & 0 \\ \vdots & \vdots & \vdots & \ddots & \vdots \\ 0 & 0 & 0 & \cdots & 1 \\ -\frac{p_{0,\tau}}{p_{N,\tau}} & -\frac{p_{1,\tau}}{p_{N,\tau}} & -\frac{p_{2,\tau}}{p_{N,\tau}} & \cdots & -\frac{p_{N-1,\tau}}{p_{N,\tau}} \end{pmatrix} \begin{pmatrix} x_\tau \\ x_{\tau+1} \\ \vdots \\ x_{\tau+N-2} \\ x_{\tau+N-1} \end{pmatrix} + \begin{pmatrix} 0 \\ 0 \\ \vdots \\ 0 \\ \frac{q_\tau}{p_{N,\tau}} \end{pmatrix}. \tag{7.17}$$

Formally, the solution to this problem is a sequence of vectors

$$\left\{ \begin{pmatrix} x_0 \\ x_1 \\ \vdots \\ x_{N-2} \\ x_{N-1} \end{pmatrix}, \ldots, \begin{pmatrix} x_\tau \\ x_{\tau+1} \\ \vdots \\ x_{\tau+N-2} \\ x_{\tau+N-1} \end{pmatrix}, \begin{pmatrix} x_{\tau+1} \\ x_{\tau+2} \\ \vdots \\ x_{\tau+N-1} \\ x_{\tau+N} \end{pmatrix}, \ldots \right\}$$

but, as the nomenclature indicates, these vectors are related in such a way that any vector in the sequence is obtained from the previous vector by discarding the first element, moving the remaining elements up one position and calculating a new last element from Eq. (7.10). In other words, the information represented by this sequence of vectors can just as well be represented by the sequence of scalars $\{x_\tau\}$. However, the formulation in Eq. (7.17) suggests the more general problem.

$$\begin{pmatrix} x_{1,\tau+1} \\ x_{2,\tau+1} \\ \vdots \\ x_{N,\tau+1} \end{pmatrix} = \begin{pmatrix} a_{1,1,\tau} & \cdots & a_{1,N,\tau} \\ a_{2,1,\tau} & \cdots & a_{2,N,\tau} \\ \vdots & \ddots & \vdots \\ a_{N,1,\tau} & \cdots & a_{N,N,\tau} \end{pmatrix} \begin{pmatrix} x_{1,\tau} \\ x_{2,\tau} \\ \vdots \\ x_{N,\tau} \end{pmatrix} + \begin{pmatrix} b_{1,\tau} \\ b_{2,\tau} \\ \vdots \\ b_{N,\tau} \end{pmatrix}$$

or

$$X_{\tau+1} = A_\tau X_\tau + B_\tau \tag{7.18}$$

where A_τ can be any matrix and B_τ any vector. This equation is homogeneous iff $B_\tau = 0$. For this problem, the solution is again a sequence of vectors and an initial condition is a specification of a vector for an initial value of the counter τ. However, these solution vectors are not related as simply as the solution vectors of Eq. (7.17).

7.4.1 *Basic theorems*

The existence and uniqueness theorem for systems of linear difference equations is trivial.

Theorem 7.2. *Let X_τ be*

$$X_\tau = \begin{pmatrix} x_\tau \\ x_{\tau+1} \\ \vdots \\ x_{\tau+N-1} \end{pmatrix}$$

and let A_τ be any $N \times N$ matrix, the elements of which may be functions of τ, and B_τ any N-dimensional vector, the elements of which may be functions of τ, then there exist one and only one solution to

$$X_{\tau+1} = A_\tau X_\tau + B_\tau$$

which satisfy the initial condition

$$X_0 = (x_0, x_1, \ldots, x_{N-1}) = (C_0, C_1, \ldots, C_{N-1})$$

Proof: By substituting the initial condition for X_τ in the matrix equation, a unique value for X_1 is obtained. The solution process can now be continued in an iterative fashion to obtain a unique solution for $\{X_\tau\}_{\tau=0}^{\tau=\infty}$. □

Notice that the simple argument in the proof above can be used whenever it is possible to solve for the highest order term, $x_{\tau+N}$, in Eq. (7.10). Thus, it can be easy to infer existence and uniqueness of the solution of even nonlinear difference equations.

We will want to find the complete or general solution to a difference equation in the form given in Eq. (7.18). By a complete or general solution, we mean an expression which all possible solutions, for all possible initial conditions, must satisfy. We will start by showing that this problem can be reduced to finding the complete solution to a homogeneous difference equation.

Claim: *The complete solution to*

$$X_{\tau+1} = A_\tau X_\tau + B_\tau$$

equals a particular solution to the problem plus the complete solution to the homogeneous problem.

Proof: Let a particular solution be denoted $X_{p,\tau}$ and let $X_\tau = X_{h,\tau} + X_{p,\tau}$. Substitute this expression for X_τ into the difference equation to get

$$X_{h,\tau} + X_{p,\tau} = A_\tau X_{h,\tau} + A_\tau X_{p,\tau} + B_\tau\,.$$

Since $X_{p,\tau}$ can be any solution, $X_{p,\tau} = A_\tau X_{p,\tau} + B_\tau$, and the expression above simplifies to

$$X_{h,\tau} = A_\tau X_{h,\tau}$$

and $X_{h,\tau}$ is the complete solution to the homogeneous problem. □

Although there is no general method for finding a homogeneous solution to a set coupled, linear difference equation, it is possible to say a great deal about the solution structure. Notice first that linearity and homogeneity of a difference equation implies that a linear combination of any two solutions is also a solution. The solutions therefore form a vector space, call it V_s, the solution space, and the complete solution can be written as a linear combination of a set of basis vectors or basis solutions to this space. Although a formula for finding basis solutions for the general case does not exist, it is possible to determine the dimension of the solution space, i.e. the number of basis solutions that are needed in order to write the general solution.

Claim: *The solution space for*

$$X_{\tau+1} = A_\tau X_\tau$$

where A_τ is an $N \times N$ matrix, is an N-dimensional vector space over $\mathbb{R}$ if A and X are real, or over $\mathbb{C}$, if A and X are complex.

Proof: We will first show that there exists a set of N linearly independent solutions. Consider the N solutions, $\{X_{n,\tau}\}$, generated by the initial conditions

$$\begin{aligned} X_{1,\tau} \text{ by } X_{1,0} &= (1, 0, \ldots, 0) \\ X_{2,\tau} \text{ by } X_{2,0} &= (0, 1, \ldots, 0) \\ &\vdots \\ X_{N,\tau} \text{ by } X_{N,0} &= (0, 0, \ldots, 0, N) \end{aligned}$$

in other words, the initial vector for $\{X_{n,\tau}\}$ has zeros everywhere, except in position n where it equals 1. Clearly, the initial vectors are linearly independent and therefore, so are the solutions.

To see that any solution can be written as a linear combination of these solutions, observe that any other solution, X_τ, will satisfy some initial condition, say

$$X_0 = (C_0, C_1, \ldots, C_{N-1})$$

and can thus be written as

$$X_\tau = \sum_{n=0}^{N-1} C_n X_{n,\tau}\,. \qquad \square$$

The results so far are summarized in Theorem 7.3.

Theorem 7.3. *The complete solution to*

$$X_{\tau+1} = A_\tau X_\tau + B_\tau$$

where A is of size $N \times N$, is the sum of a particular solution and a linear combination of N linearly independent solutions, so-called basis solutions, to the homogeneous problem.

A complete set of linearly independent basis solutions is said to form a *fundamental system* of solutions. Evidently, the trivial solution, $X = 0$, is technically speaking a basis solution, but it is of no interest and will not be considered in the following.

If a complete set of basis solutions is known, however, a particular solution can always be found using variation of parameters.

7.4.2 *Particular solution by variation of parameters*

Assume that a complete set of basis solutions of

$$X_{\tau+1} = A_\tau X_\tau + B_\tau$$

is known and arrange these as columns of a matrix Φ_τ. This matrix is known as *the fundamental matrix* of the system, a misnomer in the sense that the matrix is not unique but depends on the choice of basis functions and on how these functions are arranged in the matrix. Hence, "a fundamental matrix" would be a more appropriate nomenclature. Note that the complete homogeneous solution can be written in terms of the fundamental matrix as

$$X_{h,\tau} = \Phi_\tau C$$

where C is a vector of arbitrary constants. Assume now a particular solution of the form

$$X_{p,\tau} = \Phi_\tau F_\tau$$

where F_τ is a vector of functions f_τ to be determined. Substitute the assumed particular solution into the difference equation

$$\Phi_{\tau+1} F_{\tau+1} = A_\tau \Phi_\tau F_\tau + B_\tau\,.$$

Clearly

$$\Phi_{\tau+1} = A_\tau \Phi_\tau \;\Rightarrow\; A_\tau = \Phi_{\tau+1}\Phi_\tau^{-1}$$

so eliminating A_τ gives

$$\Phi_{\tau+1} F_{\tau+1} = \Phi_{\tau+1}\Phi_\tau^{-1}\Phi_\tau F_\tau + B_\tau \;\Rightarrow$$

$$F_{\tau+1} - F_\tau = \Phi_{\tau+1}^{-1} B_\tau\,.$$

This equation is readily solved for F_τ. To see this, let $F_\tau = (f_{1,\tau}, \ldots, f_{N,\tau})$ and write the right hand side as $\Phi_{\tau+1}^{-1} B_\tau = (\hat{b}_{1,\tau}, \ldots, \hat{b}_{N,\tau})$. Then

$$f_{n,\tau+1} - f_{n,\tau} = \hat{b}_{n,\tau}$$

and therefore

$$f_{n,\tau} = f_{n,0} + \sum_{m=0}^{\tau-1} \hat{b}_{n,m}\,.$$

Common sense tells us to use the most convenient initial condition, $f_{n,0} = 0$. With this choice, the complete solution to the difference equation can be written

$$X_\tau = \Phi_\tau C + \Phi_\tau \begin{pmatrix} \sum_{m=0}^{\tau-1} \hat{b}_{1,m} \\ \vdots \\ \sum_{m=0}^{\tau-1} \hat{b}_{N,m} \end{pmatrix}.$$

Example 7.11: Particular solution by variation of parameters.

The basis solutions for the system

$$X_{\tau+1} = \begin{pmatrix} 0 & 1 \\ -\frac{\tau+2}{\tau} & 2\frac{\tau+2}{\tau+1} \end{pmatrix} X_\tau + \begin{pmatrix} \tau+1 \\ \tau+2 \end{pmatrix}$$

can be written

$$X_{1,\tau} = \begin{pmatrix} \tau \\ \tau+1 \end{pmatrix} , \; X_{2,\tau} = \begin{pmatrix} \tau^2 \\ (\tau+1)^2 \end{pmatrix}.$$

Notice that $\tau = 0$ is a singularity of A_τ and initial conditions must therefore be given at $\tau = 1$ or some larger value of τ. The fundamental matrix is

$$\Phi_\tau = \begin{pmatrix} \tau & \tau^2 \\ \tau+1 & (\tau+1)^2 \end{pmatrix} \Rightarrow \Phi_{\tau+1}^{-1} = \begin{pmatrix} \frac{\tau+2}{\tau+1} & -\frac{\tau+1}{\tau+2} \\ -\frac{1}{\tau+1} & \frac{1}{\tau+2} \end{pmatrix}.$$

The equations for the functions $f_{1,\tau}$ and $f_{2,\tau}$ become

$$f_{1,\tau+1} = f_{1,\tau} + 1 \, , \; f_{2,\tau+1} = f_{2,\tau}$$

and with the initial conditions $f_{1,1} = f_{2,1} = 0$ the solutions are

$$f_{1,\tau} = \tau - 1 \, , \; f_{2,\tau} = 0 \, .$$

Putting everything back together, the complete solution for the difference equation is

$$X_\tau = C_1 \begin{pmatrix} \tau \\ \tau+1 \end{pmatrix} + C_2 \begin{pmatrix} \tau^2 \\ (\tau+1)^2 \end{pmatrix} + \begin{pmatrix} \tau(\tau-1) \\ (\tau+1)(\tau-1) \end{pmatrix}.$$

7.4.3 *Equations with constant coefficients*

The most important type of coupled difference equations are linear difference equation with constant coefficients, equations of the form

$$X_{\tau+1} = AX_\tau + B_\tau$$

where A is a matrix of constants. For these equations, analytical expressions can be found for both the homogeneous and particular solutions.

7.4.3.1 *Homogeneous solutions*

To find the basis solutions, determine a Jordan Canonical matrix that is similar to the matrix A, i.e. let $A = KJK^{-1}$ and substitute this into the homogeneous difference equation

$$X_{\tau+1} = KJK^{-1}X_\tau \;\Rightarrow\; K^{-1}X_{\tau+1} = JK^{-1}X_\tau$$

or, by setting $Y_\tau = K^{-1}X_\tau$,

$$Y_{\tau+1} = JY_\tau\,.$$

For a semi simple matrix A, the Jordan matrix J is a diagonal matrix and the problem now consists of N independent, linear first order equations. The solution is trivial.

$$Y_\tau = \begin{pmatrix} C_1\lambda_1^\tau \\ C_2\lambda_2^\tau \\ \vdots \\ C_N\lambda_N^\tau \end{pmatrix} = C_1\lambda_1^\tau e_1 + C_2\lambda_2^\tau e_2 + \cdots + C_N\lambda_N^\tau e_N$$

where λ_n is the eigenvalue in the nth diagonal element of J and $\{e_n\}$ is the canonical set of basis vectors. The solution in the original variable becomes

$$X_\tau = K\begin{pmatrix} C_1\lambda_1^\tau \\ C_2\lambda_2^\tau \\ \vdots \\ C_N\lambda_N^\tau \end{pmatrix}$$

or

$$X_\tau = \sum_{n=1}^{N} C_n\tilde{v}_n\lambda_n^\tau$$

where $\tilde{v}_n$ are the eigenvectors of A.

When the matrix A is degenerate, each degenerate eigenvalue gives rise to a Jordan block. The equations of this Jordan block are coupled with each other but decoupled from all the remaining equations. Each Jordan block can therefore be solved separately from the other Jordan blocks, giving problems of the form

$$Z_{\tau+1} = \begin{pmatrix} \lambda & 1 & 0 & \cdots & 0 \\ 0 & \lambda & 1 & \cdots & 0 \\ \vdots & \vdots & \vdots & \ddots & \vdots \\ 0 & 0 & 0 & \cdots & \lambda \end{pmatrix} Z_\tau\,. \tag{7.19}$$

The equation for the last element of Z_τ is solved to give $Z_{\tau,N} = C_N\lambda^\tau$. The equation for the second to last element now becomes

$$Z_{\tau+1,N-1} = \lambda Z_{\tau,N-1} + C_N\lambda^\tau$$

which is a linear, first order equation. The solution is $Z_{\tau,N-1} = C_{N-1}\lambda^\tau + C_N\tau\lambda^{\tau-1}$, where the first term in the expression is the homogeneous solution and the second term is a particular solution. The third to last element satisfy

$$Z_{\tau+1,N-2} = \lambda Z_{\tau,N-2} + C_{N-1}\lambda^\tau + C_N\tau\lambda^{\tau-1}$$

which again is a linear, first order equation with the solution

$$Z_{\tau,N-2} = C_{N-2}\lambda^\tau + C_{N-1}\tau\lambda^{\tau-1} + C_N\tau(\tau-1)\lambda^{\tau-2}/2\,.$$

It is obvious how one now proceeds to find the solutions for all the $Z_{\tau,n}$. Once the solution for $Z_{\tau,n+1}$ is obtained the difference equation for $Z_{\tau,n}$ can be written and solved.

By analogy with single higher order equations, one might expect that a repeated eigenvalue will give rise to homogeneous basis solutions of the form λ^τ, $\tau\lambda^\tau$, $\tau^2\lambda^\tau$, etc. But note that this is not necessarily the case. A semi-simple eigenvalue only gives basis solutions of the form λ^τ and these solutions are linearly independent because they are associated with different and linearly independent eigenvectors. On the other hand, a degenerate eigenvalue does give solutions with terms of the form λ^τ, $\tau\lambda^\tau$, $\tau^2\lambda^\tau$, etc. Though the solution above may appear slightly different, it can easily be rewritten to fit this form.

Example 7.12: Coupled difference equations.

Consider

$$X_{\tau+1} = \begin{pmatrix} \frac{17}{8} & \frac{3}{8} & -\frac{5}{8} \\ \frac{1}{2} & \frac{7}{2} & -\frac{5}{2} \\ \frac{1}{8} & \frac{3}{8} & \frac{11}{8} \end{pmatrix} X_\tau\,.$$

A spectral analysis yields the following result

$$\begin{pmatrix} \frac{17}{8} & \frac{3}{8} & -\frac{5}{8} \\ \frac{1}{2} & \frac{7}{2} & -\frac{5}{2} \\ \frac{1}{8} & \frac{3}{8} & \frac{11}{8} \end{pmatrix} = \begin{pmatrix} 1 & 2 & 1 \\ 3 & 1 & 4 \\ 2 & 1 & 1 \end{pmatrix} \begin{pmatrix} 2 & 0 & 0 \\ 0 & 2 & 0 \\ 0 & 0 & 3 \end{pmatrix} \begin{pmatrix} -\frac{3}{8} & -\frac{1}{8} & \frac{7}{8} \\ \frac{5}{8} & -\frac{1}{8} & -\frac{1}{8} \\ \frac{1}{8} & \frac{3}{8} & -\frac{5}{8} \end{pmatrix}.$$

Therefore, the problem for $Y_\tau = K^{-1}X_\tau$ becomes

$$Y_{\tau+1} = \begin{pmatrix} 2 & 0 & 0 \\ 0 & 2 & 0 \\ 0 & 0 & 3 \end{pmatrix} Y_\tau$$

which is solved to give

$$Y_\tau = \begin{pmatrix} C_1 2^\tau \\ C_2 2^\tau \\ C_3 3^\tau \end{pmatrix}$$

from which

$$X_\tau = \begin{pmatrix} C_1 2^\tau + 2C_2 2^\tau + C_3 3^\tau \\ 3C_1 2^\tau + C_2 2^\tau + 4C_3 3^\tau \\ 2C_1 2^\tau + C_2 2^\tau + C_3 3^\tau \end{pmatrix}.$$

Example 7.13: Coupled difference equations.

Solve the initial value problem

$$x_{\tau+2} - 2x_{\tau+1} + y_{\tau+1} + y_\tau + x_\tau = 0$$
$$y_{\tau+2} - 2x_{\tau+1} - 2y_{\tau+1} = 0$$

where $x_0 = 1$, $x_1 = 2$, $y_0 = 3$ and $y_1 = 4$. Solving each equation for the highest order term and arranging the variables in a vector, one obtains

$$\begin{pmatrix} x_{\tau+1} \\ x_{\tau+2} \\ y_{\tau+1} \\ y_{\tau+2} \end{pmatrix} = \begin{pmatrix} 0 & 1 & 0 & 0 \\ -1 & 2 & -1 & -1 \\ 0 & 0 & 0 & 1 \\ 0 & 2 & 0 & 2 \end{pmatrix} \begin{pmatrix} x_\tau \\ x_{\tau+1} \\ y_\tau \\ y_{\tau+1} \end{pmatrix}.$$

A spectral analysis of the matrix shows it to be similar to

$$J = \begin{pmatrix} 0 & 1 & 0 & 0 \\ 0 & 0 & 0 & 0 \\ 0 & 0 & 2+i\sqrt{3} & 0 \\ 0 & 0 & 0 & 2-i\sqrt{3} \end{pmatrix}$$

with the transition matrix

$$K = \begin{pmatrix} -1 & -3 & i\frac{\sqrt{3}}{2} & -i\frac{\sqrt{3}}{2} \\ 0 & -1 & -\frac{3}{2} + i\sqrt{3} & -\frac{3}{2} - i\sqrt{3} \\ 1 & 0 & 1 & 1 \\ 0 & 1 & 2 + i\sqrt{3} & 2 - i\sqrt{3} \end{pmatrix}.$$

Let

$$Z_\tau = K^{-1} \begin{pmatrix} x_\tau \\ x_{\tau+1} \\ y_\tau \\ y_{\tau+1} \end{pmatrix}$$

giving

$$Z_{\tau+1} = \begin{pmatrix} 0 & 1 & 0 & 0 \\ 0 & 0 & 0 & 0 \\ 0 & 0 & 2 + i\sqrt{3} & 0 \\ 0 & 0 & 0 & 2 - i\sqrt{3} \end{pmatrix} Z_\tau .$$

The zero eigenvalue is in some respects special because it gives rise to solutions that are equal to zero after τ exceeds some value. In this case, the solutions for z_1 and z_2 are $\{z_{1,\tau=0}, z_{2,\tau=0}, 0, 0, \ldots\}$ and $\{z_{2,\tau=0}, 0, 0, \ldots\}$ so the two elements are identically zero for $\tau > 1$. One can formally write the solutions for z_1 and z_2 as functions of τ and carry these terms in the manipulations that follow. However, this seems a waste of effort for terms that are identically equal to zero for any $\tau > 1$. It is more sensible to restrict ourselves to developing the solution for $\tau > 1$ only. In this case, z_1 and z_2 are identically zero and the solution becomes

$$Z_\tau = \begin{pmatrix} 0 \\ 0 \\ C_1(2 + i\sqrt{3})^\tau \\ C_2(2 - i\sqrt{3})^\tau \end{pmatrix}, \quad \tau > 1 .$$

As always when working with complex numbers raised to powers, the algebraic manipulations are much easier when polar notation is used. Doing so gives

$$Z_\tau = \begin{pmatrix} 0 \\ 0 \\ C_1(\sqrt{7}e^{i\theta_1})^\tau \\ C_2(\sqrt{7}e^{-i\theta_1})^\tau \end{pmatrix} = \begin{pmatrix} 0 \\ 0 \\ C_1\sqrt{7^\tau}e^{i\tau\theta_1} \\ C_2\sqrt{7^\tau}e^{-i\tau\theta_1} \end{pmatrix}$$

where $\theta_1 = \arctan\sqrt{3}/2$. Also

$$K = \begin{pmatrix} -1 & -3 & i\frac{\sqrt{3}}{2} & -i\frac{\sqrt{3}}{2} \\ 0 & -1 & -\frac{3}{2}+i\sqrt{3} & -\frac{3}{2}-i\sqrt{3} \\ 1 & 0 & 1 & 1 \\ 0 & 1 & 2+i\sqrt{3} & 2-i\sqrt{3} \end{pmatrix} = \begin{pmatrix} -1 & -3 & i\frac{\sqrt{3}}{2} & -i\frac{\sqrt{3}}{2} \\ 0 & -1 & \sqrt{21/4}e^{i\theta_2} & \sqrt{21/4}e^{-i\theta_2} \\ 1 & 0 & 1 & 1 \\ 0 & 1 & \sqrt{7}e^{i\theta_1} & \sqrt{7}e^{-i\theta_1} \end{pmatrix}$$

where $\theta_2 = \pi - \arctan(2/\sqrt{3})$. The final result is found from

$$\begin{pmatrix} x_\tau \\ x_{\tau+1} \\ y_\tau \\ y_{\tau+1} \end{pmatrix} = KZ_\tau = \begin{pmatrix} -1 & -3 & \frac{\sqrt{3}}{2}e^{i\pi/2} & \frac{\sqrt{3}}{2}e^{-i\pi/2} \\ 0 & -1 & \sqrt{21/4}e^{i\theta_2} & \sqrt{21/4}e^{-i\theta_2} \\ 1 & 0 & 1 & 1 \\ 0 & 1 & \sqrt{7}e^{i\theta_1} & \sqrt{7}e^{-i\theta_1} \end{pmatrix} \begin{pmatrix} 0 \\ 0 \\ C_1\sqrt{7^\tau}e^{i\tau\theta_1} \\ C_2\sqrt{7^\tau}e^{-i\tau\theta_1} \end{pmatrix}$$

which after simplifications become

$$\begin{pmatrix} x_\tau \\ x_{\tau+1} \\ y_\tau \\ y_{\tau+1} \end{pmatrix} = \begin{pmatrix} \frac{\sqrt{3}}{2}\sqrt{7^\tau}i(C_1e^{i\tau\theta_1} - C_2e^{-i\tau\theta_1}) \\ \frac{\sqrt{3}}{2}\sqrt{7(\tau+1)}e^{i(\theta_2+\tau\theta_1)}C_1 + \frac{\sqrt{3}}{2}\sqrt{7^{\tau+1}}e^{-i(\theta_2+\tau\theta_1)}C_2 \\ C_1\sqrt{7^\tau}e^{i\tau\theta_1} + C_2\sqrt{7^\tau}e^{-i\tau\theta_1} \\ \sqrt{7^{\tau+1}}e^{i\theta_1(1+\tau)}C_1 + \sqrt{7^{\tau+1}}e^{-i\theta_1(1+\tau)}C_2 \end{pmatrix}.$$

To extract the real part of this solution, let $C_1 = C_3 + iC_4$ and $C_2 = C_3 - iC_4$, where C_3 and C_4 are real. After simplifications

$$\begin{pmatrix} x_\tau \\ x_{\tau+1} \\ y_\tau \\ y_{\tau+1} \end{pmatrix} = \begin{pmatrix} -\sqrt{3}\sqrt{7^\tau}C_4\cos(\tau\theta_1) + C_3\sin(\tau\theta_1)) \\ \sqrt{3}\sqrt{7^{\tau+1}}(C_3\cos(\theta_2+\tau\theta_1)) - C_4\sin(\theta_2+\tau\theta_1)) \\ 2\sqrt{7^\tau}(C_3\cos(\tau\theta_1) - C_4\sin(\tau\theta_1)) \\ 2\sqrt{7^{\tau+1}}(C_3\cos(\theta_1(1+\tau)) - C_4\sin(\theta_1(1+\tau)) \end{pmatrix}, \ \tau > 1.$$

When calculating the values of the arbitrary constants, C_3 and C_4, it is important to remember that this result is only valid for $\tau > 1$. Initial conditions are typically given for $\tau = 0$, so one must first work with the initial conditions and the difference equations directly to calculate the solution at a τ-value for which the solution above is valid. One finds that for $\tau = 2$

$$\begin{pmatrix} x_2 \\ x_3 \\ y_2 \\ y_3 \end{pmatrix} = \begin{pmatrix} -4 \\ -26 \\ 12 \\ 16 \end{pmatrix}$$

giving the following set of equations for the two arbitrary constant

$$\begin{pmatrix} -7\sqrt{3}\sin(2\theta_1) & -7\sqrt{3}\cos(2\theta_1) \\ 7\sqrt{3}\sqrt{7}\cos(\theta_2+2\theta_1) & -7\sqrt{3}\sqrt{7}\sin(\theta_2+2\theta_1) \\ 14\cos(2\theta_1) & -14\sin(2\theta_1) \\ 14\sqrt{7}\cos(3\theta_1) & -14\sqrt{7}\sin(3\theta_1) \end{pmatrix} \begin{pmatrix} C_3 \\ C_4 \end{pmatrix} = \begin{pmatrix} -4 \\ -26 \\ 12 \\ 16 \end{pmatrix}.$$

This system is overdetermined, but if the result on which it is based is correct, then any two equations can be used to solve for the arbitrary constants. Solving the first two equations numerically gives

$$C_3 = 0.448979592\ldots,\ C_4 = -0.801220782\ldots$$

As a check on the result, these two values are substituted back into the last two equations

$$\begin{pmatrix} 14\cos(2\theta_1) & -14\sin(2\theta_1) \\ 14\sqrt{7}\cos(3\theta_1) & -14\sqrt{7}\sin(3\theta_1) \end{pmatrix} \begin{pmatrix} C_3 \\ C_4 \end{pmatrix} = \begin{pmatrix} 12 \\ 16 \end{pmatrix}$$

where 10 digits were used in the calculations.

Clearly, the stability result given previously for single homogeneous higher order equations still apply. The solution is stable if all the eigenvalues are inside the unit circle and unstable if any eigenvalue is outside the unit circle. We will use this result to determine the values of the controller gain for which the system previously considered is stable.

Example 7.14: Stability of PID controller.

The model is given by Eqs. (7.6) and (7.7). These are first written in vector-matrix notation as

$$\begin{pmatrix} e_{\mathcal{T}-1} \\ e_{\mathcal{T}} \\ q_{\mathcal{T}} \end{pmatrix} = \begin{pmatrix} 0 & 1 & 0 \\ -\frac{\tau_D K_c \tau_I}{\text{Denom}} & \frac{\tau_I(2\tau_D K_c + K_c\Delta t + 1)}{\text{Denom}} & \frac{\Delta t \tau_I}{\text{Denom}} \\ -\frac{(1/\Delta t + R)\tau_D \tau_I K_c}{\text{Denom}} & \frac{(\tau_D\tau_I/\Delta t + 2\tau_I R \tau_D + \tau_I R\Delta t - \Delta t)K_c}{\text{Denom}} & \frac{(R\Delta t + 1)\tau_I}{\text{Denom}} \end{pmatrix} \begin{pmatrix} e_{\mathcal{T}-2} \\ e_{\mathcal{T}-1} \\ q_{\mathcal{T}-1} \end{pmatrix}$$

where Denom $= K_c\tau_I\Delta t + K_c\Delta t^2 + K_c\tau_D\tau_I + \tau_I + \tau_I R\Delta t$. Notice that the disturbance terms, the $U_{\mathcal{T}}$-terms, are not included as we are interested in determining when the system is stable after a disturbance has passed. The eigenvalues of the system matrix can be found to be

$$\lambda_{1,2} = \frac{\Delta t \tau_I (K_c + R) + 2\tau_I(1 + K_c \tau_D) \pm \Delta t \sqrt{\tau_I^2 (K_c + R)^2 - 4K_c^2 \tau_D \tau_I - 4\tau_I K_c}}{2(\Delta t \tau_I (K_c + R) + \tau_I (1 + K_c \tau_D) + K_c \Delta t^2)}.$$

$$\lambda_3 = 0$$

Only the non-zero eigenvalues are of interest and as they are a complex conjugate pair, they are both inside or both outside the unit circle and we need therefore only consider one of them.

The expression above is messy and difficult to evaluate, though one thing should be noted: when $K_c = 0$ the eigenvalue equals 1, suggesting that $K_c = 0$ is a stability boundary. This is confirmed by plots of the solutions of $|\lambda_1(K_c)| = 1$. Such plots are shown in Fig. 7.6. The unstable region is the region between the vertical axis, corresponding to $K_c = 0$ and the curves.

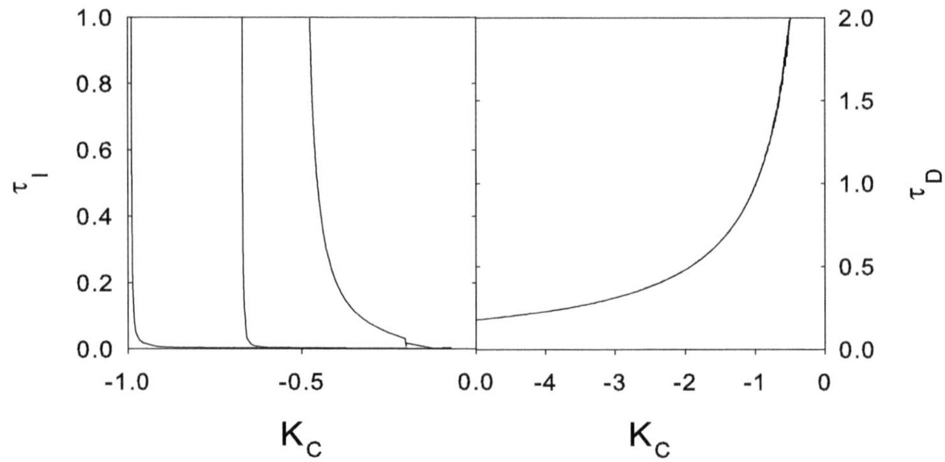

Fig. 7.6 Stability limits of digital Proportional-Integial-Derivative (PID) controller. Left figure: Variation with change in τ_I. The three curves, from left to right, correspond to $\tau_D = 1, 1.5, 5$. The unstable region lies between the curves and the vertical axis. Right figure: Variation with change in τ_D. The three curves correspond to $\tau_I = 0.1, 5, 10$ and are effectively superimposed. The unstable region lies between the curves and the vertical axis and all curves are for $R = 0.5$.

7.4.3.2 *Particular solutions for constant inhomogeneous term*

Of course, particular solutions can always be found using variation of parameters, but the special case where the inhomogeneous term is a constant vector is so common and so easy to solve that it deserves to be mentioned as a special case. The problem is

$$X_{\tau+1} = AX_\tau + B$$

and, as long as 1 is not an eigenvalue of A and a constant vector therefore a homogeneous solution, a particular solution is a constant vector v which can be found by substitution into the difference equation.

$$v = Av + B \Rightarrow$$
$$v = (I - A)^{-1}B$$

When $I - A$ is a semi simple matrix, the complete solution can therefore be written explicitely as

$$X_\tau = \sum_{n=1}^{N} C_n \tilde{v}_n \lambda_n^\tau + (I - A)^{-1}B$$

where the C_ns are arbitrary constants and $\tilde{v}_n$ the eigenvectors of A. When an initial vector, X_0, is given at $\tau = 0$, the arbitrary constants can be found by solving the system of equations

$$X_0 = \sum_{n=1}^{N} C_n \tilde{v}_n + (I - A)^{-1}B.$$

Rather than go through the trouble of solving a set of linear equations to find the arbitrary constants, the eigenrows can be used to this purpose. Consider again the equation obtained after application of the initial conditions to the complete solution and write this in the form

$$X_0 - (I - A)^{-1}B = \sum_{n=1}^{N} C_n \tilde{v}_n.$$

Take the inner product with an eigenrow, $\tilde{w}_m$, of the matrix A.

$$\langle X_0 - (I - A)^{-1}B, \tilde{w}_m\rangle = \sum_{n=1}^{N} C_n \langle \tilde{v}_n, \tilde{w}_m\rangle$$

and, since eigenrows and eigenvectors that do not belong to the same eigenvalue are orthogonal, all the terms in the sum, except the mth term, are zero giving

$$\langle X_0 - (I - A)^{-1}B, \tilde{w}_m\rangle = C_m \langle \tilde{v}_m, \tilde{w}_m\rangle \Rightarrow$$
$$C_m = \frac{\langle X_0 - (I - A)^{-1}B, \tilde{w}_m\rangle}{\langle \tilde{v}_m, \tilde{w}_m\rangle}$$

or the closed form solution

$$X_\tau = \sum_{n=1}^{N} \frac{\langle X_0 - (I-A)^{-1}B, \tilde{w}_n \rangle}{\langle \tilde{v}_n, \tilde{w}_n \rangle} \tilde{v}_n {\lambda_n}^\tau + (I-A)^{-1}B\,.$$

Do not forget that this result was derived under the assumption that the matrix A is semi-simple. The method just described is particularly convenient when the matrix A is symmetric since, in this case, the eigenvectors and eigenrows are identical.

Example 7.15: Initial value problem with symmetric matrix.

Find the solution of

$$X_{\tau+1} = \begin{pmatrix} 9 & -3 & 0 \\ -3 & 12 & -3 \\ 0 & -3 & 9 \end{pmatrix} X_\tau + \begin{pmatrix} 30 \\ 0 \\ 0 \end{pmatrix}$$

which satisfy the initial condition $\tau = 0 \Rightarrow X_1 = (3, 1, -1)$.

The system matrix A is symmetric and a spectral analysis yields the following three simple eigenvalues and associated eigenvectors

$$\lambda_1 = 6\,,\ \tilde{v}_1 = \begin{pmatrix} 1 \\ 1 \\ 1 \end{pmatrix}\ ;\ \lambda_2 = 9\,,\ \tilde{v}_2 = \begin{pmatrix} 1 \\ 0 \\ -1 \end{pmatrix}\ ;\ \lambda_3 = 15\,,\ \tilde{v}_3 = \begin{pmatrix} 1 \\ -2 \\ 1 \end{pmatrix}$$

from which the complete homogeneous solution is obtained as

$$X_{h,\tau} = C_1 \begin{pmatrix} 1 \\ 1 \\ 1 \end{pmatrix} 6^\tau + C_2 \begin{pmatrix} 1 \\ 0 \\ -1 \end{pmatrix} 9^\tau + C_3 \begin{pmatrix} 1 \\ -2 \\ 1 \end{pmatrix} 15^\tau.$$

A particular solution is $(I-A)^{-1}B$. Rather than finding $(I-A)^{-1}$ and post multiplying it by B, use the computationally more efficient method of finding $(I - A)^{-1}B$ as the solution of $(I-A)x = B$.

$$\begin{pmatrix} -8 & 3 & 0 & 30 \\ 3 & -11 & 3 & 0 \\ 0 & 3 & -8 & 0 \end{pmatrix} \Rightarrow \begin{pmatrix} 1\,0\,0 & -\frac{237}{56} \\ 0\,1\,0 & -\frac{9}{7} \\ 0\,0\,1 & -\frac{27}{56} \end{pmatrix}$$

giving the complete solution

$$X_\tau = C_1 \begin{pmatrix} 1 \\ 1 \\ 1 \end{pmatrix} 6^\tau + C_2 \begin{pmatrix} 1 \\ 0 \\ -1 \end{pmatrix} 9^\tau + C_3 \begin{pmatrix} 1 \\ -2 \\ 1 \end{pmatrix} 15^\tau - \begin{pmatrix} \frac{237}{56} \\ \frac{9}{7} \\ \frac{27}{56} \end{pmatrix}.$$

Application of the initial condition gives

$$\begin{pmatrix} 3 \\ 1 \\ -1 \end{pmatrix} + \begin{pmatrix} \frac{237}{56} \\ \frac{9}{7} \\ \frac{27}{56} \end{pmatrix} = C_1 \begin{pmatrix} 1 \\ 1 \\ 1 \end{pmatrix} + C_2 \begin{pmatrix} 1 \\ 0 \\ -1 \end{pmatrix} + C_3 \begin{pmatrix} 1 \\ -2 \\ 1 \end{pmatrix}.$$

The arbitrary constants are now determined by taking the inner product of this equation with each of the eigenvectors. Because the eigenvectors of a symmetric matrix are orthogonal, all terms on the right hand side, except the term with the nth eigenvector, will vanish when taking the inner product with the nth eigenvector. For instance, taking the inner product with $\tilde{v}_1$ gives

$$\left\langle \begin{pmatrix} 1 \\ 1 \\ 1 \end{pmatrix}, \begin{pmatrix} \frac{405}{56} \\ \frac{16}{7} \\ -\frac{29}{56} \end{pmatrix} \right\rangle = C_1 \left\langle \begin{pmatrix} 1 \\ 1 \\ 1 \end{pmatrix}, \begin{pmatrix} 1 \\ 1 \\ 1 \end{pmatrix} \right\rangle \Rightarrow$$

$$9 = 3C_1 \Rightarrow$$

$$C_1 - 3.$$

In a similar manner, one finds

$$C_2 = \frac{31}{8}$$

and

$$C_3 = \frac{5}{14}.$$

7.5 Nonlinear equations

Nonlinear equations will only concern us here in as far as they can be changed to linear equations through variable transformations.

7.5.1 *Riccati's equation*

An example of solution by variable transformation is the Riccati equation, difference equations of the form

$$x_{\tau+1}x_\tau + ax_{\tau+1} + bx_\tau + C = 0\,.$$

The first step is to make the equation homogeneous using $x_\tau = y_\tau + \delta$.

$$(y_{\tau+1} + \delta)(y_\tau + \delta) + a(y_{\tau+1} + \delta) + b(y_\tau + \delta) + c = 0 \;\Rightarrow$$

$$y_{\tau+1}y_\tau + \delta^2 + (a + \delta)y_{\tau+1} + (b + \delta)y_\tau + a\delta + b\delta + c = 0$$

This equation is homogeneous if

$$\delta^2 + (a + b)\delta + c = 0$$

from which the appropriate value of δ can be found. The resulting homogeneous equation

$$y_{\tau+1}y_\tau + (a + \delta)y_{\tau+1} + (b + \delta)y_\tau = 0$$

is first rewritten in the form

$$1 + (a + \delta)\frac{1}{y_\tau} + (b + \delta)\frac{1}{y_{\tau+1}} = 0$$

and then converted to a linear equation using $z_\tau = 1/y_\tau$

$$1 + (a + \delta)z_\tau + (b + \delta)z_{\tau+1} = 0\,.$$

Chapter 8

Linear Differential Equations

An ODEs of Nth order is an equation of the form

$$F\left(t, x, \frac{dx}{dt}, \ldots, \frac{d^N x}{dt^N}\right) = 0$$

where F is some function, t is called the free variable and $x(t)$ is the dependent variable. A given problem may consist of coupled equations of this type.

If F is not an explicit function of the free variable t, then the equation is said to be *autonomous.* For autonomous equations, the origin or zero of t is arbitrary in the sense that it can be shifted by any amount, t_0, by the variable transformation $\tau = t - t_0$ to produce the equivalent formulation

$$F\left(x, \frac{dx}{d\tau}, \ldots, \frac{d^N x}{d\tau^N}\right) = 0\,.$$

Any non-autonomous system of differential equations can be rendered autonomous by adding an equation of the form

$$\frac{dy}{dt} = 1\ ,\ y(t_0) = t_0$$

which has the solution $y = t$, and substituting $y(t)$ for t in the remaining equations.

Most differential equation models in chemical engineering are either initial value problems, such as problems in process dynamics, or two point boundary value problems. For the former, the side conditions or initial conditions specify the values of the dependent variable x and its derivatives up to order $N-1$, all at the same value of the free variable t. Differential balances in transport problems typically result in second order, two point boundary value problems with the free variable being a spatial coordinate and the side or boundary conditions given at two different values of the free variable. In this case, the free variable is usually not designated t but x, y, r or whatever letter signifies the relevant spatial coordinate. The notation used for the free variable in this chapter will vary, depending on which notation is most appropriate for a given problem.

Linear, ODEs occur most often in either of two different formulations. As a single equation of order N or as a set of coupled first order equations. Single equations will be covered first.

8.1 Linear equations with constant coefficients

A linear ODE of order N, with constant coefficients is an equation of the form

$$a_N \frac{d^N x}{dt^N} + a_{N-1} \frac{d^{N-1} x}{dt^{N-1}} + \cdots + a_1 \frac{dx}{dt} + a_0 x = q(t) \tag{8.1}$$

where the a_n are constants. The equation is *homogeneous* iff $q(t) = 0$. We will later prove a theorem which states that the complete solution to this equation is a linear combination of N linearly independent basis solutions to the homogeneous equation plus a particular solution to the complete equation. ("A particular solution" is the standard term for any solution.)

8.1.1 *Homogeneous solutions*

The complete homogeneous solution is a linear combination of N linearly independent basis solutions. The basis solutions are found by assuming a solution of the form $x = \exp(\lambda t)$. Substituting this guess into the homogeneous equivalent of Eq. (8.1) and dividing out the common factor $\exp(\lambda t)$ gives

$$a_N \lambda^N + a_{N-1} \lambda^{N-1} + \cdots + a_1 \lambda + a_0 = 0$$

which is the *characteristic polynomial* of the ODE. If all the roots of this polynomial, say λ_1 through λ_N, have multiplicity one, then they provide the N linearly independent basis solutions needed and the complete homogeneous solution becomes

$$x(t) = \sum_{n=1}^{M} C_n \exp(\lambda_n t)$$

where the C_ns are arbitrary constants which can be determined from the initial conditions.

If the characteristic polynomial has roots with multiplicity greater than 1, additional solutions must be found to obtain a complete set of basis solutions. To find these, suppose that λ_n is a root with multiplicity m_n and claim that m_n linearly independent basis solutions are $C_{n,1} \exp(\lambda_n t)$, $C_{n,2} \exp(\lambda_n t)t, \ldots,$ $C_{n,m_n} \exp(\lambda_n t) t^{m_n - 1}$. These can be seen to be solutions by substitution into the differential equation. The simplifications that follow the substitution use the fact that a polynomial and its derivatives up to order m_n are equal to zero at a root with multiplicity m_n. The complete homogeneous solution to Eq. (8.1) can therefore be written

$$x(t) = \sum_{n=1}^{M} \left(\sum_{m=1}^{m_n} C_{n,m} \exp(\lambda_n t) t^{m-1} \right).$$

The stability of this solution is easily inferred. As t goes to ∞, $\exp(\lambda t)$ goes to zero or ∞ depending on whether the real part of λ is negative or positive. Thus, the stability of the solutions of these equations depend exclusively on the location of the λs. The initial conditions are irrelevant in this context.

Theorem 8.1. *The solution of a linear, homogeneous, constant coefficient differential equation is stable if all the roots of the characteristic polynomial are located in the left half plane. Or equivalently*

$$\textit{Solution is stable if } Re(\lambda) < 0\,,\ \forall\, \lambda\,.$$

A solution is unstable if there is a single root in the right half plane. Or equivalently

$$\textit{Solution is unstable if } \exists\, \lambda \ni Re(\lambda) > 0\,.$$

A complex solution is obtained if any of the roots of the characteristic polynomial are complex. Often the solution is known from physical arguments to be real or, for whatever reason, only the real part of the solution may be of interest. A purely real solution can be obtained by appropriate choice of the arbitrary constants. Assuming the coefficients in the ODE are real, complex roots of the characteristic polynomial will appear as complex conjugate pairs giving solutions of the form

$$x_n(t) = C_1 t^k e^{(\alpha+i\beta)} + C_2 t^k e^{(\alpha-i\beta)t}$$

which can be rewritten using Euler's formula, $e^{i\beta} = \cos\beta + i\sin\beta$, as

$$x_n(t) = (C_1 + C_2)t^k e^{\alpha t}\cos(\beta t) + (C_1 - C_2)t^k e^{\alpha t} i \sin(\beta t).$$

If now C_1 and C_2 are complex conjugate and $C_1 = C_3 + iC_4$, the expression becomes

$$x_n(t) = 2C_3 t^k e^{\alpha t}\cos(\beta t) - 2C_4 t^k e^{\alpha t}\sin(\beta t)$$

where $2C_3$ and $2C_4$ can be regarded as arbitrary constants. We can conclude that real solutions corresponding to two complex conjugate roots, $\alpha \pm i\beta$, of the characteristic polynomial are

$$x_1(t) = t^k e^{\alpha t} \cos(\beta t)$$

and

$$x_2(t) = t^k e^{\alpha t} \sin(\beta t).$$

Example 8.1: Homogeneous, linear ODE with constant coefficients.
Consider the problem

$$\frac{d^4x}{dt^4} - 7\frac{d^3x}{dt^3} + 12\frac{d^2x}{dt^2} + 4\frac{dx}{dt} - 16x = 0$$

with the initial condition $t = 0 \Rightarrow x = 2, x' = 3, x'' = 25, x''' = 127$.
The characteristic polynomial is

$$\lambda^4 - 7\lambda^3 + 12\lambda^2 + 4\lambda - 16 = (\lambda - 2)(\lambda - 2)(\lambda + 1)(\lambda - 4)$$

and the complete solution thus is

$$x(t) = C_1 e^{-t} + C_2 e^{4t} + C_3 e^{2t} + C_4 e^{2t} t\,.$$

Applying the initial conditions give the following set of equations.

$$\begin{aligned} 2 =& C_1 + C_2 + C_3 \\ 3 =& -C_1 + 4C_2 + 2C_3 + C_4 \\ 25 =& C_1 + 16C_2 + 4C_3 + 4C_4 \\ 127 =& -C_1 + 64C_2 + 8C_3 + 12C_4 \end{aligned}$$

which are solved to

$$C_1 = 1\,,\ C_2 = 3\,,\ C_3 = -2\,,\ C_4 = -4$$

giving the solution

$$y(t) = e^{-t} + 3e^{4t} - 2e^{2t} - 4e^{2t} t\,.$$

8.1.2 *Method of matched coefficients*

In Section 8.2.2 we will describe a general method, variation of parameters, for finding a particular solution to any linear ODE. However, a simpler, easier to use method, called the method of matched coefficients or the annihilator method, can be used for the special case of linear ODEs with constant coefficients for which the right hand side is of the same form as the basis solutions to these ODEs.

The annihilator of the right hand side of Eq. (8.1), $q(t)$, is a linear homogeneous ODE with constant coefficients such that $q(t)$ solves this equation. I.e.

$$\sum_{m=0}^{M} b_m \frac{d^m q}{dt^m} = 0\,. \tag{8.2}$$

Let x_p be a particular solution to Eq. (8.1). Then

$$\sum_{n=0}^{N} a_n \frac{d^n x_p}{dt^n} = q(t)\,. \tag{8.3}$$

Now substitute this expression for $q(t)$ into the annihilator equation, Eq. (8.2).

$$\sum_{m=0}^{M} b_m \frac{d^m}{dt^m} \left(\sum_{n=0}^{N} a_n \frac{d^n x_p}{dt^n} \right) = 0\,.$$

This is a homogeneous linear ODE with constant coefficients and a solution to the equation is the particular solution x_p. Therefore, to find an expression for x_p, we solve this equation. There is no need to carry out the nested differentiations and simplifying since solving the equation only requires that we know the characteristic polynomial. This we can find by assuming a solution of the form $x_p = e^{\lambda t}$.

$$\begin{aligned}
\sum_{m=0}^{M} b_m \frac{d^m}{dt^m} \left(\sum_{n=0}^{N} a_n \frac{d^n e^{\lambda t}}{dt^n} \right) &= \sum_{m=0}^{M} b_m \frac{d^m}{dt^m} \left(p_n(\lambda) e^{\lambda t} \right) \\
&= p_n(\lambda) \sum_{m=0}^{M} b_m \frac{d^m e^{\lambda t}}{dt^m} \\
&= p_n(\lambda) p_m(\lambda) e^{\lambda t}
\end{aligned}$$

where $p_n(\lambda)$ and $p_m(\lambda)$ are the characteristic polynomials of Eqs. (8.2) and (8.3) respectively. It should now be obvious that the particular solution, x_p, must have the form

$$x_p(t) = \sum_{n=1}^{N} C_n \Phi_n(t) + \sum_{m=1}^{M} C_{N+m} \Psi_m(t)$$

where the $\Phi_n(t)$ are the basis solutions generated by the roots of $p_n(\lambda)$ and the $\Psi_m(t)$ are the additional basis solutions generated by the roots of $p_m(\lambda)$. It is important to remember that the form of these additional basis solutions will depend on whether or not the roots of $p_m(\lambda)$ are simple roots in the polynomial $p_n(\lambda)p_m(\lambda)$ or whether they have multiplicities greater than 1.

The arbitrary constants in the expression for x_p can be determined by substituting the above expression for x_p into the differential equation. However, the first sum in the expression correspond to the homogeneous solution and can be dropped giving an expression of the form

$$x_p(t) = \sum_{m=1}^{M} C_m \Psi_m(t).$$

When this expression is substituted into Eq. (8.3) one obtains an equation in which the coefficients of linearly independent basis solutions on the left and right hand side can be compared to extract a set of linear algebraic equations that can be solved for the coefficients.

Example 8.2: Particular solution by method of matched coefficients.
Consider the equation

$$\frac{d^2x}{dt^2} - 4\frac{dx}{dt} + 4x = 2e^{2t} + 1\,.$$

An annihilator of the right hand side is $\frac{d^2q}{dt^2} - 2\frac{dq}{dt}$, but all we really need to know is that the annihilator has the characteristic polynomial $\lambda^2 - 2\lambda$. The characteristic polynomial of the ODE itself is $\lambda^2 - 4\lambda + 4 = (\lambda - 2)^2$. A particular solution therefore has the form

$$x_p(t) = (C_1e^{2t} + C_2te^{2t}) + (C_3t^2e^{2t} + C_4).$$

The first parenthesis equals the homogeneous solution and can be dropped. We therefore will use the simpler expression

$$x_p(t) = C_3t^2e^{2t} + C_4\,.$$

Substituting this expression into the ODE gives

$$(2C_3e^{2t} + 8C_3te^{2t} + 4C_3t^2e^{2t}) - 4(2C_3te^{2t} + 2C_3t^2e^{2t}) + 4(C_3t^2e^{2t} + C_4)$$
$$= 2e^{2t} + 1 \Rightarrow 2C_3e^{2t} + 4C_4 = 2e^{2t} + 1\,.$$

Matching coefficients on the left and right hand side shows that

$$C_3 = 1\ ,\ C_4 = \frac{1}{4}\,.$$

So a particular solution is

$$x_p(t) = t^2 e^{2t} + \frac{1}{4}\,.$$

And the complete solution is

$$x(t) = C_1 e^{2t} + C_2 t e^{2t} + t^2 e^{2t} + \frac{1}{4}\,.$$

Expressions for the particular solution obtained using the annihilator method has been tabulated for all the possible right hand sides for which the method works. The results are shown in Table 8.1.

Importantly, the results in the table all assume that the roots of the characteristic polynomial of the annihilator are different from the roots of the ODE itself. This is not the case if the right hand side, $q(t)$, is a basis solution or a linear combination of basis solutions. Then the suggested particular solution is also a solution to the homogeneous problem and therefore cannot also be a particular solution. In this case, the appropriate form for the guessed particular solution is found by considering the sequence of basis solutions that are generated by multiple roots, the sequence $C_{n,1}\exp(\lambda_n t)$, $C_{n,2}\exp(\lambda_n t)t, \ldots, C_{n,m_n}\exp(\lambda_n t)t^{m_n-1}$. The particular solution is guessed as the next element in this sequence, multiplied by a parameter to be determined, i.e. $A\exp(\lambda_n t)t^{m_n}$.

Table 8.1 Particular solutions corresponding to different inhomogeneous terms. The parameters to be matched are given by capital Latin letters.

$q(t)$	$x_p(t)$
$a_n t^n + \cdots + a_1 t + a_0$	$A_n t^n + \cdots + A_1 t + A_0$
$a\exp(bt)$	$A\exp(bt)$
$\sin(at)$ or $\cos(at)$	$A\sin(at) + B\cos(at)$
$a\exp(bt)\sin(ct)$ or $a\exp(bt)\cos(ct)$	$\exp(bt)(A\sin(ct) + B\cos(ct))$

Example 8.3: Particular solution by method of matched coefficients.
Consider

$$\frac{d^2x}{dt^2} + \frac{dx}{dt} - 2x = t^2.$$

The homogeneous solution is

$$x_h(t) = C_1 \exp(t) + C_2 \exp(-2t)$$

and, consulting Table 8.1, a guess of a particular solution is

$$x_p(t) = At^2 + Bt + C.$$

Substitution of the guess into the differential equation gives

$$2A + 2At + B - 2At^2 - 2Bt - 2C = t^2 \;\Rightarrow$$

$$(2A + 1)t^2 + (2B - 2A)t + 2C - B - 2A = 0$$

from which one can extract the equations

$$\left.\begin{array}{r} 2A + 1 = 0 \\ 2B - 2A = 0 \\ 2C - B - 2A = 0 \end{array}\right\} \Rightarrow \begin{array}{l} A = -\frac{1}{2} \\ B = -\frac{1}{2} \\ C = -\frac{3}{4} \end{array}$$

and conclude that the complete solution is

$$x(t) = C_1 \exp(t) + C_2 \exp(-2t) - \frac{3}{4} - \frac{1}{2}t - \frac{1}{2}t^2.$$

Example 8.4: Particular solution by method of matched coefficients.
Find the complete solution to

$$\frac{d^3x}{dt^3} - 3\frac{d^2x}{dt^2} + 9\frac{dx}{dt} + 13x = q(t)$$

for the following right hand sides,

$$q_1(t) = 4t^3 - 3t \;,\; q_2(t) = \sin(t) + e^t.$$

The homogeneous solution is found first. The characteristic polynomial is

$$\lambda^3 - 3\lambda^2 + 9\lambda + 13 = (\lambda + 1)(\lambda - 2 + 3i)(\lambda - 2 - 3i)$$

and the homogeneous solution therefore is

$$x_h(t) = C_1 e^{-t} + C_2 e^{(2-3i)t} + C_3 e^{(2+3i)t}.$$

The real part of this solution is

$$x_h(t) = C_1 e^{-t} + C_2 e^{2t}\cos(3t) + C_3 e^{2t}\sin(3t)$$

where C_1, C_2 and C_3 are real arbitrary constants.

Consider first the right hand side $q_1(t) = 4t^3 - 3t$. The particular solution takes the form $At^3 + Bt^2 + Ct + D$. Substituting this expression into the differential equation yields

$$6A - 3(6At + 2B) + 9(3At^2 + 2Bt + C) + 13(At^3 + Bt^2 + Ct + D) = 4t^3 - 3t$$

simplifying the left hand side

$$13At^3 + (27A + 13B)t^2 + (-18A + 18B + 13C)t + (6A - 6B + 9C + 13D) = 4t^3 - 3t$$

and equating coefficients of terms of equal power gives the following equations for the unknown coefficients

$$\begin{aligned} 13A &= 4 \\ 27A + 13B &= 0 \\ -18A + 18B + 13C &= 0 \\ 6A - 6B + 9C + 13D &= 0 \end{aligned}$$

which are solved to give

$$A = \frac{4}{13}\,,\ B = -\frac{108}{169}\,,\ C = \frac{2373}{2197}\,,\ D = -\frac{33837}{28561}$$

and the complete real solution thus is

$$x(t) = C_1 e^{-t} + C_2 e^{2t}\cos(3t) + C_3 e^{2t}\sin(3t) + \frac{4}{13}t^3 - \frac{108}{169}t^2 + \frac{2373}{2197}t - \frac{33837}{28561}.$$

For the right hand side $q_2(t) = \sin(t) + e^t$, the particular solution has the form $x_p(t) = A\sin(t) + B\cos(t) + Ce^t$. Substituting this into the differential equation and equating coefficients of equal terms gives

$$A + 2B = 0\ ,\ 16A - 8B = 1\ ,\ 20C = 1$$

from which the particular solutions is obtained as

$$x_p(t) = \frac{1}{20}\sin(t) - \frac{1}{40}\cos(t) + \frac{1}{20}e^t.$$

Hence, the complete solution is

$$x(t) = C_1e^{-t} + C_2e^{2t}\cos(3t) + C_3e^{2t}\sin(3t) + frac120\sin(t) - \frac{1}{40}\cos(t) + \frac{1}{20}e^t.$$

Example 8.5: Particular solution by method of matched coefficients.
Find the complete solution to

$$\frac{d^2x}{dt^2} + 2\frac{dx}{dt} + x = \exp(-t).$$

The homogeneous solution is found to be

$$x_h(t) = C_1e^{-t} + C_2te^{-t}.$$

According to Table 8.1, a particular solution must be of the form Ae^{-t}. However, this is a basis solution so it must be discarded as a possible particular solution. Instead, the particular solution must be of the form At^me^{-t} where the value of the power m is as small as it can possibly be without at the same time being a basis solution. Considering the basis solutions in the homogeneous solution, it is obvious that m must equal 2. Thus a guess of the particular solution is

$$x_p(t) = At^2e^{-t}$$

which, when substituted into the differential equation, gives

$$A\left(-4te^{-t} + 2e^{-t} + t^2e^{-t} + 4te^{-t} - t^2e^{-t} + t^2e^{-t}\right) = e^{-t} \Rightarrow$$

$$A = \frac{1}{2}$$

so the complete solution is

$$x(t) = C_1e^{-t} + C_2te^{-t} + \frac{1}{2}t^2e^{-t}.$$

8.2 Single, higher order equations

We will now consider the problem of linear ODEs of the form

$$p_N(t)\frac{d^N x}{dt^N} + p_{N-1}(t)\frac{d^{N-1}x}{dt^{N-1}} + \cdots + p_1(t)\frac{dx}{dt} + p_0(t)x = \sum_{n=0}^{N} p_n(t)\frac{d^n x}{dt^n} = q(t). \quad (8.4)$$

The results regarding the solution structure can be summarized in a single theorem.

Theorem 8.2. *The complete solution, $x_h(t)$, of the Nth order linear, homogeneous problem $\sum_{n=0}^{N} p_n(t)\frac{d^n x}{dt^n} = 0$ is a linear combination of N linearly independent basis solutions.*

$$x_h(t) = C_1x_1(t) + C_2x_2(t) + \cdots + C_Nx_N(t)$$

The complete solution, $x(t)$, of the inhomogeneous problem $\sum_{n=0}^{N} p_n(t)\frac{d^n x}{dt^n} = q(t)$ is the sum of the complete solution to the homogeneous problem plus a particular solution, $x_p(t)$, of the inhomogeneous problem.

$$x(t) = C_1x_1(t) + C_2x_2(t) + \cdots + C_Nx_N(t) + x_p(t)$$

If a complete set of basis solutions is known, then a particular solution can always be found using variation of parameters. Unfortunately, there is no method for finding a complete set of basis solutions for the general case. However, it is sometimes possible to find a basis solution to a linear differential equation by guessing or by trial and error. This known basis solution can be used to reduce the order of the problem for the remaining solutions, much like a known root of a polynomial can be divided out to obtain a lower order polynomial for the remaining roots. The method used is a straight forward application of the variation of parameters concept.

8.2.1 *Reduction of order*

Let $x_1(t)$ be a known basis solution of an Nth order, linear differential equation. Construct a solution, linearly independent of $x_1(t)$, as $x_2(t) = f(t)x_1(t)$ and substitute this expression in to the differential equation. After simplification an $(N-1)$th order linear differential equation in $f'(t)$ is obtained.

Example 8.6: Reduction of order by variation of parameters.

Having guessed and confirmed that $x_1(t) = t$ is a solution of the differential equation

$$t^3\frac{d^3x}{dt^3} - 3t^2\frac{d^2x}{dt^2} + t(6-t^2)\frac{dx}{dt} - (6-t^2)x = 0$$

find the other two basis solutions.

From the substitution $x = f(t)t$ one obtains

$$x' = tf' + f$$

$$x^{(2)} = tf^{(2)} + 2f'$$

$$x^{(3)} = tf^{(3)} + 3f^{(2)}$$

where $x^{(k)}$ is the kth derivative of x. When substituted into the differential equation, this yields upon simplification

$$(f')^{(2)} - f' = 0 \text{ or } \frac{d^2f'}{dt^2} - f' = 0$$

which has two basis solutions, f_1' and f_2'.

$$f_1' = e^t \ , \ f_2' = e^{-t} \ \Rightarrow$$

$$f_1 = e^t \ , \ f_2 = -e^{-t}$$

so

$$x_2(t) = te^t \ , \ x_3(t) = -te^{-t} .$$

Second order, linear differential equations appear so frequently that it makes sense to state the relevant result explicitly for these equations.

Theorem 8.3. *Given the ODE*

$$\frac{d^2x}{dt^2} + p_1(t)\frac{dx}{dt} + p_0(t)x = 0 .$$

If a basis solution $x_1(t)$ to this equation is known, and $x_1(t) \neq 0$ over the interval of interest, then another basis solution to the ODE, linearly independent of $x_1(t)$ is

$$x_2(t) = x_1(t)\int \frac{1}{x_1^2(t)} e^{-\int p_1(t)\,dt}\,dt .$$

8.2.2 *Particular solution by variation of parameters*

Assume a complete set of basis solutions of a linear differential equation is known. Call these known solutions x_n, where n goes from 1 to N, the order of the differential equation. Construct a particular solution, $x_p(t)$, as

$$x_p(t) = \sum_{n=1}^{N} f_n(t) x_n(t).$$

In order to determine the N unknown $f_n(t)$ functions, N equations will be needed. One of these is obtained by the requirement that the particular solution must satisfy the differential equation. The remaining equations are obtained as follows. Find the first derivative of $x_p(t)$ as

$$x_p' = \sum_{n=1}^{N} f_n' x_n + \sum_{n=1}^{N} f_n x_n' .$$

In order to obtain the simplest possible equations for the unknown functions, $f_n(t)$, we seek conditions such that higher order derivatives of these functions do not appear in the expressions for the derivatives of x_p. Thus, to avoid the appearance of second order derivatives of the f_ns in the expression for the second order derivative of x_p, we demand that

$$\sum_{n=1}^{N} f_n' x_n = 0 .$$

Continuing in this way, calculating progressively higher order derivatives of x_p while demanding that sums which contain factors of first derivatives in f_ns are zero results in the following set of equations

$$\sum_{n=1}^{N} f_n' x_n^{(k)} = 0 \ , \ 0 \leq k \leq N-2$$

$$x_p^{(k)} = \sum_{n=1}^{N} f_n x_n^k \ , \ 0 \leq k \leq N-1$$

$$x_p^{(N)} = \sum_{n=1}^{N} f_n' x_n^{(N-1)} + \sum_{n=1}^{N} f_n x_n^{(N)} .$$

The parameter k only takes values less than N, the order of the differential equation, using larger values of k would result in an over specified system for the functions f_n, i.e. more equations than unknowns. Substituting the expressions for the derivatives of x_p into the differential equation gives

$$\sum_{n=0}^{N} p_n x_p^{(n)} = \sum_{n=0}^{N} \left\{ p_n \sum_{m=1}^{N} f_m x_m^{(n)} \right\} + p_N \sum_{n=1}^{N} f_n' x_n^{(N-1)}$$

$$= \sum_{m=1}^{N} \left\{ f_m \sum_{n=0}^{N} p_n x_m^{(n)} \right\} + p_N \sum_{n=1}^{N} f_n' x_n^{(N-1)} = q(t).$$

Since the x_ms are solutions to the homogeneous differential equation, the sum $\sum_{n=0}^{N} p_n x_m^{(n)}$ must equal zero and one obtains

$$\sum_{n=1}^{N} f_n' x_n^{(N-1)} = q(t)/p_N(t)$$

which is the last equation needed to determine the unknown functions f_n. These equations can be written in matrix notation as

$$\begin{pmatrix} x_1 & x_2 & \cdots & x_N \\ x_1' & x_2' & \cdots & x_N' \\ \vdots & \vdots & \ddots & \vdots \\ x_1^{(N-1)} & x_2^{(N-1)} & \cdots & x_N^{(N-1)} \end{pmatrix} \begin{pmatrix} f_1' \\ f_2' \\ \vdots \\ f_N' \end{pmatrix} = \begin{pmatrix} 0 \\ 0 \\ \vdots \\ q(t)/p_N(t) \end{pmatrix}.$$

The matrix of this set of equations is the Wronskian matrix of the basis solutions, and since the basis solutions are linearly independent, the determinant of the Wronskian is different from zero and the system of equations for the f_n's does have a solution. This solution is a set of first order differential equations in each of the unknown functions, f_n. These equations can, at least in principle, be solved and from the solution for the f_ns the particular solution can be found.

Example 8.7: Particular solution by variation of parameters.
Solve

$$\frac{d^3x}{dt^3} - 6\frac{d^2x}{dt^2} + 11\frac{dx}{dt} - 6x = \sin(t^2).$$

The basis solutions are found to be e^t, e^{2t} and e^{3t}. The equations for the f_ns thus become

$$\begin{pmatrix} e^t & e^{2t} & e^{3t} \\ e^t & 2e^{2t} & 3e^{3t} \\ e^t & 4e^{2t} & 9e^{3t} \end{pmatrix} \begin{pmatrix} f_1' \\ f_2' \\ f_3' \end{pmatrix} = \begin{pmatrix} 0 \\ 0 \\ \sin(t^2) \end{pmatrix}$$

which is solved to give

$$f_1' = \frac{1}{2}\sin(t^2)e^{-t}$$

$$f_2' = -\sin(t^2)e^{-2t}$$

$$f_3' = \frac{1}{2}\sin(t^2)e^{-3t}$$

and the complete solution therefore is

$$x(t) = C_1e^t + C_2e^{2t} + C_3e^{3t}$$
$$+\frac{e^t}{2}\int e^{-t}\sin(t^2)\,dt - e^{2t}\int e^{-2t}\sin(t^2)\,dt + \frac{e^{3t}}{2}\int e^{-3t}\sin(t^2)\,dt\,.$$

For the special case of second order equations, the general result is summarized in the following theorem.

Theorem 8.4. *Given the differential equation*

$$\frac{d^2x}{dt^2} + p_1(t)\frac{dx}{dt} + p_0(t)x = q(t)$$

and a basis, $x_1(t)$, $x_2(t)$, for the homogeneous solution, a particular solution is

$$x_p(t) = -x_1(t)\int \frac{x_2(t)}{W(t)}q(t)\,dt + x_2(t)\int \frac{x_1(t)}{W(t)}q(t)\,dt$$

where $W(t)$ is the Wronskian determinant

$$W(t) = \begin{vmatrix} x_1 & x_2 \\ x_1' & x_2' \end{vmatrix}.$$

8.3 Systems of linear differential equations

A single, linear higher order ODE takes the form

$$p_N(t)\frac{d^Nx}{dt^N} + p_{N-1}(t)\frac{d^{N-1}x}{dt^{N-1}} + \cdots + p_1(t)\frac{dx}{dt} + p_0(t)x = q(t).$$

This formulation can be converted to a set of coupled first order equations if the coefficient function $p_N(t) \neq 0$. If this is not the case we will consider the problem in each of the intervals between the zeros of $p_N(t)$. Define a vector x of new unknowns by

$$\begin{aligned} x_1(t) &= x(t) \\ x_2(t) &= \tfrac{dx}{dt} \\ &\vdots \\ x_N(t) &= \tfrac{d^{N-1}x}{dt^{N-1}} \,. \end{aligned}$$

Clearly

$$\frac{d}{dt}\begin{pmatrix} x_1 \\ x_2 \\ \vdots \\ x_N \end{pmatrix} = \begin{pmatrix} x_2 \\ x_3 \\ \vdots \\ (q(t) - (p_{N-1}(t)x_N + \cdots + p_1(t)x_2 + p_0(t)x_1))/p_N(t) \end{pmatrix}$$

or, in vector-matrix notation

$$\frac{dX}{dt} = \begin{pmatrix} 0 & 1 & 0 & \cdots & 0 \\ 0 & 0 & 1 & \cdots & 0 \\ \vdots & \vdots & \vdots & \ddots & \vdots \\ -\frac{p_0(t)}{p_N(t)} & -\frac{p_1(t)}{p_N(t)} & -\frac{p_2(t)}{p_N(t)} & \cdots & -\frac{p_{N-1}(t)}{p_N(t)} \end{pmatrix} \begin{pmatrix} x_1 \\ x_2 \\ \vdots \\ x_N \end{pmatrix} + \begin{pmatrix} 0 \\ 0 \\ \vdots \\ \frac{q(t)}{p_N(t)} \end{pmatrix} .$$

This formulation suggests the more general problem

$$\frac{dX}{dt} = A(t)X + B(t)$$

where X is a vector of N unknown functions, $x_n(t)$, of the free variable t, A is an $N \times N$ matrix in which each element can be a function of t (but not of the x_n's) and $B(t)$ a vector of N specified functions, $b_n(t)$. The equations are *homogeneous* if $B(t) = 0$. An initial condition for this problem is stated by specifying the values $x_n(t_0)$, the values of the elements of the vector of unknowns at some given time. We will assume that the initial time is $t_0 = 0$ unless stated otherwise. This formulation, a set of coupled first order equations, is the most common formulation for dynamic models in which case t is time.

There is no easy way to reverse the process and write a set of coupled linear equations as a single higher order equation. The matrix formulation is more general than the single equation formulation and, by extension, results that apply to the matrix formulation are true for the single equation formulation. For instance, existence and uniqueness of the solution to the general initial value problem in matrix formulation assures existence and uniqueness of the solution of all single equation formulations. We will therefore discuss the theory of the solution structure for the matrix formulation in a little more detail than for single equations. However, many types of practical problems are traditionally written as single higher order equations and are most easily recognized and solved without first converting them to matrix formulation. Two point boundary value problems belong to this class.

8.3.1 *Basic theorems*

As mentioned, the basic theorems regarding the solution structure of linear differential equations are very similar to those for linear difference equations. Some of the proofs are also very similar.

Uniqueness and existence of the solutions to coupled linear first order equations follow from the existence and uniqueness of coupled, though not necessarily linear, first order equations in general. We refer the reader to the comments about the proof for a single first order equations, page 124 and note that, in order for existence and uniqueness to hold, the coupled differential equations must satisfy a Lipschitz condition for vector valued functions.

Taking existence and uniqueness for granted, the theorem regarding the solutions structure is

Theorem 8.5. *The complete solution to*

$$\frac{dX}{dt} = A(t)X + B(t)$$

where A is an $N \times N$ matrix is the sum of a particular solution and a linear combination of N linearly independent solutions, so-called basis solutions, to the homogeneous problem.

Proof: Same as for linear difference equations, page 329. □

A complete set of linearly independent basis solutions is said to form a fundamental system of solutions or a fundamental set. A complete set of basis solutions can be arranged into a matrix $\Phi(t)$ such that the nth column of the matrix equals the nth solution vector. This matrix is *the fundamental matrix* of the system and, as for difference equations, it is not unique in spite of the name. The determinant of $\Phi(t)$ is called the *Wronsky determinant.*[1] The Wronsky determinant has great theoretical importance and linear independence of the basis solutions require that the

[1]Named after the Polish mathematician Hoene Wronsky, 1778–1853.

Wronsky determinant is different from zero and that an inverse of the fundamental matrix therefore exists. However, one is rarely, if ever, faced with the problem of evaluating the Wronsky determinant in order to determine whether or not a particular set of solutions are linearly independent or not. Note that if $\Phi(t)$ is known, then the complete homogeneous solution is a linear combination of the rows of $\Phi(t)$ and can be written

$$X_h(t) = \Phi(t)C$$

where C is a vector of arbitrary constants.

8.3.2 *Particular solution by variation of parameters*

When a fundamental matrix is known, a particular solution can be found by variation of parameters. The method is essentially the same as for difference equations. Assume a particular solution of the form

$$X_p(t) = \Phi(t)F(t)$$

and substitute this expression into the differential equation, giving

$$\Phi'(t)F(t) + \Phi(t)F'(t) = A(t)\Phi(t)F(t) + B(t) \;\Rightarrow$$

$$\{\Phi'(t) - A(t)\Phi(t)\}F(t) + \Phi(t)F'(t) = B(t).$$

The term inside the brackets is zero, so

$$F'(t) = \Phi^{-1}(t)B(t) \;\Rightarrow$$

$$F(t) = \int_{t_0}^{t} \Phi^{-1}(\tau)B(\tau)\,d\tau$$

assume that $F(t_0) = 0$, the most convenient initial condition for our purpose. Reassembling everything, the complete solution is

$$X(t) = \Phi(t)C + \Phi(t)\int_{t_0}^{t} \Phi^{-1}(\tau)B(\tau)\,d\tau$$

where C is a vector of arbitrary constants. This appears to be a powerful result, but its use is hampered by the fact that finding a fundamental matrix is difficult or impossible in many cases.

Example 8.8: Variation of parameters for coupled ODEs.
Consider

$$\begin{aligned}
\frac{dx}{dt} &= y(t) + e^t(t+1) \\
\frac{dy}{dt} &= -\frac{t+2}{t}x(t) + 2\frac{t+2}{t+1}y(t) + e^t(t+1)^2.
\end{aligned}$$

A fundamental matrix is

$$\Phi(t) = \begin{pmatrix} te^t & e^t(t^2-1+\ln(t)t) \\ e^t(t+1) & e^t(t+1)(t+1+2\ln(t)) \end{pmatrix} \Rightarrow$$

$$\Phi^{-1} = \begin{pmatrix} \frac{t+2\ln(t)+1}{e^t(t+1)} & -\frac{t^2-1+2\ln(t)t}{e^t(2t+t^2+1)} \\ -\frac{1}{e^t(t+1)} & \frac{t}{e^t(2t+t^2+1)} \end{pmatrix} \Rightarrow$$

$$\Phi^{-1}B(t) = \begin{pmatrix} -t^2+t-2\ln(t)t+2\ln(t)+2 \\ t-1 \end{pmatrix}.$$

The two differential equations for the elements of $F(t)$ become

$$\frac{df_1}{dt} = -t^2+t-2\ln(t)t+2\ln(t)+2 \Rightarrow f_1(t) = -\frac{1}{3}t^3 - t^2\ln(t) + t^2 + 2\ln(t)t$$

$$\frac{df_2}{dt} = t-1 \Rightarrow f_2(t) = \frac{1}{2}t^2 - t$$

where the constants of integration have been set equal to zero.
The complete solution can now be written

$$\begin{aligned}
\begin{pmatrix} x(t) \\ y(t) \end{pmatrix} = \begin{pmatrix} te^t & e^t(t^2-1+\ln(t)t) \\ e^t(t+1) & e^t(t+1)(t+1+2\ln(t)) \end{pmatrix} \Bigg(\begin{pmatrix} C_1 \\ C_2 \end{pmatrix} \\
+ \begin{pmatrix} -\frac{1}{3}t^3 - t^2\ln(t) + t^2 + 2\ln(t)t \\ \frac{1}{2}t^2 - t \end{pmatrix} \Bigg).
\end{aligned}$$

8.3.3 *Equations with constant coefficients*

These are equations for which the matrix A is independent of t.

$$\frac{dX}{dt} = AX + B(t)\,.$$

Analytical expressions can be found for both the homogeneous and particular solutions.

8.3.3.1 *Homogeneous solutions*

The matrix A, like any square matrix, is similar to a matrix in Jordan canonical form. Let K be a matrix of eigenvectors and generalized eigenvectors in a similarity transformation between A and a similar Jordan canonical form matrix J. Then

$$\frac{dX}{dt} = AX = KJK^{-1}X\,.$$

Premultiplying both sides of this equation by K^{-1} gives

$$\frac{d}{dt}(K^{-1}X) = J(K^{-1}X) \;\Rightarrow$$

$$\frac{dY}{dt} = JY \tag{8.5}$$

where $Y = K^{-1}X$. Each Jordan block now represents a group of equations that are decoupled from the remaining equations and which have the general form

$$\frac{dZ}{dt} = \begin{pmatrix} \lambda & 1 & 0 & \cdots & 0 & 0 \\ 0 & \lambda & 1 & \cdots & 0 & 0 \\ \vdots & \vdots & \vdots & \ddots & \vdots & \vdots \\ 0 & 0 & 0 & \cdots & \lambda & 1 \\ 0 & 0 & 0 & \cdots & 0 & \lambda \end{pmatrix} Z\,. \tag{8.6}$$

In the special case where A is semi-simple, all Jordan blocks have size 1, and Eq. (8.5) consists of N, decoupled equations of the general form

$$\frac{dy_n}{dt} = \lambda_n y_n \;\Rightarrow\; y_n = C_n e^{\lambda_n t}$$

from which the solution vector $Y(t)$ is obtained as

$$Y(t) = \begin{pmatrix} C_1 e^{\lambda_1 t} \\ \vdots \\ C_N e^{\lambda_N t} \end{pmatrix}$$

and the solution $X(t)$ is found as

$$X(t) = KY(t)$$

or

$$X(t) = \sum_{n=1}^{N} C_n \tilde{x}_n e^{\lambda_n t}$$

where $\tilde{x}_n$ are the eigenvectors of A. The sum is taken over all N eigenvectors of the eigenspace of A and when A is simple, this is equivalent to taking the sum over all eigenvalues. When A is semi-simple, the sum will contain terms with identical values of λ_n for each of the semi-simple eigenvalues. The C_ns are arbitrary constants.

Example 8.9: Coupled ODEs, semi-simple matrix.

Consider

$$\begin{aligned}\frac{dx_1}{dt} &= \frac{17}{8}x_1 + \frac{3}{8}x_2 - \frac{5}{8}x_3\\ \frac{dx_2}{dt} &= \frac{1}{2}x_1 + \frac{7}{2}x_2 - \frac{5}{2}x_3\\ \frac{dx_3}{dt} &= \frac{1}{8}x_1 + \frac{3}{8}x_2 + \frac{11}{8}x_3\,.\end{aligned}$$

A spectral analysis yields the following result for the matrix A.

$$A = \begin{pmatrix} \frac{17}{8} & \frac{3}{8} & -\frac{5}{8} \\ \frac{1}{2} & \frac{7}{2} & -\frac{5}{2} \\ \frac{1}{8} & \frac{3}{8} & \frac{11}{8} \end{pmatrix} = \begin{pmatrix} 1 & 2 & 1 \\ 3 & 1 & 4 \\ 2 & 1 & 1 \end{pmatrix} \begin{pmatrix} 2 & 0 & 0 \\ 0 & 2 & 0 \\ 0 & 0 & 3 \end{pmatrix} \begin{pmatrix} -\frac{3}{8} & -\frac{1}{8} & \frac{7}{8} \\ \frac{5}{8} & -\frac{1}{8} & -\frac{1}{8} \\ \frac{1}{8} & \frac{3}{8} & -\frac{5}{8} \end{pmatrix}.$$

Therefore, the problem for $Y = K^{-1}X$ becomes

$$\frac{dY}{dt} = \begin{pmatrix} 2 & 0 & 0 \\ 0 & 2 & 0 \\ 0 & 0 & 3 \end{pmatrix} Y$$

which is solved to give

$$\begin{pmatrix} y_1 \\ y_2 \\ y_3 \end{pmatrix} = \begin{pmatrix} C_1 c^{2t} \\ C_2 e^{2t} \\ C_3 e^{3t} \end{pmatrix}$$

from which

$$\begin{pmatrix} x_1 \\ x_2 \\ x_3 \end{pmatrix} = \begin{pmatrix} C_1 e^{2t} + 2C_2 e^{2t} + C_3 e^{3t} \\ 3C_1 e^{2t} + C_2 e^{2t} + 4C_3 e^{3t} \\ 2C_1 e^{2t} + C_2 e^{2t} + C_3 e^{3t} \end{pmatrix}$$

or in terms of the fundamental matrix

$$X(t) = \begin{pmatrix} e^{2t} & 2e^{2t} & e^{3t} \\ 3e^{2t} & e^{2t} & 4e^{3t} \\ 2e^{2t} & e^{2t} & e^{3t} \end{pmatrix}$$

or in a more familiar form

$$\begin{pmatrix} x_1 \\ x_2 \\ x_3 \end{pmatrix} = C_1 \begin{pmatrix} 1 \\ 3 \\ 2 \end{pmatrix} e^{2t} + C_2 \begin{pmatrix} 2 \\ 1 \\ 1 \end{pmatrix} e^{2t} + C_3 \begin{pmatrix} 1 \\ 4 \\ 1 \end{pmatrix} e^{3t}.$$

When the matrix A is degenerate, a set of equations similar to Eq. (8.6) must be solved for each degenerate eigenvalue. Although the equations are coupled, they can be solved by starting with the bottom equation, which is not coupled to any of the other equations, and working up to the top equation. Let Z be of size N, then the equation for z_N is

$$\frac{dz_N}{dt} = \lambda z_N \;\Rightarrow$$

$$z_N = C_N e^{\lambda t}.$$

The equation for z_{N-1} becomes

$$\frac{dz_{N-1}}{dt} = \lambda z_{N-1} + C_N e^{\lambda t}$$

which is solved to give

$$z_{N-1} = C_{N-1} e^{\lambda t} + C_N t e^{\lambda t}$$

from which

$$z_{N-2} = C_{N-2} e^{\lambda t} + C_{N-1} t e^{\lambda t} + C_N \frac{t^2}{2!} e^{\lambda t}.$$

Rather than continuing mindlessly, let us postulate a general solution.

Claim. *The general solution for the dependent variables of Eq. (8.6) is*

$$z_n = \sum_{m=n}^{N} C_m e^{\lambda t} \frac{t^{m-n}}{(m-n)!}.$$

Proof: The claim is true for $n = N$ and $n = N - 1$. Use induction to show that if the claim is true for n, then the claim is true for $n - 1$. Assuming that the claim is true, then the differential equation for z_{n-1} is

$$\frac{dz_{n-1}}{dt} = \lambda z_{n-1} + \sum_{m=n}^{N} C_m e^{\lambda t} \frac{t^{m-n}}{(m-n)!} \quad \Rightarrow$$

$$\frac{d}{dt}(e^{-\lambda t}z_{n-1}) = \sum_{m=n}^{N} C_m \frac{t^{m-n}}{(m-n)!} \quad \Rightarrow$$

$$e^{-\lambda t}z_{n-1} = C_{n-1} + \sum_{m=n}^{N} C_m \frac{t^{m-n+1}}{(m-n+1)!} \quad \Rightarrow$$

$$z_{n-1} = \sum_{m=n-1}^{N} C_m e^{\lambda t} \frac{t^{m-(n-1)}}{(m-(n-1))!}$$

but this is precisely the form that the claim takes for $n-1$. □

Once the solution is known for Eq. (8.6), the solution to the original problem can be reconstructed. A comment, similar to that made for coupled difference equations, can now be made that although one might suspect, based on the results for single higher order equations, that repeated eigenvalues will give basis solutions for $t^m e^{\lambda_n t}$, this is not always the case. Semi-simple eigenvalues only give solutions of the form $e^{\lambda_n t}$ and these are linearly independent because they are associated with different and therefore linearly independent eigenvectors. Degenerate eigenvalues do give basis solutions of the form $t^m e^{\lambda_n t}$.

Example 8.10: Homogeneous solution, degenerate matrix.

Consider

$$\frac{dX}{dt} = AX = \begin{pmatrix} \frac{75}{28} & -\frac{23}{56} & \frac{9}{14} & -\frac{1}{56} \\ \frac{6}{7} & \frac{24}{7} & \frac{9}{7} & -\frac{9}{7} \\ \frac{4}{7} & -\frac{3}{14} & \frac{27}{7} & -\frac{5}{14} \\ \frac{9}{14} & \frac{23}{28} & \frac{12}{7} & \frac{29}{28} \end{pmatrix} X.$$

A spectral analysis shows that

$$A = KJK^{-1}$$

$$= \begin{pmatrix} 1 & 0 & 2 & -1 \\ 3 & 3 & 0 & -1 \\ 1 & 2 & 1 & 2 \\ 5 & 3 & 0 & 1 \end{pmatrix} \begin{pmatrix} 2 & 0 & 0 & 0 \\ 0 & 3 & 1 & 0 \\ 0 & 0 & 3 & 1 \\ 0 & 0 & 0 & 3 \end{pmatrix} \begin{pmatrix} \frac{3}{28} & -\frac{11}{56} & -\frac{3}{14} & \frac{19}{56} \\ -\frac{1}{7} & \frac{3}{7} & \frac{2}{7} & -\frac{2}{7} \\ \frac{11}{28} & -\frac{3}{56} & \frac{3}{14} & -\frac{5}{56} \\ -\frac{3}{28} & -\frac{17}{56} & \frac{3}{14} & \frac{9}{56} \end{pmatrix}.$$

In this result we explicitly state K^{-1}. However, calculating an explicit expression for K^{-1} is actually a waste of effort as K^{-1} is not needed in the calculations that follow. After diagonalization the problem in the variable $Y = K^{-1}X$ becomes

$$\frac{dY}{dt} = \begin{pmatrix} 2 & 0 & 0 & 0 \\ 0 & 3 & 1 & 0 \\ 0 & 0 & 3 & 1 \\ 0 & 0 & 0 & 3 \end{pmatrix} Y$$

or

$$\frac{dy_1}{dt} = 2y_1$$

and

$$\frac{d}{dt}\begin{pmatrix} y_2 \\ y_3 \\ y_4 \end{pmatrix} = \begin{pmatrix} 3 & 1 & 0 \\ 0 & 3 & 1 \\ 0 & 0 & 3 \end{pmatrix}\begin{pmatrix} y_2 \\ y_3 \\ y_4 \end{pmatrix}.$$

The solutions are

$$\begin{aligned} y_1 &= C_1 e^{2t} \\ y_2 &= C_2 e^{3t} + C_3 t e^{3t} + C_4 \frac{t^2}{2} e^{3t} \\ y_3 &= C_3 e^{3t} + C_4 t e^{3t} \\ y_4 &= C_4 e^{3t}. \end{aligned}$$

The vector $X = KY$ is obtained as

$$\begin{aligned} \begin{pmatrix} x_1 \\ x_2 \\ x_3 \\ x_4 \end{pmatrix} &= \begin{pmatrix} 1 & 0 & 2 & -1 \\ 3 & 3 & 0 & -1 \\ 1 & 2 & 1 & 2 \\ 5 & 3 & 0 & 1 \end{pmatrix} \begin{pmatrix} C_1 e^{2t} \\ C_2 e^{3t} + C_3 t e^{3t} + C_4 \frac{t^2}{2} e^{3t} \\ C_3 e^{3t} + C_4 t e^{3t} \\ C_4 e^{3t} \end{pmatrix} \\ &= \begin{pmatrix} C_1 e^{2t} + 2C_3 e^{3t} + 2C_4 t e^{3t} - C_4 e^{3t} \\ 3C_1 e^{2t} + 3C_2 e^{3t} + 3C_3 t e^{3t} + 3C_4 \frac{t^2}{2} e^{3t} - C_4 e3t \\ C_1 e^{2t} + 2C_2 e^{3t} + 2C_3 t e^{3t} + 2C_4 \frac{t^2}{2} e^{3t} + C_3 e^{3t} + 2C_4 t e^{3t} + C_4 e^{3t} \\ 5C_1 e^{2t} + 3C_2 e^{3t} + 3C_3 t e^{3t} + 3C_4 \frac{t^2}{2} e^{3t} + C_4 e^{3t} \end{pmatrix}. \end{aligned}$$

So the fundamental matrix for this system is

$$\Phi(t) = \begin{pmatrix} e^{2t} & 0 & 2e^{3t} & 2te^{3t} - e^{3t} \\ 3e^{2t} & 3e^{3t} & 3te^{3t} & 3\frac{t^2}{2} e^{3t} - e3t \\ e^{2t} & 2e^{3t} & 2te^{3t} + e^{3t} & t^2 e^{3t} + 2te^{3t} + e^{3t} \\ 5e^{2t} & 3e^{3t} & 3te^{3t} & 3\frac{t^2}{2} e^{3t} + e^{3t} \end{pmatrix}.$$

8.3.3.2 *Particular solutions for constant inhomogeneous term*

The special case where the inhomogeneous term is a constant vector is so common and so easy to solve that it deserves to be mentioned as a special case. The problem is

$$\frac{dX}{dt} = AX + B$$

and if a constant solution is not a homogeneous solution, i.e. if none of the eigenvalues equal zero, then a particular solution is a constant vector k which can be found by substitution into the differential equation

$$\begin{aligned} 0 &= Ak + B \Rightarrow \\ k &= -A^{-1}B. \end{aligned}$$

When A is a semi-simple matrix, the complete solution can therefore be written explicitly as

$$X(t) = \sum_{n=1}^{N} C_n \tilde{x}_n e^{\lambda_n t} - A^{-1}B$$

where the C_ns are arbitrary constants. When an initial condition of the form $X(0) = x_0$ is given, the arbitrary constants can be found by solving the system of equations

$$X_0 = \sum_{n=1}^{N} C_n \tilde{x}_n - A^{-1}B.$$

Rather than go through the trouble of solving a set of linear equations to find the arbitrary constants, the eigenrows can be used to this purpose. Consider again the equation obtained after application of the initial conditions to the complete solution and write this in the form

$$X_0 + A^{-1}B = \sum_{n=1}^{N} C_n \tilde{x}_n.$$

Take the inner product with an eigenrow, $\tilde{y}_m$ of the matrix A.

$$\langle \tilde{y}_m, X_0 + A^{-1}B \rangle = \sum_{n=1}^{N} C_n \langle \tilde{y}_m, \tilde{x}_n \rangle$$

and, since eigenrows and eigenvectors are orthogonal unless they belong to the same eigenvalue, all the terms in the sum, except the mth term, are zero giving

$$\langle \tilde{y}_m, X_0 + A^{-1}B \rangle = C_m \langle \tilde{y}_m, \tilde{x}_m \rangle \ \Rightarrow$$

$$C_m = \frac{\langle \tilde{y}_m, X_0 + A^{-1}B \rangle}{\langle \tilde{y}_m, \tilde{x}_m \rangle}$$

or the closed form solution

$$X(t) = \sum_{n=1}^{N} \frac{\langle \tilde{y}_n, X_0 + A^{-1}B \rangle}{\langle \tilde{y}_n, \tilde{x}_n \rangle} \tilde{x}_n e^{\lambda_n t} - A^{-1}B.$$

Do not forget that this result was derived under the assumption that the matrix A is semi-simple.

The method just described is particularly convenient when the matrix A is symmetric since, in this case, the eigenvectors and eigenrows are identical.

Example 8.11: Initial value problem with symmetric matrix.

Find the solution of

$$\frac{dx_1}{dt} = 9x_1 - 3x_2 + 30$$

$$\frac{dx_2}{dt} = -3x_1 + 12x_2 - 3x_3$$

$$\frac{dx_3}{dt} = -3x_2 + 9x_3$$

which satisfy the initial condition $t = 0 \Rightarrow (x_1, x_2, x_3) = (3, 1, -1)$.

The system matrix A is symmetric

$$A = \begin{pmatrix} 9 & -3 & 0 \\ -3 & 12 & -3 \\ 0 & -3 & 9 \end{pmatrix}$$

and a spectral analysis yields the following three simple eigenvalues and associated eigenvectors

$$\lambda_1 = 6 \ , \ \tilde{x}_1 = \begin{pmatrix} 1 \\ 1 \\ 1 \end{pmatrix} \ ; \ \lambda_2 = 9 \ , \ \tilde{x}_2 = \begin{pmatrix} 1 \\ 0 \\ -1 \end{pmatrix} \ ; \ \lambda_3 = 15 \ , \ \tilde{x}_3 = \begin{pmatrix} 1 \\ -2 \\ 1 \end{pmatrix}$$

from which the complete homogeneous solution is obtained as

$$x_h(t) = C_1 \begin{pmatrix} 1 \\ 1 \\ 1 \end{pmatrix} e^{6t} + C_2 \begin{pmatrix} 1 \\ 0 \\ -1 \end{pmatrix} e^{9t} + C_3 \begin{pmatrix} 1 \\ -2 \\ 1 \end{pmatrix} e^{15t}.$$

A particular solution is $A^{-1}b$. Rather than finding A^{-1} and post multiplying it by b use the computationally more efficient method of finding $A^{-1}b$ as the solution of $Ax = b$.

$$\begin{pmatrix} 9 & -3 & 0 & 30 \\ -3 & 12 & -3 & 0 \\ 0 & -3 & 9 & 0 \end{pmatrix} \Rightarrow \begin{pmatrix} 1 & 0 & 0 & \frac{11}{3} \\ 0 & 1 & 0 & 1 \\ 0 & 0 & 1 & \frac{1}{3} \end{pmatrix}$$

giving the complete solution

$$x(t) = C_1 \begin{pmatrix} 1 \\ 1 \\ 1 \end{pmatrix} e^{6t} + C_2 \begin{pmatrix} 1 \\ 0 \\ -1 \end{pmatrix} e^{9t} + C_3 \begin{pmatrix} 1 \\ -2 \\ 1 \end{pmatrix} e^{15t} - \begin{pmatrix} \frac{11}{3} \\ 1 \\ \frac{1}{3} \end{pmatrix}.$$

Application of the initial condition gives

$$\begin{pmatrix} 3 \\ 1 \\ -1 \end{pmatrix} + \begin{pmatrix} \frac{11}{3} \\ 1 \\ \frac{1}{3} \end{pmatrix} = C_1 \begin{pmatrix} 1 \\ 1 \\ 1 \end{pmatrix} + C_2 \begin{pmatrix} 1 \\ 0 \\ -1 \end{pmatrix} + C_3 \begin{pmatrix} 1 \\ -2 \\ 1 \end{pmatrix}.$$

The arbitrary constants are now determined by taking the inner product of this equation with each of the eigenvectors. Because the eigenvectors of a symmetric matrix are orthogonal, all terms on the right hand side, except the term with the nth eigenvector, will vanish when taking the inner product with the nth eigenvector. For instance, taking the inner product with $\tilde{x}_1$ gives

$$\left\langle \begin{pmatrix} 1 \\ 1 \\ 1 \end{pmatrix}, \begin{pmatrix} \frac{20}{3} \\ 2 \\ -\frac{2}{3} \end{pmatrix} \right\rangle = C_1 \left\langle \begin{pmatrix} 1 \\ 1 \\ 1 \end{pmatrix}, \begin{pmatrix} 1 \\ 1 \\ 1 \end{pmatrix} \right\rangle \Rightarrow$$

$$8 = 3C_1 \Rightarrow$$

$$C_1 = \frac{8}{3}.$$

In a similar manner, one finds

$$C_2 = \frac{11}{3}$$

and

$$C_3 = \frac{1}{3}.$$

8.3.3.3 *Dealing with complex eigenvalues*

Complex eigenvalues can be a great nuisance when it is known a priori that the desired solution is real. One can avoid having to work with complex arithmetic if one is willing to relax the requirement that the matrix of the similarity transformed system is diagonal or in Jordan form. The idea is to instead represent a pair of complex conjugate eigenvalues of the form $\lambda = a + bi$ as a 2-by-2 sub matrix of the form $\begin{pmatrix} a & -b \\ b & a \end{pmatrix}$ along the diagonal of the transformed matrix. This gives a set of two coupled equations instead of two separate equations, but the two coupled equations are real and easy to solve without having to resort to complex arithmetic.

Let the matrix A have a pair of complex conjugate eigenvalues $\lambda_1 = a + bi$ and $\lambda_2 = a - bi$ and let the associated eigenvectors be $\tilde{x}_1$ and $\tilde{x}_2$. The two eigenvectors are complex conjugates of one another. Instead of using the two complex eigenvectors as columns in the matrix for the similarity transformation, define two real basis vectors of the eigenspace as

$$x_1 = \frac{1}{2}(\tilde{x}_1 + \tilde{x}_2) \text{ and } x_2 = -\frac{i}{2}(\tilde{x}_1 - \tilde{x}_2).$$

Clearly, x_1 is the real part and x_2 the imaginary part of the eigenvector $\tilde{x}_1$. It is now trivial to calculate that

$$Ax_1 = ax_1 - bx_2$$

and

$$Ax_2 = bx_1 + ax_2\,.$$

Using x_1 and x_2 as adjacent columns in the matrix for the similarity transformation in lieu of the complex eigenvectors will therefore give a matrix with the structure claimed above.

Example 8.12: Complex eigenvalues.

Consider

$$\frac{dx}{dt} = \begin{pmatrix} -11 & -20 & 5 \\ 15 & 24 & -10 \\ 20 & 30 & -16 \end{pmatrix} x\,.$$

Formally, one would find the eigenvectors of this matrix and arrange them as columns in the similarity transformation matrix to get

$$\begin{pmatrix} 1 & 2 & -1 \\ -\frac{1}{2}-\frac{3}{2}i & -\frac{1}{2}-\frac{5}{2}i & \frac{1}{2}+i \\ -\frac{1}{2}+\frac{3}{2}i & -\frac{1}{2}+\frac{5}{2}i & \frac{1}{2}-i \end{pmatrix} \begin{pmatrix} -11 & -20 & 5 \\ 15 & 24 & -10 \\ 20 & 30 & -16 \end{pmatrix} \begin{pmatrix} -3 & -1+i & -1-i \\ 1 & 1 & 1 \\ -2 & 1+i & 1-i \end{pmatrix}$$
$$= \begin{pmatrix} -1 & 0 & 0 \\ 0 & -1+5i & 0 \\ 0 & 0 & -1-5i \end{pmatrix}$$

and end up with a set of decoupled but complex ODEs. Instead, following the suggestion above, do a similarity transformation with the matrix

$$K = \begin{pmatrix} -3 & -1 & -1 \\ 1 & 1 & 0 \\ -2 & 1 & -1 \end{pmatrix}$$

to get the transformed system

$$\frac{dz}{dt} = \begin{pmatrix} -1 & 0 & 0 \\ 0 & -1 & -5 \\ 0 & 5 & -1 \end{pmatrix} z$$

which is readily solved to get

$$\begin{pmatrix} z_1 \\ z_2 \\ z_3 \end{pmatrix} = \begin{pmatrix} C_1 e^{-t} \\ (C_2 \sin(5t) + C_3 \cos(5t))e^{-t} \\ (C_3 \sin(5t) - C_2 \cos(5t))e^{-t} \end{pmatrix}.$$

8.3.3.4 *Classification of steady states*

It is clear from the foregoing analysis that systems of coupled first order ODEs with constant coefficients have a steady state at zero and that this steady state is stable if all eigenvalues have negative real parts and it is unstable if just a single eigenvalue has a positive real part.

A special case is that for which none of the eigenvalues have positive real parts but some of the eigenvalues have real parts that equal zero. These systems are stable in the sense that they have solutions that stay constant but these solutions are not stable in the sense that they return to the previous constant value if perturbed. For instance, the system

$$\frac{dx}{dt} = \begin{pmatrix} -1 & 1 \\ 1 & -1 \end{pmatrix} x$$

has the solution

$$\begin{pmatrix} x_1 \\ x_2 \end{pmatrix} = C_1 \begin{pmatrix} 1 \\ 1 \end{pmatrix} + C_2 \begin{pmatrix} e^{-2t} \\ -e^{-2t} \end{pmatrix}$$

so any solution of the form $(x_1, x_2) = (C_1, C_1)$ is stable in the sense that it remains constant as t goes to infinity. However, if such a state, e.g. $(x_1, x_2) = (1, 1)$, is perturbed to e.g. $(x_1, x_2) = (1, 1.01)$, then the solution is

$$\begin{pmatrix} x_1 \\ x_2 \end{pmatrix} = \begin{pmatrix} 1.005 - 0.005_2 e^{-2t} \\ 1.005 + 0.0052 e^{-2t} \end{pmatrix}$$

and the system will approach the different steady state solution $(x_1, x_2) = (1.005, 1.005)$ as t goes to infinity. Solutions that are stable in this sense, with one or more eigenvalues that have a zero real part (and no eigenvalues with positive real parts) are called *critically* or *neutrally stable.*

It is often desirable to have a more detailed description of the dynamics close to a steady state than simply stable, unstable or critically stable. This is obtained by classifying the dynamics based on the eigenvalue structure and the different types of dynamics are usually illustrated through phase space plots.

A phase space plot is a plot of the dependent variables against one another, giving rise to a family of curves called trajectories that are parameterized by the free variable. This family of curves defines the *flow* in the phase space. Each point in the phase space can be thought of as an initial condition for the differential equations and, as the solution corresponding to any initial condition is unique, the trajectories cannot intersect one another, except at a steady state.

The phase space concept is most easily illustrated in two dimensions where the space is usually called the phase plane.

Nodes: Nodes are steady states for which all eigenvalues are simple or semi-simple, real and all have the same sign. If the eigenvalues are all negative, the node is stable; if they are all positive, then the node is unstable. For instance, the system

$$\frac{dx}{dt} = \begin{pmatrix} -3 & \frac{1}{2} \\ 2 & -3 \end{pmatrix} x \tag{8.7}$$

which has the solution

$$\begin{pmatrix} x_1 \\ x_2 \end{pmatrix} = C_1 \begin{pmatrix} e^{-2t} \\ 2e^{-2t} \end{pmatrix} + C_2 \begin{pmatrix} e^{-4t} \\ -2e^{-4t} \end{pmatrix}$$

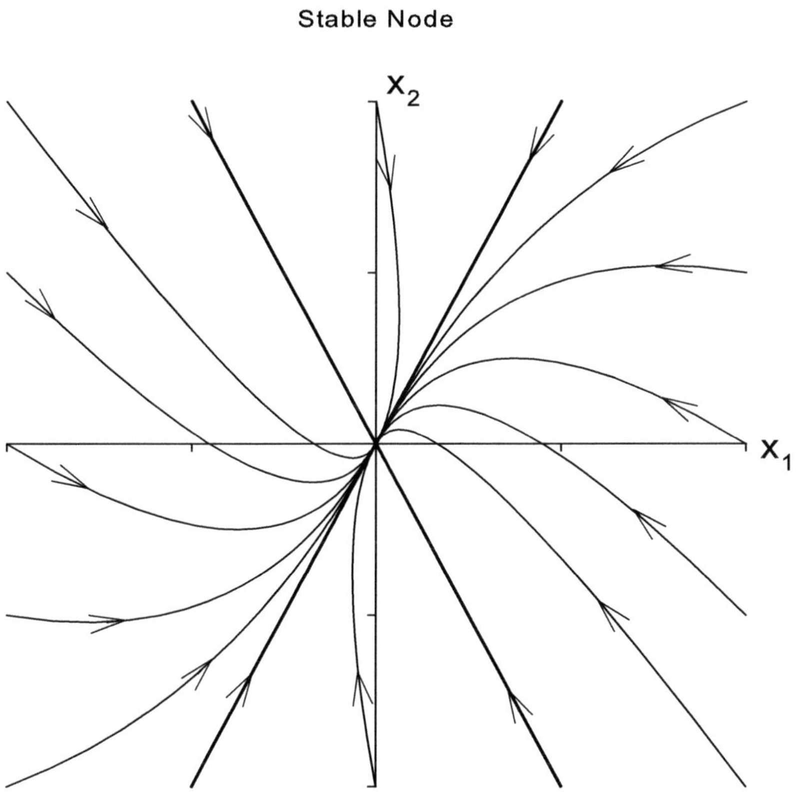

Fig. 8.1 Phase plane plot of the system in Eq. (8.7). The two eigendirections are shown by the thick lines.

has a steady state which is a stable node. Picking a variety of initial conditions and plotting the corresponding solutions in the (x_1, x_2)-plane gives the phase plane plot for this system, Fig. 8.1.

The arrows indicate the direction of the flow, that is, the direction along the trajectory in which the dependent variables change when the free variable increases. There are two, and only two, straight line trajectories in the flow, corresponding to the two eigendirections, (1,2) for $\lambda = -2$ and (1,-2) corresponding to $\lambda = -4$. It is immediately obvious from the figure which eigendirection corresponds to the largest and which to the smallest negative eigenvalue. Since the solution component corresponding to the largest negative eigenvalue goes to zero the fastest all the trajectories must approach the steady state along the eigendirection corresponding to the smallest negative eigenvalue. The trajectories are therefore tangent to the eigendirection of the smallest negative eigenvalue at the steady state (except, of course for the two trajectories that correspond to the eigendirection of the largest negative eigenvalue.)

If the system in Eq. (8.7) is similarity transformed to diagonalize the matrix, it becomes

$$\frac{dz}{dt} = \begin{pmatrix} -2 & 0 \\ 0 & -4 \end{pmatrix} z \tag{8.8}$$

and its phase plane plot is shown in Fig. 8.2.

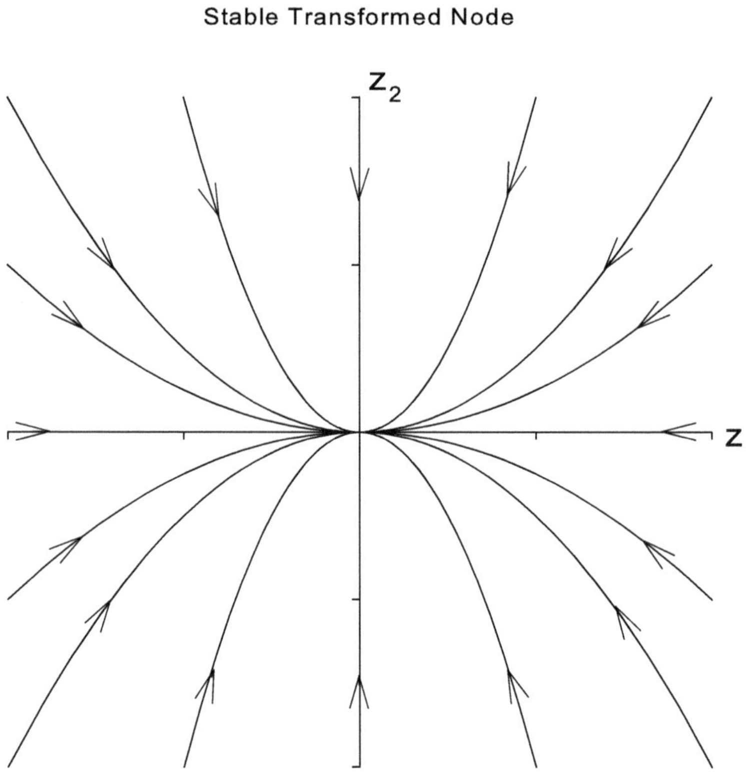

Fig. 8.2 Phase plane plot of the system in Eq. (8.8). The two eigendirections are the two axis.

The phase plane plot of the transformed system is considerably simpler looking than for the original system and shows the advantage using the similarity transformation. The eigendirections are now orthogonal and form the axis of the coordinate system and, as the flow is tangent to the z_1 axis, it is immediately clear that the eigenvalues are ordered as $\lambda_1 > \lambda_2$.

The phase plane plots for an unstable node are obtained simply by reversing the direction of the flow.

A special case of a node is a star. A star occurs when the eigenvalues are real, equal and semi-simple.

Saddle points: Saddle points are steady states for which the eigenvalues are simple or semi-simple, real and of opposite signs. For instance, the steady state of the system

$$\frac{dx}{dt} = \begin{pmatrix} 5 & -4 \\ 8 & -7 \end{pmatrix} x \tag{8.9}$$

which has the solution

$$\begin{pmatrix} x_1 \\ x_2 \end{pmatrix} = C_1 \begin{pmatrix} e^t \\ e^t \end{pmatrix} + C_2 \begin{pmatrix} e^{-3t} \\ 2e^{-4t} \end{pmatrix}$$

is a saddle point. The phase plane plot is shown in Fig. 8.3.

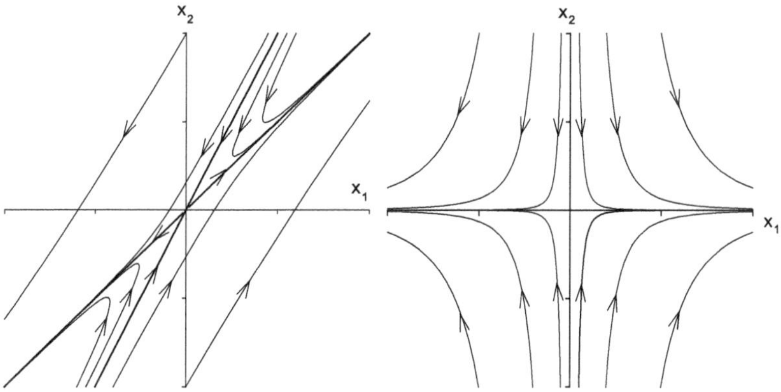

Fig. 8.3 Phase plane plot of a saddle point. On the left, the plot of the system in Eq. (8.9). On the right, the plot for the similarity transformed system.

The two straight lines are the trajectories that correspond to the stable and unstable basis solutions, e^{-3t} and e^t. These lines are known as the stable and unstable manifolds, respectively. Trajectories that start close to the stable manifold initially approach the steady state but then bend towards the unstable manifold as the unstable term in the solution becomes more significant and move away from the steady state as they approach the unstable manifold asymptotically.

Nodes and saddle points are the most important types of steady states. Examples of phase plane plots of the remaining types of steady states are shown in Figs. 8.4 and 8.5. All are shown for a stable steady state. The unsteady plots are obtained by reversal of the flow direction.

The star has already been mentioned as the phase plot for two identical real eigenvalues. It has the same appearance for all values of the eigenvalues, Fig. 8.4.

The focus: The phase plane is a focus when the two eigenvalues are complex conjugates. The trajectories are spirals that approach or leave from the steady

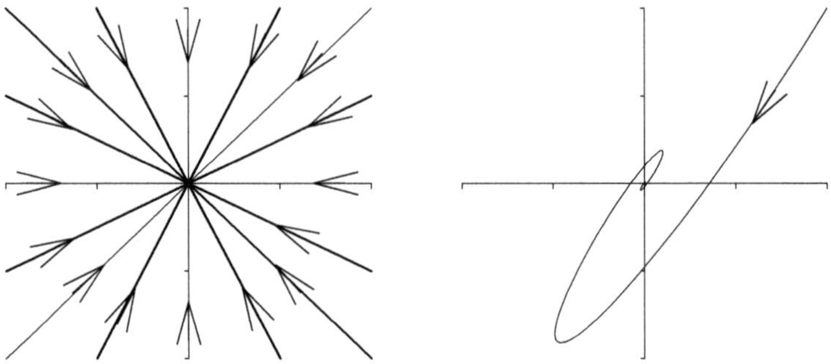

Fig. 8.4 Various phase plane plots. On the left, a stable star, obtained for two identical and real simple or semi-simple eigenvalues. On the right, a stable focus, obtained for a pair of complex conjugate eigenvalues. The plot is for the system $x_1' = -7x_1 + 4x_2$, $x_2' = -10x_1 + 5x_2$.

state, depending on whether the focus is stable or unstable. The plot shown in Fig. 8.4 is for the system

$$\frac{dx}{dt} = \begin{pmatrix} -7 & 4 \\ -10 & 5 \end{pmatrix} x \Rightarrow$$

$$\begin{pmatrix} x_1 \\ x_2 \end{pmatrix} = C_1 \begin{pmatrix} \sin(2t)e^{-t} \\ \frac{1}{2}(3\sin(2t) + \cos(2t))e^{-t} \end{pmatrix} + C_2 \begin{pmatrix} \cos(2t)e^{-t} \\ \frac{1}{2}(3\cos(2t) - \sin(2t))e^{-t} \end{pmatrix}.$$

The focus is stable if the real part of the eigenvalues is negative and unstable if the real parts are positive. If the real parts are zero, the trajectories become closed curves, a family of ellipses nested around the steady state, and the phase plane plot is then called a center.

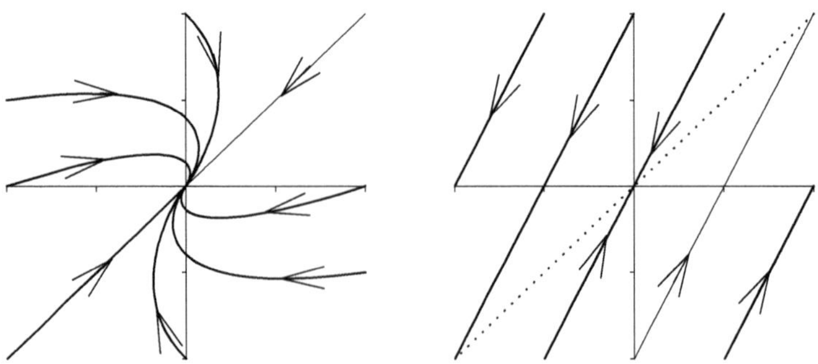

Fig. 8.5 Various phase plane plots. On the left, a degenerate system, $x_1' = -3x_1 + x_2$, $x_2' = -x_1 + x_2$. On the right, a system with an eigenvalue equal to zero,$x_1' = 2x_1 - x_2$, $x_2' = 4x_1 - 4x_2$.

When the system matrix is degenerate, a plot looking like a distorted node, as seen on the left in Fig. 8.5, is obtained. The plot is shown for the system

$$\frac{dx}{dt} = \begin{pmatrix} -3 & 1 \\ -1 & -1 \end{pmatrix} x \Rightarrow$$

$$\begin{pmatrix} x_1 \\ x_2 \end{pmatrix} = C_1 \begin{pmatrix} e^{-2t} \\ e^{-2t} \end{pmatrix} + C_2 \begin{pmatrix} te^{-2t} \\ (1+t)e^{-2t} \end{pmatrix}$$

which has the eigenvalue, $\lambda = -2$ and a one-dimensional eigenspace spanned by $(x_1, x_2) = (1, 1)$. The one-dimensional degenerate eigenspace shows up as the single straight line in the figure and trajectories on opposite sides of the eigendirection line approach the steady state from opposite directions.

The phase plane plot for a critically stable system is shown on the right in Fig. 8.5. The system shown is

$$\frac{dx}{dt} = \begin{pmatrix} 2 & -2 \\ 4 & -4 \end{pmatrix} x \Rightarrow$$

$$\begin{pmatrix} x_1 \\ x_2 \end{pmatrix} = C_1 \begin{pmatrix} 1 \\ 1 \end{pmatrix} + C_2 \begin{pmatrix} e^{-2t} \\ 2e^{-2t} \end{pmatrix}$$

and the dotted line indicates the critically stable steady states, $x_2 = x_1$. For the system shown, the trajectories all intersect the line of critically stable steady states. However, it is also possible for the trajectories to be parallel to the line of steady states. The reader is urged to try and figure out when this will happen.

These ideas are easily extended to higher dimensions though actual plots are obviously limited to three or fewer dimensions. Phase space plots for the system

$$\frac{dx}{dt} = \begin{pmatrix} -11 & -20 & 5 \\ 15 & 24 & -10 \\ 20 & 30 & -16 \end{pmatrix} x \tag{8.10}$$

are shown in Fig. 8.6 on the left.

This system can be similarity transformed to

$$\frac{dz}{dt} = \begin{pmatrix} -1 & 0 & 0 \\ 0 & -1 & -5 \\ 0 & 5 & -1 \end{pmatrix} z \tag{8.11}$$

which is shown in the plot on the right in Fig. 8.6. Neither plot is particularly easy to interpret and one can usually gain a better understanding of the dynamics by

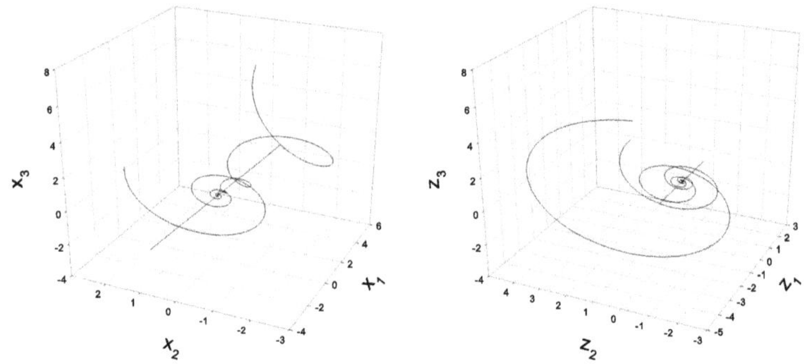

Fig. 8.6 Phase space plots for a three-dimensional system. On the left, the plot for Eq. (8.10). On the right, the plot for the transformed system, Eq. (8.11).

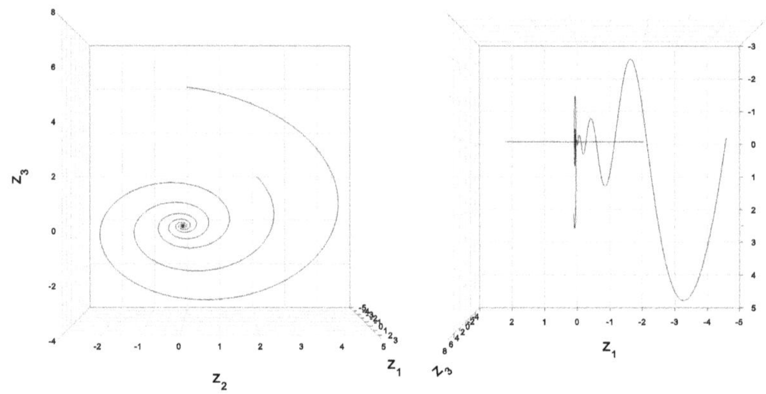

Fig. 8.7 Phase space plots for a three-dimensional system. On the left, the plot for Eq. (8.10). On the right, the plot for the transformed system, Eq. (8.11).

turning the plot in space until a better, more comprehensible orientation is obtained. Such plots are shown in Fig. 8.7.

The plot on the left in Fig. 8.7 shows the phase space plot as it appears when looking down the eigendirection of the real eigenvalue. Clearly, the remaining two eigenvalues are a complex conjugate pair forming a focus like structure in the (z_2, z_3) plane. The plot on the right shows a view at a right angle to the eigendirection of the real eigenvalue but with the same three trajectories as in the figure on the left. One of the spiral trajectories is viewed from the side and has no component or dynamics in the z_1 direction. The other spiral trajectory captures both the oscillations with decreasing amplitude of the two complex eigenvalues and the monotone decrease towards the steady state in the z_1 direction.

8.4 Stability of nonlinear ODEs

Nonlinear ODEs is a gargantuan topic and all we will mention here is the issue of stability because it is closely linked to the issue of the dynamics of linear differential equations with constant coefficients. The stability analysis that is presented here is a so-called local analysis and it determines whether a system will return to a steady state if perturbed by a "small amount". It does not tell us what constitutes a "small" perturbation. Results from this type of analysis must therefore be viewed with caution. More sophisticated types of stability analysis provide stronger conclusions but are generally far more difficult to carry out. In fact, simply talking about stability, as we will do in this cursory discussion, turns out to be inadequate for a deeper understanding and different types of stability concepts, asymptotic stability, conditional stability, global stability etc., are needed.

One can extend the stability analysis for homogeneous, linear ODEs with constant coefficients to nonlinear differential equations by linearizing these equations about the steady states. Linearization creates a system of coupled, linear differential equations with constant coefficients and the local stability of the steady state of the nonlinear equations can, in most cases, be inferred from the stability of the linear system. The linearized system of equations will provide an approximate description of the dynamics of the nonlinear system close to the steady state around which the nonlinear system was linearized. In fact, in some engineering models, the approximate model obtained by linearization of the full nonlinear model may be adequate for engineering purposes such as process control.

Consider an autonomous set of N coupled ODEs of the form

$$\frac{dx}{dt} = F(x).$$

The steady states (or equilibrium points or critical points or fixed points) of this system are the solutions of

$$F(x) = 0\,.$$

In engineering applications, the dependent variable, x, is almost always real. We are therefore *only* interested in real roots of $F(x)$ and complex roots are ignored. Let one of these equilibrium points be x_e. The set of differential equations can now be linearized around this point by expanding $F(x)$ in Taylor series around x_e.

$$F_n(x) = F_n(x_e) + \left.\frac{dF_n}{dx_1}\right|_{x=x_e}(x_1 - x_{1,e}) + \ldots + \left.\frac{dF_n}{dx_N}\right|_{x=x_e}(x_N - x_{N,e}) + \text{HOT}$$

where F_n is the nth element of $F(x)$ and $X_e = (x_{1,e}, x_{2,e}, \ldots, x_{N,e})$. Neglecting higher order terms (HOT) and noting that $F_n(x_e) = 0$, the approximate linear system becomes

$$\frac{dX}{dt} \approx \begin{pmatrix} \frac{\partial F_1}{\partial x_1} & \frac{\partial F_1}{\partial x_2} & \cdots & \frac{\partial F_1}{\partial x_N} \\ \frac{\partial F_2}{\partial x_1} & \frac{\partial F_2}{\partial x_2} & \cdots & \frac{\partial F_2}{\partial x_N} \\ \vdots & \vdots & \ddots & \vdots \\ \frac{\partial F_N}{\partial x_1} & \frac{\partial F_N}{\partial x_2} & \cdots & \frac{\partial F_N}{\partial x_N} \end{pmatrix}_{x=x_e} \begin{pmatrix} x_1 - x_{1,e} \\ x_2 - x_{2,e} \\ \vdots \\ x_N - x_{N,e} \end{pmatrix} = J(X_e)(X - X_e).$$

The matrix $J(x)$ is called the Jacobian matrix[2] and represents the first derivative of a vector valued function of a vectorial argument. The notation, J, should not lead one to confuse it with a Jordan matrix. It is convenient to define deviation variables, $\bar{x}_n = x_n - x_{n,e}$ or $\bar{X} = X - X_e$, giving

$$\frac{d\bar{X}}{dt} = J(X_e)\bar{X}.$$

The stability of the original nonlinear set of equations at the equilibrium point is now identical to the stability of the point $X = 0$ in the above equations. Obviously, the stability of $X = 0$ is determined by the eigenvalues of the Jacobian matrix evaluated at X_e. Stability requires that all eigenvalues have negative real parts. If any eigenvalue has a positive real part, then the equilibrium point is unstable. If the real part of the eigenvalue with the largest real part equals zero, one cannot conclude anything about the stability from this type of analysis.

The last point, that one cannot conclude anything about the stability of an equilibrium point if the largest real part of the eigenvalues equal zero, is often forgotten. After all, a zero eigenvalue implies that the associated eigenfunction is a constant which indicates stability. This is incorrect. A zero real part means that the stability is determined by the higher order terms in the Taylor expansion and the linear analysis presented here is therefore insufficient to conclude anything about stability.

To see this, consider the ordinary differential equation

$$\frac{dx}{dt} = \alpha x^3.$$

The only equilibrium point is $x = 0$ and linearizing the system around this point gives

$$\frac{d\bar{x}}{dt} = 0$$

but the equilibrium point is stable iff $\alpha < 0$ and unstable if $\alpha > 0$. (Convince yourself of this. You can do so easily by solving the differential equation).

[2]After the German mathematician C.G.J. Jacobi, 1804–1851.

Example 8.13: Stability of the logistic equation.

We previously encountered the logistic equation, page 126, Example 4.2.

$$\frac{dx}{dt} = ax - bx^2$$

where the two parameters a and b are positive. We found the the equation had two steady states

$$x_1 = \frac{a}{b} \text{ and } x_2 = 0$$

and argued from graphing a few solutions that the x_1-solution is stable while the x_2-solution is unstable. We can now show this rigorously. Expanding the right hand side in a Taylor series around x_e gives

$$\frac{dx}{dt} = (ax_e - bx_e^2) + (a - 2bx_e)(x - x_e) - 2b(x - x_e)^2.$$

Discarding higher order terms, simplifying and introducing the deviation variable gives

$$\frac{d\bar{x}}{dt} = (a - 2bx_e)\bar{x}.$$

Consider first $x_1 = \frac{a}{b}$. At this point, the equation becomes

$$\frac{d\bar{x}}{dt} = -a\bar{x}$$

and since $a > 0$ the single eigenvalue is negative and the steady state is stable. The other steady state, $x_2 = 0$ gives

$$\frac{d\bar{x}}{dt} = a\bar{x}$$

so the single eigenvalue is positive and the steady state is unstable.

Example 8.14: Stability of two coupled ODEs.

Consider the two coupled equations

$$\frac{dx}{dt} = x^2 + y^2 - 1$$

$$\frac{dy}{dt} = x^2 + 1 - y.$$

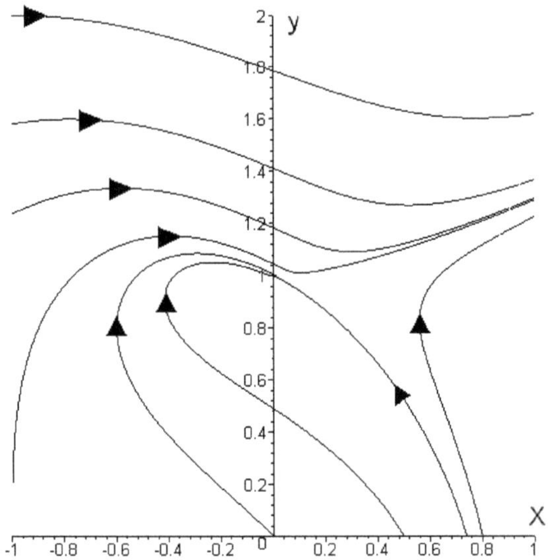

Fig. 8.8 Phase plane plot for $\frac{dx}{dt} = x^2 + y^2 - 1$, $\frac{dy}{dt} = x^2 + 1 - y$.

The fixed points are found by solving

$$x^2 + y^2 = 1\ ,\ y = x^2 + 1\,.$$

Eliminating x^2 gives

$$y - 1 + y^2 = 1 \Rightarrow y = 1 \text{ or } y = -2\,.$$

The solution $y = -2$ gives $x = \pm\sqrt{3}i$ which is complex and this solution is discarded. The only real valued fixed point therefore is $(x, y) = (0, 1)$. The numerically obtained phase plane plot in the vicinity of this point is shown in Fig. 8.8.

Linearizing around the fixed point gives

$$\frac{dx}{dt} = 2(y-1) \text{ or } \frac{d\overline{x}}{dt} = 2\overline{y}$$

$$\frac{dy}{dt} = -(y-1) \text{ or } \frac{d\overline{y}}{dt} = -\overline{y}$$

or

$$\begin{pmatrix} \frac{d\overline{x}}{dt} \\ \frac{d\overline{y}}{dt} \end{pmatrix} = \begin{pmatrix} 0 & 2 \\ 0 & -1 \end{pmatrix} \begin{pmatrix} \overline{x} \\ \overline{y} \end{pmatrix}.$$

Since one of the eigenvalues is zero the system is critically stable and it is not possible to conclude anything about the stability of the fixed point. Moreover, since one of the eigenvalues is zero, the system is not even approximately linear and

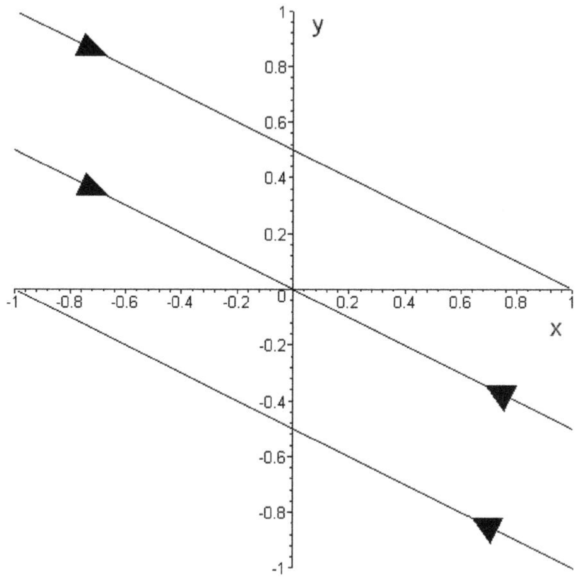

Fig. 8.9 Phase plane plot for $\frac{dx}{dt} = 2y$, $\frac{dy}{dt} = -y$.

dynamics of the system in the vicinity of the fixed point cannot be approximated by the linear system. This is clearly seen by comparing the phase plane plots of the two systems. The phase plane plot of the linearized system is shown in Fig. 8.9.

The two eigenvectors are $\tilde{v}_1 = (1, 0)$ for $\lambda_1 = 0$ and $\tilde{v}_2 = (2, -1)$ for $\lambda_2 = -1$. The flow is always parallel to the second eigenvector, in the direction towards the x-axis and all trajectories terminate on the x-axis. Of course, the phase plane plot looks like this at any "magnification".

A close-up of the phase plane plot around the fixed point of the nonlinear system is shown in Fig. 8.10.

Superficially, this may look like the phase plane plot for the linear system, with trajectories that are parallel to the eigenvector $\tilde{v}_2 = (2, -1)$ and a flow that is always towards the x-axis. However, the trajectories all bend towards the right as they approach the x-axis and either reach the fixed point or proceed towards infinity. This dynamics is fundamentally different from the dynamics of the linearized system and the difference will remain no matter how small a neighborhood around the fixed point one considers.

Example 8.15: Stability of bioreactor.

Consider a CSTR in which a microorganism grows on a single substrate. The rate of biomass formation, R_X, relative to the rate of substrate consumption, R_S, is called the yield, Y, and will be assumed constant. The steady state balances on

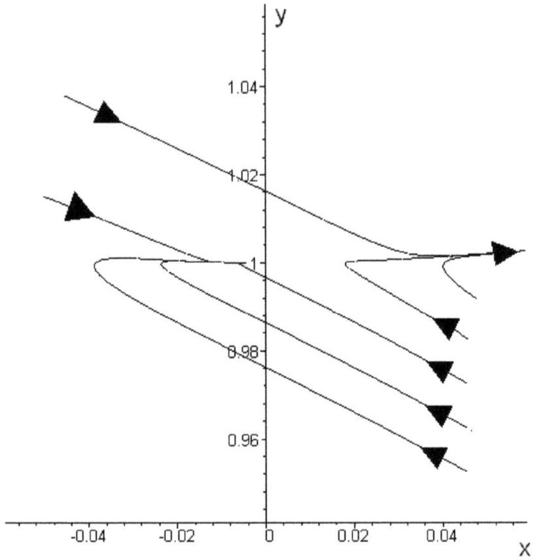

Fig. 8.10 Phase plane plot for $\frac{dx}{dt} = x^2 + y^2 - 1$, $\frac{dy}{dt} = x^2 + 1 - y$.

biomass and substrate can thus be written

$$V\frac{dX}{dt} = R_X V - FX$$

$$V\frac{dS}{dt} = F(S_f - S) - \frac{1}{Y}R_X V$$

where X is the biomass concentration, S the substrate concentration, V the reactor volume, F the volumetric flow rate through the reactor and S_f the substrate concentration in the feed stream. The feed stream is assumed sterile.

The rate of biomass growth will be described by the Monod model.[3]

$$R_X = \frac{\mu_m S}{K + S} X$$

where μ_m is the maximum specific growth rate and K is a constant simply called the Monod constant. Substituting this expression into the balances and introducing the dilution rate $D = F/V$ gives

$$\frac{dX}{dt} = \frac{\mu_m S}{K + S} X - DX$$

[3]Monod, J. [1949] "The Growth of Bacterial Cultures," *Ann. Rev. Microbiol.* **3**, 371; Monod, J. [1950] "La technique de culture continue theorie at applacations," *Ann. Inst. Pasteur, Paris* **79**, 390.

$$\frac{dS}{dt} = D(S_f - S) - \frac{1}{Y}\frac{\mu_m S}{K+S}X.$$

Before proceeding with the linearization of the equations, it is often a good idea to simplify them by introducing dimensionless variables. Define

$$\kappa \equiv \frac{K}{S_f}\ ,\ y \equiv \frac{X}{YS_f}\ ,\ z \equiv \frac{S}{S_f}\ ,\ \mathcal{D} \equiv \frac{D}{\mu_m}\ ,\ \tau \equiv t\mu_m$$

which gives

$$\frac{dy}{d\tau} = \frac{zy}{\kappa + z} - \mathcal{D}y$$

$$\frac{dz}{d\tau} = \mathcal{D}(1-z) - \frac{zy}{\kappa+z}\,.$$

Two equilibrium points can be found, the sterile steady state

$$y = 0\ ,\ z = 1$$

and the growth steady state

$$y = 1 - \frac{\kappa\mathcal{D}}{1-\mathcal{D}}\ ,\ z = \frac{\kappa\mathcal{D}}{1-\mathcal{D}}\,.$$

Expanding the right hand sides in Taylor series around either equilibrium point, (y_e, z_e), discarding higher order terms and introducing deviation variables gives

$$\begin{pmatrix} \frac{d\bar{y}}{d\tau} \\ \frac{d\bar{z}}{d\tau} \end{pmatrix} = \begin{pmatrix} \left(\frac{z_e}{\kappa+z_e} - \mathcal{D}\right) & \frac{\kappa y_e}{(\kappa+z_e)^2} \\ -\frac{z_e}{\kappa+z_e} & -\left(\mathcal{D} + \frac{\kappa y_e}{(\kappa+z_e)^2}\right) \end{pmatrix} \begin{pmatrix} \bar{y} \\ \bar{z} \end{pmatrix}.$$

At the sterile steady state, the Jacobian becomes

$$J = \begin{pmatrix} \left(\frac{1}{\kappa+1} - \mathcal{D}\right) & 0 \\ -\frac{1}{\kappa+1} & -\mathcal{D} \end{pmatrix}.$$

The eigenvalues are

$$\lambda_1 = -\mathcal{D} = -\frac{D}{\mu_m}$$

$$\lambda_2 = -\mathcal{D} + \frac{1}{\kappa+1} = -\frac{D}{u_m} + \frac{S_f}{K+S_f}\,.$$

Both are real. From physical arguments, all the dimensioned parameters, D, μ_m, K and S_f, are positive. Therefore, λ_1 is negative and λ_2 is negative iff $D >$

$\mu_m S_f/(K + S_f)$. The quantity, $\mu_m S_f/(K + S_f)$ is called the washout dilution rate and the result shows that the sterile steady state is stable when the reactor is operated above the washout dilution rate and unstable when operated below the washout dilution rate. As long as the reactor is operated below the washout dilution rate, it can therefore be started by placing even just one single cell in the reactor and it will eventually reach the steady state. Of course, using only a single cell to start the reactor is not likely to be the most economical way of operating the process.

At the growth steady state, the Jacobian equals

$$J = \begin{pmatrix} \frac{\kappa\mathcal{D}}{(1-\mathcal{D})(\kappa+\frac{\kappa\mathcal{D}}{1-\mathcal{D}})} - \mathcal{D} & \frac{\kappa(1-\frac{\kappa\mathcal{D}}{1-\mathcal{D}})}{(\kappa+\frac{\kappa\mathcal{D}}{1-\mathcal{D}})^2} \\ -\frac{\kappa\mathcal{D}}{(1-\mathcal{D})(\kappa+\frac{\kappa\mathcal{D}}{1-\mathcal{D}})} & -\mathcal{D} - \frac{\kappa(1-\frac{\kappa\mathcal{D}}{1-\mathcal{D}})}{(\kappa+\frac{\kappa\mathcal{D}}{1-\mathcal{D}})^2} \end{pmatrix}$$

which has the eigenvalues

$$\lambda_1 = -\mathcal{D}$$

$$\lambda_2 = -\frac{1 - 2\mathcal{D} + \mathcal{D}^2 - \kappa\mathcal{D} + \kappa\mathcal{D}^2}{\kappa}.$$

Both are real and λ_1 is negative. λ_2 is seen to be positive only when $1/(1+\kappa) < \mathcal{D} < 1$, (when $\mathcal{D}$ is between the roots of the polynomial in the numerator). In dimensioned variables, this is equivalent to

$$\frac{\mu_m S_f}{\kappa + S_f} < D < \mu_m \,.$$

The lower limit is expected, the growth steady state is stable if the dilution rate is less than the washout dilution rate. The upper limit is puzzling. It implies that the growth steady state is stable if the dilution rate is greater than the maximum specific growth rate of the organism. This conclusion, of course, makes no sense biologically but the problem is resolved once it is realized that the steady state solutions for $D > \mu_m$ correspond to negative biomass concentrations, i.e. steady state solutions that are not physically meaningful.

8.5 Series solutions

In general, it is not possible to find closed form expressions for the basis solutions of linear equations of order higher than 1, only infinite series solutions are possible. The most common series solutions, such as the exponential and trigonometric functions, are so familiar that most of us give little thought to the fact that these functions cannot be evaluated algebraically or exactly, but only approximately by evaluations of some finite number of terms in an infinite series. In this section we will present a general method for finding infinite series representation of the basis solutions of linear differential equations. The technique for obtaining infinite series solutions dates to 1873 to the Mathematician Frobenius[4] and is often just called Frobenius' method. Discussion of the method will be limited to second order equations of the form

$$\frac{d^2x}{dt^2} + p_1(t)\frac{dx}{dt} + p_0(t)x = 0$$

and to development of series solutions around $t = 0$. If a series solution is desired around another point, $t = t_0$, a translation along the t-axis, or equivalently, the variable transformation $\tau = t - t_0$, will transform the problem to one of finding the series solution around 0. The last restriction does therefore not result in any loss of applicability of the method. The two coefficient functions, $p_1(t)$ and $p_0(t)$ will be assumed to be polynomials in t. If this is not the case, they must be represented instead by the Taylor series expansions.

Occasionally, one needs to consider equations in the form

$$p_2(t)\frac{d^2y}{dt^2} + p_1(t)\frac{dx}{dt} + p_0(t)x = q(t)$$

where $p_2(t)$ equals zero at some points in the interval of interest and $q(t)$ is continuous. If these zeros are isolated, i.e. if they occur as discrete points instead of over an interval, one must subdivide the interval of interest by the zeros of $p_2(t)$ and find a solution in each subinterval. A twice differentiable function $x(t)$ which satisfies the ODE in each subinterval not containing a zero of $p_2(t)$ will then be a solution over the whole interval of interest. This follows from the fact that by substituting $x(t)$ into the left hand side of the ODE, one obtains a continuous function over the interval of interest which equals $q(t)$ in each point which is not a zero of $p_2(t)$. But this function must equal $q(t)$ even at the zeros of $p_2(t)$ because $q(t)$ is continuous.

An initial value problem presents a special case which is somewhat simpler to solve than the problem of finding series representations for the basis solutions. This type of problem will therefore be considered first. Let the initial condition be $(t, x, x') = (0, a, b)$ and seek a series solution around $t = 0$. Note that the initial condition assures that derivatives of any order exists and are finite (because these can be obtained by repeated differentiation of the ODE) so the solution can be written in terms of a Taylor series as

[4]Frobenius, G. [1873] "Ueber die Integration der linearen Differentialgleichungen durch Reihen," *J. für Math.* **76**, 214–235. More recent descriptions of the series solution method can be found, with varying amounts of detail, in a large number of books. The two books suggested here are therefore somewhat arbitrary picks.
Coddington, E.A. [1961] *An Introduction to Ordinary Differential Equations* (Prentice-Hall, Englewood Cliffs, NJ).
Boyce, W.E. and DiPrima, R.C. [1969] *Elementary Differential Equations and Boundary Value Problems* (Wiley, New York).

$$x(t) = \sum_{n=0}^{\infty} a_n t^n .$$

It follows from the initial condition that $a_0 = a$ and $a_1 = b$. The rest of the coefficients are found by differentiating the series term by term and substituting the results into the differential equation. After appropriate simplifications one will obtain a series which is identically equal to zero, implying that all the coefficients are zero. This last result is then used to obtain a difference equation for the a_ns. This difference equation is sufficient for evaluating the series on a computer but a closed form expression for a_n is usually obtainable from the equation without too much effort if such an expression is desired. The method is illustrated in the next example.

Example 8.16: Series solution of initial value problem.

Consider

$$\frac{d^2x}{dt^2} + (t^2+1)\frac{dx}{dt} + tx = 0$$

subject to the initial condition $(t, x, x') = (0, 1, 2)$. The solution can be represented by a Taylor series of the form

$$x(t) = \sum_{n=0}^{\infty} a_n t^n$$

where the initial conditions dictate that $a_0 = 1$ and $a_1 = 2$. Substituting the series into the differential equation gives

$$\sum_{n=2}^{\infty} a_n n(n-1)t^{n-2} + \sum_{n=1}^{\infty} a_n n t^{n+1} + \sum_{n=1}^{\infty} a_n n t^{n-1} + \sum_{n=0}^{\infty} a_n t^{n+1} = 0 .$$

Notice that the lower limit on the sums were changed in the differentiations so that terms equal to zero do not appear in the sums. Doing so is not necessary, but it tends to prevent errors later in the analysis. In the expression above, the counter n is now shifted in each sum such that the expressions for powers of t are identical in all the sums

$$\sum_{n=0}^{\infty} a_{n+2}(n+2)(n+1)t^n + \sum_{n=2}^{\infty} a_{n-1}(n-1)t^n + \sum_{n=0}^{\infty} a_{n+1}(n+1)t^n + \sum_{n=1}^{\infty} a_{n-1}t^n = 0 .$$

Terms for $n \geq 2$ appear in all the sums. Collect these terms as a single summation and write the remaining terms, terms for $n = 0$ and $n = 1$, separate from this sum

$$2a_2 + 6a_3t + a_1 + 2a_2t + a_0t$$

$$+\sum_{n=2}^{\infty}\{a_{n+2}(n+2)(n+1) + a_{n-1}(n-1) + a_{n+1}(n+1) + a_{n-1}\}t^n = 0$$

and, since this infinite sum can only equal zero if all the coefficients of powers of t equal zero, it immediately follows that

$$2a_2 + a_1 = 0 \Rightarrow a_2 = -1$$

$$6a_3 + 2a_2 + a_0 = 0 \Rightarrow a_3 = \frac{1}{6}$$

$$a_{n+2}(n+2)(n+1) + a_{n-1}(n-1) + a_{n+1}(n+1) + a_{n-1} = 0 \Rightarrow$$

$$a_{n+2} = \frac{n}{(n+2)(n+1)}a_{n-1} + \frac{n+1}{(n+2)(n+1)}a_{n+1}\,.$$

All the coefficients in the Taylor series for the solution can now be found from this difference equation combined with the initial condition given by the values of a_0 through a_3.

In the case where an initial condition is not given, the problem becomes one of finding two series representations for the two basis solutions. This problem is complicated by the fact that the basis solutions may not have a Taylor series representation. For instance, the differential equation

$$t^2\frac{d^2x}{dt^2} + t\frac{dx}{dt} - 3x = 0$$

has the solution

$$x = C_1t^{\sqrt{3}} + C_2t^{-\sqrt{3}}$$

which does not have a Taylor series expansion around $t = 0$.

Before stating the relevant theorem about the series representation of the basis solutions, a few terms must be defined.

Definitions:

1) A function is said to be analytical at a point, iff it has a Taylor series at this point that represents the function in some neighborhood of the point.

Now consider a second order, linear differential equation of the form

$$\frac{d^2x}{dt^2} + p_1(t)\frac{dx}{dt} + p_0(t)x = 0\,.$$

2) If the coefficient functions $p_1(t)$ and $p_0(t)$ are both analytic at t_0 then t_0 is said to be an ordinary point of the equation.

3) If at least one of the coefficient functions, $p_1(t)$ or $p_0(t)$ is not analytic at t_0, but if the functions defined by $(t-t_0)p_1(t)$ and $(t-t_0)^2p_0(t)$ are analytic at t_0, then t_0 is said to be a regular singular point of the equation.

4) If at least one of the products, $(t-t_0)p_1(t)$ or $(t-t_0)^2p_0(t)$, is not analytic at t_0, then t_0 is said to be an irregular singular point of the equation.

We can now state the theorems.

Theorem 8.6. *At an ordinary point t_0 of the ODE given above, every solution is analytical. In other words, every solution can be represented by a series of the form*

$$x(t) = a_0 + a_1(t-t_0) + a_2(t-t_0)^2 + \cdots$$

Moreover, the radius of convergence of each series solution is not less than the distance from t_0 to the nearest singular point of the equation.

Theorem 8.7. *At a regular singular point t_0 of the ODE, there is at least one solution which possesses an expansion of the form*

$$x(t) = (|t-t_0|)^c(a_0 + a_1(t-t_0) + a_2(t-t_0)^2 + \cdots$$

and this series will converge for $(|t-t_0|) < R$ where R is not less than the distance from t_0 to the nearest other singular point of the equation. The other linearly independent solution may or may not involve fractional powers of $t-t_0$.

Theorem 8.8. *At an irregular singular point of the ODE, there are in general no solutions with expansions consisting solely of powers of $t-t_0$.*

Series representations of the basis solutions can only be found at ordinary points and regular singular points. In both cases, the series are found by substituting the appropriate form of the series into the differential equation and simplifying the result to obtain a difference equation for the coefficients. The process is illustrated in the next examples.

Example 8.17: Series solution at an ordinary point.

Consider the well known problem

$$\frac{d^2x}{dt^2} - x = 0$$

and find a series representation of the basis solution. It is already known that the basis solutions can be written several ways, as (e^t, e^{-t}) or as $(\cosh(t), \sinh(t))$, so one must expect that different series can be obtained depending on some choice made during the derivation of the series. Obviously, $t = 0$ is an ordinary point so the theorem states that the solutions can be written as

$$x(t) = \sum_{n=0}^{\infty} a_n t^n$$

which, when substituted into the differential equation gives

$$\sum_{n=0}^{\infty} \{a_{n+2}(n+2)(n+1) - a_n\} t^n = 0$$

from which we obtain the difference equation

$$a_n = \frac{a_{n-2}}{n(n-1)}.$$

It is not difficult to see that this difference equation is solved by the following expression for a_n

$$a_n = \begin{cases} \frac{a_0}{n!} \; n \text{ even} \\ \\ \frac{a_1}{n!} \; n \text{ odd} \end{cases}.$$

Since initial conditions are not given, the two coefficients, a_0 and a_1 are arbitrary. To get linearly independent basis solutions, pick a set of linearly independent initial conditions, $(a_0, a_1)_1$ and $(a_0, a_1)_2$. Different choices of values for these coefficients will give different choices of basis functions. For instance, if the basis solution x_1 is picked as the solution for which $a_0 = 1$ and $a_1 = 0$ while the other basis solution, x_2 is picked as the solution for which $a_0 = 0$ and $a_1 = 1$, then

$$x_1(t) = 1 + \frac{1}{2!}t^2 + \frac{1}{4!}t^4 + \cdots = \cosh(t)$$

$$x_2(t) = t + \frac{1}{3!}t^3 + \frac{1}{5!}t^5 + \cdots = \sinh(t)$$

while the choice that $a_0 = 1$ for both basis solutions and $a_1 = 1$ for y_1 and $a_1 = -1$ for y_2 gives

$$x_1(t) = 1 + t + \frac{1}{2!}t^2 + \frac{1}{3!}t^3 + \cdots = e^t$$

$$x_2(t) = 1 - t + \frac{1}{2!}t^2 - \frac{1}{3!}t^3 + \cdots = e^{-t}$$

an equally valid set of basis solutions.

Example 8.18: Series solution at a regular, singular point.

Consider

$$\frac{d^2x}{dt^2} + 2\frac{dx}{dt} - \frac{x}{t^2} = 0$$

for which the point $t = 0$ is a regular, singular point. The theorem claims that a solution can be written as

$$x(t) = \sum_{n=0}^{\infty} a_n t^{n+c}$$

where t has been assumed positive such that the absolute value bars around t can be dropped. Multiplying the differential equation through by t^2 and substituting the infinite series into the equation gives

$$\sum_{n=0}^{\infty} a_n(n+c)(n+c-1)t^{n+c} + 2\sum_{n=0}^{\infty} a_n(n+c)t^{n+c+1} - \sum_{n=0}^{\infty} a_n t^{n+c} = 0\,.$$

Notice that the lower limits on this sum cannot be changed upon differentiation because the value of c is unknown and the factors which multiply the powers of t may therefore be different from zero for all values of n. Matching powers of t and collecting terms gives

$$a_0c(c-1)t^c - a_0t^c + \sum_{n=1}^{\infty}\{a_n(n+c)(n+c-1) + 2a_{n-1}(n+c-1) - a_n\}t^{n+c} = 0$$

and an equation for c is obtained by setting the term outside the summation equal to zero. This equation is called the *indicial equation.*

$$c(c-1) - 1 = 0 \;\Rightarrow\; c = \frac{1 \pm \sqrt{5}}{2}\,.$$

The two roots of the indicial equation give rise to the two different basis solutions. The difference equation for the coefficients a_n is found to be

$$a_n = -\frac{2(n+c-1)}{(n+c)(n+c-1) - 1}a_{n-1}\,.$$

It is obvious, in this case, that one cannot use different choices of a_0 and a_1 to generate two different solutions because a_1 is determined by the difference equation once a_0 is known. From the difference equation, one finds that

$$a_n = (-2)^n \frac{c(c+1)(c+2)\cdots(c+n-1)}{\{(c+1)c-1\}\{(c+2)(c+1)-1\}\cdots\{(n+c)(n+c-1)-1\}} a_0 .$$

The two basis solutions can therefore be written formally as

$$x_1(t) =$$

$$a_{01} \sum_{n=0}^{\infty} \frac{(-2)^n c_1(c_1+1)(c_1+2)\cdots(c_1+n-1)t^{n+c_1}}{\{(c_1+1)c_1-1\}\{(c_1+2)(c_1+1)-1\}\cdots\{(n+c_1)(n+c_1-1)-1\}}$$

and

$$x_2(t) =$$

$$a_{02} \sum_{n=0}^{\infty} \frac{(-2)^n c_2(c_2+1)(c_2+2)\cdots(c_2+n-1)t^{n+c_2}}{\{(c_2+1)c_2-1\}\{(c_2+2)(c_2+1)-1\}\cdots\{(n+c_2)(n+c_2-1)-1\}}$$

where $c_1 = \frac{1+\sqrt{5}}{2}$ and $c_2 = \frac{1-\sqrt{5}}{2}$. The two unknown coefficients a_{01} and a_{02} now take the place of the two arbitrary constant which multiply the basis solutions in the expression for the complete solution to the ODE.

It is obvious that this method fails if the indicial equation has a double root, since, in this case, it will only provide a single basis solution. Less obvious is the fact that the method also fails if the roots of the indicial equation differ by an integer. In this case it turns out that one of the roots gives division by zero in the difference equation for the coefficients. However, these pathological cases can be handled using the following theorem which provides the appropriate form of the series representations of the solutions.

Theorem 8.9. *Consider the differential equation above where $t = 0$ is a regular singular point. Let the roots of the indicial equation be c_1 and c_2 with $Re(c_1) \geq Re(c_2)$. If $c_1 = c_2$ then there are two linearly independent solutions y_1 and y_2 of the form*

$$x_1 = |t|^{c_1} \sum_{n=0}^{\infty} a_n t^n \ , \ a_0 = 1$$

$$x_2 = x_1 \ln|t| + |t|^{c_1+1} \sum_{n=0}^{\infty} b_n t^n$$

whose coefficients may be determined by direct substitution into the differential equation.

If the two indicial roots differ by an integer, there are two linearly independent solutions of the form

$$x_1 = |t|^{c_1} \sum_{n=0}^{\infty} a_n t^n \ , \ a_0 = 1$$

$$x_2 = A x_1 \ln|t| + |t|^{c_2} \sum_{n=0}^{\infty} b_n t^n \ , \ b_0 \neq 0$$

where A is a constant (possibly zero) and the coefficients may be determined by direct substitution into the differential equation.

8.6 Some common functions defined by ODEs

Series solutions to equations that occur with any reasonable frequency in the sciences have commonly accepted names such as exponential functions, trigonometric functions, Bessel functions, Legendre functions, etc. An equation is considered solved if it can be identified as having solutions which can be written in terms of such known transcendental functions. The relevant properties of the transcendental functions in question can then be found in tables or in the literature. In this section, some of the more common and well known transcendental functions are discussed.

8.6.1 *Exponential and trigonometric functions*

The exponential and trigonometric functions are solutions to the differential equation

$$\frac{d^2x}{dt^2} - \lambda x = 0 \,.$$

There are several common ways of writing the basis for this solution space.

$$B_{exp} = \{e^{\sqrt{\lambda}t}, e^{-\sqrt{\lambda}t}\}$$

or

$$B_{hyp} = \{\cosh\sqrt{\lambda}t, \sinh\sqrt{\lambda}t\}$$

or

$$B_{trig} = \{\cos(\sqrt{-\lambda}t), \sin(\sqrt{-\lambda}t)\}.$$

If λ is real and positive the basis solutions are usually picked as the exponential or hyperbolic functions, and the hyperbolic functions are used when symmetry indicates that the solution is odd or even. When λ is real and negative, the basis solutions are usually written using the trigonometric functions. The different choices of basis functions are related through

$$\cosh(t) = \frac{e^t + e^{-t}}{2}\,, \ \sinh(t) = \frac{e^t - e^{-t}}{2}$$

and Euler's formula

$$e^{(\alpha+i\beta)t} = e^{\alpha t}(\cos(\beta t) + i\sin(\beta t)).$$

These functions are all so well known that we will not insult the readers' intelligence by discussing their properties any further.

8.6.2 *Bessel functions*

Bessel equations are differential equations which typically arise in differential balances in cylindrical coordinates. In this section, the free variable will therefore be indicated r. Bessel's equation is a differential equation of the form

$$r^2\frac{d^2x}{dr^2} + r\frac{dx}{dr} + (r^2 - k^2)x = 0\,.$$

Specifically, this is the Bessel equation of order k. The complete solution is a linear combination of the Bessel functions of first and second kind of order k.

$$x(r) = C_1 J_k(r) + C_2 Y_k(r)$$

where $J_k(r)$ is the Bessel function of first kind of order k and $Y_k(r)$ is the Bessel function of second kind of order k. In some texts, the notation $J_{-k}(r)$ is used for the Bessel function of second kind if k is not an integer. The Bessel functions of second kind are also known as Neumann functions.

The modified Bessel equation of order k is a differential equation of the form

$$r^2\frac{d^2x}{dr^2} + r\frac{dx}{dr} - (r^2 + k^2)x = 0$$

and its complete solution is a linear combination of the modified Bessel functions of order k.

$$x(r) = C_1 I_k(r) + C_2 K_k(r)$$

where $I_k(r)$ is the modified Bessel function of first kind of order k and $K_k(r)$ is the modified Bessel function of second kind of order k. As for the Bessel function of second kind, the notation $I_{-k}(r)$ is used in some texts for the modified Bessel function of second kind if k is not an integer. Plots of some of the Bessel functions are shown in Figs. 8.11, 8.12 and 8.13.

Of course, these expressions do not in any way define what the Bessel functions and modified Bessel functions are. They can be any pair of functions which form a basis for the solution spaces of the Bessel equations. An unambiguous definition is

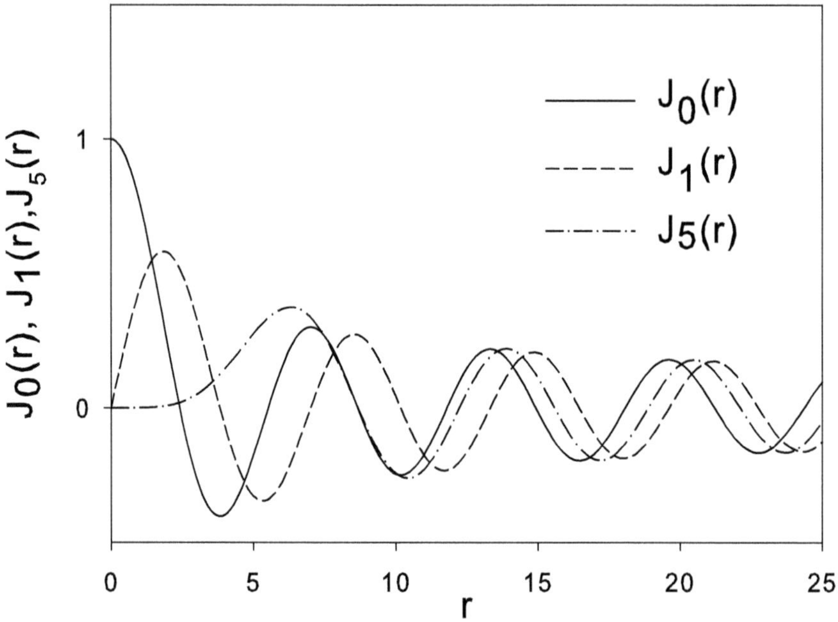

Fig. 8.11 Bessel functions of the first kind.

obtained by a series representations of the functions. These series can be obtained by the methods already described but they are not particularly nice. The formula given below for Y_k is only valid for integer orders.

$$J_k(r) = \sum_{n=0}^{\infty} \frac{(-1)^n \left(\frac{r}{2}\right)^{2n+k}}{n!\Gamma(k+n+1)}$$

$$Y_k(r) = \frac{2}{\pi} \ln\left(\frac{r}{2}\right) J_k(r) - \frac{1}{\pi}\sum_{n=0}^{k-1} \frac{(k-n-1)! \left(\frac{r}{2}\right)^{2n-k}}{n!}$$

$$- \frac{1}{\pi}\sum_{n=0}^{\infty} \frac{(-1)^n \{\psi(n+1) + \psi(n+k+1)\} \left(\frac{r}{2}\right)^{2n+k}}{n!(n+k)!}$$

$$I_k(r) = \sum_{n=0}^{\infty} \frac{\left(\frac{r}{2}\right)^{2n+k}}{n!\Gamma(n+k+1)}$$

$$K_k(r) = (-1)^{k+1} \left(\ln\left(\frac{r}{2}\right) + \gamma\right) I_k(r) + \frac{1}{2}\sum_{n=0}^{k-1} \frac{(-1)^n (k-n-1)! \left(\frac{r}{2}\right)^{2n-k}}{n!}$$

$$+ \frac{1}{2}\sum_{n=0}^{\infty} \frac{(-1)^k \left(\frac{r}{2}\right)^{2n+k} \left\{ \sum_{m=1}^{n} \frac{1}{m} + \sum_{m=1}^{k+n} \frac{1}{m} \right\}}{n!(n+k)!}$$

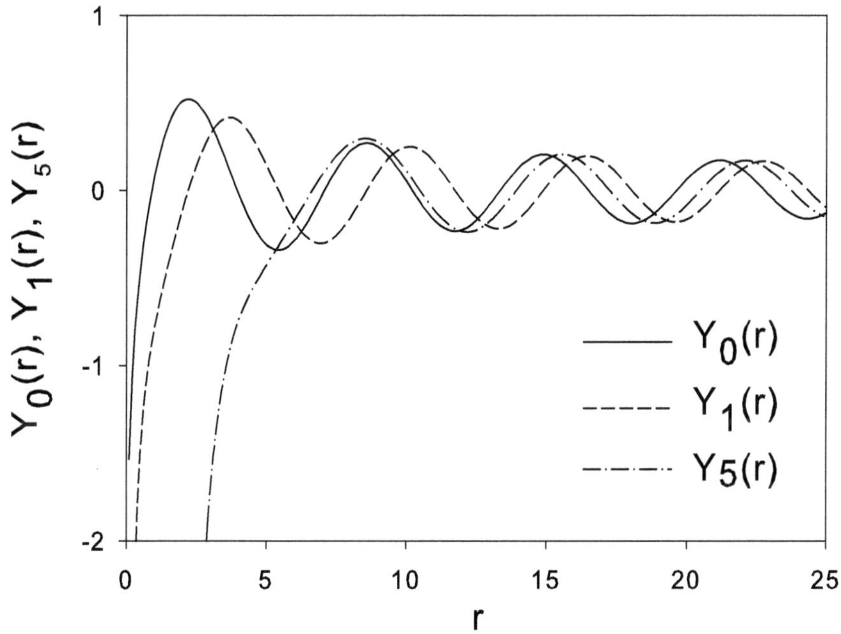

Fig. 8.12 Bessel functions of the second kind.

where $\Gamma(r)$ is the gamma function, $\psi(r)$ is the Psi or Digamma function, $\psi(r) = \Gamma'(r)/\Gamma(r)$ and $\gamma = 0.5772156649015...$ is Euler's constant.

As can be seen from the graphs, the Bessel functions oscillate like the trigonometric functions while the two modified Bessel functions go monotonically to respectively zero and infinity for large values of the argument, just like the two exponential function solutions of the exponential differential equation. In order to remember what the graphs of the Bessel functions look like, it is often convenient to think of the Bessel functions as somewhat similar to the trigonometric functions and the modified Bessel functions as somewhat similar to the exponential functions. In fact, this similarity becomes clearer at large values of the argument when the approximations following become valid.

$$J_k(r) \simeq \sqrt{\frac{2}{\pi t}} \cos\left(r - \frac{\pi}{4} - \frac{k\pi}{2}\right)$$

$$Y_k(r) \simeq \sqrt{\frac{2}{\pi r}} \sin\left(r - \frac{\pi}{4} - \frac{k\pi}{2}\right)$$

$$I_k(r) \simeq \frac{1}{\sqrt{2\pi r}} e^r$$

$$K_k(r) \simeq \sqrt{\frac{\pi}{2r}} e^{-r}.$$

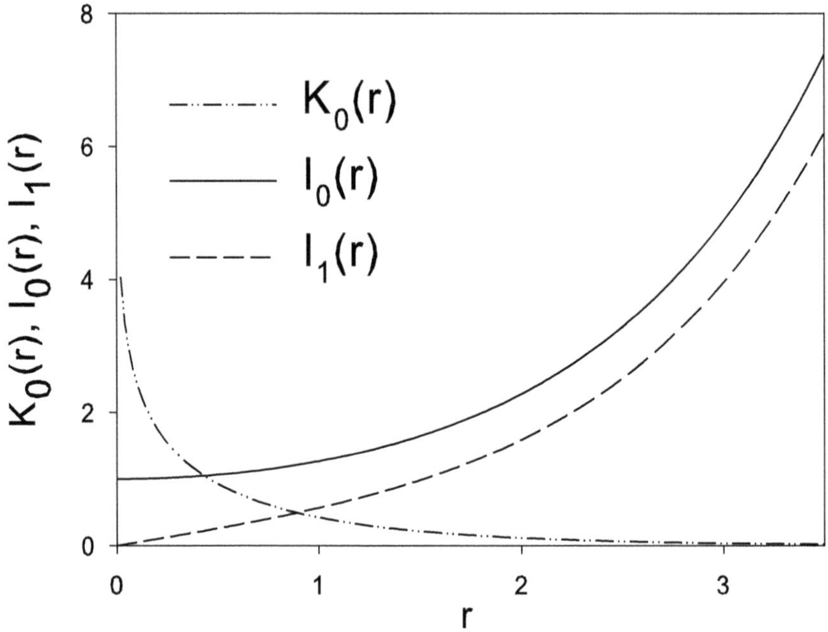

Fig. 8.13 Modified Bessel functions.

The trigonometric functions and the exponential functions are related through Euler's formula and an analogous formula can be derived for Bessel functions. Making the substitution $\xi = ir$ in the Bessel equation results in the modified Bessel equation.

$$\xi^2 \frac{d^2x}{d\xi^2} + \xi \frac{dx}{d\xi} - (\xi^2 + k^2)x = 0$$

making it clear that the modified Bessel functions can be written as a complex linear combination of the Bessel functions of complex arguments. The specific equations are

$$I_k(r) = e^{-\frac{k\pi i}{2}} J_k(ir)$$

$$K_k(r) = e^{\frac{(k+1)\pi i}{2}} \{J_k(ir) + iY_k(ir)\}.$$

Some of the important differential and integral properties of the Bessel functions are given in Table 8.2.

Problems that can be transformed to one of the Bessel equations occur so frequently that various equations, more general than the Bessel equations themselves, have occasionally been published. The most commonly seen version in the chemical

Table 8.2 Some differential and integral properties of the Bessel functions.

$$r\frac{d}{dr}J_k(\alpha r) = kJ_k(\alpha r) - \alpha r J_{k+1}(\alpha r) = \alpha r J_{k-1}(\alpha r) - kJ_k(\alpha r)$$

$$r\frac{d}{dr}Y_k(\alpha r) = kY_k(\alpha r) - \alpha r Y_{k+1}(\alpha r) = \alpha r Y_{k-1}(\alpha r) - kY_k(\alpha r)$$

$$r\frac{d}{dr}I_k(\alpha r) = kI_k(\alpha r) + \alpha r I_{k+1}(\alpha r) = \alpha r I_{k-1}(\alpha r) - kI_k(\alpha r)$$

$$r\frac{d}{dr}K_k(\alpha r) = kK_k(\alpha r) - \alpha r K_{k+1}(\alpha r) = -\alpha r K_{k-1}(\alpha r) - kK_k(\alpha r)$$

$$\int \alpha r^k J_{k-1}(\alpha r)\, dr = r^k J_k(\alpha r)$$

$$\int \alpha r^k Y_{k-1}(\alpha r)\, dr = r^k Y_k(\alpha r)$$

$$\int \alpha r^k I_{k-1}(\alpha r)\, dr = r^k I_k(\alpha r)$$

$$\int \alpha r^k K_{k-1}(\alpha r)\, dr = -r^k K_k(\alpha r)$$

engineering community comes originally from an old chemical engineering textbook[5] and is frequently called the generalized Bessel equations. However, it obviously does not cover all possible equations which have Bessel functions as part of their solutions. This equation has the form

$$r^2\frac{d^2x}{dr^2} + r(a + 2br^q)\frac{dx}{dr} + (c + dr^{2s} - b(1 - a - q)r^q + b^2r^{2q})x = 0\,.$$

Through various variable transformations, this equation can be transformed to one of the two Bessel equations and it can be shown that its complete solution is

$$x(r) = r^{\frac{1-a}{2}} \exp\left(-\frac{b}{q}r^q\right)\{C_1 Z_p(u) + C_2 Z_{-p}(u)\}$$

where

$$u = \frac{1}{s}\sqrt{|d|}\, r^s$$

$$p = \frac{1}{s}\sqrt{\left(\frac{1-a}{2}\right)^2 - c}$$

and the functions Z_p and Z_{-p} depend on d and p as follows

[5]Mickley, H.S., Sherwood, T.K. and Reed, C.E. [1957] *Applied Mathematics in Chemical Engineering* (McGraw-Hill, New York).

p	d	Z_p	Z_{-p}
Non integer	≥ 0	J_p	J_{-p}
Integer	≥ 0	J_p	Y_p
Non integer	< 0	I_p	I_{-p}
Integer	< 0	I_p	K_p

One should not give in to the temptation to use the generalized Bessel equation or some similar tool to solve any differential equation one encounters. Although many differential equations can be solved using Bessel functions, Bessel functions should only be used to report a solution if simpler expressions suffice. The unwritten but generally accepted rule is that solutions should be reported using the simplest possible forms and that algebraic expressions are regarded as the simplest, basic transcendental functions such as trigonometric, hyperbolic and exponentials as the next simplest and Bessel functions and other less common transcendental functions as the least simple. For instance, balances in spherical coordinates often give rise to the differential equation

$$\frac{d}{dr}\left(r^2\frac{dx}{dr}\right) + r^2x(r) = 0$$

where r is the radial coordinate. This equation is obtained from the generalized Bessel equation when $a = 2$, $b = c = 0$ and $d = s = 1$ which gives the solution

$$x(r) = \frac{1}{\sqrt{r}}\left(C_1J_{1/2}(r) + C_2J_{-1/2}(r)\right)$$

but this can also be written as (not the same C_ns)

$$x(r) = \frac{1}{r}(C_1\sin(r) + C_2\cos(r))$$

which most people would regard as a simpler expression. In fact, Bessel function of integer and a half order can always be written in terms of trigonometric or hyperbolic functions and they occur so frequently that they have provided the basis for definition of another set of functions, the *spherical Bessel functions*. Specifically, the spherical Bessel function of the first kind is defined as

$$j_\nu(r) = \sqrt{\frac{\pi}{2r}}\,J_{\nu+1/2}(r)$$

and one can show that

$$j_\nu(r) = (-r)^\nu \left(\frac{1}{r}\frac{d}{dr}\right)^\nu \frac{\sin(r)}{r}.$$

Thus

$$j_0(r) = \frac{\sin(r)}{r}$$

$$j_1(r) = (-r)\frac{1}{r}\frac{d\sin(r)/r}{dr} = \frac{\sin(r)}{r^2} - \frac{\cos(r)}{r}$$

$$j_2(r) = (-r)^2\frac{1}{r}\frac{d}{dr}\left(\frac{1}{r}\frac{d\sin(r)/r)}{dr}\right) = \frac{(3-r^2)\sin(r)}{r^3} - 3\,\frac{\cos(r)}{r^2}.$$

The spherical Bessel function of order zero, $j_0(r)$ is also know as the *sinc function*. Similarly, the spherical Bessel function of second kind is

$$y_\nu(r) = \sqrt{\frac{\pi}{2r}}\,Y_{\nu+1/2}(r)$$

and

$$y_\nu(r) = -(-r)\left(\frac{1}{r}\frac{d}{dr}\right)^\nu \frac{\cos(r)}{r}$$

$$y_0(r) - \frac{\cos(r)}{r}$$

$$y_1(r) = -\frac{\cos(r)}{r} - \frac{\sin(r)}{r}$$

$$y_2(r) = \frac{(r^2-3)\cos(r)}{r^3} - 3\,\frac{\sin(r)}{r^2}$$

and the modified spherical Bessel functions are

$$i_\nu(r) = \sqrt{\frac{\pi}{2r}}\,I_{\nu+1/2}(r)$$

$$k_\nu(r) = \sqrt{\frac{\pi}{2r}}\,K_{\nu+1/2}(r)$$

where

$$i_\nu(r) = r^\nu\left(\frac{1}{r}\frac{d}{dr}\right)^\nu \frac{\sinh(r)}{r}$$

$$k_\nu(r) = (-1)^\nu r^\nu\left(\frac{1}{r}\frac{d}{dr}\right)^\nu \frac{\exp(r)}{r}$$

and one finds that

$$i_0(r) = \frac{\sinh(r)}{r}\,, \ i_1(r) = \frac{r\cosh(r) - \sinh(r)}{r^2}\,, \ i_2(r) = \frac{(r^2+3)\sinh(r) - 3r\cosh(r)}{r^3}$$

$$k_0(r) = \frac{e^{-r}}{r}\,, \ k_1(r) = \frac{e^{-r}(r+1)}{r^2}\,, \ k_2(r) = \frac{e^{-r}(r^2+3r+3)}{r^3}\,.$$

8.6.3 *Legendre functions*

Legendre functions typically occur in spherical coordinates when the system exhibits latitudinal (θ) dependence. In spherical coordinates, the Laplace operator contains a term of the form

$$\frac{1}{r^2\sin\theta}\frac{\partial}{\partial\theta}\left(\sin\theta\frac{\partial y}{\partial\theta}\right).$$

For reasons that will become clear when solutions of partial differential equations are discussed, operator often give rise to problems of the form

$$\frac{d^2y}{d\theta^2} + \frac{\cos\theta}{\sin\theta}\frac{dy}{d\theta} + \lambda y = 0$$

where λ is a parameter. Let $x = \cos\theta$ to get

$$\frac{dy}{d\theta} = -\sin\theta\frac{dy}{dx}$$

and

$$\begin{aligned}\frac{d^2y}{d\theta^2} &= -\sin\theta\frac{d}{dx}\left(-\sin\theta\frac{dy}{dx}\right)\\ &= \sin^2\theta\frac{d^2y}{dx^2} + \sin\theta\frac{dy}{dx}\frac{d\sin\theta}{dx}\\ &= (1-x^2)\frac{d^2y}{dx^2} + \sqrt{1-x^2}\frac{dy}{dx}\frac{(-2x)}{2\sqrt{1-x^2}}\end{aligned}$$

where we have used that $\sin\theta = \sqrt{1-\cos^2\theta} = \sqrt{1-x^2}$. Substitution into the differential equation gives Legendre's equation, usually written as

$$\frac{d}{dx}\left((1-x^2)\frac{dy}{dx}\right) + \lambda y = 0\,.$$

Because θ, as a coordinate in a spherical system, runs between 0 and π, the variable x is usually restricted to the interval from -1 to 1. But $x = \pm 1$ are regular singular points of the differential equation, so one expects solutions that have singularities at these points. However, $x = 0$ is a regular point and the solutions therefore have Taylor series representations around this point. It is easy to show that the Taylor series around 0 have a finite number of terms iff $\lambda = n(n+1)$ where n is a non-negative integer and Legendre's differential equation is therefore often written

$$\frac{d}{dx}\left((1-x^2)\frac{dy}{dx}\right) + n(n+1)y = 0\,.$$

The two basis solutions to this equation are known as Legendre functions of the first and second kind and are indicated $P_n(x)$ and $Q_n(x)$ respectively. When n is a non-negative integer, the Legendre functions of first kind are polynomials called Legendre polynomials. It can be shown by series expansions around $x = \pm 1$, that when n is not an integer, both $P_n(x)$ and $Q_n(x)$ are unbounded at $x = \pm 1$ and, as this is ruled out by most boundary conditions in physical systems, the only solutions of interest are often the Legendre polynomials. The first few Legendre polynomials are

$$P_0(x) = 1$$

$$P_1(x) = x$$

$$P_2(x) = \frac{1}{2}(3x^2 - 1)$$

$$P_3(x) = \frac{1}{2}(5x^3 - 3x)$$

and the graphs of some low order Legendre polynomials are plotted in Fig. 8.14.

For the curious, we can mention that the Legendre functions of the second kind have simple expressions when n is an integer. The first few are

$$Q_0(x) = \frac{1}{2}\ln\left(\frac{1+x}{1-x}\right)$$

$$Q_1(x) = \frac{x}{2}\ln\left(\frac{1+x}{1-x}\right) - 1$$

$$Q_2(x) = \frac{3x^2-1}{2}\ln\left(\frac{1+x}{1-x}\right) - \frac{3x}{2}$$

$$Q_3(x) = \frac{5x^3-3x}{4}\ln\left(\frac{1+x}{1-x}\right) - \frac{5x^2}{2} + \frac{2}{3}\,.$$

Legendre polynomials of any order can of course be found from the Taylor series solution to the differential equation which gives the following recursive relationship for the coefficients in the series

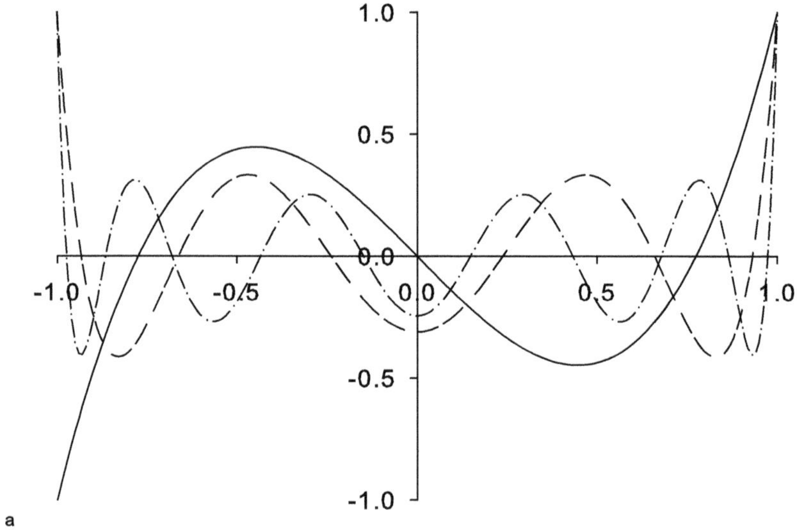

Fig. 8.14 Graphs of the Legendre polynomials of order 3, 6 and 10. It is hopefully obvious to the reader which graph corresponds to which order.

$$a_{k+2} = -\frac{(n-k)(n+k+1)}{(k+1)(k+2)} a_k \,.$$

Notice that n is the parameter in Legendre's differential equation and not the counter in the Taylor series. However, rather than using this result to find explicit expressions for Legendre polynomials, it is often easier to use either of the two relationships below which are given without proofs

$$(n+1)P_{n+1}(x) = (2n+1)xP_n(x) - nP_{n-1}(x)$$

$$P_n(x) = \frac{1}{2^n n!}\frac{d^n}{dx^n}(x^2-1)^n.$$

The Legendre polynomials can be shown to satisfy the relationship

$$\int_{-1}^{1} P_n(x)P_m(x)\,dx = \begin{cases} 0\,, & \text{if } n \neq m \\ \frac{2}{2n+1}\,, & \text{if } n = m \end{cases}$$

and are said to be orthogonal over the interval $[-1,1]$ with the weight function 1. This property will turn out to be important in applications with these polynomials.

Chapter 9

Hilbert Spaces

Most of the *algebraic* concepts and insights obtained from the development of finite dimensional spaces transfer over to infinite dimensional spaces. However, there are important differences between finite dimensional and infinite dimensional vector spaces. Arguably the most important difference is that the Hamel basis, as defined previously for finite dimensional spaces, is of little use in infinite dimensional spaces. The reason for this is that the Hamel basis is an algebraic concept and therefore does not permit infinite sums which is a concept that belongs to calculus. The way out of this dilemma is to impose additional structure on the vector space so that infinite sums can be well-defined, and for this we need the *topological* notion of convergence. This is less important in finite dimensions because finite dimensional vector spaces of the same dimension are isomorphic with the same topology. The isomorphism property between infinite dimensional spaces is only meaningful if it also incorporates the topology.[1] There may be a variety of topologies that can be used in a given context, but the one with the most useful properties is the main topic of this chapter, and is called a *Hilbert space*. We first loosely describe some of the issues and problems that complicate the study of infinite dimensional spaces, and then introduce normed spaces. The concept of a Hilbert space is a type of normed space that is suitable for many applications, mainly because it admits easy-to-use bases. A particularly important example of a Hilbert space involves Fourier series, which is discussed in some detail.

[1] Although many readers may already have encountered the concept of topology and will have an understanding of what the concept involves, most engineering students are not in that situation. Even a short introduction to topology would take us too far away from the main topic so it will have to suffice to say that topology describes which elements in a set are "close" to other elements without actually imposing a distance measure on the set. Thus, the notion of convergence is a topological concept because it states that the distance between two objects becomes arbitrarily small. This notion becomes much less abstract once one defines a measure of distance on the set. The measure that is of relevance here is the norm, defined in Section 9.1.2. Fortunately, this is the only topological concept that will be required for the material in this chapter.

9.1 Infinite dimensional vector spaces

The property required for a vector space to be infinite dimensional was specified in Chapter 5, and recall this means that there are arbitrarily large sets of linearly independent vectors. For example, any collection of polynomials whose elements have different degree is an independent set, and so it follows that any vector space that contains all polynomials is infinite dimensional.

Another collection of independent functions defined on $[-1,1]$ are the sine functions $\{\sin(n\pi x)\}_{n=1,\infty}$. To see these as independent, suppose there are scalars $C_1,\ldots,C_k$ so that

$$C_1\sin(\pi x)+C_2\sin(2\pi x)+\cdots+C_k\sin(k\pi x)=0\,. \tag{9.1}$$

We will show that all the C_ns must be zero, and therefore the functions are independent. We multiply both sides of Eq. (9.1) by $\sin(m\pi x)$, where m is a positive integer such that $m\le k$, and integrate the resulting equation from $x=-1$ to $x=1$ to get

$$\sum_{n=1}^{k}C_n\int_{-1}^{1}\sin(m\pi x)\sin(n\pi x)\,dx=0\,.$$

The integrals are evaluated as

$$\int_{-1}^{1}\sin(m\pi x)\sin(n\pi x)\,dx=\begin{cases}0 & \text{if } m\ne n\\ 1 & \text{if } m=n\end{cases}$$

leaving only a single term C_m in the sum. Consequently, C_m must be 0.

We have seen in finite dimensions the usefulness of finding a good basis for a vector space. Any basis gives a unique way to represent the elements of the space as a linear combination. If one is restricted to using only the algebraic property of finite sums in infinite dimensional vector spaces, it generally precludes the use of the most natural bases in function spaces. The idea is to introduce additional structure that defines a notion of convergence so that infinite sums can be defined.

First, recall from elementary calculus how the convergence of a series of numbers was defined. If $\alpha_n\in\mathbb{F}$, then the series $\sum_{n=0}^{\infty}\alpha_n$ converges to S provided the partial sums $S_N=\sum_{n=0}^{N}$ satisfy $S_N\to S$ as $N\to\infty$. The same definition can be extended to vector spaces provided that (1) the vector space has a structure that defines the meaning of convergence, and (2) the limiting object will always remain in the space. The structure required for (1) is called a norm, and is introduced in Section 9.1.2. The property (2) depends on the norm, and if it holds, then the space has the desirable property of being *complete*.

9.1.1 *Countable and uncountable infinities*

In the following discussion, it will be useful to have a firmer understanding of the concept of infinity. Basically, we want to give rigor to the notion of infinitely large sets having the same or different number of elements.

The number of elements in a set is called the cardinality of the set, and two sets are said to have the same cardinality or the same cardinal number if there is a one-to-one correspondence between the elements of the sets. This may seem like an unduly convoluted way of determining if two sets have the same number of elements. Is it not simpler to just count the number of elements in each set and see if the counts are the same for the two sets? Actually, it is not simpler because the counting process itself is a fairly advanced concept while the notion of a one-to-one correspondence is more fundamental.

Sets that can be counted (all finite sets are like this) can be distinguished from sets that cannot be counted. Infinite sets are said to be *countably infinite* if there is a one-to-one correspondence between the members of the set and the natural numbers. For instance, the set of all even numbers is countably infinite as is the set of all odd numbers. The cardinality of such sets are indicated $\aleph_0$ (The first letter in the Hebrew alphabet, pronounced "aleph zero".)

The set of rational numbers, which seems so much larger than the set of natural numbers, is nonetheless countable. This can be seen by arranging the rationale numbers in a table such as

$$
\begin{array}{ccccccccc}
\frac{1}{1} & \rightarrow & \frac{1}{2} & & \frac{1}{3} & \rightarrow & \frac{1}{4} & \cdots \\
 & \swarrow & & \nearrow & & \swarrow & & \\
\frac{2}{1} & & \frac{2}{2} & & \frac{2}{3} & & \frac{2}{4} & \cdots \\
\downarrow & \nearrow & & \swarrow & & \nearrow & & \\
\frac{3}{1} & & \frac{3}{2} & & \frac{3}{3} & & \frac{3}{4} & \cdots \\
 & \swarrow & & \nearrow & & \swarrow & & \\
\frac{4}{1} & & \frac{4}{2} & & \frac{4}{3} & & \frac{4}{4} & \cdots \\
\vdots & & \vdots & & \vdots & & \vdots &
\end{array}
$$

and defining a one-to-one correspondence to the natural numbers by following the path indicated (or any of many other possible paths through the table) and letting the Nth rational number encountered when moving along the path correspond to the natural number N.

Other important countable sets are the algebraic numbers which are numbers that can be roots of polynomials with rational coefficients, and the set of all polynomials with rational coefficients. It can be shown that the union of a countable number of countable sets is also countable.

The set of real numbers in the interval $[0, 1]$ is not countable. To see this, assume that they are and write them in decimal form in the order in which they are enumerated

$$\begin{aligned} a &= 0.a_1a_2a_3a_4\cdots \\ b &= 0.b_1b_2b_3b_4\cdots \\ c &= 0.c_1c_2c_3c_4\cdots \\ &\text{and so on}. \end{aligned}$$

Any such list cannot contain all real numbers between 0 and 1 because one can find a real number not in the list by picking the first decimal as different from a_1, the second decimal as different from b_2 and so forth. The cardinality of the set of real numbers in $[0,1]$ is said to be an *uncountable infinity* and is indicated by c. The uncountable infinity is also called the *power of the continuum.* A one-to-one correspondence between $\mathbb{R}$ and the real numbers in $[0,1]$ is easily set up, e.g. $\tan(\pi x)$, and it follows that the cardinality of all the real numbers is also c.

9.1.2 *Normed spaces*

A norm is a non-zero number associated to each vector of a vector space, and can be thought of as an abstract or generalized concept of length. The formal definition follows.

Definition: *Suppose V is a vector space. A norm $\|\cdot\|$ defined on V is a map from V into the non-zero real numbers that satisfies the following list of axioms:*

(1) Positive definiteness:

$$\|v\| = 0 \text{ if and only if } v = 0\,.$$

(2) Homogeneity:

$$\|\alpha v\| = |\alpha|\|v\| \;\forall\; \alpha \in \mathbb{F} \text{ and } v \in V.$$

(3) Triangle inequality:

$$\|v + w\| \leq \|v\| + \|w\| \;\forall\; v, w \in V.$$

A vector space on which a norm has been defined is called a *normed space.* The norm of a vector can be interpreted as its length, or as its distance from the origin. The distance between any two vectors is then the norm of their difference.

There may be many norms defined on a given vector space. For instance, on $\mathbb{R}^N$ or $\mathbb{C}^N$, for $1 \leq p < \infty$, the *p-norm* is defined as

$$\|v\|_p = \left\{ \sum_{n=1}^{N} |v_n|^p \right\}^{1/p}.$$

The particular case $p = 2$ is usually referred to as the Euclidean norm. It is easy to see a p-norm as positive definite and homogeneous, but the triangle inequality is considerably more difficult to verify, and is called the Minkowski inequality. It says that for finite sums

$$\left(\sum_{n=1}^{N} |v_n \pm w_n|^p\right)^{1/p} \le \left(\sum_{n=1}^{N} |v_n|^p\right)^{1/p} + \left(\sum_{n=1}^{N} |w_n|^p\right)^{1/p}.$$

The case $p = \infty$ is called either the *infinity-norm* or the *sup-norm*, and is defined as

$$\| v \|_\infty = max\{|v_1|, |v_2|, \ldots, |v_N|\}.$$

It can be easily checked that it in fact is a norm, and is obtained by taking the limit of the p-norm as $p \to \infty$.

The infinite dimensional vector spaces that will primarily concern us are spaces of functions defined over a closed interval. We assume the interval to be $[0, 1]$ for simplicity. The following norms are defined in terms of integrals, and so the functions must be at least integrable. It actually suffices to consider the norms defined only for polynomials $\mathcal{P}$, and we shall see later how the space of polynomials can be completed to include additional functions.

For each $f \in \mathcal{P}$, the p-norm of f is defined by

$$\| f(t) \|_p = \left\{\int_0^1 |f(t)|^p \, dt\right\}^{1/p}.$$

It is again easy to verify that positive definiteness and homogeneity both hold; the triangle inequality is considerably more difficult to verify (we shall not do so), but it is nonetheless true and is referred to as the Minkowski inequality for integrals

$$\left(\int_0^1 |f(t) + g(t)|^p \, dt\right)^{1/p} \le \left(\int_0^1 |f(t)|^p \, dt\right)^{1/p} + \left(\int_0^1 |g(t)|^p \, dt\right)^{1/p}.$$

The ∞-norm or sup-norm on a function space is defined as

$$\| f(t) \|_\infty = \sup\{|f(t)| : t \in [0, 1]\}$$

where "sup" indicates the supremum or least upper bound.

As mentioned above, it is desirable that a normed space be complete. Formally, a normed space V is said to be complete if every *Cauchy sequence* converges to an element in V. A sequence $\{v_n\} \subseteq V$ is a Cauchy sequence if for all $\varepsilon > 0$, there exists n_0 so that $m, n \ge n_0$ implies that $\|v_m - v_n\| < \varepsilon$. It is easy to see that every convergent sequence is a Cauchy sequence, but the property of being complete is the converse statement: for every Cauchy sequence $\{v_n\}$, there exists a vector $v \in V$ for which $\|v_n - v\| \to 0$ as $n \to \infty$.

We emphasize that the property of being complete depends on the norm as well as the vector space — the same vector space as — be complete in one norm but not in another. For example, consider $V = C[0, 1]$, the space of continuous functions defined on $[0, 1]$. We shall see that $C[0, 1]$ is complete in the infinity norm, but first we show that it is not complete in the 1-norm (or for that matter, in any p-norm with $1 \le p < \infty$). Define a sequence and a function f by

$$f_n(t) = \begin{cases} (2t)^n & \text{if } 0 \le t < \frac{1}{2} \\ 1 & \text{if } \frac{1}{2} \le t \le 1 \end{cases} \quad \text{and} \quad f(t) = \begin{cases} 0 & \text{if } 0 \le t < \frac{1}{2} \\ 1 & \text{if } \frac{1}{2} \le t \le 1 \end{cases}.$$

Then $\|f_n - f\|_1 = \int_0^{\frac{1}{2}} (2t)^n \, dt = \frac{2}{n+1} \to 0$ as $n \to \infty$, and so $\{f_n\}$ is a Cauchy sequence in the 1-norm, but it does not converge to an element in $C[0,1]$ since f is not continuous. The sequence $\{f_n\}$ is not Cauchy in the sup-norm. Indeed, for any n_0, there exists $0 < t < \frac{1}{2}$ for which $f_{n_0}(t) > \frac{3}{4}$, and then there exists $m > n_0$ with $f_m(t) < \frac{1}{4}$. Consequently $\|f_{n_0} - f_m\| \ge \frac{1}{2}$, and so $\{f_n\}$ cannot be Cauchy in the sup-norm. To summarize, convergence in infinite dimensional spaces can be quite subtle and is highly dependent on the norm.

Theorem 9.1. *$C[0,1]$ is complete in the sup-norm.*

Proof: Suppose $\{f_n\} \subset C[0,1]$ is a Cauchy sequence with respect to the sup-norm. For each $t \in [0,1]$, the sequence $\{f_n(t)\}$ is a Cauchy sequence of scalars, and so converges to a scalar that we call $f(t)$. This property can be described as saying the sequence f_n converges *pointwise* to f (which specifically means that for all $t \in [0,1]$, we have $f_n(t) \to f(t)$ as $n \to \infty$), while the conclusion of the theorem asserts that the convergence is uniform over $t \in [0,1]$ (the latter means that $\|f_n(t) - f(t)\|$ is small for *all* $t \in [0,1]$ when n is large). For the function f defined in this manner, we shall show firstly that $\|f_n - f\|_\infty \to 0$ as $n \to \infty$, and secondly, f is continuous.

Let $\varepsilon > 0$, and choose n_0 so that $m, n \ge n_0$ implies $\|f_m - f_n\|_\infty < \varepsilon$. For any $n \ge n_0$ and any $t \in [0,1]$, we have

$$|f_n(t) - f(t)| = \lim_{m\to\infty} |f_n(t) - f_m(t)| \le \sup_{m \ge n_0} \|f_n - f_m\| \le \varepsilon,$$

and hence $\|f_n - f\|_\infty \to 0$ as $n \to \infty$. To show f is continuous, let $t \in [0,1]$ and $\varepsilon > 0$. By the previous assertion, there exists n_0 so that n_0 implies $\|f_{n_0} - f\|_\infty < \varepsilon$. Since f_{n_0} is continuous, there exists $\delta > 0$ so that $|s - t| < \delta \Rightarrow |f_{n_0}(t) - f_{n_0}(s)| < \varepsilon$. Now suppose $|s - t| < \delta$. We have

$$\begin{aligned} |f(t) - f(s)| &\le |f(t) - f_{n_0}(t)| + |f_{n_0}(t) - f_{n_0}(s)| + |f_{n_0}(s) - f(s)| \\ &\le 2\|f_{n_0} - f\|_\infty + |f_{n_0}(t) - f_{n_0}(s)| \\ &\le 3\varepsilon. \end{aligned}$$

This shows that f is continuous at t, and hence $f \in C[0,1]$. □

We have just seen that the completeness of an infinite dimensional vector space depends on the norm chosen for the space, for $C[0,1]$ is complete in the sup-norm but not with any of the other p-norms. Different norms in finite dimensional spaces are somewhat less important, for the following reason. It can be shown that if $\|\cdot\|$

and $\|\cdot\|'$ are any two norms defined on $\mathbb{F}^N$, then there are positive constants c_1 and c_2 so that

$$c_1\|v\| \le \|v\|' \le c_2\|v\| \quad \text{for all } v \in \mathbb{F}^N.$$

For example, considering the p-norms, for any $1 \le p_1 < p_2 \le \infty$, one has

$$\|v\|_\infty \le \|v\|_{p_2} \le \|v\|_{p_1} \le \|v\|_1 \le N\|v\|_\infty.$$

Hence in finite dimensional spaces, a sequence is Cauchy in one norm if and only if it is Cauchy in any other norm.

Theorem 9.2. *For each N, $\mathbb{F}^N$ is complete in any norm.*

The Euclidean norm (that is, the $p = 2$-norm) is usually used almost exclusively because it reflects the geometry of physical space, and as we will soon see, it has another advantage that makes it convenient to work with.

9.1.3 *Bases in infinite dimensional spaces*

Recall that a subset $\mathcal{B}$ of a vector space V is a basis if (1) $\mathcal{B}$ consists of independent vectors, and (2) $\mathcal{B}$ spans V. One of the main reasons for introducing norms in infinite dimensional spaces is that the definition of span in (2) can be extended to include infinite sums, and thus have a more efficient and useful definition of a basis. Specifically, suppose $\mathcal{B}$ is an independent set. Let W be the span of $\mathcal{B}$, which consists of all possible *finite* linear combinations of elements of $\mathcal{B}$. That is, $v \in W$ provided there exist a positive integer M, scalars $\{\alpha_n\}_{n=1}^M$, and vectors $\{v_n\}_{n=1}^M \subseteq \mathcal{B}$ so that $v = \sum_{n=1}^M \alpha_n v_n$. If $W = V$, then $\mathcal{B}$ is often called a *Hamel* basis, although in finite dimensions the adjective "Hamel" is usually dropped because there are no other kind of bases. As we shall see, a Hamel basis is generally not useful in infinite dimensional spaces.

Thus far, we are still considering only vector space operations, but now suppose V is also a normed spaced with norm $\|\cdot\|$. We say that a subset U is *dense* in V provided that for every $v \in V$ and $\varepsilon > 0$, there exists $u \in U$ so that $\|v - u\| < \varepsilon$. In other words, for every element $v \in V$, there are elements in the dense set that are arbitrarily close to v. For example, the rational numbers are dense in the set of real numbers. With $\mathcal{B}$ and W given as in the previous paragraph, if W is dense in V, then $\mathcal{B}$ is called a *basis* for V.

As an example, consider the infinite dimensional space of polynomials $\mathcal{P}$. The set of monomials $\mathcal{B} = \{t^n : n = 0, 1, 2\dots\}$ is a Hamel basis for $\mathcal{P}$, since every element in $\mathcal{P}$ has the form $p(t) = a_0 + a_1 t + \cdots + a_n t^n$. Let us now consider $C[0, 1]$ with the sup-norm. The elements of $\mathcal{B}$ restricted to $[0, 1]$ form an independent subset, and its span is $W = \mathcal{P}$. The next theorem, which we shall not prove, is called the Weierstrass approximation theorem, and in effect says that $\mathcal{B}$ is a basis for $C[0, 1]$.

Theorem 9.3. *The polynomials $\mathcal{P}$ are dense in $C[0,1]$ with respect to the sup-norm. That is, for every $f \in C[0,1]$ and $\varepsilon > 0$, there exists $p \in \mathcal{P}$ so that*

$$\|f - p\|_\infty = \sup_{t\in[0,1]} \big|f(t) - p(t)\big| < \varepsilon.$$

Of course $\mathcal{B}$ is not a Hamel basis for $C[0,1]$ because there are continuous functions that are not polynomials. Nonetheless, one may still seek to find a Hamel basis for $C[0,1]$. By so doing, the need to consider infinite sums in order to efficiently represent elements of $C[0,1]$ would be considerably diminished. However, it is a general fact that if V is a complete infinite dimensional normed space, then any Hamel basis of V is necessarily uncountable and impossible to construct, and consequently, offers no practical use. This is why we do not seek a Hamel basis for $C[0,1]$, but instead, extend the notion of a basis to allow for infinite sums. The tradeoff is that we must be able to say how the infinite sums converge.

Heretofore, we have assumed that a normed space V is given, and we defined a basis as an independent set whose span is dense in V. It is often the case, however, that the space V is created as the "completion" from a more familiar space that is endowed with a norm. Suppose we begin with an independent set $\mathcal{B}$, and consider its span W that is endowed with a norm $\|\cdot\|$. An obvious question that arises is whether there is a complete space for which $\mathcal{B}$ is a basis. And if so, how can it be found? In other words, can W be extended to a larger vector space in some manner so that it includes enough elements in order to be complete?

There is an analogy with the space of rational numbers, which we explain first. For any sequence $q_0, q_1, q_2, \ldots$ of rational numbers, each finite sum $s_N = \sum_{n=0}^{N} q_n$ is also rational. If the sequence $\{s_N\}_N$ of partial sums is a Cauchy sequence, the sequence may not converge in the space of rational numbers. For example, with $q_n = \frac{1}{n!}$, the sequence $s_N = \sum_{n=1}^{N} \frac{1}{n!}$ converges to the infinite sum $\sum_{n=0}^{\infty} 1/n!$, which equals the irrational number e. In general, suppose x is any real number with a decimal expansion $x = d_0.d_1d_2\ldots$, where d_n is one of the digits $0, 1, 2\ldots, 9$. Then x equals the infinite sum $\sum_{n=0}^{\infty} q_n$ with $q_n = \frac{d_n}{10^n}$, and is the limit of the rational partial sum $\sum_{n=0}^{N} q_n$. Thus, although the rational numbers are not complete, if all the infinite sums of rational numbers whose partial sums are Cauchy are appended to the rational numbers, the result is the set of real numbers, which is complete.

A similar procedure can be carried out more generally. Suppose $\mathcal{B} = \{v_1, v_2, \ldots\}$ is an independent set, and W is its span endowed with a norm $\|\cdot\|$. Suppose $\alpha_1, \alpha_2, \ldots$ are scalars, and for each N, consider the finite sum $s_N = \sum_{n=1}^{N} \alpha_n v_n$, which is an element of W. If the sequence $\{s_N\}_N$ is a Cauchy sequence, then the infinite sum $\sum_{n=1}^{\infty} \alpha_n v_n$ can be used to represent a new element that does not necessarily belong to W. The collection of all such convergent sums, called the completion of $\mathcal{B}$ (under the norm $\|\cdot\|$), is a complete normed space for which $\mathcal{B}$ is a basis.

In summary, if V is complete and $\mathcal{B} = \{v_1, v_2, \ldots\}$ is a basis, then every $v \in V$ can be written as the infinite sum $v = \sum_{n=1}^{\infty} \alpha_n v_n$. As in finite dimensions, it is natural to consider the uniqueness of this expression, and to have a convenient method to obtain the coordinates $\alpha_1, \alpha_2, \ldots$. Bases in general infinite dimensional spaces is a complicated issue, but simplifies greatly if the additional structure of an inner product is also present. Such spaces are called Hilbert spaces, and will be introduced formally in the next section.

9.1.4 *The function spaces $\mathcal{L}^p[0,1]$*

We have seen that $\mathcal{P}$ is dense in $C[0,1]$ under the sup-norm. The completion of $\mathcal{P}$ under the p-norm, $1 \leq p < \infty$, is denoted by $\mathcal{L}^p[0,1]$. We will describe this completion in further detail.

The elements of $\mathcal{L}^p[0,1]$ are, by definition, represented as infinite sums whose partial sums belong to $\mathcal{P}$ and are Cauchy under the p-norm. This is of course very abstract, and a more concrete description is given next. A complicating factor, however, is that elements of $\mathcal{L}^p[0,1]$ are not functions in the usual sense of that term, but rather are equivalence classes with respect to the equivalence relation of being equal "almost everywhere". To understand what this means requires some concepts from measure theory.

An open set O of $\mathbb{R}$ is any set consisting of the union of disjoint open intervals, and so has the form $O = \cup_{n=1}^{\infty}(a_n, b_n)$ where $a_n < b_n$ for all n. The length of O, denoted by $|O|$, is the sum of the length of the intervals: $|O| = \sum_{n=1}^{\infty}(b_n - a_n)$. Any subset N of $\mathbb{R}$ is said to have measure zero (or be called a null set) provided that for every $\varepsilon > 0$, there is an open set O that contains N and has length less than ε; that is, $N \subseteq O$ and $|O| < \varepsilon$. Any property (P) is said to hold "almost everywhere" (abbreviated to ae) if there exists a null set N so that (P) holds for all points not in N. For example, two functions f and g are equal ae provided there exists a null set N so that $f(x) = g(x)$ for all $x \notin N$.

We make a few elementary observations about null sets: (i) if $N_1 \subseteq N_2$ and N_2 is a null set, then N_1 is a null set. (ii) A set N consisting of a single element $N = \{a\}$ is a null set, since for $\varepsilon > 0$, the open set $O = \left(a - \frac{\varepsilon}{3}, a + \frac{\varepsilon}{3}\right)$ contains N and has length $|O| = \frac{2\varepsilon}{3} < \varepsilon$. (iii) If $N_1, N_2, \ldots$ is a sequence of null sets, then the union $N = \cup_{n=1}^{\infty} N_n$ is also a null set. To see this, let $\varepsilon > 0$, and for each $n = 1, 2, \ldots$, let O_n be an open set that contains N_n and has length $|O_n| < \frac{\varepsilon}{2^n}$. Then $N \subseteq O = \cup_{n=1}^{\infty} O_n$ and $|O| \leq \sum_{n=1}^{\infty} |O_n| < \sum_{n=1}^{\infty} \frac{\varepsilon}{2^n} = \varepsilon$. In particular, every countable set (that is, a set that can be consecutively enumerated as $\{a_1, a_2, \ldots\}$) is a null set.

A function $f : [0,1] \to \mathbb{F}$ is said to be measurable on $[0,1]$ provided it is the ae limit of a sequence of polynomials. More precisely, f is measurable on $[0,1]$ when there exists a sequence of polynomials $\{p_n\}$ and a null set $N \subset [0,1]$ so that for all $x \in [0,1]/N$, one has $p_n(x) \to f(x)$ as $n \to \infty$. The property of two functions being equal ae is an equivalence relation, and by a measurable function, we precisely mean any element of an equivalence class rather than a particular function that is defined pointwise. This abuse of terminology should cause no confusion, but one should keep in mind there is no distinction between measurable functions if they are equal ae.

Let $1 \leq p < \infty$. We can now describe the elements of $\mathcal{L}^p[0,1]$. Suppose f is a measurable function. Then $f \in \mathcal{L}^p[0,1]$ provided there exists a sequence $\{p_n\} \subseteq \mathcal{P}$ that converges pointwise ae to f and is Cauchy in the p-norm. It is easy to see that $\mathcal{L}^p[0,1]$ is a vector space. Since

$$\big|\|p_n\|_p - \|p_m\|_p\big| \leq \|p_n - p_m\|_p \to 0 \text{ as } m, n \to \infty,$$

the sequence $\{\|p_n\|_p\}_n$ is a Cauchy sequence of real numbers and so has a limit. We define $\|f\|_p = \lim_{n\to\infty} \|p_n\|_p$, and this defines a norm for each $f \in \mathcal{L}^p[0,1]$. By construction, $\mathcal{L}^p[0,1]$ is complete under the norm $\|\cdot\|_p$ for each $1 \leq p < \infty$, and the monomials form a basis.

The case $p = \infty$ needs a special definition, since it is not the completion of $\mathcal{P}$ with respect to the sup-norm (we have already seen that $C[0,1]$ is that completion). A measurable function f belongs to $\mathcal{L}^\infty[0,1]$ provided there is a constant $M \geq 0$ so that $|f(x)| \leq M$ ae. If the least M that has this property is denoted by $\|f\|_\infty$, then $\|\cdot\|_\infty$ is a norm on $\mathcal{L}^\infty[0,1]$ that agrees with our previous sup-norm that was earlier defined just on $\mathcal{P}$. Another (equivalent) way to define elements in $\mathcal{L}^\infty[0,1]$ is to say they are measurable functions that are pointwise ae limits of polynomials $\{p_n\}$ that satisfy $\|p_n\|_\infty \leq M$ for some M and all n.

As an illustration, the function f defined by

$$f(x) = \begin{cases} -1 & \text{if } 0 \leq x \leq \frac{1}{2} \\ 1 & \text{if } \frac{1}{2} < x \leq 1 \end{cases}$$

belongs to $\mathcal{L}^\infty[0,1]$. One can show that $\|f - p\|_\infty \geq 1$ for every $p \in \mathcal{P}$, which demonstrates that the set $\mathcal{B}$ of monomials is not a basis for $\mathcal{L}^\infty[0,1]$. In fact, it is very difficult to find any basis for $\mathcal{L}^\infty[0,1]$, since such a basis must be *uncountable.* Fortunately, the space that arises most frequently in applications is $\mathcal{L}^2[0,1]$, which not only has a countable base, but have bases with other desirable and useful properties.

9.2 Hilbert spaces

Hilbert spaces are complete normed spaces with the additional structure of an inner product that determines the norm. An inner product was introduced earlier for finite dimensional spaces, and the same definition is recalled here for any vector space.

9.2.1 *Inner products*

Definition *Suppose V is a vector space over the field $\mathbb{F}$. An inner product is a binary operation which maps two elements of V into $\mathbb{F}$, and which has the following properties.*

(1) Additivity:

$$\langle v + w, q \rangle = \langle v, q \rangle + \langle w, q \rangle.$$

(2) Hermitian symmetry:

$$\langle v, w \rangle = \overline{\langle w, v \rangle}.$$

(3) Homogeneity:

$$\langle \alpha v, w \rangle = \alpha \langle v, w \rangle.$$

where α is a scalar.

(4) Positive definiteness

$$\langle v, v \rangle > 0 \;,\; \textit{if } v \neq 0\,.$$

A vector space V with an inner product is called an inner product space.

Notice that for $\mathbb{F} = \mathbb{R}$, the complex conjugate in (2) does not appear and the inner product is symmetric. It also follows from the symmetry property (2) that the inner product of a vector with itself is real, even if the associated field is the complex numbers. It makes sense then to define

$$\|v\| = \sqrt{\langle v, v \rangle}, \tag{9.2}$$

and we will next see that this defines a norm on V. The properties of positivity, positive definiteness, and homogeneity are immediate. The triangle inequality needs more significant justification, for which we need the following inequality known as the *Schwarz inequality.*

Lemma 9.1. *Suppose V is a vector space with an inner product. Then*

$$|\langle v, w \rangle| \leq \|v\| \; \|w\| \tag{9.3}$$

for each $v, w \in V$.

Proof: The inequality (9.3) is trivial if $w = 0$, so now suppose $w \neq 0$. For any $\alpha \in \mathbb{F}$, we have

$$\begin{aligned} 0 \leq & \langle v - \alpha w, v - \alpha w \rangle \\ = & \langle v, v \rangle - \alpha \langle w, v \rangle - \bar{\alpha} \langle v, w \rangle + |\alpha|^2 \langle w, w \rangle \\ = & \|v\|^2 - 2\text{Re } \alpha \langle w, v \rangle + |\alpha|^2 \|w\|^2. \end{aligned}$$

Now set $\alpha = \frac{\langle v, w \rangle}{\|w\|^2}$, and then

$$0 \leq \|v\|^2 - 2\frac{|\langle v, w \rangle|^2}{\|w\|^2} + \frac{|\langle v, w \rangle|^2}{\|w\|^4} \|w\|^2,$$

which reduces to Eq. (9.3) after simplifying and rearranging terms. $\square$

We now turn to verifying the triangle inequality for the quantity $\|v\|$ defined in Eq. (9.2).

$$0 \le \| v+w \|^2 = \langle v+w, v+w \rangle = \| v \|^2 + \langle v,w \rangle + \overline{\langle v,w \rangle} + \| w \|^2$$
$$= \| v \|^2 + 2Re(\langle v,w \rangle) + \| w \|^2 \le \| v \|^2 + 2|\langle v,w \rangle| + \| w \|^2$$

Using the Cauchy-Schwarz inequality on the middle term gives

$$\| v+w \|^2 \le \| v \|^2 + 2 \| v \| \cdot \| w \| + \| w \|^2 = (\| v \| + \| w \|)^2$$

from which the triangle inequality immediately follows. □

Definition: *A Hilbert space is an inner product space in which the norm given by (9.2) is complete.*

The choice of α in the proof of the Cauchy-Schwarz inequality has a geometrical interpretation when $\mathbb{F} = \mathbb{R}$. Suppose w is a non-zero element of a Hilbert space V, and let W be the one-dimensional subspace spanned by w; that is, $W = \{\alpha w : \alpha \in \mathbb{R}\}$. Now let $v \in V$, and consider the problem of finding the point in W that is closest to v. This (unique) point is called the *projection* of v into W, and can be found by first considering the problem of minimizing the quantity $g(\alpha) = \|v - \alpha w\|^2$ over $\alpha \in \mathbb{R}$. The minimum will occur precisely when the derivative of g is equal to zero. We calculate $\frac{d}{d\alpha} g(\alpha) = \langle v - \alpha w, v - \alpha w \rangle = -2\langle v - \alpha w, w \rangle = -2\langle v, w \rangle + 2\alpha \|w\|^2$, which equals zero when $\alpha = \frac{\langle v,w \rangle}{\|w\|^2}$. We will soon see that one can project into any subspace in a similar manner.

For each fixed $w \in V$, we observe that the map $v \mapsto \langle v, w \rangle$ is continuous. The reason is that if $v_n \to v$ as $n \to \infty$, then

$$\big|\langle v_n, w \rangle - \langle v, w \rangle\big| = \big|\langle v_n - v, w \rangle\big| \le \|v_n - v\| \, \|w\| \to 0$$

as $n \to \infty$, where the inequality holds due to the Schwarz inequality (9.3). It follows that if the $\sum_{n=1}^{\infty} \alpha_n v_n$ converges, then for any w, one has

$$\left\langle \sum_{n=1}^{\infty} \alpha_n v_n, w \right\rangle = \sum_{n=1}^{\infty} \alpha_n \langle v_n, w \rangle.$$

That is, not only does the additive property (1) for inner products hold for finite sums, but it also holds for convergent infinite sums.

9.2.2 *Examples*

We have already seen several examples of Hilbert spaces, although we have not called them that. The finite dimensional Euclidean space $\mathbb{F}^N$ is a Hilbert space, where the inner product is given by the classical scalar product. There are many other inner products that could be defined in $\mathbb{F}^N$. For example, let $\langle v, w \rangle$ denote the classical scalar product, and suppose A is any $N \times N$ matrix that satisfies $\langle Av, v \rangle > 0$ for every $v \neq 0$ (such a matrix is called *positive definite*). Recall from the earlier discussion on inner products from page 240 that a new inner product is defined via

$$\langle v, w \rangle_A = \langle Av, w \rangle.$$

Recall from page 150 the sequence space ℓ_2 consisting of all infinite sequences that are square summable. If $v = \{\alpha_1, \alpha_2, \ldots, \alpha_n, \ldots\}$ and $w = \{\beta_1, \beta_2, \ldots, \beta_n, \ldots\}$ belong to ℓ_2, then an inner product can be defined by

$$\langle v, w \rangle = \sum_{n=1}^{\infty} \alpha_n \bar{\beta}_n,$$

in which case, ℓ_2 is a Hilbert space that has the norm given by $\|v\|_2$. The other ℓ_p spaces with $1 \leq p \leq \infty$ and $\neq 2$ do not have an inner product that induces the norm, and so are not Hilbert spaces.

Consider the function spaces $\mathcal{L}^p[0,1]$. As for sequence spaces the only Hilbert space is also the case where $p = 2$, and the inner product is given by

$$\langle f, g \rangle = \int_0^1 f(t)\bar{g}(t)\, dt,$$

for $f, g \in \mathcal{L}^2[0,1]$.

As in $\mathbb{F}^N$, there are many other inner products that could be imposed on a function space. Let $w(t)$ be any real-valued measurable function defined on $[0,1]$ with $0 < w(t) \leq M$ ae for some constant $M > 0$. The function w should be interpreted as a "weight" function that gives greater or lesser importance to areas of the interval $[0,1]$ depending on the values of w in that area. Define

$$\langle f, g \rangle_w = \int_0^1 f(t)\bar{g}(t)w(t)\, dt$$

for $f, g \in \mathcal{P}$. One can check that this is indeed an inner product, and it induces the norm $\|f\|_w = \sqrt{\langle f, f \rangle_w}$. The completion of the norm is the space denoted by $\mathcal{L}^2_w[0,1]$. One can verify that if there are constants $0 < m \leq M < \infty$ with $m \leq w(t) \leq M$ ae $t \in [0,1]$, then the spaces $\mathcal{L}^2[0,1]$ and $\mathcal{L}^2_w[0,1]$ have the same elements, although if $w(t) \neq 1$ ae, they have different inner products.

9.2.3 *Orthogonality*

Hilbert spaces are attractive to use in applications largely because they share many of the same properties that one is familiar with in finite dimensions. This is mainly because the concept of orthogonality can be defined, which gives a Hilbert space a familiar sense of geometry that is in general more complicated in other infinite dimensional spaces. Furthermore, orthogonality will be used to construct bases that provide a coordinate representation of a vector that is reminiscent to the finite dimensional case.

Two non-zero vectors v and w in a Hilbert space V are *orthogonal* provided $\langle v, w\rangle = 0$. An arbitrary subset $\mathcal{B}$ of non-zero vectors is called orthogonal if any two its members are orthogonal. Such a set is called *orthonormal* if in addition all its members have norm equal to one. That is, $\mathcal{B}$ is orthormal if and only if for any v, $w \in \mathcal{B}$, we have

$$\langle v, w\rangle = \begin{cases} 0 & \text{if } v \neq w \\ 1 & \text{if } v = w. \end{cases}$$

The set $\mathcal{B}$ is a *maximal orthonormal* set if it is orthonormal and not properly contained in a strictly larger orthonormal set. We record some of the basic properties of orthogonal and orthonormal sets in the following theorem.

Theorem 9.4.

(1) The Pythagorean theorem holds in any Hilbert space. That is, if v and w are orthogonal, then

$$\|v + w\|^2 = \|v\|^2 + \|w\|^2.$$

More generally, if $\mathcal{B}$ is an orthogonal subset and $v_1, \ldots, v_N$ belong to $\mathcal{B}$, then

$$\left\|\sum_{n=1}^{N} v_n\right\|^2 = \sum_{n=1}^{N} \|v_n\|^2.$$

(2) Any collection of orthogonal vectors is an independent set.
(3) Suppose $\mathcal{B} = \{v_n\}$ is any orthonormal set and $v \in V$. Then $\sum_{n=1}^{\infty} \left|\langle v, v_n\rangle\right|^2 \leq \|v\|^2$, and $\sum_{n=1}^{\infty} \langle v, v_n\rangle v_n$ converges in V.

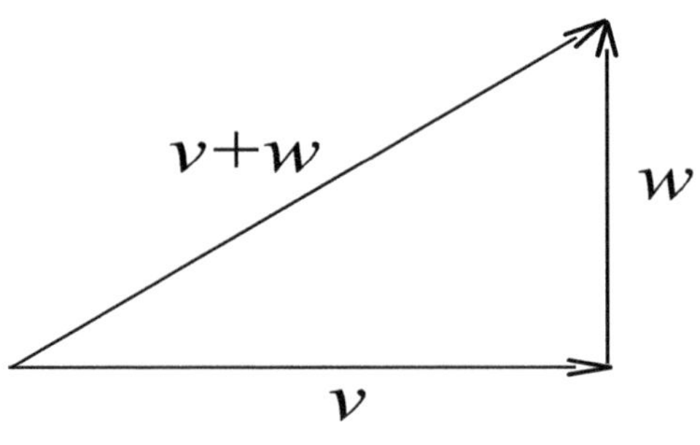

Fig. 9.1 Pythagorean theorem.

Proof: (1) Let v and w be two orthogonal vectors, corresponding to the two sides of a right angled triangle. Then

$$\begin{aligned} \|v + w\|^2 &= \langle v + w, v + w\rangle \\ &= \langle v, v\rangle + \langle v, w\rangle + \langle w, v\rangle + \langle w, w\rangle \end{aligned}$$

and, since v and w are orthogonal, their inner product equals zero and so

$$\|v+w\|^2 = \|v\|^2 + \|w\|^2$$

which is the Pythagorean theorem. The general case of N orthogonal vectors is proved similarly.
(2) Suppose $\mathcal{B}$ is orthogonal, $\{v_1, \dots, v_N\} \subseteq \mathcal{B}$, and $\sum_{n=1}^N \alpha_n v_n = 0$ for some choice of scalars $\alpha_1, \dots, \alpha_N$. Then for each $k = 1, \dots, N$,

$$0 = \left\langle \sum_{n=1}^N \alpha_n v_n, v_k \right\rangle = \sum_{n=1}^N \alpha_n \langle v_n, v_k \rangle = \alpha_k \|v_k\|^2,$$

and hence $\alpha_k = 0$.
(3) Suppose $\mathcal{B} = \{v_n\}$ is orthonormal and $v \in V$. Then for each N, we have

$$\begin{aligned}
0 &\le \left\langle v - \sum_{n=1}^N \langle v, v_n\rangle v_n, v - \sum_{m=1}^N \langle v, v_m\rangle v_m \right\rangle \\
&= \langle v, v\rangle - \left\langle v, \sum_{m=1}^N \langle v, v_m\rangle v_m \right\rangle - \left\langle \sum_{n=1}^N \langle v, v_n\rangle v_n, v \right\rangle + \left\langle \sum_{n=1}^N \langle v, v_n\rangle v_n, \sum_{m=1}^N \langle v, v_m\rangle v_m \right\rangle \\
&= \langle v, v\rangle - \sum_{m=1}^N \overline{\langle v, v_m\rangle}\, \langle v, v_m\rangle - \sum_{n=1}^N \langle v, v_n\rangle \langle v_n, v\rangle + \sum_{n,m=1}^N \langle v, v_n\rangle \overline{\langle v, v_m\rangle} \langle v_n, v_m\rangle \\
&= \langle v, v\rangle - \sum_{n=1}^N \left|\langle v, v_n\rangle\right|^2.
\end{aligned}$$

Hence $\sum_{n=1}^N \left|\langle v, v_n\rangle\right|^2 \le \|v\|^2$, for all N, and letting $N \to \infty$ we conclude that $\sum_{n=1}^\infty \left|\langle v, v_n\rangle\right|^2 \le \|v\|^2$.

To see that $\sum_{n=1}^\infty \langle v, v_n\rangle v_n$ converges, let $s_N = \sum_{n=1}^N \langle v, v_n\rangle v_n$. For $N < M$, note that

$$\left\|s_N - s_M\right\|^2 = \left\| \sum_{n=M+1}^N \langle v, v_n\rangle v_n \right\|^2 = \sum_{n=M+1}^N \left|\langle v, v_n\rangle\right|^2,$$

where we see the last equality holds by the Pythagorean Theorem. The last term approaches 0 as $M, N \to \infty$, and therefore the sequence $\{s_n\}$ is Cauchy and the series $\sum_{n=1}^\infty \langle v, v_n\rangle v_n$ converges. □

Recall that a set $\{v_n\}_n$ is a basis in an infinite dimensional space V provided its span is dense in V. We now see the great advantage of orthogonality by considering a basis consisting of an orthonormal set. Part (4) of the following theorem gives an attractive way to represent a vector in terms of an orthonormal basis expansion, and is usually called the Fourier series of v.

Theorem 9.5. *Suppose $\mathcal{B} = \{v_n\}$ is an orthonormal set of vectors in the Hilbert space V. Then the following statements are equivalent.*

(1) $\mathcal{B}$ is a basis for V.
(2) $\mathcal{B}$ is a maximal orthonormal set.
(3) If $\langle w, v_n \rangle = 0$ for all n, then $w = 0$.
(4) For every $v \in V$, we have $v = \sum_{n=1}^{\infty} \langle v, v_n \rangle v_n$.

Proof: (1)⇒(2): Suppose there exists w so that $\mathcal{B} \cup \{w\}$ is orthonormal. Then for any set of scalars $\alpha_1, \ldots, \alpha_N$, the Pythagorean Theorem implies

$$\left\| w - \sum_{n=1}^{N} \alpha_n v_n \right\|^2 = \|w\|^2 + \sum_{n=1}^{N} \left\| \alpha_n v_n \right\|^2 = 1 + \sum_{n=1}^{N} \left| \alpha_n \right|^2 \geq 1,$$

and so w is not arbitrarily close to elements in the span of $\mathcal{B}$. Hence $\mathcal{B}$ is not a basis.

(2)⇒(3) Suppose $w \in V$ that satisfies $\langle w, v_n \rangle = 0$ for all n. If $w \neq 0$, then $\tilde{w} = w/\|w\|$ is a vector of norm one that is orthogonal to every element of $\mathcal{B}$. It follows that $\mathcal{B} \cup \{\tilde{w}\}$ is an orthonormal set that is strictly larger than $\mathcal{B}$.

(3)⇒(4): By Theorem 9.4(3), the series $\sum_{n=1}^{\infty} \langle v, v_n \rangle v_n$ converges. Let $w = v - \sum_{n=1}^{\infty} \langle v, v_n \rangle v_n$, and observe that for each m,

$$\langle w, v_m \rangle = \left\langle v - \sum_{n=1}^{\infty} \langle v, v_n \rangle v_n, v_m \right\rangle = \langle v, v_m \rangle - \sum_{n=1}^{\infty} \langle v, v_n \rangle \langle v_n, v_m \rangle = 0.$$

The assumption (3) implies $w = 0$, and so (4) holds.

(4)⇒(1): Let $v \in V$, and suppose $v = \sum_{n=1}^{\infty} \langle v, v_n \rangle v_n$. Since $s_N = \sum_{n=1}^{N} \langle v, v_n \rangle v_n$ belongs to span of $\mathcal{B}$ and converges to v, it follows that the span of $\mathcal{B}$ is dense in V. Therefore $\mathcal{B}$ is a basis of V. □

We have yet to address the issue of whether a given Hilbert space has an orthonormal basis. This is not a trivial concern in abstract spaces, where a basis may even be uncountable. We will not go into much detail, but rather just assert the following.

Theorem 9.6. *Every Hilbert space contains an orthonormal basis.*

The idea of proving this theorem is to build a maximal orthonormal set by starting with any normalized vector, and keep adding vectors to the set that are not in the span of those previously chosen. Since in finite dimensions, this process eventually stops, the Gram–Schmidt procedure (introduced earlier in finite-dimensions) can then be utilized to produce an orthonormal basis. In infinite dimensions, this type of construction will go on indefinitely, but even after infinitely many additions, there is still no guarantee that the final result will produce a basis. For example, suppose we chose only monomials of even degree at each step. These would not form a basis of $\mathcal{L}^2[0, 1]$ even though there are infinitely many of them. The correct and complete proof of the theorem requires the mathematical device of "tranfinite induction", and we will not go into that.

But now suppose a basis of V is known. For example, the set of all monomials form a basis for $\mathcal{L}^2[0,1]$, but they are not orthonormal. The Gram–Schmidt procedure can still be utilized to produce an orthonormal basis from a given basis in the following way. Suppose $\{v_n\}_n$ is a basis, which means that its span is dense in V. Let $\hat{w}_1 = v_1/\|v_1\|$, which is of norm one. Let $w_2 = v_2 - \langle v_2, \hat{w}_1\rangle\hat{w}_1 \neq 0$. Note that $\langle w_2, \hat{w}_1\rangle = 0$, and so normalizing by letting $\hat{w}_2 = w_2/\|w_2\|$ produces an orthonormal set $\{\hat{w}_1, \hat{w}_2\}$ that has the same span as $\{v_1, v_2\}$. One can continue inductively by assuming that $\{\hat{w}_1, \ldots, \hat{w}_n\}$ have been chosen so that it is orthonormal and spans the same subspace as $\{v_1, \ldots, v_n\}$. One proceeds by choosing

$$w_{n+1} = v_{n+1} - \sum_{m=1}^{n} \langle v_{n+1}, \hat{w}_m\rangle\hat{w}_m \quad \text{and} \quad \hat{w}_{n+1} = \frac{w_{n+1}}{\|w_{n+1}\|}.$$

It can be easily seen that $\{\hat{w}_1, \ldots, \hat{w}_{n+1}\}$ is orthonormal and spans the same subspace that is spanned by $\{v_1, \ldots, v_{n+1}\}$. The entire collection $\{\hat{w}_1, \hat{w}_2 \ldots\}$ is an orthonormal basis for V.

9.2.4 *Orthogonal projections*

A *closed* subspace W of a normed space is vector subspace that is also complete. Recall this means that

- for $v, w \in W$ and $\alpha \in \mathbb{F}$, we have $\alpha v + w \in W$, and
- if $\{w_n\}$ is a Cauchy sequence belonging to W, then its limit v (which is known to exist in V since V is complete) also belongs to W.

Suppose $v \notin W$ where W is a closed subspace. A *projection* of v into W is a vector $P(v)$ in W that is closest to v. This means that $P(v) \in W$ and

$$\|v - P(v)\| = \inf\{\|v - w\| : w \in W\}. \tag{9.4}$$

In general infinite dimensional normed spaces, $P(v)$ may not exist, be unique, or be easy to find, but in Hilbert spaces, all of these issues are resolved easily and in a natural way. The next theorem will show that $P(v)$ exists and is unique, and also that $v - P(v)$ is orthogonal to every element in W, and hence is also called the *orthogonal projection.*

Since every closed subspace of a Hilbert space is a Hilbert space in its own right, by Theorem 9.6 there exists an orthonormal basis for W. We shall also need to invoke the so-called parallelogram law, which is valid in any inner product space. This says that

$$\|x + y\|^2 + \|x - y\|^2 = 2(\|x\|^2 + \|y\|)^2 \tag{9.5}$$

for all $x, y \in V$, and can be easily verified by just expanding the terms on the left.

Theorem 9.7. *Suppose W is a closed subspace and $v \notin W$. Then there exists a unique vector $P(v) \in W$ satisfying Eq. (9.4). Furthermore, if $\{w_n\}$ is an orthonormal basis for W, then*

$$P(v) = \sum_n \langle v, w_n \rangle w_n, \tag{9.6}$$

and $v - P(v)$ is orthogonal to every element of W.

Proof: Let $r = \inf\{\|v-w\| : w \in W\}$, and let $v_N \in W$ so that $\|v - v_N\| \leq \inf\{\|v - w\| : w \in W\} + \frac{1}{N}$. We will first show $\{v_N\}$ is Cauchy. Note that $\frac{1}{2}(v_N + v_M) \in W$, and so that $r \leq \left\|v - \frac{1}{2}(v_N + v_M)\right\|$. Using this observation, we have

$$\begin{aligned} r^2 &\leq \left\|v - \frac{1}{2}(v_N + v_M)\right\|^2 \\ &= \frac{1}{4}\|2v - (v_N + v_M)\|^2 \\ &= \frac{1}{2}\|v - v_N\|^2 + \frac{1}{2}\|v - v_M\|^2 - \frac{1}{4}\|v_N - v_M\|^2, \end{aligned}$$

where the last equality follows from the parallelogram law in Eq. (9.5) with $x = v - v_N$ and $y = v - v_m$. Rearranging terms produces the inequality

$$\|v_N - v_M\|^2 \leq 2\|v - v_N\|^2 + 2\|v - v_M\|^2 - 4r^2$$

which approaches 0 as $M, N \to \infty$. Hence $\{v_N\}$ is Cauchy, and therefore has a limit w that belongs to W since W is closed.

It is not yet clear that w is unique, but we shall assume that $w = P(v)$ is unique and has the form in Eq. (9.6). Recall that $\{w_n\}$ is an orthonormal basis, and by Theorem 9.5(4), the representation $w = \sum_{n=1}^{\infty} \langle w, w_n \rangle w_n$ holds. Fix an index k of the orthonormal basis. For $\alpha \in \mathbb{F}$, let $w^\alpha = \sum_{n \neq k} \langle w, w_n \rangle w_n + \alpha w_k \in W$, and consider the function $g_k : \mathbb{F} \to \mathbb{R}$ defined by

$$g_k(\alpha) = \|v - w^\alpha\|^2.$$

Note that $\langle w_k, w^\alpha \rangle = \alpha$, and so one calculates

$$g_k'(\alpha) = -2 \text{ Re } \langle w_k, v - w^\alpha \rangle = -2 \text{ Re } \big[\langle w_k, v \rangle - \alpha\big].$$

Assume first that $\mathbb{F} = \mathbb{R}$. Since $w^\alpha \in W$ for all α and w is a projection of v into W, it follows that g_k attains a minimum when $\alpha = \langle w, w_k \rangle$. Therefore $g_k'(\alpha) = 0$ for $\alpha = \langle w, w_k \rangle$, and so $\langle v, w_k \rangle = \langle w, w_k \rangle$. If $\mathbb{F} = \mathbb{C}$, then similar arguments can be employed to the real and imaginary parts, and the conclusion is again $\langle v, w_k \rangle = \langle w, w_k \rangle$. Since this holds for each index k, We see that $P(v)$ is indeed unique and satisfies Eq. (9.6).

To prove the last statement, let $w \in W$. Note that $w = \sum_m \langle w, w_m \rangle w_m$ by Theorem 9.5, and so by Eq. (9.6)

$$\begin{aligned} \langle v - P(v), w \rangle &= \left\langle v - \sum_n \langle v, w_n \rangle w_n, \sum_m \langle w, w_m \rangle w_m \right\rangle \\ &= \sum_m \langle w, w_m \rangle \langle v, w_m \rangle - \sum_{n,m} \langle v, w_n \rangle \langle w, w_m \rangle \langle w_n, w_m \rangle \\ &= \sum_m \langle w, w_m \rangle \langle v, w_m \rangle - \sum_m \langle v, w_m \rangle \langle w, w_m \rangle = 0. \end{aligned}$$

Hence $v - P(v)$ is orthogonal to every $w \in W$, as asserted. □

9.2.5 *Orthogonal complements*

Direct sums play an important role in the finite dimensional theory, and provide the framework to describe how vectors can be decomposed into components with desirable properties. Orthogonal subspaces provide a convenient mechanism to work similarly in Hilbert spaces.

The *orthogonal complement* of a closed subspace W is defined by

$$W^{\perp} = \{u : \langle u, w\rangle = 0 \text{ for all } w \in W\}.$$

Theorem 9.8. *Suppose W is a closed subspace of a Hilbert space V. Then the following statements hold.*

(1) $W^{\perp}$ is a closed subspace and $W \cap W^{\perp} = \{0\}$.
(2) For $v \in V$, there exists unique elements $w \in W$ and $u \in W^{\perp}$ so that $v = w + u$.
(3) $(W^{\perp})^{\perp} = W$.

Proof: (1) If $u_1, u_2 \in W^{\perp}$ and $\alpha \in \mathbb{F}$, then $\langle \alpha u_1 + u_2, w\rangle = \alpha\langle u_1, w\rangle + \langle u_2, w\rangle = 0$ for any $w \in W$, and so $\alpha u_1 + u_2 \in W^{\perp}$ and $W^{\perp}$ is a subspace. If $v \in W \cap W^{\perp}$ then $\|v\|^2 = \langle v, v\rangle = 0$, and so $v = 0$.
(2) Let $v \in V$ and let $w = P(v)$ be the orthogonal projection of v onto W. Then $u = v - P(v)$ belongs to $W^{\perp}$ by Theorem 9.7, and clearly $v = w + u$. If in addition $v = w_1 + u_1$ with $w_1 \in W$ and $u_1 \in W^{\perp}$, then $w + u = v = w_1 + u_1$ implies $w - w_1 = u_1 - u \in W \cap W^{\perp}$. Since the latter set is $\{0\}$ by part (1), we conclude $w = w_1$ and $u = u_1$ or that the decomposition is unique.
(3) Let $w \in W$. For any $u \in W^{\perp}$, we have $\langle w, u\rangle = 0$, and so $w \in (W^{\perp})^{\perp}$. Therefore $W \subseteq (W^{\perp})^{\perp}$. To prove the reverse inclusion, let $v \in (W^{\perp})^{\perp}$. By part (2), there exist $w \in W$ and $u \in W^{\perp}$ so that $v = w + u$. Now $w \in W \subseteq (W^{\perp})^{\perp}$ and $v \in (W^{\perp})^{\perp}$, so $u = v - w \in (W^{\perp})^{\perp}$. But also $u \in W^{\perp}$, and so $\langle u, u\rangle = 0$, or that $u = 0$. Therefore $v = w \in W$. □

9.3 Linear operators in Hilbert spaces

Suppose V and W are Hilbert spaces over the same field $\mathbb{F}$. Recall that a linear operator is a map $L : V \to W$ that satisfies $L(\alpha v_1 + v_2) = \alpha L(v_1) + L(v_2)$ for all $v_1, v_2 \in V$ and $\alpha \in \mathbb{F}$. If V and W are finite dimensional with given bases, then a linear operator can be represented as a matrix that operates by matrix multiplication. A linear operator defined on infinite dimensional spaces can be considerably more complicated. We shall not go into much generality, but consider only the most important linear operators that arise in applications. Although we say that a linear operator L will map from $\mathcal{L}^2_w[0,1]$ into $\mathcal{L}^2_w[0,1]$ (for some weight function $w(t)$), it is usually the case that $L(f)$ is only defined for certain $f \in \mathcal{L}^2_w[0,1]$, and the set of all these f is called the domain of L and is denoted by $\mathcal{D}_L$. The operators we shall study are the so-called self-adjoint second order *differential operators* that are often accompanied with boundary conditions that affect its domain and other properties of the operator. An example of such an operator is $Lf(t) = \frac{d^2}{dt^2}f(t)$, in which the domain consists of the twice differentiable functions. The domain $\mathcal{D}_L$ can be further restricted by requiring its elements to satisfy boundary conditions of the form $\alpha_1 f(0)+\alpha_2 f(1)+\alpha_3 f'(0)+\alpha_4 f'(1) =$ and $\beta_1 f(0)+\beta_2 f(1)+\beta_3 f'(0)+\beta_4 f'(1) = 0$. An initial value problem is where $f(0)$ and $f'(0)$ is specified, and there is always a unique solution, but boundary value problem have a more complicated structure.

Eigenvalues and eigenvectors of a linear operator are defined similarly as they are in finite dimensions, although there is no analog of a determinant that can be used to characterize and find eigenvalues. A scalar λ is an eigenvalue with an associated eigenvector v provided $v \in \mathcal{D}_L$, $v \neq 0$, and $Lv = \lambda v$.

9.3.1 *The adjoint operator*

We consider the abstract spectral properties of so-called self-adjoint operators in this section, and then specialize them to function spaces in the rest of the chapter.

Suppose $\mathcal{D}_L \subseteq V$ is a dense subspace of a Hilbert space V, and $L : \mathcal{D}_L \to V$ is a linear operator. The adjoint operator of L is the operator L^* that satisfies

$$\langle Lv, w\rangle = \langle v, L^* w\rangle \tag{9.7}$$

for $v \in \mathcal{D}_L$. More precisely, L^* is another linear operator that is defined on its domain $\mathcal{D}_{L^*}$ consisting of all those w for which there exists an element $z \in V$ so that $\langle Lv, w\rangle = \langle v, z\rangle$ for all $v \in \mathcal{D}(L)$. Since $\mathcal{D}_L$ is dense, if such a vector z exists, then it is unique, and we write $L^* w = z$. This is the long-winded explanation of Eq. (9.7). For linear operators defined on a finite dimensional space with a matrix representation, the adjoint operator is the operator whose representation is given by the complex conjugate matrix transpose. If it should happen that $Lv = L^* v$ for all $v \in \mathcal{D}_L$, then the operator is said to be *formally self-adjoint*. If in addition one has $\mathcal{D}_L = \mathcal{D}_{L^*}$, then L is called *self-adjoint*.

We next see, as in finite dimensions, self-adjoint operators have real eigenvalues which, when distinct, admit orthogonal eigenvectors.

Theorem 9.9. *Suppose L is a self-adjoint linear operator. Then*

(1) All the eigenvalues of L are real.
(2) Distinct eigenvalues of L have associated eigenvectors that are orthogonal.

Proof: (1) Suppose λ is an eigenvalue of the self-adjoint, linear operator L, and let v be the associated eigenvector. Then

$$\langle Lv, v\rangle = \langle \lambda v, v\rangle = \lambda\langle v, v\rangle.$$

Since L is self-adjoint, we also must have that

$$\langle Lv, v\rangle = \langle v, L^*v\rangle = \langle v, Lv\rangle = \langle v, \lambda v\rangle = \overline{\lambda}\langle v, v\rangle,$$

and consequently

$$\lambda\langle v, v\rangle = \overline{\lambda}\langle v, v\rangle \quad \Rightarrow \quad \lambda = \overline{\lambda}.$$

Therefore any eigenvalue equals its own complex conjugate, and so must be real.
(2) Let λ and μ be distinct eigenvalues of L with associated eigenvectors v and w, respectively. Then

$$\lambda\langle v, w\rangle = \langle \lambda v, w\rangle = \langle Lv, w\rangle = \langle v, Lw\rangle = \langle v, \mu w\rangle = \mu\langle v, w\rangle$$

(since μ is real), and so $(\lambda - \mu)\langle v, w\rangle = 0$. Since $\lambda \neq \mu$, we conclude that v and w are orthogonal. □

9.3.2 *Examples*

For a given weight function $w(t)$, we are mainly interested in the Hilbert space $\mathcal{L}^2_w[0, 1]$, which recall has the inner product

$$\langle f(t), g(t)\rangle_w = \int_0^1 f(t)\overline{g(t)}w(t)\, dt\,.$$

If the weight function is $w(t) = 1$, then the subscript for the inner product is dropped. If L is a differential operator, then its adjoint is another differential operator that can be found directly from the definition of the adjoint, using integration by parts to rearrange the inner product as will be illustrated in the examples that follow. This process determines both the adjoint operator and the adjoint boundary conditions. When the adjoint operator has the same form as the operator but has different boundary conditions, then the operator is formally self-adjoint. If the boundary conditions also coincide, then the operator is self-adjoint.

Example 9.1: A formally self-adjoint operator.

Consider the operator $L_1 f = \frac{d^2f}{dt^2}$ whose domain consists of the twice differentiable functions that satisfy the boundary conditions $f(0)+f(1) = 0$ and $f'(0) = 0$. The adjoint is calculated by integration by parts twice and then substituting the boundary conditions into the boundary terms.

$$\begin{aligned}
\langle L_1 f, g\rangle &= \int_0^1 \frac{d^2f}{dt^2} g(t)\, dt \\
&= \left[\frac{df}{dt}g(t)\right]_0^1 - \int_0^1 \frac{df}{dt}\frac{dg}{dt}\, dt \\
&= \left[\frac{df}{dt}g(t)\right]_0^1 - \left[f(t)\frac{dg}{dt}\right]_0^1 + \int_0^1 f(t)\frac{d^2g}{dt^2}\, dt \\
&= f'(1)g(1) + f(0)\{g'(1) + g'(0)\} + \left\langle f, \frac{d^2g}{dt^2}\right\rangle.
\end{aligned}$$

The boundary terms in the last line vanish when $g(1) = 0$ and $g'(0) + g'(1) = 0$, and these are the adjoint boundary conditions that determine the domain of L^*. Thus L_1 is formally self-adjoint since it agrees with L_1^* as a differentiable operator, but it is not self-adjoint because the boundary conditions differ and they have different domains.

Example 9.2: A self-adjoint operator.

Let L_2 be the same operator as the previous example but with boundary conditions given by $f(0) = f'(0)$ and $f(1) = f'(1)$. The adjoint operator is found by

$$\begin{aligned}\langle L_2 f, g\rangle &= \int_0^1 \frac{d^2 f}{dt^2} g(t)\, dt \\ &= \left[\frac{df}{dt} g(t)\right]_0^1 - \left[f(t)\frac{dg}{dt}\right]_0^1 + \int_0^1 f(t)\frac{d^2 g}{dt^2}\, dt \\ &= f(1)\{g(1) - g'(1)\} + f(0)\{g'(0) - g(0)\} + \left\langle f, \frac{d^2 g}{dt^2}\right\rangle .\end{aligned}$$

Hence the adjoint boundary conditions are seen to be $g(0) = g'(0)$ and $g(1) = g'(1)$, which are identical to those of L_2. Thus $L_2 = L_2^*$ and is self-adjoint.

Example 9.3: An operator under different inner products.

Consider the operator $L_3 f = \frac{1}{t}\frac{d}{dt}\left(t\frac{df}{dt}\right)$ with boundary conditions $f(0) = f'(0)$ and $f(1) = 2f'(1)$. We first evaluate its adjoint using the inner product with weight $w(t) = 1$.

$$\begin{aligned}\langle L_3 f, g\rangle &= \int_0^1 \frac{1}{t}\frac{d}{dt}\left(t\frac{df}{dt}\right) g(t)\, dt \\ &= \left[\frac{df}{dt} g(t)\right]_0^1 - \left[f(t) t\frac{d}{dt}\left(\frac{g}{t}\right)\right]_0^1 + \int_0^1 f(t)\frac{d}{dt}\left(t\frac{d}{dt}\left(\frac{g}{t}\right)\right) dt \\ &= f'(1)g(1) - f'(0)g(0) - 2f'(1)\left.\frac{d}{dt}\left(\frac{g}{t}\right)\right|_{t=1} + \langle f, L_3^* g\rangle .\end{aligned}$$

The boundary terms vanish precisely when

$$g(0) = 0\,, \quad \frac{1}{2}g(1) = \left.\frac{d}{dt}\left(\frac{g}{t}\right)\right|_{t=1} ,$$

and these are the boundary conditions of L_3^*. The operator is clearly not self-adjoint.

However, changing the inner product by using the weight function $w(t) = t$, the operator will have a much simpler presentation. We obtain

$$\begin{aligned}\langle L_3 f, g\rangle_w &= \int_0^1 \frac{1}{t}\frac{d}{dt}\left(t\frac{df}{dt}\right) g(t) t\, dt = \int_0^1 \frac{d}{dt}\left(t\frac{df}{dt}\right) g(t)\, dt \\ &= \left[t\frac{df}{dt} g(t)\right]_0^1 - \left[f(t) t\frac{dg}{dt}\right]_0^1 + \int_0^1 f(t)\frac{d}{dt}\left(t\frac{dg}{dt}\right) dt \\ &= f'(1)g(1) - f(1)g'(1) + \langle f, L_3^* g\rangle.\end{aligned}$$

The adjoint boundary condition becomes

$$g(1) = 2g'(1)$$

which is identical to one of the boundary conditions on L_3. The other boundary on L_3 can be applied to L_3^* as $g(0) = g'(0)$ without affecting the result of the inner product calculation. The operator is therefore self-adjoint with respect to this inner product.

As these examples show, the adjoint operator depends on both the boundary conditions of the operator, and at least in some cases, can be made self-adjoint with a favorable choice for the inner product.

9.3.3 *Sturm–Liouville operators*

The last example above is an example of a general class of operators that occur so frequently in applications that they deserve special attention. We shall do this only for the field $\mathbb{F} = \mathbb{R}$.

Definition: *A Sturm–Liouville operator is a second order differential operator of the form*

$$Lf = \frac{1}{w(t)}\left\{\frac{d}{dt}\left(p(t)\frac{df}{dt}\right) + q(t)f(t)\right\} \tag{9.8}$$

which is defined for all $f \in \mathcal{C}^2[0,1]$, the set of all twice differentiable functions. The coefficient data functions are real-valued, and are such that $p(t)$ is differentiable, $w(t)$ and $q(t)$ are continuous, and $w(t) > 0$ and $p(t) > 0$ for all $t \in (0,1)$, $w(t)$ and $p(t)$ may take the value 0 at $t = 0$ and $t = 1$.

A regular *Sturm–Liouville operator is an operator of the form Eq. (9.8) whose domain $\mathcal{D}_L$ consists of those $f \in \mathcal{C}^2[0,1]$ that satisfy boundary conditions of the form*

$$\alpha_1 f'(0) - \alpha_2 f(0) = 0 \quad \textit{and} \quad \beta_1 f'(1) + \beta_2 f(1) = 0,$$

where α_1, α_2, β_1, and β_2 are real constants. A periodic *Sturm–Liouville operator is similarly defined but with boundary conditions*

$$f(0) = f(1) \quad and \quad p(0)f'(0) = p(1)f'(1).$$

Both regular and periodic Sturm–Liouville operators L are self-adjoint with the inner product $\langle f, g\rangle_w = \int_0^1 f(t)g(t)w(t)\,dx$, where the function $w(t)$ that appears in the definition of the operator is the weight function for the inner product. We provide the details only for the regular case

$$\begin{aligned}\langle Lf, g\rangle_w &= \left\langle \frac{1}{w(t)}\left\{\frac{d}{dt}\left(p(t)\frac{df}{dt}\right) + q(t)f\right\}, g\right\rangle_w \\ &= \int_0^1 \frac{w(t)}{w(t)}\left\{\frac{d}{dt}\left(p(t)\frac{df}{dt}\right) + q(t)f(t)\right\} g(t)\,dt \\ &= [pf'g]_0^1 - [pfg']_0^1 + \left\langle f, \frac{1}{w(t)}\left\{\frac{d}{dt}\left(p(t)\frac{dg}{dt}\right) + q(t)g\right\}\right\rangle_w .\end{aligned}$$

Setting the boundary terms equal to zero gives

$$p(1)f'(1)g(1) - p(0)f'(0)g(0) - p(1)f(1)g'(1) + p(0)f(0)g'(0) = 0\,.$$

Applying the boundary conditions on $f(t)$ and simplifying yields

$$p(1)f(1)\left\{-g'(1) - \frac{\beta_2}{\beta_1}g(1)\right\} + p(0)f(0)\left\{g'(0) - g(0)\frac{\alpha_2}{\alpha_1}\right\} = 0$$

and the adjoint boundary conditions reduce to

$$\beta_1 g'(1) + \beta_2 g(1) = 0 \quad \text{and} \quad \alpha_1 g'(0) - \alpha_2 g(0) = 0.$$

The conclusion is that the operator L is self-adjoint.

9.4 Eigenvalue problems

There is no standard way of writing the eigenvalue problem for differential operators, since both forms $Lf = \lambda f$ and $Lf + \lambda f = 0$ are used in the literature. The two forms are identical except for a change in signs. One must be careful when using theorems regarding the signs of eigenvalues since such theorems depend on the precise way the problem is written. Nonetheless, eigenvalue problems can be formulated for differential operators in much the same way as they are formulated for linear operators on finite dimensional vector spaces. The eigenvalue problems of differential operators are linear, homogeneous differential equations with the added complication that they contain a parameter λ. The values of λ for which there exists a non-zero solution is called an eigenvalue, and these are in general not known in advance. The eigenvectors are the non-zero solutions to the differential equation, and are called *eigenfunctions* rather than eigenvectors. The eigenspace of an eigenvalue consists of all the eigenfunctions, and its dimension is limited by the order of the differential operator. We know that an nth order, homogeneous, linear differential equation has a general solution which is a linear combination of n basis solutions. The eigenspace can therefore at most have dimension n. In many practical problems, the boundary conditions of the operator limit the dimension of the eigenspaces even further.

Note that the sum of self-adjoint operators with the same boundary conditions is self-adjoint, and also that the operator given by $Lf = \lambda f$ is self-adjoint. Hence if L is self-adjoint, the operator $Lf + \lambda f$ is also self-adjoint.

Example 9.4: A self-adjoint eigenvalue problem.

Consider the eigenvalue problem $\frac{d^2f}{dt^2} + \lambda f(t) = 0$ with boundary conditions $f(0) = f'(0)$ and $f(1) = f'(1)$. The right hand side of the equation is an example of a Sturm–Liouville operator with $w(t) = 1$, $p(t) = 1$, $q(t) = \lambda$, $\alpha_1 = \alpha_2 = 1$ and $\beta_1 = -\beta_2 = 1$, and so is self-adjoint. It is also similar to the operator L_2 in the example on page 426, and can be shown to be self-adjoint directly in the same manner as that example. The eigenvalues are therefore all real by Theorem 9.9(1), and the easiest way to proceed is to consider the three separate possibilities $\lambda = 0$, $\lambda < 0$ and $\lambda > 0$.

The case $\lambda = 0$ is when $\frac{d^2f}{dt^2} = 0$, and this has only the linear solution of form $f(t) = at + b$. The boundary conditions imply $a = b = 0$, and so $\lambda = 0$ is not an eigenvalue.

Suppose $\lambda < 0$, and the second order equation $\frac{d^2f}{dt^2} = -\lambda f(t)$ has the general solution

$$f(t) = C_1 e^{\sqrt{-\lambda}t} + C_2 e^{-\sqrt{-\lambda}t}.$$

The boundary conditions give the two algebraic equations

$$\begin{aligned} C_1 + C_2 &= C_1\sqrt{-\lambda} - C_2\sqrt{-\lambda} \\ C_1 e^{\sqrt{-\lambda}} + C_2 e^{-\sqrt{-\lambda}} &= C_1\sqrt{-\lambda}e^{\sqrt{-\lambda}} - C_2\sqrt{-\lambda}e^{-\sqrt{-\lambda}}, \end{aligned}$$

and now the complication of having the parameter λ appear in the differential equation comes to the forefront: we have two equations but three unknowns. However, the differential equation and the boundary conditions are all homogenous, and the solution can only be determined up to a constant multiplier. In other words, C_1 and C_2 cannot both be determined, only their ratio. First, suppose $C_2 \neq 0$ and let $C_3 = C_1/C_2$. The algebraic equations simplify to

$$\begin{aligned} C_3 + 1 &= C_3\sqrt{-\lambda} - \sqrt{-\lambda} \\ C_3 e^{\sqrt{-\lambda}} + e^{-\sqrt{-\lambda}} &= C_3\sqrt{-\lambda}e^{\sqrt{-\lambda}} - \sqrt{-\lambda}e^{-\sqrt{-\lambda}}, \end{aligned}$$

which is now a system with the same number of equations and unknowns. Nonetheless, the equations are nonlinear, and in general can be very difficult to solve. But we also know that the problem is self-adjoint, and so the roots are all real (Theorem 9.9). In this case, and the only real solutions are

$$\{C_3 = -1, \sqrt{-\lambda} = 0\} \quad \text{and} \quad \{C_3 = 0, \sqrt{-\lambda} = -1\}.$$

The first of these has already been treated with $\lambda = 0$ and given the trivial solution so it can be discarded. The second solution can also be discarded since a real, negative value of λ cannot satisfy $\sqrt{-\lambda} = -1$. When we consider the case $C_2 = 0$, we get

$$C_1 = C_1\sqrt{-\lambda} \text{ and } C_1 e^{\sqrt{-\lambda}} = C_1\sqrt{-\lambda}e^{\sqrt{-\lambda}}.$$

The only non-trivial solutions is $\lambda = -1$. Hence $\lambda = -1$ is the only negative eigenvalue, and its eigenspace is one-dimensional and spanned by the eigenfunction $\{e^t\}$. Finally, consider the case when $\lambda > 0$. The general solution has the form

$$f(t) = C_1 \cos\left(\sqrt{\lambda}t\right) + C_2 \sin\left(\sqrt{\lambda}t\right),$$

and in a manner analogous to the above, let $C_4 = C_1/C_2$. The boundary conditions reduce to $C_4 = \sqrt{\lambda}$ and $(1+\lambda)\sin\sqrt{\lambda} = 0$. Hence the eigenvalues are $\lambda_n = (n\pi)^2$ for $n = 1, 2, \ldots$, and for each n, the associated eigenfunction for λ_n is of the form $C_2(n\pi\cos(n\pi t) + \sin(n\pi t))$.

This completes the search for eigenvalues and eigenfunctions. It is left as an exercise to verify the conclusion of Theorem 9.9(2), and show that all the eigenfunctions are mutually orthogonal with respect to the inner product in which the operator is self-adjoint.

The previous example shows that care must be taken when solving the algebraic equations that arise when applying the boundary conditions to the general solution of a differential equation eigenvalue problem. In general, the equations are nonlinear and may be difficult to solve and analyze, yet all the solutions must be found to solve the eigenvalue problem. Often this is not possible to do analytically and the best one can hope for is to determine an equation for the eigenvalues, an equation that must then be solved numerically. However, eigenvalue problems in infinite dimensional Hilbert spaces have infinitely many eigenvalues so all the eigenvalues cannot be found numerically. Most of them must be determined from an approximate analytical expression. Self-adjoint problems are in general far simpler to solve than problems that are not self-adjoint. The latter may have complex eigenvalue making the algebraic problem of finding these much more difficult. The next example is for a less simple self-adjoint case.

Example 9.5: Another self-adjoint eigenvalue problem.

Consider the eigenvalue problem $\frac{d^2f}{dt^2} + \lambda f(t) = 0$ with boundary conditions $f(0) = -f'(1)$ and $f'(0) = f(1) + f'(1)$. The boundary conditions keep the associated operator from being Sturm–Liouville, and so we cannot immediately conclude that it is self-adjoint. However, we can find the adjoint directly.

$$\begin{aligned}\left\langle \frac{d^2f}{dt^2}, g\right\rangle &= \left[\frac{df}{dt}g(t)\right]_0^1 - \left[f(t)\frac{dg}{dt}\right]_0^1 + \int_0^1 f(t)\frac{d^2g}{dt^2}\,dt \\ &= f'(1)(g(1) - g(0) - g'(0)) - f(1)(g(0) + g'(1)) + \left\langle f, \frac{d^2g}{dt^2}\right\rangle\end{aligned}$$

The operator is formally self-adjoint, but the adjoint boundary conditions are $g'(1) = -g(0)$ and $g'(0) = -g(0) + g(1)$ and the second boundary condition is not identical to the boundary condition on $f(t)$. However, using the first boundary conditions, the second can be rewritten as $g'(0) = g'(1) + g(1)$, and the problem is seen to be self-adjoint.

Now we find the eigenvalues. The case $\lambda = 0$ is easily found to give the trivial solution and so is not an eigenvalue. For negative values of λ, the solution can be written

$$f(t) = C_1 \cosh(\sqrt{-\lambda}t) + C_2 \sinh(\sqrt{-\lambda}t).$$

Applying the boundary conditions give the two algebraic equations

$$\begin{aligned}C_1 &= -C_1\sqrt{-\lambda}\sinh(\sqrt{-\lambda}) - C_2\sqrt{-\lambda}\cosh(\sqrt{-\lambda}) \\ C_2\sqrt{-\lambda} &= C_1\cosh(\sqrt{-\lambda}) + C_2\sinh(\sqrt{-\lambda}) + C_1\sqrt{-\lambda}\sinh(\sqrt{-\lambda}) + C_2\sqrt{-\lambda}\cosh(\sqrt{-\lambda}).\end{aligned}$$

In matrix notation, this becomes

$$\begin{pmatrix} 1 + \sqrt{-\lambda}\sinh(\sqrt{-\lambda}) & \sqrt{-\lambda}\cosh(\sqrt{-\lambda}) \\ \cosh(\sqrt{-\lambda}) + \sqrt{-\lambda}\sinh(\sqrt{-\lambda}) & \sinh(\sqrt{-\lambda}) + \sqrt{-\lambda}\cosh(\sqrt{-\lambda}) - \sqrt{-\lambda}\end{pmatrix}\begin{pmatrix} C_1 \\ C_2\end{pmatrix} = \begin{pmatrix} 0 \\ 0\end{pmatrix}.$$

A non-trivial solution exists only if the determinant of the matrix is zero, which provides the desired equation for the eigenvalues.

$$\begin{aligned}0 = {} & (1 + \sqrt{-\lambda}\sinh(\sqrt{-\lambda}))(\sinh(\sqrt{-\lambda}) + \sqrt{-\lambda}\cosh(\sqrt{-\lambda}) - \sqrt{\lambda}) \\ & -\sqrt{-\lambda}\cosh(\sqrt{-\lambda})(\cosh(\sqrt{-\lambda}) + \sqrt{-\lambda}\sinh(\sqrt{-\lambda}))\end{aligned}$$

Expanding the expression and using the properties of the hyperbolic functions, this can be simplified down to

$$\sinh(\sqrt{-\lambda}) + \sqrt{-\lambda}\cosh(\sqrt{-\lambda}) - 2\sqrt{-\lambda} + \lambda\sinh(\sqrt{-\lambda}) = 0.$$

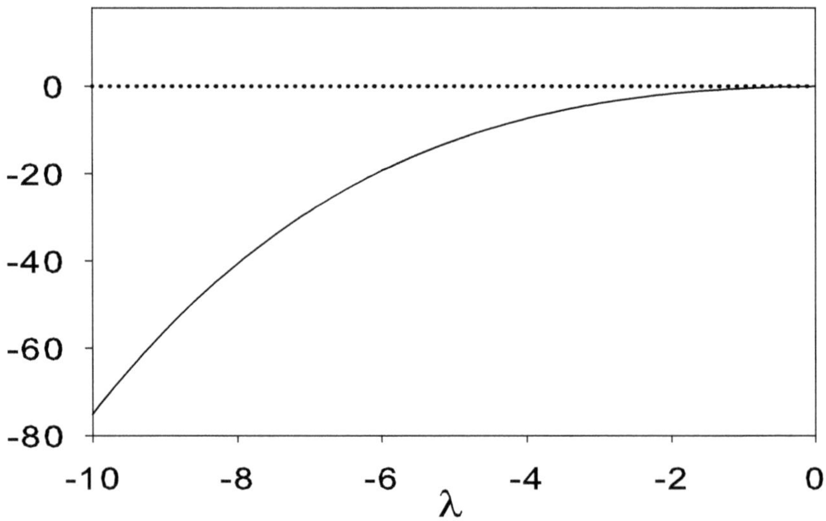

Fig. 9.2 Plot of the function $\sinh(\sqrt{-\lambda}) + \sqrt{-\lambda}\cosh(\sqrt{-\lambda}) - 2\sqrt{-\lambda} + \lambda\sinh(\sqrt{-\lambda})$.

There are no negative real roots. The easiest, though non rigorous, way to convince yourself of this is to plot the function on the left hand side for positive arguments, Fig. 9.2.

For positive values of λ, the solution is written

$$f(t) = C_1\cos(\sqrt{\lambda}t) + C_2\sin(\sqrt{\lambda}t).$$

Applying the boundary conditions gives the matrix equation

$$\begin{pmatrix} 1 - \sqrt{\lambda}\sin(\sqrt{\lambda}) & \sqrt{\lambda}\cos(\sqrt{\lambda}) \\ -\cos(\sqrt{\lambda}) + \sqrt{\lambda}\sin(\sqrt{\lambda}) & -\sin(\sqrt{\lambda}) - \sqrt{\lambda}\cos(\sqrt{\lambda}) + \sqrt{\lambda} \end{pmatrix} \begin{pmatrix} C_1 \\ C_2 \end{pmatrix} = \begin{pmatrix} 0 \\ 0 \end{pmatrix}$$

and the requirement of a zero determinant gives, after simplifications

$$\sin(\sqrt{\lambda}) + \cos(\sqrt{\lambda})\sqrt{\lambda} - 2\sqrt{\lambda} + \sin(\sqrt{\lambda})\lambda = 0\,.$$

A plot of the left hand side, Fig. 9.3, indicates that this equation has an infinity of negative roots. Their values will have to be determined numerically.

The problem of finding all the eigenvalues and of convincing oneself that none of these are positive is rather tedious. The point to note here, however, is that this problem is simplified tremendously by the knowledge that the operator is self-adjoint. This means the search can be a priori restricted to finding only real eigenvalues, even though that can be difficult enough. The difficulty is considerably heightened when trying to find all the potentially complex roots of transcendental equations like those above that arise from operators that are not self-adjoint.

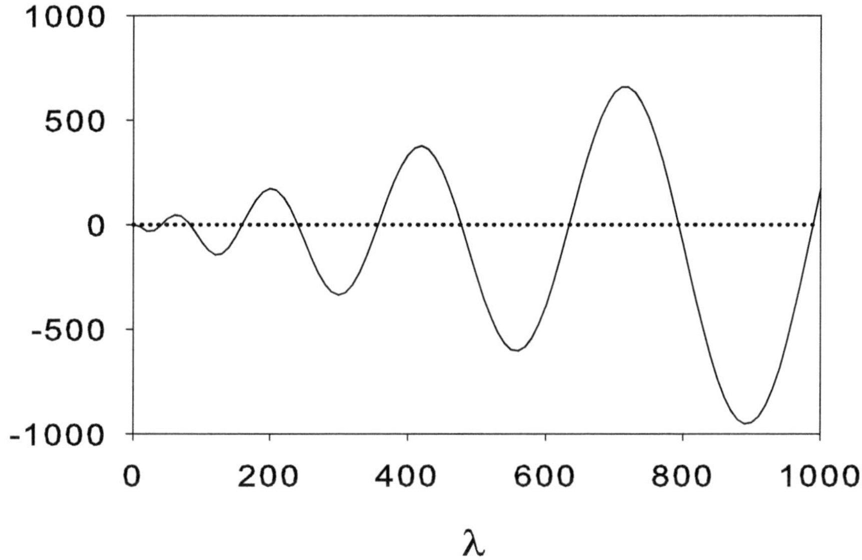

Fig. 9.3 Plot of the function $\sin(\sqrt{\lambda}) + \cos(\sqrt{\lambda})\sqrt{\lambda} - 2\sqrt{\lambda} + \sin(\sqrt{\lambda})\lambda$.

9.4.1 *Sturm–Liouville problems*

The eigenvalue problem associated with the Sturm–Liouville operator is called a Sturm–Liouville Problem, or SLP for short. In the engineering literature it is usually written

$$\frac{d}{dt}\left(p(t)\frac{df}{dt}\right) + \left(q(t) + \lambda w(t)\right)f(t) = 0 \tag{9.9}$$

with the same boundary conditions and restrictions on $p(t)$ and $w(t)$ as for the Sturm–Liouville operator. The values of λ for which a solution exists are the eigenvalues and the corresponding solutions are the eigenfunctions or eigenvectors. In the mathematical literature, the SLP is usually written with the $\lambda w(t)$-term on the right hand side, or equivalently, the SLP is defined with a "−" sign in front of the $\lambda w(t)$ terms instead of the "+" used in Eq. (9.9). Unfortunately, as a consequence of this lack of agreement on how to write a SLP, there is no agreement on the precise nomenclature for a SLP. The term eigenvalue is used casually and some people refer to $\sqrt{\lambda}$ or to $-\lambda$ as the eigenvalue. Obviously, these alternative nomenclature choices will give different versions of the theorems which follow.

If $p(t)$ and $w(t)$ are strictly positive, the SLP is said to be regular. If $p(t)$ and/or $w(t)$ equal zero at any of the interval endpoints 0 or 1, the problem is said to be a singular SLP.

Because the Sturm–Liouville operator is self-adjoint, it is already known that all eigenvalues are real. The following theorem states that all the eigenvalues of regular SLP are simple, which means that each associated eigenspace is one-dimensional.

Theorem 9.10. *All eigenvalues of the regular SLP are simple.*

Proof: The regular SLP is a homogeneous, second order, linear ODE, so the complete solution is a linear combination of two basis solutions. The eigenspace is therefore at most two-dimensional. Suppose λ is an eigenvalue and $u(t)$ and $v(t)$ are both eigenfunctions. First note that $u(0) \neq 0$, since otherwise the boundary condition implies $u'(0) = 0$ also, and so $u \equiv 0$ by the uniqueness theorem of ODEs. Similarly $v(0) \neq 0$. Let $g(t) = \frac{1}{u(0)}u(t) - \frac{1}{v(0)}v(t)$, which satisfies a second order linear differential equation with $g(0) = g'(0) = 0$. Therefore $g(t)$ must be identically 0. The conclusion is that $u(t) = \frac{u(0)}{v(0)}v(t)$, and the eigenspace is one-dimensional, and λ is simple. $\square$

There are two other important theorems that describe the structure of the SLP. Both of these theorems are true for regular and periodic SLP.

Theorem 9.11 (Eigenvalue theorem). *Consider the regular or periodic SLP in Eq. (9.9). Then there exist a countably infinite number of eigenvalues that can be arranged as*

$$\lambda_0 < \lambda_1 < \lambda_2 < \cdots < \lambda_n < \lambda_{n+1} < \cdots$$

where

$$\lambda_n \to \infty \ \ \textit{for} \ \ n \to \infty.$$

Furthermore, the zeros of the eigenfunction f_n, associated with the eigenvalue λ_n partitions the interval $[0,1]$ into $n+1$ subintervals.

Proof: It is really not that difficult, but very long, and so will be omitted.

The last statement of the theorem says the roots of the nth eigenfunction partitions the interval $[0,1]$ into $n+1$ subintervals, and illustrates something about the appearance of the eigenfunctions: They must be oscillatory functions over the interval of the SLP and exhibit an increasing number of oscillations as n increases. Appreciating this fact can be an aide in solving SLPs. For instance, the solutions to the problem

$$\frac{d^2 f}{dt^2} + \lambda f(t) = 0$$

are the hyperbolic (exponential) functions and the trigonometric functions, depending on the sign of λ. Clearly, the hyperbolic functions cannot generate a family of oscillating functions and the family of eigenfunctions should therefore be sought among the trigonometric functions which arise for positive values of λ. Yet, one cannot conclude that none of the eigenfunctions are hyperbolic or exponential functions. We saw this in Example 9.4 on page 429. Figure 9.4 contains the graphs of the first four eigenfunctions.

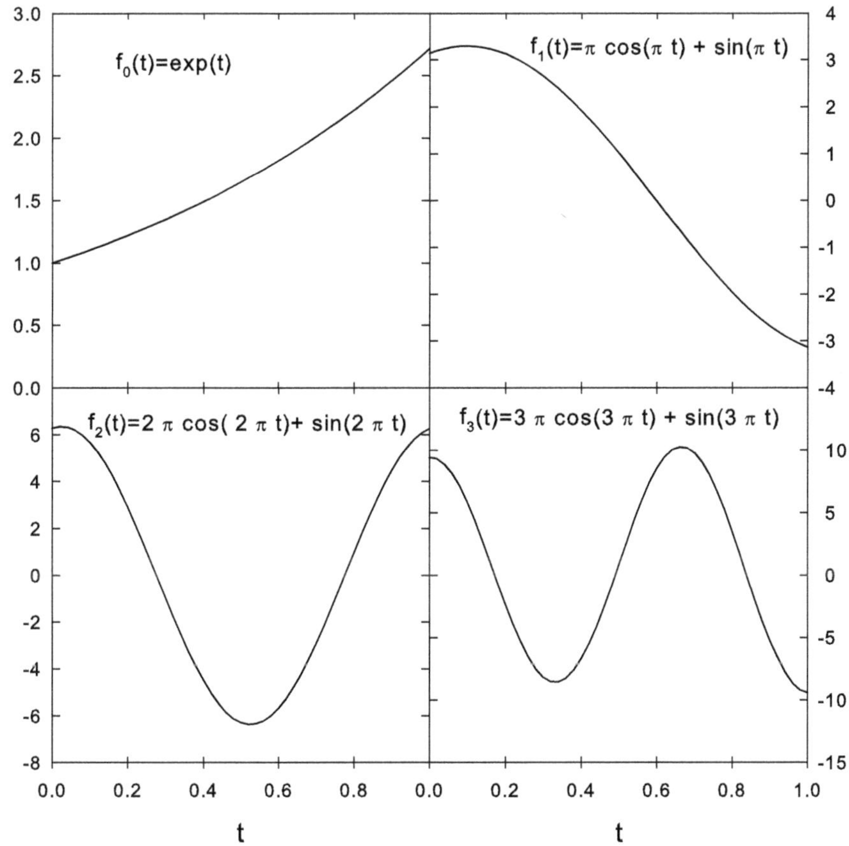

Fig. 9.4 Plots of the first four eigenfunctions of $f'' + \lambda f = 0$, $f(0) = f'(0)$, $f(1) = f'(1)$.

Similar comments hold for eigenvalue problems based on the Bessel equations and other types of equations. For instance, for Bessel equation problems (Section 8.6.2), one should expect that the oscillating Bessel functions $J_k(t)$ and $Y_k(t)$ to form an infinite family of oscillating eigenfunctions while the non-oscillating, modified Bessel functions, $I_k(t)$ and $K_k(t)$, only appear as eigenfunctions for a few eigenvalues, if at all.

Theorem 9.12 (Orthonormal basis theorem). *Consider either the regular or periodic SLP given in Eq. (9.9). Suppose $\lambda_0 < \lambda_1 < \ldots$ is the collection of eigenvalues as in Theorem 9.11, and f_n the associated eigenfunction to λ_n normalized with respect to the norm induced by the inner product*

$$\langle f, g\rangle_w = \int_0^1 f(t)\overline{g(t)}w(t)\,dt. \tag{9.10}$$

Then $\{f_n\}$ is an orthonormal basis of $\mathcal{L}^2_w[0,1]$, and so every $f(t) \in \mathcal{L}^2_w[0,1]$ can be expanded in a unique series of the form

$$f = \sum_{n=1}^{\infty}\langle f, f_n\rangle f_n. \tag{9.11}$$

Moreover, if $f(t)$ is piecewise continuous and t_0 is a point of discontinuity, then the partial sum $s_N(t) = \sum_{n=1}^{N}\langle f, f_n\rangle f_n(t)$ evaluated at $t = t_0$ converges to the average value $(f(t_0-) + f(t_0+))/2$ of the two sides of the discontinuity. At the interval endpoints 0 and 1, the series converges to $(f(0+) + f(1-))/2$.

Proof: Many of the assertions follow from the above development. We leave unproven that $\{f_n\}$ is a basis and the validity of the convergence statements, since these are difficult to verify and are beyond this text. □

A relatively minor theorem, but at times useful in applications to physical systems, is the following.

Theorem 9.13. *Suppose $q(t) \le 0$ for $t \in [0,1]$, $\alpha_2/\alpha_1 \ge 0$, and $\beta_2/\beta_1 \ge 0$. Then all the eigenvalues are positive or zero.*[2] *Furthermore, zero is an eigenvalue if and only if $q(t) \equiv 0$ and the boundary conditions are of the form $f'(0) = f'(1) = 0$. An eigenfunction associated with a zero eigenvalue is a constant.*

9.4.2 *Conversion of linear equations to SLP*

Suppose one is given a second order, linear differential equation in the usual form

$$p_2(t)\frac{d^2 f}{dt^2} + p_1(t)\frac{df}{dt} + (p_0(t) + \lambda w_0(t))f(t) = 0. \tag{9.12}$$

[2] Again, this result depends on how the SLP is written. If a "$-$" sign is used in front of the $\lambda w(t)$ term, the eigenvalues are negative.

It first appears that the SLP theory cannot be applied here, but multiplying Eq. (9.12) by an appropriate integrating factor can convert the equation into a standard SLP. Assuming $p_2(t) \neq 0$, the integrating factor is

$$\frac{e^{\int \frac{p_1(t)}{p_2(t)}\,dt}}{p_2(t)}$$

which after multiplying Eq. (9.12) and simplifying gives

$$\frac{d}{dt}\left(e^{\int \frac{p_1(t)}{p_2(t)}\,dt}\frac{df}{dt}\right) + (p_0(t) + \lambda w_0(t))\frac{e^{\int \frac{p_1(t)}{p_2(t)}\,dt}}{p_2(t)} f(t) = 0\,.$$

This is in the form of an SLP with

$$p(t) = e^{\int \frac{p_1(t)}{p_2(t)}\,dt},\; q(t) = \frac{p_0(t)}{p_2(t)}e^{\int \frac{p_1(t)}{p_2(t)}\,dt},\text{ and } w(t) = \frac{w_0(t)}{p_2(t)}e^{\int \frac{p_1(t)}{p_2(t)}\,dt}.$$

Notice that this method can only be applied when $p_2(t) \neq 0$.

Example 9.6: Bessel's equation.

The well known Bessel's equation has the form

$$t^2\frac{d^2 f}{dt^2} + t\frac{df}{dt} - (k^2 + \lambda t^2)f(t) = 0\,. \tag{9.13}$$

The integrating factor for converting this equation to SLP form is

$$\frac{1}{t^2}e^{\int \frac{t}{t^2}\,dt} = \frac{1}{t},$$

which gives

$$\frac{d}{dt}\left(t\frac{df}{dt}\right) - \left(\frac{k^2}{t} + \lambda t\right) f(t) = 0.$$

Thus $p(t) = t$, $q(t) = -\frac{k^2}{t}$, and $w(t) = t$.

One must be careful when solving the Bessel equation in the form of an eigenvalue problem because the sign of the eigenvalue determines whether it is a Bessel or a modified Bessel equation. When $\lambda < 0$, Eq. (9.13) can be rewritten as

$$t^2\frac{d^2 f}{dt^2} + t\frac{df}{dt} - (k^2 - \sqrt{-\lambda}^2 t^2)f(t) = 0\,.$$

Letting $\tau = \sqrt{-\lambda}t$ gives an expression that is recognized as the usual Bessel's equation

$$\tau^2\frac{d^2 f}{d\tau^2} + \tau\frac{df}{d\tau} + (\tau^2 - k^2)f(t) = 0$$

with solutions

$$f_k(t) = C_1 J_k(\sqrt{-\lambda}t) + C_2 Y_k(\sqrt{-\lambda}t).$$

Similarly, for $\lambda > 0$, Eq. (9.13) can be rewritten as

$$t^2\frac{d^2 f}{dt^2} + t\frac{df}{dt} - (k^2 + \sqrt{\lambda}^2 t^2)f(t) = 0$$

which is the modified Bessel equation with the solution

$$f_k(t) = C_1 I_k(\sqrt{\lambda}t) + C_2 K_k(\sqrt{\lambda}t).$$

Finally, for $\lambda = 0$,

$$f_k(t) = C_1 t^k + C_2 t^{-k}.$$

9.5 Fourier series

The orthogonal basis theorem is the foundation of Fourier series, expansion of functions in infinite series of trigonometric functions. Three types of Fourier series are commonly used, sine, cosine and complete Fourier series. However, other functions that are eigenfunctions of a SLP, such as Bessel functions, can be used as basis vectors in the expansion. Such expansions are sometimes called generalized Fourier series to distinguish them from the trigonometric series.

9.5.1 *Fourier sine series*

These are expansions over the interval $[0, l]$ in the functions

$$f_n(t) = \sin\left(\frac{\pi n}{l} t\right).$$

We can see that this is a complete set simply by noting that the expansion functions are the eigenfunctions of the SLP

$$\frac{d^2 f}{dt^2} + \lambda f(t) = 0\,,\ f(0) = 0\,,\ f(l) = 0\,.$$

Thus a function over the interval $[0, l]$ can be expanded as

$$f(t) = \sum_{n=1}^{\infty} \frac{\int_0^l f(t) \sin(\frac{\pi n}{l} t)\, dt}{\int_0^l \sin^2(\frac{\pi n}{l} t)\, dt} \sin\left(\frac{\pi n}{l} t\right).$$

9.5.2 *Fourier cosine series*

These are expansions over the interval $[0, l]$ in the functions

$$f_n(t) = \cos\left(\frac{\pi n}{l} t\right).$$

We can see that this is a complete set simply by noting that the expansion functions are the eigenfunctions of the SLP

$$\frac{d^2 f}{dt^2} + \lambda f(t) = 0\,,\ f'(0) = 0\,,\ f'(l) = 0\,.$$

Note that for this problem, zero is an eigenvalue and the corresponding eigenfunction is a constant. Thus a function over the interval $[0, l]$ can be expanded as

$$f(t) = \frac{\int_0^l f(t)\, dt}{l} + \sum_{n=1}^{\infty} \frac{\int_0^l f(t) \cos(\frac{\pi n}{l} t)\, dt}{\int_0^l \cos^2(\frac{\pi n}{l} t)\, dt} \cos\left(\frac{\pi n}{l} t\right)$$

where the first term corresponds to the zero eigenvalue.

9.5.3 *Complete Fourier series*

These are extensions of the Fourier sine and cosine series to the interval $[-l, l]$ and are of the form,

$$f(t) = a_0 + \sum_{n=1}^{\infty} \left(a_n \cos\left(\frac{\pi n}{l} t\right) + b_n \sin\left(\frac{\pi n}{l} t\right) \right).$$

This type of expansion arise from eigenfunctions of the SLP with periodic boundary conditions

$$\frac{d^2 f}{dt^2} + \lambda f(t) = 0 \ , \ f(-l) = f(l) \ , \ f'(-l) = f'(l).$$

As already noted, the difference between this problem and previous two SLPs is that the eigenvalues are not all simple. In fact, zero is the only simple eigenvalue. Each positive eigenvalue has two eigenfunctions associated with it, the sine and the cosine. Thus there are two terms in the summation.

Example 9.7: Fourier series.

Find the sine and cosine Fourier series of $f(t) = t(1-t)$ over the interval $[0, 1]$ and the complete Fourier series over the interval $[-1, 1]$.

Sine series: The coefficients are all given by the expression

$$C_n = \frac{\int_0^1 t(1-t)\sin(\pi n t)\, dt}{\int_0^1 \sin^2(\pi n t)\, dt} = -\frac{2}{\pi^2 n^2} \frac{\pi n \sin(\pi n) + 2\cos(\pi n) - 2}{\pi n - \cos(\pi n)\sin(\pi n)}.$$

Since n is an integer, this simplifies further to

$$C_n = -2\frac{2(-1)^n - 2}{\pi^3 n^3}$$

giving the sine series

$$t(1-t) = \sum_{n=1}^{\infty} -2\frac{2(-1)^n - 2}{\pi^3 n^3} \sin(\pi n t) \ , \ 0 \le t \le 1 \, .$$

The series, together with the function $t(1-t)$, is plotted in Fig. 9.5.

Some obvious observations are that the Fourier series only matches the function well over the interval from 0 to 1 and that the series, when evaluated outside of this interval, is an odd function and periodic with the period 2. This must be since the series is a sum of sines function that are odd with period 2.

Cosine series: The coefficient of the zero eigenvalue is

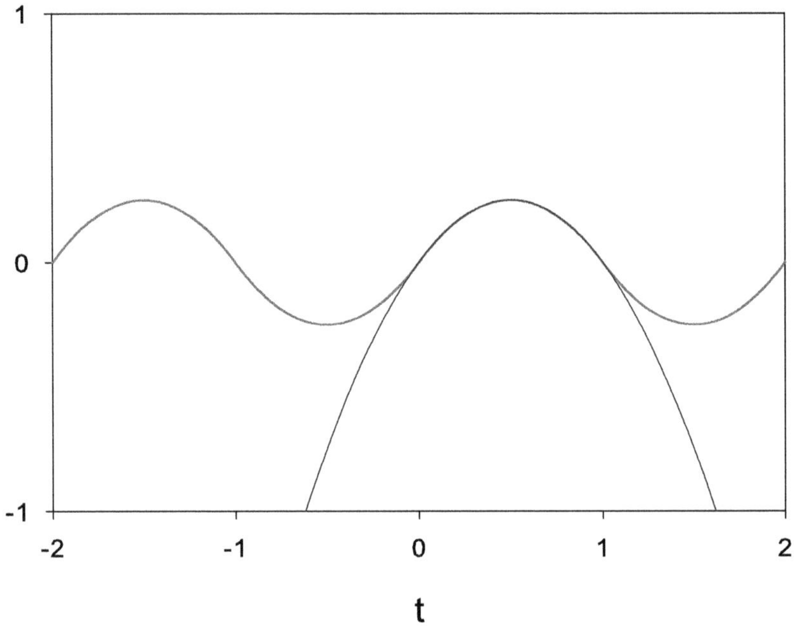

Fig. 9.5 Plots of the function $t(1-t)$ and its Fourier sine series. The series was calculated using 30 terms.

$$C_0 = \frac{\int_0^1 t(1-t)\,dt}{\int_0^1 1\,dt} = \frac{1}{6}\,.$$

The remaining coefficients are given by

$$C_n = \frac{\int_0^1 t(1-t)\cos(\pi n t)\,dt}{\int_0^1 \cos^2(\pi n t)\,dt} = -2\frac{(-1)^n+1}{\pi^2 n^2}$$

giving the cosine series

$$t(1-t) = \frac{1}{6} + \sum_{n=1}^{\infty} \frac{\int_0^1 t(1-t)\cos(\pi n t)\,dt}{\int_0^1 \cos^2(\pi n t)\,dt} = \frac{1}{6} - 2\sum_{n=1}^{\infty} \frac{(-1)^n+1}{\pi^2 n^2}\cos(\pi n t)\ ,\ 0 \le t \le 1\,.$$

The series, together with the function $t(1-t)$, is plotted in Fig. 9.6.

Complete Fourier series: The coefficient of the zero eigenvalue is

$$C_0 = \frac{\int_{-1}^1 t(1-t)\,dt}{\int_{-1}^1 1\,dt} = -\frac{1}{3}\,.$$

The remaining coefficients of the cosine terms are

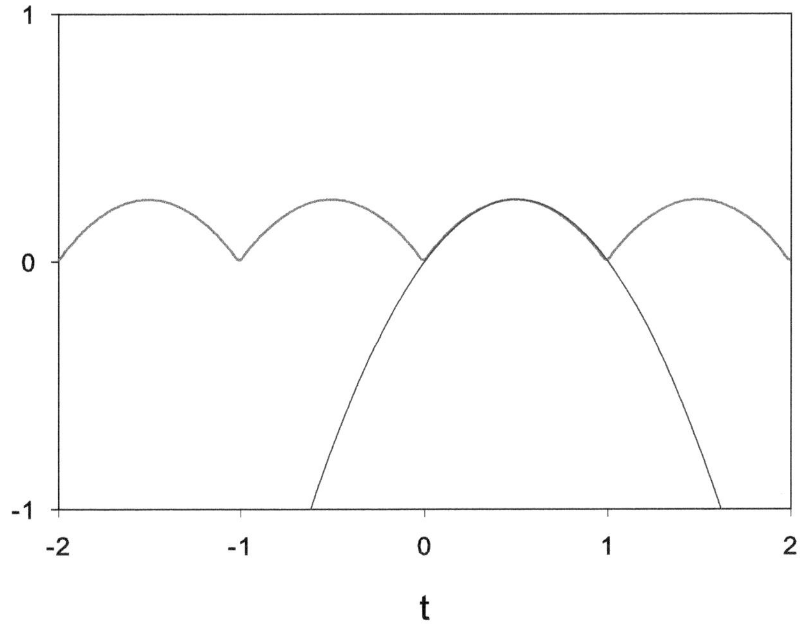

Fig. 9.6 Plots of the function $t(1-t)$ and its Fourier cosine series. The series was calculated using 30 terms.

$$C_n = \frac{\int_{-1}^{1} t(1-t)\cos(\pi n t)\,dt}{\int_{-1}^{1} \cos^2(\pi n t)\,dt} = (-1)^{n+1}\frac{4}{\pi^2 n^2}\,.$$

Finally, the coefficients of the sine terms are

$$S_n = \frac{\int_{-1}^{1} t(1-t)\sin(\pi n t)\,dt}{\int_{-1}^{1} \sin^2(\pi n t)\,dt} = (-1)^{n+1}\frac{2}{\pi n}\,.$$

Assembling the results for the coefficients gives the series

$$t(1-t) = -\frac{1}{3} + \sum_{n=1}^{\infty}(-1)^{n+1}\left\{\frac{4}{\pi^2 n^2}\cos(\pi n t) + \frac{2}{\pi n}\sin(\pi n t)\right\}\,, \quad -1 \le t \le 1\,.$$

The series, together with the function $t(1-t)$, is plotted in Fig. 9.7.

9.5.4 *Gibb's phenomena*

Figure 9.7 shows that, at a discontinuity, the Fourier series overshoots the value of the function in both directions. This overshoot and the accompanying oscillations

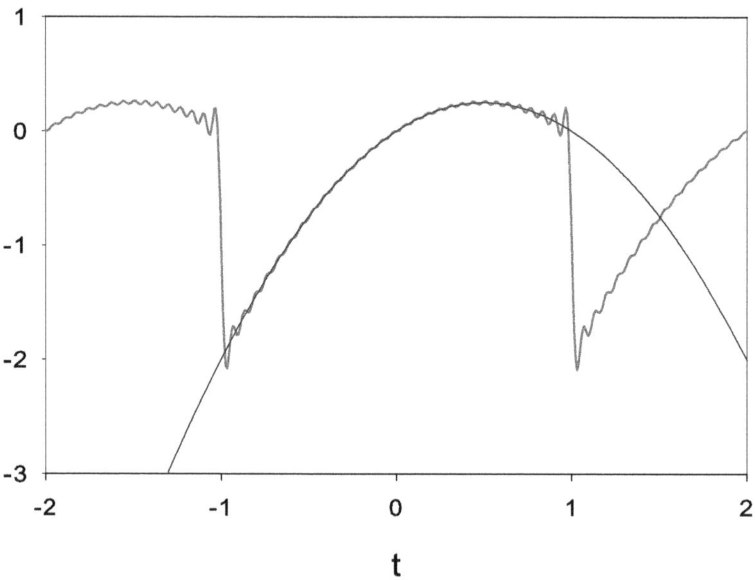

Fig. 9.7 Plots of the function $t(1-t)$ and its Fourier sine-cosine series. The series was calculated using 30 terms.

are called Gibb's phenomena and occur at all discontinuities, not just those at the end of the interval over which the series is calculated, and for all three types of Fourier series, sine, cosine and complete series. This is illustrated by considering the sine, cosine and complete Fourier series for the function

$$f(t) = \begin{cases} 3/2 \text{ if } t < 1/4 \\ 2 \text{ if } 1/4 < t < 1/2 \\ 0 \text{ if } 1/2 < t \end{cases}.$$

The sine, cosine and complete Fourier series are, respectively

$$f_{sin}(t) = 2\sum_{n=1}^{\infty}\left(\frac{3}{2}\left(1-\cos\left(\frac{n\pi}{4}\right)\right) - 2\left(\cos\left(\frac{n\pi}{2}\right) - \cos\left(\frac{n\pi}{4}\right)\right)\right)\frac{\sin(n\pi t)}{n\pi}$$

$$f_{cos}(t) = \frac{7}{8} + 2\sum_{n=1}^{\infty}\left(\frac{3}{2}\sin\left(\frac{n\pi}{4}\right) - 2\left(\sin\left(\frac{n\pi}{4}\right) - \sin\left(\frac{n\pi}{2}\right)\right)\right)\frac{\cos(n\pi t)}{n\pi}$$

$$f_{sin,cos}(t) = \frac{19}{16} + \sum_{n=1}^{\infty}\left(\frac{3}{2}\sin\left(\frac{n\pi}{4}\right) + 2\left(\sin\left(\frac{n\pi}{2}\right) - \sin\left(\frac{n\pi}{4}\right)\right)\right)\frac{\cos(n\pi t)}{n\pi}$$
$$+ \sum_{n=1}^{\infty}\left(\frac{3}{2}\left((-1)^n - \cos\left(\frac{n\pi}{4}\right)\right) + 2\left(\cos\left(\frac{n\pi}{4}\right) - \cos\left(\frac{n\pi}{2}\right)\right)\right)\frac{\sin(n\pi t)}{n\pi}.$$

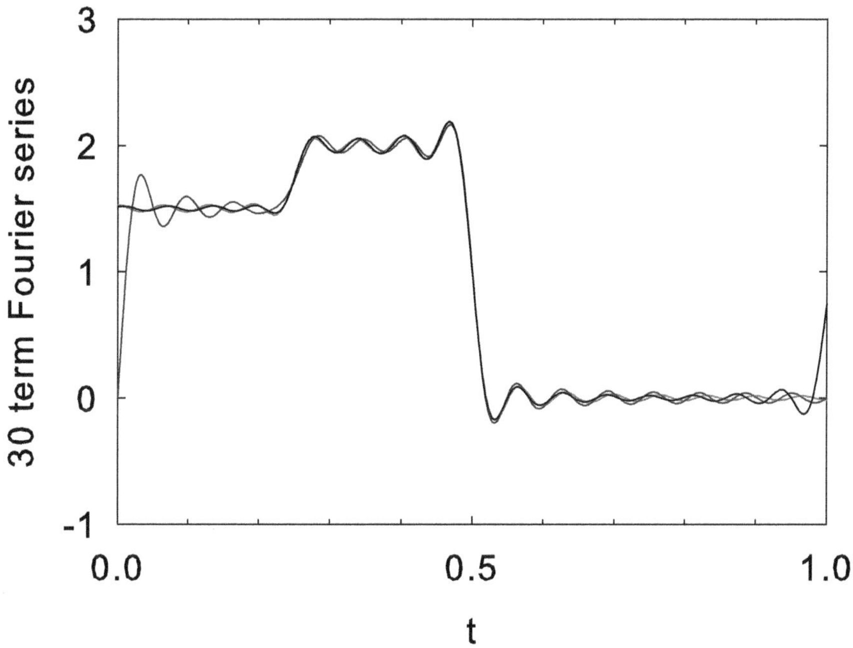

Fig. 9.8 Cosine, sine and complete Fourier series. All were calculated using an upper limit of $n = 30$.

The sine and cosine series are over the interval $[0, 1]$, while the complete series is over $[-1, 1]$. The three series are plotted in Fig. 9.8 using an upper limit in the truncated sums of $n = 30$.

The three series are very similar but, as expected, the sine-series exhibit an overshoot at $t = 0$ because the series is discontinuous at this point when it is extended beyond the $[0, 1]$-interval. Similarly, the overshoot and oscillations that can be observed at $t = 1$ are those of the complete Fourier series.

The Gibb's phenomenon makes it hard to use a Fourier series to accurately evaluate a function close to a discontinuity. One might hope that the overshoot would decrease as the number of terms in the Fourier series is increased, but this is not the case. The oscillations close to the discontinuity persist with a higher frequency but with an overshoot that remains proportional to the magnitude of the discontinuity. It can be shown that the overshoot equals approximately 9% of the magnitude of the discontinuity. Plots of the cosine series between $t = 0.2$ and $t = 0.6$, using upper limits in the truncated sums of $n = 50$, $n = 100$ and $n = 500$ are shown in Fig. 9.9. The overshoot is clearly smaller at the smaller discontinuity at $t = 0.25$ than at the discontinuity at $t = 0.5$. The approximately fixed value of the overshoot is also clearly visible.

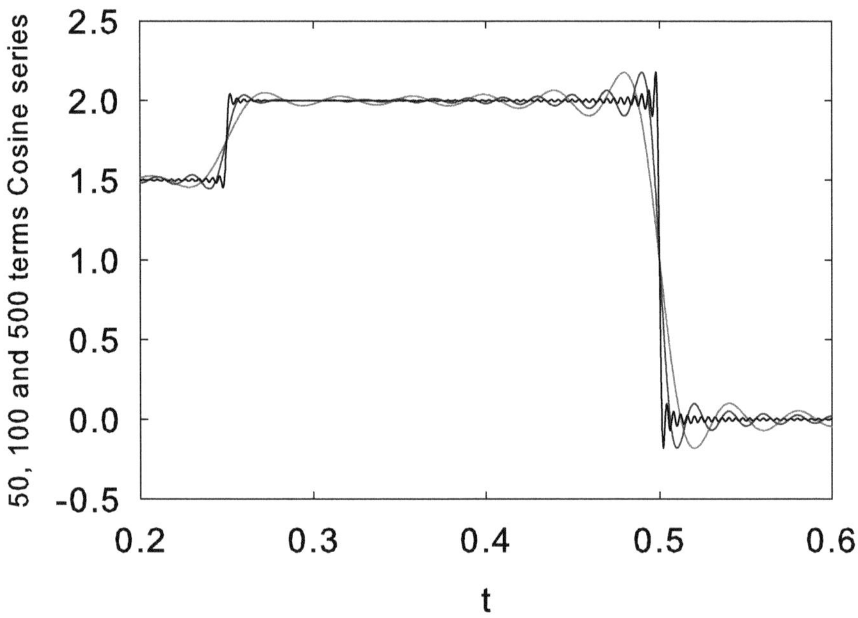

Fig. 9.9 Cosine Fourier series using 50, 100 and 500 terms.

9.5.5 *Generalized Fourier series*

As an example of a generalized Fourier series, consider the eigenvalue problem

$$\frac{d}{dt}\left(t\frac{df}{dt}\right) + \lambda t f(t) = 0\ ,\ f(0) = f(1) = 0$$

which is recognized as a Sturm–Liouville problem with the weight function $r(t) = t$. It has the solution

$$f(t) = C_1 J_0(\sqrt{\lambda}t) + C_2 Y_0(\sqrt{\lambda}t).$$

The boundary condition at $t = 0$ indicates that $C_2 = 0$. The boundary conditions at $t = 1$ provides the equation for the eigenvalues.

$$J_0(\sqrt{\lambda}) = 0$$

In other words, the eigenvalues are the squares of the zeros of the Bessel function of first kind of order zero. The eigenfunctions are

$$f_n(t) = J_0(\sqrt{\lambda_n}t).$$

The Fourier series in this set of eigenfunctions can be written

$$f(t) = \sum_{n=1}^{\infty} \frac{\langle f(t), J_0(\sqrt{\lambda_n} t)\rangle_t}{\langle J_0(\sqrt{\lambda_n} t), J_0(\sqrt{\lambda_n} t)\rangle_t} J_0(\sqrt{\lambda_n} t)$$

where the inner product is

$$\langle f(t), g(t)\rangle_t = \int_0^1 t f(t) g(t)\, dt\,.$$

Example 9.8: Fourier series in Bessel functions.

Find the generalized Fourier series in the basis $\{J_0(\sqrt{\lambda_n} t)\}$, $J_0(\sqrt{\lambda_n}) = 0$ for

$$f(t) = \begin{cases} 1\ , & t < \frac{1}{2} \\ \frac{1}{2}\ , & t > \frac{1}{2} \end{cases}\,.$$

The two required inner products evaluate as follows

$$\begin{aligned} \langle f(t), J_0(\sqrt{\lambda_n} t)\rangle_t &= \int_0^{1/2} t J_0(\sqrt{\lambda_n} t)\, dt + \int_{1/2}^1 \frac{t}{2} J_0(\sqrt{\lambda_n} t)\, dt \\ &= \frac{1}{2\sqrt{\lambda_n}} J_1(\sqrt{\lambda_n}/2) + \frac{1}{4\sqrt{\lambda_n}} (J_1(\sqrt{\lambda_n}) - J_1(\sqrt{\lambda_n}/2)) \end{aligned}$$

$$\begin{aligned} \langle J_0(\sqrt{\lambda_n} t), J_0(\sqrt{\lambda_n} t)\rangle_t &= \int_0^1 t J_0(\sqrt{\lambda_n} t)^2\, dt \\ &= \frac{1}{2}(J_0(\sqrt{\lambda_n})^2 + J_1(\sqrt{\lambda_n})^2). \end{aligned}$$

Substitution of these expressions into the Fourier series gives, upon simplification

$$f(t) = \sum_{n=1}^{\infty} \frac{J_1(\sqrt{\lambda_n}) + J_1(\sqrt{\lambda_n}/2)/2}{\sqrt{\lambda_n}(J_0(\sqrt{\lambda_n})^2 + J_1(\sqrt{\lambda_n})^2)} J_0(\sqrt{\lambda_n} t).$$

The main practical problem when using an expression such as this is to write a robust algorithm for finding all the zeros of the Bessel function. When doing this, it pays to make use of the approximate formula given earlier. For Bessel functions of first kind of order 0, this is

$$J_0(t) \simeq \sqrt{\frac{2}{\pi t}} \cos\left(t - \frac{\pi}{4}\right).$$

A plot of this approximate expression and $J_0(t)$ itself is shown in Fig. 9.10.

As expected, the approximate formula is increasingly accurate for increasing values of the argument, but even for arguments less than the smallest or first root, the agreement is quite good. This suggests that a robust, fast algorithm for finding the nth root of $J_0(t)$ can be based on an initial guess of the root of the approximate

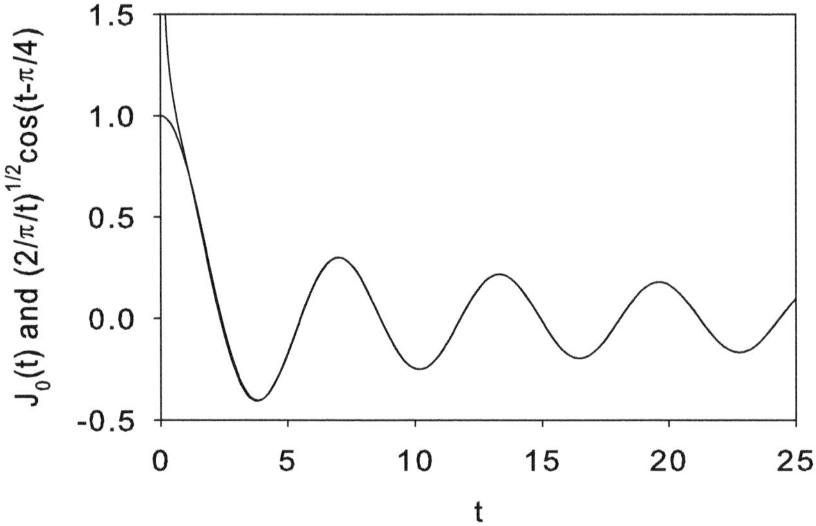

Fig. 9.10 Graphs of the Bessel function of first kind, order zero and the approximation formula for this function.

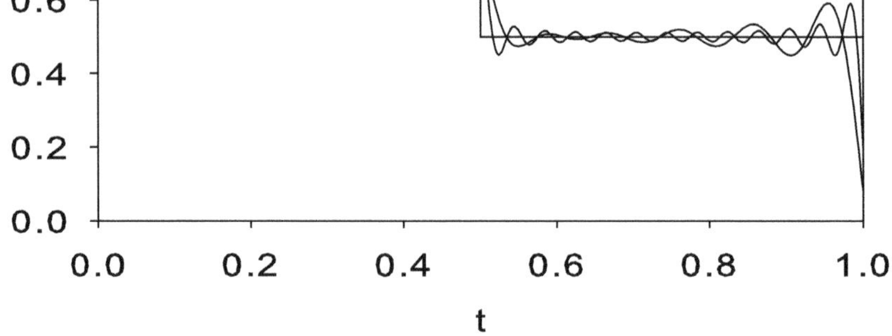

Fig. 9.11 Plots of the Bessel–Fourier series of $f(t)$ using 20 and 50 terms in the series.

formula, $3\pi/4 + n\pi$, and a search interval of $[\pi/4 + (n-1)\pi, \pi/4 + n\pi]$. A plot of the Fourier series with 20 and 50 terms are shown in Fig. 9.11.

Finally, note that Legendre's differential equation

$$\frac{d}{dt}\left((1-t^2)\frac{df}{dt}\right)+n(n+1)f=0$$

is a Sturm–Liouville problem with a weight function equal to unity. When boundary conditions stipulate that the solution is finite at $t=\pm 1$, the only possible solutions are those for which n is a positive integer and the eigenfunctions are the Legendre polynomials. Thus, the eigenvalues equal $n(n+1)$ for $n \in \mathbb{Z}_+$ and the inner product in which the eigenfunction/Legendre polynomials are orthogonal is

$$\langle f, g\rangle = \int_{-1}^{1} f(t)g(t)\,dt\,.$$

Legendre polynomials form an orthogonal basis for functions defined over the interval from -1 to 1 and a Legendre Fourier series is given by

$$f(t)=\sum_{n=0}^{\infty}\frac{\langle f(t), P_n(t)\rangle}{\langle P_n(t), P_n(t)\rangle}P_n(t)=\frac{1}{2}\sum_{n=0}^{\infty}\langle f(t), P_n(t)\rangle(2n+1)P_n(t).$$

For instance, for the function $\sqrt{t+1}$, one obtains

$$\langle\sqrt{t+1}, P_n(t)\rangle = \int_{-1}^{1}\sqrt{t+1}\cdot P_n(t)\,dt\,.$$

It is not easy to find a closed form expression for this integral in terms of n, but for given, small values of n, the integral is readily obtained

$$\langle\sqrt{t+1}, P_0(t)\rangle = \frac{4}{3}\sqrt{2}$$

$$\langle\sqrt{t+1}, P_1(t)\rangle = \frac{4}{15}\sqrt{2}$$

$$\langle\sqrt{t+1}, P_2(t)\rangle = -\frac{4}{105}\sqrt{2}$$

$$\langle\sqrt{t+1}, P_3(t)\rangle = \frac{4}{315}\sqrt{2}$$

and so on. A plot of Legendre Fourier series of $\sqrt{t+1}$ is shown in Fig. 9.12. The series representation is remarkably good even when only a few terms are used in the series.

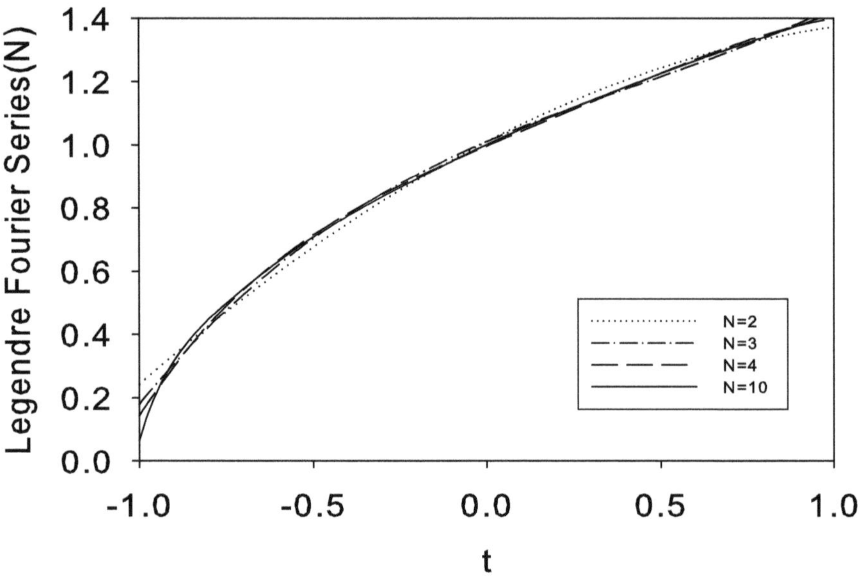

Fig. 9.12 Plots of Legendre Fourier series of $\sqrt{t+1}$ using 2, 3, 4 and 10 terms in the series.

Chapter 10

PDEs with Self-Adjoint Operators

Partial Differential Equations (PDEs) are almost always solved by converting the problem to solution of one or more ODEs. This chapter describes a solution method for solving various types of partial differential equations by summing solutions of self-adjoint ODEs. These sums are Fourier series and the solution method rely on identifying a linear operator in the PDE and on solving the eigenvalue problem associated with this operator. The method is primarily used when the operators are self-adjoint and the associated eigenvalue problem therefore is easy to solve. Typically, this means that Fourier series methods can be used for PDEs of even order over a finite domain.

10.1 Fourier series methods

As the name implies, these methods find a Fourier series expression for the solution to a PDE. They will do so, mindlessly one might say, even in the special cases where the solution to the PDE can be written in terms of simple, finite expressions. Fourier series methods are used primarily for second order PDEs and this section starts with a brief classification of these types of equations followed by an explanation of the inner product method. This is followed by a series of solved examples that illustrate how the method is applied and how to deal with various types of problems that may arise during the solution process. Finally, an alternative method, the finite Fourier transform, is described. This method is so similar to the inner product method that once the inner product method has been mastered, it is but a small step to understanding finite Fourier transforms and a single example will therefore suffice to illustrate the method.

10.1.1 *Classification of second order PDEs*

Two dimensional problems occur so often that these equations have been classified into several recognizable types. The most general second order linear PDE over two dimensions can be written

$$F_{xx}(x,t)\frac{\partial^2\Theta}{\partial x^2}+2F_{xt}(x,t)\frac{\partial^2\Theta}{\partial x\partial t}+F_{tt}(x,t)\frac{\partial^2\Theta}{\partial t^2}$$
$$+F_x(x,t)\frac{\partial\Theta}{\partial x}+F_t(x,t)\frac{\partial\Theta}{\partial t}+F(x,t)\Theta=G(x,t)$$

where the free variables have been named t and x which suggests time and a spacial dimension although the free variables could both be spatial variables without in any way affecting the classification of the PDE. The subscripts on the coefficient functions, the $F_n(x,t)$s, are part of the name and should not be construed as indicating partial derivative as is sometimes done.

This general equation will be recast in a standard form by expanding the terms with second order derivatives into two terms of first order derivatives. The process is similar to the process that can be used to recast a second order ordinary differential equation as two coupled first order equations, a particularly useful trick when the dependent variable does not occur explicitly. While this is trivial to do for ordinary differential equations, it is a bit more involved for PDEs. Start by considering the operator

$$L(\Theta)=L_1(L_2(\Theta))=\left(\frac{\partial}{\partial x}-f_+(x,t)\frac{\partial}{\partial t}\right)\left(\frac{\partial\Theta}{\partial x}-f_-(x,t)\frac{\partial\Theta}{\partial t}\right).$$

It is important to understand that the expression does not indicate multiplication in the usual algebraic sense of the terms in the two parenthesis. Instead, the first set of parenthesis contains an operator, L_1, which operates on everything inside the second set of parenthesis. In other words, the second set of parenthesis contains the argument of the operator enclosed by the first set of parenthesis. Notice that both L_1 and L_2 are first order. Expanding this expression

$$L(\Theta)=\frac{\partial^2\Theta}{\partial x^2}-\frac{\partial f_-}{\partial x}\frac{\partial\Theta}{\partial t}-f_-(x,t)\frac{\partial^2\Theta}{\partial x\partial t}-f_+(x,t)\frac{\partial^2\Theta}{\partial x\partial t}$$
$$+f_+(x,t)\frac{\partial f_-}{\partial t}\frac{\partial\Theta}{\partial t}+f_+(x,t)f_-(x,t)\frac{\partial^2\Theta}{\partial t^2}$$

and rearranging

$$L(\Theta)=\frac{\partial^2\Theta}{\partial x^2}-(f_-(x,t)+f_+(x,t))\frac{\partial\Theta}{\partial x\partial t}+f_+(x,t)f_-(x,t)\frac{\partial^2\Theta}{\partial t^2}$$
$$-\frac{\partial f_-}{\partial x}\frac{\partial\Theta}{\partial t}+f_+(x,t)\frac{\partial f_-}{\partial t}\frac{\partial\Theta}{\partial t}.$$

Comparing this result to the general second order linear PDE, it becomes clear that one can write

$$\frac{\partial^2\Theta}{\partial x^2}+\frac{2F_{xt}}{F_{xx}}\frac{\partial^2\Theta}{\partial x\partial t}+\frac{F_{tt}}{F_{xx}}\frac{\partial^2\Theta}{\partial t^2}=\left(\frac{\partial}{\partial x}-f_+(x,t)\frac{\partial}{\partial t}\right)\left(\frac{\partial\Theta}{\partial x}-f_-(x,t)\frac{\partial\Theta}{\partial t}\right)$$
$$+\left(\frac{\partial f_-}{\partial x}-f_+(x,t)\frac{\partial f_-}{\partial t}\right)\frac{\partial\Theta}{\partial t}$$

where the two coefficient functions, f_+ and f_- are determined from

$$f_+(x,t) + f_-(x,t) = -\frac{2F_{xt}}{F_{xx}} \quad \text{and} \quad f_+(x,t)f_-(x,t) = \frac{F_{tt}}{F_{xx}}$$

or, as roots of a second order polynomial

$$f_\pm(x,t) = \frac{-F_{xt} \pm \sqrt{F_{xt}^2 - 4F_{xx}F_{tt}}}{F_{xx}}\,.$$

The PDE is now classified based on nature of these roots. If

$$F_{xt}^2 - 4F_{xx}F_{tt} > 0$$

then the PDE is called *hyperbolic*. The roots, and thus f_+ and f_-, are real and so the equation can be factored into two first order PDE and is often best solved this way. The classical example of a hyperbolic problem is the wave equation

$$\frac{\partial^2\Theta}{\partial t^2} - c^2\frac{\partial^2\Theta}{\partial x^2} = 0\,.$$

If

$$F_{xt}^2 - 4F_{xx}F_{tt} = 0$$

then the PDE is *parabolic*. Transient transport processes such as transient diffusion and conduction, are governed by parabolic equations. For instance, in one spatial dimension, the mass or energy balance, using Fick's law for diffusion and Fourier's law for conduction and assuming no fluid flow, is

$$\frac{\partial\Theta}{\partial t} = \alpha\frac{\partial^2\Theta}{\partial x^2}\,.$$

Finally, if

$$F_{xt}^2 - 4F_{xx}F_{tt} < 0$$

then the PDE is said to be *elliptic*. Elliptic equations describe the steady state temperature or concentration profile inside a solid or a stagnant fluid, for instance. The steady state mass/energy balance is called Laplace's equation and in two spatial dimensions it has the form

$$\frac{\partial^2\Theta}{\partial x^2} + \frac{\partial^2\Theta}{\partial y^2} = 0\,.$$

While hyperbolic problems are often solved by viewing them as coupled first order problems, this is not possible for parabolic and elliptical problems. Over finite domains, these problems can often be solved by a method we will simply call "the inner product method" which will be described and illustrated in the remainder of this section.

By a finite domain we mean that all the boundary conditions are given at points located at finite distances from one another as opposed to problems in which a boundary condition is given at plus or minus infinity. For these problems, Fourier series representations of the solutions can be found when the partial differential equation contain self-adjoint or formally self-adjoint operators. The two methods that will be described here, the inner product method and the finite Fourier transform, are in principle quite simple. The devil is usually in the details of finding the eigenfunctions and eigenvalues that are needed in order to find the complete solution. The bulk of this chapter is a collection of examples that illustrate the inner product method and shows how to handle a variety of problems that one may encounter in the solution process.

Readers may already be familiar with two other methods for solving certain PDEs: Laplace transformation and separation of variables. Before describing the inner product method, it behooves us to discuss why this method is superior to Laplace transformation and separation of variables for many problems.

The Laplace transform method only works for problems with constant coefficients and which have one spatial dimension and a time dimension. Transformation of higher dimensional problems gives a PDE in the Laplace domain which usually cannot be further simplified by Laplace transformation. Furthermore, even when the Laplace transformation method does work, it still requires back transformation from the Laplace domain. In general, this is not possible using tables of Laplace transforms, but can only be accomplished by integration in the complex domain using residue calculus, a technique that may be unfamiliar to many readers.

The separation of variables method makes use of the assumption that a particular solution to the PDE can be written as a product of functions, each of which depend on only a single free variable. Standard manipulations then give rise to an infinity of particular solutions that are summed in a Fourier series to give the final, complete solution. The manipulations required are only possible when the PDE as well as the boundary conditions (except one of them) are homogeneous. Separation of variables is not useless for inhomogeneous problems because such problems can often be rendered homogeneous by well chosen variable transformations, a trick we will need when using the inner product method. However, it does not handle inhomogeneities as elegantly as the inner product method and is difficult to use with time dependent boundary conditions, a complication that is easily handled by the inner product method.

10.1.2 *Inner product method*

This solution method is quite general and can be used to solve a wide range of problems that involve linear operators. Let the problem be written

$$Lx = b$$

where L is a linear operator, x the unknown and b a known vector. Assume for now that L is a matrix, that it is self-adjoint and that all its eigenvectors, $\tilde{x}_n$, and associated eigenvalues, λ_n, are known. Then take the inner product of the equation above with an arbitrary eigenvector.

$$\langle Lx, \tilde{x}_n \rangle = \langle x, L\tilde{x}_n \rangle = \lambda_n \langle x, \tilde{x}_n \rangle = \langle b, \tilde{x}_n \rangle$$

but, by Fourier's theorem, the unknown x can be written as

$$x = \sum_{n=1}^{N} \frac{\langle x, \tilde{x}_n \rangle}{\langle \tilde{x}_n, \tilde{x}_n \rangle} \tilde{x}_n$$

where the upper limit, N, may be infinity. Combining the last two results give

$$x = \sum_{n=1}^{N} \frac{\langle b, \tilde{x}_n \rangle}{\lambda_n \langle \tilde{x}_n, \tilde{x}_n \rangle} \tilde{x}_n$$

which is a Fourier series solution for the unknown. The method works even if the matrix is not self-adjoint. If this is the case, take the inner product of the equation with a generalized eigenrow, $\tilde{y}_n$, of the matrix

$$\langle Lx, \tilde{y}_n \rangle = \langle x, L^* \tilde{y}_n \rangle = \langle b, \tilde{y}_n \rangle \, .$$

Since the eigenrows are the eigenvectors of the adjoint matrix, $L^* \tilde{y}_n = \overline{\lambda}_n \tilde{y}_n$, giving

$$\lambda_n \langle x, \tilde{y}_n \rangle = \langle b, \tilde{y}_n \rangle \;\Rightarrow\; \langle x, \tilde{y}_n \rangle = \frac{\langle b, \tilde{y}_n \rangle}{\lambda_n} \, .$$

Now, write the unknown, x, in terms of the basis given by the eigenvectors of L.

$$x = \sum_{m=1}^{N} C_m \tilde{x}_m$$

and take the inner product of this equation with the eigenrow $\tilde{y}_n$.

$$\langle x, \tilde{y}_n \rangle = \sum_{m=1}^{N} C_m \langle \tilde{x}_m, \tilde{y}_n \rangle .$$

The eigenvectors and eigenrows are biorthogonal sets. Thus, the only term remaining on the right hand side is $\langle \tilde{x}_n, \tilde{y}_n \rangle$ and

$$C_n = \frac{\langle x, \tilde{y}_n \rangle}{\langle \tilde{x}_n, \tilde{y}_n \rangle} .$$

Substituting in the expression found for $\langle x, \tilde{y}_n \rangle$ gives the solution as

$$x = \sum_{n=1}^{N} \frac{1}{\lambda_n} \frac{\langle b, \tilde{y}_n \rangle}{\langle \tilde{x}_n, \tilde{y}_n \rangle} \tilde{x}_n .$$

Example 10.1: Fourier solution of coupled, linear equations.

Consider the problem

$$\begin{pmatrix} 1 & 6 & 0 & 2 \\ 3 & 2 & -1 & 0 \\ 0 & -1 & 3 & 5 \\ 2 & 0 & 5 & 4 \end{pmatrix} x = \begin{pmatrix} 3 \\ 2 \\ 1 \\ 0 \end{pmatrix} .$$

This system matrix is not self-adjoint but its adjoint equals its transpose. The eigenvalues are simple and the eigenvectors, $\tilde{x}_n$, and rows, $\tilde{y}_n$, are, with 9 decimals accuracy

$$\lambda_1 = 5.989808032 \, , \; \tilde{x}_1 = \begin{pmatrix} 0.718298331 \\ 0.618412243 \\ -0.31245114 \\ -0.063151337 \end{pmatrix} , \; \tilde{y}_1 = \begin{pmatrix} -0.521343604 \\ -0.830852094 \\ 0.186878735 \\ -0.05442411 \end{pmatrix}$$

$$\lambda_2 = -3.140071903 \, , \; \tilde{x}_2 = \begin{pmatrix} -0.779672420 \\ 0.404358499 \\ -0.260585502 \\ 0.400874443 \end{pmatrix} , \; \tilde{y}_2 = \begin{pmatrix} -0.638999264 \\ 0.734039443 \\ -0.060980063 \\ 0.221692283 \end{pmatrix}$$

$$\lambda_3 = -1.654543556 \, , \; \tilde{x}_3 = \begin{pmatrix} -0.031255643 \\ 0.22669897 \\ 0.734714331 \\ -0.638612177 \end{pmatrix} , \; \tilde{y}_3 = \begin{pmatrix} -0.208424839 \\ 0.521437351 \\ 0.655066477 \\ -0.50551962 \end{pmatrix}$$

$$\lambda_4 = 8.804807426\,,\ \tilde{x}_4 = \begin{pmatrix} 0.178424977 \\ -0.015925705 \\ 0.643646287 \\ 0.744063409 \end{pmatrix}\,,\ \tilde{y}_4 = \begin{pmatrix} 0.234428275 \\ 0.115826709 \\ 0.618391199 \\ 0.741093707 \end{pmatrix}.$$

The various inner products needed for the solution can now be evaluated.

$$C_1 = \frac{\left\langle \begin{pmatrix} 3 \\ 2 \\ 1 \\ 0 \end{pmatrix}, \begin{pmatrix} -0.208424839 \\ 0.521437351 \\ 0.655066477 \\ -0.50551962 \end{pmatrix} \right\rangle}{\left\langle \begin{pmatrix} 0.718298331 \\ 0.618412243 \\ -0.31245114 \\ -0.063151337 \end{pmatrix}, \begin{pmatrix} -0.208424839 \\ 0.521437351 \\ 0.655066477 \\ -0.50551962 \end{pmatrix} \right\rangle} = 0.537865546$$

etc.

$$C_2 = 0.180470024$$

$$C_3 = -0.697983215$$

$$C_4 = 0.178303149$$

and the solution is

$$\begin{aligned} x =& 0.537865546 \begin{pmatrix} 0.718298331 \\ 0.618412243 \\ -0.31245114 \\ -0.063151337 \end{pmatrix} + 0.180470024 \begin{pmatrix} -0.779672420 \\ 0.404358499 \\ -0.260585502 \\ 0.400874443 \end{pmatrix} \\ & - 0.697983215 \begin{pmatrix} -0.031255643 \\ 0.22669897 \\ 0.734714331 \\ -0.638612177 \end{pmatrix} + 0.178303149 \begin{pmatrix} 0.178424977 \\ -0.015925705 \\ 0.643646287 \\ 0.744063409 \end{pmatrix} \\ =& \begin{pmatrix} 0.299270073 \\ 0.244525547 \\ -0.613138686 \\ 0.616788321 \end{pmatrix}. \end{aligned}$$

It is easy to verify that this is the solution.

Evidently, solving coupled linear equations with this method is inefficient compared to solution by Gauss elimination. However, in infinite dimensional vector spaces, where there is no equivalent to Gauss elimination, this solution method becomes useful.

As mentioned, the practical difficulties one encounters when solving problems by this method are those of finding all the eigenvectors and eigenvalues. This is particularly difficult when the operator is not self-adjoint because the eigenvectors and eigenvalues can then be complex. An important first step is therefore always to try and make the operator self-adjoint. This can be an easy matter of finding a simple variable transformation or may be a very difficult problem of finding an inner product in which the operator is self-adjoint. The only problems we will consider here are those that can be made self-adjoint.

It is an open question whether this solution method is valid for linear operators in infinite dimensional spaces. For finite dimensional operators, i.e. matrices, we have shown that the set of generalized eigenvectors is a basis for the appropriate vector space and we have postulated the orthogonal basis theorem for SLPs. However, we have not shown that the (generalized) eigenvectors or eigenrows of all self-adjoint operators form an orthogonal basis for the appropriate Hilbert space. This is a difficult problem in infinite dimensions and we will ignore it and simply accept on faith that the solution method works.

Inhomogeneous, ODEs can also be solved with this method, although it is mainly a curiosity and not a particularly usefully technique for ODEs. However, the method is illustrated for an ODE in the next example.

Example 10.2: Fourier solution of ODE.

Consider

$$Ly = \frac{d^2y}{dt^2} = e^t \ , \ y(0) = y(1) = 0 \, .$$

First note that the left hand side is a Sturm–Liouville operator and writing the eigenvalue problem for this operator as $Ly_n + \lambda_n y_n = 0$, the eigenvalues and eigenfunctions are

$$y_n(t) = \sin(n\pi t) \ , \ \lambda_n = (n\pi)^2 \ , \ n = 1, 2, 3, \cdots$$

Take the inner product of the ODE with an eigenfunction

$$\langle Ly, y_n \rangle = \langle e^t, y_n \rangle$$

and make use of the self-adjoint property of the operator to rewrite this as

$$\langle y, Ly_n \rangle = \langle e^t, y_n \rangle \;\Rightarrow$$

$$-\lambda_n \langle y, y_n \rangle = \langle e^t, y_n \rangle \;\Rightarrow$$

$$\langle y, y_n \rangle = -\frac{1}{\lambda_n}\langle e^t, y_n \rangle \;\Rightarrow \;.$$

The Fourier series representation of the solution is

$$y(t) = \sum_{n=1}^{\infty} \frac{\langle y, y_n \rangle}{\langle y_n, y_n \rangle} y_n = -\sum_{n=1}^{\infty} \frac{1}{\lambda_n} \frac{\langle e^t, y_n \rangle}{\langle y_n, y_n \rangle} y_n$$

which, after evaluation of the various integrals, becomes

$$y(t) = \sum_{n=1}^{\infty} \frac{-2}{n^2\pi^2} \left\{ \frac{n\pi}{1+n^2\pi^2} - \frac{n\pi e(-1)^n}{1+n^2\pi^2} \right\} \sin(n\pi t).$$

One can check that this is the Fourier series of $e^t - 1 + (1-e)t$, the solution obtained by variation of parameters.

10.1.3 *PDEs with Sturm–Liouville operators*

Partial Differential Equations with Sturm–Liouville operators are simple to solve because the appropriate inner product can be found from inspection of the SLP operator. The only problem may be that of making the boundary conditions homogeneous such that the operator becomes self-adjoint.

10.1.3.1 *Homogeneous problem*

This is the simplest possible case and many problems can be reduced to this case with the appropriate variable transformations.

Example 10.3: Transient conduction in wall.

The energy balance for thermal conduction in a flat wall is

$$\frac{\partial T}{\partial t} = \alpha \frac{\partial^2 T}{\partial x^2}$$

where T is the temperature, x is the coordinate orthogonal to the wall surface, t is time and α is the thermal conductivity, $\alpha = k/(\rho c_p)$. Place the origin of the x-axis at the center plane of the wall and let the boundary and initial conditions be

$$x = \pm L \;\Rightarrow\; T = T_\infty \;, \quad \text{all } t$$

$$x = 0 \Rightarrow \frac{\partial T}{\partial x} = 0 \,, \quad \text{all } t$$

$$t = 0 \Rightarrow T = T_0 \,, \quad \text{all } x \,.$$

Now, introduce dimensionless variables. This not only simplifies the problem by reducing the number of parameters, but render the boundary condition at the surface homogeneous. This is an essential step since a Sturm–Liouville operator requires homogeneous boundary conditions. Define

$$\Theta = \frac{T - T_\infty}{T_0 - T_\infty} \,, \; \xi = \frac{x}{L} \,, \; \tau = \frac{\alpha t}{L^2}$$

which gives

$$\frac{\partial \Theta}{\partial \tau} = \frac{\partial^2 \Theta}{\partial \xi^2}$$

with the dimensionless boundary and initial conditions

$$\xi = \pm 1 \Rightarrow \Theta = 0 \,, \quad \text{all } \tau$$

$$\xi = 0 \Rightarrow \frac{\partial \Theta}{\partial \xi} = 0 \,, \quad \text{all } \tau$$

$$\tau = 0 \Rightarrow \Theta = 1 \,, \quad \text{all } \xi \,.$$

This problem is homogeneous in both the PDE and the boundary conditions.

Recognize that the right hand side of the dimensionless energy balance, together with the boundary conditions in ξ, are a Sturm–Liouville operator with an associated SLP which can be written

$$\frac{d^2 \Theta}{d\xi^2} + \lambda \Theta = 0$$

$$\xi = \pm 1 \Rightarrow \Theta = 0$$

$$\xi = 0 \Rightarrow \frac{d\Theta}{d\xi} = 0 \,.$$

The eigenvalues and eigenfunctions are

$$\lambda_n = \left(\frac{\pi}{2} + n\pi\right)^2 \,, \quad n = 0, 1, 2, \cdots$$

$$\Theta_n(\xi) = \cos\left(\left(\frac{\pi}{2} + n\pi\right)\xi\right) \,, \quad n = 0, 1, 2, \cdots$$

and the inner product is over $[0, 1]$ with the weight function 1. Now take the inner product of the energy balance with an eigenfunction

$$\left\langle \Theta_n(\xi), \frac{\partial \Theta}{\partial \tau} \right\rangle = \left\langle \Theta_n(\xi), \frac{\partial^2 \Theta}{\partial \xi^2} \right\rangle .$$

Since the operator is self-adjoint, it can be moved back and forth between the two factors of the inner product and, since the inner product is an integration with respect to ξ, the differentiation with respect to τ can be taken outside the inner product.

$$\frac{d}{d\tau}\langle \Theta_n, \Theta \rangle = \left\langle \frac{d^2 \Theta_n}{d\xi^2}, \Theta \right\rangle = -\lambda_n \langle \Theta_n, \Theta \rangle$$

This is an ODE in $\langle \Theta_n, \Theta \rangle$ with the initial condition

$$\tau = 0 \ \Rightarrow \ \langle \Theta_n, \Theta \rangle = \langle \Theta_n, 1 \rangle$$

which is solved to give

$$\langle \Theta_n, \Theta \rangle = \langle \Theta_n, 1 \rangle e^{-\lambda_n \tau}.$$

What remains to be done is substituting this result into the Fourier series for Θ

$$\Theta(\xi, \tau) = \sum_{n=0}^{\infty} \frac{\langle \Theta_n, \Theta \rangle}{\langle \Theta_n, \Theta_n \rangle} \Theta_n(\xi) = \sum_{n=0}^{\infty} e^{-\lambda_n \tau} \frac{\langle \Theta_n, 1 \rangle}{\langle \Theta_n, \Theta_n \rangle} \Theta_n(\xi)$$

and evaluating the inner products

$$\Theta(\xi, \tau) = 2 \sum_{n=0}^{\infty} \frac{(-1)^n}{\frac{\pi}{2} + n\pi} e^{-\left(\frac{\pi}{2} + n\pi\right)^2 \tau} \cos\left(\left(\frac{\pi}{2} + n\pi\right) \xi\right) . \tag{10.1}$$

The solution is plotted in Fig. 10.1 for several values of τ and using 2, 11 and 101 terms in the series. At time 0, all graphs show Gibbs oscillations but at $\tau = 0.01$ the graph plotted using 101 terms in the Fourier series is perfectly smooth while oscillations still dominate the two other graphs. As the dimensionless time increases, the exponential terms in the Fourier series that correspond to large values of n go rapidly towards zero and at $\tau = 0.1$ the graphs show that 10 terms are sufficient for a smooth curve. At $\tau = 0.5$ the three graphs are virtually superimposed so all terms but the first two have become insignificant and only the first two terms in the Fourier series are required.

So, although a Fourier series solution may appear to be cumbersome to work with and computationally expensive, only the first few terms may be needed in calculations with the solution except for calculations done for very small times.

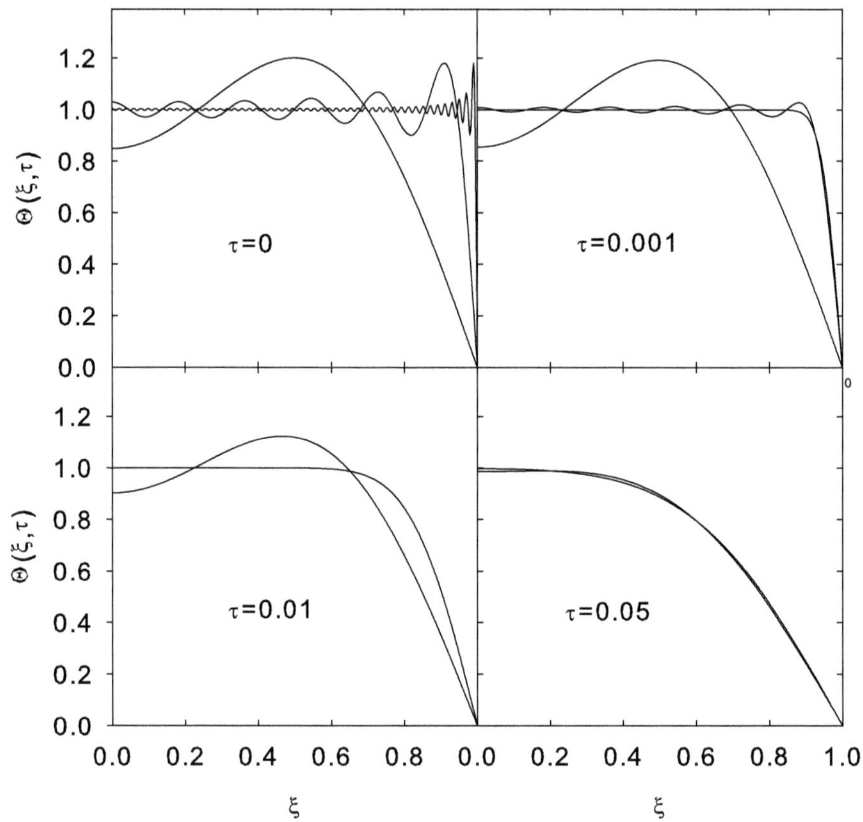

Fig. 10.1 Dimensionless temperature profile in wall, Eq. (10.1), using 2, 11 and 101 terms in the Fourier series.

10.1.3.2 *Homogeneous problem with transcendental equation for eigenvalues*

The next example is still conduction in a wall, but with a film resistance at the surface and an arbitrary initial temperature profile. The film condition gives rise to an eigenvalue problem that does not permit an analytical solution for the eigenvalues. However, the eigenvalues can still be indexed by the natural numbers and arranged in increasing order.

Example 10.4: Asymmetric, transient conduction in wall with film resistance.

Consider again the problem of transient conduction in a semi-infinite wall, but this time assume that the initial temperature distribution can be any function and that the boundary condition at the surface is a film condition. Place the x-axis at a right angle to the wall surface with the origin at the center of the wall. The energy balance and side conditions are

$$\frac{\partial T}{\partial t} = \alpha \frac{\partial^2 T}{\partial x^2}$$

$$x = L \Rightarrow -k \left.\frac{dT}{dx}\right|_{x=L} = h(T(L) - T_\infty) \ , \ \text{all } t$$

$$x = -L \Rightarrow -k \left.\frac{dT}{dx}\right|_{x=-L} = h(T_\infty - T(-L)) \ , \ \text{all } t$$

$$t = 0 \Rightarrow T = T_0(x)$$

where h is the heat transfer coefficient. All other variables have the same meaning as in the previous example. Note that one cannot use the non-flux condition at $x = 0$ because the initial temperature distribution may a asymmetric. Define dimensionless variables as follows

$$\Theta = \frac{T - T_\infty}{T_\infty} \ , \ \xi = \frac{x}{L} \ , \ \tau = \frac{\alpha t}{L^2} \ , \ Bi = \frac{hL}{k}$$

where Bi is the Biot number. In dimensionless form, the problem becomes

$$\frac{\partial \Theta}{\partial \tau} = \frac{\partial^2 \Theta}{\partial \xi^2}$$

$$\xi = 1 \Rightarrow \frac{\partial \Theta}{\partial \xi} + Bi\Theta = 0 \ , \ \text{all } \tau$$

$$\xi = -1 \Rightarrow \frac{\partial \Theta}{\partial \xi} - Bi\Theta = 0 \ , \ \text{all } \tau$$

$$\tau = 0 \Rightarrow \Theta = \Theta_0(\xi).$$

The associated SLP is

$$\frac{d^2 \Theta_n}{d\xi^2} + \lambda_n \Theta_n = 0$$

$$\xi = 1 \Rightarrow \frac{d\Theta_n}{d\xi} + Bi\Theta_n = 0$$

$$\xi = -1 \Rightarrow \frac{d\Theta_n}{d\xi} - Bi\Theta_n = 0 \, .$$

Solve this by considering the cases $\lambda = 0$, $\lambda < 0$ and $\lambda > 0$.

$\lambda_n = 0 \Rightarrow$

$$\Theta_n = C_1 \xi + C_2 \, .$$

Apply boundary conditions

$$\left.\begin{aligned} C_1 + Bi(C_1 + C_2) = 0 \\ C_1 - Bi(-C_1 + C_2) = 0 \end{aligned}\right\} \Rightarrow \begin{aligned} C_1 = 0 \\ C_2 = 0 \end{aligned}$$

i.e. the trivial solution. So 0 is not an eigenvalue.

$\lambda_n < 0 \Rightarrow$

$$\Theta = C_1 e^{\sqrt{-\lambda_n}\xi} + C_2 e^{-\sqrt{-\lambda_n}\xi}\,.$$

Applying the boundary conditions give $C_1 = C_2 = 0$, once again the trivial solution. The eigenvalues must all be positive.

$\lambda_n > 0 \Rightarrow$

$$\Theta_n = C_1 \cos(\sqrt{\lambda_n}\xi) + C_2 \sin(\sqrt{\lambda_n}\xi).$$

Applying boundary conditions give the two equations

$$-C_1\sqrt{\lambda_n}\sin(\sqrt{\lambda_n}) + C_2\sqrt{\lambda_n}\cos(\sqrt{\lambda_n}) + BiC_1\cos(\sqrt{\lambda_n}) + BiC_2\sin(\sqrt{\lambda_n}) = 0$$

$$C_1\sqrt{\lambda_n}\sin(\sqrt{\lambda_n}) + C_2\sqrt{\lambda_n}\cos(\sqrt{\lambda_n}) - BiC_1\cos(\sqrt{\lambda_n}) + BiC_2\sin(\sqrt{\lambda_n}) = 0\,.$$

By adding and subtracting these two equations from one another, they are simplified to

$$C_2(\sqrt{\lambda_n}\cos(\sqrt{\lambda_n}) + Bi\sin(\sqrt{\lambda_n})) = 0$$

$$C_1(-\sqrt{\lambda_n}\sin(\sqrt{\lambda_n}) + Bi\cos(\sqrt{\lambda_n})) = 0\,.$$

Obviously, C_1 and C_2 cannot both be zero since this will give the trivial solution. Additionally, if both C_1 and C_2 are different from zero, then there is no solution (convince yourself of this). We must therefore consider the two cases, $C_1 = 0$, $C_2 \neq 0$ and $C_1 \neq 0$, $C_2 = 0$. The following solutions are found.

$$C_2 = 0 \Rightarrow \Theta_n(\xi) = C_n \cos(\sqrt{\lambda_n}\xi)\,,\ \tan(\sqrt{\lambda_n}) = \frac{Bi}{\sqrt{\lambda_n}}$$

$$C_1 = 0 \Rightarrow \Theta_m(\xi) = C_m \sin(\sqrt{\lambda_m}\xi)\,,\ \tan(\sqrt{\lambda_m}) = -\frac{\sqrt{\lambda_m}}{Bi}\,.$$

Notice that two counters are used, n and m. This notation helps to clarify or avoid confusing the eigenvalues corresponding to the two kinds of eigenfunctions, the cosine and sine eigenfunctions. The inner product is over $[0, 1]$ with the weight function 1.

There is no way of expressing this family of eigenfunctions by a single, simple expression. A sensible way of writing the Fourier series solution is therefore to write it as the sum of two series, a cosine-series and a sine-series.

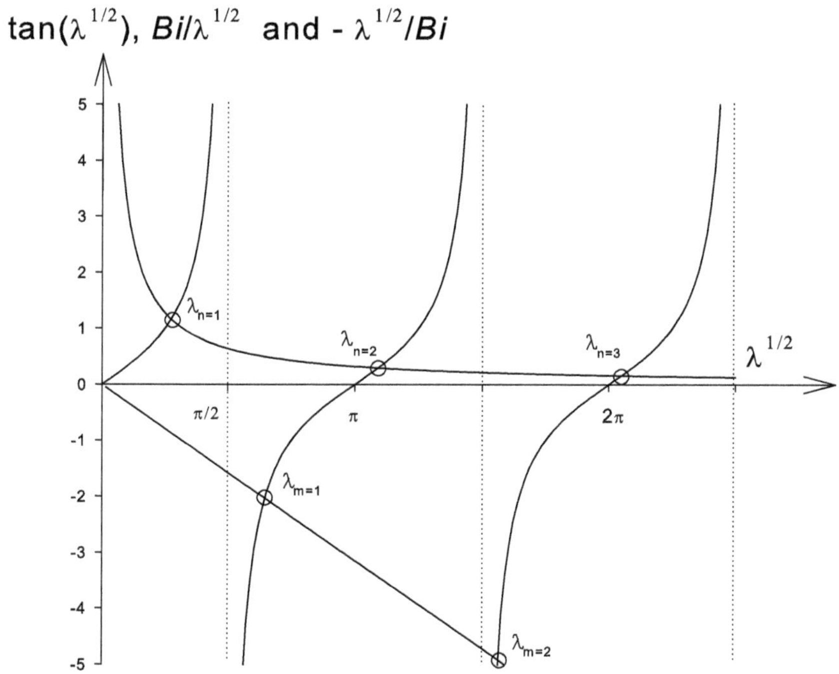

Fig. 10.2 Graphs of $\tan(\sqrt{\lambda})$, $Bi/\sqrt{\lambda}$ and $-\sqrt{\lambda}/Bi$. The eigenvalues are found at the intersection of the tan-graph and the two other graphs.

The two eigenvalue equations are plotted in Fig. 10.2.

Plots such as this are helpful because the eigenvalues must be found numerically. It does not matter much that one can only find a finite number of eigenvalues this way, since calculations with the solution always use but a finite number of terms in the Fourier series. However, it is important that the numerical search does not miss any of the eigenvalues that correspond to terms of the Fourier series that are used in calculations. To guard against this possibility, it is often helpful to try and locate the eigenvalues in disjoint search intervals and to determine the limiting value of the eigenvalues as the index, n or m, becomes large.

The eigenvalues are found where the graphs in Fig. 10.2 intersect. The intersections of the hyperbola $Bi/\sqrt{\lambda_n}$ with the tangent graph gives the eigenvalues associated with the cosine eigenfunctions. A single eigenvalue is found in each interval over which the tangent is positive, so if the eigenvalues are indexed starting with 1, the bounds are

$$(n-1)\pi < \lambda_n < n\pi - \pi/2 \ , \ n = 1, 2, 3, \cdots$$

Similarly, the eigenvalues associated with the sine eigenfunctions are located at the intersection of the tangent graph with the line $-\sqrt{\lambda_m}/Bi$, with a single eigenvalue in each interval over which tangent is negative. Thus

$$m\pi - \pi/2 < \lambda_m < m\pi \ , \ m = 1, 2, 3, \cdots$$

Now that the SLP associated with the PDE has been solved, take the inner product of the PDE with an eigenfunction

$$\left\langle \Theta_k, \frac{\partial^2 \Theta}{\partial \xi^2} \right\rangle = \left\langle \Theta_k, \frac{\partial \Theta}{\partial \tau} \right\rangle$$

where Θ_k is any eigenfunction. Rearrange this as usual to get

$$-\lambda_k \langle \Theta_k, \Theta \rangle = \frac{d}{d\tau} \langle \Theta_k, \Theta \rangle$$

which must be solved subject to the initial condition

$$\tau = 0 \Rightarrow \langle \Theta_k, \Theta \rangle = \langle \Theta_k, \Theta_0(\xi) \rangle$$

giving

$$\langle \Theta_k, \Theta \rangle = \langle \Theta_k, \Theta_0(\xi) \rangle e^{-\lambda_k \tau}.$$

The Fourier series for the solution can now be assembled, keeping in mind that it will be written as two series. One for the cosine eigenfunctions and one for the sine eigenfunctions

$$\Theta(\xi, \tau) = \sum_{n=1}^{\infty} e^{-\lambda_n \tau} \frac{\langle \cos(\sqrt{\lambda_n} \xi), \Theta_0(\xi) \rangle}{\langle \cos(\sqrt{\lambda_n} \xi), \cos(\sqrt{\lambda_n} \xi) \rangle} \cos(\sqrt{\lambda_n} \xi)$$

$$+ \sum_{m=1}^{\infty} e^{-\lambda_m \tau} \frac{\langle \sin(\sqrt{\lambda_m} \xi), \Theta_0(\xi) \rangle}{\langle \sin(\sqrt{\lambda_m} \xi), \sin(\sqrt{\lambda_m} \xi) \rangle} \sin(\sqrt{\lambda_m} \xi).$$

The inner products must now be evaluated

$$\langle \cos(\sqrt{\lambda_n} \xi), \cos(\sqrt{\lambda_n} \xi) \rangle = \int_{-1}^{1} \cos^2(\sqrt{\lambda_n} \xi) \, d\xi = \frac{\cos(\sqrt{\lambda_n}) \sin(\sqrt{\lambda_n}) + \sqrt{\lambda_n}}{\sqrt{\lambda_n}}.$$

The last result can be simplified using the equation for the λ_n eigenvalues

$$\tan(\sqrt{\lambda_n}) = \frac{Bi}{\sqrt{\lambda_n}} \Rightarrow \sin(\sqrt{\lambda_n}) = \frac{Bi}{\sqrt{\lambda_n}} \cos(\sqrt{\lambda_n}) \Rightarrow$$

$$\langle \cos(\sqrt{\lambda_n} \xi), \cos(\sqrt{\lambda_n} \xi) \rangle = 1 + \frac{Bi}{\lambda_n} \cos^2(\sqrt{\lambda_n}).$$

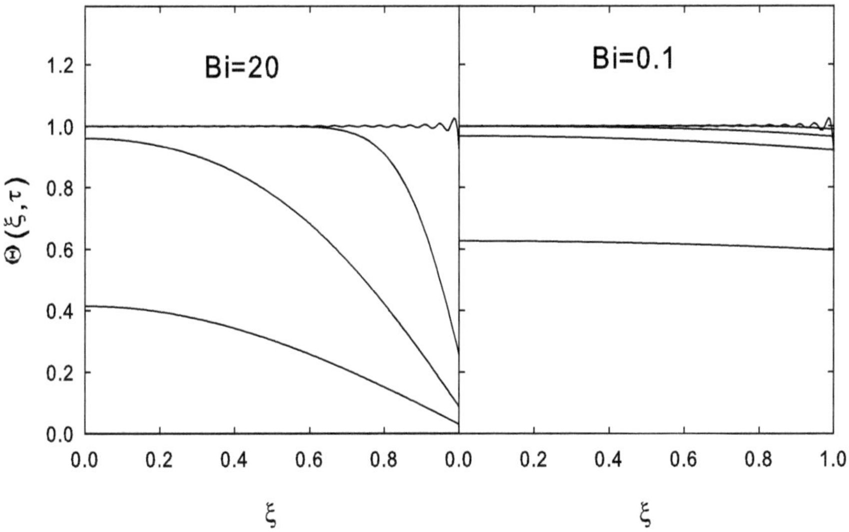

Fig. 10.3 Dimensionless temperature profile in a wall with film resistance, Eq. (10.3). The initial condition used is $\Theta_0(\xi) = 1$ and the profiles are plotted for $\tau = 0, 0.01, 0.1, 0.5$ and 5.

Similarly

$$\langle \sin(\sqrt{\lambda_m}\xi), \sin(\sqrt{\lambda_m}\xi)\rangle = 1 + \frac{1}{Bi}\cos^2(\sqrt{\lambda_m}).$$

In summary, the solution becomes

$$\Theta(\xi,\tau) = \sum_{n=1}^{\infty} e^{-\lambda_n \tau} \frac{\langle \cos(\sqrt{\lambda_n}\xi), \Theta_0(\xi)\rangle}{1 + \frac{Bi}{\lambda_n}\cos^2(\sqrt{\lambda_n})} \cos(\sqrt{\lambda_n}\xi) \tag{10.2}$$

$$+ \sum_{m=1}^{\infty} e^{-\lambda_m \tau} \frac{\langle \sin(\sqrt{\lambda_m}\xi), \Theta_0(\xi)\rangle}{1 + \frac{1}{Bi}\cos^2(\sqrt{\lambda_m})} \sin(\sqrt{\lambda_m}\xi) \tag{10.3}$$

where

$$\tan(\sqrt{\lambda_n}) = \frac{Bi}{\sqrt{\lambda_n}}\,, \ \tan(\sqrt{\lambda_m}) = -\frac{\sqrt{\lambda_m}}{Bi}\,.$$

Graphs of the solution are plotted in Figs. 10.3 and 10.4. Figure 10.3 shows the solution for the symmetric initial condition $\Theta_0(\xi) = 1$ and for two different values of the Biot number, 20 and 0.1. At large values of the Biot number, the external film resistance is negligible and the situation is very similar to the case with no film resistance. At low values of the Biot number, on the other hand, the external resistance is the controlling resistance and conduction through the solid

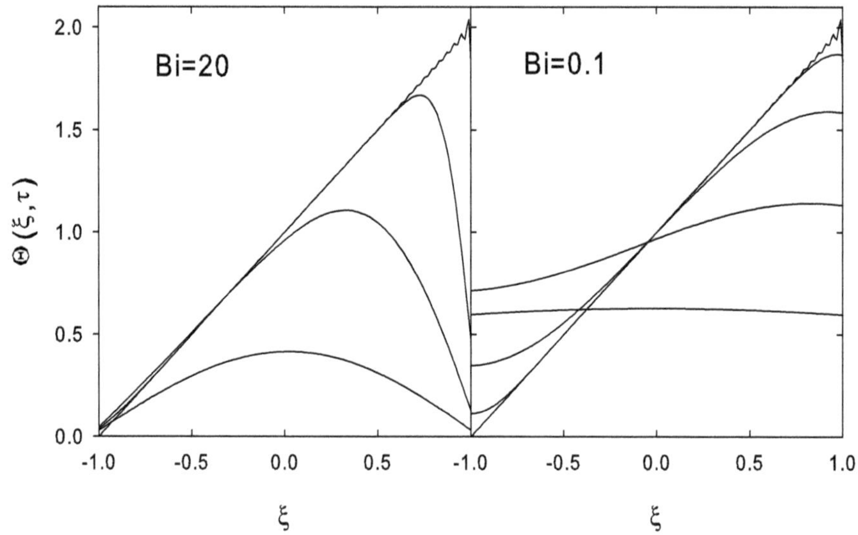

Fig. 10.4 Dimensionless temperature profile in a wall with film resistance, Eq. (10.3). The initial condition used is $\Theta_0(\xi) = 1 + \xi$ and the profiles are plotted for $\tau = 0$, 0.01, 0.1, 0.5 and 5.

itself is fast compared to loss of thermal energy through the film. Consequently, the temperature profile is in a quasi-steady state and stays close to flat at all times.

Figure 10.4 shows the solution for the asymmetric initial condition $\Theta_0(\xi) = 1 + \xi$. For both values of the Biot number, the temperature profiles become more symmetric with time. For high values of the Biot number, symmetry is approached as the excess thermal energy at high values of ξ is lost through the film while, at low values of the Biot number, symmetry is attained as the temperature profile approaches the steady state profile faster than thermal energy is lost through the film.

The next example does not introduce any novel aspects of the solution method per se, but illustrates the method for an elliptic PDE.

Example 10.5: Steady state temperature profile in a square with heat generation.

The dimensionless energy balance an a solid square in which heat evolves at a constant rate can be written.

$$\frac{\partial^2 \Theta}{\partial x^2} + \frac{\partial^2 \Theta}{\partial y^2} + q = 0$$

We will solve this problem subject to the boundary conditions

$$\Theta(x = 0, y) = \Theta(x = 1, y) = \Theta(x, y = 0) = 0$$

$$\Theta(x, y = 1) = 1\,.$$

The associated eigenvalue problem is

$$\frac{d^2\Theta_n}{dx^2} + \lambda_n\Theta_n = 0\;,\;\; \Theta_n(0) = \Theta_n(1) = 0$$

and it is left as an exercise to show that it has no negative eigenvalues. We still need to investigate $\lambda = 0$ and positive values of λ. For $\lambda = 0$ we have

$$\frac{d^2\Theta_n}{dx^2} = 0 \;\Rightarrow$$

$$\Theta_n = C_1 x + C_2\,.$$

Applying the boundary conditions give $\Theta = 0$, the trivial solution. Thus zero is not an eigenvalue. For $\lambda > 0$ we have

$$\Theta_n(x) = C_1\cos(\sqrt{\lambda_n}x) + C_2\sin(\sqrt{\lambda_n}x).$$

Applying the boundary conditions give

$$0 = C_1$$

$$0 = C_2\sin(\sqrt{\lambda_n}) \;\Rightarrow \lambda_n = (n\pi)^2\;,\;\; n \in \mathbb{N}\,.$$

So

$$\Theta_n(x) = \sin(n\pi x)$$

and the eigenfunctions are orthogonal in the inner product

$$\langle f, g\rangle = \int_0^1 f(x)g(x)dx\,.$$

Now take the inner product of the PDE with $\Theta_n(x)$.

$$-\lambda_n\langle\Theta_n, \Theta\rangle + \frac{d^2}{dy^2}\langle\Theta_n, \Theta\rangle + \langle\Theta_n, q\rangle = 0\,.$$

Clearly

$$\langle\Theta_n, \Theta\rangle = C_{1,n}\cosh(\sqrt{\lambda_n}y) + C_{2,n}\sinh(\sqrt{\lambda_n}y) + \langle\Theta_n, q\rangle/\lambda_n\,.$$

So the Fourier series for Θ is

$$\Theta(x,y)=\sum_{n=1}^{\infty}\frac{C_{1,n}\cosh(\sqrt{\lambda_n}y)+C_{2,n}\sinh(\sqrt{\lambda_n}y)+\langle\Theta_n,q\rangle/\lambda_n}{\langle\Theta_n(x),\Theta_n(x)\rangle}\Theta_n(x).$$

Apply the boundary condition at $y=0$

$$y=0\ \Rightarrow\ \Theta=0$$

which gives

$$0=\sum_{n=1}^{\infty}\frac{C_{1,n}+\langle\Theta_n,q\rangle/\lambda_n}{\langle\Theta_n(x),\Theta_n(x)\rangle}\Theta_n(x).$$

The series can only equal 0 if all the coefficients equal 0, hence

$$C_{1,n}=-\langle\Theta_n,q\rangle/\lambda_n\ ,\ \forall n\,.$$

Now apply the boundary condition at $y=1$.

$$y=1\ \Rightarrow\ \Theta=1\ \Rightarrow$$

$$1=\sum_{n=1}^{\infty}\frac{\langle\Theta_n,q\rangle/\lambda_n(1-\cosh(\sqrt{\lambda_n}))+C_{2,n}\sinh(\sqrt{\lambda_n})}{\langle\Theta_n(x),\Theta_n(x)\rangle}\Theta_n(x)\,.$$

Take the inner product of this equation with $\Theta_m(x)$.

$$\langle\Theta_m(x),1\rangle=\sum_{n=1}^{\infty}\frac{\langle\Theta_n,q\rangle/\lambda_n(1-\cosh(\sqrt{\lambda_n}))+C_{2,n}\sinh(\sqrt{\lambda_n})}{\langle\Theta_n(x),\Theta_n(x)\rangle}\langle\Theta_m(x),\Theta_n(x)\rangle\,.$$

Due to the orthogonality of the eigenfunctions, the only term left in the series is the m'th term.

$$\langle\Theta_m(x),1\rangle=\langle\Theta_m,q\rangle/\lambda_m(1-\cosh(\sqrt{\lambda_m}))+C_{2,m}\sinh(\sqrt{\lambda_m})\ \Rightarrow$$

$$C_{2,n}=\frac{\langle\Theta_n(x),1\rangle-\langle\Theta_n,q\rangle/\lambda_n(1-\cosh(\sqrt{\lambda_n}))}{\sinh(\sqrt{\lambda_n})}\,.$$

Putting everything together, the Fourier series for Θ is

$$\Theta(x,y)=\sum_{n=1}^{\infty}\frac{\begin{array}{l}\langle\theta_n,q\rangle/\lambda_n(1-\cosh(\sqrt{\lambda_n}y))\\ \quad+\left(\frac{\langle\theta_n(x),1\rangle-\langle\theta_n,q\rangle/\lambda_n(1-\cosh(\sqrt{\lambda_n}))}{\sinh(\sqrt{\lambda_n})}\right)\sinh(\sqrt{\lambda_n}y)\end{array}}{\langle\theta_n(x),\theta_n(x)\rangle}\theta_n(x).$$

One finds that

$$\langle \Theta_n, \Theta_n \rangle = \int_0^1 \sin^2(n\pi x)dx = \frac{1}{2}$$

$$\langle \Theta_n, 1 \rangle = \int_0^1 \sin(n\pi x)dx = \frac{1-(-1)^n}{n\pi}$$

$$\langle \Theta_n, q \rangle = q\frac{1-(-1)^n}{n\pi}$$

which "simplifies" the result to

$$\begin{aligned}\Theta(x,y) =& 2\sum_{n=1}^{\infty}\left\{q\frac{1-(-1)^n}{n^3\pi^3}(1-\cosh(n\pi y))\right.\\ &+\left.\left(\frac{1-(-1)^n}{n\pi} - q\frac{1-(-1)^n}{n^3\pi^3}(1-\cosh(n\pi))\right)\frac{\sinh(n\pi y)}{\sinh(n\pi)}\right\}\sin(n\pi x).\end{aligned}$$

The next example shows the use of Legendre polynomials in a Fourier series solution.

Example 10.6: Concentration profile in a spherical catalytic pellet with latitudinally varying surface concentration.

If the distance over which concentrations change in a packed bed reactor are small compared to the diameter of the catalytic particles, then the reactant surface concentration that is seen by the particles will change from front to back. To model this, consider a spherical catalytic particle placed in a steady fluid flow and use a spherical coordinate system for the pellet with the $\theta = 0$ direction placed opposite the direction of flow, Fig. 10.5.

Assume that the reaction inside the particle is first order with respect to the reactant and that the surface concentration of the reactant is only a function of θ. The concentration inside the pellet is therefore a function of r and θ only, giving the dimensionless steady state balance equation

$$\frac{1}{r^2}\frac{\partial}{\partial r}\left(r^2\frac{\partial C}{\partial r}\right) + \frac{1}{r^2\sin\theta}\frac{\partial}{\partial\theta}\left(\sin\theta\frac{\partial C}{\partial\theta}\right) = \Phi^2 C$$

where Φ is the Thiele modulus, and with the boundary conditions

$$r = 1 \Rightarrow C = C_s(\theta)$$

and C finite for all $r \in [0,1]$ and all $\theta \in [0,\pi]$. There are two possible associated eigenvalue problems, one in terms of r, another in terms of θ. Choosing the latter, we must consider

$$\frac{1}{\sin\theta}\frac{d}{d\theta}\left(\sin\theta\frac{dC_n}{d\theta}\right)+\lambda_n C_n=0$$

subject to the boundary condition that C_n is finite for $\theta \in [0,\pi]$. A recognizable problem can be obtained using the variable transformation $x=\cos\theta$ which gives, (see page 402).

$$(1-x^2)\frac{d^2C_n}{dx^2}-2x\frac{dC_n}{dx}+\lambda_n C_n=0$$

which is Legendre's differential equation. The boundary conditions require that the solutions are finite for all $x \in [-1,1]$, i.e. the eigenfunctions are the Legendre polynomials, $P_n(x)$, and the eigenvalues are

$$\lambda_n = n(n+1),$$

the appropriate inner product is

$$\langle f,g\rangle = \int_{-1}^{1} fg\,dx$$

and the eigenfunctions satisfy

$$\int_{-1}^{1} P_n(x)P_m(x)\,dx = \begin{cases} 0\,, & \text{if } n \neq m \\ \frac{2}{2n+1}\,, & \text{if } n = m \end{cases}.$$

At this point it is probably a good idea to also use the variable transformation $x=\cos\theta$ in the PDE. One obtains

$$\frac{1}{r^2}\frac{\partial}{\partial r}\left(r^2\frac{\partial C}{\partial r}\right)+\frac{1}{r^2}\frac{\partial}{\partial x}\left((1-x^2)\frac{\partial C}{\partial x}\right)=\Phi^2 C\,.$$

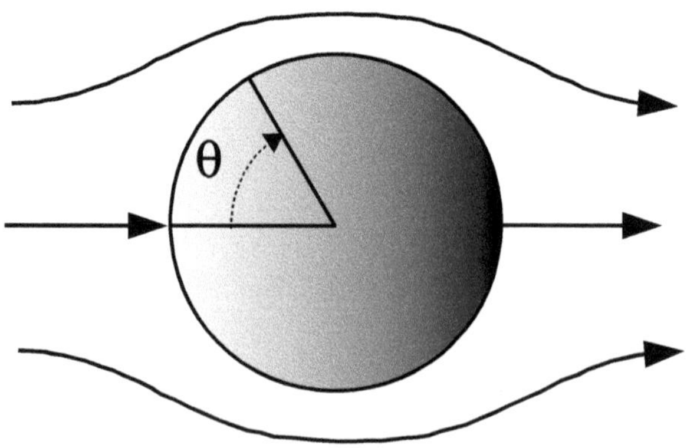

Fig. 10.5 Definition of θ direction in the spherical coordinate system for a catalytic particle in a flow.

Take the inner product with a Legendre polynomial, P_n, to get

$$\frac{1}{r^2}\frac{d}{dr}\left(r^2\frac{d\langle P_n,C\rangle}{dr}\right)+\frac{1}{r^2}\left\langle P_n,\frac{\partial}{\partial x}\left((1-x^2)\frac{\partial C}{\partial x}\right)\right\rangle=\Phi^2\langle P_n,C\rangle \Rightarrow$$
$$\frac{1}{r^2}\frac{d}{dr}\left(r^2\frac{d\langle P_n,C\rangle}{dr}\right)+\frac{1}{r^2}\left\langle \frac{d}{dx}\left((1-x^2)\frac{dP_n}{dx}\right),C\right\rangle=\Phi^2\langle P_n,C\rangle \Rightarrow$$
$$\frac{1}{r^2}\frac{d}{dr}\left(r^2\frac{d\langle P_n,C\rangle}{dr}\right)-\frac{n(n+1)}{r^2}\langle P_n,C\rangle=\Phi^2\langle P_n,C\rangle$$

which is subject to the boundary conditions

$$r=0 \Rightarrow \langle P_n,C\rangle \text{ finite}$$
$$r=1 \Rightarrow \langle P_n,C\rangle=\langle P_n,C_S(\theta(x))\rangle$$

and therefore has the solution

$$\langle P_n,C\rangle=C_1\frac{I_{n+1/2}(\Phi r)}{\sqrt{r}}+C_2\frac{K_{n+1/2}(\Phi r)}{\sqrt{r}}.$$

To evaluate the two arbitrary constants, determine the limiting behavior of the two basis functions as r goes to zero. It is easy to show from the Taylor series that $K_{n+1/2}(\Phi r)$ goes to infinity while $I_{n+1/2}(\Phi r)$ is bounded. Thus, $C_2=0$. The boundary conditions at $r=1$ now gives

$$\langle P_n,C_S(\theta(x))\rangle=C_1 I_{n+1/2}(\Phi) \Rightarrow$$
$$\langle P_n,c\rangle=\langle P_n,C_S(\theta(x))\rangle\frac{I_{n+1/2}(\Phi r)}{\sqrt{r}I_{n+1/2}(\Phi)}.$$

Substitution into the Fourier series finally gives

$$C(r,\theta)=\sum_{n=0}^{\infty}\frac{\langle P_n,C\rangle}{\langle P_n,P_n\rangle}P_n(\cos\theta)$$

or

$$C(r,\theta)=2\sum_{n=0}^{\infty}\frac{\langle P_n,C_S(\theta(x))\rangle}{2n+1}\frac{I_{n+1/2}(\Phi r)}{\sqrt{r}I_{n+1/2}(\Phi)}P_n(\cos\theta). \tag{10.4}$$

The inner product $\langle P_n,C_S(\theta(x))\rangle$ can be hard to evaluate analytically unless the surface concentration is a simple function of x or, equivalently, of $\cos\theta$. A simple function that model a decrease in surface concentration as θ increases is $C_S=1+\cos\theta=1+x$. The solution is plotted for this surface concentration in Figs. 10.6 and 10.7. Figure 10.6 shows the concentration profile for $\Phi=1$, a relatively small value of the Thiele modulus at which the rate of diffusion is high relative to the rate of reaction. The profile is therefore almost flat throughout the pellet.

Figure 10.7 shows the profile for $\Phi=10$. At this value of the Thiele modulus the reaction rate is fast relative to the rate of transport and the reaction occurs predominantly in a narrow region close to the pellet surface. The reactant concentration in the interior of the pellet is effectively zero.

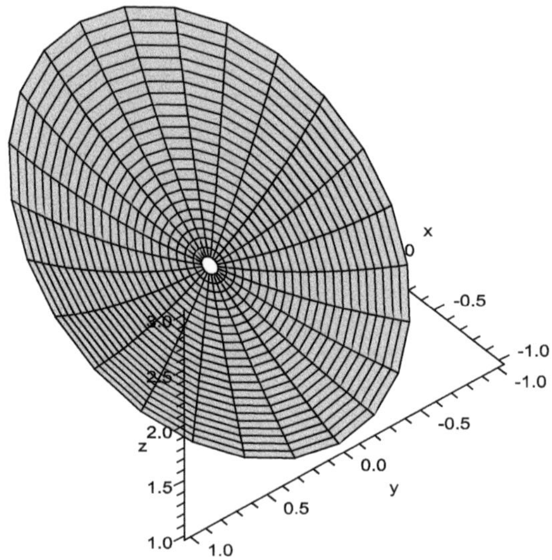

Fig. 10.6 Dimensionless concentration profile in spherical catalytic pellet for $\Phi = 1$, Eq. (10.4). The dimensionless concentration is plotted in the z-direction and the $\theta = 0$ direction correspond to the positive x direction. Eleven terms were used in the Fourier series.

10.1.3.3 *Inhomogeneous PDE*

An inhomogeneous term in the PDE may initially appear to be a problem for the solution method, but the inner product solution method works without any problems.

Example 10.7: Transient conduction in wall with heat generation.
Consider the problem

$$\frac{\partial \Theta}{\partial \tau} = \frac{\partial^2 \Theta}{\partial \xi^2} + Q(\xi, \tau)$$

$$\tau = 0 \Rightarrow \Theta = 1$$

$$\xi = 0 \Rightarrow \frac{\partial \Theta}{\partial \xi} = 0$$

$$\xi = 1 \Rightarrow \Theta = 0$$

which can be the dimensionless energy balance, in rectangular coordinates, for a wall with heat generation. For instance a flat catalytic pellet in which the heat of reaction depends on position and time. The associated SLP is identified by simply ignoring the inhomogeneous term in the PDE.

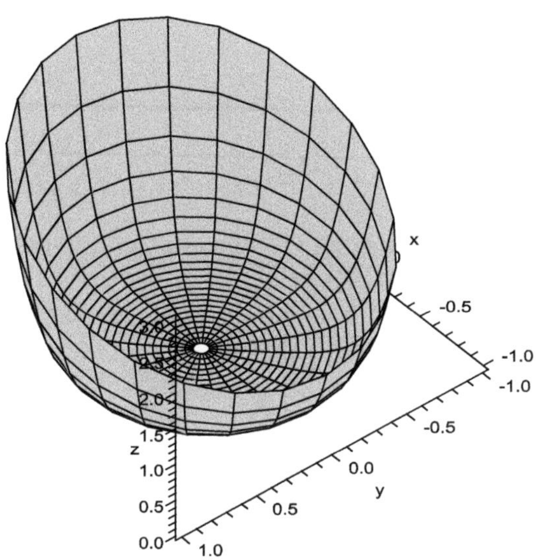

Fig. 10.7 Dimensionless concentration profile in spherical catalytic pellet for $\Phi = 10$, Eq. (10.4). The dimensionless coincentration is plotted in the z-direction and the $\theta = 0$ direction correspond to the positive x direction. Eleven terms were used in the Fourier series.

$$\frac{d^2\Theta_n}{d\xi^2} + \lambda_n \Theta_n = 0 \ , \ \Theta_n(1) = 0 \ , \ \left.\frac{d\Theta_n}{d\xi}\right|_{\xi=0} = 0$$

which has the solution

$$\Theta_n(\xi) = \cos(\sqrt{\lambda_n}\xi) \ , \ \sqrt{\lambda_n} = \frac{\pi}{2} + n\pi \ , \ n = 0, 1, 2, \cdots$$

and the inner product is over $[0, 1]$ with the weight function 1. Take the inner product of the PDE with an eigenfunction

$$\left\langle \Theta_n, \frac{\partial \Theta}{\partial \tau} \right\rangle = \left\langle \Theta_n, \frac{\partial^2 \Theta}{\partial \xi^2} \right\rangle + \langle \Theta_n, Q(\xi, \tau) \rangle.$$

Rearrange as usual to get

$$\frac{d}{d\tau}\langle \Theta_n, \Theta \rangle = -\lambda_n \langle \Theta_n, \Theta \rangle + \langle \Theta_n, Q(\xi, \tau) \rangle.$$

Notice that in the inner product $\langle \Theta_n, Q(\xi, \tau) \rangle$, the inner product that arise from the inhomogeneous term in the PDE, all the ξ dependence is eliminated since the inner product is an integration with respect to ξ. The equation above is therefore an ODE in the free variable τ, subject to the initial conditions

$$\tau = 0 \Rightarrow \langle \Theta_n, \Theta \rangle = \langle \Theta_n, 1 \rangle.$$

The solution to this ODE is

$$\langle \Theta_n, \Theta \rangle = e^{-\lambda_n \tau} \int_0^\tau e^{\lambda_n s} \langle \Theta_n, Q(\xi, s) \rangle \, ds + e^{-\lambda_n \tau} \langle \Theta_n, 1 \rangle.$$

Substituting these results back into the Fourier series for Θ, we get

$$\Theta(\xi, \tau) = \sum_{n=0}^{\infty} \frac{\langle \cos((n\pi + \frac{\pi}{2})\xi), 1 \rangle + \int_0^\tau e^{(n\pi + \frac{\pi}{2})^2 s} \langle \cos((n\pi + \frac{\pi}{2})\xi), Q(\xi, s) \rangle \, ds}{\langle \cos((n\pi + \frac{\pi}{2})\xi), \cos((n\pi + \frac{\pi}{2})\xi) \rangle} \cdot e^{-(n\pi + \frac{\pi}{2})^2 \tau} \cos\left(\left(n\pi + \frac{\pi}{2}\right)\xi\right).$$

This can be written as two Fourier series, one arising from the initial condition, another arising from the inhomogeneous term in the PDE.

$$\Theta(\xi, \tau) = \sum_{n=0}^{\infty} \frac{\langle \cos((n\pi + \frac{\pi}{2})\xi), 1 \rangle}{\langle \cos((n\pi + \frac{\pi}{2})\xi), \cos((n\pi + \frac{\pi}{2})\xi) \rangle} e^{-(n\pi + \frac{\pi}{2})^2 \tau} \cos\left(\left(n\pi + \frac{\pi}{2}\right)\xi\right)$$

$$+ \sum_{n=0}^{\infty} \frac{\int_0^\tau e^{(n\pi + \frac{\pi}{2})^2 s} \langle \cos((n\pi + \frac{\pi}{2})\xi), Q(\xi, s) \rangle \, ds}{\langle \cos((n\pi + \frac{\pi}{2})\xi), \cos((n\pi + \frac{\pi}{2})\xi) \rangle} e^{-(n\pi + \frac{\pi}{2})^2 \tau} \cos\left(\left(n\pi + \frac{\pi}{2}\right)\xi\right).$$

The reason for doing this is to separate out the transient terms and the terms that correspond to a steady state solution (assuming one exists). Doing so results in a simpler expression. To illustrate this, let $Q = 1$. Then the solution for Θ becomes

$$\Theta(\xi, \tau) = \sum_{n=0}^{\infty} \frac{2(-1)^n}{(n\pi + \pi/2)^3} \cos((n\pi + \pi/2)\xi) + \sum_{n=0}^{\infty} \frac{2(-1)^n}{(n\pi + \pi/2)} \left(1 - \frac{1}{(n\pi + \pi/2)^2}\right) e^{-(n\pi + \pi/2)^2 \tau} \cos((n\pi + \pi/2)\xi).$$

Clearly, the second sum contains all the time dependence and, because of the appearance of the exponential functions in each term of this sum, the sum goes to zero as τ goes to infinity. The steady state solution, $\Theta_{s.s}$, must therefore be given by the Fourier series

$$\Theta_{s.s.}(\xi) = \sum_{n=0}^{\infty} \frac{2(-1)^n}{(n\pi + \pi/2)^3} \cos((n\pi + \pi/2)\xi) \tag{10.5}$$

but this solution can also be found from the steady state balance equation which is

$$\frac{d^2\Theta_{s.s.}}{d\xi^2} = -1\ ,\ \Theta_{s.s.}(1) = 0\ ,\ \frac{d\Theta_{s.s.}}{d\xi} = 0$$

$$\Theta_{s.s} = \frac{1}{2}(1-\xi^2).$$

The Fourier series for this function is in fact the series in Eq. (10.5). The solution for Θ can therefore be simplified to

$$\Theta(\xi,\tau) = \frac{1}{2}(1-\xi^2)$$
$$+\sum_{n=0}^{\infty}\frac{2(-1)^n}{(n\pi+\pi/2)}\left(1-\frac{1}{(n\pi+\pi/2)^2}\right)e^{-(n\pi+\pi/2)^2\tau}\cos((n\pi+\pi/2)\xi)\,.$$

This expression is computationally less expensive than the expression using a Fourier series to represent the entire solution.

It often pays to recognize that a given problem has a solution which can be written as the sum of two terms, a steady state solution plus a transient solution. When this is the case, the steady state solution, which can often be written in very simple terms, can be subtracted from the complete solution to obtain a model of just the transient. This mode is often simpler to solve than the full model. For instance, when the steady state solution for the model just solved (with $Q = 1$) is subtracted from the general solution, it is essentially equivalent to using the variable substitution

$$\Psi = \Theta - \frac{1}{2}(1-\xi^2)$$

and, when applied to the PDE, it gives the simpler homogeneous PDE

$$\frac{\partial\Psi}{\partial\tau} = \frac{\partial^2\Psi}{\partial\xi^2}$$

$$\tau = 0\ \Rightarrow\ \Psi = \frac{1}{2}(1+\xi^2)$$

$$\xi = 0\ \Rightarrow\ \frac{\partial\Psi}{\partial\xi} = 0$$

$$\xi = 1\ \Rightarrow\ \Psi = 0\,.$$

10.1.3.4 *Inhomogeneous, time varying boundary conditions*

Solving partial differential equations with the inner product method requires identifying one or more self-adjoint operators in the equation, usually Sturm–Liouville operators, and these operators require homogeneous boundary conditions. Inhomogeneous boundary conditions can often be eliminated by introducing dimensionless variables as was done in the previous examples but in some cases this is insufficient. In these cases removing the inhomogeneities in the boundary conditions becomes a key difficulty when using the inner product method. Generally, one must devise a variable transformation which removes the inhomogeneities from the boundary conditions and moves them to the PDE itself where they are easily handled by the inner product method. This is illustrated in the next example.

Example 10.8: Transient conduction with time dependent surface conditions.

As the last variation on the problem of conduction in a wall, consider a case in which the temperature of the right hand surface of the wall changes in some known way with time, say $f(\tau)$ and the flux at the left hand surface changes with time as $g(\tau)$. This gives rise to inhomogeneous and time dependent boundary conditions. Inhomogeneous boundary conditions do present a problem for the inner product solution method because the operator associated with the inhomogeneous conditions cannot be a Sturm–Liouville operator. The problem is solved by using a variable substitution to move the inhomogeneity from the boundary conditions to the PDE.

In dimensionless form, the problem can be stated as

$$\frac{\partial \Theta}{\partial \tau} = \frac{\partial^2 \Theta}{\partial \xi^2}$$

$$\xi = 0 \Rightarrow \frac{\partial \Theta}{\partial \xi} = g(\tau)\,, \quad \text{all } \tau$$

$$\xi = 1 \Rightarrow \Theta = f(\tau)\,, \quad \text{all } \tau$$

$$\tau = 0 \Rightarrow \Theta = \Theta_0(\xi)\,, \quad \text{all } \xi\,.$$

In order to make the boundary conditions of this problem homogeneous, one must subtract some function from Θ such that the boundary condition at both $\xi = 1$ and $\xi = 0$ become homogeneous. There is an infinity of possible choices for this transformation and it is usually not too hard to find a simple function that will suffice. To render the boundary condition at $\xi = 1$ homogeneous without interfering with the condition at $\xi = 0$, pick a function that has a zero derivative with respect to ξ at $\xi = 0$, e.g. $C(\tau)$ (or $C(\tau)\xi^n$ where n is even). The function $C(\tau)$ is determined by the requirement that $\Theta(\xi, \tau) - C(\tau)$ must equal zero at $\xi = 1$. Clearly $C(\tau) = f(\tau)$.

Similarly, make the boundary condition at $\xi = 0$ homogeneous by subtracting a function from Θ that does not interfere with the condition at $\xi = 1$. For instance, $g(\tau)(\xi - 1)$.

Thus, define

$$\Psi(\xi,\tau) = \Theta(\xi,\tau) - f(\tau) - g(\tau)(\xi - 1)$$

and the problem in Ψ becomes

$$\frac{\partial \Psi}{\partial \tau} + \frac{df}{d\tau} + \frac{dg}{d\tau}(\xi - 1) = \frac{\partial^2 \Psi}{\partial \xi^2}$$

$$\xi = 0 \;\Rightarrow\; \frac{\partial \Psi}{\partial \xi} = \frac{\partial \Theta}{\partial \xi} - g(\tau) = 0\,, \quad \text{all } \tau$$

$$\xi = 1 \;\Rightarrow\; \Psi = 0\,, \quad \text{all } \tau$$

$$\tau = 0 \;\Rightarrow\; \Psi = \Theta_0(\xi) - f(0) - g(0)(\xi - 1).$$

This is an inhomogeneous PDE with homogeneous boundary conditions, so we are on familiar turf. The operator in ξ is a Sturm–Liouville operator and the problem is now almost identical to the problem in the previous example. The inner product is over $[0,1]$ with the weght function 1 and the eigenfunctions and eigenvalues are

$$\Psi_n(\xi) = \cos(\sqrt{\lambda_n}\xi)\,,\ \sqrt{\lambda_n} = n\pi + \pi/2\,,\ n = 0,1,2,\cdots$$

Proceeding now as usual

$$\left\langle \Psi_n, \frac{\partial \Psi}{\partial \tau} \right\rangle + \left\langle \Psi_n, \frac{df}{d\tau} + \frac{dg}{d\tau}(\xi - 1) \right\rangle - \left\langle \Psi_n, \frac{\partial^2 \Psi}{\partial \xi^2} \right\rangle \Rightarrow$$

$$\frac{d}{d\tau}\langle \Psi_n, \Psi \rangle + F_n(\tau) = -\lambda_n \langle \Psi_n, \Psi \rangle$$

where

$$F_n(\tau) = \left\langle \Psi_n, \frac{df}{d\tau} + \frac{dg}{d\tau}(\xi - 1) \right\rangle = \left(\frac{df}{d\tau} - \frac{dg}{d\tau} \right) \langle \Psi_n, 1 \rangle + \frac{dg}{d\tau} \langle \Psi_n, \xi \rangle.$$

The initial condition is

$$\tau = 0 \;\Rightarrow\; \langle \Psi_n, \Psi \rangle = \langle \Psi_n, \Theta_0(\xi) - f(0) - g(0)(\xi - 1) \rangle$$

and the solution for the Fourier coefficients is

$$\langle \Psi_n, \Psi \rangle = \left(\langle \Psi_n, \Theta_0(\xi) - f(0) - g(0)(\xi - 1) \rangle - \int_0^\tau e^{\lambda_n s} F_n(s)\, ds \right) e^{-\lambda_n \tau}$$

and the solution for $\Psi(\xi,\tau)$ is

$$\Psi(\xi,\tau) = \sum_{n=0}^{\infty} \frac{\langle \Psi_n, \Theta_0(\xi) - f(0) - g(0)(\xi - 1)\rangle - \int_0^\tau e^{\lambda_n s} F_n(s)\, ds}{\langle \cos(\sqrt{\lambda_n}\xi), \cos(\sqrt{\lambda_n}\xi)\rangle} e^{-\lambda_n \tau} \cos(\sqrt{\lambda_n}\xi).$$

Evaluating the inner product of the eigenfunctions with themselves, ($\langle \cos(\sqrt{\lambda_n}\xi), \cos(\sqrt{\lambda_n}\xi)\rangle = 1/2$) and splitting the sum into two gives

$$\Psi(\xi,\tau) = 2\sum_{n=0}^{\infty}(\langle \Psi_n, \Theta_0(\xi) - f(0) - g(0)(\xi - 1)\rangle) e^{-\lambda_n \tau} \cos(\sqrt{\lambda_n}\xi)$$

$$- 2\sum_{n=0}^{\infty} \int_0^\tau e^{\lambda_n s} F_n(s)\, ds e^{-\lambda_n \tau} \cos(\sqrt{\lambda_n}\xi).$$

The reason for writing the result as two sums is purely computational. The first sum represents the transient temperature profile that results from the initial conditions and the sum goes to zero as τ goes to infinity. But often, one is not interested in this transient and one only wants to determine the dynamic behavior which is left after all the effects of the initial condition has died out. This dynamics is represented by the second Fourier series in the solution which is all that needs to be evaluated when one is not interested in the effect of specific initial conditions.

A set of results are plotted in Fig. 10.8.

The initial condition used in all the graphs shown in Fig. 10.8 is $\Theta_0(\xi) = \xi(1-\xi)$. However, all the graphs are for times that are so large that the initial condition is basically irrelevant because all initial transient have died out. In all cases, the left surface of the wall is assumed insulated, i.e. $g(\tau) = 0$, while the temperature at the right surface is given by $f(\tau) = \sin(\omega\tau)$. The four figures, from top to bottom, left to right, shows the temperature profile for increasing values of ω. The dynamic behavior illustrated by the plots is characteristic of linear, dynamic systems that are subject to some external, periodic forcing. When the forcing is slow relative to the rate at which the system responds (when the characteristic time of the forcing, $1/\omega$ in this case, is large relative to the characteristic time of the system, given approximately by $1/\lambda_0 = 4/\pi^2 \approx 0.4$) then the system remains in a quasi-steady state corresponding to the current value of the forcing function as evidenced by the almost flat temperature profile in the top left figure. When the forcing is very rapid ($1/\omega. << 1/\lambda_0$) then the system is unable to "keep up" with the forcing and barely responds as shown by the temperature profiles in the bottom right figure. Through most of the wall, the profiles remain close to the steady state profiles that would be seen without forcing except close to the surface where the forcing occurs. It is only at intermediate values of the forcing frequency that anything even remotely interesting happens and the temperature profile shows a response throughout the wall without being close to a quasi-steady state profile.

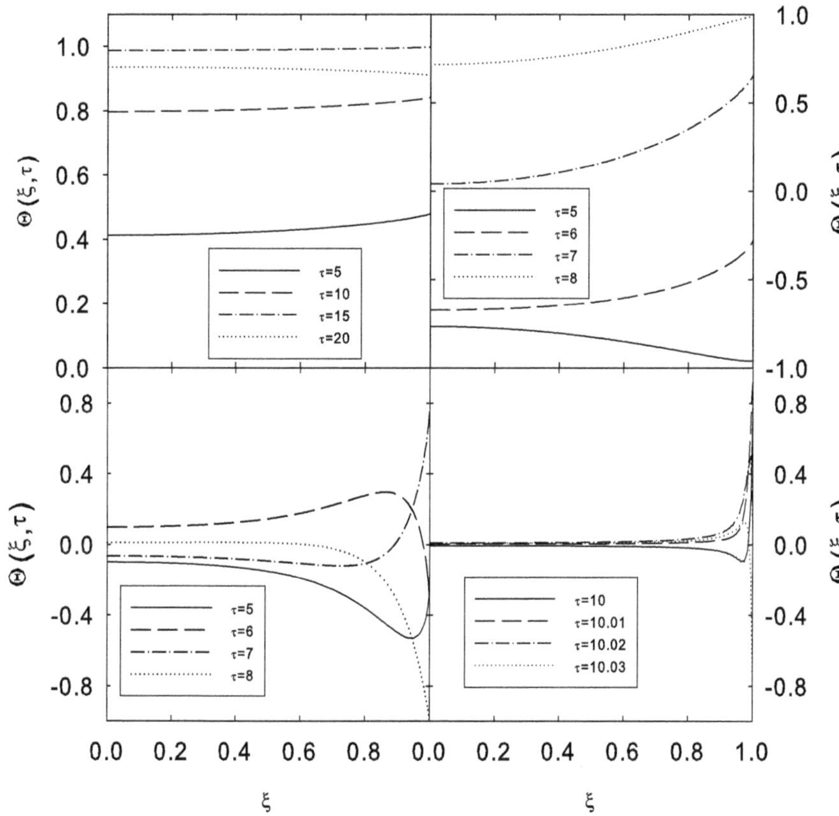

Fig. 10.8 Dimensionless temperature profile in a wall which is insulated at the left surface and with a specified time varying temperature at the right surface given by $\sin(\omega\tau)$. The value of ω changes from left to right, top to bottom as 0.1, 1, 10 and 100.

10.1.4 *Other self-adjoint PDEs*

All the previous PDE problems have been trivial in the sense that once an SLP operator had been recognized, solving the problem became routine. The far more interesting and challenging problems are those that contain other linear operators that may or may not be self-adjoint. The challenge for these problems is to find "the appropriate inner product" or Hilbert space in which the operator is self-adjoint. Once this is accomplished, the rest of the solution can be obtained through routine manipulations. There is no systematic way of finding an inner product in which an operator is self-adjoint, but with some experience it becomes possible to come up with reasonable guesses that can be tested. The rest of this section aims to build some of this insight by finding the appropriate inner product for several examples.

Example 10.9: Leaching of hazardous waste.

Consider a toxic waste site or a hazardous waste spill in which a hazardous compound is leaching from the soil into a body of water. We will assume that the solid containing the hazardous compound is positioned on top of a layer of clay or some other material through which the hazardous compound cannot travel. In other words, the compound is only removed from the soil by leaching into the water above. The water will be assumed well mixed and with a constant supply of fresh water and an identical constant rate of removal of contaminated water by runoff. A simple model of this system is a well mixed, flow through vessel in which a compound is leached from a solid inside the vessel, Fig. 10.9.

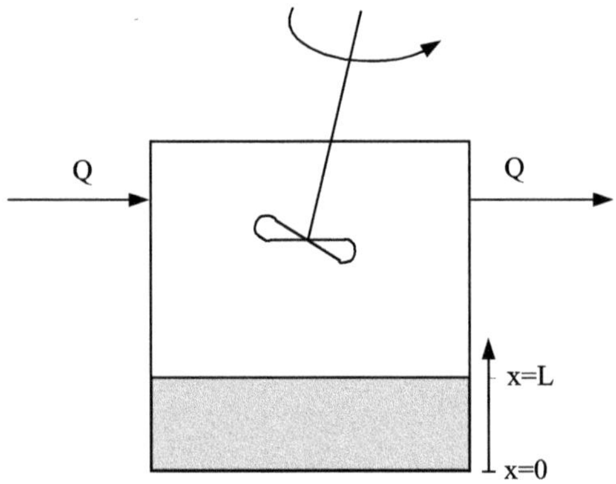

Fig. 10.9 Simple model of hazardous waste leaching into a moving body of water.

Let the volumetric flow rate through the vessel be Q and place a coordinate system in the solid as shown. Assuming that transport through the solid is by Fickian diffusion, the balance on the solid becomes

$$\frac{\partial C_s}{\partial t} = D\frac{\partial^2 C_s}{\partial x^2}$$

and on the water in the vessel

$$V\frac{d\tilde{C}_w}{dt} = -D\left.\frac{\partial C_s}{\partial x}\right|_{x=L} A - Q\tilde{C}_w$$

where C_s is the concentration in the solid, $\tilde{C}_w$ the concentration in the water, A the surface area of the solid, Q the flow rate of water through the vessel and D the diffusion coefficient of the compound in the solid. The initial and boundary conditions are

$$t = 0 \Rightarrow \tilde{C}_w = 0$$

$$t = 0 \Rightarrow C_s = C_{s0}$$

$$x = 0 \Rightarrow \frac{\partial C_s}{\partial x} = 0$$

$$x = L \Rightarrow C_s = \tilde{C}_w$$

where we have assumed no film resistance between the solid and the water. Make this model dimensionless with the variables

$$\xi = \frac{x}{L}$$

$$C = \frac{C_s}{C_{s0}}$$

$$C_w = \frac{\tilde{C}_w}{C_{s0}}$$

$$\tau = \frac{tD}{L^2}.$$

Giving the problem

$$\frac{\partial C}{\partial \tau} = \frac{\partial^2 C}{\partial \xi^2}$$

$$\frac{dC_w}{d\tau} = -\alpha \left.\frac{\partial C}{\partial \xi}\right|_{\xi=1} - \beta C_w$$

where $\alpha = AL/V$ and $\beta = QL^2/(VD)$,

$$\tau = 0 \Rightarrow C_w = 0\,, C(\xi) = 1$$

$$\xi = 0 \Rightarrow \frac{\partial C}{\partial \xi} = 0$$

$$\xi = 1 \Rightarrow C = C_w\,.$$

To write this problem so that an operator clearly appears, define a vector of the two unknowns as

$$\overline{C} = \begin{pmatrix} C(\xi, \tau) \\ C_w(\tau) \end{pmatrix}$$

and restate the problem as

$$\frac{\partial \overline{C}}{\partial \tau} = \begin{pmatrix} \frac{\partial^2 C}{\partial \xi^2} \\ -\alpha \left.\frac{\partial C}{\partial \xi}\right|_{\xi=1} - \beta C_w \end{pmatrix} = L\overline{C}.$$

This brings us to the main problem of finding an inner product for which this operator is self-adjoint. In this case, that is not too difficult to do and the train of thoughts leading to a good inner product candidate might go as follows. The part of the operator that operates on the first element of the vector of unknowns is the well known Sturm–Liouville operator in rectangular coordinates and the appropriate inner product for this is the integral from 0 to 1 with a weight function of unity. The second element of the vector of unknowns is not even a function of ξ so it is tempting to think of this as a regular element of a finite dimensional vector for which the appropriate inner product is the usual scalar product. The scalar product is a sum of products of vector elements and in a sense so is the inner product for the Sturm–Liouville operator. After all, an integral is usually defined as the limit of an infinite sum of terms. A candidate for the appropriate inner product of our operator is therefore a linear combination of the appropriate inner products of the two elements in the vector of unknowns.

$$\langle z, y\rangle = \int_0^1 z(\xi)y(\xi)\,d\xi + C_1 z_w y_w\,.$$

Of course, one would have to show that this candidate satisfies the requirements of an inner product, a step we will skip. The next logical step is to determine whether or not the operator is self-adjoint with this inner product

$$\langle Lz, y\rangle = \int_0^1 \frac{d^2z}{d\xi^2}y(\xi)\,d\xi + C_1\left(-\alpha\left.\frac{dz}{d\xi}\right|_{\xi=1} - \beta z_w\right)y_w\,.$$

Integrating by parts

$$\begin{aligned}\langle Lz, y\rangle = &\int_0^1 z(\xi)\frac{d^2y}{d\xi^2}\,d\xi + y(1)\left.\frac{dz}{d\xi}\right|_{\xi=1} - z(1)\left.\frac{dy}{d\xi}\right|_{\xi=1}\\ &-y(0)\left.\frac{dz}{d\xi}\right|_{\xi=0} + z(0)\left.\frac{dy}{d\xi}\right|_{\xi=0} - \alpha C_1\left.\frac{dz}{d\xi}\right|_{\xi=1} y_w - \beta C_1 z_w y_w\,.\end{aligned}$$

From the boundary conditions, the fourth and fifth term on the right hand side are zero and y_w in the sixth term can be replaced by $y(1)$ such that the second and sixth term can be combined. Finally, $z(1)$ equals z_w so that the third and last terms can be combined.

$$\langle Lz, y\rangle = \int_0^1 z(\xi)\frac{d^2y}{d\xi^2}\,d\xi + (1-\alpha C_1)\left.\frac{dz}{d\xi}\right|_{\xi=1} y(1) + C_1\left(-\frac{1}{C_1}\left.\frac{dy}{d\xi}\right|_{\xi=1} - \beta y_w\right)z_w\,.$$

It is now reasonably clear that if $C_1 = 1/\alpha$ this becomes

$$\langle Lz, y\rangle = \int_0^1 z(\xi)\frac{d^2y}{d\xi^2}\,d\xi + C_1\left(-\alpha\left.\frac{dy}{d\xi}\right|_{\xi=1} - \beta y_w\right)z_w = \langle z, Ly\rangle$$

and the operator is seen to be self-adjoint. The remaining steps, although tedious, are routine. First solve the eigenvalue problem

$$L\overline{C} + \lambda\overline{C} = 0$$

subject to the boundary conditions already stated. This implies solving

$$\frac{d^2C}{d\xi^2} + \lambda C = 0$$

which has the general solution

$$C = C_1 \cos(\sqrt{\lambda}\xi) + C_2 \sin(\sqrt{\lambda}\xi).$$

The boundary condition at $\xi = 0$ gives

$$C_2 = 0 \text{ or } \lambda = 0$$

but only the trivial solution is obtained from $\lambda = 0$. The same is found to hold for negative values of λ. Thus

$$C_n(\xi) = \cos(\sqrt{\lambda_n}\xi).$$

Now consider the second part of the operator equation, the scalar part.

$$-\alpha \left.\frac{\partial C}{\partial \xi}\right|_{\xi=1} - \beta C_w = -\lambda C_w$$

which after substitution of the solution for $C_n(\xi)$ gives

$$\alpha\sqrt{\lambda_n}\sin(\sqrt{\lambda_n}) - \beta C_w = -\lambda_n C_w\,.$$

Apply the boundary conditions at $\xi = 1$ to obtain the equation for the eigenvalues.

$$\alpha\sqrt{\lambda_n}\sin(\sqrt{\lambda_n}) - \beta\cos(\sqrt{\lambda_n}) = -\lambda_n\cos(\sqrt{\lambda_n}) \Rightarrow$$

$$\tan(\sqrt{\lambda_n}) = \frac{\beta - \lambda_n}{\alpha\sqrt{\lambda_n}}\,.$$

Because the operator is self-adjoint the roots of this equation are all real. This greatly simplifies the numerical search for all the roots. Graphs of the two sides of the eigenvalue equation are shown in Fig. 10.10. It is evident that the following bounds apply: $0 < \lambda_1 < \frac{\pi}{2}$ and $\frac{\pi}{2} + (n-2)\pi < \lambda_n < \frac{\pi}{2} + (n-1)\pi$.

The eigenvectors are

$$\overline{C}_n = \begin{pmatrix} \cos(\sqrt{\lambda_n}\xi) \\ \cos(\sqrt{\lambda_n}) \end{pmatrix}.$$

With the eigenvectors now known, take the inner product of the operator equation with an eigenvector.

$$\langle L\overline{C}, \overline{C}_n\rangle = \left\langle \frac{\partial \overline{C}}{\partial \tau}, \overline{C}_n \right\rangle \Rightarrow$$

$$\langle \overline{C}, L\overline{C}_n\rangle = -\lambda_n\langle \overline{C}, \overline{C}_n\rangle = \frac{d}{d\tau}\langle \overline{C}, \overline{C}_n\rangle$$

which is subject to the initial condition

$$\tau = 0 \Rightarrow \langle \overline{C}, \overline{C}_n\rangle = \left\langle \begin{pmatrix} 1 \\ 0 \end{pmatrix}, \overline{C}_n \right\rangle.$$

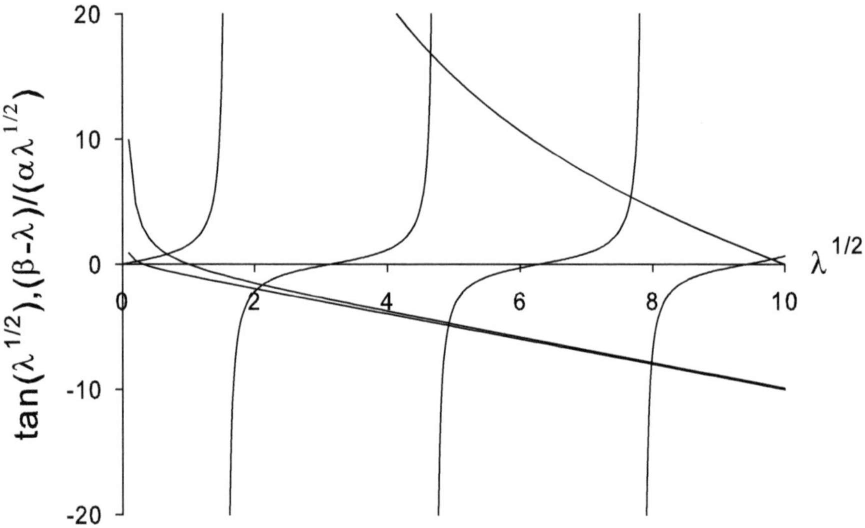

Fig. 10.10 Graphs of the eigenvalue equations for $\alpha = 1$ and $\beta = 0.1, 1, 100$.

This is solved to give

$$\langle \overline{C}, \overline{C}_n \rangle = \left\langle \begin{pmatrix} 1 \\ 0 \end{pmatrix}, \overline{C}_n \right\rangle \exp(-\lambda_n \tau)$$

and the Fourier series for the solution is

$$\overline{C} = \sum_{n=1}^{\infty} \frac{\left\langle \begin{pmatrix} 1 \\ 0 \end{pmatrix}, \overline{C}_n \right\rangle}{\langle \overline{C}_n, \overline{C}_n \rangle} \exp(-\lambda_n \tau) \overline{C}_n(\xi)$$

where

$$\begin{aligned} \langle \overline{C}_n, \overline{C}_n \rangle &= \int_0^1 \cos(\sqrt{\lambda_n}\xi)^2 \, d\xi + \cos(\sqrt{\lambda_n})^2 \\ &= \frac{1}{2} + \cos(\sqrt{\lambda_n})^2 \left(1 + \frac{\tan(\sqrt{\lambda_n})}{2\sqrt{\lambda_n}}\right) \end{aligned}$$

and

$$\left\langle \begin{pmatrix} 1 \\ 0 \end{pmatrix}, \overline{C}_n \right\rangle = \int_0^1 \cos(\sqrt{\lambda_n}\xi) \, d\xi + 0 \cos(\sqrt{\lambda_n}) = \frac{\sin(\sqrt{\lambda_n})}{\sqrt{\lambda_n}}$$

or, in its full glory

$$\begin{pmatrix} C(\xi, \tau) \\ C_w(\tau) \end{pmatrix} = \sum_{n=1}^{\infty} \frac{\sin(\sqrt{\lambda_n}) \exp(-\lambda_n \tau)}{\sqrt{\lambda_n} \left(\frac{1}{2} + \cos(\sqrt{\lambda_n})^2 \left(1 + \frac{\tan(\sqrt{\lambda_n})}{2\sqrt{\lambda_n}}\right)\right)} \begin{pmatrix} \cos(\sqrt{\lambda_n}\xi) \\ \cos(\sqrt{\lambda_n}) \end{pmatrix}.$$

Graphs of the solution are plotted in Fig. 10.11.

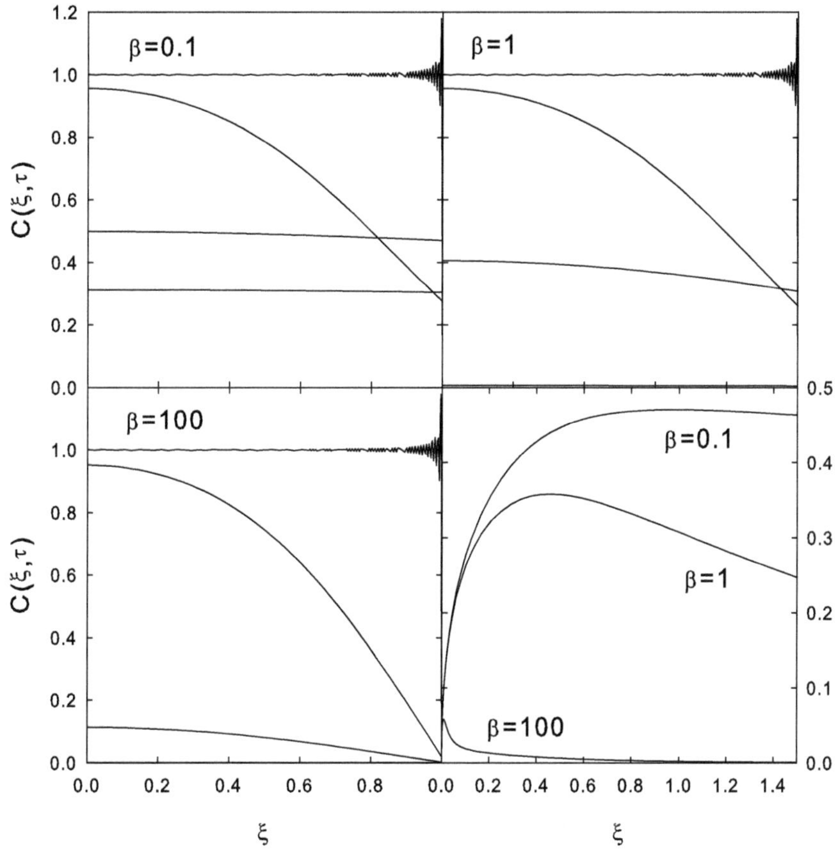

Fig. 10.11 Dimensionless concentrations in the solid and fluid during leaching. The solid profiles are shown for times $\tau = 0, 0.1, 1, 10$.

Small values of β (upper left figure) correspond to situations where the fluid flow rate through the system, Q is small and limits the rate of removal of the hazardous compound. When this is the case, diffusion through the solid is fast compared to the rate of removal so the solid profile first undergoes a rapid transition to an almost flat, quasi-steady state profile in which the solid and liquid concentrations are in equilibrium. This is followed by a much slower transient during which the hazardous compound is slowly washed out of the fluid phase while the solid concentration profile remains almost flat, in a quasi-steady state, as it slowly decreases towards zero.

Large values of β (lower left figure) correspond to large fluid flow rates and the fluid phase concentration therefore remains low throughout. The transient of the solid profile is roughly similar to the transient seen when the surface concentration is suddenly set to zero.

The concentrations in the fluid phase versus time are plotted in the figure at the lower right. They clearly exhibit the transitions just described. For $\beta = 0.1$ the concentration initially increases rapidly as the fluid and solid concentrations equilibrate. This is followed by the slow transient washout of the hazardous compound as the fluid is slowly replaced. The transient fluid concentration for $\beta = 100$ is low throughout as the hazardous compound cannot diffuse out of the solid fast enough to raise the concentration in the fluid relative to the rate of removal by the fluid flow through the system.

We conclude this section by solving the classical Graetz–Nusselt problem, the problem of finding the temperature profile in a fluid in laminar flow when the otherwise constant wall temperature or wall heat flux changes to a new value at some point. This problem has been discussed in many textbooks but with the assumption that the rate of conduction in the axial direction is negligible compared to the rate of convection. However, in 1980, the full problem without this simplifying assumption yielded to the inner product method and solutions were found for both constant wall temperature.[1] and constant wall flux.[2] The example describes the solution for constant wall temperature and closely follows the first of the papers just cited. The purpose of this example is not to just solve another problem by the inner product method, but to show how problems that look unsuited for this method because they appear not to be self-adjoint, may still be solved if one can figure out some way of rendering the problem self-adjoint in some appropriate Hilbert space. We will therefore only pursue the solution up to the point where a self-adjoint eigenvalue problem and a set of ordinary differential equations for the Fourier coefficients are obtained. The remaining steps are left as an exercise (admittedly not a trivial exercise, though). Readers may want to consult both papers cited above for additional details.

Example 10.10: The Graetz–Nusselt problem for constant wall temperature.

Consider a pipe through which a liquid is moving in laminar flow with the velocity profile

$$v_z(r) = v_m\left(1 - \left(\frac{r}{R}\right)^2\right)$$

where R is the radius of the pipe. The temperature of the wall is assumed constant and equal to T_0 for $z \leq 0$ and constant and equal to T_1 for $z > 0$. The fluid is assumed to enter the pipe at $z = -\infty$ with the constant temperature T_0. A cylindrical coordinate system is placed with the z-axis in the direction of flow and an energy balance is done on an annular region defined by dz and dr, yielding the

[1] Papoutsakis, E., Ramkrishna, D. and Lim, H.C. [1980] "The Extended Graetz Problem with Dirichlet wall Boundary Conditions," *Appl. Sci. Res.* **36**, 13–34.

[2] Papoutsakis, E., Ramkrishna, D. and Lim, H.C. [1980] "The Extended Graetz Problem with Prescribed Wall flux," *AIChE J.* **26**(5), 779–787.

model

$$\rho c_p v_m \left(1 - \left(\frac{r}{R}\right)^2\right) \frac{\partial T}{\partial z} = \frac{k}{r}\frac{\partial}{\partial r}\left(r\frac{\partial T}{\partial r}\right) + k\frac{\partial^2 T}{\partial z^2}$$

subject to

$$r = 0 \Rightarrow \frac{\partial T}{\partial r} = 0 \text{ or } T \text{ finite}$$

$$r = R \Rightarrow \begin{cases} T = T_0 \text{ for } z \leq 0 \\ T = T_1 \text{ for } z > 0 \end{cases}$$

$$\lim_{z \to -\infty} T = T_0 \,.$$

The following dimensionless variables are introduced

$$\Theta = \frac{T - T_1}{T_0 - T_1}\,, \; \zeta = \frac{z}{RPe}\,, \; \eta = \frac{r}{R}$$

where the Peclet number is $Pe = \frac{\rho c_p v_m R}{k}$. This gives the dimensionless problem

$$\left(1 - \eta^2\right) \frac{\partial \Theta}{\partial \zeta} = \frac{1}{\eta}\frac{\partial}{\partial \eta}\left(\eta \frac{\partial \Theta}{\partial \eta}\right) + \frac{1}{Pe^2}\frac{\partial^2 \Theta}{\partial \zeta^2}$$

$$\eta = 0 \Rightarrow \frac{\partial \Theta}{\partial \eta} = 0 \text{ or } \Theta \text{ finite}$$

$$\eta = 1 \Rightarrow \begin{cases} \Theta = 1 \text{ for } \zeta \leq 0 \\ \Theta = 0 \text{ for } \zeta > 0 \end{cases} \tag{10.6}$$

$$\lim_{\zeta \to -\infty} \Theta = 0 \,.$$

The central difficulty of this problem is figuring out how to rewrite the problem so that it becomes self-adjoint. This means both a recasting of the problem itself and determination of a Hilbert space in which the recast problem is self-adjoint. This is far from a trivial task and there is no recipe one can follow. Only physical and mathematical insight will work together with a willingness to try out ideas.

The idea presented in the above cited papers was to introduce a new variable, the *axial energy flow*, the energy flux from both conduction and convection

$$S(z, r) = \int_0^r \left(-k\frac{\partial T}{\partial z} + \rho c_p v_m \left(1 - \left(\frac{\tilde{r}}{R}\right)^2\right) T\right) 2\pi \tilde{r} d\tilde{r} \,.$$

Differentiation of this expression with respect to r gives

$$\frac{\partial S}{\partial r} = \left(-k\frac{\partial T}{\partial z} + \rho c_p v_m \left(1 - \left(\frac{\tilde{r}}{R}\right)^2\right) T\right) 2\pi r \,. \tag{10.7}$$

We will also need an expression for $\partial S/\partial z$. This is obtained from an energy balance on a disk of radius r, located between z and $z+dz$.

$$S(z,r) - S(z+dz,r) = -k\frac{\partial T}{\partial r}2\pi r dr \ \Rightarrow$$

$$\frac{\partial S}{\partial z} = k\frac{\partial T}{\partial r}2\pi r\,. \tag{10.8}$$

The dimensionless axial energy flow is defined as

$$\Sigma(\zeta,\eta) = \frac{S(z,r) - \int_0^r \rho c_p v_m \left(1-\left(\frac{\tilde{r}}{R}\right)^2\right) T 2\pi\tilde{r}d\tilde{r}}{\pi R^2 \rho c_p v_m (T_0 - T_1)} \ \Rightarrow$$

$$\Sigma(\zeta,\eta) = \int_0^\eta \left(-\frac{1}{Pe^2}\frac{\partial\Theta}{\partial\zeta} + (1-\tilde{\eta}^2)\Theta\right) 2\tilde{\eta}d\tilde{\eta}\,.$$

Recasting Eq. (10.7) as

$$\frac{\partial T}{\partial z} = \rho c_p v_m \left(1-\left(\frac{r}{R}\right)^2\right) T - \frac{1}{2\pi r}\frac{\partial S}{\partial r}$$

and making it dimensionless gives

$$\frac{\partial\Theta}{\partial\zeta} = Pe^2(1-\eta^2)\Theta - \frac{Pe^2}{2\eta}\frac{\partial\Sigma}{\partial\eta}$$

and the dimensionless version of Eq. (10.8) becomes

$$\frac{\partial\Sigma}{\partial\zeta} = 2\eta\frac{\partial\Theta}{\partial\eta}\,.$$

The two dimensionless equations can be written in vector-matrix notation as

$$\frac{\partial}{\partial\zeta}\begin{pmatrix}\Theta\\ \Sigma\end{pmatrix} = \begin{pmatrix}Pe^2(1-\eta^2) & -\frac{Pe^2}{2\eta}\frac{\partial}{\partial\eta}\\ 2\eta\frac{\partial}{\partial\eta} & 0\end{pmatrix}\begin{pmatrix}\Theta\\ \Sigma\end{pmatrix}. \tag{10.9}$$

The boundary conditions on Θ have already been stated. The boundary conditions on Σ are (convince yourself of this)

$$\Sigma(\zeta,0) = 0\ ,\ \forall\zeta \tag{10.10}$$

$$\lim_{\zeta\to\infty}\Sigma = 0\ ,\ 0<\eta<1$$

$$\lim_{\zeta\to-\infty}\Sigma = \int_0^\eta 2\tilde{\eta}(1-\tilde{\eta}^2)d\tilde{\eta}\,.$$

This concludes the recasting of the problem and what remains is to find a Hilbert space, i.e. an inner product, in which the problem is self-adjoint. The following inner product is proposed

$$\left\langle \begin{pmatrix}\Phi_1(\eta)\\ \Phi_2(\eta)\end{pmatrix}, \begin{pmatrix}\Psi_1(\eta)\\ \Psi_2(\eta)\end{pmatrix}\right\rangle = \int_0^1 \left(\frac{4}{Pe^2}\Phi_1(\eta)\Psi_1(\eta)\eta + \Phi_2(\eta)\Psi_2(\eta)\frac{1}{\eta}\right) d\eta\,.$$

A boundary condition $\Phi_2(0) = 0$ is obtained from the boundary condition on Σ, Eq. (10.10). A boundary condition on $\Phi_1(1)$ is formally obtained form the boundary condition on Θ, Eq. (10.6), and $\Phi_1(1)$ therefore equals 0 for $\zeta > 0$ and equals 1 for $\zeta \leq 0$. We will ignore this difficulty for the moment and use the boundary conditions $\Phi_1(1) = \Phi_2(0) = 0$ for our inner product. First check that the matrix operator in Eq. (10.9) is self-adjoint in this inner product.

$$\left\langle \begin{pmatrix} Pe^2(1-\eta^2) & -\frac{Pe^2}{2\eta}\frac{\partial}{\partial\eta} \\ 2\eta\frac{\partial}{\partial\eta} & 0 \end{pmatrix} \begin{pmatrix} \Phi_1(\eta) \\ \Phi_2(\eta) \end{pmatrix}, \begin{pmatrix} \Psi_1(\eta) \\ \Psi_2(\eta) \end{pmatrix} \right\rangle$$

$$= \int_0^1 \left(\frac{4}{Pe^2}\left(Pe^2(1-\eta^2)\Phi_1 - \frac{Pe^2}{2\eta}\frac{\partial \Phi_2}{\partial\eta} \right) \Psi_1\eta + 2\eta\frac{\partial\Phi_1}{\partial\eta}\Psi_2\frac{1}{\eta} \right) d\eta$$

$$= \int_0^1 \left(\frac{4}{Pe^2}\left(Pe^2(1-\eta^2)\Psi_1 - \frac{Pe^2}{2\eta}\frac{\partial \Psi_2}{\partial\eta} \right) \Phi_1\eta + 2\eta\frac{\partial\Psi_1}{\partial\eta}\Phi_2\frac{1}{\eta} \right) d\eta$$

$$- 2(\Phi_2(1)\Psi_1(1) - \Phi_2(0)\Psi_1(0) - \Phi_1(1)\Psi_2(1) + \Phi_1(0)\Psi_2(0))$$

$$= \left\langle \begin{pmatrix} \Phi_1(\eta) \\ \Phi_2(\eta) \end{pmatrix}, \begin{pmatrix} Pe^2(1-\eta^2) & -\frac{Pe^2}{2\eta}\frac{\partial}{\partial\eta} \\ 2\eta\frac{\partial}{\partial\eta} & 0 \end{pmatrix} \begin{pmatrix} \Psi_1(\eta) \\ \Psi_2(\eta) \end{pmatrix} \right\rangle .$$

Thus, the associated eigenvalue problem is

$$\begin{pmatrix} Pe^2(1-\eta^2) & -\frac{Pe^2}{2\eta}\frac{\partial}{\partial\eta} \\ 2\eta\frac{\partial}{\partial\eta} & 0 \end{pmatrix} \begin{pmatrix} \Theta_n(\eta) \\ \Sigma_n(\eta) \end{pmatrix} + \lambda_n \begin{pmatrix} \Theta_n(\eta) \\ \Sigma_n(\eta) \end{pmatrix} = 0$$

subject to boundary conditions already given. We will not pursue the solution to this eigenvalue problem here but proceed straight to the step of taking the inner product of Eq. (10.9) with an eigenfunction (Θ_n, Σ_n). As pointed out above, when $\zeta > 0$, the boundary condition on Φ is $\Phi(1) = 0$, the same as used in the definition of the inner product, so one obtains

$$\frac{d}{d\zeta}\left\langle \begin{pmatrix} \Theta \\ \Sigma \end{pmatrix}, \begin{pmatrix} \Theta_n \\ \Sigma_n \end{pmatrix} \right\rangle + \lambda_n \left\langle \begin{pmatrix} \Theta \\ \Sigma \end{pmatrix}, \begin{pmatrix} \Theta_n \\ \Sigma_n \end{pmatrix} \right\rangle = 0\ , \ \zeta > 0\ .$$

However, when $\zeta \leq 0$, taking the inner product gives rise to boundary terms giving

$$\frac{d}{d\zeta}\left\langle \begin{pmatrix} \Theta \\ \Sigma \end{pmatrix}, \begin{pmatrix} \Theta_n \\ \Sigma_n \end{pmatrix} \right\rangle + \lambda_n \left\langle \begin{pmatrix} \Theta \\ \Sigma \end{pmatrix}, \begin{pmatrix} \Theta_n \\ \Sigma_n \end{pmatrix} \right\rangle = 2\Phi_{n2}(1)\ , \ \zeta \leq 0\ .$$

At this stage, the problem is reduced to protocol, solving an eigenvalue problem and two ODEs for the coefficients in the Fourier series.

10.2 Finite Fourier transform

In the finite Fourier transform method or FFT,[3] a Fourier series solution is assumed for the PDE. The basis functions are chosen such that they are solutions to a formally self-adjoint operator in the PDE, but homogeneous boundary conditions are not required. At this point, one may be tempted to substitute the series into the PDE as a first step in determining the Fourier coefficients. However, Fourier series are generally not term by term differentiable, so this is not a permissible operation. Instead, one takes the inner product of the PDE with a basis function. Since the relevant operator is only required to be formally self-adjoint, carrying out the integration steps of the inner product usually results in creation of boundary terms in the same manner as when adjoint boundary conditions are calculated. These boundary terms are unknowns in the sense that they depend on the boundary values of the solution to the PDE, but they can be eliminated by an appropriate choice of boundary conditions for the basis functions. The method is perhaps best explained by an example.

Example 10.11: Conduction in a slab with heat generation.

Consider a semi-infinite slab of fixed thickness in which heat is generated as a function of time and position. In dimensionless form, the energy balance for the slab can be written

$$\frac{\partial \Theta}{\partial \tau} = \frac{\partial^2 \Theta}{\partial \xi^2} + F(\xi, \tau)$$

subject to the boundary and initial conditions

$$\xi = 0 \Rightarrow \Theta = \Theta_1(\tau)$$

$$\xi = 1 \Rightarrow \Theta = \Theta_2(\tau)$$

$$\tau = 0 \Rightarrow \Theta = \Theta_0(\xi)$$

where it is assumed that the surface temperatures can be different at the two sides of the slab.

The operator $\frac{\partial^2}{\partial \xi^2}$ is formally self-adjoint with the inner product

$$\langle f, g \rangle = \int_0^1 f(\xi) g(\xi)\, d\xi$$

so given an appropriate set of boundary conditions, the eigenfunctions of this operator will form an orthogonal basis that can be used in a Fourier series representation of Θ. Call these yet to be specified eigenfunctions $y_n(\xi)$ and take the inner product of y_n and the PDE

$$\left\langle \frac{\partial \Theta}{\partial \tau}, y_n(\xi) \right\rangle = \left\langle \frac{\partial^2 \Theta}{\partial \xi^2}, y_n(\xi) \right\rangle + \langle F(\xi), y_n(\xi) \rangle . \tag{10.11}$$

[3]The abbreviation FFT is also used for the Fast Fourier Transform, a method for data analysis.

The first inner product on the right hand side is evaluated using integration by parts, viz.

$$\begin{aligned}\left\langle \frac{\partial^2\Theta}{\partial\xi^2}, y_n(\xi)\right\rangle &= \int_0^1 \frac{\partial^2\Theta}{\partial\xi^2} y_n(\xi)\, d\xi \\ &= \left[\frac{\partial\Theta}{\partial\xi} y_n(\xi)\right]_0^1 - \left[\Theta\frac{dy_n}{d\xi}\right]_0^1 + \int_0^1 \Theta\frac{d^2y_n}{d\xi^2}\, d\xi \\ &= \left.\frac{\partial\Theta}{\partial\xi}\right|_{\xi=1} y_n(1) - \left.\frac{\partial\Theta}{\partial\xi}\right|_{\xi=0} y_n(0) \\ &\quad - \Theta_2(\tau)y_n'(1) + \Theta_1(\tau)y_n'(0) + \left\langle \Theta, \frac{d^2y_n}{d\xi^2}\right\rangle .\end{aligned}$$

The two first terms on the right hand side are not known. However, one can get rid of these terms simply by demanding that $y_n(1)$ and $y_n(0)$ both equal zero. This demand provides the boundary conditions that permits one to find the $y_n(\xi)$'s and the eigenvalue problem for $y_n(\xi)$ thus becomes

$$\frac{d^2y_n}{d\xi^2} + \lambda_n y_n = 0\ ,\ y_n(0) = y_n(1) = 0$$

giving the basis functions

$$y_n(\xi) = \sin(\sqrt{\lambda_n}\xi)\ ,\ \lambda_n = (n\pi)^2\ ,\ n \in \mathbb{Z}_+$$

which then permits evaluation of the inner product $\left\langle \frac{\partial^2\Theta}{\partial\xi^2}, y_n(\xi)\right\rangle$.

$$\left\langle \frac{\partial^2\Theta}{\partial\xi^2}, y_n(\xi)\right\rangle = -\Theta_2(\tau)\sqrt{\lambda_n}\cos(\sqrt{\lambda_n}) + \Theta_1(\tau)\sqrt{\lambda_n} - \lambda_n\langle\Theta, y_n\rangle$$

and Eq. (10.11) becomes

$$\frac{d}{d\tau}\langle\Theta, y_n\rangle = \langle F(\xi,\tau), y_n\rangle - (n\pi)^2\langle\Theta, y_n\rangle + \Theta_1(\tau)n\pi - \Theta_2(\tau)n\pi(-1)^n$$

which is an ODE in the inner product $\langle\Theta, y_n\rangle$. The solution to this ODE is

$$\begin{aligned}&\langle\Theta, y_n\rangle = \langle\Theta_0, y_n\rangle e^{-(n\pi)^2\tau} \\ &\quad + e^{-(n\pi)^2\tau}\int_0^\tau e^{(n\pi)^2 t}\left\{\langle F(\xi,t), y_n\rangle + \Theta_1(t)n\pi - \Theta_2(t)n\pi(-1)^n\right\} dt\,.\end{aligned}$$

Substitution of this result into the Fourier series for Θ gives

$$\begin{aligned}\Theta(\xi,\tau) &= \sum_{n=1}^\infty \frac{\langle\Theta_0(\xi), y_n(\xi)\rangle e^{-(n\pi)^2\tau}}{\langle y_n(\xi), y_n(\xi)\rangle} y_n(\xi) \\ &+ \sum_{n=1}^\infty \frac{y_n(\xi)e^{-(n\pi)^2\tau}}{\langle y_n(\xi), y_n(\xi)\rangle}\int_0^\tau e^{(n\pi)^2 t}\left\{\langle F(\xi,t), y_n\rangle + \Theta_1(t)n\pi - \Theta_2(t)n\pi(-1)^n\right\} dt\,.\end{aligned} \tag{10.12}$$

The first sum represents the transient that results from the initial condition. The terms in this series, and therefore the series itself, all go to zero as τ increases and in many situations this part of the solution can be ignored as irrelevant.

The second series represents the part of the solution that results from the forcing of the plate, the internal heating and the imposed surface temperatures.

Some graphs of the solutions are shown in Fig. 10.12.

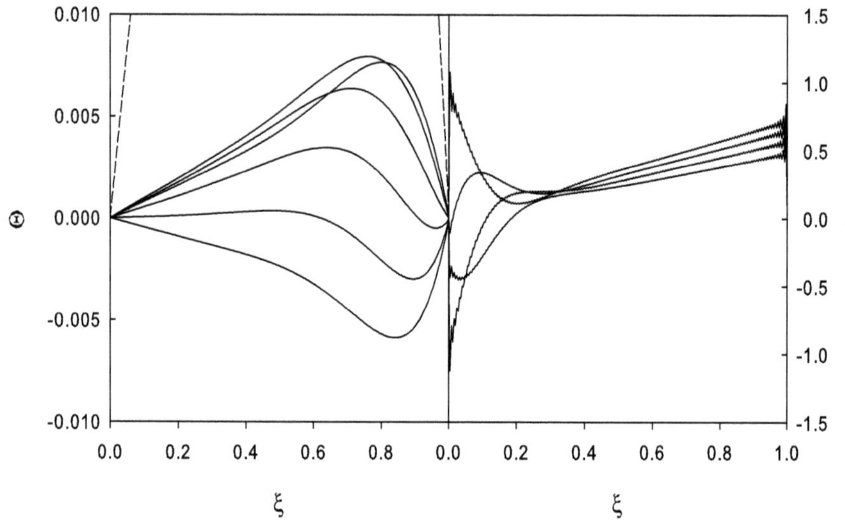

Fig. 10.12 Graphs of the solution in Eq. (10.12). On the left are plotted solutions at $\tau = 5, 6, 7, 8, 9$ and 10 when there is no forcing of the surface temperature, Θ_1 and Θ_2 are both 0, but when the internal heating rate is given by $F(\xi, \tau) = \xi \cos(100\tau)$. The frequency of this forcing is rapid compared to the characteristic time of conduction and the temperature profiles are therefore close to zero, the average heating rate, at all times. The steady state temperature profile when $F(\xi, \tau) = \xi$ is shown for comparison purposes by the broken line. On the right are graphs of the solution for $\tau = 5, 6, 7$ and 8 when there is no internal heating but when the surface temperatures are forced as $\Theta_1(\tau) = \sin(250\tau)$ and $\Theta_2(\tau) = \sin(0.1\tau)$. As expected, the fast forcing on the left surface does not penetrate far into the slab and has only an effect close to the surface. The forcing on the right hand side is slow enough to leave the system in quasi-steady state and the transient profiles, except for the region close to the rapidly forced left side, are close to straight lines corresponding to the steady state solutions for an imposed fixed temperature on the right surface.

Chapter 11

First Order PDEs

First order linear partial differential equations (PDEs) are equations of the form

$$\sum_{n=1}^{N} p_n(x) \cdot \frac{\partial C}{\partial x_n} = q(x) \cdot C(x) + r(x)$$

where x is a vector of the free variables, x_n. If the coefficient functions, p_ns are also functions of the dependent variable, C, then the equation is called semi-linear or quasi-linear. If any of the partial derivatives appear in a nonlinear fashion, then the equation is nonlinear.

Linear first order PDEs are solved using Cauchy's method or the method of characteristics. This is essentially a variable transformation which converts a PDE into a set of coupled ODEs. Cauchy's method is also the starting point for semi-linear and nonlinear equations. However, only single linear equations will be considered here.

11.1 Models that give rise to first order PDEs

First order PDEs occur in transport problems where transport is predominantly convective instead of dissipative, i.e. where transport is by fluid flow rather than by diffusion or conduction. For instance, chromatographic columns are designed to minimize mixing between fluid elements and are therefore typically modelled assuming only convective transport. First order PDEs are also used in modelling traffic flow and first order partial differential operators occur in models of particulate systems, so called population balance equations, such as models of aerosols, crystallizers or cell populations.

We start by giving some examples of models that give rise to first order PDEs.

Example 11.1: Chromatographic column.

To work well, chromatographic columns must be designed to minimize mixing or dispersion and it is therefore reasonable to model them assuming only convective transport. In a chromatographic column, there is a mobile fluid phase that transports the compounds of interest through the column and a stationary phase, the packing. Compounds partition themselves between these two phases and when they have different partition coefficients they are separated from one another as they move through the column, Fig. 11.1.

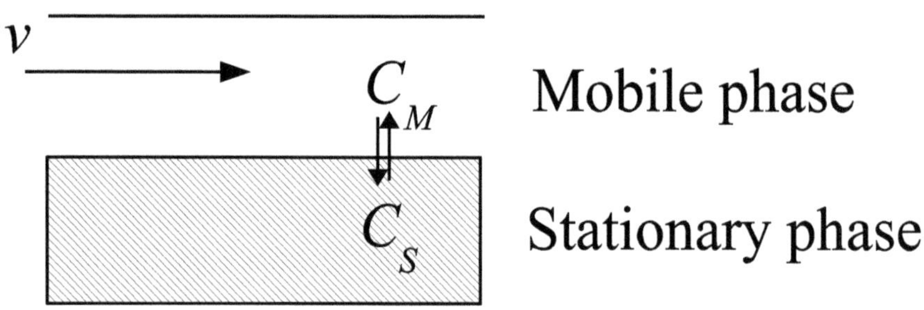

Fig. 11.1 Schematic of chromatographic column. The superficial velocity of and concentration in the mobile phase is v and C_M respectively. The concentration in the stationary phase is C_S.

Placing an x-axis in the direction of the flow, a balance on a compound gives

$$\epsilon v \frac{\partial C_M}{\partial x} + \epsilon \frac{\partial C_M}{\partial t} + (1-\epsilon)\frac{\partial C_S}{\partial t} = 0 \qquad (11.1)$$

where ϵ is the void fraction in the bed, v the interstitial velocity of the mobile phase and C_M and C_S the concentrations in the mobile and stationary phase respectively. The model assumes that the concentrations in the two phases are in equilibrium and that this equilibrium is not affected by the presence of other compounds in the column. If the equilibrium relationship is given by $C_S = f(C_M)$, then the balance becomes

$$\epsilon v \frac{\partial C_M}{\partial x} + \epsilon \frac{\partial C_M}{\partial t} + (1-\epsilon)\frac{df}{dC_M}\frac{\partial C_M}{\partial t} = 0$$

or, as it is commonly written

$$\epsilon v \frac{\partial C_M}{\partial x} + \left(\epsilon + (1-\epsilon)\frac{df}{dC_M}\right)\frac{\partial C_M}{\partial t} = 0$$

which is quasi-linear except in the special case where f' is not a function of C_M. To complete the model, the balance needs two side conditions, an initial condition specifying the concentration profile in the column at time zero and a condition specifying the inlet concentration versus time

$$C_M(t=0) = C_0(x)\,,\ C_M(x=0) = C_{\text{in}}(t).$$

The problem becomes particularly simple for dilute solutions for which we can assume that $C_S = KC_M$, giving

$$\epsilon v \frac{\partial C_M}{\partial x} + (\epsilon + (1-\epsilon)K)\frac{\partial C_M}{\partial t} = 0$$

which is not only linear but also has constant coefficient functions.

Anyone familiar with chromatography knows that, at least for dilute solutions, different compounds travel through the column at different constant speeds. It is therefore tempting to guess that the solution to the balance equation is such that the initial and inlet concentration profiles simply move through the column unchanged in shape and with some velocity less than the velocity of the mobile phase. Call this velocity v_c, then the assumed solution can be stated as

$$C_M(x,t) = C_0(x - v_c t)\ ,\ x > v_c t$$

and

$$C_M(x,t) = C_{\text{in}}\left(t - \frac{x}{v_c}\right)\ ,\ x < v_c t\,.$$

Notice that we are forced to work with two expressions for the solution, depending on whether the appropriate side condition is along the t- or x-axis. This is typical of the type of problems we will encounter. Substitution of either of these expressions in the balance equation confirms that the solutions are correct. E.g. for the first expression

$$vC_0' + (\epsilon + (1-\epsilon)K)C_0'(-v_c) \ \Rightarrow$$

$$v_c = \frac{v}{\epsilon + (1-\epsilon)K}\,.$$

The velocity v_c is called the *concentration wave velocity*. Remember, however, that this result is only valid for dilute solutions and under the assumption that the solutes do not interact.

The next examples are population balance equations, PBEs for short, first introduced in the example on page 68. These are models of particles such as droplets in an emulsion or an aerosol, living cells in a culture, crystals, polymers or even cars. Although population balance models describe a wide range of phenomena, all these models share a common mathematical structure. In all cases, what one seeks to find is the distribution of states, also referred to as the frequency. This can be the size distribution, the mass distribution or the age distribution. Distributions can be multi-dimensional such as the DNA-protein-carbohydrate concentration distribution in a cell culture or the length-width-height distribution of crystals.

Example 11.2: Traffic flow.

Consider a section of a road, no stop signs or lights and no side street, and place an x-axis along the road in the direction of the flow of traffic. The density of traffic can be described using the number distribution of cars

$$C(x,t) = \text{Number of cars per unit length of road at position } x \text{ at time } t\,.$$

Notice that $C(x,t)$ is essentially a concentration measure; number of cars per volume (which in this case is one-dimensional) at position x at time t. The velocity of a car, v, will generally depend on this density. Let

$$v(x,t,C) = \text{Velocity of cars at } (x,t) \text{ given the density } C.$$

A balance on the total number of cars between positions x_1 and x_2 gives

$$\frac{d}{dt}\int_{x_1}^{x_2} C(x,t)\,dx = v(x_1,t,C(x_1,t)) - v(x_2,t,C(x_2,t)).$$

The term on the left hand side is the accumulation term and the two terms on the right hand side represent the flux of cars in and out of the control volume. The expressions are typical of the way fluxes are expressed in these balances, as the product of the frequency and a convective term, in this case the velocity of the cars. The balance is easily rewritten as

$$\int_{x_1}^{x_2} \frac{\partial C}{\partial t}\,dx = -\int_{x_1}^{x_2} \frac{\partial vC}{\partial x}\,dx \;\Rightarrow$$

$$\int_{x_1}^{x_2} \left\{\frac{\partial C}{\partial t} + \frac{\partial vC}{\partial x}\right\} dx = 0\,.$$

Since the CV is completely arbitrary, this last equation can only be true if the integrand is zero

$$\frac{\partial C}{\partial t} + \frac{\partial vC}{\partial x} = 0$$

or, in the standard form

$$\frac{\partial C}{\partial t} + v\frac{\partial C}{\partial x} = -\frac{\partial v}{\partial x}C.$$

This is the population balance equation for this simple traffic flow situation. The solution properties depend strongly on the kind of dependence of v on the density C. As for the chromatographic column, the side conditions for this model has two parts. The initial density on the road at $C(x,0)$ and the density of cars as they enter the road $C(0,t)$.

Example 11.3: The age population balance in a cell culture.

Growing cells pass through a cell cycle in which larger cells divide to form two smaller cells which in turn grow and divide following the same pattern. To derive a model of the age distribution in a growing culture, we will therefore need to introduce some function which describes this division process. We define the age distribution as

$$C(a,t) = \text{Frequency of cells with age } a \in [a, a+da] \text{ per reactor volume}.$$

A cell number balance will be done on ages between a_1 and $a_2 > a_1$ for a culture in a CSTR with constant volume V and volumetric flow rate Q. The feed to the CSTR will be assumed sterile

$$\frac{d}{dt}\int_{a_1}^{a_2} VC(a,t)\,da = 1{\cdot}VC(a_1,t) - 1{\cdot}VC(a_2,t) - Q\int_{a_1}^{a_2} C(a,t)\,da - \int_{a_1}^{a_2} \Gamma(a)VC\,da\,.$$

The term on the left hand side represents accumulation of cells. The two first terms on the right hand side represent the cell fluxes due to growth. As always, these equal the distribution multiplied by the convective rate, in this case the age growth rate which equals unity and, since the balance is done over the entire reactor, a factor equal to the reactor volume V. The third term on the right hand side represents removal of cells by the exit stream and the last term represents removal of cells from the age interval $[a_1, a_2]$ by division of cells. The division rate is modelled by the function $\Gamma(a)$ which equals the rate at which cells with the age a divides. More formally, this function is defined as

$$\Gamma(a)dt = \text{Probability that a cell with age } a \text{ will divide next } dt \text{ time step.}$$

$\Gamma(a)$ is know as the division intensity or modulus or, for particles that are not living cells, the breakage intensity. Rearranging the balance equation and introducing the dilution rate $D = Q/V$, one obtains

$$\int_{a_1}^{a_2} \left(\frac{\partial C}{\partial t} + \frac{\partial C}{\partial a} + DC + \Gamma(a)C\right) da = 0 \;\Rightarrow$$

$$\frac{\partial C}{\partial t} + \frac{\partial C}{\partial a} = -(D + \Gamma(a))C.$$

The side conditions for this balance are the specified initial distribution $C(a,0)$ and the value of the distribution at age zero, $C(0,t)$. The latter must be calculated from a cell balance. Taking the point $a = 0$ as a control volume, a balance states that the cell flux out of the point, $1 \cdot C(0,t)$, equals twice (since each cell division gives rise to two new cells) the total rate of cell division. The latter equals twice the rate of cell division of cells with age a integrated over all ages. Thus

$$C(0,t) = 2\int_0^\infty \Gamma(a)C(a,t)\,da\,.$$

This boundary condition is usually referred to as the *renewal equation.*

11.2 First order PDE and Cauchy's method

Cauchy's method uses a variable transformation to convert the PDE into a set of ODEs and in some cases the variable transformation is so obvious that it is carried out almost automatically, without much thought. We will start by presenting such a case, a plug flow reactor with a constant volume reaction. A PDE model is derived using a stationary CV and the solution to this model is obtained by writing an equally valid model over a control volume that moves with the fluid, thereby obtaining an ODE model. This transformation from a stationary CV to a moving CV is at the heart of Cauchy's method.

A plug flow reactor is a flow through reactor in which the fluid moves with a flat velocity profile and without any mixing or exchange of matter between fluid elements. It is a reasonable model to use when convective transport dominates over diffusive or dispersive transport mechanisms.

To derive a model for a plug flow reactor we will first use a CV fixed in space, i.e. an Eulerian description. Place a x-axis along the reactor axis with an origin at the reactor inlet and pointing in the direction of the fluid flow. Let the fluid velocity be v and let the specific rate of the reaction that consumes the reactant be $r(C)$, where C is the reactant concentration. Do a balance on a thin slice of thickness dx and after the usual manipulations obtain the model

$$\frac{\partial C}{\partial t} + v\frac{\partial C}{\partial x} = -r(C). \tag{11.2}$$

Alternatively, one can write a balance on a differential control volume which moves with the fluid flow, i.e. a Lagrangian description. In this case, the balance on the CV is effectively equivalent to a balance on a differentially small bath reactor. The result is an ordinary differential equation,

$$\frac{dC}{dt} = -r(C)$$

where t now refers to the time elapsed since the fluid element entered the reactor. It is trivial to solve this model. Assuming that the fluid element enters the reactor at time $t_{in} > 0$, the solution is

$$C(t) = C_{inlet}(t_{in})e^{-k(t-t_{in})}.$$

We would of course like to eliminate t_{in} and write the solution in terms of t and x only. To do that, note that the position, x, of the fluid element is related to the time at which the fluid element entered the reactor, t_{in}, by

$$x = v(t - t_{in}) \Rightarrow t_{in} = t - \frac{x}{v}$$

and therefore

$$C(t, x) = C_{inlet}(t - x/v)e^{-\frac{kx}{v}}.$$

If, on the other hand, the fluid element in question was already inside the reactor at time zero, the solution to the Lagrangian balance is

$$C(t) = C_0(x_{initial})e^{-kt}$$

where $C_0(x)$ is the concentration profile in the reactor at time zero. And, since

$$x_{initial} = x - vt$$

this solution can be written as

$$C(t, x) = C_0(x - vt)e^{-kt}.$$

The two solutions are summarized as

$$C(t,x) = \begin{cases} C_{inlet}(t - x/v)e^{-\frac{kx}{v}} \ , \ x < vt \\ C_0(x - vt)e^{-kt} \ , \ x > vt \end{cases}.$$

This, one can easily confirm, is a solution to the Eulerian model, Eq. (11.2) and it was obtained by switching to a CV that moved with the fluid elements. In this case, it was trivial to find these trajectories because the fluid velocity was constant, independent of both time, position and reactant concentration.

For this example, the ODE which was solved along the fluid trajectories, as well as the fluid trajectories themselves, were obtained from physical insight into the problem. In the general case a very similar solution procedure is used: an ODE is found which must be solved along a family of curves in the state space. For some first order PDEs, one may be able to reformulate the problem in this way using physical insight, after all, the reformulation is only a change from an Eulerian description to a Lagrangian description. However, since we cannot rely on being able to do this all the time, we will now look at a systematic way of converting a first order PDE problem to a problem involving only ODEs. In a sense, we will try and abstract the process which converts an Eulerian model to a Lagrangian model.

11.2.1 Cauchy's method for linear equations

The general form of a two-dimensional, first order, linear PDE can be written

$$p_t(t,x)\frac{\partial c}{\partial t} + p_x(t,x)\frac{\partial c}{\partial x} = q(t,x)c + r(t,x). \tag{11.3}$$

If one interprets the free variables t as time and x as space, it is tempting to divide through by $p_t(t,x)$ and interpret the term p_x/p_t as the velocity that a CV must have in a Lagrangian description in which case the position of the control volume is the solution to $dx/dt = p_x/p_t$. However, such an interpretation is not possible in all cases and an explicit solution for $x(t)$ may not even exist in the general case. A different approach is therefore called for. We will still convert the PDE to an ODE which is solved along some set of curves in (t,x) space, but these curves will be given by a parametrically as opposed to an explicit expression for $t(x)$ or $x(t)$. For instance, for the plug flow reactor, the CV that moves with the flow follows trajectories given by

$$t = s \ , \ x = vs + x_0$$

where s is a parameter and x_0 is the position of the CV when $s = 0$, i.e. when $t = 0$. Parametric representations of curves are not unique and one can equally well write

$$t = \frac{s}{v} + t_0 \ , \ x = s$$

where t_0 is the time at which the CV is in position $x = 0$. For the plug flow reactor, the parametric representation of the CV trajectory is formally obtained by solving the two coupled ODEs

$$\frac{dt}{ds} = 1 \ \text{ and } \ \frac{dx}{ds} = v$$

or for the general equation by solving

$$\frac{dt}{ds} = p_t(t,x) \ \text{ and } \ \frac{dx}{ds} = p_x(t,x).$$

The solutions to these equations are called the *characteristic base curves* and the boundary conditions that the differential equations must satisfy are obtained form the initial and/or boundary conditions of the PDE. Initial and boundary conditions are treated in an identical fashion and usually they are both referred to as initial conditions. Let these be specified along a curve in the (t,x)-plane. This curve is called the *initial manifold*, it will be indicated $\mathcal{M}$ and it will be specified parametrically. Again, parametric representations of curves are not unique and one should simply pick the representation which is the most convenient. As an example consider the entire x-axis for which a possible parameterization is

$$\mathcal{M}:\ x = \tau\ ,\ t = 0\ ,\ \tau \in \mathbb{R}$$

where τ is now the parameter. Notice that in this parameterization τ is the distance along the x-axis and parameterizations for which the parameter is a measure of the distance along the curve are called *natural parameterizations.* However, an equally valid parameterization of the x-axis is

$$\mathcal{M}:\ x = \tau^3\ ,\ t = 0\ ,\ \tau \in \mathbb{R}\,.$$

while an invalid parameterization is

$$\mathcal{M}:\ x = \tau^2\ ,\ t = 0\ ,\ \tau \in \mathbb{R}\,.$$

This parameterization is invalid because, as τ runs through all the real numbers, x only runs through the non-negative real numbers, not the entire x-axis. However, it is a valid parameterization of the positive x-axis even if τ is restricted to $\mathbb{R}_+$. As another example, let the initial manifold be the branch of the hyperbola $x = 1/t$ for which both x and t are positive. This can be parameterized as

$$\mathcal{M}:\ x = \tau\ ,\ t = 1/\tau\ ,\ \tau \in \mathbb{R}_+$$

or as

$$\mathcal{M}:\ x = \tau^2\ ,\ t = 1/\tau^2\ ,\ \tau \in \mathbb{R}_+\,.$$

Once a parameterization of the initial manifold has been chosen, the initial condition in the dependent variable can formally be written as

$$C(t(\tau), x(\tau)) = C_0(\tau)$$

along $\mathcal{M}$. The initial conditions for the two coupled differential equations for the characteristic base curves are given at $s = 0$ as

$$\frac{dt}{ds} = p_t(t, x)\ ,\ t(0) = t(\tau)$$

$$\frac{dx}{ds} = p_x(t, x)\ ,\ x(0) = x(\tau).$$

In other words, the parameter s is set equal to zero along the initial manifold where the values of t and x are given in terms of the parameter τ. A solution for the characteristic base curves is therefore a family of s-parameterized curves where τ is used to specify a particular member of this family of curves. The parameterized problem can be viewed as a variable transformation from the (t, x) coordinate system to a (τ, s) coordinate system, Fig. 11.2.

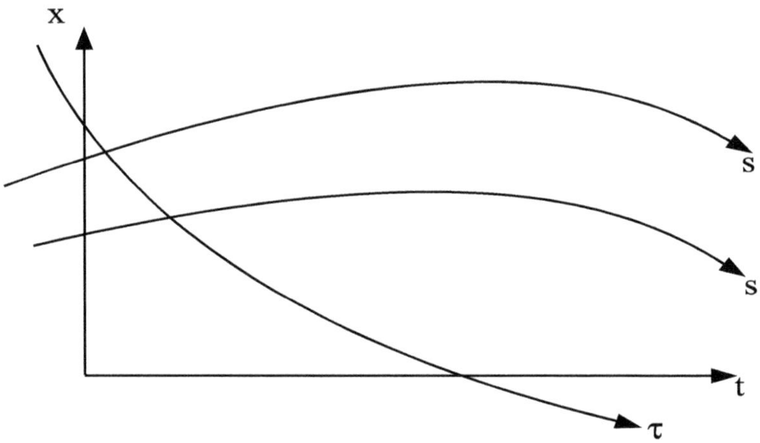

Fig. 11.2 The (τ, s) coordinate system superimposed on the original (t, x) coordinate system.

To solve the problem in the (τ, s) coordinates, one must integrate a differential equation along the characteristic base curves (parameterized by s) using the value of the dependent variable along the initial manifold as the initial condition. Looking at Fig. 11.2, a few things become immediately obvious. First of all, a solution is only possible in the domain that is traversed by characteristic base curves which intersect the initial manifold. Characteristic base curves that do not intersect the initial manifold, are not associated with an initial condition and therefore have no solution. Secondly, if the initial manifold intersects the same characteristic base curves multiple times, the problem may not be well posed. It may not be well posed because the solution at a second intersection point can be found two ways, from the specified initial condition on the initial manifold and from the solution obtained by integration along the characteristic curve starting at a different intersection point. Only in the special cases where these solutions agree is the solution well defined. Finally, the solution may not be well defined at points where the characteristic curves intersect because solutions along different intersecting characteristics are different. When this happens, a unique solution can sometimes be found by introducing discontinuities or shocks in the solution.

The characteristic base curves are solutions to the first order differential equation

$$\frac{dx}{dt} = \frac{p_x(t,x)}{p_t(t,x)}$$

and the fundamental existence theorem of these equations state that a unique solution through a point (t_0, x_0) exists if the function on the right hand side is bounded and satisfies a Lipschitz condition. Most functions in engineering models satisfy these requirements so the models will have unique solutions.

After this qualitative description of the solution method, we can now state Cauchy's method rigorously.

Define the three characteristic equations

$$\frac{dt}{ds} = p_t(t(s), x(s)) \ , \ t(s=0) = t(\tau)$$

$$\frac{dx}{ds} = p_x(t(s), x(s)) \ , \ x(s=0) = x(\tau)$$

$$\frac{dC}{ds} = r(t(s), x(s))C(s) + f(t(s), x(s)) \ , \ C(0) = C_0(\tau)$$

where the functions p_t, p_x, r and f are those appearing in Eq. (11.3). The first two characteristic equations are the base equations. The last characteristic equation is slightly different from the base equations in that the right hand side of this equation cannot be written out explicitly until an explicit solution for the variables $t(s, \tau)$ and $x(s, \tau)$ has been found. The solution to Eq. (11.3) is obtained by solving these three equations and eliminating τ and s.

The proof that this does indeed give the solution to the PDE is straight forward and is simply a check that the solution of the PDE satisfies the characteristic equations. Write this solution as $C(t(s), x(s)) = C(s)$ and use the chain rule to get

$$\frac{dC}{ds} = \frac{\partial C}{\partial t}\frac{dt}{ds} + \frac{\partial C}{\partial x}\frac{dx}{ds} .$$

Then, using the base equations, one obtains

$$\frac{dC}{ds} = \frac{\partial C}{\partial t} p_t(t(s), x(s)) + \frac{\partial C}{\partial x} p_x(t(s), x(s))$$

but since $C(t, x)$ is a solution to the PDE, the right hand side can be rewritten to obtain

$$\frac{dC}{ds} = r(t(s), x(s))c + f(t(s), x(s))$$

which is the third characteristic equation, the defining equation for $C(s)$. Thus, $C(t(s), x(s)) = C(s)$ is a solution to the PDE initial value problem, provided that $C(t(0), x(0)) = C_0(\tau)$.

To eliminate s and τ from the problem, one must solve the characteristic base equations to obtain t and x as functions of s and τ. I.e. find

$$x = x(s, \tau) \ , \ t = t(s, \tau).$$

Invert these equations to obtain

$$\tau = \tau(t, x)\ ,\ s = s(t, x)$$

and substitute this result into the expression for $C(s, \tau)$

$$C(t, x) = C(s(t, x), \tau(t, x))$$

which concludes the proof.

Cauchy's method is easily generalized to problems of any dimensionality. We will state this general procedure before considering specific examples.

Given

$$\sum_{n=1}^{N} p_n(x) \frac{\partial C}{\partial x_n} = r(x)C(x) + f(x)$$

where x is a vector of the free variables, the x_ns. The initial values of $C(x)$ are given along the initial manifold $\mathcal{M}$. The solution can be obtained as follows.

i) Parameterize the initial manifold and initial condition by the parameters $\tau_1, \tau_2, \ldots, \tau_{N-1}$. We can formally write this as

$$\mathcal{M}:\ (x_1(\tau), x_2(\tau), \ldots, x_N(\tau))\ ,\ C_0 = C_0(\tau)$$

where τ is the vector of parameters, $\tau = (\tau_1, \tau_2, \ldots, \tau_{N-1})$.

ii) Write and solve the N coupled ODEs which constitute the characteristic base equations using the initial conditions given along the initial manifold

$$\frac{dx_n}{ds} = p_n(X(s))\ ,\ x_n(s = 0, \tau) = x_n(\tau).$$

iii) Invert these equations to obtain

$$s = s(x)\ ,\ \tau = \tau(x).$$

Inverting the solutions for the x_ns means that one must solve N coupled equations for the N unknowns, s and τ_1 through τ_{N-1}. If the dimensionality of the problem is large, this can be a very difficult problem and may in fact be the hardest step in obtaining the solution to the PDE.

iv) Solve the ODE

$$\frac{dC}{ds} = r(x(s,\tau))C(s,\tau) + f(X(s,\tau))$$

subject to the initial condition

$$C(s=0,\tau) = C_0(\tau).$$

v) The solution of the PDE then is

$$C(x) = C(s(x),\tau(x)).$$

Example 11.4: Plug flow reactor by Cauchy's method.

For first order reaction kinetics, the concentration of a reactant in a plug flow reactor is given by

$$\frac{\partial C}{\partial t} + v\frac{\partial C}{\partial x} = -kC.$$

The initial and boundary conditions are

$$t = 0 \Rightarrow C(0,x) = C_0(x)$$

$$x = 0 \Rightarrow C(t,0) = C_{in}(t).$$

The initial manifold is the union of the positive x-axis and positive t-axis. This is most easily parameterized in two steps, parameterizing each axis separately. Starting with the x-axis, we have the initial manifold and initial condition

$$\mathcal{M} : t = 0 \ , \ x = \tau \ , \ \tau \in \mathbb{R}_+ \ , \ C = C_0(\tau).$$

The characteristic equations are

$$\frac{dt}{ds} = 1 \ , \ t(0) = 0$$

$$\frac{dx}{ds} = v \ , \ x(0) = \tau$$

$$\frac{dC}{ds} = -kc \ , \ C(0) = C_0(\tau).$$

These are solved to give

$$t = s\ ,\ x = vs + \tau\ ,\ C(s,\tau) = C_0(\tau)e^{-ks}.$$

Invert the first two solutions to find that

$$s = t\ ,\ \tau = x - vt$$

and substitute this into the expression for $C(s,\tau)$

$$C(x,t) = C_0(x - vt)e^{-kt}.$$

Repeat this procedure for the initial condition given at the reactor inlet. Parametrize the initial condition as

$$\mathcal{M} : x = 0\ ,\ t = \tau\ ,\ \tau \geq 0\ ,\ C = C_{in}(\tau).$$

Write and solve the characteristic equations

$$\frac{dx}{ds} = v\ ,\ \frac{dt}{ds} = 1\ ,\ \frac{dC}{ds} = -kC \Rightarrow$$

$$x = vs\ ,\ t = s + \tau\ ,\ C(s,\tau) = C_{in}(\tau)e^{-ks}.$$

Solve for s and τ and substitute into the expression for $C(s,\tau)$

$$s = \frac{x}{v}\ ,\ \tau = t - \frac{x}{v}\ ,\ C(x,t) = C_{in}\left(t - \frac{x}{v}\right)e^{-k\frac{x}{v}}.$$

Example 11.5: Particle growth without breakage and nucleation.

When a solid precipitates out of solution under gentle conditions, breakage of the crystals does not occur and the mass population balance in a batch crystallizer can be written

$$\frac{\partial C}{\partial t} + \frac{\partial r(m)C}{\partial m} = 0$$

where $C(t,m)$ is the particle number distribution. I.e. $C(t,m)dm$ is the concentration of particles with mass between m and $m + dm$ at time t. The function $r(m)$ is the rate of growth of a particle of mass m. We will find the transient solution for three different models of $r(m)$; zero order kinetics for which the characteristic curves are parallel lines, first order kinetics for which the characteristics diverge and a rate expression of the form $r = e^{-m}$ for which the characteristics converge. In all cases, we assume the initial condition the $C(0,m) = C_0(m)$.

Case I: $r = k$. The model becomes

$$\frac{\partial C}{\partial t} + k\frac{\partial C}{\partial m} = 0\,.$$

Parameterize the positive m-axis as

$$\mathcal{M} : m = \tau\;,\; t = 0\,.$$

The base equations are

$$\left.\begin{array}{l} \frac{dt}{ds} = 1\;,\; t(0) = 0 \;\Rightarrow\; t = s \\ \frac{dm}{ds} = k\;,\; m(0) = \tau \;\Rightarrow\; m = ks + \tau \end{array}\right\} \Rightarrow \begin{array}{l} s = t \\ \tau = m - kt \end{array}.$$

The last characteristic equation is

$$\frac{dC}{ds} = 0\; C(0) = C_0(\tau) \;\Rightarrow\; C = C_0(\tau).$$

Substituting in the expressions for s and τ gives

$$C(t, m) = C_0(m - kt)\;,\; m > kt\,. \tag{11.4}$$

Note that this solution is only valid along the characteristics that intersect the positive m axis. Thus the constraint that $m > kt$. The solution for $m < kt$ can be found from the characteristics that intersect the t-axis if an initial condition along this axis is specified, in other words, if the nucleation rate is specified. The solution in Eq. (11.4) is plotted for several values of kt in Fig. 11.3.

The dynamics is quite dull. The initial distribution simply translates down the m-axis with the constant speed k without changing shape.

Case II: $r(m) = \nu m$.

The population balance becomes

$$\frac{\partial C}{\partial t} + \nu m\frac{\partial C}{\partial m} = -\nu C.$$

Parameterization of the positive m-axis is as above and the base equations are

$$\left.\begin{array}{l} \frac{dt}{ds} = 1\;,\; t(0) = 0 \;\Rightarrow\; t = s \\ \frac{dm}{ds} = \nu m\;,\; m(0) = \tau \;\Rightarrow\; m = \tau e^{\nu s} \end{array}\right\} \Rightarrow \begin{array}{l} s = t \\ \tau = me^{-\nu t} \end{array}.$$

Last characteristic equation

$$\frac{dC}{ds} = -\nu C\;,\; C(0) = C_0(\tau) \;\Rightarrow\; C(s, \tau) = C_0(\tau)e^{-\nu s}$$

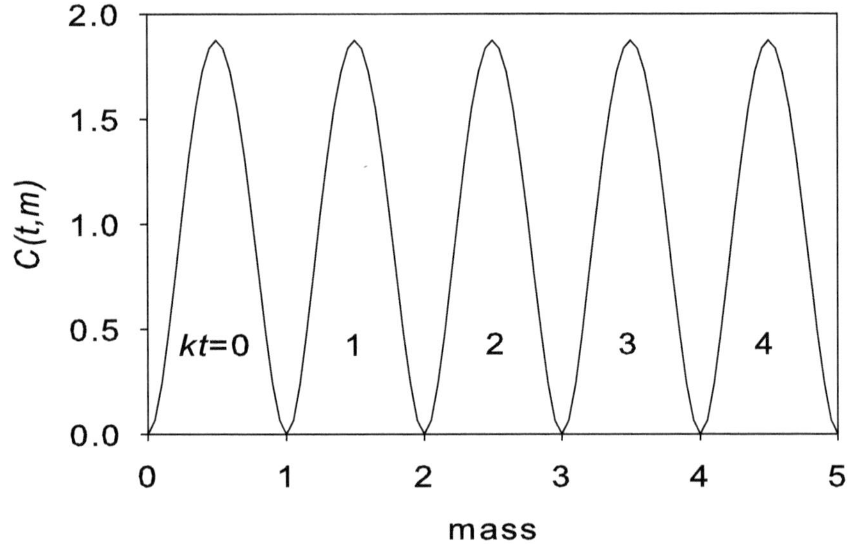

Fig. 11.3 Plots of the solution to Eq. (11.4) for $kt = 0, 1, 2, 3, 4$. The initial distribution is $C_0(m) = 30m^2(1-m)^2$.

or

$$C(t,m) = C_0(me^{-\nu t})e^{-\nu t}. \tag{11.5}$$

Notice that this solution is valid for all positive times and masses because the characteristics that intersect the initial manifold, the positive t-axis, pass through all points in this domain. The characteristics diverge and the distribution therefore broadens with time. This broadening "stretches" the initial distribution and results in a decrease in the magnitude of the frequency as shown in Fig. 11.4.

Case III: $r(m) = e^{-m}$.

The population balance becomes

$$\frac{\partial C}{\partial t} + e^{-m}\frac{\partial C}{\partial m} = e^{-m}C.$$

Parameterizing the m-axis as before the base equations are

$$\left.\begin{array}{l} \frac{dt}{ds} = 1\ ,\ t(0) = 0\ \Rightarrow\ t = s \\ \frac{dm}{ds} = e^{-m}\ ,\ m(0) = \tau\ \Rightarrow\ m = \ln(s + e^{\tau}) \end{array}\right\} \Rightarrow \begin{array}{c} s = t \\ \tau = \ln(e^m - t) \end{array}.$$

The last characteristic equation is

$$\frac{dC}{ds} = e^{-m}C\ ,\ C(0) = C_0(\tau).$$

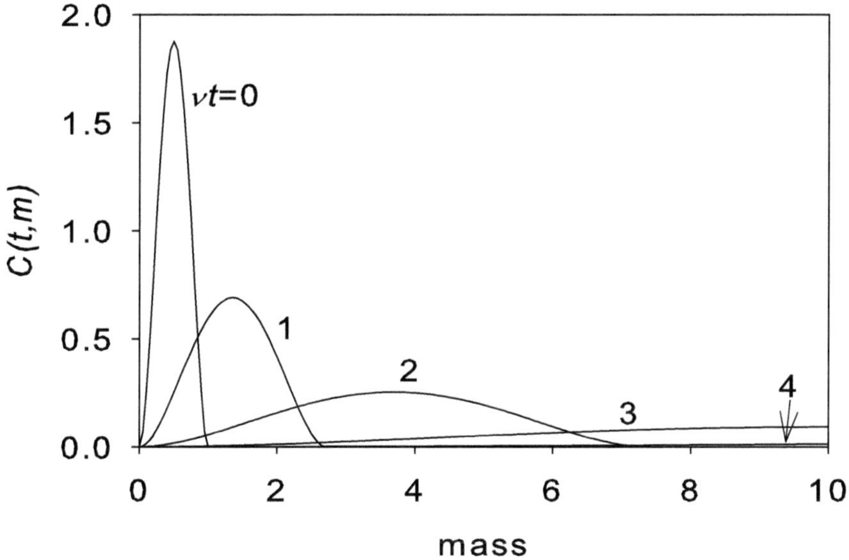

Fig. 11.4 Plots of the solution to Eq. (11.4) for $\nu t = 0, 1, 2, 3, 4$. The initial distribution is $C_0(m) = 30m^2(1-m)^2$.

This equation has a right hand side that is a function of m. We must therefore substitute in the expression for $m(s)$ found in the solution of the base equations.

$$\frac{dC}{ds} = e^{-\ln(s+e^\tau)}C = \frac{C}{s+e^\tau} \Rightarrow C(s,\tau) = C_0(\tau)\frac{s+e^\tau}{e^\tau}$$

or

$$C(t,m) = C_0(\ln(e^m - t))\frac{e^m}{e^m - t}\ ,\ m > \ln(1+t). \tag{11.6}$$

Notice the restriction of the solution to $m > \ln(t+1)$, the domain covered by the characteristics that intersect the positive m-axis. As before, the solution in the rest of the domain is zero as we assume no nucleation. This solution is plotted in Fig. 11.5. Notice how the converging characteristics lead to a narrowing of the distribution which causes an increase in the magnitude of the distribution.

The next example is not just another illustration of Cauchy's method, but also introduces the concept of regularity conditions and substrate balances.

Example 11.6: Dissolution of particles.

Consider a population of spherical particles that dissolve into a well mixed solvent. Assuming that dissolution is mass transfer limited, the rate of dissolution in

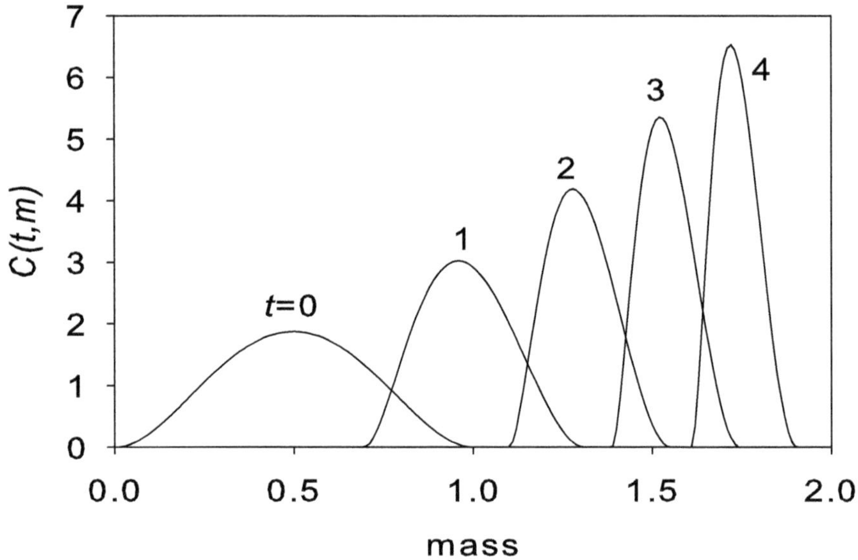

Fig. 11.5 Plots of the solution to Eq. (11.6) for $t = 0, 1, 2, 3, 4$. The initial distribution is $C_0(m) = 30m^2(1-m)^2$.

terms of particle mass can be written as

$$r(m) = -\alpha m^{2/3}(C_{\text{sat}} - C_{\text{liquid}})$$

where C_{sat} is the saturation concentration of the solute and C_{liquid} the concentration in the bulk liquid. Assuming that the particle concentration is low, the bulk concentration of the solute will not change significantly as the particles dissolve and C_{liquid} can then be assumed constant. We will also assume that C_{liquid} equals zero, giving the population balance

$$\frac{\partial C}{\partial t} - \alpha C_{\text{sat}} \frac{\partial m^{2/3} C}{\partial m} = 0$$

or

$$\frac{\partial C}{\partial t} - \alpha C_{\text{sat}} m^{2/3} \frac{\partial C}{\partial m} = \frac{2\alpha C_{\text{sat}}}{3} m^{-1/3} C$$

subject to the initial condition $C(0, m) = C_0(m)$. The initial manifold is the positive m-axis. The characteristics are the curves in the (t, m)-plane that particles follow as they dissolve and no initial condition is therefore needed along the time axis because all the characteristics that intersect the positive m-axis also intersect the positive time axis at the time when the particle in question disappears. Parameterize the initial manifold as

$$\mathcal{M}: \; m = \tau \,, \; t = 0$$

and the base equations become

$$\left.\begin{array}{l}\frac{dt}{ds}=1,\ t(0)=0\Rightarrow t=s\\ \frac{dm}{ds}=-\alpha C_{\rm sat}m^{2/3},\ m(0)=\tau\Rightarrow m=\left(\tau^{1/3}-\frac{\alpha C_{\rm sat}s}{3}\right)^3\end{array}\right\}\Rightarrow\begin{array}{c}s=t\\ \tau=\left(m^{1/3}+\frac{\alpha C_{\rm sat}t}{3}\right)^3\end{array}.$$

Finally

$$\begin{aligned}\frac{dC}{ds}&=\frac{2\alpha C_{\rm sat}}{3}m^{-1/3}C\\ &=\frac{2\alpha C_{\rm sat}C}{3(\tau^{1/3}-\alpha C_{\rm sat}s/3)}\,,\ C(0)=C_0(\tau)\ \Rightarrow\\ C(s,\tau)&=\frac{9\tau^{2/3}C_0(\tau)}{(\alpha C_{\rm sat}s-3\tau^{1/3})^2}\ \Rightarrow\\ C(t,m)&=C_0\left(\left(m^{1/3}+\frac{\alpha C_{\rm sat}t}{3}\right)^3\right)\frac{(m^{1/3}+\frac{\alpha C_{\rm sat}t}{3})^2}{m^{2/3}}\,.\end{aligned}$$

Notice that, for $t>0$, this solution has a singularity at $m=0$. Thus, the number density is unbounded and can take any value for m close enough to zero. This is not the problem it may appear to be because an unbounded distribution does not imply an unbounded particle number concentration (which, for physical reasons is unacceptable). The particle number concentration is obtained as an integral of the distribution and integrals of functions with singularities are not necessarily unbounded. An equation for the particle number concentration can be obtained from the zeroth moment of the population balance.

$$\begin{aligned}&\int_0^\infty\frac{\partial C}{\partial t}-\alpha C_{\rm sat}\int_0^\infty\frac{\partial m^{2/3}C}{\partial m}=0\Rightarrow\\ \frac{dN}{dt}&=\alpha C_{\rm sat}\lim_{m\to\infty}(m^{2/3}C)-\alpha C_{\rm sat}\lim_{m\to0}(m^{2/3}C)\,.\end{aligned}$$

The first term on the left hand side represents a convective flux of particles towards infinite size. From physical arguments, there can be no "sink" at infinity so this flux must be zero. In the general case, we say that a solution of a population balance must satisfy

$$\lim_{m\to\infty}r(m)C(t,m)=0$$

and this is called a *regularity condition* at infinity. Similarly, in a large number of situations, particles cannot form from nothing, only from other particles and a convective flux out of $m=0$ is therefore not physically meaningful. This is certainly true for cell cultures and holds for any system where nucleation is absent. These systems therefore satisfy a regularity condition at zero.

$$\lim_{m\to 0} r(m)C(t,m) = 0\,.$$

Notice that the two regularity conditions state that the convective flux, the product rC, equals zero, not the distribution itself.

The dissolution problem is interesting in that it is one of the systems that do not satisfy the regularity condition at zero. There is a non-zero flux into $m = 0$ because particles that dissolve eventually disappear completely as exhibited by the moment equation

$$\frac{dN}{dt} = -\alpha C_{\text{sat}} \lim_{m\to 0}(m^{2/3}C).$$

We can again see from this equation that the distribution must have a singularity at zero. Since the particles do disappear, dN/dt must be negative, not zero, and the only way that the limit at zero of $m^{2/3}W$ can be non-zero is for C to go to infinity at zero.

Plots of the transient mass distribution during dissolution are shown in Fig. 11.6. The singularity is barely visible and only at $t = 1.2$ because the large values of C only appear very close to zero.

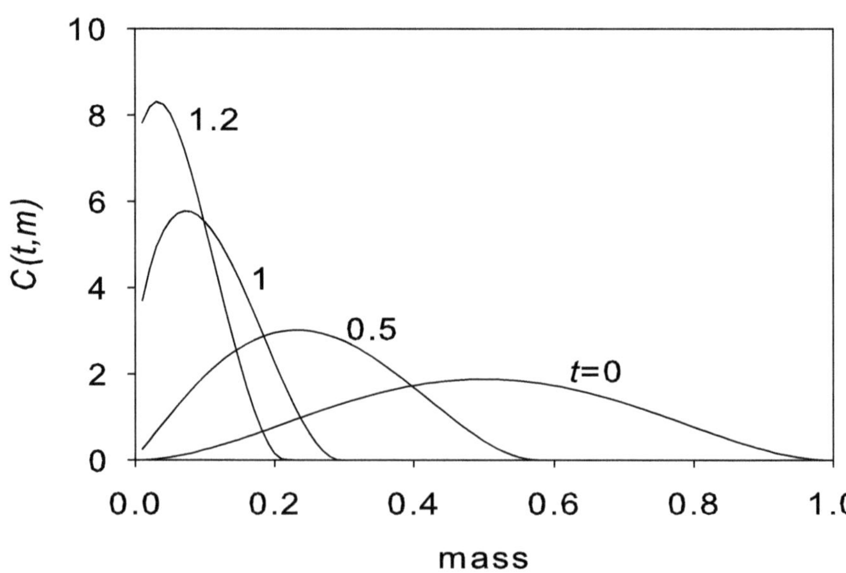

Fig. 11.6 Mass distribution during dissolution when $\alpha C_{\text{sat}} = 1$ for $t = 0, 0.5, 1, 1.2$. The initial distribution is $C_0(m) = 30m^2(1-m)^2$.

11.2.2 *Cauchy's method for semi-linear equations*

Cauchy's method for semi-linear equations is no different from the method for linear equations. However, for semi-linear equations, the two base equations are coupled to the last characteristic equation and the three coupled equations must be solved together. It is not possible to solve for the characteristic curves independently of the last characteristic equation. The characteristic equations are

$$\begin{aligned}
\frac{dt}{ds} &= p_t(t(s), x(s), C(s,\tau)) \ , \ t(s=0) = t(\tau) \\
\frac{dx}{ds} &= p_x(t(s), x(s), C(s,\tau)) \ , \ x(s=0) = x(\tau) \\
\frac{dC}{ds} &= r(t(s), x(s), C(s,\tau))C(s) + f(t(s), x(s), C(s,\tau)) \ , \ C(0) = C_0(\tau).
\end{aligned}$$

When these equations can be solved, the solution for $C(t, x)$ can be obtained as it was for linear equations. Doing this can be difficult and the method is illustrated for a rather simple case, elution of a chromatographic column.

Example 11.7: Elution of a chromatographic column.

The governing mass balance for a chromatographic column has already been found, see Eq. (11.1), repeated below

$$\epsilon v \frac{\partial C_M}{\partial x} + \epsilon \frac{\partial C_M}{\partial t} + (1-\epsilon)\frac{\partial C_S}{\partial t} = 0\,.$$

Recall that this balance equation is for a single compound so when applied to a column separating multiple compounds, the balance is only valid if one can assume that there is no interaction between the compounds. Recall also that ϵ is the void fraction in the bed, v the interstitial velocity of the mobile phase and C_M and C_S the concentrations in the mobile and stationary phase respectively.

The model assumes that the concentrations in the two phases are in equilibrium and we will assume that the equilibrium is given by a Langmuir-type isotherm

$$C_S = \frac{aC_M}{1 + bC_M}$$

where a and b are constants. Hence

$$\frac{\partial C_S}{\partial t} = \frac{a}{(1+bC_M)^2}\frac{\partial C_M}{\partial t}\,.$$

And the balance equation becomes

$$\epsilon v \frac{\partial C_M}{\partial x} + \epsilon \frac{\partial C_M}{\partial t} + (1-\epsilon)\frac{a}{(1+bC_M)^2}\frac{\partial C_M}{\partial t} = 0\,.$$

We will write this as

$$v\frac{\partial C_M}{\partial x} + \left(1 + \frac{(1-\epsilon)}{\epsilon}\frac{a}{(1+bC_M)^2}\right)\frac{\partial C_M}{\partial t} = 0\,.$$

The balance equation does not contain a characteristic length so there is no obvious way to define a dimensionless length coordinate. However, we will consider the problem of column elution, when the column contains some specified initial concentration profile and the feed to the column has zero concentration of the compound in question. We will model the initial condition as a constant concentration through the column except at the inlet where the concentration is assumed to increase linearly from zero to the specified constant concentration $C_{M,0}$ over the distance Δx

$$C_M(x=0,t) = 0\;,\; C_M(x,t=0) = \begin{cases} \frac{C_{M,0}x}{\Delta x}\;,\; 0 \le x \le \Delta x \\ C_{M,0}\;,\; \Delta x \le x \end{cases}.$$

The new model parameter Δx allows us to define the following dimensionless variables

$$C = bC_M\;,\; \tau = \frac{tv}{\Delta x}\;,\; y = \frac{x}{\Delta x}\;,\; \alpha = a\frac{1-\epsilon}{\epsilon}\,.$$

This gives the model

$$\frac{\partial C}{\partial y} + \left(1 + \frac{\alpha}{(1+C)^2}\right)\frac{\partial C}{\partial \tau} = 0$$

with the initial condition

$$C(y=0,\tau) = 0\;,\; C(y,\tau=0) = \begin{cases} C_0 y\;,\; 0 \le y \le 1 \\ C_0\;,\; 1 \le x \end{cases}.$$

The initial manifold is the union of the positive y- and τ-axis. Consider first the positive τ-axis. Since τ has already been used for the dimensionless time, the characteristic equations are written in terms of κ in lieu of τ.

$$\begin{aligned}
\frac{dy}{ds} &= 1\;,\; y(s=0) = 0 \\
\frac{d\tau}{ds} &= 1 + \frac{\alpha}{(1+C)^2}\;,\; \tau(s=0) = \kappa \\
\frac{dC}{ds} &= 0\;,\; C(0) = 0\,.
\end{aligned}$$

These are solved to give

$$
\begin{aligned}
y =& s \\
\tau =& (1+\alpha)s + \kappa \\
C =& 0\,.
\end{aligned}
$$

The characteristic curves are straight lines with a slope of $1+\alpha$. Since $C=0$, there is no point in solving for s and κ. The solution is easily obtained as

$$
C(y,\tau) = 0\ ,\ \tau > (1+\alpha)y\,.
$$

Consider now the initial manifold $0 < y < 1$. The characteristic equations are

$$
\begin{aligned}
\frac{dy}{ds} =& 1\ ,\ y(s=0) = \kappa \\
\frac{d\tau}{ds} =& 1 + \frac{\alpha}{(1+C)^2}\ ,\ \tau(s=0) = 0 \\
\frac{dC}{ds} =& 0\ ,\ C(\kappa) = C_0\kappa\,.
\end{aligned}
$$

This gives

$$
\begin{aligned}
y =& s + \kappa \\
\tau =& \left(1 + \frac{\alpha}{(1+C)^2}\right) s \\
C =& C_0\kappa\,.
\end{aligned}
$$

This is not easily inverted to solve for $C(y,\tau)$. However, it is easy to see that the characteristic curves are straight lines of the form

$$
\tau = \left(1 + \frac{\alpha}{(1+C)^2}\right)(y-\kappa)
$$

and that the concentration $C = C_0\kappa$ along this line. Thus we can eliminate κ in the equation above for the characteristic curve and solve for C. This requires finding the real root of a third order polynomial.

Finally, we need to consider the initial manifold $y > 1$. It is a trivial matter to see that the characteristic curves are straight lines with the slope $1 + \frac{\alpha}{(1+C_0)^2}$ and that the concentration is constant and equal to C_0 along these curves. In summary, we can write the solution as

$$C(y,\tau) = \begin{cases} 0, & \tau > (1+\alpha)y \\ \text{Root of } \tau = \left(1+\frac{\alpha}{(1+C)^2}\right)\left(y - \frac{C}{C_0}\right), & \left(1+\frac{\alpha}{(1+C_0)^2}\right)(y-1) < \tau < (1+\alpha)y\,. \\ C_0, & \tau < \left(1+\frac{\alpha}{(1+C_0)^2}\right)(y-1) \end{cases}$$

The solution is plotted in Fig. 11.7. It is evident that the tail of the distribution becomes longer as the elution proceeds.

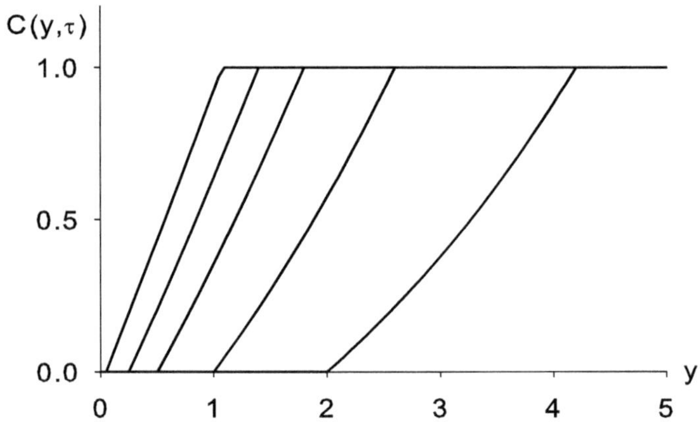

Fig. 11.7 Concentration profiles during elution of chromatographic column. The graphs are plotted for $\alpha = 1$, $C_0 = 1$ and for $\tau = 0.1, 0.5, 1, 2$ and 4.

11.3 Shocks

It was mentioned previously that Cauchy's method may fail to give a unique solution if the characteristic curves intersect and that this problem can be solved by introducing discontinuities or shocks in the solution. Typically, the model written for a given problems fails at a shock and an additional model must be written just for the shock itself. A complete solution to a problem then requires that one finds the point at which the shock forms and the path that the shock follows. The shock forms at the point where the characteristic curves first intersect and follows a path through state space such that the characteristic curves on the two sides of the shock never intersect. The solution on each side of the shock is therefore well defined and the magnitude of the shock is found by solving the original model for both sides of the shock.

The mathematics of shock formation and dynamics is involved and all we will attempt to do here is present an illustrative example that provides some of the flavors of the methods used.

An example of a shock phenomenon familiar to most people is the abrupt change from traffic that flows smoothly to traffic that is at a standstill. Shock waves in compressible gasses can be modelled as shocks in the mathematical sense used here as can the sharp fronts seen in peaks exiting a chromatographic columns. The chromatographic column is the example most relevant to chemical engineering and is used in the next example to illustrate how a shock solution is obtained.

Example 11.8: Shock solution for loading of a chromatographic column.

The component balance for a chromatographic column with a Langmuir isotherm was found in the previous example to be

$$v\frac{\partial C_M}{\partial x} + \left(1 + \frac{(1-\epsilon)}{\epsilon}\frac{a}{(1+bC_M)^2}\right)\frac{\partial C_M}{\partial t} = 0\,.$$

We will model the column as initially free of the compound in question and with a feed concentration that increases linearly with time until a constant feed concentration is reached at time Δt. Hence

$$C_M(x,t=0) = 0\,,\; C_M(x=0,t) = \begin{cases} \frac{C_{M,0}t}{\Delta t}\,, & 0 \le t \le \Delta t \\ C_{M,0}\,, & \Delta \le t \end{cases}.$$

The model parameter Δt allows us to define the following dimensionless variables

$$C = bC_M\,,\; \tau = \frac{t}{\Delta t}\,,\; y = \frac{x}{v\Delta t}\,,\; \alpha = a\frac{1-\epsilon}{\epsilon}\,.$$

This gives the model

$$\frac{\partial C}{\partial y} + \left(1 + \frac{\alpha}{(1+C)^2}\right)\frac{\partial C}{\partial \tau} = 0$$

with the initial condition

$$C(y,\tau=0) = 0\,,\; C(y=0,\tau) = \begin{cases} \frac{bC_{M,0}t}{\Delta t} = C_0\tau\,, & 0 \le \tau \le 1 \\ bC_{M,0} = C_0\,, & 1 \le \tau \end{cases}.$$

The initial manifold is the positive y- and τ-axis and we can formally parametrize this by κ and obtain the characteristic equations

$$\frac{dy}{ds} = 1\,,\; y(0) = y(\kappa)$$
$$\frac{d\tau}{ds} = \left(1 + \frac{\alpha}{(1+C)^2}\right)\,,\; \tau(0) = \tau(\kappa)$$
$$\frac{dC}{ds} = 0\,,\; C(0) = C(\kappa).$$

The characteristic curves are straight lines with slopes of $1+\frac{\alpha}{(1+C)^2}$ and the concentration is constant along these lines. For the purpose of clarifying the discussion that follows, it is convenient to partition the characteristic lines into three types. Type I are the lines that intersect the positive y-axis. They all carry the concentration 0. Type II are the lines that intersect the τ-axis while the feed concentration is increasing with time and since the slope of the lines decrease as the concentration increases, the slope of these lines decreases as the intersection point with the τ-axis increases. Type III are the lines that intersect the τ-axis after the feed concentration has become constant and they all carry the constant concentration C_0. The situation is depicted in Fig. 11.8. It is evident that the characteristic lines do intersect, and to determine when the shock forms, the first point of intersection must be found.

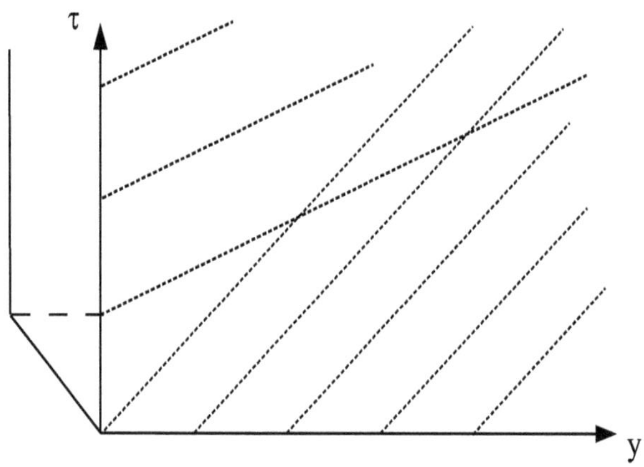

Fig. 11.8 The characteristic curves in the (y, τ) plane. The feed concentration versus τ is graphed along the τ-axis.

Shock formation:

The two characteristics, starting at $(0, 0)$ and at $(0, 1)$ obviously do intersect, so the first intersection point must happen for the characteristics that start along the τ-axis and for $0 \leq \tau \leq 1$, the type II characteristics. We start by parametrizing these characteristics by η

$$\tau = \eta + \left(1 + \frac{\alpha}{(1+\eta C_0)^2}\right) y \,.$$

Pick two lines, given by say η_1 and η_2, and solve for their point of intersection. Then take the limit as $\eta_1 \to \eta_2$ to determine when adjacent lines intersect. The set of all these intersection points determines the so called *envelope* of the family of parametrized curves. Formally, the envelope is obtained by writing the equations for the parametrized characteristic curves as

$$f(y,\tau,\eta)=0$$

which is the first equation that must be satisfied by the points on the envelope. Then let $\eta_2=\eta_1+\delta$ and write the equation for the point of intersection as

$$f(y,\tau,\eta_1)-f(y,\tau,\eta_1+\delta)=0\,.$$

Letting $\delta\to 0$ gives

$$\frac{\partial f(y,\tau,\eta)}{\partial\eta}=0$$

which is the second equation that must be satisfied by the points on the envelope. The points on the envelope are found by eliminating η from these two coupled equations. For the model under consideration, the two equations are

$$f(y,\tau,\eta)=\tau-\eta-\left(1+\frac{\alpha}{(1+\eta C_0)^2}\right)y=0$$

and

$$\frac{\partial f}{\partial\eta}=-1+\frac{2\alpha C_0}{(1+\eta C_0)^3}y=0\,.$$

Though it is possible to eliminate η, it is a cumbersome problem and we are not interested in the envelope per se but only in the first intersection point. To find this point, note first that that the two equations above can be rearranged to give a parametric representation of y and τ in terms of η

$$y=\frac{(1+\eta C_0)^3}{2\alpha C_0}$$

$$\tau=\eta+\frac{(1+\eta C_0)^3}{2\alpha C_0}+\frac{(1+\eta C_0)}{2C_0}\,.$$

Both of these expressions are monotonically increasing with η, so the first point of intersection, the point with the smallest y and τ value must be the point given by the smallest value of η. But η is limited to the interval $0\le\eta\le 1$ so the first point of intersection must be given by $\eta=0$. Substituting this value into the equations above give the point at which the shock first forms

$$y_{\text{formation}}=\frac{1}{2\alpha C_0}$$

$$\tau_{\text{formation}}=\frac{1+\alpha}{2\alpha C_0}\,.$$

Note that this point is on the characteristic line that passes through the origin, the first type I characteristic, so to speak. Therefore, the shock path which starts at this point is intersected from below by only type I characteristics and hence the concentration in front of the shock is always zero.

Shock path:
To find the path of the shock, we will need to write a new balance equation, a conservation balance over the shock. The balance will be written in a coordinate system that moves with the velocity of the shock. The situation is sketched in Fig. 11.9.

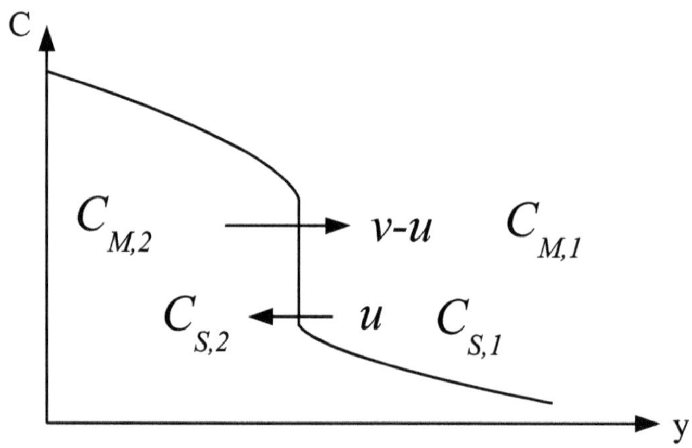

Fig. 11.9 The relevant velocities and concentrations used in writing the conservation balance over the shock.

As before, let C_M and C_S indicate the concentrations in the mobile and stationary phases. The lower concentrations are in the region that the shock is moving towards, the region in front of the shock. Let these concentrations be $C_{M,1}$ and $C_{S,1}$ respectively. Let the concentrations behind the shock be $C_{m,2}$ and $C_{S,2}$. The mobile phase moves with the velocity v and if the velocity of the shock is u, then, in a coordinate system moving with the shock, the stationary phase moves with a velocity $-u$ and the mobile phase with a velocity $v - u$. Consequently, the conservation balance over the shock is

$$(v - u)\epsilon C_{M,2} - (v - u)\epsilon C_{M,1} + u(1 - \epsilon)C_{S,1} - u(1 - \epsilon)C_{S,2} = 0\,.$$

Eliminate C_S using the Langmuir isotherm, $C_S = \frac{aC_M}{1+bC_M}$, to get

$$(v - u)\epsilon C_{M,2} - (v - u)\epsilon C_{M,1} + u(1 - \epsilon)\frac{aC_{M,1}}{1 + bC_{M,1}} - u(1 - \epsilon)\frac{aC_{M,2}}{1 + bC_{M,2}} = 0\,.$$

Solve for v/u

$$\begin{aligned}\frac{v}{u} =& \frac{(C_{M,1}C_{M,2}b^2\epsilon + C_{M,1}b\epsilon + C_{M,2}b\epsilon + \epsilon - a\epsilon + a)}{\epsilon(1 + bC_{M,1})(1 + bC_{M,2})} \\ =& 1 + \frac{a\frac{1-\epsilon}{\epsilon}}{(1 + bC_{M,1})(1 + bC_{M,2})}.\end{aligned}$$

Using the dimensionless variables already defined, this is written

$$\frac{v}{u} = 1 + \frac{\alpha}{(1 + C_1)(1 + C_2)}.$$

This is an equation for the shock velocity u in terms of the interstitial fluid velocity v and the dimensionless model parameters.

To find a solution for u, i.e. to find the shock path, one needs to know the values of the dimensionless concentrations before and after the shock, C_1 and C_2. A diagram of the path and the characteristics is helpful here, Fig. 11.10.

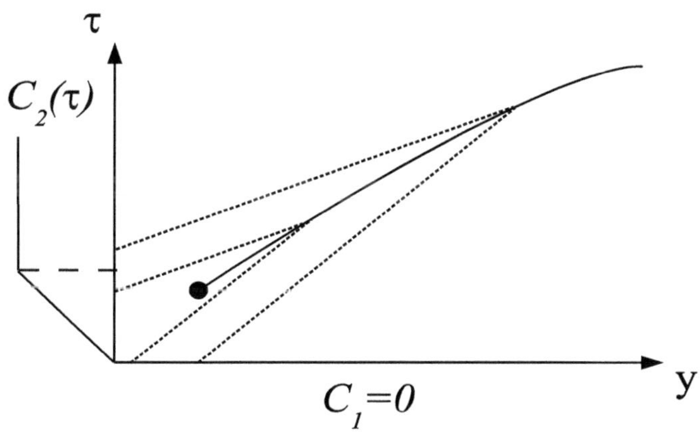

Fig. 11.10 The shock path and the characteristics that intersect the shock path. See text for a detailed explanation.

The shock first forms at the point indicated by the dot and its path is indicated by the solid curve. The characteristic curves are indicated by dotted lines. The characteristics all carry a constant concentration which is determined by following the characteristic back to the initial condition. Those that approach the shock path from below represent the concentration ahead of the shock and carry the concentration initially present in the column, thus $C_1 = 0$. Those that approach the shock path from above represent the concentration behind the shock and the concentration they carry equals the feed concentration at a point in time.

The zero concentration before the shock immediately gives the simpler expression

$$\frac{v}{u} = 1 + \frac{\alpha}{1 + C_2} \,. \tag{11.7}$$

Although the right hand side of this equation has been made dimensionless, the two velocities v and u are still dimensioned. Let a point on the shock path be indicated x_s, then

$$u = \frac{dx_s}{dt} = \frac{y_s v \Delta t}{d\tau \Delta t} = v \frac{dy_s}{d\tau}$$

where y_s is the dimensionless point on the shock path. Hence

$$\frac{d\tau}{dy_s} = 1 + \frac{\alpha}{1 + C_2} \,. \tag{11.8}$$

We shall need this result shortly.

To find the concentration behind the shock, C_2, requires a bit more work. There are two kinds of characteristic lines that represents conditions behind the shock, type II and III characteristics. The latter are easy to deal with since they all carry the constant concentration C_0 and they therefore give a constant shock velocity of $u = v/(1 + \frac{\alpha}{1+bC_0})$. The former constitute a parametrized family of straight lines that we represented as

$$\begin{aligned} \tau &= \eta + \left(1 + \frac{\alpha}{(1 + \eta C_0)^2}\right) y \,, \; 0 \leq \eta \leq 1 \\ &= \frac{C_2}{C_0} + \left(1 + \frac{\alpha}{(1 + C_2)^2}\right) y \,, \; 0 \leq C_2 \leq C_0 \,. \end{aligned} \tag{11.9}$$

Formally, one can now find the shock path by solving the above equation for the parameter C_2 and substitute the result into Eq. (11.8) to obtain an equation for the shock velocity. This equation can then be integrated to obtain the shock path. The important word here is "formally" since carrying out these steps is difficult, even using software for symbolic manipulation. However, there is a clever trick that can be used to find a parametric representation for the shock path.

In Eq. (11.9), let $y = y_s$, i.e. a point on the shock path, and differentiate with respect to C_2 keeping in mind that y_s is also a function of C_2.

$$\frac{d\tau}{dC_2} = \frac{1}{C_0} + \frac{dy_s}{dC_2}\left(1 + \frac{\alpha}{(1 + C_2)^2}\right) - \frac{2y_s\alpha}{(1 + C_2)^3} \,.$$

Turning now to Eq. (11.8), $\frac{d\tau}{dy_s} = 1 + \frac{\alpha}{1+C_2}$, we obtain

$$\frac{d\tau}{dC_2} = \frac{d\tau}{dy_s}\frac{dy_s}{dC_2} = \left(1 + \frac{\alpha}{1 + C_2}\right)\frac{dy_s}{dC_2} \,.$$

Eliminating $\frac{d\tau}{dC_2}$ gives a differential equation in $\frac{dy_s}{dC_2}$

$$\frac{1}{C_0}+\frac{dy_s}{dC_2}\left(1+\frac{\alpha}{(1+C_2)^2}\right)-\frac{2y_s\alpha}{(1+C_2)^3}=\left(1+\frac{\alpha}{1+C_2}\right)\frac{dy_s}{dC_2}\;\Rightarrow$$

$$\frac{C_2^2}{(1+C_2)^2}\frac{dy_s}{dC_2}+\frac{2y_sC_2}{(1+C_2)^3}=\frac{C_2}{\alpha C_0}\;\Rightarrow$$

$$\frac{d}{dC_2}\left(\frac{C_2^2y_s}{(1+C_2)^2}\right)=\frac{C_2}{\alpha C_0}\;\Rightarrow$$

$$\frac{C_2^2y_s}{(1+C_2)^2}=\frac{C_2^2}{2\alpha C_0}+C.$$

To determine the value of the arbitrary constant C, use the initial condition that at the point of shock formation, $C_2=0$ and $y_s=y_{\text{formation}}=1/(2\alpha C_0)$. This gives $C=0$ and hence

$$y_s=\frac{(1+C_2)^2}{2\alpha C_0}$$

which is a parametric representation for the point on the shock path. To find a representation for τ_s, the value of τ for a point on the path, use Eq. (11.9), the parametric representation for the characteristic curves

$$\begin{aligned}\tau_s&=\frac{C_2}{C_0}+\left(1+\frac{\alpha}{(1+C_2)^2}\right)y_s\\&=\frac{C_2}{C_0}+\frac{(1+C_2)^2}{2\alpha C_0}+\frac{1}{2C_0},\quad 0\le C_2\le C_0\,.\end{aligned}$$

Eliminating C_2 from these two parametric equations is not overly cumbersome and gives the parabola

$$\tau_s=y_s+\sqrt{\frac{2y_s\alpha}{C_0}}-\frac{1}{2C_0}\,.$$

This expression for the shock path was derived assuming that the shock intersected type II characteristics. The expression is no longer valid once the shock has travelled to a point where it intersects type III characteristics. The transition point is therefore determined by setting $C_2=C_0$ in the parametric expressions for y_s and τ_s

$$\begin{aligned}y_{\text{transition}}&=\frac{(1+C_0)^2}{2\alpha C_0}\\\tau_{\text{transition}}&=1+\frac{(1+C_0)^2}{2\alpha C_0}+\frac{1}{2C_0}\,.\end{aligned}$$

After this transition point, the concentration on the back of the shock is constant and equal to C_0. So, from Eq. (11.8)

$$\tau = y_s\left(1 + \frac{\alpha}{1+C_0}\right) + C.$$

The arbitrary constant is determined to equal $\frac{1}{2}$ using the transition values found above, so the shock path is

$$\tau_s = y_s\left(1 + \frac{\alpha}{1+C_0}\right) + \frac{1}{2}.$$

To sum up, the equation for the shock path can be written

$$\tau_s(y_s) = \begin{cases} \text{NULL}\,, & y_s < \frac{1}{2\alpha C_0} \\ y_s + \sqrt{\frac{2y_s\alpha}{C_0}} - \frac{1}{2C_0}\,, & \frac{1}{2\alpha C_0} < y_s < \frac{(1+C_0)^2}{2\alpha C_0} \\ y_s\left(1 + \frac{\alpha}{1+C_0}\right) + \frac{1}{2}\,, & y_s > \frac{(1+C_0)^2}{2\alpha C_0} \end{cases}$$

where NULL indicates the absence of the path.

The expression for the complete solution can now be written. It is so cumbersome that it is convenient to break it into several parts. The solution before the shock forms is

$$C(y,\tau) = \begin{cases} 0\,, & \tau < (1+\alpha)y \\ \text{Root of } \tau = \frac{C_2}{C_0} + \left(1 + \frac{\alpha}{(1+C_2)^2}\right)y\,, & (1+\alpha)y < \tau < 1 + (1 + \frac{\alpha}{(1+C_0)^2})y. \\ C_0\,, & \tau > 1 + (1 + \frac{\alpha}{(1+C_0)^2})y \end{cases}$$

By "Root of" we mean the real root of the equation, a third order polynomial. The closed form solution for this is[1]

$$\begin{aligned} C_2 = \frac{1}{6}\Bigg(& 8 + C_0[24\tau - 24y - 108\alpha y] + 24C_0^2(\tau-y)^2 + 8C_0^3\left[\tau^3 - 3\tau^2 y + 3\tau y^2 - y^3\right] \\ & + 12\sqrt{\begin{matrix} 12C_0^4\alpha[y^4 - 3y^3\tau + 3y^2\tau^2 - y\tau^3] - 36C_0^3\alpha y(\tau-y)^2 + 9C_0^2\alpha[9\alpha y^2 - 4\tau y + 4y^2] \\ -12C_0\alpha y \end{matrix}}\ \Bigg)^{1/3} \\ & + \frac{2}{3}\left(2\tau C_0 - 2yC_0 + 1 + \tau^2C_0^2 - 2\tau C_0^2 y + y^2C_0^2\right)/\{8 + C_0[24\tau - 24y - 108\alpha y] \\ & + 24C_0^2(\tau-y)^2 + 8C_0^3[\tau^3 - 3\tau^2 y + 3\tau y^2 - y^3] \\ & + 12\sqrt{\begin{matrix} 12C_0^4\alpha[y^4 - 3y^3\tau + 3y^2\tau^2 - y\tau^3] - 36C_0^3\alpha y(\tau-y)^2 + 9C_0^2\alpha[9\alpha y^2 - 4\tau y + 4y^2] \\ -12C_0\alpha y \end{matrix}}\ \Bigg\}^{1/3} \\ & + \frac{(\tau-y)C_0 - 2}{3}. \end{aligned}$$

[1]A highly motivated reader who tries to work with this expression may discover that the argument under the square root becomes negative for some parameter values. This turns out not to be a problem because the two imaginary numbers that arise from each square root cancel one another such that the final result, the real root, is in fact real.

The solution after the shock has formed can be written

$$C(y,\tau)=\begin{cases} 0\,, & \tau<\tau_s \\ \left\{\begin{array}{ll}\text{Root of } \tau=\frac{C_2}{C_0}+\left(1+\frac{\alpha}{(1+C_2)^2}\right)y\,, & \tau<1+(1+\frac{\alpha}{(1+C_0)^2})y \\ C_0\,, & \tau>1+(1+\frac{\alpha}{(1+C_0)^2})y\end{array}\right\}, & \tau>\tau_s\,.\end{cases}$$

The solution is plotted in Fig. 11.11.

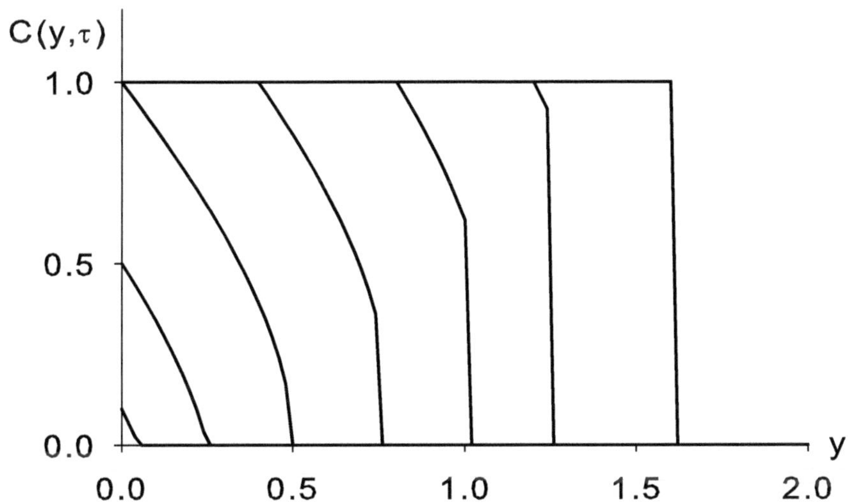

Fig. 11.11 Shock formation in chromatographic column. The graphs are plotted for $\alpha=1$, $C_0=1$ and for $\tau=0.1, 0.5, 1, 1.5, 2, 2.5, 3$.

11.4 Reduction of higher order PDEs

We close this chapter by discussing the use of first order linear PDEs as a tool for solving higher order PDEs. Just like some Nth order ODEs can be solved by recasting them as N coupled first order ODEs that are then solved, some higher order PDEs can be solved by recasting them as coupled first order PDEs. An elegant example of this technique is d'Alembert's[2] solution of the wave equation.

Example 11.9: The wave equation.

A simple derivation of the wave equation can be done by a force balance on an ideal, elastic string that is undergoing small deflections in the plane. Ideal in this

[2] Jean le Rond d'Alembert, French mathematician, 1717–1783.

case means that the tension in the string is constant and in the direction of the tangent to the string. Let the deflection at point x at time t be $C(t,x)$ and let the constant tension be τ. For small deflections, the force on a piece of string between x and $x+dx$ is

$$-\tau\frac{\partial C}{\partial x}\bigg|_{x}+\tau\frac{\partial C}{\partial x}\bigg|_{x+dx}=\tau\frac{\partial^2 C}{\partial x^2}dx\,.$$

By Newton's second law of motion, this equals the acceleration of the piece

$$\tau\frac{\partial^2 C}{\partial x^2}dx=\rho dx\frac{\partial^2 C}{\partial t^2}$$

where ρ is the mass per length of the string. The result is usually written

$$\frac{\partial^2 C}{\partial t^2}-v^2\frac{\partial^2 C}{\partial x^2}=0$$

where $v=\sqrt{\tau/\rho}$ and equals (as the solution will show) the velocity at which waves travel along the string. Initial conditions are usually given as initial displacement and initial displacement velocity

$$C(0,x)=f(x)\,,\quad \frac{\partial C}{\partial t}\bigg|_{(0,x)}=g(x).$$

We can write the PDE as two first order differential operators, operating sequentially on C

$$\left(\frac{\partial}{\partial t}+v\frac{\partial}{\partial x}\right)\left(\frac{\partial}{\partial t}-v\frac{\partial}{\partial x}\right)C=0\,.$$

Remember that the expressions in the parentheses are operators and the parentheses are not the common algebraic kind indicating multiplication of the terms inside the parenthesis. Instead, the expression above means that all the derivatives inside a pair of parentheses, operate on the expression to the right of the pair. Thus all the derivatives inside the first pair operate on everything inside the second pair and the derivatives inside the second pair operate on C. If we call the argument of the outer operator U, we have

$$\frac{\partial U}{\partial t}+v\frac{\partial U}{\partial x}=0$$

and

$$U=\frac{\partial C}{\partial t}-v\frac{\partial C}{\partial x}\,.$$

The initial condition on U becomes

$$U(0,x) = g(x) - vf'(x).$$

We start by solving for U. We first parametrize the initial manifold and initial conditions by

$$\mathcal{M} : x = \tau \;,\; t = 0 \;,\; U = g(\tau) - vf'(\tau).$$

The characteristic equations become

$$\frac{dx}{ds} = v \;,\; x(0) = \tau$$

$$\frac{dt}{ds} = 1 \;,\; t(0) = 0$$

$$\frac{dU}{ds} = 0 \;,\; U(0) = g(\tau) - vf'(\tau).$$

Solving these subject to the initial conditions gives

$$x = vs + \tau \;,\; t = s \;,\; U(s,\tau) = g(\tau) - vf'(\tau).$$

Now solve for s and τ and substitute into the result for $U(s,\tau)$

$$s = t \;,\; \tau = x - vt \;\Rightarrow$$

$$U(t,x) = g(x - vt) - vf'(x - vt).$$

From the definition of U one then obtains a new PDE problem.

$$\frac{\partial C}{\partial t} - v\frac{\partial C}{\partial x} = g(x - vt) - vf'(x - vt) \;,\; C(0,x) = f(x)\,.$$

The initial conditions are parameterized as

$$\mathcal{M} : x = \tau \;,\; t = 0 \;,\; C = f(\tau).$$

The characteristic equations are

$$\frac{dx}{ds} = -v \;,\; x(0) = \tau$$

$$\frac{dt}{ds} = 1 \;,\; t(0) = 0$$

$$\frac{dC}{ds} = g(x(s,\tau) - vt(s,\tau)) - vf'(x(s,\tau) - vt(s,\tau)) \ , \ C(0) = f(x(s,\tau))$$

and their solutions are

$$x = -vs + \tau \ , \ t = s$$

$$C = \int_0^s (g(x(s,\tau) - vt(s,\tau)) - vf'(x(s,\tau) - vt(s,\tau)))ds + f(\tau).$$

Both g and f' are functions of only one argument, $x - vt$. From the solution for x and t we find that this argument can be written as, $x - vt = -2vs + \tau$. We can therefore write the solution for $C(s,\tau)$ as

$$\begin{aligned} C(s,\tau) &= \int_0^s (g(-2vs + \tau) - vf'(-2vs + \tau))ds + f(\tau) \\ &= -\frac{1}{2v}\int_\tau^{-2vs+\tau} g(\lambda)d\lambda + \frac{1}{2}f(-2vs + \tau) + \frac{1}{2}f(\tau) \end{aligned}$$

where the variable transformation, $\lambda = -2vs + \tau$, has been used to obtain the last integral. We can now proceed as usual, solve for s and τ in terms of x and t and substitute this result into the expression for $C(s,\tau)$ to get d'Alembert's solution

$$C(t,x) = \frac{1}{2}(f(x + vt) + f(x - vt)) + \frac{1}{2v}\int_{x-vt}^{x+vt} g(\lambda)d\lambda \,.$$

Chapter 12

Similarity Transformations

A similarity transformations in finite dimensional vector spaces is essentially a variable transformation resulting from a change of basis. In particular, the transformation can change a problem in N dimensions to N uncoupled problems, each in one dimension. When the term is applied to PDEs, it indicates a similar idea; the idea that the dimensionality of the problem can be reduced by some clever variable transformation. Most often, that PDE in two free variables can be simplified to an ODE.

12.1 Semi-infinite media

The Gibbs oscillations that can occur in Fourier series solutions render these solutions virtually useless for numerical evaluation of the solution because the number of terms required in the Fourier series for an accurate evaluation becomes astronomical. Usually, though not always, the oscillations are only significant at small values of the dimensionless time, τ, when a change in boundary value has not yet penetrated far into the domain and the solution is very steep at the boundary. For instance, this is the case when the surface temperature of an object is changed suddenly. These situations can be approximated by assuming that the domain is semi-infinite since the actual extent of the domain is irrelevant when the change at the surface has not penetrated a significant distance into the domain.

Modelling surface transients, at short exposure times, by assuming the substance to be semi-infinite is the basis of penetration theory. Penetration theory is a good bit more involved than this brief description implies and here we shall only be concerned with the method of solutions for models with semi-infinite media. Similarity transformation or combinations of variables reduce the dimensionality of a problem by combining several free variables into a new independent variable called the *similarity variable*. The method will be illustrated using the classical example of transient conduction in an semi-infinite slab.

Example 12.1: Conduction in a semi-infinite slab.

The transient energy balance for a solid with constant thermal conductivity is

$$\frac{\partial T}{\partial t} = \alpha \frac{\partial^2 T}{\partial x^2}$$

where α is the thermal diffusivity, $\alpha = \frac{k}{\rho C_p}$. For a semi-infinity solid which is initially at a uniform temperature but which at time zero is subjected to a different surface temperature, the boundary and initial conditions can be written

$$x = 0 \Rightarrow T = T_1$$
$$x = \infty \Rightarrow T = T_0$$
$$t = 0 \Rightarrow T = T_0$$

where the x-axis has been placed with the origin at the solid surface and the positive direction pointing into the solid. This problem has no characteristic length, so length cannot be made dimensionless in any physically reasonable manner. Temperature, however, can be made dimensionless by defining

$$\Phi \equiv \frac{T - T_0}{T_1 - T_0}$$

giving the problem

$$\frac{\partial \Phi}{\partial t} = \alpha \frac{\partial^2 \Phi}{\partial x^2}$$
$$x = 0 \Rightarrow \Phi = 1$$
$$x = \infty \Rightarrow \Phi = 0$$
$$t = 0 \Rightarrow \Phi = 0\,.$$

In this form, the problem has two free variables and one model parameter, so the solution must be of the form

$$\Phi = \Phi(\alpha, x, t)$$

but Φ is dimensionless, so in the solution, α, x and t must somehow combine into a dimensionless group. Dimensional analysis shows this variable to be of the form

$$\eta = A \left(\frac{x^2}{\alpha t}\right)^B .$$

Any value of A and B could be used in the analysis, but some values are more convenient to work with than others. To find these values, substitute the expression for η into the dimensionless energy balance. This variable substitution is a bit tricky and should be done with care. Define a new dependent variable $\Theta(\eta)$ such that $\Theta(\eta(x,t)) = \Phi(x,t)$. Note that Θ is not an explicit function of either x or t. Using the chain rule, one now obtains

$$\frac{\partial \Phi}{\partial t} = \frac{d\Theta}{d\eta}\frac{\partial \eta}{\partial t} = -AB\left(\frac{x^2}{\alpha t}\right)^B \frac{1}{t}\frac{d\Theta}{d\eta}\,.$$

Similarly, one obtains

$$\frac{\partial \Phi}{\partial x} = \frac{d\Theta}{d\eta}\frac{\partial \eta}{\partial x} = 2AB\frac{x^{2B-1}}{(\alpha t)^B}\frac{d\Theta}{d\eta}\,.$$

Note that $\frac{\partial \Phi}{\partial x}$ is an explicit function of both x and η. Thus, the chain rule gives

$$\begin{aligned}\frac{\partial^2 \Phi}{\partial x^2} &= \frac{\partial}{\partial x}\left(\frac{d\Theta}{d\eta}\frac{\partial \eta}{\partial x}\right) + \frac{\partial}{\partial \eta}\left(\frac{d\Theta}{d\eta}\frac{\partial \eta}{\partial x}\right)\frac{\partial \eta}{\partial x} \\ &= \frac{d\Theta}{d\eta}\frac{\partial^2 \eta}{\partial x^2} + \frac{\partial \eta}{\partial x}\frac{\partial}{\partial x}\left(\frac{d\Theta}{d\eta}\right) + \frac{\partial \eta}{\partial x}\left(\frac{d^2\Theta}{d\eta^2}\frac{\partial \eta}{\partial x} + \frac{d\Theta}{d\eta}\frac{\partial}{\partial \eta}\left(\frac{\partial \eta}{\partial x}\right)\right).\end{aligned}$$

Consider the term $\frac{\partial}{\partial x}\left(\frac{d\Theta}{d\eta}\right)$. This term is equal to zero, as can be seen by changing the order of integration

$$\frac{\partial}{\partial x}\left(\frac{d\Theta}{d\eta}\right) = \frac{d}{d\eta}\left(\frac{\partial \Theta}{\partial x}\right) = 0\,.$$

The last equality was obtained using the fact that Θ is not an explicit function of x. The term $\frac{\partial}{\partial \eta}\left(\frac{\partial \eta}{\partial x}\right)$ is evaluated the same way

$$\frac{\partial}{\partial \eta}\left(\frac{\partial \eta}{\partial x}\right) = \frac{\partial}{\partial x}\left(\frac{\partial \eta}{\partial \eta}\right) = \frac{\partial 1}{\partial x} = 0\,.$$

The transformed equation now becomes

$$\frac{\partial^2 \Phi}{\partial x^2} = \frac{d\Theta}{d\eta}\frac{\partial^2 \eta}{\partial x^2} + \left(\frac{\partial \eta}{\partial x}\right)^2 \frac{d^2\Theta}{d\eta^2}\,.$$

Consider the term

$$\frac{\partial^2 \eta}{\partial x^2} = A(\alpha t)^{-B} 2B(2B-1)x^{2B-2}.$$

It appears that $B = 1/2$ will be a particularly good choice because it renders the above term zero. With this choice, the transformed equation becomes

$$\alpha A^2 \left(\frac{1}{\sqrt{\alpha t}}\right)^2 \frac{d^2\Theta}{d\eta^2} + \frac{A}{2}\frac{x}{\sqrt{\alpha t}}\frac{1}{t}\frac{d\Theta}{d\eta} = 0 \Rightarrow$$
$$\frac{d^2\Theta}{d\eta^2} + \frac{1}{2A}\frac{x}{\sqrt{\alpha t}}\frac{d\Theta}{d\eta} = 0\,.$$

The choice of A is not critical but $A = 1/2$ is a traditional choice giving the final form

$$\frac{d^2\Theta}{d\eta^2} + \frac{x}{\sqrt{\alpha t}}\frac{d\Theta}{d\eta} = 0 \Rightarrow$$
$$\frac{d^2\Theta}{d\eta^2} + 2\eta\frac{d\Theta}{d\eta} = 0\,.$$

The transformed boundary and initial conditions become

$$x = 0 \Rightarrow \Phi = 1 \text{ becomes } \eta = 0 \Rightarrow \Theta = 1$$
$$x = \infty \Rightarrow \Phi = 0 \text{ becomes } \eta = \infty \Rightarrow \Theta = 0$$
$$t = 0 \Rightarrow \Phi = 0 \text{ becomes } \eta = \infty \Rightarrow \Theta = 0\,.$$

The solution for Θ is now obtained as

$$\frac{d\Theta}{d\eta} = C_1 e^{-\eta^2} \Rightarrow$$
$$\Theta = C_2 + C_1 \int_0^\eta e^{-\tau^2}\, d\tau\,.$$

The boundary condition at $\eta = 0$ gives

$$\Theta = 1 + C_1 \int_0^\eta e^{-\tau^2}\, d\tau\,.$$

It can be shown that the remaining integral equals $\sqrt{\pi}/2$ when η equals ∞, so from the remaining boundary condition it follows that

$$\Theta = 1 - \frac{2}{\sqrt{\pi}} \int_0^\eta e^{-\tau^2}\, d\tau\,.$$

Which is usually written using the error function as

$$\Theta = 1 - \text{erf}(\eta) = 1 - \text{erf}\left(\frac{x}{2\sqrt{\alpha t}}\right) \tag{12.1}$$

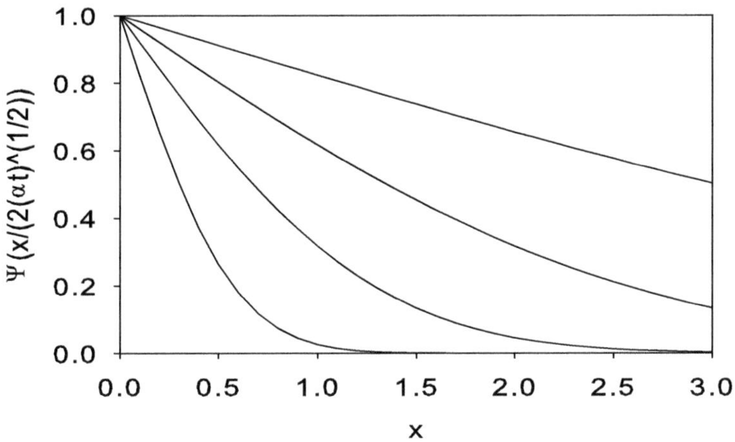

Fig. 12.1 Graphs of the solution in Eq. (12.1). The solutions are for $\alpha = 1$ and are shown for $t = 0.1$, 0.5, 2 and 10.

where the error function is

$$\operatorname{erf}(\eta) \equiv \frac{2}{\sqrt{\pi}} \int_0^{\eta} e^{-\tau^2}\, d\tau\,.$$

The solution is plotted in Fig. 12.1.

In the next example, we model diffusion in a tube and find and compare the solution for both a tube of finite length and for a tube that is assumed infinitely long. The infinitely long tube solution, the penetration solution, is well behaved at short times and superior to the solution for a finite tube which exhibits Gibbs phenomena at short times.

Example 12.2: The Loschmidt apparatus.

Diffusion coefficients in gases can be measured in an instrument known as a Loschmidt apparatus, shown in Fig. 12.1 Two identical hollow cylinders are each filled with a pure gas at the same pressure. Say A in the top cylinder and B in the bottom cylinder. At time zero, the two cylinders are placed end-to-end and the gases are allowed to diffuse into one another. After some time t, the experiment is terminated. The cylinders are moved away from each other and the mole fractions of the two gases in each are determined. From knowledge of the mole fractions and the duration of the experiment, it is possible to calculate the diffusion coefficient.

The problem is symmetric in the sense that it is immaterial which gas is labeled A and which is labeled B. It holds that $x_A(z) = x_B(-z)$ and, since A and B are the only gases in the system, $x_A(z) + x_B(z) = 1$. It is therefore sufficient to find, for instance, $x_B(z)$ for positive values of z because the remaining mole fractions can be found from this result using the two algebraic equations just given.

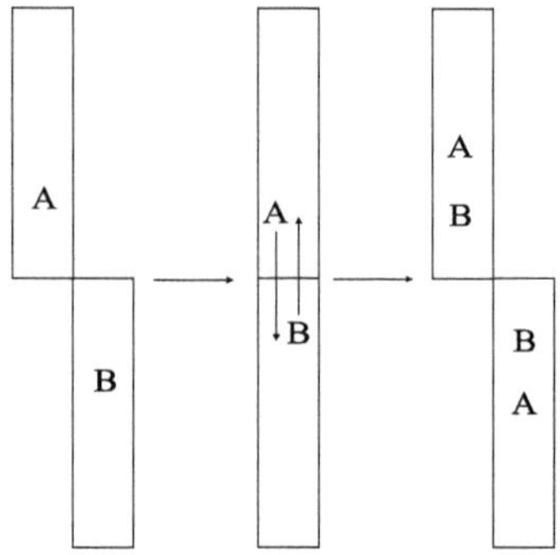

Fig. 12.2 Schematic of a Loschmidt apparatus.

Solution for finite cylinders:
The balance on component B is

$$\frac{\partial x_B}{\partial t} = D\frac{\partial^2 x_B}{\partial z^2}$$

and the initial and boundary conditions are

$$t = 0 \Rightarrow x_B = \begin{cases} 0\,, & z > 0 \\ 1\,, & z < 0 \end{cases}$$

$$z = \pm L \Rightarrow \frac{\partial x_B}{\partial z} = 0\,.$$

Introduce the dimensionless variables

$$\xi = \frac{z}{L}\,,\quad \tau = \frac{tD}{L^2}$$

to get

$$\frac{\partial x_B}{\partial \tau} = \frac{\partial^2 x_B}{\partial \xi^2}$$

$$\tau = 0 \Rightarrow x_B = \begin{cases} 0\,, & \xi > 0 \\ 1\,, & \xi < 0 \end{cases}$$

$$\xi = \pm 1 \Rightarrow \frac{\partial x_B}{\partial \xi} = 0\,.$$

The Fourier series solution is found to be

$$x_B(\xi,\tau) = \frac{1}{2} - 2\sum_{m=0}^{\infty} \frac{e^{-(\pi/2+m\pi)^2\tau}}{\pi(1+2m)} \sin((\pi/2+m\pi)\xi).$$

Solution for infinite cylinders:
The problem becomes

$$\frac{\partial x_B}{\partial t} = D\frac{\partial^2 x_B}{\partial z^2}$$

$$z = 0 \Rightarrow x_B = 1/2$$

$$z = \infty \Rightarrow x_B = 0$$

$$t = 0 \Rightarrow x_B = 0\,.$$

Define $\eta = z/(2\sqrt{Dt})$ and carry out the variable transformation to get

$$\frac{d^2 x_B}{d\eta^2} + 2\eta\frac{dx_B}{d\eta} = 0$$

$$\eta = 0 \Rightarrow x_B = 1/2$$

$$\eta = \infty \Rightarrow x_B = 0$$

giving

$$x_B(\xi,\tau) = \frac{1}{2} - \frac{1}{2}\,\mathrm{erf}\left(\frac{\xi}{2\sqrt{\tau}}\right).$$

We can express all our results in terms of the dimensionless variables

$$\xi = \frac{z}{L}\,,\ \tau = \frac{tD}{L^2}$$

and compare the solutions by plotting $x(\tau,\xi)$ for all solutions for ξ between 0 and 0.1 and for τ equal to 0.00001. Graphs of the two solutions are shown in Fig. 12.3.

The agreement between the graphs in Fig. 12.3 is poor and the Gibbs phenomena suggests that the penetration theory solution is superior to the Fourier series solution. This is confirmed by the graphs in Fig. 12.4 which show the two solutions for different values of τ using 250 terms in the Fourier series solution. To the naked eye, the two solutions are identical but the Fourier series solutions obviously require substantially more computations to evaluate.

Finally, Fig. 12.5 shows the two solutions for later times when the penetration theory solution increasingly fail to provide an accurate answer.

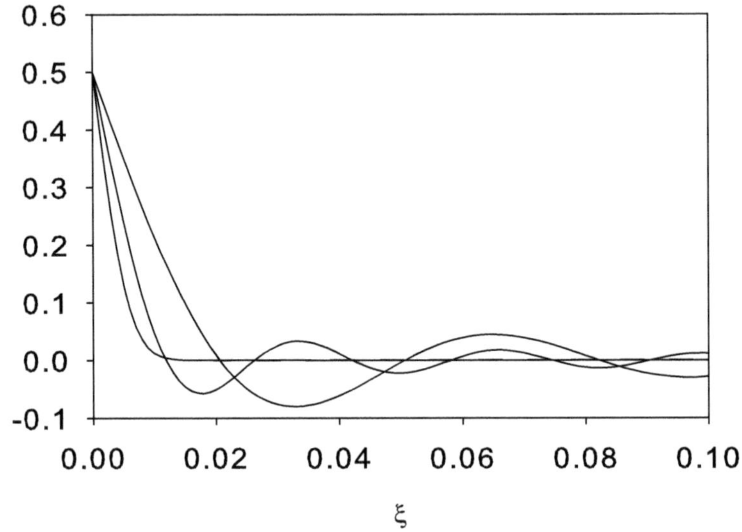

Fig. 12.3 Graphs of the Fourier series solution and the penetration theory solution for the Loschmidt apparatus for $\tau = 0.00001$. The graphs are, left to right, penetration theory solution, Fourier series solution with 30 terms and Fourier solution with 60 terms. A graph of the Fourier series solution with 250 terms is also plotted. It is indistinguishable from the penetration theory solution at resolution used in the figure.

To demonstrate that similarity transformation can be used for other problems than the simple problems in the above examples, let us deviate from the main theme of this text and consider a nonlinear problem. Assume the same physical situation, but with the added and physically reasonable assumption that the thermal conductivity is temperature dependent. With this assumption, the energy balance becomes

$$\frac{\partial \Phi}{\partial t} = \frac{\partial}{\partial x}\left(\alpha(\Phi)\frac{\partial \Phi}{\partial x}\right)$$

$$x = 0 \Rightarrow \Phi = 1$$

$$x = \infty \Rightarrow \Phi = 0$$

$$t = 0 \Rightarrow \Phi = 0$$

and one can, it turns out, use the same similarity variable, but now without the constant factor α, in this problem as for the constant conductivity problem. So, set $\eta = \frac{x}{2\sqrt{t}}$ and $\Theta(\eta) = \Phi(x,t)$ to obtain

$$-2\eta\frac{d\Theta}{d\eta} = \frac{d}{d\eta}\left(\alpha(\Theta)\frac{d\Theta}{d\eta}\right) \tag{12.2}$$

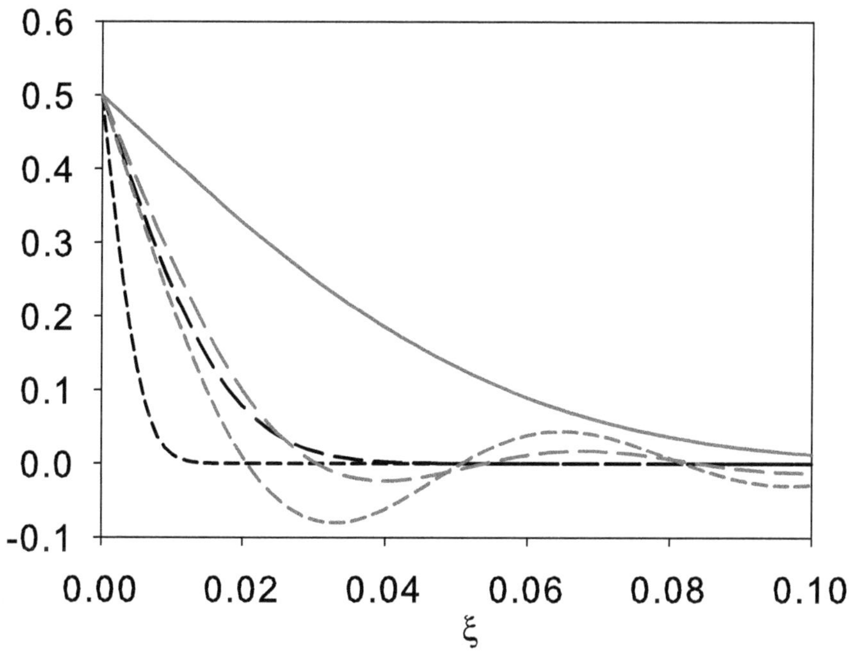

Fig. 12.4 Graphs of the Fourier series solution and the penetration theory solution for the Loschmidt apparatus for $\tau = 0.00001$ (short dash), $\tau = 0.0001$ (long dash) and $\tau = 0.001$ (solid). All Fourier series solutions are evaluated using 30 terms. For $\tau = 0.001$, the graphs of the penetration theory solution and the Fourier series solution are so similar that they cannot be distinguished in this plot.

with the boundary conditions

$$\eta = 0 \Rightarrow \Theta = 1 \text{ and } \eta \to \infty \Rightarrow \Theta = 0\,.$$

Although this equation cannot, in general, be solved analytically, a numerical solution is trivial to obtain and certainly much easier to obtain as well as more compact than a numerical solution to the partial differential equation. Numerical solutions are plotted in Fig. 12.6 for the case of $\alpha(\Theta) = 1 + \beta\Theta$.

In the next example, similarity transformation is used to solve the equations for the boundary layer over a flat plate.

Example 12.3: Boundary layer solution.

Most readers have probably been introduced to the ideas of hydrodynamic boundary layer theory, so we will go straight to the boundary layer equations and show how these can be solved by similarity transformation for the laminar boundary layer over a flat plate.

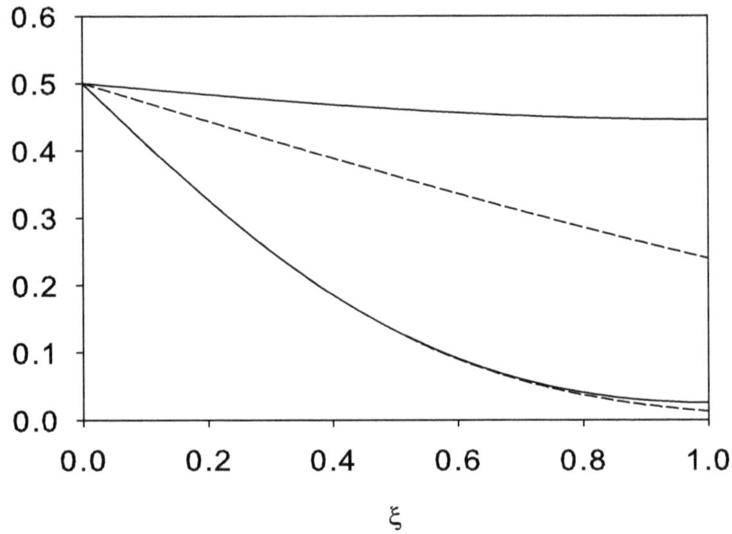

Fig. 12.5 Graphs of the Fourier series solution (solid curves) and the penetration theory solution (dashed curves) for the Loschmidt apparatus for $\tau = 0.1$ and $\tau = 1$. All Fourier series solutions are evaluated using 30 terms. The top graphs are the solutions for $\tau = 1$, the bottom graphs for $\tau = 0.1$.

The boundary layer equations are[1]

$$\frac{\partial v_x}{\partial x} + \frac{\partial v_y}{\partial y} = 0$$

$$v_x \frac{\partial v_x}{\partial x} + v_y \frac{\partial v_x}{\partial y} = \nu \frac{\partial^2 v_x}{\partial y^2}$$

where $\nu = \mu/\rho$, the viscosity divided by the density or the so called kinematic viscosity and v_x and v_y are the velocity components in the x and y directions, respectively. These equations are subject to the boundary conditions

$$v_x(y=0) = 0 \ , \ v_y(y=0) = 0 \text{ and } v_x(y=\infty) = V$$

where the first condition is the no-slip condition at the plate surface and the second condition states that far away from the surface of the plate, the velocity is constant and given as V. The coordinate system used here is defined in Fig. 12.7.

Physically, the notion of similarity implies that the shape of the velocity profile changes by stretching in the y-direction as the flow moves down the plate. This stretching is indicated by the dashed curve in Fig. 12.7. This suggests a similarity

[1] A derivation of these equations can be found in virtually any advanced text on fluid mechanics. The classical reference is Schlichting, H. and Gersten, K. [2000] *Boundary-Layer Theory*, 8th edition (Springer, Berlin).

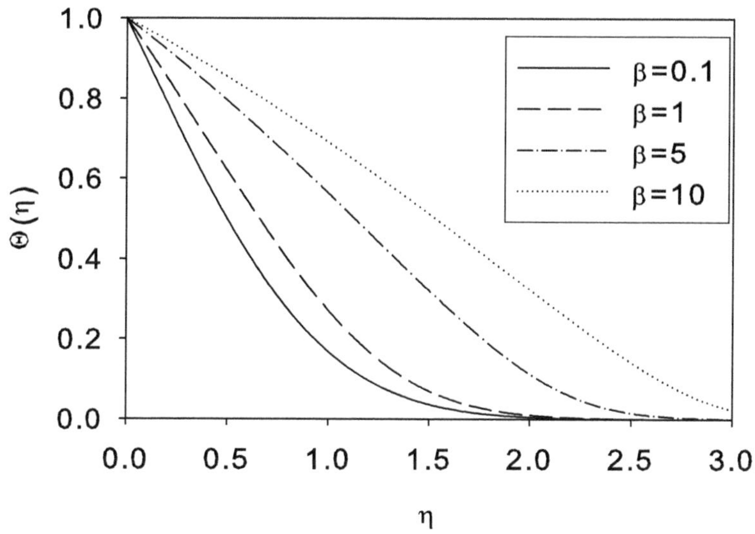

Fig. 12.6 Numerical solutions of Eq. (12.2).

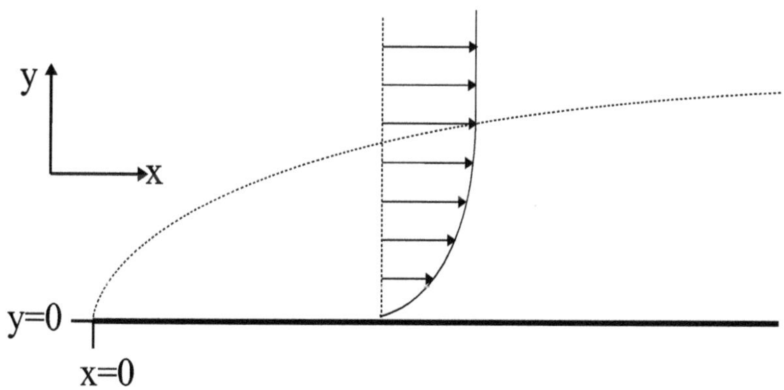

Fig. 12.7 Coordinate system for laminar boundary layer problem.

variable of the form

$$\eta = \frac{y}{\delta}$$

where δ must be a function of the x-coordinate, increasing with x as indicated in the figure. A dimensional analysis shows that δ must be given by

$$\delta = \sqrt{\frac{\nu x}{V}} \Rightarrow \eta = y\sqrt{\frac{V}{\nu x}}$$

or be proportional to this variable. δ is of course a measure of the boundary layer thickness. We now obtain the following

$$\frac{\partial v_x}{\partial x} = \frac{dv_x}{d\eta}\frac{\partial \eta}{\partial x} = -\eta\frac{1}{2x}\frac{dv_x}{d\eta}$$

$$\frac{\partial v_x}{\partial y} = \frac{dv_x}{d\eta}\frac{\partial \eta}{\partial y} = \sqrt{\frac{V}{\nu x}}\frac{dv_x}{d\eta}$$

$$\frac{\partial^2 v_x}{\partial y^2} = \left(\frac{\partial \eta}{\partial y}\right)^2\frac{d^2 v_x}{d\eta^2} = \frac{V}{\nu x}\frac{d^2 v_x}{dy^2}$$

which, upon substitution into the boundary layer equations give

$$-\eta\frac{1}{2x}\frac{dv_x}{d\eta} + \sqrt{\frac{V}{\nu x}}\frac{dv_y}{d\eta} = 0$$

$$\frac{dv_x}{d\eta}\left(-v_x\eta\frac{1}{2x} + v_y\sqrt{\frac{V}{\nu x}}\right) = \frac{V}{x}\frac{d^2 v_x}{d\eta^2}$$

which perhaps does not look promising. However, define the variables

$$V_x = \frac{v_x}{V}\,,\; V_y(\eta) = \sqrt{\frac{x}{V\nu}}v_y$$

and the equations become

$$-\eta\frac{1}{2}\frac{dV_x}{d\eta} + \frac{dV_y}{d\eta} = 0$$

$$\frac{dV_x}{d\eta}\left(-V_x\eta\frac{1}{2} + V_y\right) = \frac{d^2 V_x}{d\eta^2}$$

i.e. two coupled ordinary differential equations. The boundary conditions become

$$\eta = 0 \Rightarrow V_x = V_y = 0 \text{ and } \eta = \infty \Rightarrow V_x = 1\,.$$

These two equations can be solved numerically as is or can be recast as a single third order equation. A single equation can also be obtained by introducing the stream function, a mathematical trick that will not be covered here. Plots of the numerical solution of the two equations are shown in Fig. 12.8.

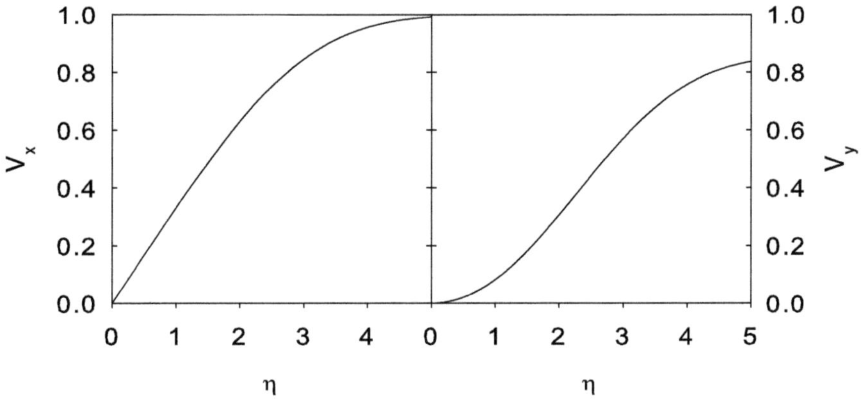

Fig. 12.8 Numerical solution of the two ODEs obtained by similarity transformation of the boundary layer equations.

Chapter 13

Problems

Model formulation

Macroscopic balances

1. Set up a mathematical model for a stirred, flow-through tank in which a solid powder is dissolving in the liquid that flows through the tank. All the particles are present initially in the tank and do not wash out. You can make the following simplifying assumptions:

- The liquid is well mixed.
- The total volume in the tank, liquid plus solid is constant.
- The solid is in the shape of perfect spheres, all of the same size.
- No volume change is associated with the dissolution of the solid (ideal mixture).

Explain your model derivation and define all symbols used in the model. Also make sure that the model has the same number of equations as unknowns.

2. 40 mole of a benzene/toluene mixture must be cleaned of a non-volatile impurity. The mole fraction of benzene in the mixture is $x_f = 0.32$. The purification is done in a simple still, where the complete vapor stream is condensed as the purified product. The boiler holds 20 moles of the mixture and is controlled in such a way that 10 mole of the mixture boils off every hour. The remainder of the mixture is added to the boiler with the constant rate of 10 moles per hour. The equilibrium relationship between the mole fraction of benzene in the vapor and in the liquid can be described adequately by

$$y = \begin{cases} 2.07x \ , \ x < 0.15 \\ 1.4x + 0.1 \ , \ 0.15 < x < 0.35 \, . \end{cases}$$

a) Find the composition in the boiler, as given by x, the mole fraction of benzene, for times shorter than 2 hours.

b) Sketch a curve of y, the mole fraction of benzene in the product, as a function of time for $t < 2$hrs. What would y converge to if the feed stream to the boiler was not stopped after 2 hours?
c) To smooth the time dependence of y, the condensate is collected into a well mixed tank. After 10 mole of condensate has collected, a stream of 10 moles per hour is removed as product. What is the initial composition of this product?
d) How does x change with time in the product stream for $t < 2$ hrs?
e) Find the composition in the boiler during the last half of the process ($2 < t < 4$), while the boiler slowly empties.

3. Consider two well mixed tanks that are in thermal contact with one another but perfectly insulated from the rest of the world. The overall heat transfer coefficient between the tanks is U and the area for energy transfer is a. Write a model for the transient temperatures in the tanks assuming constant physical properties. You cannot assume that the tanks have the same volume or that they contain the same substance.

Assume now that the tanks do contain the same substance and that energy transfer also occurs via two recycle streams of volumetric flow rate Q. I.e., one stream goes from tank 1 to tank 2 and another stream goes from tank 2 to tank 1.

4. An Othmer or equilibrium still is a device used in determining compositions in binary liquid-vapor systems. The still consists of a boiler and a trap. Vapor from the boiler is condensed and collects in the trap. When the liquid level in the trap reaches a preset level, liquid is returned to the boiler via an overflow. After a sufficient amount of time, a steady state is reached. Consider now the following case

- At time zero, the boiler of the still is charged with M_0 moles of a benzene/toluene mixture while the trap is left empty.
- Liquid is boiled at the constant molar rate F and the mole fraction of benzene in the liquid x is related to the mole fraction of benzene in the vapor y by

$$y = 2.07x\,.$$

- The trap can hold up to M_t moles of liquid before liquid starts returning to the boiler.
- Both boiler and trap can be considered well mixed and there is no hold-up of material in either the vapor or liquid return lines.
- The initial mole fraction of benzene in the boiler is x_0.

i) Determine the time, t_1, at which the trap is first full.

ii) Determine the composition in the boiler during this time period.

iii) Write a complete model of the composition in the boiler and trap for times greater than t_1.

iv) Find the composition in the boiler at steady state.

5. A mixing tank is supplied with two liquid feed streams, streams 1 and 2, and a single exit stream. Stream 1 with volumetric flow rate Q_1 is pure compound A. Stream 2 with volumetric flow rate Q_2 is pure compound B. The volumetric flow rate of the exit stream is controlled such that the liquid volume inside the tank, V, is constant.

The tank can be considered well mixed, i.e. the composition of the exit stream is identical to the composition in the tank. There is no volume change associated with the mixing.

The tank is open to the atmosphere and compound B evaporates from the surface of the liquid with the know rate $N(t)$, (moles/time). Compound A is non-volatile and consequently does not evaporate.

This is essentially an exercise in converting between volumetric terms such as the flow rates of the process stream and molar terms such as the evaporation rate.

Let the concentrations of A and B in the tank be C_A and C_B respectively. Write a transient model for the volumetric flow rate of the exit stream, Q_{exit}, and the composition of the exit stream. Be sure the model has the same number of equations and unknowns and define all the symbols that you use which are not defined in the problem statement.

6. Two liquid feed streams of the same substance feed a well mixed tank of constant volume V. The two streams have volumetric flow rates Q_1 and Q_2 and temperatures T_1 and T_2. All four are given functions of time.

The tank is heated with a jacket and the overall heat transfer coefficient between the tank and the substance in the jacket is U and the area for heat transfer is A. The temperature of the heating fluid in the jacket can be assumed constant and equal to T_j.

i) Assuming constant physical properties (heat capacity c_P and density ρ) and assuming that the exit stream is liquid, write a model for the time varying temperature in the exit stream, $T_e(t)$. You must clearly specify the control volume(s) used to formulate the model and the variable on which your balance(s) are done.

ii) How does the model from part i) change if the density, ρ, is a known function of temperature?

iii) How does the model from part i) change if the exit stream is saturated vapor and the heat of vaporization is ΔH (energy per mole)?

7. An impurity is removed from an organic phase by extraction into water. The extraction takes place in a well mixed tank with an organic holdup volume of V_O and an aqueous holdup of V_A. The specific interfacial area is a and the overall mass transfer coefficient based on the organic phase is K_m (i.e., the flux is given by $N = K_m(C_O - K \cdot C_A)$ where K is the partition coefficient.) The organic feed stream to the extractor has the volumetric flow rate Q_O and the aqueous volumetric flow rate is Q_A. The raffinate, the organic exit stream, is recycled to the feed reservoir which contains a total liquid volume of V_R and which can be considered well mixed. The extract, the aqueous exit stream, is not recycled. Write a model for the transient concentration of the impurity in the feed reservoir.

Suppose now that the reservoir is open and that the organic solvent evaporates at the rate of F moles per hour. How does this affect the model?

8. Consider the following system.

- The feed stream to a process has the volumetric flow rate Q and carries the reactant concentration C_f and an inert at concentration $C_{I,f}$.
- The feed stream is mixed with a recycle stream (see next bullet) before entering a reactor in which a first order reaction with rate constant k consumes the reactant.
- The exit stream from the reactor enters a separator which produces two streams of equal volumetric flow rates, one enriched in unconsumed reactant which is recycled and mixed with the feed stream, another which is the exit stream.
- The reactant concentration in the recycle stream is 10 times the concentration in the product stream and all the inert is in the recycle steam.
- To prevent buildup of the inert in the recycle loop the recycle stream is split into two equal streams, one is recycled to the reactor, the other stream is purged.
- There is no volume change in the reaction and the reactor volume is a constant V.

Write a steady state model of this process. The model must be complete, i.e. it must have as many equations as unknowns. State the model in vector matrix notation with the vector of unknowns being $(C, C_e, C_I, C_{I,p})$ where C and C_I are the reactant and inert concentrations in the reactor, C_e the reactant concentration in the exit stream and $C_{I,p}$ the inert concentration in the purge.

9. A Newtonian fluid is moving in laminar flow between two parallel plates, separated by the distance $2L$. Given an x-axis located in the center of the gap between the plates, with a positive direction in the direction of the flow and a y-axis at a right angle to the plates, the velocity profile is

$$v_x(y) = v_{\max}\left(1 - \left(\frac{y}{L}\right)^2\right).$$

The fluid flow generates heat through viscous dissipation (fluid friction). It can be shown that the rate of viscous heat generation (energy per time per volume) in rectangular coordinates is given by

$$\begin{aligned}\dot{q} =& 2\mu\left(\left(\frac{\partial v_x}{\partial x}\right)^2+\left(\frac{\partial v_y}{\partial y}\right)^2+\left(\frac{\partial v_z}{\partial z}\right)^2\right)\\ &+\mu\left(\left(\frac{\partial v_x}{\partial y}+\frac{\partial v_y}{\partial x}\right)^2+\left(\frac{\partial v_x}{\partial z}+\frac{\partial v_z}{\partial x}\right)^2+\left(\frac{\partial v_z}{\partial y}+\frac{\partial v_y}{\partial z}\right)^2\right).\end{aligned}$$

Use an energy balance to derive a model of the steady state temperature in the fluid. Thermal conduction is governed by Fourier's law with constant conductivity k and the physical properties of the fluid can be assumed constant. Note that you are not asked for the boundary conditions to the energy balance.

10. It is desired to find the transient temperature distribution in a hemi-spherical solid object. The spherical surface of the object is in contact with a fluid of temperature T_f and the base is kept at a fixed temperature T_b. The object is initially at a uniform temperature T_0. Write the differential energy balance and the boundary conditions for this object, assuming a film resistance in the fluid with heat transfer coefficient h. Do an energy balance in spherical coordinates. Describe your control volume and account carefully for all fluxes in and out of the volume. Remove all terms that must equal zero due to symmetry considerations.

11. One design of a pressurized water nuclear reactor uses spherical fuel pellet suspended in water. A simple model of this reactor assumes N identical pellet of radius R in which heat is generated at the constant rate $\dot{q}$ (energy/(time volume)). The water is assumed well mixed and of a constant volume V and is passed through the reactor with the volumetric flow rate F. The heat transfer coefficient between the fuel pellets and the water is h, assumed constant and all physical properties can be assumed constant.

a) Write a complete transient (time dependent) model for the temperature of this reactor.

b) Find the steady state water outlet temperature and steady state temperature profile in the fuel pellets.

Macroscopic balances with reaction

12. Growth of microorganisms (X) on a substrate (S) may be modelled with kinetic expressions of the form

$$r_X = \mu S X$$

$$r_S = -\frac{1}{Y} r_X$$

where r_X is the specific rate of biomass formation (biomass per time per volume), r_S the specific rate of substrate formation, X is the biomass concentration, S the substrate concentration, μ a rate constant and Y is a constant known as the yield.

a) Using these rate expressions, write the transient biomass and substrate balances for a CSTR with constant volume V, and constant volumetric feed stream Q. The feed stream is sterile, i.e. $X_{feed} = 0$, but has a constant substrate concentration S_f.

The exit stream from the reactor in part a) is now equipped with a separator that splits the stream into two streams with the same volumetric flow rate and same substrate concentration. However, one stream is enriched in biomass and contains 90% of the mass that enters the separator and is recycled to the reactor while the other stream, the product stream, contains the remaining cell mass.

b) Write a complete steady state model for this system and solve for the biomass and substrate concentrations in the product stream.

13. A compound in liquid solution is oxidized by passing the solution through a CSTR while bubbling an oxygen containing gas mixture through the CSTR. The liquid and gas volumes in the CSTR are V and V_g respectively and both can be assumed constant and well mixed.

The volumetric flow rates of the solution and the gas are $Q(t)$ and $Q_g(t)$ and their reactant and oxygen concentrations are $C_f(t)$ and $C_{g,f}(t)$ respectively.

Transfer of oxygen from the gas phase into the solution is controlled by a liquid side mass transfer coefficient k_m and the gas-liquid interfacial area in the reactor can be assumed constant and equal to A. (The mass transfer resistance in the gas phase is insignificant and can be ignored.)

The equilibrium relationship between gas and liquid concentration of oxygen is given by $C_{O_2} = KC_g$ where C_{O_2} is the liquid oxygen concentration, C_g the gas phase oxygen concentration and K an equilibrium constant.

The specific reaction rate of the reaction is kCC_{O_2} where k is a reaction rate constant and C the liquid concentration of the compound being oxidized.

Derive a transient (time dependent) model for the concentrations in the exit streams. (Note: there are three exit concentrations, the compound in the liquid and the oxygen in the liquid and gas streams.)

14. Consider a CSTR of volume V in which the gas phase reaction $A \rightarrow B$ takes place as a first order reaction in the reactant. We will assume that the gas phase is an ideal gas, that the reactor operates at constant pressure and that the pressure difference between the feed stream and exit stream is insignificant and that there are no significant heat effects of the reactions.

The feed to the reactor consists of pure A at the molar concentration C_{Af} and has the volumetric flow rate Q_f.

i) Derive a complete transient model for the properties of the exit stream, i.e. for the exit concentrations C_A and C_B and the volumetric flow rate Q.

ii) Repeat if the stoichiometry of the reaction is $A \rightarrow 2B$.

15. The feed stream to a CSTR with volume V has a constant volumetric flow rate Q and carries a constant reactant concentration C_f. In the reactor, the reactant is consumed in a first order reaction with reaction rate constant k. The exit stream from the reactor is fed to a separator which splits it into two streams with the same volumetric flow rate. One stream, which is returned to the reactor, is enriched in unconverted reactant, the other stream is depleted in the reactant and is the process exit stream. This process exit stream carries the reactant concentration C_{Exit}. The separator splits the two streams in such a way that the enriched recycle stream has a reactant concentration equal to 10 times the reactant concentration of the process exit stream.

Write a complete steady state model for this process and solve for C_{Exit} as a function of V, Q, C_f and k.

16. Consider the reactor depicted in Fig. 16.1. In this reactor a membrane separates the vessel into two chambers of equal volume V. Both chambers are well mixed and the membrane serves to contain a catalyst inside the reactor. The catalyst itself cannot diffuse across the membrane while the reactant and product can. The catalyst catalyzes the reaction $C \rightarrow P$ which is first order in reactant concentration C and has rate constant k.

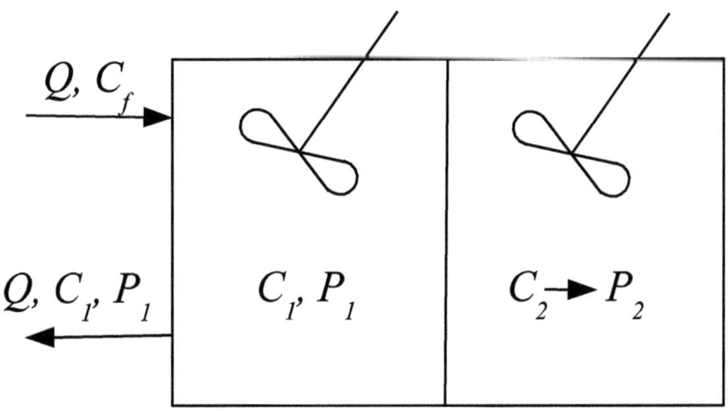

Fig. 16.1 Membrane reactor nomenclature.

The feed and product streams, with constant volumetric flow rates Q, enter and leave the left hand chamber, chamber "1". The feed stream contains the reactant concentration C_f but no product. The reactant can diffuse across the membrane into the right hand chamber, chamber "2", which contains the catalyst and where the reaction occurs. Similarly, the product formed in chamber 2 can diffuse across the membrane to chamber 1. No reaction occurs in chamber 1. The membrane has a surface area A and the overall mass transfer coefficient for transport across the membrane is K_m.

i) Write a complete steady state model from which the product concentration in the exit stream can be found.

ii) Write your model in vector-matrix form or, if this is not possible, clearly state why it cannot be done.

17. A common way of increasing the concentration of biomass in a well mixed flow through reactor is to recycle some of the cells in the exit stream back to the reactor. Consider such a system. Let the reactor volume be V and the volumetric flow rate through the system Q. The feed stream is sterile but contains the substrate at a concentration of S_f. The substrate is consumed by the cells in the reactor and the rate of biomass growth can be modeled by the Monod model,

$$r_X = \frac{\mu_m S}{K+S} X$$

where X is the biomass concentration and S the substrate concentration. The constant model parameter μ_m is called the maximum specific growth rate and K the Monod constant. The rate of substrate consumption is given by

$$r_S = \frac{1}{Y} r_X$$

where Y is the yield, assumed constant. The exit stream from the reactor is passed through a separator which produces a product stream and a recycle stream with the biomass concentration βX. The recycle and product streams have the same substrate concentration. Let the volumetric flow rate of the recycle stream be αQ where α is called the recycle ration and write a steady state model that can be solved for the biomass concentration in the product stream as a function of the model parameters and feed concentration.

18. Consider the following system.

- The feed stream to a process has the volumetric flow rate Q and carries the reactant concentration C_f.
- The feed stream is mixed with a recycle stream (see next bullet) before entering a reactor in which a first order reaction with rate constant k consumes the reactant.
- The exit stream from the reactor enters a separator which produces two streams of equal volumetric flow rates, one enriched in unconsumed reactant which is recycled and mixed with the feed stream, another which is the product stream.

- The reactant concentration in the recycle stream is 10 times the concentration in the product stream.
- There is no volume change associated with the reaction and the reactor volume is a constant V.

Write a steady state model of this process. The model must be complete, i.e. it must have as many equations as unknowns. Verify that your model is complete by solving for the reactant concentration in the reactor as a function of Q, C_f, V and k.

If you need to make additional assumptions when you write your model, state these clearly.

19. In a reactive distillation process, the boiler of a still is initially charged with the volume V_0 of pure A which is then consumed and produced in the reaction

$$A \leftrightarrow 2B\,.$$

The specific reaction rate of the forward reaction, $A \rightarrow 2B$, is given as moles of A consumed per time per volume and equals $k_A C_A$ where k_A is a reaction rate constant and C_A is the concentration of A in moles per volume. Similarly, the reverse reaction, $2B \rightarrow A$ has the specific reaction rate $k_B C_B^2$ in terms of moles of B consumed per time per volume. There are no heat effects or any volume change associated with either reaction.

Distillation removes vapor at the constant volumetric flow rate Q. The vapor can be modeled as an ideal gas at a known pressure P and temperature T.

Over the operating conditions relevant to the process, the composition of the vapor is determined by the equilibrium relationship for B, given by

$$y_B = 10 x_B$$

where y_B is the mole fraction of B in the vapor and x_B is the mole fraction of B in the liquid.

Write a complete model for the composition in the boiler as a function of time and provide an explicit list of the unknowns in the model. Define the symbols in the model that are not defined in the problem statement above. Note:

- To be complete, the model must have the same number of equations and unknowns.
- The model equations must be independent in the sense that some of the equations cannot be derived by manipulations of other equations.
- You are not asked to simplify or solve your model.

20. Derive a model for a two-phase CSTR with constant dilution rate D. ($D = Q/V$ or volumetric flow rate divided by reactor volume.) You can assume that total reactor volume is constant and that the inlet volumetric flow rates, $Q_{I,f}$ and $Q_{II,f}$ are known. You can also assume that the composition of the inlet streams are known and call these concentrations $C_{n,f}$ where $n = 1, 2, \cdots, N$, N being the number of compounds in the phase. Let the specific rate of formation of compound m in reaction x be $r_{x,m}$(moles/(time volume)) and assume these rate expressions are all known. Model transfer between the two phases using an overall mass transfer coefficient, K_m. Due to symmetry, you need only derive the balance equations for one of the phases. The balance equations for the other phase are identical. If you make any additional assumptions, clearly state these.

Clearly state all the dependent unknowns and make sure your model is complete (has as many equations as unknowns).

21. A mixing tank with imperfect mixing can be modelled as two perfectly mixed and interacting tanks.

a) Consider first a tank of volume V which is modeled as two tanks, tank 1 and tank 2 with volumes V_1 and V_2 respectively. ($V = V_1 + V_2$). A process feed stream with volumetric flow rate Q and concentration C_f of some compound enters tank 1. Similarly, a product stream with volumetric flow rate Q and concentration C_1 leaves tank 1. The two tanks are connected with a recycle stream with volumetric flow rate Q_r. One recycle stream goes from tank 1 to tank two and a return stream goes in the opposite direction. Let the concentration in tank 2 be C_2 and write the transient component balances on the two tanks. Define dimensionless concentrations as

$$\theta_1 = \frac{C_1}{C_f}\ , \ \theta_2 = \frac{C_2}{C_f}$$

and a dimensionless time as

$$\tau = Dt = \frac{Q}{V}t$$

where D is the dilution rate for the entire tank, and state the dimensionless form of the balance equations.

b) Repeat if the transfer of matter between the tanks is modelled as a film resistance with overall mass transfer coefficient K_m and total transfer area A.

22. An impurity is removed from a gas stream in a counter current staged adsorption tower. The gas enters at the bottom of the tower with a molar flow rate of G and a mole fraction of the impurity of y_f. Pure water enters at the top with a molar flow

rate of L. The impurity is assumed to be present at such a low concentration that the molar flow rates of the two streams can be assumed constant.

Let the mole fraction of the impurity on stage n be y_n in the gas phase and x_n in the aqueous phase. Stages are numbered in ascending order from 1 to N starting at the top of the tower and all stages are equilibrium stages with equilibrium given by

$$y_n = Kx_n$$

where K is constant.

a) Write a steady state balance on the impurity on stage n and eliminate the gas phase mole fraction to obtain an equation solely in xs.

b) Arrange the unknown aqueous phase mole fractions in a vector as $(x_1, x_2, \cdots, x_N)$ and state the equations in vector-matrix- notation. Be sure to clearly state the expressions for the elements of the matrix and the elements of the right hand side.

Differential Balances

23. Derive the steady state energy balance for a wall of thickness $2L$ in which heat is generated with the constant rate $\dot{q}$. In your derivation, place the origin of the x-axis in the center (the symmetry plane) of the wall and assume constant physical properties.

Assume a film resistance at the wall surface with a heat transfer coefficient of h and an ambient temperature of T_0 and state the two boundary conditions at the surface.

Repeat for a pipe wall with inner and outer radii R_1 and R_2 respectively.

24. A nuclear fuel pellet is shaped as a cylinder with radius R and length L. Fission reactions in the pellet generate thermal energy at the rate $\dot{q}(r, z)$ (energy per time per volume).

Assuming perfect cylindrical geometry (i.e. no θ-dependence) derive the transient (time dependent) energy balance for the pellet. You can assume that the density of the pellet ρ is constant but that the thermal conductivity k and heat capacity C_P are known functions of temperature. Simplify the equation to the extent possible.

25. In a process known as friction welding, two identical, metal rods are fused by placing them end-to-end and rotating them in opposite directions around their common axis. Friction at the interphase generates heat at a not necessarily constant rate of Q energy/time/area. The heat generated by the friction eventually causes the metal at the interphase to melt such that the rods fuse.

i) Derive the transient energy balance for a rod, assuming cylinder symmetry and constant physical properties.

ii) What boundary condition(s) on temperature hold at the metal-metal interphase?

iii) Once the metal melts, the gap between the rods will be filled with liquid metal in laminar flow. Simplify the Navier–Stokes equation as much as possible for this flow. You can assume constant physical properties and steady flow.

iv) Assuming that the two rods rotate with angular velocities ω (in opposite directions) and that the fluid containing gap has width 2δ, find the fluid velocities in the gap. You can neglect the effect of gravity.

v) Consider now friction welding of a rod to a plate as shown in Fig. 25.1. The depth of the indentation in the plate is such that the axis of rotation is in the same plane as the plate surface. Derive the transient energy balance for the rod assuming constant physical properties.

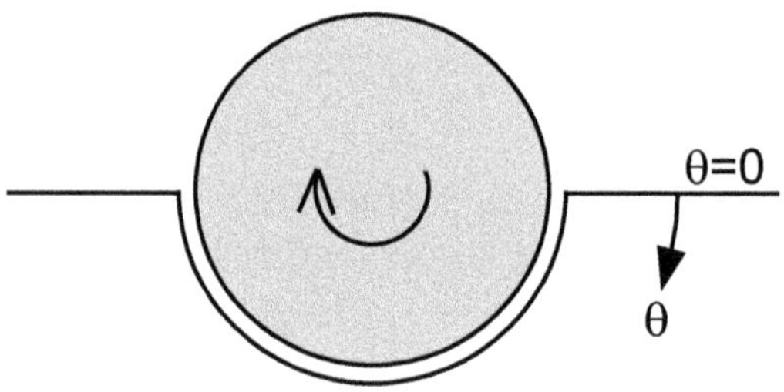

Fig. 25.1 Friction welding of a rod to a plate. The indentation in the plate is of just size that the axis of rotation is in the same plane as the surface of the plate.

vi) If the $\theta = 0$-direction is defined as being in the plane of the plate surface and pointing to the right in the figure and with θ increasing in the same direction as the direction of rotation, what is the boundary condition at the metal-metal surface?

26. A spherical nuclear fuel pellet is made by covering a sphere of radioactive material with a non-radioactive coating. In the radioactive core, heat is continuously generated at the constant rate $\dot{q}$ (energy/(time volume)). No heat is generated in the coating. The radioactive core has radius R_1 and the entire pellet, radioactive core plus coating, has radius R_2. The thermal conductivities of the core and coating are k_1 and k_2 respectively and all physical properties can be assumed constant.

Assuming prefect spherical symmetry, derive the transient (time dependent) energy balances for the temperature in the pellet, including all necessary boundary conditions. The ambient temperature is T_0 and there is a film resistance at the pellet surface with heat transfer coefficient h. For notational convenience, please use T_1 to indicate the temperature in the core and T_2 to indicate the temperature in the coating.

27. A possible way to measure the thermal conductivity of a solid is to heat (or cool) a lump of it to a uniform temperature; then transfer the lump to a well stirred and well insulated bath containing a liquid and measure the liquid temperature as a function of time. Assuming that the lump is spherical with a radius R and thermal conductivity k, write a complete model for this system assuming a film resistance around the solid sphere. Repeat assuming no film resistance. Be careful to clearly define all additional model parameters. You may assume constant physical properties.

Simplify the model assuming that the thermal conductivity of the solid is much greater than that of the liquid. This model can only be written for the film resistance case. Why?

Differential balances with reaction

28. The two reactions $A \to B$ and $B + C \to P$ are carried out in the two-reactor system shown in Fig. 28.1. The following information is available

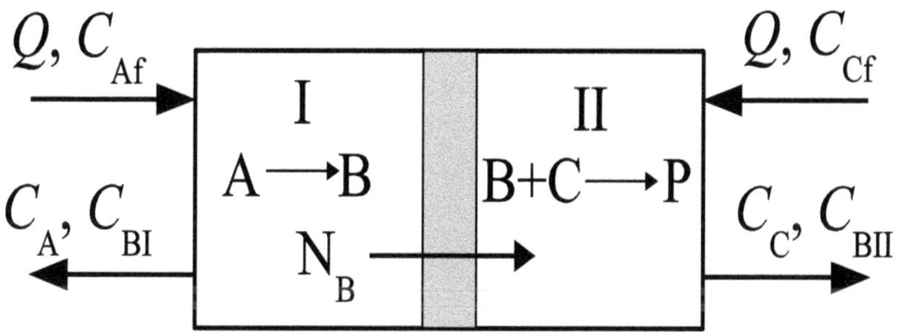

Fig. 28.1 Two identical CSTRs exchanging compound B through a membrane.

- The two vessels are identical, both with constant volume V, and they are well mixed.
- Both vessels are supplied by a feed stream with volumetric flow rate Q.
- The feed stream to vessel I contains A with the concentration C_{Af} but no B or C.
- The feed stream to vessel II contains C with concentration C_{Cf} but no A or B.
- Compound B is exchanged between the vessels by diffusion through a shared membrane with surface area a and the flux of B (moles per time per area) through the membrane is N_B. No other compounds are exchanged through the membrane.
- The specific reaction rate of $A \to B$ is $k_I C_A$ where C_A is the concentration of A. The specific reaction rate of $B + C \to P$ is $k_{II} C_B C_C$ where C_B and C_C are the concentrations of B and C respectively. There is no volume change associated with either reaction.

a) Write the steady state balances for A, B and C for both vessels.

b) The thickness of the membrane is L, the effective diffusion coefficient of B in the membrane is D and there is no film resistance between the well mixed fluids and the membrane surface. Find the flux N_B as a function of the process variables. (Your expression for the flux should close the model written in part a.)

29. A simplifying assumption, frequently used in reactor design, is the assumption that flow through a tubular reactor is plug flow, i.e. that the linear velocity in the tube is constant across a cross section of the tube. In reality, diffusion, turbulent mixing and other factors will result in a back mixing of the reactants and products. This back mixing can significantly affect the performance of the reactor.

The back mixing flux, N, can be modelled by assuming that it is proportional to the concentration gradient in the axial direction. Thus an expression similar to Fick's law is obtained

$$N(mol/hr.) = -DA\frac{dC}{dz}.$$

However, D is not a diffusion coefficient but a so-called axial dispersion coefficient. (Radial dispersion coefficients are also used in more elaborate models.) In general, D will depend strongly on the fluid flow field.

a) Use a differential mass balance to show that the differential equation model for the tubular reactor at steady state is

$$D\frac{d^2C}{dz^2} - v\frac{dC}{dz} - r(C) = 0.$$

Where, $r(C)$ is the reaction rate, mole/(time volume), D is the axial dispersion coefficient and v is the constant linear velocity through the reactor.

b) Dedimensionalize the differential equation using the reactor length L and the inlet concentration C_0 and obtain an expression of the form

$$\frac{d^2y}{dx^2} - Pe\frac{dy}{dx} - \frac{\tau Pe}{C_0}\tilde{r}(y) = 0\,.$$

Where, $y = C/C_0$, $x = z/L$ and $\tau = L/v$, the mean residence time. What is Pe (Peclet's number) and $\tilde{r}$?

c) Assume that no conversion or mixing takes place in the section upstream of the reactor, ($y = 1$ for all $x < 0$). Assume further, that the same holds for the section downstream of the reactor, ($y = y(1)$ for all $x > 1$). Using mass balances over $x = 0$ and $x = 1$ derive the boundary conditions below[1]

[1] Several different boundary conditions have been suggested for this problem. None seem quite right. For instance, the conditions used here allow a discontinuity in y at $x = 0$. A discussion of boundary conditions for tubular reactors can be found in: Wehner and Wilhelm [1956] *Chem. Eng. Sci.*, **6**, 89.

$$x = 0 \Rightarrow \frac{dy}{dx} - Pe \cdot y + Pe = 0$$

$$x = 1 \Rightarrow \frac{dy}{dx} = 0\,.$$

d) Find the exit concentration from the reactor, $y(1)$, when the reaction is first order, $r(C) = kC$.
e) Find $y(1)$ when $Pe \to \infty$ (plug flow) and sketch $y(x)$.
f) Repeat when $Pe \to 0$ (CSTR) and sketch $y(x)$.

30. A spherical catalytic pellet is made by coating an inert core of radius R_1 with a catalytic layer to give a total radius of R_2. The reactant diffuses through the catalytic layer with an effective constant diffusion coefficient D but **cannot** diffuse through the inert core. In the catalytic layer the reactant is consumed in a first order reaction with reaction rate constant k and the concentration of the reactant at the catalytic surface is known to be C_0.

a) Write a complete steady state model for the reactant concentration in the catalytic layer, differential equation and boundary conditions.

b) The inert core is changed to a different material through which the reactant **can** diffuse with the same diffusion coefficient as in the catalytic layer. Will this change increase or decrease the effectiveness factor of the pellet? Explain your reasoning.

31. A catalytic porous membrane is used to separate a reactant stream and a product stream as shown in Fig. 31.1.

The reactor is a rectangular channel with the membrane placed in the center or symmetry plane of the channel. The reactant stream flows through the channel above the membrane and the product stream flows through the channel below the membrane. The two streams flow in the same direction, indicated as the x-direction on the figure, and have known laminar velocity profiles, $v(y)$. The reactant diffuses

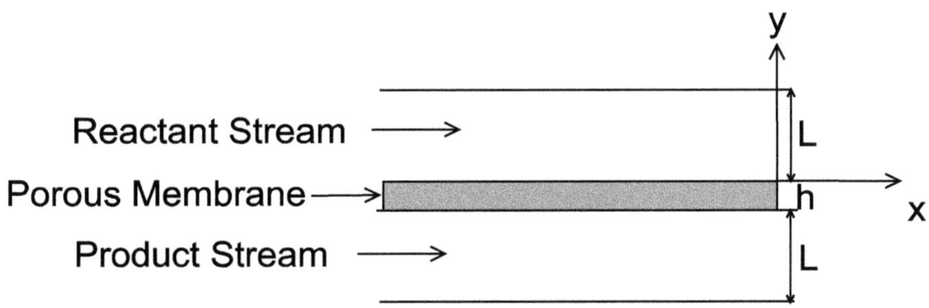

Fig. 31.1 Porous catalytic membrane separating a reactant and a product stream.

through the membrane with an effective diffusion coefficient of D_e while being converted in a first order reaction with reaction rate constant k. The diffusion coefficient of the reactant, D, is the same in the two fluid streams. The thickness of the membrane is h and the height of each channel is L. The channels can be considered infinite in extent in the z-direction and the top and bottom of the reactor are impermeable to the reactant.

i) Derive the steady state reactant balance for the reactant and product streams.

ii) Derive the steady state reactant balance in the membrane.

iii) Place the origin of the y-axis at the top membrane surface and state all the boundary conditions in the y-direction in this coordinate system.

32. A reactant is converted in a flat catalytic particle of thickness $2L$. The reaction is third order with respect to reactant concentration and D, the effective diffusion coefficient of the reactant, is constant. The reactant concentration at the catalytic surface is C_∞.

a) Assuming that heat effects and volume changes in the reaction can be ignored, write a steady state model, boundary conditions included, for the concentration of reactant in the particle.

b) Derive an equation between position, x, concentration $C(x)$ and the (unknown) center concentration $C(0)$.

c) If the reaction is diffusion limited, it only occurs in a region close to the catalyst surface. In that case, the center concentration can be assumed zero. Using the assumption, derive an equation for the effectiveness factor of the catalyst.

33. A gas phase reaction is carried out in a tubular reactor with constant radius R. The feed to the reactor is a mixture of the reactant A and inert I, both with mole fractions equal to 1/2. Write the reactant balances under the modelling assumptions below and in each case, clearly identify the control volume used in writing the balance.

a) Plug flow with fluid velocity v and a homogeneous reaction in the gas phase with rate kC_A and the stoichiometry $A \rightarrow B$. C_A is the concentration of the reactant A and diffusional transport can be neglected.

b) Same reaction as above, but with parabolic velocity profile: $v(r) = v_M(1 - \left(\frac{r}{R}\right)^2)$ and non-negligible diffusional transport. The diffusion coefficient can be assumed constant. Also state the boundary conditions in the radial direction.

c) Same velocity profile and reaction stoichiometry as above, but reaction is now a surface catalyzed reaction at the reactor wall with the rate kC_A in moles per time per area. State both the reactant balance in the gas and the boundary conditions in the radial direction.

d) Repeat part a) changing only the stoichiometry of the reaction to $A \to 2B$. You can assume ideal gas and constant pressure through the reactor. Write as many additional model equations as is needed to give a closed model. (Equal number of equations and unknowns.)

34. In a spherical catalytic particle of radius R, the following two reactions occur. An irreversible reaction $A \to B$ with rate constant k_1, and a reversible reaction $B \leftrightarrow P$ with forward rate constant k_2 and reverse rate constant k_{-2}. All reactions are first order. At the catalyst surface, the concentration of A is C_{A0}, the other reactant concentrations are zero.

Derive the steady state balances and their boundary conditions for all three reactants in pellet. You may assume that the effective diffusion coefficient is constant and the same for all compounds.

35. Some catalytic converters are made of a series of rectangular channels through which the engine exhaust passes, Fig. 35.1. The channel walls are covered with catalyst but reaction takes place both in the fluid and on the catalytic wall. Model this reactor by placing an x-axis in the center of a channel, pointing in the direction of the flow, a y-axis at a right angle to the channel wall and ignore changes in the z-direction.

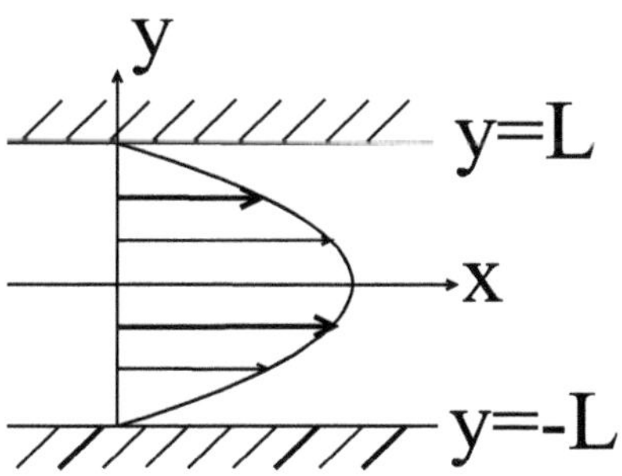

Fig. 35.1 Coordinate system for catalytic converter channel.

Flow can be assumed laminar, i.e. $v(y) = v_m \left(1 - \left(\frac{y}{L}\right)^2\right)$ and transport of reactant is by convection and diffusion. The reaction in the fluid is first order with reaction rate constant k.

a) Write a transient balance on the reactant in the fluid.

b) Assuming the reaction on the catalyst is instantaneous, write all the boundary conditions in the y-direction.

36. Consider a flow-through reactor made by coating the wall of a cylindrical pipe with a layer of solid catalyst. The pipe has a radius of R and the catalyst layer has a constant thickness ΔR. A fluid containing the reactant is passed through the pipe with the linear velocity v. Transport in the catalyst is by diffusion with effective diffusion coefficient D_e and the reactant is consumed by a first order reaction. No reaction occurs in the fluid phase and transport in the fluid is by convection and dispersion with dispersion coefficient D. The dispersion coefficient can be assumed the same for the axial and radial directions. There is no film resistance between the fluid and the catalyst.

a) Write the balance equation for the reactant in the fluid. Clearly indicate the control volume you use.

b) Write the balance equation for the reactant in the catalyst. Clearly indicate the control volume you use.

c) Write the boundary conditions in the radial direction.

37. Enzyme catalyzed reactions have the rate $r = kC/(K + C)$, where C is the concentration of the reactant and k and K are constants. Thus, a shell balance on a flat immobilized enzyme pellet gives the model

$$\frac{D}{k}\frac{d^2C}{dx^2} = \frac{C}{K+C}\,, \quad C(1) = C_s\,, \quad \left.\frac{dC}{dx}\right|_{x=0} = 0\,.$$

a) Show that the reactant concentration, $C(x)$, and position x in the pellet are related through

$$x = 1 - \int_C^{C_s} \left\{ \frac{2k}{D}\left(C - C(0) - K\ln\left(\frac{K+C}{K+C(0)}\right)\right)\right\}^{-1/2} dC.$$

b) Use a total pellet balance to derive an expression for the effectiveness factor of the pellet, η, when the reaction is diffusion limited and you can assume that $C(0) = 0$. η is defined as

$$\eta = \frac{\text{observed rate}}{\text{rate at surface}}\,.$$

38. A spherical catalytic particle of radius R is placed in a laminar fluid flow, Fig. 38.1. The fluid transports reactant over the surface of the pellet such that the reactant concentration in the fluid, adjacent to the pellet surface, is a known function of θ and θ only. Call this function $f(\theta)$, where θ indicates the latitudinal coordinate in the spherical coordinate system indicated in the figure below. Reaction in the pellet is first order and transport through the pellet is by Fickian diffusion with effective diffusion coefficient D.

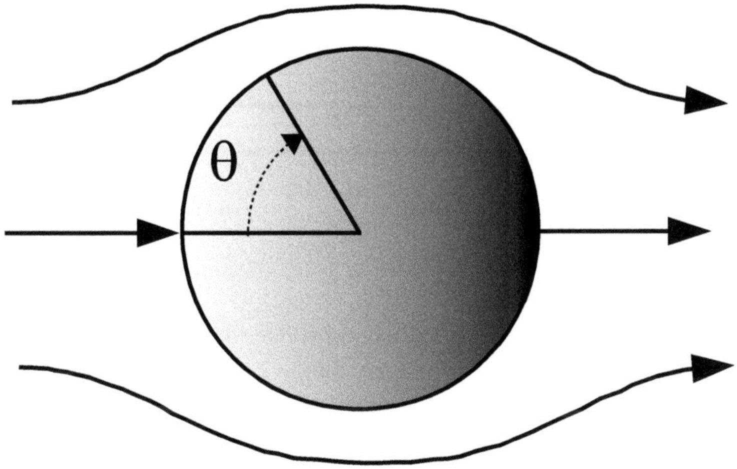

Fig. 38.1 Coordinate system for spherical catalytic particle.

a) Derive the transient conservation equation for the reactant in the pellet, in its simplest possible form and in the indicated spherical coordinate system.

b) State the boundary conditions in the θ direction.

39. In a flat catalytic pellet of thickness $2L$ a reactant is converted in two parallel reactions, one first order the other zero order. The reaction rates are k_1C and k_0, respectively.

a) Show that the steady state model for the reactant concentration in the pellet can be written as

$$\frac{d^2\theta}{d\zeta^2} = \Phi_1^2\theta + \Phi_0^2 \ , \ \theta(1) = 1$$

and identify θ, ζ, Φ_1 and Φ_0.

b) Assuming that the reactant is present throughout the pellet, find the steady state concentration profile.

40. Porous, catalytic pellets can be modelled by assuming that the pores are straight cylindrical cavities with constant diameters. When the reaction is a constant volume reaction, the fluid in the pore is stagnant, but reactants can move through the pore by diffusion. Let the reactant be converted at the pore wall by a first order reaction with reaction rate coefficient k_s. The subscript s indicates that the rate constant is for a surface reaction and the rate k_xC therefore has units of mole/(area time). The concentration of the reactant will strictly speaking be a function of both the axial and radial position in the pore, however, it is reasonable to assume that the radial variation can be ignored because the length of a typical pore is orders of magnitudes greater than its diameter.

Part I:
a) Place a z-axis with the origin at the pore mouth, pointing down the pore and derive a model for the concentration profile in the pore.
b) Let the concentration at the pore mouth be C_s, the length of the pore L and for the boundary condition at the bottom of the pore, assume zero diffusion flux through the bottom and neglect the reaction that takes place at bottom face of the pore. State these boundary conditions mathematically.
c) Solve for the concentration profile through the pore.

Part II:

Repeat the previous problem for a second order reaction. The balance equation cannot be solved explicitly for the concentration but a partial analysis of the differential equation is still possible. To carry this out, note that the free variable, the axial coordinate, does not appear explicitly in the differential equation. The standard trick for such equations is to use the variable substitution $p = dC/dz$ to obtain a differential equation in $p(C)$. In other words, the variable substitution turns the dependent variable into a free variable.

41. Set up a mathematical model (**not** including boundary conditions for the bed itself) for the transient concentration profile of a reactant in a packed, catalytic bed. The catalytic particles can be modelled as porous spheres with some effective diffusion coefficient, D_e and radius ρ_0. The reaction rate in the particles is given by a known function r_C. It can be assumed that no reaction takes place in the fluid phase, that the catalytic particles are small relative to the distance over which reactant concentration changes in the bed and that there is a film resistance between the fluid and the particles. Furthermore, mixing in the fluid phase can be modelled using axial and radial dispersion. To avoid confusion in the nomenclature, use r to indicate the radial coordinate in the bed and ρ to indicate radial coordinate in the catalytic particles.

Repeat the first part of this problem for solid catalytic particles, where the reaction occurs only on the surface of the particle. The reaction is thus a surface reaction and the surface concentration and fluid concentration at the solid surface, i.e. inside the film, can be assumed to be in equilibrium according to a Langmuir type expression

$$C_{\text{surface}} = \frac{AC_{\text{fluid}}}{B + C_{\text{fluid}}}$$

where A and B are constants.

42. Set up a mathematical model, differential equations plus boundary conditions, for the steady state temperature and concentration profiles in a spherical catalytic particle in which an exothermic reaction takes place. The model should include film resistance around the particle. Define all symbols used in the model.

43. A spherical catalytic pellet is made by coating a spherical core of one substance with a layer of a different substance. Neither substance is catalytic by itself, but the interface between the two substances is catalytic.

Let the pellet core of the first substance have radius R_1 and let the coated particle have radius R_2. At the interface, a first order reaction takes place with rate $r = kC$, where k is a reaction rate constant and C the ambient concentration of the reactant. (Note: Since this is a surface reaction, kC has units of moles per time per area.)

The reactant concentration outside the pellet is C_0 and there is no film resistance at the pellet surface. Transport of the reactant through the pellet is by Fickian diffusion with constant diffusion coefficient.

i) Derive a complete model for the reactant concentration profile in the pellet, balance equations and boundary equations.

ii) Define a dimensionless length coordinate as $\xi = r/R_1$ and a dimensionless concentration as $\theta = C/C_0$ and make the model, balance equations and boundary conditions, dimensionless.

44. Consider a cylindrical catalyst pellet of radius R and length L which catalyze the reaction

$$A + B \rightarrow \text{Products.}$$

The reaction is first order in the concentration of both A and B with rate constant k and diffusion of the reactants through the pellet is Fickian with constant diffusion coefficient D. The concentration of the reactants at the surface is C_{A0} and C_{B0} respectively, both of which are constant and do not change with position on the surface.

Derive a complete steady state model for the two concentration profiles in the pellet. Note that the concentrations will be functions of both the radial and axial coordinate. State the relevant boundary conditions assuming a film resistance at the surface with mass transfer coefficient k_m.

45. A catalytic pellet is shaped like a long cylinder with a cylindrical channel through the center. The radius of the channel is R_1 and the radius of the pellet is R_2. A first order reaction with reaction rate constant k occurs inside the pellet and transport of reactant through the pellet is by Fickian diffusion with constant diffusion coefficient D.

i) Derive a transient model, including boundary conditions, for the reactant concentration profile in the pellet. Clearly define the control volume used in the derivation and assume that the pellet is so long that end effects can be neglected and that there is a film resistance at the pellet surface with mass transfer coefficient k_m. The reactant concentration outside the film is C_0.

ii) Make the model and boundary conditions dimensionless by defining a dimensionless concentration as $\theta = C/C_0$ and a dimensionless radial distance as $\xi = r/\Delta R$ where $\Delta R = R_2 - R_1$. Introduce additional dimensionless variables as needed and collect the parameters in the model equations into dimensionless groups.

46. A laminar jet apparatus is a device that has been used in experiments that investigate gas-liquid reactions. In the device, a liquid laminar jet moves vertically downward through a gas containing the reactant. The fluid flow field in the jet is cylinder symmetric but the diameter of the jet and the velocity profile in the jet changes with distance downwards. The fluid flow therefore has two components, $v_z(z, r)$ and $v_r(z, r)$.

A reactant is absorbed from the surrounding gas into the liquid jet and is consumed in an nth order reaction. Reactant concentration can be assumed cylinder symmetric and diffusion obeys Fick's law with a constant diffusion coefficient.

Derive the steady state conservation equation for the reactant in the liquid. Clearly state the control volume that you use. What are the boundary conditions in the radial direction?

47. Write a complete transient model (balances and boundary conditions) for a tubular reactor of radius R through which a solution of the reactant passes in steady flow. There is no reaction in the fluid, only on the wall of the reactor and the inlet and initial conditions can be assumed cylinder symmetric, i.e. no θ-dependence. Model the following cases

(1) Assume laminar flow and that the reactant is consumed "infinitely fast" at the reactor wall.
(2) Assume laminar flow and that the reactant is consumed in a first order surface reaction at the wall. In other words, the reaction rate is kC in units of mole per area per time.
(3) Assume laminar flow and that the wall of the reactor is covered with a porous catalytic layer of thickness ΔR through which the reactant must diffuse with effective diffusion coefficient D_e while being consumed in a first order reaction.
(4) Assume turbulent flow with a flat velocity profile, that mixing in the fluid phase occurs by dispersion, an "infinitely fast" reaction at the reactor wall and a mass transfer coefficient k_m at the reactor wall.

You may ignore the boundary conditions in the axial, z-direction.

How can the result in (3) be simplified if $\Delta R << R$?

Given a solution to the model, $C(t, r, z)$, find the reactant concentration in the exit stream, $C_{\text{exit}}(t)$.

48. Immobilized enzyme particles are sometimes made by micro encapsulation. In the process, a solution of the enzymes is enclosed inside a non-catalytic membrane

and the reactant must diffuse through this membrane before reaching the enzymes. Let the particles created this way be spherical with radius $R + \Delta R$ and a membrane thickness of ΔR. The enzyme solution inside the particles can be considered stagnant and the diffusion coefficient of the reactant in the membrane is D_m while it is D in the enzyme solution. The concentration of the reactant at the surface of the particles is C_0. The specific rate at which the reactant is converted in the enzyme solution is given by

$$r = \frac{r_m C}{K + C}$$

where r_m and K are constant model parameters and C is the reactant concentration. Write a steady state model for the reactant concentration through the entire particle. Solve the model for the limiting case where $C << K$.

49. A liquid film runs down a vertical wall in laminar flow. In the coordinate system shown in Fig. 49.1, the velocity profile in the liquid is

$$v_y = v_{\max}\left(1 - (x/L)^2\right)$$

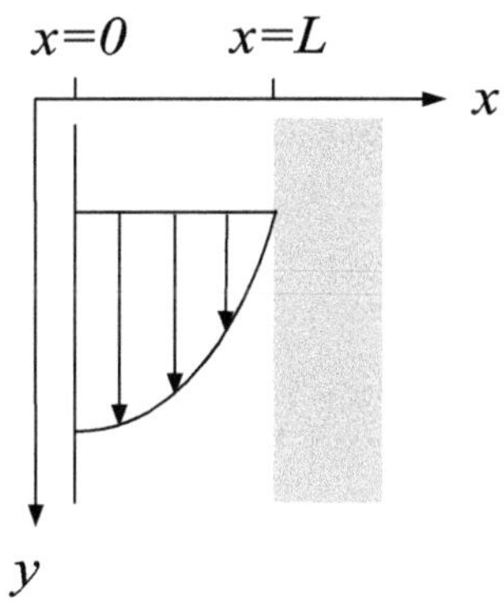

Fig. 49.1 Coordinate system for problem 2.

A compound in the adjacent gas is adsorbed in the liquid and diffuses with constant diffusion coefficient D through the liquid film while being consumed in a first order reaction with reaction rate constant k.

i) Write a conservation equation for the steady state concentration of this compound in the liquid. You must clearly specify the control volume(s) used to formulate the equation and the variable on which your balance(s) are done.

ii) Assuming a gas phase concentration of the compound of C_g, no film resistance and an equilibrium relationship between liquid and gas concentrations of the form $C_{\text{liquid}} = KC_g$ where K is a partition coefficient, what are the boundary conditions in the x-direction?

iii) What are the boundary conditions in the x-direction if there is a film resistance in the gas phase with mass transfer coefficient k_m?

iv) What form does the balance equation take and what are the boundary conditions in the x-direction if the reaction is zero order? You can assume that there is no film resistance.

50. A solution of a reactant is passed through a reactor made of a cylindrical tube with a radius that decreases slightly with distance down the tube. Fluid flow is laminar, but the decreasing tube radius induces a radial component to the velocity which therefore has two components, $v_x(x,r)$ and $v_r(x,r)$, respectively where x is the distance in the direction of flow. These velocity components are assumed known.

The reactant is converted in a first order reaction with the rate constant k and diffuses through the solvent with the constant diffusion coefficient D.

i) Write the balance equation on the reactant in the tube (note: you are not asked for the boundary conditions).

If the rate of change in tube radius with distance is small, the velocity profile can be found using the lubrication approximation which states that the profile at x is the same as the laminar velocity profile at the radius at x. In other words, the velocity profile can be approximated by

$$v_x = v_{\max}(x)\left(1 - \left(\frac{r}{R(x)}\right)^2\right).$$

ii) Assuming that the fluid is incompressible, that $R(x)$ is known and that the volumetric flow rate through the tube is Q, find $v_{\max}(x)$.

51. A first order reaction occurs in a spherical catalytic pellet of radius R with a fixed surface concentration of the reactant, C_S and a fixed surface temperature T_S. The problem satisfies spherical symmetry and all variables depend on the radial coordinate only. The reaction is exothermic and has the heat of reaction ΔH and the reaction rate constants dependence on temperature is given by Arrhenius law, $k = k_0 \exp(-E/RT)$. Conduction of heat in the pellet is given by Fourier's law

$$q_r = -k_T \frac{dT}{dr}$$

where q_r is the energy flux in the positive radial direction (energy per time per area) and k_T is the thermal conductivity in the pellet.

Write the steady state energy balance and associated boundary conditions for the pellet. Simplify the balance as much as possible assuming that the thermal conductivity is constant.

52. The equations below model the reactant concentration in a plug flow reactor.

$$\frac{\partial C}{\partial t} + v\frac{\partial C}{\partial x} = -kC^n \ , \ C(x=0,t) = C_{\text{feed}}(t).$$

Here, $C(x,t)$ is the molar concentration of a reactant being consumed in an nth order reaction. The fluid velocity through the reactor is v, time is t, x is the distance down the reactor from the reactor inlet and k is the reaction rate constant. The feed concentration is specified as $C_{\text{feed}}(t)$.

i) Let $n = 1$ and make this model dimensionless using the initial feed concentration, $C_{\text{feed}}(0)$, to define a dimensionless concentration, $\theta(\tau,\xi)$, and let the dimensionless time and distances be τ and ξ respectively. How are these two dimensionless variables defined?

ii) Now let $n = 2$. What are the units of k? Make the model dimensionless as before.

53. Write a model for the temperature and concentration of reactant in a tubular reactor with radius R in which the fluid flow is laminar. The reactant has a diffusion coefficient D and it is converted by a first order reaction with reaction rate constant k. The heat of reaction is ΔH and heat removal at the reactor wall can be modelled using a constant overall heat transfer coefficient U. The physical properties of the fluid can be assumed constant. The model must include boundary conditions in the radial direction but conditions in the axial direction are not required.

54. A fluid reaction mixture is passed through a cylindrical tubular reactor with radius R. Fluid flow through the reactor is laminar with a known velocity profile $v_z(r)$ and a reactant A in the fluid is consumed in a first order reaction. There is no volume change associated with the reaction.

i) Place a cylindrical coordinate system with an axial coordinate, z, in the center of the cylindrical reactor and a positive direction in the direction of the fluid flow. Derive the steady state mass balance/conservation equation for the reactant assuming that Fick's law applies for diffusional transport and that diffusional transport in the axial direction can be ignored.

ii) Derive the steady state energy balance for the fluid assuming that the heat of reaction is ΔH and that the reaction rate constant obeys Arrhenius law ($k = k_0 e^{-\frac{E}{RT}}$).

iii) State the boundary conditions for the reactant concentration in the radial direction.

iv) Restate the boundary conditions for the reactant concentration in the radial direction assuming that the reactant is converted infinitely fast at the pipe wall.

Navier–Stoke's type problems

55. A volatile compound is stripped from a liquid solution which runs in laminar flow downwards along the wall in a cylindrical tube. The gas which takes up the compound passes up the tube in turbulent flow. Let the radius of the tube be R and the radius at the gas-liquid interface be R_i. Find the velocity profile in the fluid. Write a model for the concentration of the volatile in the liquid and in the gas.

56. Consider steady laminar flow of an incompressible Newtonian fluid in a cylindrical conduit for which the radius decreases with distance in the direction of flow. I.e.

$$r(z) = R - \alpha z$$

where α is a given constant. Ignoring end effects, simplify the continuity equation and the Navier–Stokes equation for this flow using a cylindrical coordinate system. You can ignore the effect of gravity.

Repeat using a spherical coordinate system oriented such that the origin is located at the tip of the cone formed by the wall of the conduit and the positive $\theta = 0$ direction is along the axis of the conduit pointing in the direction of increasing radius.

57. Consider laminar flow of two immiscible, Newtonian fluids between two stationary, parallel, horizontal plates, the denser fluid, fluid I, flowing below the less dense fluid, fluid II. Place an x-axis in the direction of the flow and a y-axis at a right angle to the plates with the origin at the fluid interface and the positive direction pointing up. The z-axis is defined assuming a normal, right-handed, orthogonal coordinate system. The plates are assumed infinite in the x and z directions and the lower plate is at the y coordinate L_I and the upper plate at the y-coordinate L_{II}. The flow is driven by a known pressure gradient $\partial p/\partial x = C$.

a) State which dependent variables, velocity components and pressure are zero and which free variables the non-zero dependent variables are functions of.
b) Simplify the Navier–Stokes equation as much as possible for this problem.
c) State the relevant boundary conditions.

d) Solve for the velocity profile in both fluids.
e) Solve for the pressure distribution.

58. A two phase, laminar, gravity driven, liquid film is flowing down a cylindrical, vertical rod. Call the liquid adjacent to the rod "I" and the liquid on the outside "II". The radius of the rod is r_1, the interphase of the two liquids is at r_2 and the surface of the outer liquid is at r_3. The surrounding gas can be considered to be at a constant pressure. See Fig. 58.1.

Simplify the Navier–Stokes and continuity equation for the two liquids for steady flow and state all the boundary conditions needed for a complete model.

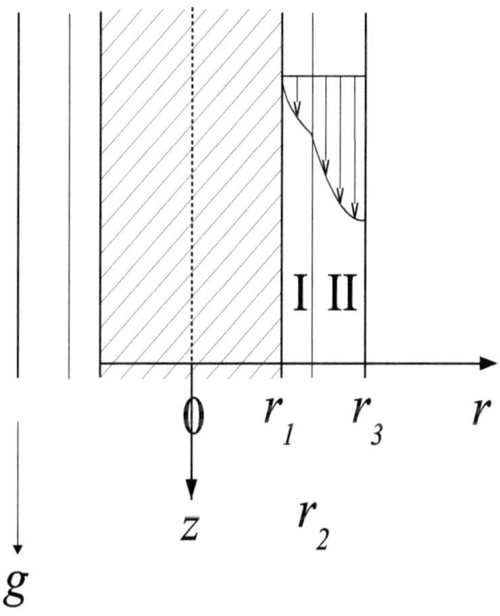

Fig. 58.1 Vertical, two phase, gravity driven flow in cylindrical coordinates.

59. A plate-and-cone viscometer consists of a horizontal stationary flat plate of radius R and a rotating cone which is placed with its tip on the center of the flat plate. Let the angle between the flat plate and the sides of the cone be θ_0 and let the cone rotate with the angular velocity ω. The surface of the liquid in the viscometer can be assumed to lie on the surface of a sphere of radius R. Simplify the Navier–Stokes equation as much as possible for laminar, stationary flow in this viscometer and state the boundary conditions on the velocity. Assume that the viscometer is oriented such that the flat plate is horizontal and below the cone and that the axis of rotation of the cone is vertical and the cone is pointing down.

60. A rotating bearing is cooled by forcing the lubricant through the bearing with a fixed and known pressure drop of $\Delta P/L$.

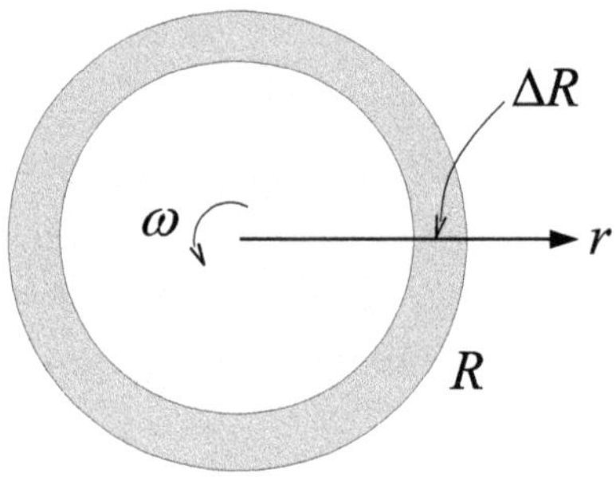

Fig. 60.1 Geometry of rotating bearing. The inner bearing rotates with a constant angular velocity of ω, the outer bearing is stationary and has a radius of R. The gap between the inner and outer bearing is ΔR.

a) Given the geometry shown in Fig. 60.1 and assuming laminar flow with constant physical properties state which velocity components are non-zero and the free coordinates that these depend on. What simplified form does the steady state Navier–Stoke and continuity equations take for this flow? You can ignore the effect of gravity. The Navier–Stoke and continuity equations in cylindrical coordinates is given on the last page.

b) Heat is generated in the lubricant by viscous dissipation according to some known function $q(r, z)$, where q has units of energy per volume per time. Derive the steady state energy equation (the equation for the temperature distribution) for the lubricant. Clearly define the CV used in the derivation of the balance. In the derivation, you may assume that the fluid velocity components are known functions; you are not required to solve the differential equations derived in part a).

ODEs

61. Find the complete solution to

$$t\frac{dx}{dt} = 4t^3 - x\,.$$

62. Solve

$$\frac{dx}{dt} + \sin(t) = \sin(t)x\ ,\ x(0) = 0\,.$$

63. Solve the following

a)

$$(x^3 + y^3)dx = 2xy^2dy \ , \ x = 1, y = 0 \, .$$

b)

$$(3x^2 - 6xy)dx - (3x^2 + 2y)dy = 0 \, .$$

c)

$$y' + y = e^{-x} \ , \ x = 0, y = 3 \, .$$

64. Solve

$$\frac{d^2x}{dt^2} = \exp(x) \ , \ x(0) = 0 \ , \ x'(0) = \sqrt{2} \, .$$

65. Find the **complete** solution (i.e. all solutions) to the ODE

$$\left(\frac{d^2x}{dt^2}\right)^2 = \frac{dx}{dt} + 1 \, .$$

66. Find the complete solution of

$$\frac{dx}{dt} + x = x \cdot t + t - 1 \, .$$

67. Solve

$$\frac{d^2x}{dt^2} + \frac{dx}{dt} = \exp(t).$$

68. Find the complete solution to

$$t\frac{d^3x}{dt^3} - \frac{d^2x}{dt^2} = 0 \, .$$

69. Find all the solutions to the set of two coupled ODEs below.

$$\frac{dx}{dt} = xye^{-t}$$
$$\frac{dy}{dt} = y + e^t.$$

70. Find the general solution of the homogeneous Riccati equation

$$\frac{dx}{dt} = p(t)x^2 + q(t)x$$

by using of the variable transformation $x = 1/y$.

71. Solve

$$\frac{d^2x}{dt^2} + \left(\frac{dx}{dt}\right)^2 + 1 = 0\,.$$

72. Solve

$$\frac{d^2x}{dt^2} + \frac{dx}{dt}\left(x + \frac{dx}{dt}\right) = 0$$

subject to

$$t = 0 \;\Rightarrow\; x = \frac{1}{2}\,,\; \frac{dx}{dt} = \frac{1}{2}\,.$$

73. A runaway chemical reaction can be modeled by the explosion equation

$$\frac{dx}{dt} = kx^2\,,\; x(0) = x_0$$

where k is a constant. Solve this equation and show that the solution has a singularity (it goes to $\pm\infty$). Find the location of the singularity as a function of k and x_0.

74. Find the **complete** solution of

$$t\frac{dx}{dt} = 2x\,,\; x(-1) = 1\,.$$

(Warning: This is a pathological problem.)

Finite dimensional vector spaces

75. Determine, directly from the definition, if the functions below are linearly independent or dependent over the interval $I = [0, 2\pi]$.

a)

$$\sin(t), \sin(2t), \cos(t)$$

b)

$$\sin(t), \cos(t), \sin(t + \pi/2)\,.$$

76. Let V_1 and V_2 be subspaces of $\mathbb{R}^5$, given by

$$V_1 = \{(\alpha, \beta, 0, \alpha + \beta, 2\alpha)\}\ ,\ \alpha, \beta \epsilon \mathbb{R}$$

and

$$V_2 = \{(0, \gamma + \delta, \delta + 2\gamma, \delta, 0)\}\ ,\ \gamma, \delta \epsilon \mathbb{R}\,.$$

Determine the dimension, N, of $V_1 + V_2$.

77. Given the two subspaces of $\mathbb{R}^6$ below, V_1 and V_2, determine their intersection. In other words, determine the subspace of vectors that are in both V_1 and V_2.

$$V_1 = \begin{pmatrix} s_1 \\ s_2 \\ 0 \\ s_1 + s_2 \\ s_1 - s_2 \\ 0 \end{pmatrix}\ ,\ V_2 = \begin{pmatrix} 0 \\ \tilde{s}_1 \\ \tilde{s}_2 \\ \tilde{s}_3 \\ \tilde{s}_1 + \tilde{s}_2 - 2\tilde{s}_3 \\ 0 \end{pmatrix}.$$

78. Below are given parametric representations of two subspaces of $\mathbb{R}^5$. Determine the parametric representation of the sum of these two subspaces and report the result in standard form.

$$V_1 = \begin{pmatrix} s_1 + s_2 \\ 0 \\ s_1 \\ -s_1 \\ s_2 \end{pmatrix}\ ,\ V_2 = \begin{pmatrix} s_1 + s_2 \\ s_2 \\ 0 \\ 0 \\ s_1 \end{pmatrix}.$$

79. Given the spanning set below, find the dimension of the vector space spanned by this set and determine a basis for this space.

$$\left\{ \begin{pmatrix} 1 \\ 1 \\ 1 \\ 1 \end{pmatrix}, \begin{pmatrix} -3 \\ 0 \\ 8 \\ 5 \end{pmatrix}, \begin{pmatrix} 4 \\ 3 \\ 1 \\ 2 \end{pmatrix}, \begin{pmatrix} 5 \\ 3 \\ 1 \\ 3 \end{pmatrix}, \begin{pmatrix} 7 \\ 10 \\ 2 \\ -1 \end{pmatrix} \right\}.$$

80. Consider the subspace V_1 of $\mathbb{R}^4$ defined by the spanning set below.

$$V_1 \text{ , spanned by } \left\{ \begin{pmatrix} 1 \\ 1 \\ 0 \\ 2 \end{pmatrix}, \begin{pmatrix} 0 \\ 1 \\ 1 \\ -1 \end{pmatrix}, \begin{pmatrix} 1 \\ 1 \\ 1 \\ 0 \end{pmatrix}, \begin{pmatrix} 2 \\ 0 \\ 1 \\ 0 \end{pmatrix} \right\}.$$

a) Determine a basis for V_1.

Let the subspace V_2 of $\mathbb{R}^4$ be defined parametrically as

$$V_2 = \begin{pmatrix} s_1 \\ s_2 \\ s_3 \\ s_1 + s_2 - 2s_3 \end{pmatrix}.$$

b) Determine whether V_1 and V_2 are the same or different subspaces. Explain your reasoning!

81. Given below are expressions for two subspaces of $\mathbb{R}^4$.

i) Put the representation of V_1 in standard form and state the dimension of this subspace.

ii) Find a standard representation of $V_1 + V_2$ and state its dimension.

iii) Are V_1 and V_2 disjoint or not? Justify your answer.

$$V_1 = \begin{pmatrix} s_1 + s_2 \\ s_2 + s_3 \\ s_3 - s_1 \\ s_1 - s_3 \end{pmatrix}, \quad V_2 = \begin{pmatrix} 0 \\ 0 \\ s_1 \\ -s_1 \end{pmatrix}$$

82. Prove or disprove (by counterexample) that if V_1, V_2 and W are subspaces of Q such that

$$V_1 + W = V_2 + W$$

then $V_1 = V_2$.

83. Prove or disprove (by counterexample) that if V_1, V_2 and W are subspaces of Q such that

$$V_1 \oplus W = Q \quad \text{and} \quad V_2 \oplus W = Q$$

then $V_1 = V_2$.

84. For each of the subsets of $\mathbb{R}^3$ given below, determine if it is a subspace of $\mathbb{R}^3$ and, if it is a subspace, provide a geometric interpretation of the subspace as a subspace of three-dimensional Euclidean space. Feel free to make use of drawings or sketches for this purpose.

i) $\{(x_1, x_2, x_3)\epsilon\mathbb{R}^3 | x_1 + x_2 + x_3 = 0\}$
ii) $\{(x_1, x_2, x_3)\epsilon\mathbb{R}^3 | x_1 + x_2 + x_3 = 1\}$
iii) $\{(x_1, x_2, x_3)\epsilon\mathbb{R}^3 | x_1 + x_2 = 0\}$

85. Consider the two vector spaces V_1 and V_2 given parametrically by

$$V_1 = \begin{pmatrix} s_1 + s_4 \\ -s_1 + s_2 \\ s_1 + 2s_3 + s_4 \\ 2s_1 - s_2 + s_4 \\ s_2 + 2s_3 + s_4 \end{pmatrix}, \quad V_2 = \begin{pmatrix} s_1 \\ 0 \\ 0 \\ s_2 \\ 0 \end{pmatrix}.$$

a) What is the dimension of V_1?
b) What is the dimension of $V_1 + V_2$?
c) What is the dimension of $V_1 \oplus V_2$?

86. Find the determinant and the rank of the matrix

$$\begin{pmatrix} -2 & -4 & -4 & 0 \\ 1 & 4 & 0 & 0 \\ 0 & 0 & 2 & -5 \\ 1 & 2 & 1 & 2 \end{pmatrix}.$$

87. Find the rank r of the set of vectors below and identify a subset of r linearly independent vectors.

$$\begin{pmatrix} 10 \\ 2 \\ 8 \\ 4 \\ 4 \\ 6 \end{pmatrix}, \begin{pmatrix} 7 \\ 8 \\ 3 \\ -1 \\ 1 \\ 12 \end{pmatrix}, \begin{pmatrix} 3 \\ -6 \\ 5 \\ 5 \\ 3 \\ -6 \end{pmatrix}, \begin{pmatrix} 2 \\ 7 \\ -1 \\ -3 \\ -1 \\ 9 \end{pmatrix}, \begin{pmatrix} -8 \\ 4 \\ 3 \\ -2 \\ -7 \\ 0 \end{pmatrix}, \begin{pmatrix} -5 \\ -2 \\ 8 \\ 3 \\ -4 \\ -6 \end{pmatrix}, \begin{pmatrix} 0 \\ -1 \\ 12 \\ 5 \\ -2 \\ -3 \end{pmatrix}.$$

88. For a mixture that contains $SO_2, SO_3, H_2O, O_2, H_2$ and H_2S, determine the number of equilibrium constants needed to calculate the equilibrium composition. Identify a set of independent reactions which determine the equilibrium composition.

89. Determine the inverse of the matrix below.

$$A = \begin{pmatrix} 1 & 2 & -1 \\ 0 & 3 & 2 \\ 1 & -1 & 1 \end{pmatrix}.$$

90. Determine the real values of A for which the equations below have non-trivial solutions and find the complete solution for these values of A. (Non-trivial means that the solution vector is not the zero vector.)

$$\begin{aligned} x + 2y - (A+1)z &= 0 \\ (2+A)x + 4y - 2z &= 0 \\ 3x + (6+2A)y - 3z &= 0\,. \end{aligned}$$

91. Determine the values of a for which the determinant of the matrix below equals zero and document the detailed steps in your calculation.

$$A = \begin{pmatrix} 1 & -1 & 1+a & -a \\ 0 & -a-1 & a & -1 \\ 2 & 1+a & a & 0 \\ 1 & 1 & a & 0 \end{pmatrix}.$$

For the largest value of a, determine the value of b_4 for which the system

$$Ax = \begin{pmatrix} 2 \\ 1 \\ 5 \\ b_4 \end{pmatrix}$$

has solutions and find these.

92. i) Using Gauss elimination, obtain a matrix in row echelon form from the matrix below. Indicate clearly in your solution which row operations you perform.

ii) From the echelon form, identify a set of columns that form a basis for the space spanned by all the columns of the row echelon matrix.

$$\begin{pmatrix} 1 & 1 & 0 & 2 & 3 & 4 & 1 \\ 2 & 2 & 1 & 5 & 6 & 9 & 2 \\ 1 & 1 & 0 & 2 & 4 & 6 & 0 \\ 3 & 3 & 1 & 7 & 10 & 15 & 2 \end{pmatrix}.$$

93. Find the range and null space of

$$A = \begin{pmatrix} 1 & 1 & 1 & 1 \\ 1 & 2 & 2 & 2 \\ 1 & 4 & 4 & 4 \\ 0 & 0 & 0 & 1 \end{pmatrix}.$$

94. Determine the rank, the range and the null space (kernel) of

$$A = \begin{pmatrix} 1 & 2 & 0 & 5 & 3 \\ 2 & 6 & 0 & -1 & 0 \\ 3 & 0 & -2 & 4 & 1 \\ 0 & 4 & 1 & 0 & 1 \end{pmatrix}.$$

What is the dimension of the range and the kernel?

95. Find the null space of

$$A = \begin{pmatrix} 1 & 2 & 1 & 2 \\ 0 & 0 & 0 & \alpha \\ 1 & 1 & 2 & 0 \\ 2 & 3 & 3 & 2 \end{pmatrix}$$

where α can be any real number.

96. Find the null space of the matrix A below

$$A = \begin{pmatrix} 1 & -1 & 3 & 2 & -1 \\ 1 & -1 & 4 & 3 & -2 \\ 2 & -2 & 7 & 6 & -3 \\ 1 & -1 & 5 & 5 & -3 \end{pmatrix}.$$

97. Find the null space of the matrices below. α can be any real number.

$$A = \begin{pmatrix} 0 & 0 & 0 & 0 & 0 \\ 0 & 0 & 0 & 0 & 0 \\ 1 & 1 & 1 & 0 & 0 \\ 1 & -1 & 1 & 0 & 0 \\ 1 & 0 & 1 & 0 & 0 \end{pmatrix}, \; B = \begin{pmatrix} 1 & 1 & 1 & 1 & 1 & 1 & 1 \\ 0 & 1 & 1 & 1 & 1 & 1 & 1 \\ 0 & 0 & 0 & 0 & 1 & 1 & 1 \\ 0 & 0 & 0 & 0 & 0 & 1 & \alpha \\ 0 & 0 & 0 & 0 & 0 & 1 & 1 \\ 0 & 0 & 0 & 0 & 0 & 0 & 1 \\ 0 & 0 & 0 & 0 & 0 & 0 & 0 \end{pmatrix}, \; C = \begin{pmatrix} 1 & 1 & 1 & 1 & 1 & 1 & 1 \\ 0 & 1 & 1 & 1 & 1 & 1 & 1 \\ 0 & 0 & 0 & 0 & 1 & 1 & 1 \\ 0 & 0 & 0 & 0 & 0 & 1 & \alpha \\ 0 & 0 & 0 & 0 & 0 & 1 & 1 \\ 0 & 0 & 0 & 0 & 0 & 0 & 0 \\ 0 & 0 & 0 & 0 & 0 & 0 & 0 \end{pmatrix}.$$

98. Determine the null space and range of the matrix A below

$$A = \begin{pmatrix} 1 & 2 & 3 & 4 \\ 0 & -1 & 1 & 6 \\ 1 & 2 & 4 & 4 \end{pmatrix}.$$

99. Find the range and null space of

$$A = \begin{pmatrix} 1 & 2 & 1 & 2 \\ 3 & 0 & 4 & 1 \\ 4 & 2 & 5 & 3 \\ 2 & -2 & 3 & -1 \end{pmatrix}.$$

Show that, except for the zero vector, a vector in the null space of A cannot be in the range. In other words, show that the range and null space of A are disjoint vector spaces.

100. Consider the matrix A below

$$A = \begin{pmatrix} 1 & 2 & -4 & -1 \\ 4 & 2 & -4 & -2 \\ 0 & 4 & -4 & 3 \\ -1 & 2 & 0 & 4 \end{pmatrix}.$$

a) Find the null space of A.

b) Find the range of A.

c) Show that the null space and range found in a) and b) are disjoint (i.e. show that any non-zero vector in the null space is not a vector in the range or vice versa.)

101. Determine the eigenvalues and eigenvectors of

$$A = \begin{pmatrix} a & -b \\ b & a \end{pmatrix}$$

and provide a similarity transformation between A and a diagonal matrix. Clearly show the diagonal matrix for the given similarity transformation.

102. Find all eigenvalues and their algebraic and geometric multiplicities of the matrix

$$A = \begin{pmatrix} 1 & 0 & 1 & -5 \\ 1 & 1 & 1 & 0 \\ 0 & 0 & 0 & 5 \\ 0 & 0 & -5 & 0 \end{pmatrix}.$$

103. Find the eigenvalues and associated eigenspaces of the matrices below. State the algebraic and geometric multiplicity for each of the eigenvalues.

$$A = \begin{pmatrix} 1 & -3 & 1 \\ 0 & -1 & 0 \\ 0 & -3 & 2 \end{pmatrix}, \quad B = \begin{pmatrix} 2 & 1 & -2 & -1 \\ 2 & 3 & -4 & -2 \\ 1 & 1 & -1 & -1 \\ 0 & 0 & 0 & 1 \end{pmatrix}, \quad C = \begin{pmatrix} 3 & -2 & 2 \\ 4 & -2 & 3 \\ 0 & -1 & 2 \end{pmatrix}.$$

104. Consider the matrix below

$$A = \begin{pmatrix} 0 & 0 & a \\ 0 & 1 & 0 \\ b & 0 & 0 \end{pmatrix}$$

where a and b are real. Determine the values of a and b for which

i) All eigenvalues are real.

ii) The matrix has a degenerate eigenvalue.

105. In the matrix A defined below, the parameter x is a real number.

$$A = \begin{pmatrix} 1 & x^2 \\ 1 & 1 \end{pmatrix}.$$

Determine the values of x for which

i) The eigenvalues of A are real and simple and find the eigenvectors.

ii) The eigenvalues are real and semi-simple and find the eigenvectors.

iii) The eigenvalues are real and degenerate and find the generalized eigenspace.

106. The matrix below has two eigenvalues $\lambda_1 = 1$ and $\lambda_2 = 0$. Both have algebraic multiplicity 2. Knowing this, find a similar upper Canonical form Jordan matrix.

$$A = \begin{pmatrix} 5 & 1 & -2 & 1 \\ -8 & -1 & 4 & -2 \\ 4 & 1 & -1 & 1 \\ -5 & -1 & 2 & -1 \end{pmatrix}.$$

107. Determine a similar Jordan matrix to the matrix A given below and state the associated similarity transform.

$$A = \begin{pmatrix} 1 & 1 & 0 \\ 0 & 1 & 0 \\ 0 & 2 & 1 \end{pmatrix}.$$

108. Find a similar diagonal or Jordan form matrix (whichever the case may be) of the matrix below. State the matrix K used in the similarity transformation.

$$A = \begin{pmatrix} 1 & 0 & 0 \\ 0 & 2 & -1 \\ 0 & 1 & 0 \end{pmatrix}.$$

109. Determine a matrix K_U which defines a similarity transformation that maps the matrix A below into a matrix in **upper** Jordan canonical form. Clearly state the matrix K_U and the Jordan canonical matrix that results from using K_U in the similarity transformation.

Repeat for a matrix K_L which defines a similarity transformation that maps A to a matrix in **lower** Jordan canonical form.

$$A = \begin{pmatrix} 2 & 0 & 1 & 0 \\ 0 & 2 & 1 & 0 \\ 0 & 0 & 2 & 0 \\ 0 & 0 & -1 & 2 \end{pmatrix}.$$

110. The matrices A and J below are similar. Determine a matrix K such that $A = KJK^{-1}$.

$$A = \begin{pmatrix} 2 & 0 & 0 \\ 2 & 4 & -1 \\ 4 & 4 & 0 \end{pmatrix}, \quad J = \begin{pmatrix} 2 & 0 & 0 \\ 0 & 2 & 1 \\ 0 & 0 & 2 \end{pmatrix}.$$

111. Find the eigenvalues, their algebraic and geometric multiplicities and their eigenspaces for the matrix below.

$$\begin{pmatrix} 1 & 3 & 5 & 9 \\ 0 & 1 & 3 & 5 \\ 0 & 0 & 4 & 0 \\ 0 & 0 & 0 & 4 \end{pmatrix}.$$

Give an example of a similar Jordan matrix.

112. Do a complete spectral analysis of the matrix below. Identify a similarity transformation between A and a matrix in Jordan canonical form and state the Jordan canonical form matrix for this transformation.

$$A = \begin{pmatrix} 1 & 1 & 2 \\ 0 & 1 & 1 \\ 0 & 0 & 1 \end{pmatrix}.$$

113. For the matrix A below,

$$A = \begin{pmatrix} 3 & 1 & -1 & 0 \\ 7 & 9 & -3 & -4 \\ 8 & 8 & -2 & -4 \\ 16 & 16 & -8 & -6 \end{pmatrix}$$

the characteristic polynomial is $(\lambda + 2)(\lambda - 2)^3$.

Determine a similarity transformation between A and a Jordan canonical matrix and state the matrices used in the transformation as well as the Jordan matrix itself.

114. Find a similar upper Jordan matrix to the following matrix A and clearly state the matrix K used in the similarity transformation, $A = KJK^{-1}$.

$$A = \begin{pmatrix} 1 & 1 & 0 & -2 \\ 0 & 1 & 0 & 1 \\ 0 & 1 & 1 & -2 \\ 0 & -1 & 0 & 3 \end{pmatrix}.$$

115. Do a complete spectral analysis of the given matrix. In other words, find all eigenvalues, their algebraic and geometric multiplicities and their associated eigenvectors. Determine the dimension of the eigenvector space of each eigenvalue and find a basis for the space.

For all eigenvalues for which the dimension of the eigenspace is less than the algebraic multiplicity of the eigenvalue, determine the index of the eigenvalue and a chain of linearly independent generalized eigenvectors. Find a basis for the generalized eigenspace of these eigenvalues. The dimension of the generalized eigenspace must equal the algebraic multiplicity of the eigenvalues.

Determine a similarity transformation between A and a matrix in Jordan canonical form and state the Jordan canonical form.

$$A = \begin{pmatrix} \frac{282}{17} & -\frac{244}{17} & 20 & \frac{105}{17} & -\frac{145}{17} & \frac{28}{17} \\ \frac{585}{17} & -\frac{487}{17} & 35 & \frac{218}{17} & -\frac{263}{17} & \frac{108}{17} \\ \frac{363}{17} & -\frac{262}{17} & 20 & \frac{114}{17} & -\frac{172}{17} & \frac{10}{17} \\ \frac{65}{17} & -\frac{56}{17} & 6 & \frac{28}{17} & -\frac{67}{17} & -\frac{56}{17} \\ \frac{548}{17} & -\frac{390}{17} & 37 & \frac{161}{17} & -\frac{330}{17} & -\frac{118}{17} \\ \frac{209}{17} & -\frac{156}{17} & 11 & \frac{78}{17} & -\frac{81}{17} & \frac{65}{17} \end{pmatrix}.$$

You do not want to do this problem by hand (unless you are insane), so use Maple, Mathematica or some similar package to do your calculations. Hand in an adequately commented printout of your document, an explanation of the calculations done in the document and a summary of the results. As always, the grader must be able to follow your work.

116. Determine an expression for A^n where

$$A = \begin{pmatrix} 1 & 0 & 0 \\ 0 & 1 & -1 \\ 1 & -4 & 1 \end{pmatrix}.$$

You may leave the expression in a form that contains matrix products as long as the matrices in the product are defined.

117. Find $\exp(A)$ where

$$A = \begin{pmatrix} 5 & 4 \\ -2 & -1 \end{pmatrix}.$$

118. Find $\exp(B)$ where

$$B = \begin{pmatrix} 1 & 1 \\ 0 & 0 \end{pmatrix}.$$

119. Evaluate e^A when

$$A = \begin{pmatrix} -1 & 2 \\ -1 & 1 \end{pmatrix}.$$

120. Derive a general expression for A^P where

$$A = \begin{pmatrix} 1 & 1 \\ 0 & 2 \end{pmatrix}.$$

121. Do a spectral analysis of matrix A. i.e. find eigenvalues, eigenvectors, generalized eigenvectors, a similar Jordan matrix and the similarity transform between the Jordan matrix and A. Then evaluate $\exp(A \cdot t)$.

$$A = \begin{pmatrix} \frac{1}{2} & \frac{1}{2} & 1 \\ -\frac{1}{2} & \frac{3}{2} & 3 \\ 0 & 0 & 2 \end{pmatrix}.$$

122. Use Gram–Schmidt orthogonalization to determine an orthogonal basis from the vectors

$$v_1 = \begin{pmatrix} 0 \\ -2 \\ 2 \end{pmatrix}, \; v_2 = \begin{pmatrix} 2 \\ -1 \\ 1 \end{pmatrix}, \; v_3 = \begin{pmatrix} -2 \\ 2 \\ -4 \end{pmatrix}.$$

123. Consider the space of first order polynomials in x over $[0, 1]$.

$$p(x) = a + bx \;, \; x \in [0, 1]$$

and let a basis of this space be B,

$$B = \{1, x\}.$$

Define an inner product on this space as

$$\langle p_1(x), p_2(x) \rangle = \int_0^1 p_1(x) p_2(x) dx$$

and Use Gram–Schmidt orthogonalization to obtain an orthogonal basis from the given basis B.

Tensors

124. Consider the function space defined by

$$y = C_1 \sin(x) + C_2 \cos(x)$$

where $\sin(x)$ and $\cos(x)$ are the basis vectors of the space and C_1 and C_2 are the coordinates of a vector in this basis. Using this basis, derive the matrix representation of the linear operator for differentiation.

125. Let the linear operator

$$LP_N = x^2 \frac{d^2 P_N}{dx^2} - x \frac{dP_N}{dx} + P_N$$

map elements of the space of Nth order polynomials in x into the same space. If the basis of the polynomial space is

$$B = \{1, x, x^2, \cdots, x^N\}$$

then find the matrix representation of L for $N = 0$, $N = 1$, $N = 2$ and for arbitrary values of N.

What is the dimension of the space spanned by LP_N? Explain your answer!

126. Elements of the vector space of second order polynomials in x, polynomials of the form $a_0 + a_1 x + a_2 x^2$, are mapped into this same vector space by the linear operator

$$Lv = x^2 \frac{d^2 v}{dx^2} + v\,.$$

If the bases of both the domain and image space is $B = \{1, x, x^2\}$, what is the matrix representation of the operator?

127. Let $p(x)$ be the space of third order polynomials, i.e.

$$p(x) = \{a + b \cdot x + c \cdot x^2 + d \cdot x^3\}.$$

The linear operator $L : p(x) \to p(x)$ is defined as

$$Lp = \frac{d^2 p}{dx^2} + x \frac{dp}{dx} + 3p\,.$$

a) What is the matrix representation of the operator in the basis $\{1, x, x^2, x^3\}$?

b) Confirm your result by evaluating $L(a + b \cdot x + c \cdot x^2 + d \cdot x^3)$ using both the differential operator and the matrix representation of the operator.

c) Find a basis for $p(x)$ such that this operator is represented by a diagonal matrix. Find the matrix representation of L in this basis?

128. Consider the vector space $\mathcal{P}_2(x)$, the space of real 2nd order polynomials in x. I.e. polynomials of the form $a_2 x^2 + a_1 x + a_0$ where a_2, a_1 and a_0 are real. Let the basis of the space be $B = \{1, x, x^2\}$.

The operator L maps elements of $\mathcal{P}_2(x)$ into $\mathcal{P}_2(x)$ by

$$L = \frac{d^2}{dx^2} + \frac{d}{dx}\,.$$

i) Find the matrix representation for L.

ii) Suppose that L maps from the space of real Nth order polynomials, polynomials of the form $a_N x^N + \cdots + a_1 x + a_0$, into the same space and assume further that the basis is $B = \{1, x, \cdots, x^N\}$. What is the matrix representation for L? (Justify your answer!)

129. Call the vector space of all Nth order polynomial P_N, i.e.P_N is the set of all polynomials of the form

$$P_N(x) = a_0 + a_1 x + a_2 x^2 + \cdots + a_N x^N.$$

Differentiation of these polynomials with respect to x maps P_N into P_{N-1}. Once a basis has been chosen for P_N and P_{N-1} this mapping can be represented by a matrix. Determine this matrix relative to the basis B, where B is

$$B = \{1, x, x^2, \cdots, x^N\}.$$

130. Let V be the vector space spanned by the basis $B = \{1, \cos(t), \sin(t)\}$ and let L be the operator given by

$$L : V \to V$$

where

$$Lv = \int_0^\pi v(s)(1 + \cos(t - s))\, ds$$

show that L is a linear operator and find the matrix representation of L using the basis B in both the domain and image space. The following formulas may be useful:

$$\sin(x \pm y) = \sin(x)\cos(y) \pm \cos(x)\sin(y)\ , \ \cos(x \pm y) = \cos(x)\cos(y) \mp \sin(x)\sin(y)$$

$$\int \cos(x)^2\, dx = \cos(x)\sin(x)/2 + x/2\ ,\ \int \sin(x)^2\, dx = -\cos(x)\sin(x)/2 + x/2$$

$$\int \cos(x)\sin(x)\, dx = -\frac{1}{2}\cos(x)^2.$$

131. Consider the vector space of points in three-dimensional space given by their coordinates in the usual rectangular (x, y, z) coordinate system. What is the matrix of the operator which projects points into the (x, y) plane?

Consider points in two-dimensional space given by coordinates in the usual (x, y) coordinate system. What is the matrix of the operator that rotates all points an angle $\pi/2$ around the origin?

132. Consider the vector space of second order polynomials of the form $P_2(x) = a_0 + a_1x + a_2x^2$ with the basis $B_2 = \{1, x, x^2\}$. Integration maps these polynomials into third order polynomials of the form $P_3(x) = C + a_0x + \frac{a_1}{2}x^2 + \frac{a_2}{3}x^3$ where C is an arbitrary constant. Let the basis for the space of third order polynomials be $B_3 = \{1, x, x^2, x^3\}$. What is the matrix of the integration operator?

Change the basis for the space of second order polynomials to $\overline{B_2} = \{1, 1+x, 1+x+x^2\}$ but keep the basis for the space of third order polynomials the same. With this choice of basis, what is the matrix for the integration operation?

133. Consider functions of the form $f(x) = \alpha_1 e^x + \alpha_2 ln(|x|) + \alpha_3 x$, where the αs are real parameters. Clearly these functions form a vector space. What is the coordinate representation of $f(x)$ in the basis $B_1 = \{e^x, ln(|x|), x\}$?

What is the coordinate representation in the basis $B_2 = \{x + e^x, ln(|x|) + x, ln(|x|)\}$?

The differentiation operator D maps the functions $f(x)$ into the space of functions $g(x) = \alpha_1 e^x + \frac{\alpha_2}{x} + \alpha_3$. If the basis for the space of $g(x)$-functions is $B_3 = \{e^x, \frac{1}{x}, 1\}$ and the basis for the $f(x)$ functions is B_1, what is the matrix of the operator?

If the basis for the $f(x)$-functions is B_2, what is the matrix of the operator?

134. Find the matrix of transition between the basis B_V and $B_{\hat{V}}$ where

$$B_V = \left\{ \begin{pmatrix} 1 \\ 3 \\ 1 \end{pmatrix}, \begin{pmatrix} 2 \\ 0 \\ 1 \end{pmatrix}, \begin{pmatrix} 1 \\ 1 \\ 4 \end{pmatrix} \right\}$$

$$B_{\hat{V}} = \left\{ \begin{pmatrix} 1 \\ 1 \\ 4 \end{pmatrix}, \begin{pmatrix} 4 \\ 2 \\ 1 \end{pmatrix}, \begin{pmatrix} 3 \\ 4 \\ 1 \end{pmatrix} \right\}.$$

Let the matrix of transition between the basis B_V and another basis, B_W, be

$$A = \begin{pmatrix} 1\,4\,0 \\ 2\,2\,1 \\ 4\,0\,1 \end{pmatrix}.$$

Find B_W.

135. An (x, y, z) coordinate system is rotated an angle θ, in the positive direction, about the z-axis. What is the rotation matrix Θ for this rotation?

In the unrotated coordinate system, a tensor has the coordinate representation

$$T = \begin{pmatrix} 0 & 1 & 0 \\ 1 & 0 & 0 \\ 0 & 0 & 1 \end{pmatrix}.$$

What is the coordinate representation of T in the rotated coordinate system?

136. What is the physical interpretation of the rotation represented by the rotation matrix

$$\Theta = \begin{pmatrix} 1 & 0 & 0 \\ 0 & \cos\alpha & \sin\alpha \\ 0 & -\sin\alpha & \cos\alpha \end{pmatrix}.$$

137. An (x, y, z) coordinate system is rotated an angle p about the z-axis, then an angle q about the x-axis. What is the rotation matrix for the combined rotation? What is the combined rotation matrix if rotation about the x-axis is done before rotation about the z-axis?

138. The stress tensor in a continuum is given by

$$\tau = \begin{pmatrix} 3xy & 5y^2 & 0 \\ 5y^2 & 0 & 2z \\ 0 & 2z & 0 \end{pmatrix}$$

in some suitable system of units. Determine the force at the point $(x, y, z) = (2, 1, \sqrt{3})$ on the tangent plane to the cylinder given by $y^2 + z^2 = 4$.

Determine the principal stresses (the stresses on the planes with pure normal stress), and their directions, as given by unit normals, for any point on the z-axis.

139. For two-dimensional solid conduction, Fourier's law takes the form

$$q = \begin{pmatrix} k_{11} & k_{12} \\ k_{21} & k_{22} \end{pmatrix} \begin{pmatrix} \frac{\partial T}{\partial x} \\ \frac{\partial T}{\partial y} \end{pmatrix}.$$

Show that the conduction tensor can be simplified to contain only one material constant for a material with a checker-board symmetry, Fig. 139.1.

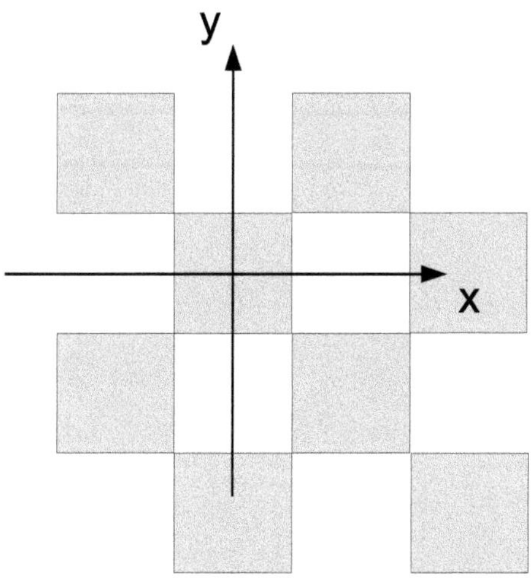

Fig. 139.1 Material with checker-board symmetry. Grey and white squares are made of material with different conductivity. The indicated pattern continues ad infinitum in both x and y-directions.

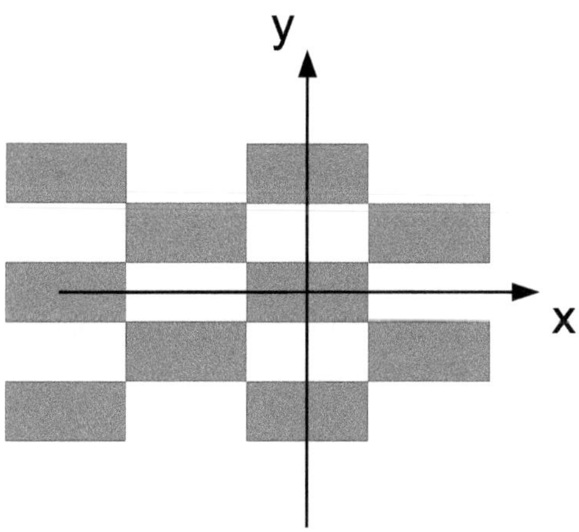

Fig. 140.1 Material with checker-board symmetry.

140. For two-dimensional solid conduction, Fourier's law takes the form

$$q = \begin{pmatrix} k_{11} & k_{12} \\ k_{21} & k_{22} \end{pmatrix} \begin{pmatrix} \frac{\partial T}{\partial x} \\ \frac{\partial T}{\partial y} \end{pmatrix} .$$

Simplify the conduction tensor as much as possible for a material with the checker-board symmetry indicated in Fig. 140.1. Grey and white rectangles are made of isotropic materials with different conductivity and the width of each rectangle equals twice the height. The indicated pattern continues ad infinitum in both x and y-directions.

Linear difference equations

Single equations

141. Find the complete solutions to the difference equation

$$X_{\tau+2} - 3X_{\tau+1} + 2X_{\tau} = 1\,.$$

142. Find the complete solution to

$$y_{\tau+2} - 4y_{\tau+1} + 4y_{\tau} = \sin(\tau).$$

143. Find the general solution of

$$(\tau+2)(\tau+1)y_{\tau+2} - (\tau+1)y_{\tau+1} - 2y_{\tau} = \frac{1}{(\tau-1)!}\,.$$

Hint: Use the variable substitution $z_{\tau} = \tau! y_{\tau}$.

144. Show that the initial value problem below has an infinity of solutions.

$$y_{\tau+1}y_{\tau} + y_{\tau+1} + y_{\tau} + 1 = 0\,,\ y_0 \text{ given.}$$

Hint: Use the variable transformation $y_{\tau} = z_{\tau} + \delta$ and pick a value δ that makes the problem easy to analyze.

145. Find the complete solution to the following difference equation. Is the homogeneous solution stable or not? Explain.

$$x_{\tau+2} + 3x_{\tau+1} + 2x_{\tau} = (-1)^{\tau}.$$

146. The equation

$$x_{\tau+1} = (1-\alpha)x_{\tau} + \alpha y_{\tau}$$

represents a digital low pass filter, a method used for reducing high frequency noise in data series. The input or raw measurements are represented by the sequence, y_{τ}, while the filtered output is represented by x_{τ}. α is a tuning parameter called the smoothing factor and it must be between 0 and 1.

To tune the filter, i.e. decide on a value for α, you want to investigate how it performs under a unit step change in the input, i.e. $y_{\tau} = 1$ for $\tau \geq 0$ and $x_0 = 0$. Find x_{τ} for this situation and determine a bound on α such that $|x_{10} - x_{\infty}| < 0.1$.

Coupled equations

147. Consider the two coupled difference equations

$$x_{\tau+2} + 2x_{\tau+1} - 3y_{\tau+1} + y_\tau = 0$$

$$y_{\tau+2} + x_{\tau+1} + 2y_{\tau+1} + x_\tau - y_\tau = 0\,.$$

Recast these as a system of first order, linear difference equations. Clearly identify the matrix of the system and the components of the vector variable in terms of the variables given here.

148. Rewrite the following two coupled, linear difference equations in vector-matrix form. Solve by doing a spectral analysis of the matrix in the vector-matrix problem.

$$x_{\tau+2} - x_{\tau+1} + 2y_{\tau+1} + 2y_\tau - 2x_\tau = 0$$

$$y_{\tau+2} + y_{\tau+1} + x_{\tau+1} + 2y_\tau - 2x_\tau = 0\,.$$

149. The following two difference equations describe (somewhat simplified) the dynamics of a first order system subject to proportional control. The first equation represents the first order system with e_τ being the deviation variable or error, the second equation represent a digital proportional controller with p_τ being the controller output and K the controller gain.

$$2e_\tau - e_{\tau-1} = p_\tau$$

$$p_\tau = p_{\tau-1} - K(e_\tau - e_{\tau-1}).$$

a) Write these equations in vector-matrix notation.

b) Determine the values of the controller gain K over which the system is unstable.

c) Determine the values of K over which the system can exhibit an oscillatory response.

150. Fibonacci's model of rabbit populations assumed that rabbits are immortal. This is a poor assumption and it can be disposed of by introducing the concept of age classes. A population model can have any number of age classes, but the simplest model has only two, pairs of young individuals Y_τ and pairs of old individuals X_τ. In one month, both young and old pairs breed to form more young pairs, i.e.

$$Y_{\tau+1} = fY_\tau + fX_\tau\,.$$

The model parameter f is known as the fertility rate and $f \geq 0$.

Old pairs simply die but the population of old pairs is replenished because young pairs grow old, i.e.

$$X_{\tau+1} = sY_{\tau}\,.$$

The model parameter s is known as the survival rate and $0 \leq s \leq 1$.

i) Write this model in vector-matrix notation.

ii) Determine a set of constraints on s and f which assure that the population will become extinct with time. I.e. such that $\lim_{\tau\to\infty} Y_{\tau} = \lim_{\tau\to\infty} X_{\tau} = 0\,.$

151. Solve

$$\begin{pmatrix} x \\ y \\ z \end{pmatrix}_{\tau+1} = \begin{pmatrix} 0 & 1 & 1 \\ 0 & 0 & 1 \\ 0 & 1 & 0 \end{pmatrix} \begin{pmatrix} x \\ y \\ z \end{pmatrix}_{\tau}\,.$$

What do you conclude about the stability of this system?

Linear ODEs

Single equations

152. Find the complete homogeneous solution and a particular solution to

$$\frac{d^4x}{dt^4} + 2\frac{d^3x}{dt^3} + \frac{d^2x}{dt^2} = t\,.$$

153. Find the complete solution of

$$\frac{d^3x}{dt^3} - 4\frac{d^2x}{dt^2} + 5\frac{dx}{dt} - 2x = e^t.$$

154. Consider

$$\frac{d^3x}{dt^3} - 3\frac{d^2x}{dt^2} + 4\frac{dx}{dt} - 2x = t\,.$$

a) Find the complete, complex homogeneous solution.

b) Determine the real part of the complete homogeneous solution.

c) Find a particular solution.

155. Solve

$$\frac{d^4x}{dt^4} - 2\frac{d^3x}{dt^3} - 3\frac{d^2x}{dt^2} + 4\frac{dx}{dt} + 4x = t\,.$$

156. Find the complete solution to

$$\frac{d^2x}{dt^2} - \frac{dx}{dt} = 2te^t.$$

157. Find the complete solution to

$$(1-t)\frac{d^2x}{dt^2} + t\frac{dx}{dt} - x = te^{-t}.$$

Note that a possible solution to the homogeneous equation is $x(t) = t$.

158. Find the complete solution to the equation below using the method of matched coefficients to find a particular solution.

$$\frac{d^3x}{dt^3} - 3\frac{dx}{dt} + 2x = e^{2t} + t\,.$$

159. a) Find the complete homogeneous solution to

$$\frac{d^3y}{dt^3} - \frac{dy}{dt} = \sin(t)\ ,\ y(0) = 2\ ,\ y'(0) = 0\ ,\ y''(0) = 1\,.$$

b) Find a particular solution using method of matched coefficients.

c) Find the solution when the initial conditions are applied.

160. Find the real part of the complete solution to

$$\frac{d^2x}{dt^2} + \frac{dx}{dt} + 2x = t^2.$$

161. Find the complete homogeneous solution to

$$\frac{d^2x}{dt^2} - 2\frac{dx}{dt} - 3x = \phi(t).$$

Find a particular solution to the equation above when $\phi(t) = \sin(t)e^t$.
Repeat when $\phi(t) = \exp(\sin(t))$.

Repeat when $\phi(t) = e^{3t}$.

162. Find the complete, real solution to

$$x^2 y'' + xy' + 9y = \frac{1}{1+x} .$$

163. Find a particular solution to

$$\frac{d^2x}{dt^2} + \frac{dx}{dt} - 2x = te^t .$$

164. The differential equation

$$(1-t)\frac{d^2x}{dt^2} + (1-t)\frac{dx}{dt} + x = 1$$

is found by trial and error to have a basis solution

$$x_1(t) = t - 1 .$$

a) Find the other basis solution using variation of parameters.

b) Find a particular solution.

165. Upon examining the differential equation below, you immediately notice that one of the basis solutions is $x_1(t) = t$. Demonstrate how the other basis solution is found by variation of parameters. ("Demonstrate" means that it is not sufficient to substitute into the formula in the text which you may or may not be able to remember correctly)

$$t^2\frac{d^2x}{dt^2} - 6t\frac{dx}{dt} + 6x = 0 .$$

166. One of the basis solutions of the ODE below is $x_1(t) = t^2$. Find the other basis solution.

$$2t^2\frac{d^2x}{dt^2} - 3t\frac{dx}{dt} + 2x = 0 .$$

Coupled equations

167. Consider the equations

$$\frac{d^2x_1}{dt^2} + 2\frac{dx_1}{dt} - 3\frac{dx_2}{dt} + x_2 = 0$$
$$\frac{d^2x_2}{dt^2} + \frac{dx_1}{dt} + 2\frac{dx_2}{dt} + x_1 - x_2 = 0\,.$$

Recast these equations as a system of coupled, linear, first order differential equations. Clearly identify the matrix of the system and the components of the vector variable in terms of the variables given here.

168. Consider

$$\frac{dx}{dt} = \begin{pmatrix} 1 & -3 & 1 \\ 0 & -1 & 0 \\ 0 & -3 & 2 \end{pmatrix} x + \begin{pmatrix} 0 \\ 0 \\ 2 \end{pmatrix}\,,\ x(0) = \begin{pmatrix} 0 \\ 1 \\ 0 \end{pmatrix}.$$

a) Find the complete homogeneous solution.

b) Find the complete solution.

c) Find the solution that satisfies the initial condition.

169. Find the complete, complex solution to the equations below, then extract the real part of this solution (i.e. write the solution in terms of real functions and real arbitrary constants.)

$$\frac{dx_1}{dt} = x_1 + x_2$$
$$\frac{dx_2}{dt} = -x_1 + x_2\,.$$

170. Given the system of differential equations

$$\frac{dx_1}{dt} = 2x_1 - 2x_2 + 3x_3$$
$$\frac{dx_2}{dt} = x_1 + x_2 + x_3$$
$$\frac{dx_3}{dt} = x_1 + 3x_2 - x_3$$

with the initial conditions

$$t = 0 \Rightarrow x_1 = 1\,,\ x_2 = 0\,,\ x_3 = 1/2$$

solve by first finding the eigenvalues, eigenvectors and eigenrows and then calculating the solution using these.

171. Find the complete solution of

$$\frac{dx_1}{dt} = 2x_1 + 3x_2 + 3x_3$$
$$\frac{dx_2}{dt} = 3x_1 - x_2$$
$$\frac{dx_3}{dt} = 3x_1 - x_3$$

then find the specific solution that satisfies the following condition at $t = 0$.

$$(x_1, x_2, x_3) = (1, -2, 0).$$

The equations are now changed by adding constant terms to the right hand side to get

$$\frac{dx_1}{dt} = 2x_1 + 3x_2 + 3x_3 + 2$$
$$\frac{dx_2}{dt} = 3x_1 - x_2 - 2$$
$$\frac{dx_3}{dt} = 3x_1 - x_3 + 1\,.$$

Find the complete solution to this set of equations.

172. Find the eigenvalues, eigenvectors and eigenrows of the matrix

$$A = \begin{pmatrix} 0 & 1 & 1 \\ 0 & 2 & 0 \\ 1 & 1 & 0 \end{pmatrix}.$$

Then find the solution to the initial value problem

$$\frac{dx}{dt} = y + z\,,\ x(0) = 1$$
$$\frac{dy}{dt} = 2y + 2\,,\ y(0) = 1$$
$$\frac{dz}{dt} = x + y + 2\,,\ z(0) = 1\,.$$

173. Find the complete solution to

$$\begin{aligned}\frac{dx_1}{dt} &= 8x_1(t) - x_2(t) + t \\ \frac{dx_2}{dt} &= -8x_1(t) + 10x_2(t) + e^{6t}.\end{aligned}$$

174. Find the complete solution to

$$\begin{aligned}\frac{dx}{dt} + \frac{dy}{dt} &= 3x \\ \frac{dx}{dt} - \frac{dy}{dt} &= -x + 4y\,.\end{aligned}$$

175. Using variation of parameters, solve

$$\begin{aligned}\frac{dx_1}{dt} &= x_1 + 2x_2 + t \\ \frac{dx_2}{dt} &= 2x_1 + x_2\,.\end{aligned}$$

Stability problems

176. Consider the following system of equations

$$\begin{aligned}\frac{dx}{dt} &= -(A + (x^2 + y^2))x - y\omega \\ \frac{dy}{dt} &= -(A + (x^2 + y^2))y + x\omega\end{aligned}$$

where $\omega > 0$.

a) Find the steady state, linearize the equations and determine the Jacobian at the steady state.
b) Determine the stability of the steady state as a function of the parameter A.
c) For the stable case: find the eigenvectors at the steady state and draw a sketch of the phase plane close to the steady state.
d) For the unstable case: show, that when the steady state is unstable, the system is still stable with respect to a circular domain around the steady state.
e) Find the minimum radius of this circular domain as a function of A.

177. Consider the set of coupled ODEs

$$\frac{dx}{dt} = x - y - 1$$

$$\frac{dy}{dt} = x^2(y-2) + 2y + 2$$

$$\frac{dz}{dt} = y - 2z - 1\,.$$

a) Find all the steady state solutions.

b) Find the Jacobian matrix of the system at the steady state $(x, y, z) = (0, -1, -1)$ and state the approximate linear model around this point.

c) Is the steady state at $(x, y, z) = (0, -1, -1)$ locally stable? Explain your answer!

d) Find the complete solution to the linearized model from part b).

178. The following system of equations has been used to describe the dynamics of certain biochemical networks.

$$\frac{dx}{dt} = 1 - xy^{\gamma}$$

$$\frac{dy}{dt} = 4xy^{\gamma} - 4y\,.$$

The parameter, γ is a positive constant.

a) Determine the Jacobian matrix for this system at an arbitrary point (x_0, y_0).
b) Determine the steady state and linearize the equations around it.
c) Determine the stability of the steady state and classify it (node, focus, etc.) as a function of γ.
d) Sketch the trajectories in the phase space close the steady state for each type of steady state.

179. One model of microbial growth in a CSTR is

$$\frac{dX}{dt} = \mu_m(1 - e^{-S/K})X - DX$$

$$\frac{dS}{dt} = D(S_f - S) - \frac{1}{Y}\mu_m(1 - e^{-S/K})X.$$

The first equation is a balance on the biomass, X, and the second equation is a balance on the substrate, S. The constant model parameters are the yield, Y, the maximum specific growth rate μ_m and a constant K. All are strictly positive.

D is the dilution rate and X and S the biomass and substrate concentrations, respectively and S_f is the substrate concentration in the feed stream.

a) The system has two steady states, a sterile and a growth steady state. Find the biomass and substrate concentrations at the growth steady state.

b) Find the washout dilution rate, D_w, the dilution rate at which the biomass concentration in the reactor becomes zero.

c) Determine the Jacobian of the system.

d) Show that the growth steady state is stable when the dilution rate is less than the washout dilution rate and unstable otherwise for the model.

180. The following equation is known as van der Pol's equation. Determine the values of ϵ for which the trivial solution, $x = 0$ is stable.

$$\frac{d^2x}{dt^2} - \epsilon(1 - x^2)\frac{dx}{dt} + x = 0\,.$$

181. Determine the steady state(s) or fixed point(s) x_e of the following system as functions of the parameter a. Then, determine the values of a for which the fixed point with the smallest value of x_e, is unstable.

$$\frac{dx}{dt} = -y - x$$

$$\frac{dy}{dt} = z - y$$

$$\frac{dz}{dt} = yz - a\,.$$

182. The following equations have been used to model the dynamics of predator-prey systems.

$$\frac{dx}{dt} = ax - bxy$$

$$\frac{dy}{dt} = cxy - dy\,.$$

i) Find the the general form of the Jacobian of this system.

ii) The model has two steady states, the trivial steady state at $(x, y) = (0, 0)$ and a second steady state away from the origin. Find this non-trivial steady state and determine what you can about its stability as a function of the model parameters a, b, c and d.

Series solutions

183. Consider the initial value problem

$$\frac{d^2x}{dt^2} + (1 - t^2)x = 0$$

$$t = 0 \Rightarrow x = 1 \,, \; \frac{dx}{dt} = 1 \,.$$

Find a series solution for $x(t)$ around $t = 0$.

184. Consider the differential equation

$$\frac{d^2y}{dt^2} - ty = 0 \,.$$

We seek the complete series solution around 0, and because 0 is an ordinary point, we know that y can be represented by a series of the form

$$y = \sum_{n=0}^{\infty} a_n t^n \,.$$

Substitute this series into the differential equation and determine two linearly independent solutions.

185. Find a series solution to

$$\frac{d^3x}{dt^3} - tx(t) = 0 \,, \; (t, x, x', x'') = (0, 0, 1, 2) \,.$$

Report your result as a difference equation with initial conditions in the coefficients of the series.

186. Consider the following differential equation

$$t\frac{d^2y}{dt^2} + \frac{dy}{dt} + y = 0 \,.$$

Derive the indicial equation and the difference equation between the coefficients as they appear when seeking a series solution around $t = 0$.

187. Find series representations around $t = 0$ for two basis solution of

$$\frac{d^2y}{dt^2} + \frac{t^2+1}{t}\frac{dy}{dt} + y = 0 \,.$$

Only positive values of t need to be considered in the solution.

188. Find series representations around $t = 0$ for two basis solutions of

$$t^2 \frac{d^2 x}{dt^2} + 5t \frac{dx}{dt} + (t + 4)x = 0\,.$$

The solution need only be worked out for positive values of t.

189. Find series solutions around $t = 0$ for the basis solutions of

$$\frac{d^2 x}{dt^2} + \frac{dx}{dt} - \frac{t + 1}{t^2} x = 0\,.$$

Bessel functions

190. a) Derive the reactant balance equation for a first order, irreversible reaction with rate constant k, in a cylindrical catalytic pellet of radius R, and state the relevant boundary conditions. Assume there is no film resistance at the surface and that transport in the pellet is by Fickian diffusion with an effective diffusion coefficient D. The reactant concentration outside the pellet is C_∞.
Make the balance equation and boundary conditions dimensionless using R to define a dimensionless radial coordinate and C_∞ to define a dimensionless concentration and show that the problem can be written as

$$\frac{d^2 x}{d\xi^2} + \frac{1}{\xi}\frac{dx}{d\xi} - 4\Phi^2 x = 0\,.$$

What is the Thiele modulus Φ and what are the dimensionless boundary conditions?
b) Solve for $x(\xi)$.
c) Find the effectiveness factor η as a function of Φ and plot $\log(\eta(\Phi))$ versus $\log(\Phi)$ for Φ between 0.1 and 100.

d) Repeat the problem if the center of the catalyst is an inert support material through which the reactant does not diffuse. Assume that the support has a radius R_1 and the catalytic layer goes from R_1 to R_2. Use R_2 to make the problem dimensionless and plot $\log(\eta(\Phi, R_1/R_2))$ versus $\log(\Phi)$ for R_1/R_2 equal to 0.01, 0.5 and 0.9.

e) If the Thiele modulus is small ($\Phi << 1$), the reaction is slow relative to the rate of diffusion and the reactant concentration is effectively constant through the pellet. For this special case, find the concentration profile of the product for the pellet considered in part a) assuming that diffusion of the product is slow relative to the reaction rate. You can assume that stoichiometry of the reaction is such that one mole of product is formed for each mole of reactant consumed. The product concentration outside the pellet equals zero and, for the product, there is a film resistance with mass transfer coefficient k_m between pellet surface and the surrounding fluid.

191. The differential equation

$$\frac{d^2x}{dt^2} - tx = 0$$

can be shown to have basis solutions of the form

$$x(t) = \sqrt{t} Z_p \left(\frac{2}{3} t^{\frac{3}{2}} \right)$$

where Z_p is some Bessel function. Determine which Bessel functions by carrying out the appropriate variable transformations to obtain one of Bessel's differential equations.

192. Consider the following equation

$$x^2 \frac{d^2y}{dx^2} + (2x^2 + x)\frac{dy}{dx} + (4x^3 + x^2 + x - 1)y = 0\,.$$

This equation has basis solutions of the form

$$y = \exp(-x) Z_p \left(\frac{4}{3} x^{3/2} \right)$$

where Z_p are the appropriate Bessel functions. Determine what these Bessel functions are and find the complete solution to the differential equation.

193. The motion of a flexible string, attached at one point and swinging freely under the effect of gravity, can be modeled by

$$x\frac{d^2f}{dx^2} + \frac{df}{dx} + \frac{\omega^2}{g} f = 0$$

where $f(x)$ is the deflection from equilibrium, x the upward pointing coordinate in the vertical direction, ω the frequency of the oscillations and g the acceleration of gravity.

a) Transform this equation using the variable transformation $x = \frac{1}{4} g \tau^2$ and find the complete solution to the transformed equation.

b) Simplify the solution using the fact that the deflection of the string is finite for all values of x.

The swinging string is attached at $x = L$, giving the boundary condition

$$x = L \Rightarrow f(x) = 0\,.$$

c) What can you conclude about the values of the frequency ω?

194. Transform the equation

$$x^2 \frac{d^2y}{dx^2} + 3x\frac{dy}{dx} + (x-1)y = 0$$

using the free variable transformation $t = 2\sqrt{x}$.

Transform the equation further using the dependent variable transformation $f = \frac{1}{4}t^2 y$.

Solve the resulting equation and find $y(x)$.

195. Consider a cylindrical catalytic pellet of radius R. The radius is much smaller than the length of the cylinder, so end effects can be ignored, i.e. reactant concentration is a function of the radial coordinate only. Let the concentration of the reactant at the surface be C_0, the effective diffusion coefficient of the reactant in the pellet be D and assume that the reaction is first order with respect to reactant concentration C, with a reaction rate constant of k.

a) Write the steady state balance equation for the reactant in the pellet and state the boundary conditions.

b) The Thiele modulus for this reaction and geometry is defined as $\Phi = R/2\sqrt{k/D}$. Make the balance equation dimensionless using this modulus and solve for the dimensionless reactant concentration of the pellet.

c) Find the effectiveness factor $\eta(\Phi)$ and plot $\ln(\eta)$ versus $\ln(\Phi)$.

d) Assume now that there is film resistance at the pellet surface with mass transfer coefficient h. Introduce the Sherwood number, $Sh = hR/D$, restate the problem (balance equation and boundary conditions) in dimensionless form and solve for the dimensionless concentration profile.

Hilbert spaces

196. The *Chebychef* polynomials can be defined as

$$T_n(x) = \cos(n\theta)$$

where

$$\cos(\theta) = 2x - 1\,.$$

Show that the *Chebychef* polynomials are orthogonal over $[0, 1]$ with the inner product

$$\langle T_n, T_m \rangle = \int_0^1 T_n(x) T_m(x) x^{-1/2} (1-x)^{-1/2}\, dx\,.$$

You will need to use : $\cos(n\theta) + i\sin(n\theta) = (\cos(\theta) + i\sin(\theta))^n$, (de Moivre's equation), where i is the imaginary unit.

197. Determine the adjoint operator of

$$Ly = -\frac{d^2y}{dx^2}\,,\; y(0) = 0\,,\; y(1) = y'(0).$$

198. Determine the adjoint operator and adjoint boundary conditions of the operator L

$$Lf = \frac{d^2f}{dx^2}\,,\; f(0) = f'(0) + f(1)\,,\; f(1) = f'(1) + f(0)$$

in the inner product space defined by

$$\langle f, g\rangle = \int_0^1 f(x)g(x)\,dx\,.$$

Is L self-adjoint or formally self-adjoint?

199. For the differential equation

$$(1-x^2)\frac{d^2f}{dx^2} - 4x\frac{df}{dx} + N(N+3)f = 0\,.$$

a) Find series representation around $x = 0$ of two linearly independent basis solutions. Clearly state the recursion formula for the coefficients and explain how two linearly independent solutions are obtained.

b) Show that if N is a positive integer, then the equation has polynomial solutions.

c) Determine whether or not all solutions are polynomial when N is a positive integer. Explain your reasoning!

d) Given the boundary conditions $f(-1) = -1$, $f(1) = 1$. Determine whether or not the operator on the left hand side of the differential equation is self-adjoint or not in the inner product

$$\langle f, g\rangle = \int_{-1}^1 (1-x^2)fg dx$$

and state the adjoint boundary conditions and operator if the operator of the ODE above is not self-adjoint.

200. Solve the eigenvalue problem below. I.e. find all the eigenvalues or the equation(s) for them and their associated eigenfunctions.

$$\frac{d^2\theta_n}{d\xi^2} = \lambda_n\theta_n\,,\; \theta_n(0) = 0\,,\; 3\theta_n(1) = \left.\frac{d\theta_n}{d\xi}\right|_{\xi=1}.$$

201. Show that the operator

$$Lf = \frac{1}{r}\frac{d}{dr}\left(r\frac{df}{dr}\right) \, , \; f(0) = f'(0) \, , \; f(1) = 2f'(1)$$

is self-adjoint in the Hilbert space defined by the inner product

$$\langle f, g \rangle = \int_0^1 rfg dr \, .$$

Then solve the eigenvalue problem

$$\frac{1}{r}\frac{d}{dr}\left(r\frac{df}{dr}\right) = \lambda f \, , \; f(0) = f'(0) \, , \; f(1) = 2f'(1).$$

202. Find the eigenvalues, λ, and eigenfunctions of the following Sturm–Liouville problem where Φ is a real positive parameter.

$$\frac{d^2y}{dx^2} + (\Phi^2 + \lambda)y(x) = 0 \, , \; y(0) = 0 \, , \; y(1) = 0 \, .$$

203. Solve the eigenvalue problem

$$\frac{1}{t^2}\frac{d}{dt}\left(t^2\frac{dx}{dt}\right) + \lambda x = 0$$

$x(t)$ is finite for all t

$$t = 1 \Rightarrow \frac{dx}{dt} = 0 \, .$$

Note: the variable substitution $y(t) = t \cdot x(t)$ may be useful.

204. Find all eigenvalues and all eigenvectors or eigenfunctions of

$$\frac{d^2x}{dt^2} - 2\frac{dx}{dt} + x = \lambda x \, , \; x(0) = 0 \, , \; x(1) = 0 \, .$$

205. Determine if the Sturm–Liouville operator is self-adjoint with periodic boundary conditions of the form

$$y(l) = y(-l) \, , \; y'(l) = y'(-l).$$

206. Find all the **real** eigenvalues and eigenfunctions of the following problem. Note that the boundary conditions are different from the Sturm–Liouville boundary conditions so this problem is not a Sturm–Liouville problem. (Depending on how you represent your solutions, the formula, $\cosh^2(x) - \sinh^2(x) = 1$, may come in handy.)

$$\frac{d^2 f}{dx^2} + \lambda f(x) = 0$$

$$\left.\frac{df}{dx}\right|_{x=0} = \left.\frac{df}{dx}\right|_{x=1}$$

$$f(0) = f(1).$$

207. Solve the eigenvalue problem

$$t^2\frac{d^2 x}{dt^2} + t\frac{dx}{dt} - 4x(t) = \lambda t^2 x(t)$$

$$t = 0 \Rightarrow x = 0$$

$$t = 1 \Rightarrow x = 0$$

and state the inner product for which the eigenfunctions are orthogonal.

208. Show that the set of functions, $\{J_0(\lambda_n x)\}_{n=0,\infty}$, where the λ_n are the zeros of $J_0(x)$, form a basis for the space of piecewise differentiable functions by identifying the appropriate SLP and state the inner product in which the basis functions are orthogonal.

209. Find the cosine and sine Fourier series of

$$f(x) = \begin{cases} x \ , \ x \leq \frac{1}{2} \\ 1 - x \ , \ x \geq \frac{1}{2} \end{cases}$$

over the interval $[0, 1]$ and find the complete Fourier series over $[-1, 1]$. Plot the results using 5 and 50 terms.

210. Expand the function e^x in a complete Fourier series over the interval $-\pi \leq x \leq \pi$. Determine the expressions for all the expansion coefficients and plot the result using 5 and 15 terms in the series.

211. Use the orthogonal basis given by the eigenfunctions of

$$\frac{d}{dx}\left(x\frac{dy}{dx}\right) + \lambda x y(x) = 0$$

$$y'(0) = 0\ ,\ y(1) = 0$$

to expand the function

$$f(x) = \begin{cases} x\ ,\ 0 \le x \le \frac{1}{2} \\ \frac{1}{2}\ ,\ \frac{1}{2} \le x \le 1 \end{cases}$$

in a Fourier series. Plot the series between 0 and 1 using 15 and 50 terms.

PDEs

212. Vibrations of a one-dimensional string, fixed at both ends, can be modeled by the dimensionless wave equation below

$$\frac{\partial^2 \Theta}{\partial t^2} - \frac{\partial^2 \Theta}{\partial x^2} = 0$$

where x is the deflection of the string and t is time. Since the ends of the string are fixed, the boundary conditions can be written

$$\Theta(0,t) = \Theta(1,t) = 0\,.$$

Let the initial conditions specify that there is no initial deflection but that the string is struck at $t = 0$ to impart the velocity $v(x)$. Mathematically this becomes

$$\Theta(x,0) = 0\ ,\ \left.\frac{\partial \Theta}{\partial t}\right|_{t=0} = v(x).$$

The associated Sturm–Liouville problem is

$$\frac{d^2\Theta_n}{dx^2} + \lambda_n \Theta_n = 0\ ,\ \Theta_n(0) = \Theta_n(1) = 0$$

and it has the solution

$$\lambda_n = (n\pi)^2\ ,\ n = 1, 2, 3, ...$$

$$\Theta_n(x) = \sin(n\pi x).$$

Knowing this, find $\Theta(x, t)$.

213. Solve

$$\frac{\partial \Theta}{\partial \tau} = \frac{\partial^2 \Theta}{\partial \xi^2}$$

$$\tau = 0 \Rightarrow \Theta = 1$$

$$\xi = 0 \Rightarrow \frac{\partial \Theta}{\partial \xi} = f(\tau)$$

$$\xi = 1 \Rightarrow \Theta = g(\tau).$$

214. Find the eigenfunctions and an equation for the eigenvalues of the problem below.

$$\frac{\partial \Theta}{\partial \tau} = \frac{1}{\xi^2}\frac{\partial}{\partial \xi}\left(\xi^2 \frac{\partial \Theta}{\partial \xi}\right)$$

$$\tau = 0 \Rightarrow \Theta = 1$$

$$\xi = 0 \Rightarrow \frac{\partial \Theta}{\partial \xi} = 0 \text{ or } \Theta \text{ finite}$$

$$\xi = 1 \Rightarrow \frac{\partial \Theta}{\partial \xi} = -Bi\Theta.$$

(The variable transformation $p(\xi) = f(\xi)\xi$ may be useful).
What orthogonality property holds for the eigenfunctions?
Write the solution for Θ in terms of the eigenfunctions and eigenvalues.

215. Solve

$$\frac{\partial^2 \Theta}{\partial \xi^2} = \frac{\partial \Theta}{\partial \tau}$$

$$\tau = 0 \Rightarrow \Theta = 1$$

$$\xi = 0 \Rightarrow \Theta = 0$$

$$\xi = 1 \Rightarrow \frac{\partial \Theta}{\partial \xi} + Bi\Theta = 0$$

where Bi is a positive constant. Clearly state the definition of the inner product that you use, the eigenfunctions and the eigenvalues. If you cannot find explicit solutions for the eigenvalues, you must locate the eigenvalues in disjoint search intervals. You should only evaluate the various inner products that appear in the solution if you have time available at the end.

b) Redo part a) if the differential equation is changed to

$$\frac{\partial^2 \Theta}{\partial \xi^2} - \Theta(\xi, \tau) = \frac{\partial \Theta}{\partial \tau}\,.$$

c) Change the boundary condition at $\xi = 1$ in part a) to $\Theta = f(\tau)$. Define a variable transformation such that the transformed problem has homogeneous boundary conditions and state the transformed PDE and boundary and initial conditions.

216. Consider a wire in which, at time $t = 0$, an electric current starts to generate heat at some constant rate Q (energy per time per volume). Show that a dimensionless differential equation model for this system is

$$q + \frac{1}{\xi}\frac{\partial}{\partial \xi}\left(\xi \frac{\partial \Theta}{\partial \xi}\right) = \frac{\partial \Theta}{\partial \tau}$$

$$\text{I.C. } \tau = 0 \Rightarrow \Theta(\xi) = \Theta_0(\xi)$$

$$\text{B.C.1 } \xi = 0 \Rightarrow \frac{\partial \Theta}{\partial \xi} = 0 \text{ or } \Theta \text{ is finite}$$

$$\text{B.C.2 } \xi = 1 \Rightarrow \frac{\partial \Theta}{\partial \xi} + Bi\Theta = 0\,.$$

What is the definition of the different parameters and variables? Solve for the transient temperature profile. Clearly identify the following: a) The associated eigenvalue problem. b) The equation for the eigenvalues. Is zero an eigenvalue? c) The orthogonality property of the eigenfunctions. (The appropriate inner product for this problem). d) The form of the expansion for Θ.

217. A cylinder symmetric body with height h and radius R is placed on its end on a cooling plate of fixed temperature T_0. The free surfaces of the body contacts a fluid of fixed temperature T_1. Film resistance can be neglected.

Derive a differential equation model, including boundary conditions, for the steady state temperature profile through the body.

a) Solve the model by introducing new variables (not necessarily dimensionless) in such a way that the boundary conditions in z, the axial direction, become homogeneous.

b) Repeat when new variables are defined in such a way that the boundary conditions in r become homogeneous.

218. Solve

$$\frac{\partial^2\Theta}{\partial x^2}+\frac{\partial^2\Theta}{\partial y^2}=0$$

$$x=0 \Rightarrow \Theta=0$$

$$x=1 \Rightarrow \Theta=0$$

$$y=0 \Rightarrow \Theta=0$$

$$y=1 \Rightarrow \Theta=1\,.$$

219. A viscous liquid runs down a vertical wall in a thin film of constant thickness L. A gas stream runs in the opposite direction while being adsorbed into the liquid. Place an x-axis pointing into the film with the origin at the liquid surface. Place a z-axis pointing down along the liquid surface with the origin at the liquid inlet. We want to find the liquid concentration of the adsorbed gas under the following simplifying model assumptions:
a. The gas phase concentration of the adsorbing compound is approximately constant $= C_g$ for all values of z. The liquid concentration at the interphase $x=0$ is given as $C_l = HC_g$, where H is a Henry's law constant.
b. All transport in the x-direction in the liquid is by diffusion with a constant diffusion coefficient D. Transport in the z-direction is by convection only.
c. At the liquid inlet, $z=0$, the liquid concentration of the adsorbing compound is $C(x, z=0)=C_0$.
d. The velocity profile in the liquid can be assumed constant $= v$.
a) Using these modeling assumptions, show that a dimensionless model of this system at steady state is

$$\frac{\partial^2\Theta}{\partial\xi^2}-\frac{\partial\Theta}{\partial y}=0 \text{ where } \Theta=\frac{C-C_l}{C_0-C_l}\,,\ \xi=\frac{x}{L}\,.$$

What is the equation between z and y, and what are the boundary conditions?
b) State the Sturm–Liouville problem that arises from this model.
c) Determine the eigenfunctions of this SLP and state clearly what orthogonality property must hold between the eigenfunctions.
d) Determine the eigenvalues. In particular, determine if zero is an eigenvalue.
e) Derive the solution for Θ in terms of well-defined inner products of the relevant functions and parameters. Evaluate the inner products.

Consider now a modification of the model in which the adsorbing compound in the liquid is consumed in a first order reaction, with a reaction rate constant k.
f) Show that the model now becomes

$$\frac{\partial^2 \Theta}{\partial \xi^2} - \frac{\partial \Theta}{\partial y} - K\Theta = \frac{K.C_l}{C_0 - C_l}$$

where the variables have the same definition as previously. What is K? Solve this problem.

220. Consider the problem below which describes, for instance, the concentration profile in a spherical catalytic pellet in which some compound is formed at a constant rate.

$$\frac{\partial \Theta}{\partial \tau} = Q + \frac{1}{\xi^2}\frac{\partial}{\partial \xi}\left(\xi^2 \frac{\partial \Theta}{\partial \xi}\right)$$

$$\tau = 0 \Rightarrow \Theta = \Theta_0(\xi)$$

$$\xi = 0 \Rightarrow \frac{\partial \Theta}{\partial \xi} = 0\,, \quad \text{or } \Theta \text{ finite}$$

$$\xi = 1 \Rightarrow \frac{\partial \Theta}{\partial \xi} + Bi\Theta = 0$$

where Q is a constant. Solve for $\Theta(\xi, \tau)$ and state clearly a) the equation for the eigenvalues, b) the eigenfunctions, c) the appropriate inner product and d) the expression for the expansion coefficients in the series solution in terms of inner products. You are not required to evaluate the inner products.

221. Consider a cylindrical pipe, inner radius R_1 and outer radius R_2, initially at a uniform temperature, T_2. At time zero, the temperature on the inside wall is changed to T_1 while the outside wall remains at the temperature T_2.

Formulate a model, differential equation, boundary and initial conditions, for the temperature profile in the pipe wall. Film resistance can be assumed negligible both on the inside and the outside of the pipe.

Solve for the transient temperature profile in the wall. You are not required to evaluate the integrals that appear in the equation for the expansion coefficients.

222. Solve

$$q + \frac{1}{\xi^2}\frac{\partial}{\partial \xi}\left(\xi^2 \frac{\partial \Theta}{\partial \xi}\right) = \frac{\partial \Theta}{\partial \tau}$$

$$\tau = 0 \Rightarrow \Theta = \Theta_0(\xi)$$

$$\xi = 0 \Rightarrow \Theta \text{ finite or } \frac{\partial \Theta}{\partial \xi} = 0$$

$$\xi = 1 \Rightarrow \Theta = 0\,.$$

Clearly state the eigenfunctions used in the series solution, the eigenvalues and the appropriate inner product.

223. Solve

$$\frac{\partial W}{\partial t} + \frac{\partial}{\partial l}((l - l^2)W) = 0$$

$$t = 0 \Rightarrow W = W_0(l).$$

Regarding this as a population balance of particles characterized by their length l, what is the physical interpretation of the term $(l - l^2)$? What can you infer about the solution based on this interpretation?

224. Consider the Loschmidt apparatus described in the book. The example in the book states the solution for cylinders of finite lengths and for cylinders of infinite length.

a) Give a detailed derivation of the solution for finite length cylinders.

b) Repeat for infinite length cylinders.

c) Solve for $x_B(t, z)$ for z between 0 and L. Use the no-flux boundary condition at $z = L$ and another boundary condition (which you must figure out yourself) at $z = 0$.

Appendix

A.1 Complex numbers

Complex numbers are objects of the form (or objects that can be rearranged to the form)

$$\alpha + \beta i$$

where α and β are real and the imaginary unit i satisfies $i^2 = -1$ or $i = \pm\sqrt{-1}$. In some texts, the imaginary unit is indicated j. The number α is called the *Real Part* of the complex number and β is called the *Imaginary Part*. The number that is obtained when changing the sign of the imaginary part is called the complex conjugate of the number and is indicated by an overbar

$$\overline{\alpha + \beta i} = \alpha - \beta i\,.$$

The absolute value of $\alpha + \beta i$ is designated $|\alpha + \beta i|$ and is defined as the non negative value of the square root of $\alpha^2 + \beta^2$. Thus $|\alpha + \beta i| = \sqrt{(\alpha + \beta i)(\overline{\alpha + \beta i})}$.

Complex numbers can be added, subtracted and multiplied the obvious way

$$(\alpha + \beta i) + (\gamma + \delta i) = \alpha + \gamma + (\beta + \delta)i$$

$$(\alpha + \beta i)(\gamma + \delta i) = \alpha\gamma + \beta\delta i^2 + \alpha\delta i + \beta\gamma i = \alpha\gamma - \beta\delta + (\alpha\delta + \beta\gamma)i\,.$$

Division is a bit more involved because of the steps required to bring the result to standard form. This involves multiplying both the numerator and the denominator by the complex conjugate of the denominator term

$$\frac{\alpha + \beta i}{\gamma + \delta i} = \frac{(\alpha + \beta i)(\gamma - \delta i)}{(\gamma + \delta i)(\gamma - \delta i)} = \frac{\alpha\gamma + \beta\delta + (\beta\gamma - \alpha\delta)i}{\gamma^2 + \delta^2} = \frac{\alpha\gamma + \beta\delta}{\gamma^2 + \delta^2} + \frac{(\beta\gamma - \alpha\delta)}{\gamma^2 + \delta^2}i\,.$$

Convince yourself that

$$|\alpha\beta| = |\alpha||\beta| \quad \text{and} \quad \left|\frac{\alpha}{\beta}\right| = \frac{|\alpha|}{|\beta|}\,.$$

Complex numbers are often thought of as a point in the *complex plane*, usually depicted with a horizontal real axis and a vertical imaginary axis. This point can obviously be defined by its real and imaginary coordinates but can also be defined in terms of length r of the position vector of the point and the angle θ that the position vector makes with the real axis. This is the basis for polar representation of complex numbers as

$$A = \alpha + \beta i = r(\cos\theta + i\sin\theta) = re^{\theta i}$$

where the last equality is Euler's formula. Notice that θ is only determined up to modulus of 2π. I. e. if θ_1 is a valid angle then so is $\theta_2 = \theta_1 + 2n\pi$ where n can be any integer. r is often called the *modulus* of the number and θ the *argument*.

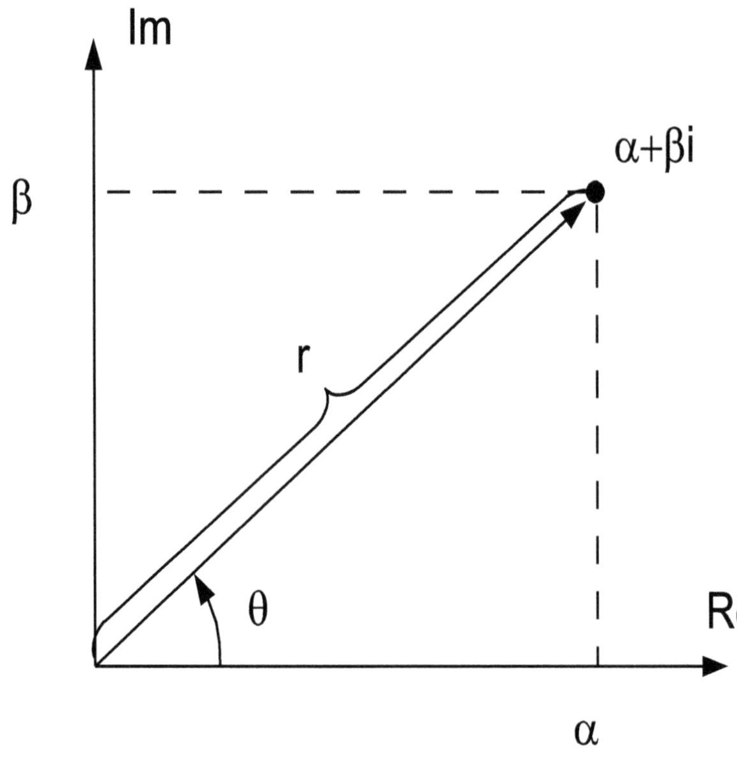

Fig. A.1 Polar representation of a complex number.

One can easily show that

$$r = |\alpha + \beta i| \text{ and } \cos\theta = \frac{\alpha}{r} \text{ or } \sin\theta = \frac{\beta}{r}.$$

Polar representation is convenient for multiplication of complex numbers

$$r_1 e^{\theta_1 i} r_2 e^{\theta_2 i} = r_1 r_2 e^{(\theta_1+\theta_2)i}$$

and makes it easy to raise complex numbers to a power

$$A = re^{\theta i} \Rightarrow A^x = r^x e^{x\theta i}.$$

Notice that if you take the Nth root of a complex number, you get N values because the argument is only defined $\pm 2n\pi$

$$A^{1/N} = r^{1/N} e^{\theta/N + 2n\pi/N}$$

and when n takes all possible values, the argument takes N possible values (keeping in mind that the argument is only defined modulus 2π). For instance

$$1^{1/3} = 1 \text{ or } 1^{1/3} = -\frac{1 \pm \sqrt{3}i}{2}$$

or

$$(-i)^{1/3} = i \text{ or } (-i)^{1/3} = -\frac{i \pm \sqrt{3}}{2}.$$

Geometrically, the n roots are n equally spaced points located on the circle in the complex plane with center at the origin and radius equal to $r^{1/n}$.

Index

$\aleph_0$, 407
l_2, 417
l_p, 150
l_∞, 150
p-norm, 408
$\mathcal{L}^2[0,1]$, 417
$\mathcal{L}^p[0,1]$, 413, 417
$\mathcal{L}^\infty[0,1]$, 414

adjoint eigenvalue problem, 248
adjoint operator, 262, 424
age population, 497
algebraic multiplicity, *204*
analytical function, 389
angle between vectors, 240, 300
annihilator, 347
Arrhenius' law, *25*, 27, 44, 87
associated vectors, 298
attractor, 309
augmented matrix, *170*
axial vector, 242

backsubstitution, 170
basis
  canonical, *154*
  Hamel, 152–156, 158, 411
  infinite, 154, *411*
  usual set, 154
basis solution
  difference equation, 314, 329
  ODE, 344, 353
batch reactor, 125
Bessel functions, *395*
  approximations for large arguments, 397
  differential and integral properties, 399
Bessel's equation, 395
bifurcation, 309
Bingham model, 40
Binnet's formula, 309
bioreactor, 383
biorthogonal sets, 249
Blackman's model, 81
boundary layer, 537
Brauer's theorem, 237
breakage intensity, 497
Briggs–Haldane kinetics, 96, 97
Buckingham's Π theorem, 190

catalytic pellet, 53, 54
  first order reaction, 41
  varying surface concentration, 469
  zero order reaction, 118
Cauchy sequence, 409
Cauchy's law, 202, 283, 285
Cauchy's method, 493, *498*
cell cycle, 305
center, 376
chain of generalized eigenvectors, 221, 222
chaotic solution, 310
characteristic equation, 203
characteristic polynomial
  difference equation, 314
  matrix, 203
  ODE, 142, 344
chemical kinetics, 24
chemical master equation, 74
chemical reactions
  independent, 187
chromatographic column, 493

closed form solution, 102
closed model, *see* complete model
closed subspace, 421
cofactor, 175
collocation, 116
column rank, 181–183
complement subspaces, 161, 162
complete model, *11*, 23
complete vector space, 406, 409
concentration wave velocity, 495
conduction
  transient, 457, 460, 472, 476, 490, 530
constitutive equation, *33*, 286
Contois' equation, 81
contravariance, *295*
contravariant degree, 264
contravariant vector, 295, 296
control volume, *2*
convergence, 406
  pointwise, 410
  uniform, 410
coordinate systems
  cylindrical, *31*, 292, 299, 302
  rectangular, *31*
  spherical, *31*, 292, 294, 299, 300, 302
coordinates, 156, 162
countable infinity, 407
countably infinite, 407
covariance, *295*
covariant degree, 264
covariant vector, 296
cross product, 241, *241*, 242–244
CSTR, *26*

degrees of freedom, 83
dense subset, 411
density gradient, 16, 99
determinant, *173*, 270
  calculation by cofactor expansion, 178
  calculation by Gauss elimination, 178
  differentiation, 179
  of diagonal matrix, 175
  of matrix product, 175
deviation variable, 380
difference equation, 75, *303*
  autonomous, 304
  basis solution, 314, 329, 332
    real part, 316
  constant coefficients, 314
    particular solution, 338
  coupled equations, 326
    constant coefficients, 331
    homogeneous, 327
  Euler's equation, 322
  existence and uniqueness theorem, 327
  first order, 321
  higher order, 322
  homogeneous, 314
  order of, 303
  particular solution
    by matched coefficients, 317
    by variation of parameters, 321, 324, 330
  Riccati equation, 342
  solution theorem, 329
differential balance, 6
diffusion coefficient, *34*
diffusion limited reaction, 93
digital process control, 311, 337
dimension, 152–155, 159, 163
dimensional analysis, 188, 190
dimensional matrix, 189, 190
dimensionless groups, 87
direct sum, 161–163
direction cosine, 266
disjoint subspaces, 161–163
dispersion, 53, 54
distributed models, 30
division intensity, 70, 497
domain of linear operator, 423
domain space, 198
dual space, 253, 259
dyad, *245*

effectiveness factor, 108
  first order reaction, 108, 109
  zero order reaction, 120
eigenbasis, 215, 217
eigenfunction, 428
eigenrow, *248*, 339, 367, 453
eigenspace, *203*, 215
eigenvalue, *202*
  as root of characteristic equation, 314
  calculation, 204
  Hilbert space, 424
  of self-adjoint operator, 262
  semi-simple, 215
  simple, 215
eigenvalue problem
  algebraic, 202

Hilbert space, 424, 428
eigenvector, *202*, 214
calculation, 204
Hilbert space, 424
of self-adjoint operator, 262
elemental matrix, 186, 187
elementary Jordan block, 225
for complex eigenvalue, 231
elliptic PDE, 451
Embden-Meyerhof, 193
endomorphism, 259
endothermic reaction, *25*
energy balance, 44, 49, 98, 277, 530
envelope, 518
enzyme, 95
enzyme kinetics, 95, 118
equivalence class, 213
equivalence relation, 158, 213, *213*
equivalence transformation, *259*, 261
error function, 533
Euclidean norm, 408, 411
Euclidean space, 290
Euler's equation
difference equation, 322
ODE, 143
Euler's formula, 316, 395
Eulerian, 30, 45
Eulerian description, 498
exothermic reaction, *25*, 27, 87
expansion by cofactors, 174, 175
exponential of a matrix, 235
extensive property, 2
extraction, 20

fed batch, 26, 29
Fibonacci equation, 308
Fibonacci number, 308
Fick's law, *34*, 41, 53, 58
film theory, 21
finite Fourier transform, 490
focus, 375, 376
formally self-adjoint, 424
forward substitution, 170
Fourier cosine series, 438
Fourier series, 438
complete, 439
generalized, 438, 444
Fourier sine series, 438
Fourier's law, *34*, 44, 50, 58, 272, 274
Frobenius' method, 387
full rank, *183*
fundamental matrix
difference equation, 330
ODE, 359
fundamental system
difference equation, 329
ODE, 359

gas constant, 64
Gauss elimination, 169
generalized Bessel equation, 399
generalized eigenbase, 221
for complex eigenvalue, 230
generalized eigenspace, 217, *218*, 219, 222
for complex eigenvalue, 230
generalized eigenvector, 206, 215, *218*
calculation, 222–224
geometric multiplicity, 206, *210*
of complex eigenvalue, 230
Gerschgorin's theorem, 237
Gibb's phenomena, 441
Graetz–Nusselt problem, 486
Gram–Schmidt orthogonalization, *245*, 247, 420
group, 280

Hamel basis, *see* basis, Hamel
heat of reaction, *25*, 27, 44, 87
Henry's law, 65
Hilbert space, *416*
homogeneous solution
difference equations, 314
ODE, 344
Hooke's law, *34*, 286–288
isotropic material, 289
non-isotropic material, 288
hyperbolic PDE, 451

ideal gas, 64
identity matrix, 167
image space, 198
implicit function theorem, 292
inconsistent system, 192
index, 218, 220, 230
indicial equation, 144, 392
infinity-norm, 409
initial manifold, 500
inner product, *239*, 240, *414*
of covariant and contravariant vectors, 300

inner product method, 452, *453*
inner product space, 239, 415
integrating factor, 127, *127*, 128, 135, 437
intensive property, 2
interstitial fluid velocity, 54
irregular singular point, 390
isomorphism, 157–159

Jacobian matrix, 380
Jordan block, 226
Jordan canonical form, 225, 227

kernel, *see* null space, 252

Lagrangian, 30, 45
Lagrangian description, 498
Langmuir isotherm, 513
Langmuir–Hinshelwood kinetics, *25*
Laplace's equation, 451
Laplacian operator, 56
left inverse, 168
Legendre functions, *402*
Legendre polynomial, 403, 404, 469, 470
Legendre's equation, 402, 447, 470
length of vector, 240, 300
linear algebraic equations
  solution structure, 191
linear combination, *151*
linear dependence, *151*
linear independence, *151*, 152, 155
  of functions, 152
linear operator, 150, 158, *252*
  Hilbert space, 423
linear transformation, *see* linear operator
Lipschitz condition, 123
loan, 305
logistic difference equation, 309
logistic equation, 126, 309, 381
Loschmidt apparatus, 533
lowering of the index, 299
lubrication approximation, 106
lumping, *92*

main diagonal, 167
mass action kinetics, 25, 66
mass balance, 26, 47, 54, 67, 118
Matrices as operators, 254
matrix, 149, 155, *166*
  adjoint, 180
  antisymmetric, 167
  diagonal, 167
  echelon form, 182, 183
  equivalent, 261
  invariant, 218
  inverse, 168, 169, 181
    calculation, 173
  lower triangular, 167
  nilpotent, 234
  orthogonal, 267
  positive, 238
  real and symmetric, 250
  self-adjoint, 262
  semi-simple, 215, 216
  similar, 211, 261, 270
  simple, 215, 216
  singular, 181
  skew-symmetric, 167
  square, 167, 168
  symmetric, 167
  transpose, 167, 168, 175
  upper triangular, 167
matrix addition, 167
matrix column, 166
matrix multiplication, 168
matrix of transition, 259
  for basis change, 260
matrix row, 166
matrix-vector multiplication, 168
maximal orthonormal set, 418
measurable function, 413
measure theory, 413
metabolic flux analysis, 193
method of characteristics, 493
method of matched coefficients
  difference equation, 317
  ODE, 347
metric tensor, 299, 300
  contravariant, 300
  covariant, 298
Michaelis–Menten kinetics, 25, 96, 97, 118
microbial growth models, 80
Minkowski inequality
  finite sums, 409
  infinite sums, 150
  integrals, 409
minor, 184
mixing
  in flow through reactor, 313
mixing point, 6
modified Bessel's equation, 395

modulus of viscosity, 40
mole balance, 21
momentum balance, 60
Monod model, *25*, 81, 384
Monte Carlo, 74
Moore-Penrose inverse, 202
Moser's equation, 81
multi-linear function, 264

Navier–Stokes equation, 60, 85
Neumann function, 395
Newton's law of cooling, 37
Newton's law of viscosity, 34, 35, 39, 60, 289, 290
Newtonian fluid, 35
node, 372
norm, *408*
normalized vector, 240
normed space, 408
null set, 413
null space, 198
  operator, 252, 271
number balance, 67

ODE, *343*
  autonomous, 343
  basis solution, 344, 362
    real part, 345
  complex eigenvalues, 370
  constant coefficients, 142, 344
    particular solution, 367
  coupled equations, 357
    constant coefficients, 361
    homogeneous, 358
  first order, 123
    Bernoulli, 134
    Clairaut, 132
    exact, 129
    homogeneous, 131
    linear, 127
    Riccati, 135
    separable, *124*, 125, 127, 128, 131, 137
  higher order, 353
  homogeneous, 344
  linearization, 379
  particular solution
    by matched coefficients, 347
    by variation of parameters, 355, 360
  second order, 136
    basis solution, 354
    dependent variable not explicit, 137
    free variable not explicit, 137
    homogeneous, 138
    particular solution, 357
  series solution, *386*
  solution theorem, 353, 359
operators
  cylindrical coordinates, 58
  differential, 56
  Laplacian, 56
  rectangular coordinates, 57
  spherical coordinates, 59
ordinary differential equation, *see* ODE
ordinary point, 390
orthogonal basis, 240
orthogonal complement, 423
orthogonal projection, 421
orthogonal vectors, 240, 418
orthonormal basis, 241
orthonormal set, 418
outer product, 245, *245*
overdetermined systems, 192, *201*

packed bed reactor, 53
parabolic PDE, 451
partial differential equation, *see* PDE
particle dissolution, 509
particle growth, 506
particular solution, 128
partition, 214
PDE
  classification, 449
  first order, 493
  quasi-linear, 493
  semi-linear, 493
period doubling, 310
permutation, 173
Perron's theorem, 238
phase plane, 372
phase plane plot
  node, 373, 374
phase space
  flow, 372
phase space plot, 372, 377, 378
physical components of a vector, 301
plug flow, 41
plug flow reactor, 44, 46, 498, 505

point balance, 6
Poisson's ratio, 289
polar notation for complex numbers, 316, 335
polar vector, 242
population balance, 68
population balance equation, 68, 71, 72, 495
power law index, 40
power law model, 40
power of the continuum, 408
powers of a matrix, 234, 235
principal stress, 284
principal stress direction, 284
projection, 416, 421
proper transformation, 292
pseudoinverse, 202

quasi-steady state, *92*, 478
quasi-steady state assumption, 102

raising of the index, 300
range, 198
  operator, 252, 271
rank, 181, 183, 184
  by Gauss elimination, 182
  calculation by determinants, 184
  generalized eigenvector, 221
  of a set of vectors, 184
  operator, 270
rank and linear independence, 184–186
Raoult's law, 65
rate limited reaction, 93
reaction order, 25
reaction propensity, 72, 73
recycle reactor, 11
reduction of order
  difference equations, 322
  ODE, 353
reflection, 266
regular singular point, 390
regularity condition, 511
renewal equation, 498
Riccati equation, 135, 342
  difference equation, 342
Riesz index, *see* index
right inverse, 168
rotation, 266, 269
  improper, 267
  proper, 267
rotation matrix, 266–268, 273
  curvilinear coordinates, 301
row rank, 181–183
row-reduced echelon form, 171

saddle, 375
scalar, *147*
scalar field, 147
scalar product, 239
scale factor, 299, 301, 302
Schwarz inequality, 415
self-adjoint operator, 262, 283, 424
series solution
  expansion theorem
    irregular singular point, 390
    ordinary point, 390
    regular singular point, 390, 393
shear strain, 287
shift operator, 304
shock, 516
SI units, 113
similarity transformation
  for matrices, 211
  for PDE, 529
similarity variable, 529
sinc function, 401
slurry reactor, 97
span, 151, 152
spanning set, 151, 156
specific reaction rate, 24
spectrum, *202*
  degenerate, 210
spherical Bessel functions, 400
stability
  critical, 372
  difference equation, 315, 337
  neutral, 372
  ODE, 345, 379
star, 374, 376
steady state
  center, 376
  critically stable, 377
  degenerate system, 377
  focus, 375, 376
  node, 372–374
  saddle, 375
  star, 374, 376
steady state classification, 371
stochastic kinetics, 72
stoichiometry, 24

strain tensor, 287
stress tensor, 282–284, 286
Sturm–Liouville operator, 427
Sturm–Liouville problem, 433
  regular, 433
  singular, 433
subspace, *160*, 161
  closed, 421
  parametric representation, 164–166
sum of subspaces, 161–163
summation convention, 299
  Cartesian systems, 268
  curvilinear systems, 297
sup-norm, *see* infinity-norm
supremum, 150
symmetry, 272
synchronous cells, 308

tanks in series model, 77
Taylor series, 379, 387, 389
TCA cycle, 193
tensile strain, 287
tensor, 263, *264*
  anti-symmetric, 271
  arbitrary order, 268
  degree, 264
  first order, 264, 267
  in curvilinear coordinates, 297
  second order, 264, 267
  skew-symmetric, 271
  symmetric, 270
  zero order, 263, 267
tensor invariant, 270
Tessier's equation, 81
thermal conductivity, *34*
thermal conductivity tensor, 272
Thiele modulus
  first order reaction, 94, 108, 109
  zero order reaction, 118
trace, 167, 271
traffic flow, 496
trajectory, 372, 373
transformation rule, 265, 268
  any tensor, 268
  first order tensor, 267
  second order tensor, 267
  zero order tensor, 267
transpose, *see* matrix, transpose
trial function, 116
triple scalar product, *242*, 244
tubular reactor, 47
two cycle, 309

uncountable infinity, 407, 408
usual basis, *see* basis, canonical

van der Waals equation, 64
variation of parameters, 321, 322, 330, 353, 360
  difference equations, 324
  ODE, 355
vector, 147
  in curvilinear coordinates, 293
vector space, *148*, 149
  complete, 406
  normed, 408
viscosity, *34*
void fraction, 53, 54
volume balance, 17

wave equation, 451, 525
Weierstrass approximation theorem, 411
weight function, 417, 425, 426, 428
well mixed, 5
Wronskian matrix, 356
Wronsky determinant, 359

Young's modulus, 34, 289

zero order reaction, 118

www.ingramcontent.com/pod-product-compliance
Ingram Content Group UK Ltd.
Pitfield, Milton Keynes, MK11 3LW, UK
UKHW051044150926
8624IPUK00006B/390

9 789813 27087